ILLUSTRATED HANDBOOK OF PHYSICAL - CHEMICAL PROPERTIES AND ENVIRONMENTAL FATE FOR ORGANIC CHEMICALS

Volume I

Monoaromatic Hydrocarbons,
Chlorobenzenes,
and PCBs

Donald Mackay
Wan Ying Shiu
Kuo Ching Ma

Library of Congress Cataloging-in-Publication Data

Mackay, Donald, Ph.D.
Illustrated handbook of physical-chemical properties and environmental fate for organic chemicals/Donald Mackay, Wan Ying Shiu, and Kuo Ching Ma.

p. cm.
Includes bibliographical references.
Contents: v. 1. Monoaromatic hydrocarbons, chlorobenzenes, and PCBs.
1. Organic compounds - Environmental aspects - Handbooks, manuals, etc.
2. Environmental chemistry - Handbooks, manuals, etc.
I. Shiu, Wan Ying. II. Ma, Kuo Ching. III. Title.
TD196.073M32 1991
628.5'2 - dc20 91-33888
ISBN 0-87371-513-6

This book represents information obtained from authentic and highly regarded sources. Reprinted material is quoted with permission, and sources are indicated. A wide variety of references are listed. Every reasonable effort has been made to give reliable data and information, but the authors and the publisher cannot assume responsibility for the validity of all materials or for the consequences of their use.

LEWIS PUBLISHERS, INC.
121 South Main Street, Chelsea, Michigan 48118

Printed in the United States of America 1 2 3 4 5 6 7 8 9 0

PREFACE

This series of Handbooks brings together physical-chemical data for similarly structured groups of chemical substances, which influence their fate in the multimedia environment of air, water, soils, sediments and their resident biota. The task of assessing chemical fate locally, regionally and globally is complicated by the large (and increasing) number of chemicals of potential concern, by uncertainties in their physical-chemical properties, and by lack of knowledge of prevailing environmental conditions such as temperature, pH and deposition rates of solid matter from the atmosphere to water, or from water to bottom sediments. Further, reported values of properties such as solubility are often in conflict. Some are measured accurately, some approximately and some are estimated by various correlation schemes from molecular structure. In some cases units or chemical identity are wrongly reported. The user of such data thus has the difficult task of selecting the "best" or "right" values. There is justifiable concern that the resulting deductions of environmental fate may be in substantial error. For example, the potential for evaporation may be greatly underestimated if an erroneously low vapor pressure is selected.

To assist the environmental scientist and engineer in such assessments, this Handbook contains compilations of physical-chemical property data for series of chemicals such as the aromatic hydrocarbons. It has long been recognized that within such series, properties vary systematically with molecular size, thus providing guidance about the properties of one substance from those of its homologs. Plots of these systematic property variations are provided to check the reported data and provide an opportunity for interpolation and even modest extrapolation to estimate unreported properties of other homologs. Most handbooks treat chemicals only on an individual basis, and do not contain this feature of chemical-to-chemical comparison which can be valuable for identifying errors and estimating properties.

The data are taken a stage further and used to estimate likely environmental partitioning tendencies, i.e., how the chemical is likely to become distributed between the various media which comprise our biosphere. The results are presented numerically and pictorially to provide a visual impression of likely environmental behavior. This will be of interest to those assessing environmental fate by confirming the general fate characteristics or behavior profile. It is, of course, only possible here to assess fate in a "typical" or "generic" or "evaluative" environment, thus no claim is made that a chemical will behave in this manner in all situations, but this assessment should reveal the broad characteristics of behavior. These evaluative fate assessments are generated using simple fugacity models which flow naturally from the compilations of data on physical-chemical properties of relevant chemicals.

It is hoped that this series of Handbooks will be of value to environmental scientists and engineers and to students and teachers of environmental science. Its aim is to contribute to better assessments of chemical fate in our multimedia environment by serving as a reference source for environmentally relevant physical-chemical property data of classes of chemicals and by illustrating the likely behavior of these chemicals as they migrate throughout our biosphere.

Donald Mackay, born and educated in Scotland, received his degrees in Chemical Engineering from The University of Glasgow. After a period of time in the petrochemical industry he joined The University of Toronto, where he is now a Professor in the Department of Chemical Engineering and Applied Chemistry, and in the Institute for Environmental Studies. Professor Mackay's primary research is the study of organic environmental contaminants, their sources, fates, effects, and control, and particularly in understanding and modeling their behavior with the aid of the fugacity concept. His work has focused especially on the Great Lakes Basin and on cold northern climates.

Wan-Ying Shiu is a Research Associate in the Department of Chemical Engineering and Applied Chemistry, and the Institute for Environmental Studies, University of Toronto. She received her Ph.D. in Physical Chemistry from the Department of Chemistry, University of Toronto, M.Sc. in Physical Chemistry from St. Francis Xavier University and B.Sc. in Chemistry from Hong Kong Baptist College. Her research interest is in the area of physical-chemical properties and thermodynamics for organic chemicals of environmental concern.

Kuo-Ching Ma obtained his Ph.D. from The Florida State University, M.Sc. from The University of Saskatchewan and B.Sc. from The National Taiwan University; all in Physical Chemistry. After working many years in the Aerospace, Battery Research, Fine Chemicals and Metal Finishing industries in Canada as Research Scientist, Technical Supervisor/Director, he is now dedicating his time and interests to Environmental Research.

TABLE OF CONTENTS

1. INTRODUCTION

1.1 THE INCENTIVE

It is alleged that there are some 60,000 chemicals in current commercial production, with approximately 1000 being added each year. Most are organic chemicals. Of these, perhaps 500 are of environmental concern because of their presence in detectable quantities in various components of the environment, their toxicity, their tendency to bioaccumulate, or their persistence. A view is emerging that some of these chemicals are of such extreme environmental concern that all production and use should be ceased, i.e., as a global society we should elect not to synthesize or use these chemicals. They should be "sunsetted". PCBs, "dioxins" and freons are examples. A second group consists of chemicals which are of concern because they are used or discharged in large quantities, or they are toxic or persistent. They are, however, of sufficient value to society that their continued use is justified, but only under conditions in which we fully understand their sources, fate and effects. This understanding is essential if society is to be assured that there are no adverse ecological or human health effects. Other groups of increasingly benign chemicals can presumably be treated with less rigor.

A key feature of this "cradle to grave" approach is that society must improve its skills in assessing chemical fate in the environment. We must better understand where chemicals originate, how they migrate in, and between, the various media of air, water, soils, sediments and their biota which comprise our biosphere. We must understand how these chemicals are transformed by chemical and biochemical processes and thus how long they will persist in the environment. We must seek a fuller understanding of the effects which they will have on the multitude of interacting organisms which occupy these media, including ourselves.

It is now clear that the fate of chemicals in the environment is controlled by a combination of two groups of factors. First are the prevailing environmental conditions such as temperatures, flows and accumulations of air, water and solid matter and the composition of these media. Second are the properties of the chemicals which influence partitioning and reaction tendencies, i.e., whether the chemical evaporates or associates with sediments, and how the chemical is eventually destroyed by conversion to other chemical species.

In recent decades there has emerged a discipline within environmental science concerned with increasing our understanding of how chemicals behave in our multimedia environment. It has been termed "chemodynamics". Practitioners of this discipline include scientists and engineers, students and teachers who attempt to measure, assess and predict how this large number of chemicals will behave in laboratory, local, regional and global environments. These individuals need data on physical-chemical and reactivity properties, as well as information on how these properties translate into environmental fate. This Handbook provides a compilation of such data and uses them to estimate the broad features of environmental fate. It does so for classes or groups of chemicals, instead of the usual approach of treating chemicals on an individual basis. This has the advantage that systematic variations in properties with molecular size can be revealed and used to check reported values, interpolate and even extrapolate to other chemicals of similar structure.

With the advent of inexpensive and rapid computation there has been a remarkable growth in interest in this general area of Quantitative-Structure-Property Relationships (QSPRs). The ultimate goal is to use information about chemical structure to deduce physical-chemical properties, environmental partitioning and reaction tendencies, and even uptake and effects on biota. The goal is far from being realized, but considerable progress has been made, as is briefly reviewed in a following section. In this Handbook we adopt a simple, and well tried, approach of using molecular structure to deduce a molar volume, which in turn is related to physical-chemical properties. Undoubtedly, other molecular descriptors such as surface area or topological indices have the potential to give more accurate correlations and will be used in the future, but at this stage we believe that the improvements in accuracy obtained by using these more complex descriptors do not justify the computational effort of generating them, at least for the purposes of routine, general assessments. In some cases the fundamental causes of the relationships remain obscure.

A major benefit of this simple QSPR analysis is that it reveals likely errors in reported data. Regrettably, the scientific literature contains a great deal of conflicting data with reported values often varying over several orders of magnitude. There are some good, but more not-so-good reasons for this lack of accuracy. Many of these properties are difficult to measure because they involve analyzing very low concentrations of 1 part in 10^9 or 10^{12}. For many purposes an approximate value, for example, that a solubility is less than 1 mg/L, is adequate. There has been a mistaken impression that if a vapor pressure is low, as is the case with DDT, it is not important. DDT evaporates appreciably from solution in water despite its low vapor pressure, because of its low solubility in water. In some cases the units are reported incorrectly or there are uncertainties about temperature or pH. In other cases the chemical is wrongly identified. One aim of this Handbook is to assist the user to identify such problems and provide guidance when selecting appropriate values.

The final aspect of chemical fate treated in this Handbook is the depiction or illustration of likely chemical fate. This is done using a series of multimedia "fugacity" models as is described in a later section. The authors' aim is to convey an impression of likely environmental partitioning and transformation characteristics, i.e., we seek to generate a "behavior profile". A fascinating feature of chemodynamics is that chemicals differ so greatly in their behavior. Some, such as chloroform, evaporate rapidly and are dissipated in the atmosphere. Others, such as DDT, partition into the organic matter of soils and sediments and the lipids of fish, birds and mammals. Phenols tend to remain in water subject to fairly rapid transformation processes such as biodegradation and photolysis. By entering the physical-chemical data into a model of chemical fate in a generic or evaluative environment, it is possible to estimate the likely general features of the chemical's behavior and fate. The output of these calculations is presented numerically and pictorially.

In total, the aim of this series of Handbooks is to provide a useful reference work for those concerned with the assessment of the fate of existing and new chemicals in the environment.

1.2 PHYSICAL-CHEMICAL PROPERTIES

1.2.1 The key physical-chemical properties

The major differences between behavior profiles of organic chemicals in the environment are attributable to physical-chemical properties. The key properties are believed to be solubility in water, vapor pressure, octanol-water partition coefficient, dissociation constant in water (when relevant) and susceptibility to degrading or transformation reactions. Other essential molecular descriptors are molecular mass and molar volume, with properties such as critical temperature and pressure and molecular area being occasionally useful for specific purposes.

Chemical identity may appear to present a trivial problem, but many chemicals have several names, and subtle differences between isomers (e.g., cis and trans) may be ignored. The most commonly accepted identifiers are the IUPAC name and the Chemical Abstracts System (CAS) number. More recently methods have been sought of expressing the structure in line notation form so that computer entry of a series of symbols can be used to define a three-dimensional structure. The Wiswesser Line Notation is quite widely used, but it appears that for environmental purposes it will be superceded by the SMILES (Simplified Molecular Identification and Line Entry System, Anderson et al. 1987).

Molecular mass is readily obtained from structure. Also of interest are molecular volume and area, which may be estimated by a variety of methods.

Solubility in water and vapor pressure are both "saturation" properties, i.e., they are measurements of the maximum capacity which a phase has for dissolved chemical. Vapor pressure P (Pa) can be viewed as a "solubility in air", the corresponding concentration C (mol/m^3) being P/RT where R is the ideal gas constant (8.314 J/mol • K) and T is absolute temperature (K). Although most chemicals are present in the environment at concentrations well below saturation, these concentrations are useful for estimating air-water partition coefficients as ratios of saturation values. It is usually assumed that the same partition coefficient applies at lower sub-saturation concentrations. Vapor pressure and solubility thus provide estimates of air-water partition coefficients K_{AW} or Henry's law constants H ($Pa \cdot m^3/mol$), and thus the relative air-water partitioning tendency.

The octanol-water partition coefficient K_{OW} provides a direct estimate of hydrophobicity or of partitioning tendency from water to organic media such as lipids, waxes and natural organic matter such as humin or humic acid. It is invaluable as a method of estimating K_{OC}, the organic carbon-water partition coefficient, the usual correlation invoked being that of Karickhoff (1981)

$$K_{OC} = 0.41\ K_{OW}$$

It is also used to estimate fish-water bioconcentration factors K_B or BCF using a correlation similar to that of Mackay (1982)

$$K_B = 0.05\ K_{OW}$$

where the term 0.05 corresponds to a 5% lipid content of the fish.

For ionizing chemicals it is essential to quantify the extent of ionization as a function of pH using the dissociation constant pKa. The parent and ionic forms behave and partition quite differently, thus pH and the presence of other ions may profoundly affect chemical fate.

Characterization of chemical reactivity presents a severe problem in Handbooks. Whereas radioisotopes have fixed half-lives, the half-life of a chemical in the environment depends not only on the intrinsic properties of the chemical, but also on the nature of the surrounding environment. Factors such as sunlight intensity, hydroxyl radical concentration and the nature of the microbial community, as well as temperature, affect the chemical's half-life so it is impossible (and misleading) to document a single reliable half-life. The compilation by Howard et al. (1991) provides an excellent review of the existing literature for a large number of chemicals. It is widely used as a source document in this work. The best that can be done is to suggest a semi-quantitative classification of half-lives into groups, assuming average environmental conditions to apply. Obviously a different class will generally apply in air and bottom sediment. In this compilation we use the following class ranges for chemical reactivity in a single medium such as water.

Class	Mean half-life (hours)	Range (hours)
1	5	< 10
2	17 (~ 1 day)	10-30
3	55 (~ 2 days)	30-100
4	170 (~ 1 week)	100-300
5	550 (~ 3 weeks)	300-1,000
6	1700 (~ 2 months)	1,000-3,000
7	5500 (~ 8 months)	3,000-10,000
8	17000 (~ 2 years)	10,000-30,000
9	55000 (~ 6 years)	> 30,000

These times are divided logarithmically with a factor of approximately 3 between adjacent classes. With the present state of knowledge it is probably misleading to divide the classes into finer groupings; indeed a single chemical may experience half-lives ranging over three classes, depending on season.

A recurring problem in compilation of this type is the criteria which should be used for selecting the "best" value of a property when several values are reported. An element of judgement is necessary. The usual considerations are:

(1) the age of the data and acknowledgment of previous conflicting or supporting values,
(2) the method of determination,

(3) the perception of the objectives of the authors, not necessarily as an indication of competence, but often as an indication of the need of the authors for accurate values.

In this Handbook we have used these considerations as well as information derived from the QSPR analyses.

It is appropriate, therefore, to review briefly the experimental methods which are commonly used for property determinations and comment on their accuracy.

1.2.2. Experimental methods

Solubility in water

The conventional method of preparing saturated solutions for the determination of solubility is batch equilibration. An excess amount of solute chemical is added to water and equilibrium is achieved by shaking gently (generally referred as the "shake flask method") or slow stirring with a magnetic stirrer. The aim is to prevent formation of emulsions or suspensions and thus avoid extra experimental procedures such as filtration or centrifuging which may be required to ensure that a true solution is obtained. Experimental difficulties can still occur because of the formation of emulsion or microcrystal suspensions with the sparingly soluble chemicals such as higher normal alkanes and polycyclic aromatic hydrocarbons (PAHs). An alternative approach is to coat a thin layer of the chemical on surface of the equilibration flask before water is added. An accurate "generator column" method has also been developed, (Weil et al. 1974, May et al. 1978a,b) in which a column is packed with an inert solid support, such as glass beads or Chromosorb, and then coated with the solute chemical. Water is pumped through the column at a controlled, known flow rate to achieve saturation.

The method of concentration measurement of the saturated solution depends on the solute solubility and its chemical properties. Some common methods used for solubility measurement are listed below.

1. Gravimetric or volumetric methods (Booth and Everson 1948)

   An excess amount of solid compound is added to a flask containing water to achieve saturation solution by shaking, stirring, centrifuging until the water is saturated with solute and undissolved solid or liquid residue appears, often as a cloudy phase. For liquids, successive known amounts of solute may be added to water and allowed to reach equilibrium, and the volume of excess undissolved solute is measured.
2. Instrumental methods
   - a. UV spectrometry (Andrews and Keffer 1950, Bohon and Claussen 1951, Yalkowsky et al. 1976);
   - b. Gas chromatographic analysis with FID, ECD or other detectors (McAuliffe 1968, Mackay et al. 1975, Chiou et al. 1982);
   - c. Fluorescence spectrophotometry (Mackay and Shiu 1977);
   - d. Interferometry (Gross and Saylor 1931);
   - e. High-pressure liquid chromatography (HPLC) with R.I., UV or fluorescence detection (May et al. 1978ab, Wasik et al. 1983, Shiu et al. 1988, Doucette and Andren 1988a);
   - f. Nephelometric methods (Davis and Parke 1942, Davis et al. 1942, Hollifield 1979).

For most organic chemicals the solubility is reported at a defined temperature in distilled water. For substances which ionize (e.g. phenols, carboxylic acids and amines) it is essential to report the pH of the determination because the extent of ionization affects the solubility. It is common to maintain the desired pH by buffering with an appropriate electrolyte mixure. This raises the complication that the presence of electrolytes modifies the water structure and changes the solubility. The effect is usually "salting-out". For example, many hydrocarbons have solubilities in seawater about 75% of their solubilities in distilled water. Care must thus be taken to interpret and use reported data properly when electrolytes are present.

The most common problem encountered with reported data is inaccuracy associated with very low solubilities, i.e., those less than 1.0 mg/L. Such solutions are difficult to prepare, handle and analyze, and reported data are often contain appreciable errors.

Octanol-water partition coefficient K_{OW}

The experimental approaches are similar to those for solubility, i.e., employing shake flask or generator-column techniques. Concentrations in both the water and octanol phases may be determinated after equilibration. Both phases can then be analyzed by the instrumental methods discussed above and the partition coefficient is calculated from the concentration ratio C_O/C_W. This is actually the ratio of solute concentration in octanol saturated with water to that in water saturated with octanol.

As with solubility, K_{OW} is a function of the presence of electrolytes and for dissociating chemicals it is a function of pH. Accurate values can generally be measured up to about 10^6, but accurate measurement beyond this requires meticulous technique. A common problem is that the presence of small quantities of emulsified octanol in the water phase could create a high concentration of chemical in that emulsion which would cause an erroneously high apparent water phase concentration.

Considerable success has been achieved by calculating K_{OW} from molecular structure; thus there has been a tendency to calculate K_{OW} rather than measure it, especially for "difficult" hydrophobic chemicals. These calculations are, in some cases, extrapolations and can be in serious error. Any calculated log K_{OW} value above 7 should be regarded as suspect, and any experimental or calculated value above 8 should be treated with extreme caution.

Details of experimental methods are described by Fujita et al. (1964), Leo et al. (1971); Hansch and Leo (1979), Rekker (1977), Chiou et al. (1977), Miller et al. (1984), Bowman and Sans (1983), Woodburn et al. (1984), Doucette and Andren (1987), and De Bruijn et al. (1989).

Vapor pressure

In principle, the determination of vapor pressure involves the measurement of the saturation concentration or pressure of the solute in a gas phase. The most reliable methods involve direct determination of these concentrations, but convenient indirect methods are also available based on evaporation rate measurement or chromatographic retention times. Some methods and approaches are listed below.

a. Comparative ebulliometry (Ambrose 1981);

b. Effusion methods (Balson 1947, Bradley and Cleasby 1953, Hamaker and Kerlinger 1969, Sinke 1974);

c. Gas saturation or transpiration methods (Spencer and Cliath 1970, 1972, Macknick and Prausnitz 1979, Westcott et al. 1981);
d. Dynamic coupled-column liquid chromatographic method - a gas saturation method (Sonnefeld et al. 1983);
e. Calculation from evaporation rates and vapor pressures of reference compound (Gückel et al. 1974, 1982, Dobbs and Grant 1980, Dobbs and Cull 1982);
f. Calculation from GC retention time data (Westcott and Bidleman 1982, Bidleman 1984, Kim et al. 1984, Foreman and Bidleman 1985, Burkhard et al. 1985a, Hinckley et al. 1990).

The greatest difficulty and uncertainty arises when determining the vapor pressure of chemicals of low volatility, i.e., those with vapor pressures below 1 Pa. Vapor pressures are strongly dependent on temperature, thus accurate temperature control is essential. Data are often regressed against temperature and reported as Antoine or Clapeyron constants. Care must be taken when using the Antoine or other equations to extropolate data beyond the temperature range specified. It must be clear if the data apply to the solid or liquid phase of the chemical.

Henry's law constant

The Henry's law constant is essentially an air-water partition coefficient which can be determined by measurement of solute concentrations in both phases. This raises the difficulty of accurate analytical determination in two very different media which require different techniques. Accordingly, some effort has been devoted to devising techniques in which concentrations are measured in only one phase and the other concentration is deduced by a mass balance. These methods are generally more accurate. The principal difficulty arises with hydrophobic, low-volatility chemicals which can establish only very small concentrations in both phases.

Henry's law constant can be regarded as a ratio of vapor pressure to solubility, thus it is subject to the same effects which electrolytes have on solubility and temperature has on both properties. Some methods are as follows:

a. Equilibrium batch stripping (Mackay et al. 1979, Dunnivant et al. 1988);
b. EPICS (Equilibrium Partioning In Closed Systems) method (Lincoff and Gossett 1984; Gossett 1987, Ashworth et al. 1988);
c. Wetted-wall column (Fendinger and Glotfelty 1988, 1990);
d. Headspace analyses (Hussam and Carr 1985);
e. Calculation from vapor pressure and solubility (Mackay and Shiu 1981).

When using vapor pressure and solubility data it is essential to ensure that both properties apply to the same chemical phase, i.e. both are of the liquid, or of the solid. Occasionally a solubility is of a solid while a vapor pressure is extrapolated from higher temperature liquid phase data.

1.3 QUANTITATIVE-STRUCTURE-PROPERTY RELATIONSHIPS (QSPRs)

1.3.1 Objectives

Because of the large number of chemicals of actual and potential concern, the difficulties and cost of experimental determinations, and scientific interest in elucidating the fundamental molecular determinants of physical-chemical properties, a considerable effort has been devoted to generating quantitative-structure-activity relationships (QSARs). This concept of structure-property relationships or structure-activity relationships is based on observations of linear free-energy relationships, and usually takes the form of a plot or regression of the property or interest as a function of an appropriate molecular descriptor which can be obtained from merely a knowledge of molecular structure.

Such relationships have been applied to solubility, vapor pressure, K_{OW}, Henry's law constant, reactivities, bioconcentration data and several other environmentally relevant partition coefficients. Of particular value are relationships involving various manifestations of toxicity, but these are beyond the scope of this Handbook. These relationships are valuable because they permit values to be checked for "reasonableness" and (with some caution) interpolation is possible to estimate undetermined values. They may be used (with extreme caution!) for extrapolation.

A large number of descriptors have been, and are being, proposed and tested. Dearden (1990) and the compilation by Karcher and Devillers (1990) give comprehensive accounts of descriptors and their applications.

Among the most commonly used molecular descriptors are molecular weight and volume, the number of specific atoms (e.g., carbon or chlorine), surface areas (which may be defined in various ways), refractivity, parachor, steric parameters, connectivities and various topological parameters. Several quantum chemical parameters can be calculated from molecular orbital calculations including charge, electron density and superdelocalizability.

It is likely that existing and new descriptors will be continued to be tested, and that eventually a generally preferred set of readily accessible parameters will be adopted of routine use for correlating purposes. From the viewpoint of developing quantitative correlations it is very desirable to seek a linear relationship between descriptor and property, but a non-linear or curvilinear relationship is quite adequate for illustrating relationships and interpolating purposes. In this Handbook we have elected to use the simple descriptor of molar volume at the normal boiling point as estimated by the LeBas method (Reid et al. 1987). This parameter is very easily calculated and proves to be adequate for the present purposes of plotting property versus relationship without seeking linearity.

The LeBas method is based on a summation of atomic volumes with adjustment for the volume decrease arising from ring formation. The full method is described by Reid et al. (1987), but for the purposes of this compilation the volumes and rules as listed in Table 1.1 are used.

Table 1.1 LeBas Molar Volume

	increment, cm^3 /mol
carbon	14.8
hydrogen	3.7
oxygen	7.4
in methyl esters and ethers	9.1
in ethyl esters and ethers	9.9
join to S, P, or N	8.3
nitrogen	
doubly bonded	15.6
in primary amines	10.5
in secondary amines	12.0
bromine	27.0
chlorine	24.7
fluorine	8.7
iodine	37.0
sulfur	25.6
ring	
three-membered	-6.0
four-membered	-8.5
five-membered	-11.5
six-membered	-15.0
naphthalene	-30.0
anthracene	-47.5

Example: The experimental molar volume of chlorobenzene is 115 cm^3/mol. From the above rules, the LeBas molar volume for chlorobenzene (C_6H_5Cl) is:

$$V = 6 \times 14.8 + 5 \times 3.7 + 24.6 - 15 = 117 \text{ cm}^3\text{/mol}$$

Accordingly, plots are presented at the end of each chapter for solubility, vapor pressure, K_{OW}, and Henry's law constant versus LeBas molar volume.

A complication arises in that two of these properties (solubility and vapor pressure) are dependent on whether the solute is in the liquid or solid state. Solid solutes have lower solubilities and vapor pressures than they would have if they had been liquids. The ratio of the (actual) solid to the (hypothetical subcooled) liquid solubility or vapor pressure is termed the fugacity ratio and can be estimated from the melting point and the entropy of fusion ΔS_{fus} as discussed by Mackay and Shiu (1981). For solid solutes, the correct property to plot is the calculated or extrapolated subcooled liquid solubility. This is calculated in this Handbook using the relationship suggested by Yalkowsky (1979) which implies an entropy of fusion of 56 J/mol K or 13.5 cal/mol • K

$$C^S_S/C^S_L = P^S_S/P^S_L = \exp\{6.79(1 - T_M/T)\}$$

where C^S is solubility, P^S is vapor pressure, subcripts S and L referring to solid and liquid phases, T_M is melting point and T is the system temperature, both in absolute (K) units. The fugacity ratio is given in the data tables at 25 °C, the usual temperature at which physical-chemical property data are reported. For liquids, the fugacity ratio is 1.0.

1.3.2 Examples

Recently, there have been efforts to extend the long established concept of Quantitative Structure-Activity Relationships (QSARs) to Quantitative Structure-Property Relationships (QSPRs) to compute all relevant environmental physical-chemical properties (e.g. aqueous solubility, vapor pressure, octanol-water partition coefficient, Henry's law constant, bioconcentration factor (BCF) and sorption coefficient from molecular structure. Examples are Burkhard (1984) and Burkhard et al. (1985a) who calculated solublity, vapor pressure, Henry's law constant, K_{OW} and K_{OC} for all PCB congeners. Hawker and Connell (1988) also calculated log K_{OW}; Abramowitz and Yalkowsky (1990) calculated melting point and solubility for all PCB congeners based on the correlation with total surface area (planar TSAs). Doucette and Andren (1988b) used six molecular descriptors to compute the K_{OW} of some chlorobenzenes, PCBs and PCDDs. Mailhot and Peters (1988) employed seven molecular descriptors to compute physical-chemical properties of some 300 compounds. Isnard and Lambert (1988, 1989) correlated solubility, K_{OW} and BCF for a large number of organic chemicals. Nirmalakhandan and Speece (1988ab, 1989) used molecular connectivity indices to predict aqueous solubility and Henry's law constants for 300 compounds over 12 logarithmic units in solubility. Kamlet and coworkers (1987, 1988) have developed the solvatochromic parameters with the intrinsic molar volume to predict solubility, log K_{OW} and toxicity of organic chemicals. Warne et al. (1990) correlated solubility and K_{OW} for lipophilic organic compound with 39 molecular descriptors and physical-chemical properties. Other correlations are reviewed by Lyman et al. (1982). As Dearden (1990) has pointed out, "new parameters are continually being devised and tested, although the necessity of that may be questioned, given the vast number already available". It must be emphasized, however, that regardless of how accurate these predicted or estimated properties are claimed to be, utimately they have to be confirmed or verified by experimental measurement.

A fundamental problem encountered in these correlations is that the molecular descriptors can be calculated with relatively high precision, usually within a few percent. The accuracy may not always be high, but for empirical correlation purposes precision is more important than accuracy. The precision and accuracy of the experimental data are often poor, frequently ranging over a factor of two or more. Certain isomers may yield identical descriptors, but have different properties. There is thus an inherent limit to the applicability of QSPRs imposed by the quality of the experimental data, and further efforts to improve descriptors, while interesting and potentially useful, are unlikely to yield demonstrably improved QSPRs.

For correlation of **solubility** the correct thermodynamic quantities for correlation are the activity coefficient γ, or the excess Gibbs free energy ΔG as discussed by Pierotti et al. (1959) and Tsonopoulos and Prausnitz (1971). Examples of such correlations are given below.

1. Carbon number or carbon plus chlorine number (Tsonopoulos and Prausnitz 1971, Mackay and Shiu 1977);
2. Molar volume cm^3/mol
   a. Liquid molar volume - from density (McAuliffe 1966, Lande and Banerjee 1981, Chiou et al. 1982, Abernethy et al. 1988);
   b. Molar volume by additive group contribution method, e.g., LeBas method, Schroeder method (Reid et al. 1987, Miller et al. 1985);
   c. Intrinsic molar volume, V_I, cm^3/mol - from van der Waals radius with solvatochromic parameters α and β (Leahy 1986, Kamlet et al. 1987, 1988);
   d. Characteristic molecular volume, m^3/mol (McGowan and Mellors 1986);
3. Molecular volume - $\text{Å}^3/mol$ (cubic Angstrom per mole)
   a. van der Waals volume (Bondi 1964);
   b. Total Molecular Volume (TMV) (Pearlman et al. 1984, Pearlman 1986);
4. Total Surface Area (TSA) - $\text{Å}^2/mol$ (Hermann 1971, Yalkowsky and Valvani 1976, Yalkowsky et al. 1979, Pearlman 1986, Andren et al. 1987, Hawker and Connell 1988);
5. Molecular Connectivity indices, χ (Kier and Hall 1976, Andren et al. 1987, Nirmalakhandan and Speece 1988b, 1989);
7. Boiling point (Almgren et al. 1979);
8. Melting point (Amidon and Williams 1982);
9. Melting point and TSA (Abramowitz and Yalkowsky 1990).

Several workers have explored the linear relationship between octanol-water partition coefficient and solubility as means of estimating solubility.

Hansch et al. (1968) established the linear free-energy relationship between aqueous and octanol-water partition of organic liquid. Others, such as Tulp and Hutzinger (1978), Yalkowsky et al. (1979), Mackay et al. (1980), Banerjee et al. (1980), Chiou et al. (1982), Bowman and Sans (1983), Miller et al. (1985), Andren et al. (1987) and Andren and Doucette (1988b) have all presented similar but modified relationships.

The UNIFAC (UNIQUAC Functional Group Activity Coefficient) group contribution (Fredenslund et al. 1975) is widely used for predicting the activity coefficient in nonelectrolytes liquid mixtures by using group-interaction parameters. This method has been used by Kabadi and Danner (1979), Banerjee (1985), Arbuckle (1983, 1986), Banerjee and Howard (1988) and Al-Sahhaf (1989).

HPLC retention time data have been used as a psuedo-molecular descriptor by Whitehouse and Cooke (1982), Hafkenscheid and Tomlinson (1981), Tomlinson and Hafkenscheid (1986) and Swann et al. (1983).

The **octanol-water partition coefficient** K_{OW} is widely used as a descriptor of hydrophobicity. Variation in K_{OW} is primarily attributable to variation in activity coefficient in the aqueous phase (Miller et al. 1985), thus, the same correlations used for solubility in water are applicable to K_{OW}. Most widely used is the Hansch-Leo compilation of data (Leo et al. 1971, Hansch and Leo 1979) and related predictive methods. Examples of K_{OW} correlations are

I. Molecular descriptors
  1. Molar volumes: LeBas method; from density; intrinsic molar volume; characteristic molecular volume (Abernethy et al. 1988, Chiou 1985, Kamlet et al. 1988, McGowan and Mellors 1986);
  2. TMV (De Bruijn and Hermens 1990);
  3. TSA (Yalkowsky et al. 1979, Yalkowsky et al. 1983, Hawker and Connell 1988);
  4. Molecular connectivity indices (Doucette and Andren 1988b);
  5. Molecular weight (Doucette and Andren 1988b).

II. Group contribution methods
  1. π-constant or hydrophobic substituent method (Hansch et al. 1968, Hansch & Leo 1979, Doucette and Andren 1988b);
  2. Fragmental constants or f-constant (Rekker 1977, Yalkowsky et al. 1983);
  3. Hansch & Leo's f-constant (Hansch & Leo 1979; Doucette and Andren 1988b)

III. From solubility - K_{OW} relationship

IV. HPLC retention data
  1. HPLC-k' capacity factor (Könemann et al. 1979, McDuffie 1981);
  2. HPLC-RT retention time (Veith et al. 1979, Rappaport and Eisenreich 1984, Doucette and Andren 1988b);
  3. HPLC-RV retention volume (Garst 1984);
  4. HPLC-RT/MS HPLC retention time with mass spectrometry (Burkhard et al. 1985c);

V. Reversed-phase thin-layer chromatography (TLV) (Bruggeman et al. 1982);

VI. Molar refractivity (Yoshida et al. 1983).

As with solubility and octanol-water partition coefficient, **vapor pressure** can be estimated with a variety of correlations as discussed in detail by Burkhard et al. (1985a) and summarized as follows:

1. Interpolation or extrapolation from equation for correlating temperature relationships, e.g., the Clausius-Clapeyron, Antoine equations (Burkhard et al. 1985a);
2. Carbon or chlorine numbers (Mackay et al. 1980, Shiu and Mackay 1986);
3. LeBas molar volume (Shiu et al. 1987, 1988);
4. Boiling point and heat of vaporization ΔH_v (Mackay et al. 1982);
5. Group contribution method (Macknick and Prausnitz 1979);

6. UNIFAC group contribution method (Burkhard et al. 1985a, Banerjee et al.1990);
7. Molecular weight and Gibbs' free energy of vaporization ΔG_v (Burkhard et al. 1985a);
8. TSA and ΔG_v (Amidon and Anik 1981, Burkhard et al. 1985a, Hawker 1989);
9. Molecular connectivity indices (Kier and Hall 1976, 1986, Burkhard et al. 1985a);
10. Melting point and GC retention index (Bidleman 1984, Burkhard et al. 1985a);
11. Solvatochromic parameters and intrinsic molar volume (Banerjee et al. 1990).

As described earlier, **Henry's law constants** can be calculated from the ratio of vapor pressure and aqueous solubility. Henry's law constants do not show a simple linear pattern as solublity, K_{OW} or vapor pressure when plotted against simple molecular descriptors, such as numbers of chlorine or LeBas molar volume, e.g., PCBs (Burkhard et al. 1985b); pesticides (Suntio et al. 1988) and chlorinated dioxins (Shiu et al. 1988). Henry's law constants can be estimated from:

1. UNIFAC derived Activity coefficients (Arbuckle 1983);
2. Molecular connectivity indices (Nirmalakhandan and Speece 1988b, Salbjic and Güsten 1989);
3. Total surface area - planar TSA (Hawker 1989).

Bioconcentration factors:

1. Correlation with K_{OW} (Neely et al. 1974, Könemann and van Leeuwen 1980, Veith et al. 1980, Chiou et al. 1977, Mackay 1982, Briggs 1981, Garten and Trabalka 1983, Davies and Dobbs 1984, Oliver and Niimi 1988, Isnard and Lambert 1988);
2. Correlation with solubility (Kenaga 1980, Kenaga and Goring 1980, Briggs 1981, Garten and Trabalka 1983, Davies and Dobbs 1984, Isnard and Lambert 1988)
3. Correlation with K_{OC} (Kenaga 1980, Kenaga and Goring 1980, Briggs 1981);
4. Calculation with HPLC retention data (Swann et al. 1983);
5. Calculation with solvatochromic parameters (Hawker 1989a, 1990).

Sorption coefficients:

1. Correlation with K_{OW} (Karickhoff et al. 1979, Schwarzenbach and Westall 1981, Mackay 1982, Oliver 1984);
2. Correlation with solubility (Karickhoff et al. 1979);
3. Molecular connectivity indices (Sabljic 1984, 1987, Sabljic et al. 1989);
4. From HPLC retention data (Swann et al. 1983, Szabo et al. 1990).

1.4 FATE MODELS

1.4.1 Evaluative Environmental Calculations

The nature of these calculations has been described in a series of papers, notably Mackay (1979), Paterson and Mackay (1985), Mackay and Paterson (1990, 1991), and a recent text (Mackay 1991). Only the salient features are presented here. Three calculations are completed for each chemical, namely the Level I, II and III fugacity calculations.

1.4.2 Level I Fugacity Calculation

The Level I calculation describes how a given amount of chemical partitions at equilibrium between six media: air, water, soil, bottom sediment, suspended sediment and fish. No account is taken of reactivity. Whereas most early evaluative environments have treated a one square kilometer region with about 70% water surface (simulating the global proportion of ocean surface), it has become apparent that a more useful approach is to treat a larger, principally terrestrial area similar to a jurisdictional region such as a U.S. state. The area selected is 100,000 km^2 or 10^{11} m^2 which is about the area of Ohio, Greece or England.

The atmospheric height is selected as a fairly arbitrary 1000 m reflecting that region of the troposphere which is most affected by local air emissions. A water surface area of 10% or 10,000 km^2 is used, with a water depth of 20 m. The water volume is thus $2x10^{11}$ m^3. The soil is viewed as being well mixed to a depth of 10 cm and is considered to be 2% organic carbon. It has a volume of $9x10^9$ m^3. The bottom sediment has the same area as the water, a depth of 1 cm and an organic carbon content of 4%. It thus has a volume of 10^8 m^3.

For the Level I calculation both the soil and sediment are treated as simple solid phases with the above volumes, i.e., the presence of air or water in the pores of these phases is ignored.

Two other phases are included for interest. Suspended matter in water is often an important medium when compared in sorbing capacity to that of water. It is treated as having 20% organic carbon and being present at a volume fraction in the water of $5x10^{-6}$, i.e., it is about 5 mg/L. Fishes are also included at an entirely arbitrary volume fraction of 10^{-6} and are assumed to contain 5% lipid, equivalent in sorbing capacity to octanol. These two phases are small in volume and rarely contain an appreciable fraction of the chemical present, but it is in these phases that the highest concentration of chemical often exists.

Another phase which is introduced later in the Level III model is aerosol particles with a volume fraction in air of $2x10^{-11}$, i.e., approximately 30 $\mu g/m^3$. Although negligible in volume, an appreciable fraction of the chemical present in the air phase may be associated with aerosols. Aerosols are not treated in Level I or II calculations because their capacity for chemical is usually negligible when compared with soil.

These dimensions and properties are summarized in Table 1.2. The user is encouraged to modify these dimensions to reflect conditions in a specific area of interest.

Table 1.2a Compartment Dimensions and Properties for Level I and II Calculations

Compartment	Air	Water	Soil	Sediment	Suspended Sediment	Fish
Volume, V (m^3)	10^{14}	$2x10^{11}$	$9x10^9$	10^8	10^6	$2x10^6$
Depth, h (m)	1000	20	0.1	0.01	-	-
Area, A (m^2)	$100x10^9$	$10x10^9$	$90x10^9$	$10x10^9$	-	-
Org. Fraction (ϕ_{OC})	-	-	0.02	0.04	0.2	-
Density, ρ (kg/m^3)	1.2	1000	2400	2400	1500	1000
Adv. Residence Time, t (hours)	100	1000	-	50,000	-	-
Adv. flow, G (m^3/h)	10^{12}	$2x10^8$	-	2000	-	-

Table 1.2b Bulk Compartment Dimensions and Volume Fraction (v) for Level III Calculations

Air	Total volume	10^{14} m^3 (as above)
	Air phase	10^{14} m^3
	Aerosol phase	2000 m^3 (v = $2x10^{-11}$)
Water	Total volume	$2x10^{11}$ m^3
	Water phase	$2x10^{11}$ m^3 (as above)
	Suspended sediment phase	$10x10^6$ m^3 (v = $5x10^{-6}$)
	Fish phase	$2x10^5$ m^3 (v = $1x10^{-6}$)
Soil	Total volume	$18x10^9$ m^3
	Air phase	$3.6x10^9$ m^3 (v = 0.2)
	Water phase	$5.4x10^9$ m^3 (v = 0.3)
	Solid phase	$9.0x10^9$ m^3 (v = 0.5) (as above)
Sediment	Total volume	$500x10^6$ m^3
	Water phase	$400x10^6$ m^3 (v = 0.8)
	Solid phase	$100x10^6$ m^3 (v = 0.2) (as above)

The amount of chemical introduced in the Level I calculation is an arbitrary 100,000 kg or 100 tonnes. If dispersed entirely in the air, this amount yields a concentration of 1 $\mu g/m^3$ which is not unusual for ubiquitous contaminants such as hydrocarbons. If dispersed entirely in the water, the concentration is a higher 500 $\mu g/m^3$ or 500 ng/L, which again is reasonable for a well-used chemical of commerce. The corresponding value in soil is about 0.0046 $\mu g/g$. It is believed that this amount is a reasonable common value for evaluative purposes. Clearly for restricted chemicals such as PCBs, this amount is too large, but it is preferable to adopt a common evaluative amount for all substances. No significance should, of course, be attached to the absolute values of the concentrations which are deduced from this arbitrary amount. Only the relative values have significance.

The Level I calculation proceeds by deducing the fugacity capacities, Z values for each medium (see Table 1.3), following the procedures described by Mackay (1991). These working equations show the necessity of having data on molecular mass, water solubility, vapor pressure, and octanol-water partition coefficient. The fugacity f (Pa) common to all media is deduced as

$$f = M/\Sigma V_i Z_i$$

where M is the total amount of chemical (mol), V_i is the medium volume (m^3) and Z_i is the corresponding fugacity capacity for the chemical in each medium.

The molar concentration C (mol/m^3) can then be deduced as Zf mol/m^3 or as WZf g/m^3 or 1000 WZf/ρ $\mu g/g$ where ρ is the phase density (kg/m^3) and W is the molecular mass (g/mol). The amount m_i in each medium is C_iV_i mol, and the total in all media is M mol. The **BASIC** computer program for undertaking this calculation is appended. For those who prefer a spreadsheet format, an identical Lotus 123® program is also provided.

The information obtained from this calculation includes the concentrations, amounts and distribution. In the figures, a pie chart illustrates the distribution between the four primary compartments of air, water, soil and sediment, the amount in fish and suspended sediment being ignored. This information is useful as an indication of the relative concentrations.

Note that this simple treatment assumes that the soil and sediment phases are entirely solid, i.e., there are no air or water phases present to "dilute" the solids. Later in the Level III calculation these phases and aerosols are included.

Table 1.3a Equations for Phase Z values used in Levels I and II and the Bulk Phase values used in Level III

Air	$Z_1 = 1/RT$
Water	$Z_2 = 1/H = C^S/P^S$
Soil	$Z_3 = Z_2 \cdot \rho_3 \cdot \phi_3 \cdot K_{OC}/1000$
Sediment	$Z_4 = Z_2 \cdot \rho_4 \cdot \phi_4 \cdot K_{OC}/1000$
Suspended Sediment	$Z_5 = Z_2 \cdot \rho_5 \cdot \phi_5 \cdot K_{OC}/1000$
Fish	$Z_6 = Z_2 \cdot \rho_6 \cdot L \cdot K_{OW}/1000$
Aerosol	$Z_7 = Z_1 \cdot 6x10^6/P^S_L$
where	R = gas constant (8.314 J/mol K)
	T = absolute temperature (K)
	C^S = solubility in water (mol/m^3)
	P^S = vapor pressure (Pa)
	H = Henry's law constant (Pa • m^3/mol)
	P^S_L = liquid vapor pressure (Pa)
	K_{OW} = octanol-water partition coefficient
	K_{OC} = organic-carbon partition coefficient (= 0.41 K_{OW})
	ρ_i = density of phase i (kg/m^3)
	ϕ_i = mass fraction organic-carbon in phase i (g/g)
	L = lipid content of fish

Note for solids $P^S_L = P^S_S/\exp\{6.79(1 - T_M/T)\}$ where T_M is melting point (K) of the solute.

Table 1.3b Bulk Phase Z values, Z_{Bi} deduced as $\Sigma\, v_iZ_i$, in which the coefficients, e.g., $2x10^{-11}$, are the volume fractions v_i of each pure phase as specified in Table 1.2b

Air	$Z_{B1} = Z_1 + 2x10^{-11}\, Z_7$	(approximately 30 μg/m^3 aerosols)
Water	$Z_{B2} = Z_2 + 5x10^{-6}\, Z_5 + 1x10^{-6}\, Z_6$	(5 ppm solids, 1 ppm fish by volume)
Soil	$Z_{B3} = 0.2\, Z_1 + 0.3\, Z_2 + 0.5\, Z_3$	(20% air, 30% water, 50% solids)
Sediment	$Z_{B4} = 0.8\, Z_2 + 0.2\, Z_4$	(80% water, 20% solids)

1.4.3 Level II Fugacity Calculation

The Level II calculation simulates a situation in which chemical is continuously discharged into the multimedia environment and achieves a steady-state equilibrium condition at which input and output rates are equal. The task is to deduce the rates of loss by reaction and advection.

The reaction rate data developed for each chemical in the tables are used to select a reactivity class as described earlier, and hence a first-order rate constant for each medium. Often these rates are in considerable doubt, thus the quantities selected should be used with extreme caution because they may not be widely applicable. The rate constants k_i h^{-1} are used to calculate reaction D values for each medium D_{Ri} as $V_iZ_ik_i$. The rate of reactive loss is then $D_{Ri}f$ mol/h.

For advection, it is necessary to select flow rates. This is conveniently done in the form of advective residence times, t h, thus the advection rate G_i is V_i/t m^3/h for each medium. For air, a residence time of 100 hours is used (approximately 4 days), which is probably too long for the geographic area considered, but shorter residence times tend to cause air advective loss to be a dominant mechanism. For water, a figure of 1000 hours (42 days) is used reflecting a mixture of rivers and lakes. For sediment burial (which is treated as an advective loss), a time of 50000 hours or 5.7 years is used. Only for very persistent, hydrophobic chemicals is this process important. No advective loss from soil is included. The D value for loss by advection D_{Ai} is G_iZ_i and the rates are $D_{Ai}f$ mol/h. These rates are listed in Table 1.2.

There may thus be losses caused by both reaction and advection D values for the four primary media. These loss processes are not included for fish or suspended matter. At steady-state, equilibrium conditions the input rate E mol/h can be equated to the sum of the output rates, from which the common fugacity can be calculated as follows

$$E = f \cdot \Sigma D_{Ai} + f \cdot \Sigma D_{Ri}$$

thus,

$$f = E/(\Sigma D_{Ai} + \Sigma D_{Ri})$$

The common assumed emission rate is 1000 kg/h or 1 tonne/h. To achieve an amount equivalent to the 100 tonnes in the Level I calculation requires an overall residence time of 100 hours. Again the concentrations and amounts m_i and Σ m_i or M can be deduced, as well as the reaction and advection rates. These rates obviously total to give the input rate E. Of particular interest are the relative rates of these loss processes, and the overall persistence or residence time which is calculated as

$$t_O = M/E$$

where M is the total amount present. It is also useful to calculate a reaction and an advection persistence t_R and t_A as

$$t_R = M/\Sigma D_{Ri}f \qquad t_A = M/\Sigma D_{Ai}f$$

Obviously $$1/t_O = 1/t_R + 1/t_A$$

These persistences indicate the likelihood of the chemical being lost by reaction as distinct from advection. The percentage distribution of chemical between phases is identical to that in Level I. A pie chart depicting the distribution of losses is presented.

1.4.4 Level III Fugacity Calculation

Whereas the Level I and II calculations assume equilibrium to prevail between all media, this is recognized as being excessively simplistic and even misleading. In the interests of algebraic simplicity only the four primary media are treated for this level. The task is to develop expressions for intermedia transport rates by the various diffusive and nondiffusive processes as described by Mackay (1991). This is done by selecting values for 12 intermedia transport velocity parameters which have dimensions of velocity (m/h or m/year), are designated as U_i m/h and are applied to all chemicals. These parameters are used to calculate seven intermedia transport D values.

It is desirable to calculate new "bulk phase" Z values for the four primary media which include the contribution of dispersed phases within each medium as described by Mackay and Paterson (1991) and as listed in Tables 1.2 and 1.3. The air is now treated as an air-aerosol mixture, water as water plus suspended particles and fish, soil as solids, air and water, and sediment as solids and porewater. The Z values thus differ from the Level I and Level II "pure phase" values. The necessity for introducing this complication arises from the fact that much of the intermedia transport of the chemicals occurs in association with the movement of chemical in these dispersed phases. To accommodate this change the same volumes of the soil solids and sediment solids are retained, but the total phase volumes are increasd. These Level III volumes are also given in Table 1.2. The reaction and advection D values employ the generally smaller bulk phase Z values but the same resisdence times, thus the G values are increased and the D values are generally larger.

Intermedia D values

The justification for each intermedia D value follows. It is noteworthy that, for example, air-to-water and water-to-air values differ because of the presence of one-way nondiffusive processes. A fuller description of the background to these calculations is given by Mackay (1991).

1. Air to water (D_{12})

Four processes are considered: diffusion (absorption), dissolution in rain of gaseous chemical, and wet and dry deposition of particle-associated chemical.

For diffusion, the conventional two-film approach is taken with water-side (k_W) and air-side (k_A) mass transfer coefficients (m/h) being defined. Values of 0.05 for k_W and 5 m/h for k_A are used. The absorption D value is then

$$D_{VW} = 1/(1/(k_A A_W Z_1) + 1/(k_W A_W Z_2))$$

where A_W is the air-water area (m^2) and Z_1 and Z_2 are the pure air and water Z values. The velocities k_A and k_W are designated as U_1 and U_2.

For rain dissolution, a rainfall rate of 0.876 m/year is used, i.e., U_R or U_3 is 10^{-4} m/h. The D value for dissolution D_{RW} is then

$$D_{RW} = U_R A_W Z_2 = U_3 A_W Z_2$$

For wet deposition, it is assumed that the rain scavenges Q (scavenging ratio) or about 200,000 times its volume of air. Using a particle concentration (volume fraction) v_Q of 2×10^{-11}, this corresponds to the removal of Qv_Q or 4×10^{-6} volumes of aerosol per volume of rain. The total rate of particle removal by wet deposition is then $Qv_Q U_R A_W$ m^3/h, thus the wet "transport velocity" $Qv_Q U_R$ is 4×10^{-10} m/h.

For dry deposition, a typical deposition velocity U_Q of 10 m/h is selected yielding a rate of particle removal of $U_Q v_Q A_W$ or $2 \times 10^{-10} A_W$ m^3/h corresponding to a transport velocity of 2×10^{-10} m/h. Thus,

$$U_4 = Qv_Q U_R + U_Q v_Q = v_Q(QU_R + U_Q)$$

The total particle transport velocity U_4 for wet and dry deposition is thus 6×10^{-10} m/h and the total D value D_{QW} is

$$D_{QW} = U_4 A_W Z_7$$

where Z_7 is the aerosol Z value.

The overall D value is given by

$$D_{12} = D_{VW} + D_{RW} + D_{QW}$$

2. Water to air (D_{21})

Evaporation is treated as the reverse of absorption thus D_{21} is simply D_{VW} as before.

3. Air to soil (D_{13})

A similar approach is adopted as for air to water transfer. Four processes are considered with rain dissolution (D_{RS}) and wet and dry deposition (D_{QS}) being treated identically except that the area term is now the air-soil area A_S.

For diffusion, the approach of Jury et al. (1983, 1984) is used as described by Mackay and Stiver (1991) and Mackay (1991) in which three diffusive processes are treated. The air boundary layer is characterized by a mass transfer coefficient k_S or U_7 of 5 m/h, equal to that of the air-water MTC coefficient k_A used in D_{12}.

For diffusion in the soil air-pores, a molecular diffusivity of 0.02 m^2/h is reduced to an effective diffusivity using a Millington-Quirk type of relationship by a factor of about 20 to 10^{-3} m^2/h. Combining this with a path length of 0.05 m gives an effective air to soil mass transfer coefficient k_{SA} of 0.02 m/h which is designated as U_5.

Similarly, for diffusion in water a molecular diffusivity of 2×10^{-6} m^2/h is reduced by a factor of 20 to an effective diffusivity of 10^{-7} m^2/h, which is combined with a path length of 0.05 m to give an effective soil to water mass transfer coefficient of k_{SW} 2×10^{-6} m/h.

It is probable that capillary flow of water contributes to transport in the soil. For example, a rate of 7 cm/year would yield an equivalent water velocity of 8×10^{-6} m/h which exceeds the water diffusion rate by a factor of four. For illustrative purposes we thus select a water transport velocity or coefficient U_6 in the soil of 10×10^{-6} m/h, recognizing that this may be in error by a substantial amount, and will vary with rainfall characteristics and soil type.

The soil processes are in parallel with boundary layer diffusion in series, so the final equation is

$$D_{VS} = 1/[1/D_S + 1/(D_{SW} + D_{SA})]$$

where

$$D_S = U_7 A_S Z_1 \qquad (U_7 = 5 \text{ m/h})$$
$$D_{SW} = U_6 A_S Z_2 \qquad (U_6 = 10\times10^{-6} \text{ m/h})$$
$$D_{SA} = U_5 A_S Z_1 \qquad (U_5 = 0.02 \text{ m/h})$$

where A_S is the soil horizontal area.

Air-soil diffusion thus appears to be much slower than air-water diffusion because of the slow migration in the soil matrix. In practice, the result will be a nonuniform composition in the soil with the surface soil (which is much more accessible to the air than the deeper soil) being closer in fugacity to the atmosphere.

The overall D value is given as

$$D_{13} = D_{VS} + D_{QS} + D_{RS}$$

4. Soil to air (D_{31})

Evaporation is treated as the reverse of absorption, thus the D value is simply D_{VS}.

5. Water to sediment (D_{24})

Two processes are treated, diffusion and deposition.

Diffusion is characterized by a mass transfer coefficient U_8 of 10^{-4} m/h which can be regarded as a molecular diffusivity of 2×10^{-6} m^2/h divided by a path length of 0.02 m. In practice, bioturbation may contribute substantially to this exchange process, and in shallow water current-induced turbulence may also increase the rate of transport. Diffusion in association with organic colloids is not included.

The D value is thus given as $U_8A_WZ_2$.

Deposition is assumed to occur at a rate of 5000 m^3/h which corresponds to addition of a depth of solids of 0.438 cm/year; thus 43.8% of the solids resident in the accessible bottom sediment is added each year. This rate is about 12 $cm^3/m^2 \cdot$ day which is high compared to values observed in large lakes. The velocity U_9, corresponding to the addition of 5000 m^3/h over the area of 10^{10} m^2, is thus 5×10^{-7} m/h.

It is assumed that of this 5000 m^3/h deposited, 2000 m^3/h or 40% is buried (yielding the advective flow rate in Table 1.2), 2000 m^3/h or 40% is resuspended (as discussed later) and the remaining 20% is mineralized organic matter. The organic carbon balance is thus only approximate.

The transport velocities are thus:

deposition U_9	5.0×10^{-7} m/h or 0.438 cm/years
resuspension U_{10}	2.0×10^{-7} m/h or 0.175 cm/year
burial U_B	2.0×10^{-7} m/h or 0.175 cm/year (included as an advective residence time of 50,000 h)

The water to sediment D value is thus

$$D_{24} = U_8A_WZ_2 + U_9A_WZ_5$$

where Z_5 is the Z value of the particles in the water column.

6. Sediment to water (D_{42})

This is treated similarly to D_{24} giving:

$$D_{42} = U_8A_WZ_2 + U_{10}A_WZ_4$$

where U_{10} is the sediment resuspension velocity of 2.0×10^{7} m/h and Z_4 is the Z value of the sediment solids.

7. Sediment advection (D_{A4})

This D value is $U_BA_WZ_4$ where U_B, the sediment burial rate, is 2.0×10^{-7} m/h. It can be viewed as G_BZ_{B4} where G_B is the total burial rate specified as V_S/t_B where t_B (residence time) is 50,000 h, and V_S (the sediment volume) is the product of sediment depth (0.01 cm) and area A_W.

Z_4, Z_{B4} are the Z values of the sediment solids and of the bulk sediment respectively. Since there are 20% solids, Z_{B4} is about 0.2 Z_4. There is a slight difference between these approaches because in the advection approach (which is used here) there is burial of water as well as solids.

8. Soil to water (D_{32})

It is assumed that there is run-off of water at a rate of 50% of the rain rate, i.e., the D value is

$$D = 0.5\ U_3 A_S Z_2 = U_{11} A_S Z_2$$

thus the transport velocity term U_{11} is $0.5U_3$ or $5x10^{-5}$ m/h.

For solids run-off it is assumed that this run-off water contains 200 parts per million by volume of solids; thus the corresponding velocity term U_{12} is $200x10^{-6}U_{11}$, i.e., 10^{-8} m/h. This corresponds to the loss of soil at a rate of about 0.1 mm per year. If these solids were completely deposited in the aquatic environment (which is about 1/10th the soil area), they would accumulate at about 0.1 cm per year, which is about a factor of four less than the deposition rate to sediments. The implication is that most of this deposition is of naturally generated organic carbon and from sources such as bank erosion.

Summary

The 12 intermedia transport parameters are listed in Table 1.4 and the equations are summarized in Table 1.5.

Table 1.4 Intermedia Transport Parameters

U		m/h	m/year
1	Air side, air-water MTC*, k_A	5	43800
2	Water side, air-water MTC, k_W	0.05	438
3	Rain rate, U_R	10^{-4}	0.876
4	Aerosol deposition	$6x10^{-10}$	$5.256x10^{-6}$
5	Soil-air phase diffusion MTC, k_{SA}	0.02	175.2
6	Soil-water phase diffusion MTC	$10x10^{-6}$	0.0876
7	Soil-air boundary layer MTC, k_S	5	43800
8	Sediment-water MTC	10^{-4}	0.876
9	Sediment deposition	$5.0x10^{-7}$	0.00438
10	Sediment resuspension	$2.0x10^{-7}$	0.00175
11	Soil-water run-off	$5x10^{-5}$	0.438
12	Soil-solids run-off	10^{-8}	0.0000876

* Mass transfer coefficient
with,
Scavenging ratio $Q = 2x10^5$
Dry deposition velocity $U_Q = 10$ m/h
Sediment burial rate $U_B = 2.0x10^{-7}$ m/h

Table 1.5 Intermedia Transport D Value Equations

Air-Water

$$D_{12} = D_{VW} + D_{RW} + D_{QW}$$
$$D_{VW} = A_W/(1/U_1Z_1 + 1/U_2Z_2)$$
$$D_{RW} = U_3A_WZ_2$$
$$D_{QW} = U_4A_WZ_7$$

Water-Air

$$D_{21} = D_{VW}$$

Air-Soil

$$D_{13} = D_{VS} + D_{RS} + D_{QS}$$
$$D_{VS} = 1/(1/D_S + 1/(D_W + D_A))$$
$$D_S = U_7A_SZ_1$$
$$D_{SA} = U_5A_SZ_1$$
$$D_{SW} = U_6A_SZ_2$$
$$D_{RS} = U_3A_SZ_2$$
$$D_{QS} = U_4A_SZ_7$$

Soil-Air

$$D_{31} = D_{VS}$$

Water-Sediment

$$D_{24} = U_8A_WZ_2 + U_9A_WZ_5$$

Sediment-Water

$$D_{42} = U_8A_WZ_2 + U_{10}A_WZ_4$$

Soil-Water

$$D_{32} = U_{11}A_SZ_2 + U_{12}A_SZ_3$$

Algebraic solution

Four mass balance equations can be written, one for each medium resulting in a total of four unknown fugacities, enabling simple algebraic solution as shown in Table 1.6. From the four fugacities, the concentration, amounts and rates of all transport and transformation processes can be deduced, yielding a complete mass balance.

The new information from the Level III calculations are the intermedia transport data, i.e., the extent to which chemical discharged into one medium tends to migrate into another. This migration pattern depends strongly on the proportions of the chemical discharged into each medium; indeed, the relative amounts in each medium are largely a reflection of the locations of discharge. It is difficult to interpret these mass balance diagrams because, for example, chemical depositing from air to water may have been discharged to air, or to soil from which it evaporated, or even to water from which it is cycling to and from air.

To simplify this interpretation, it is best to conduct three separate Level III calculations in which unit amounts (1000 kg/h) are introduced individually into air, soil and water. Direct discharges to sediment are unlikely and are not considered here. These calculations show clearly the extent to which intermedia transport occurs. If, for example, the intermedia D values are small compared to the reaction and advection values, the discharged chemical will tend to remain in the discharge or "source" medium with only a small proportion migrating to other media. Conversely, if the intermedia D values are relatively large the chemical becomes very susceptible to intermedia transport. This behavior is observed for persistent substances such as PCBs which have very low rates of reaction.

A direct assessment of multimedia behavior is thus possible by examining the proportions of chemical found at steady-state in the "source" medium and in other media. For example, when discharged to water, an appreciable fraction of the benzene is found in air, whereas for atrazine, only a negligible fraction of atrazine reaches air.

Table 1.6 Level III Solutions to Mass Balance Equations

Mass balance equations:

Air $E_1 + f_2D_{21} + f_3D_{31} = f_1D_{T1}$

Water $E_2 + f_1D_{12} + f_3D_{32} + f_4D_{42} = f_2 D_{T2}$

Soil $E_3 + f_1D_{13} = f_3D_{T3}$

Sediment $E_4 + f_2D_{24} = f_4D_{T4}$

where E_i is discharge rate, E_4 usually being zero.

$$D_{T1} = D_{R1} + D_{A1} + D_{12} + D_{13}$$

$$D_{T2} = D_{R2} + D_{A2} + D_{21} + D_{23} + D_{24}, \quad (D_{23} = 0)$$

$$D_{T3} = D_{R3} + D_{A3} + D_{31} + D_{32}, \quad (D_{A3} = 0)$$

$$D_{T4} = D_{R4} + D_{A4} + D_{42}$$

Solution:

$$f_2 = [E_2 + J_1J_4/J_3 + E_3D_{32}/D_{T3} + E_4D_{42}/D_{T4}]/(D_{T2} - J_2J_4/J_3 - D_{24} \cdot D_{42}/D_{T4})$$

$$f_1 = (J_1 + f_2J_2)/J_3$$

$$f_3 = (E_3 + f_1D_{13})/D_{T3}$$

$$f_4 = (E_4 + f_2D_{24})/D_{T4}$$

where

$$J_1 = E_1/D_{T1} + E_3D_{31}/(D_{T3} \cdot D_{T1})$$

$$J_2 = D_{21}/D_{T1}$$

$$J_3 = 1 - D_{31} \cdot D_{13}/(D_{T1} \cdot D_{T3})$$

$$J_4 = D_{12} + D_{32} \cdot D_{13}/D_{T3}$$

Linear additivity

Because these equations are entirely linear, the solutions can be scaled linearly. The concentrations resulting from a discharge of 2000 kg/h are simply twice those of 1000 kg/h. Further, if discharge of 1000 kg/h to air causes 500 kg in water and discharge of 1000 kg/h to soil causes 100 kg in water, then if both discharges occur simultaneously, there will be 600 kg in water. If the discharge to soil is increased to 3000 kg/h, the total amount in the water will rise to (500 + 300) or 800 kg. It is thus possible to deduce the amount in any medium arising from any combination of discharge rates by scaling and adding the responses from the unit inputs. This "linear additivity principle" is more fully discussed by Stiver and Mackay (1989).

In the diagrams presented later, these three-unit (1000 kg/h) responses are given. Also, an illustrative "three discharge" mass balance is given in which a total of 1000 kg/h is discharged, but in proportions judged to be typical of chemical use and discharge to the environment. For example, benzene is believed to be mostly discharged to air with minor amounts to soil and water.

Also given in the tables are the rates of reaction, advection and intermedia transport for each case.

The reader can deduce the fate of any desired discharge pattern by appropriate scaling and addition. It is important to re-emphasize that because the values of transport velocity parameters are only illustrative, actual environmental conditions may be quite different; thus, simulation of conditions in a specific region requires determination of appropriate parameter values as well as the site specific dimensions, reaction rate constants and the physical-chemical properties which prevail at the desired temperature.

In total, the aim is to convey an impression of the likely environmental behavior of the chemical in a readily assimilable form.

1.5. DATA SOURCES AND PRESENTATION

1.5.1 Data sources

Most physical properties such as molecular weight (MW, g/mol), melting point (M.P., °C), boiling point (B.P., °C), and density have been obtained from commonly used Handbooks such as the CRC Handbook of Physics and Chemistry (Weast 1972, 1984), Lange's Handbook of Chemistry (Dean 1979, 1985) and the Merck Index (1983, 1987). Other physical-chemical properties such as aqueous solubility, vapor pressure, octanol-water partition coefficient, Henry's law constant, bioconcentration factor and sorption coefficient have been obtained from scientific journals or other environmental Handbooks, notably Verschueren's Handbook of Environmental Data on Organic Chemicals (1977, 1983) and Howard et al.'s Handbook of Environmental Fate and Exposure Data, Vol. I and II (1989, 1990). Other important sources of vapor pressure are the CRC Handbook of Physics and Chemistry (Weast 1972), Lange's Handbook of Chemistry (Dean 1985), the Handbook of Vapor Pressures and Heats of Vaporization of Hydrocarbons and Related Compounds (Zwolinski and Wilhoit 1971), the Vapor Pressure of Pure Substances (Boublik et al. 1973, 1984), the Handbook of the Thermodynamics of Organic Compounds (Stephenson and Malanowski 1987). For aqueous solubilities, valuable sources include the IUPAC Solubility Data Series (1985, 1989a,b) and Horvath's Halogenated Hydrocarbons, Solubility-Miscibility with Water (Horvath 1982). Octanol-water partition coefficients are conveniently obtained from the compilation by Leo et al. (1971) and Hansch and Leo (1979), or can be calculated from molecular structure by the methods of Hansch and Leo (1979) or Rekker (1977). Lyman et al. (1982) also outline methods of estimating solubility, K_{OW}, vapor pressure, and bioconcentration factor for organic chemicals. The recent Handbook of Environmental Degradation Rates by Howard et al. (1991) is a valuable source of rate constants and half-lives for inclusion in subsequent fugacity calculations.

The most reliable sources of data are the original citations in the reviewed scientific literature. Particularly reliable are those papers which contain a critical review of data from a number of sources as well as independent experimental determinations. Calculated or correlated values are reported in the tables but are viewed as being less reliable. A recurring problem is that a value is frequently quoted, then requoted and the original paper may not be cited. The aim in this work has been to gather and list the citations, interpret them and select a "best" or "most likely" value. To assist in this process, plots are prepared of properties as a function of molar volume as the molecular descriptors. These are discussed at the end of each chapter.

1.5.2 Data Format

Each data sheet lists the following properties, although not all quantities are included for all chemicals. In all cases citations are provided.

Common Name:
Synonym:
Chemical Name:
CAS Registry No:
Molecular Formula:

Molecular Weight (g/mol):
Melting Point (°C):
Boiling Point (°C):
Density at 20 °C (g/cm^3):
Molar Volume (cm^3/mol):
Molecular Volume (Å^3):
Total Surface Area, TSA (Å^2):
Heat of Fusion, kcal/mol:
Fugacity Ratio at 25 °C:
Water Solubility (g/m^3 or mg/L at 25°C):
Vapor Pressure (Pa at 25°C):
Henry's Law Constant (Pa m^3/mol) or Air-Water Partition Coefficient:
Octanol-Water Partition Coefficient K_{OW} or log K_{OW}:
Bioconcentration Factor K_B or BCF (or log K_B):
Sorption Partition Coefficient to Organic Carbon K_{OC} or to Organic Matter K_{OM}:
Half-lives in the Environment:

Air:
Surface water:
Goundwater:
Soil:
Sediment:
Biota:

Environmental Fate Rate Constants or Half-Lives:

Volatilization/Evaporation:
Photolysis:
Oxidation or photooxidation:
Hydrolysis:
Biotransformation/Biodegradation:
Bioconcentration: uptake and elimination rate constants:

1.5.3 Explanation of Data Presentations

Example: Benzene (data sheets presented in Chapter 2)

1. Chemical Properties.

The names, formula, melting and boiling point and density data are self explanatory.

The molar volumes are in some cases at the stated temperature and in others at the normal boiling point. Certain calculated molecular volumes are also used; thus the reader is cautioned to ensure that when using a molar volume in any correlation, it is correctly selected.

The total surface areas (TSAs) are calculated in various ways and may contain the hydration shell, thus giving a much larger area. Again the reader is cautioned to ensure that values are consistent.

Heats of fusion, ΔH_{fus}, are generally expressed in kcal/mol or kJ/mol and entropies of fusion, ΔS_{fus} in cal/mol•K (e.u. or entropy unit) or J/mol•K. In the case of liquids such as benzene it is 1.0. For solids it is a fraction representing the ratio of solid to liquid solubility or vapor pressure. It is generally assumed that for a rigid organic molecule, the entropy of fusion is 13.5 e.u. or 56 J/mol•K, which is an average value of a number of organic compounds (Yalkowsky 1979, Miller et al. 1984). The fugacity ratio, F, given is calculated using ΔS_{fus} = 56 J/mol•K in the following expression

$$F = \exp\{(\Delta H_{fus}/RT)(1 - T_M/T)\} = \exp\{(\Delta S_{fus}/R)(1 - T_M/T)\}$$

where R is the ideal gas constant (8.314 J/mol•K or 1.987 cal/mol•K) and T_M is the melting point and T is the system temperature (K).

As is apparent, a wide variety of solubilities (in units of g/m³ or the equivalent mg/L) have been reported. Experimental data have the method of determinations indicated. In other compilations of data the reported value has merely been quoted from another secondary source. In some cases the value has been calculated. The abbreviations are generally self-explanatory and usually include two entries, the method of equilibration followed by the method of determination. From these values a single value is selected for inclusion in the summary data table. In the case of benzene the selected solubility value is 1780 g/m³ at 25°C. From an examination of the data it is judged that the true value almost certainly lies between 1765 and 1795 g/m³.

The vapor pressure data are treated similarly with a value of 12700 Pa being selected. The true value is judged to lie between 12500 and 12900 Pa. Vapor pressures are, of course very temperature dependent.

The Henry's law constant data are measured in some cases, but in other cases are calculated from the ratio of vapor pressure and solubility (in units of mol/m³). In this case a value of 550 Pa m³/mol is selected, the actual value probably lying between 500 and 600 Pa m³/mol. Care must be exercised when water is appreciably soluble in the chemical because the assumption that the Henry's law constant is the ratio of vapor pressure to solubility may be invalid.

The octanol-water partition coefficient data are similarly a combination of calculated and experimental values. A value of 2.13 is selected as being the most likely value of log K_{OW}, i.e., K_{OW} is 135.

A number of (log) bioconcentration factors are listed. These generally range from 0.5 to 1.6 which corresponds to BCF of 3 to 40. This range could be interpreted as lipid contents ranging from 2.2 to 30%.

The (log) organic-carbon partition coefficients listed range from 1.0 to 2.0. It is expected that K_{OC} usually lies in the range of 20 to 80% of K_{OW}, i.e., log K_{OC} will range from 1.4 to 2.0. Organic matter partition coefficients are also reported. Since organic carbon accounts for some 50 to 60% of the content of organic matter, K_{OM} is expected to be about half K_{OC}.

The reader is advised to consult the original reference when using these values of BCF, K_{OC} and K_{OM}, to ensure that conditions are as close as possible to those of specific interest.

The "Half-life in the Environment" data reflect observations of the rate of disappearance of the chemical from a medium, without necessarily identifying the cause of mechanism of loss. For example, loss from water may be a combination of evaporation, biodegradation and photolysis. Clearly these times are highly variable and depend on factors such as temperature, meteorology and the nature of the media. Again, the reader is urged to consult the original reference.

The "Environmental Fate Rate Constants" refer to specific degradation processes rather than media. As far as possible the original numerical quantities are given and thus there is a variety of time units with some expressions being rate constants and others half-lives.

The conversion is

$$k = 0.693/t_{1/2}$$

where k is the first-order rate constant (h^{-1}) and $t_{1/2}$ is the half-life (h).

From these data a set of medium-specific degradation reaction half-lives was selected for use in Level II and III calculations. Emphasis was based on the fastest and the most plausible degradation process for each of the environmental compartment considered. Instead of assuming an equal half-life for both the water and soil compartment as suggested by Howard et al. (1991), a slower active class (in the reactivity table described earlier) was assigned for soil and sediment compared to that of the water compartment. This is in part because the major degradation processes are often photolysis (or photooxidation) and biodegradation. There is an element of judgement in this selection and it may be desirable to explore the implications of selecting other values. The selected values of the monoaromatics are given in Table 2.3 at the end of Chapter 2.

In summary, the physical-chemical and environmental fate data listed result in the selection of values of solubility, vapor pressure, K_{OW} and reaction half-lives which are used in the evaluative environmental calculations.

The physical-chemical data of monoaromatics are also plotted in the appropriate QSPR plots on Figures 1.1 to 1.6 (which are the same as Figures 2.1 to 2.6 for the monoaromatic hydrocarbons). These plots show that the benzene data are relatively "well-behaved" and are consistent with data obtained for homologous chemicals. In the case of benzene this QSPR plot

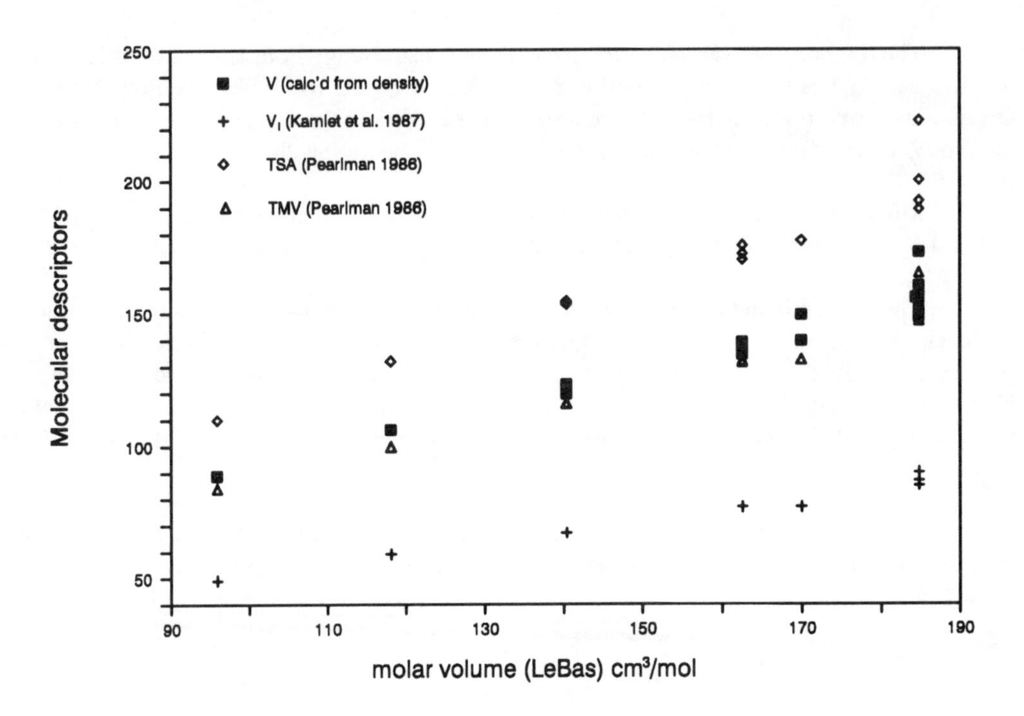

Figure 1.1 Plot of molecular descriptors versus LeBas molar volume

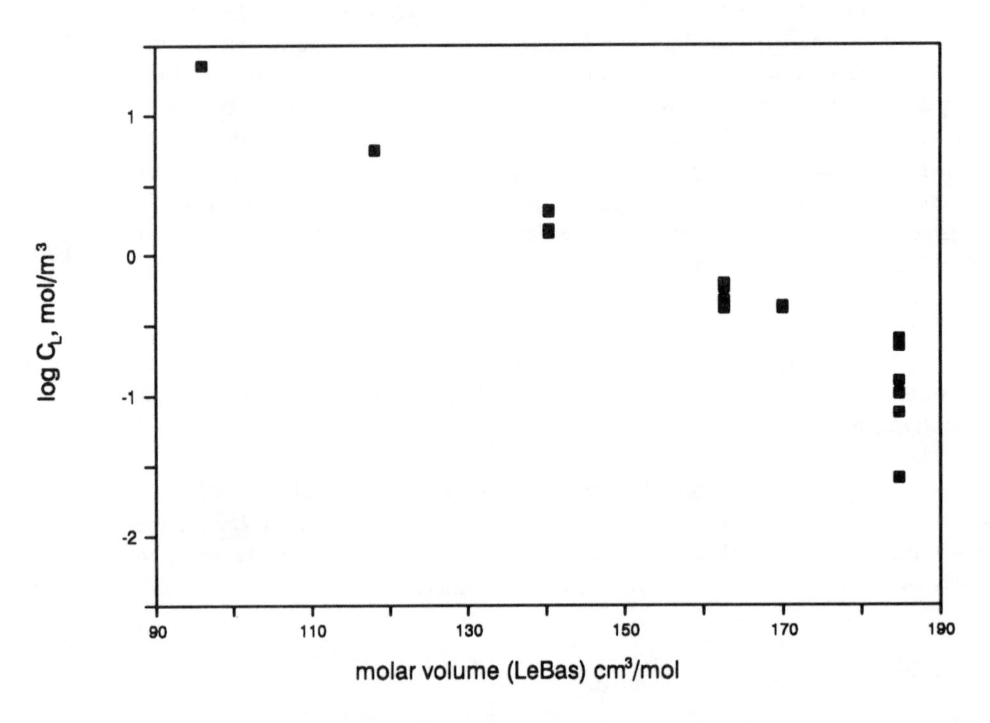

Figure 1.2 Plot of log C_L (liquid solubility) versus molar volume

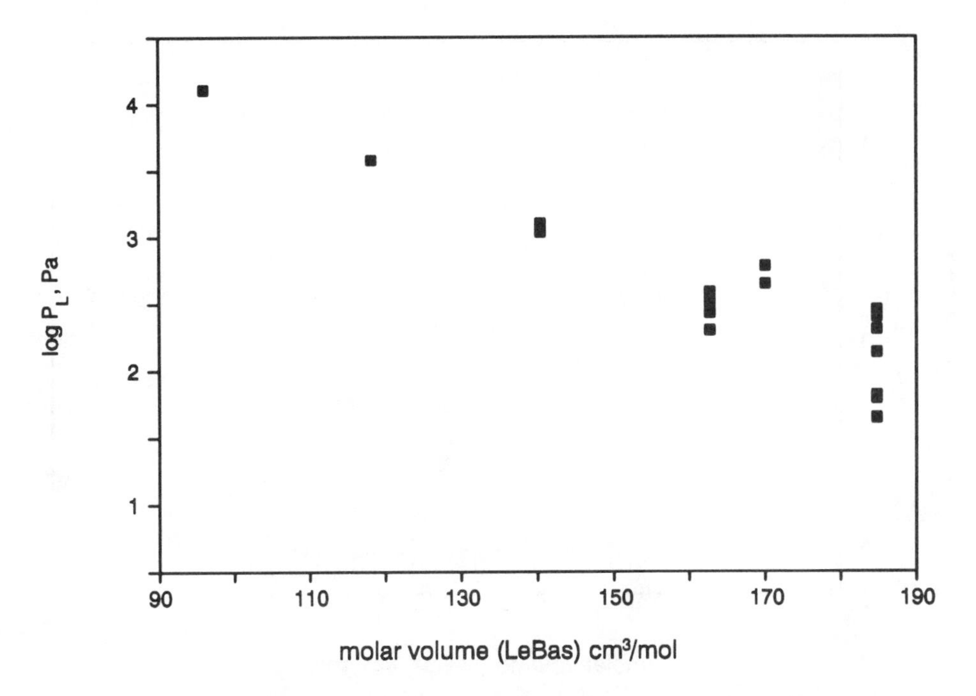

Figure 1.3 Plot of log P_L (liquid vapor pressure) versus molar volume

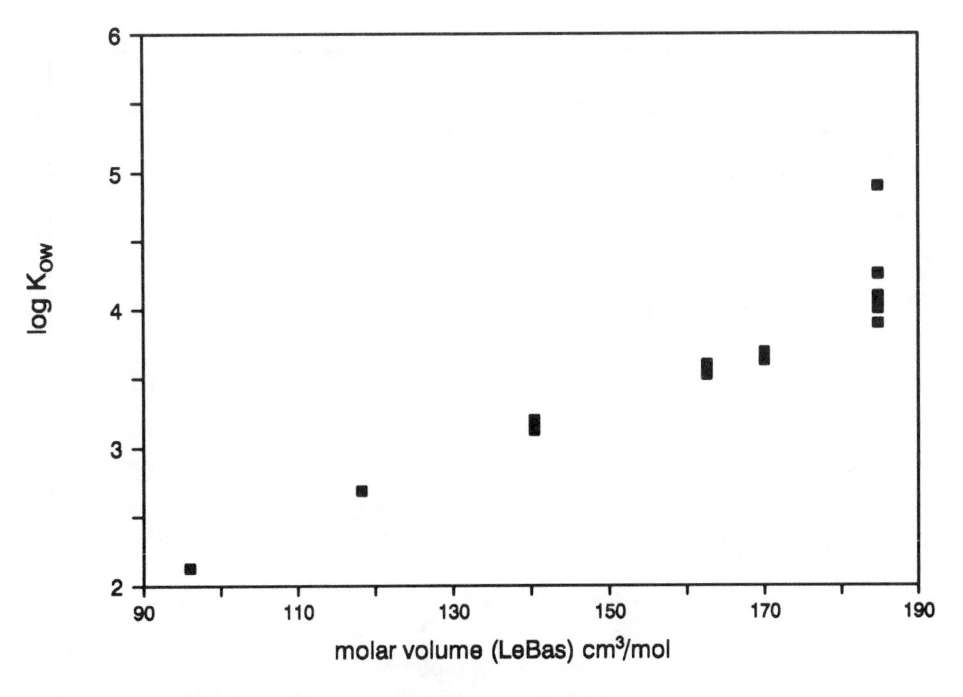

Figure 1.4 Plot of log K_{OW} versus LeBas molar volume

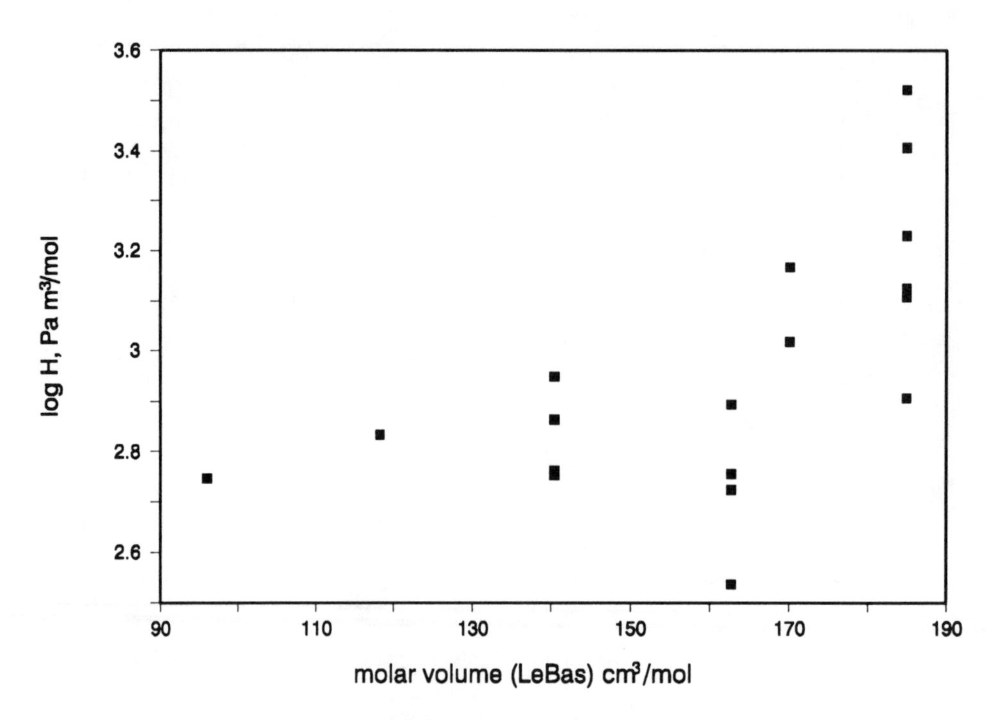

Figure 1.5 Plot log H (Henry's law constant) versus molar volume

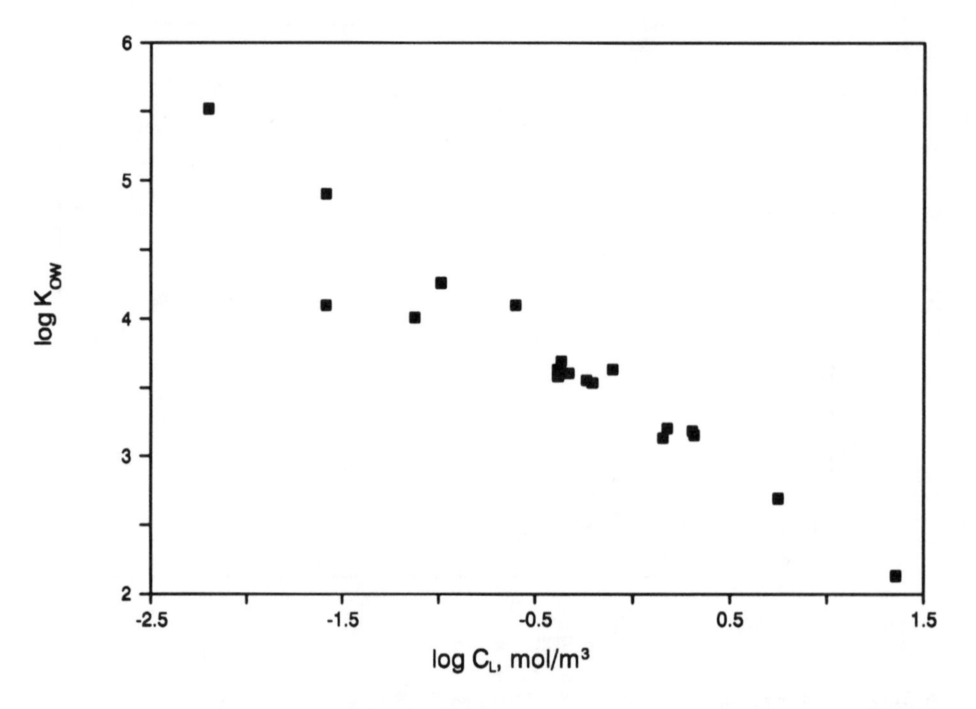

Figure 1.6 Plot of log K_{OW} versus log C_L

is of little value because this is a well-studied chemical, but for other less-studied chemicals the plots are invaluable as a means of checking the reasonableness of data. The plots can also be used, with appropriate caution, to estimate data for untested chemicals.

Figure 1.1 shows the linear relationships among various molecular descriptors. Figures 1.2 to 1.5 show the dependence of the physical-chemical properties on LeBas molar volume. Figure 1.2 shows the solubilities of the monoaormatics decrease steadily with increasing molar volume. The vapor pressure data are similar but K_{OW} increases with increasing molar volume also in a linear fashion. The plot between Henry's law constant and molar volume (Figure 1.5) is more scattered. Figure 1.6 shows the often reported inverse relationship between octanol-water partition coefficient and the subcooled liquid solubility.

The QSPR plots show that an increase in molar volume by 100 cm^3/mol generally causes:

(i) A decrease in log solubility by 2.5 units, i.e., a factor of $10^{2.5}$ or 316;

(ii) A decrease in log vapor pressure by 2.2 units, i.e., a factor of $10^{2.2}$ of 159;

(iii) An increase in log Henry's law constant of 0.3 (i.e., 2.5 - 2.2) or a factor or $10^{0.3}$ or 2.0;

(iv) An increase in log K_{OW} by 2.0 units, i.e., a factor of 100.

The plot of log K_{OW} versus log solubility thus has a slope of approximately 2.0/2.5 or 0.8. This slope of less than 1 has been verified experimentally by Chiou et al. (1982) and Bowman and Sans (1983), the theoretical basis has been discussed in detail by Miller et al. (1985).

1.5.4. Evaluative Calculations

Level I

The Level I calculation suggests that if 100,000 kg (100 tonnes) of benzene are introduced into the 100,000 km^2 environment, 99% will partition into air at a concentration of 9.9 x 10^{-7} g/m^3 or about 1 $\mu g/m^3$. The water will contain nearly 1% at a low concentration of 4 $\mu g/m^3$ or equivalently 4 ng/L. Soils would contain 5 x 10^{-6} $\mu g/g$ and sediments about 9.7 x 10^{-6} $\mu g/g$. These values would normally be undetectable as a result of the very low tendency of benzene to sorb to organic matter in these media. The fugacity is calculated to be 3.14 x 10^{-5} Pa. The dimensionless soil-water and sediment-water partition coefficients or ratios of Z values are 2.6 and 5.3 as a result of a K_{OC} of about 55 and a few percent organic carbon in these media. There is little evidence of bioconcentration with a very low fish concentration of 3 x 10^{-5} $\mu g/g$. The pie chart in Figure 1.7 (which is the same as Figure 2.7 in Chapter 2) clearly shows that air is the primary medium of accumulation.

Level II

The Level II calculation includes the half-lives of 17 h in air, 170 h in water, 550 h in soil and 1700 h in sediment. No reaction is included for suspended sediment or fish. The input of 1000 kg/h results in an overall fugacity of 6 x 10^{-6} Pa which is about 20% of the Level I value. The concentrations and amounts in each medium are thus about 20% of the Level I values. The relative mass distribution is identical to Level I. The primary loss mechanism is reaction in air which accounts for 802 kg/h or 80.2% of the input. Most of the remainder is lost by advective outflow. The water, soil and sediment loss processes are unimportant largely because so little of the benzene is present in these media, but also because of the slower reaction and advection rates. The overall residence time is 19.9 h; thus there is an inventory of benzene in the system of 19.9 x 1000 or 19900 kg. The pie chart in Figure 1.7 illustrates the dominance of air reaction and advection.

If the primary loss mechanism of atmospheric reaction is accepted as having a 17h half-life, the D value is 1.6 x 10^9 mol/Pa • h. For any other process to compete with this would require a value of at least 10^8 mol/Pa • h. This is achieved by advection (4 x 10^8) but the other processes range in D value from 19 (advection in bottom sediment) to 1.5 x 10^6 (reaction in water) and are thus a factor of over 100 or less. The implication is that the water reaction rate constant would have to be increased by 100-fold to become significant. The soil rate constant would require an increase by 10^4 and the sediment by 10^6. These are inconceivably large numbers corresponding to very short half-lives, thus the actual values of the rate constants in these media are relatively unimportant in this context. They need not be known accurately. The most sensitive quantity is clearly the atmospheric reaction rate.

The amounts in the compartments can be calculated easily from the total amount and the percentages of mass distribution in Level I. For example, the amount in water is 0.881% of 19877 kg or 175 kg.

Level III

The Level III calculation includes an estimation of intermedia transport. Examination of the magnitude of the intermedia D values given in the fate diagram (Figure 1.8, which is the same as Figure 2.8 in Chapter 2) suggests that air-water and air-soil transport are most important with water-sediment and soil-water transport being negligible in potential transfer rate. The magnitude of these larger intermedia transport D values (approximately 10^6 mol/Pa • h) compared to the atmospheric reaction and advection values of 10^8 to 10^9 suggests that reaction and advection will be very fast relative to transport.

The bulk Z values are similar for air and water to the values for the "pure" phases in Level I and II, but they are lower for soil and sediment because of the "dilution" of the solid phase with air or water.

The first row describes the condition if 1000 kg/h is emitted into the air. The result is similar to the Level II calculation with 19700 kg in air, 57 kg in water, 24 kg in soil and only 0.2 kg in sediment. It can be concluded that benzene discharged to the atmosphere has very little potential to enter other media. The rates of transfer from air to water and air to soil are both only about 0.4 kg/h. Even if the transfer coefficients were increased by a factor of 10, the rates would remain negligible. The reason for this is the value of the mass transfer coefficients which control this transport process. The overall residence time is 19.8 hours, similar to Level II.

If 1000 kg/h of benzene is discharged to water, as in the second row, there is predictably a much higher concentration in water (by a factor of over 2000). There is reaction of 546 kg/h in water, advective outflow of 134 kg/h and transfer to air of 320 kg/h with negligible loss to sediment. The amount in the water is 134000 kg, thus the residence time in the water is 134 h and the overall environmental residence time is a longer 140 hours. The key processes are thus reaction in water (half-life 170 h), evaporation (half-life 290 h) and advective outflow (residence time 1000 h). The evaporation half-life can be calculated as (0.693 x mass in water)/rate of transfer, i.e., (0.693 x 133863)/320 = 290 h. Clearly competition between reaction and evaporation in the water determines the overall fate. 95% of the benzene discharged is now found in the water and the concentration is a fairly high as 6.7×10^{-4} g/m^3 or 670 ng/L.

The third row shows the fate if discharge is into soil. The amount in soil is 67460 kg, reflecting an overall 87 h residence time. The rate of reaction in soil is only 85 kg/h, there is no advection, thus the primary loss mechanism is transfer to air (T_{31}) at a rate of 905 kg/h, with a relatively minor 10 kg/h to water by run-off. The net result is that the air concentrations are similar to those for air discharge and the soil acts only as a reservoir. The soil concentration of 3.75×10^{-3} g/m^3 or 2.5×10^{-3} μg/g or 2.5 ng/g is controlled almost entirely by the rate at which the benzene can evaporate.

The net result is that benzene behaves entirely differently when discharged to the three media. If discharged to air it reacts rapidly and advects with a residence time of 20 h with little transport to soil or water. If discharged to water it reacts and evaporates to air with a residence time of 140 h. If discharged to soil it mostly evaporates to air with a residence time in soil of 53 h.

The final scenario is a combination of discharges, 600 kg/h to air, 300 kg/h to water, and 100 kg/h to soil. The concentrations, amounts and transport and transformation rates are merely linearly combined versions of the three initial scenarios. For example, the rate of reaction in air is now 632 kg/h. This is 0.6 of the first (air emission) rate of 803 kg/h, i.e., 482 kg/h, plus 0.3 of the second (water emission) rate of 257 kg/h, i.e., 77 kg/h and 0.1 of the third (soil emission) rate of 729 kg/h, i.e., 73 kg yielding a total of (482 + 77 + 73) or 632 kg/h. It is also apparent that the amount in the air of 15500 kg causing a concentration of 0.155 μg/m^3 is attributable to emissions to air (0.6 x 0.197 or 0.118 μg/m^3), emissions to water (0.3 x 0.063 or 0.019 μg/m^3) and emissions to soil (0.1 x 0.179 or 0.018 μg/m^3). The concentration in water of 2.0×10^{-4} g/m^3 or 202 μg/m^3 or ng/L is largely attributable to the discharges to water which alone cause 0.3 x 669 or 200 μg/m^3. Although more is emitted to air it contributes less than 1 μg/m^3 to the water with soil emissions accounting for about 1 μg/m^3. Similarly, the prevailing soil concentration is controlled by the rate of discharge to the soil.

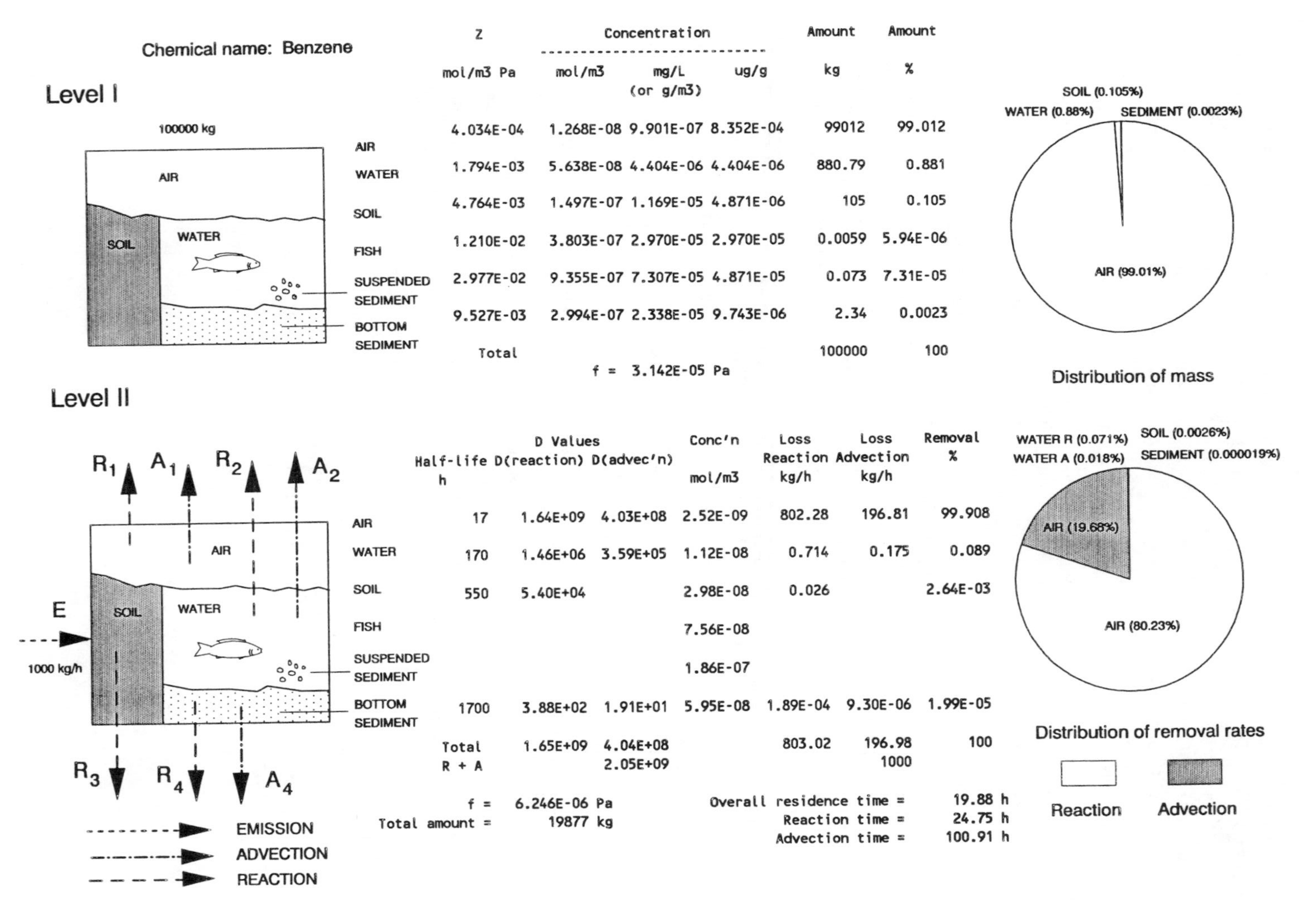

Level I

	Z mol/m3 Pa	Concentration mol/m3	mg/L (or g/m3)	ug/g	Amount kg	Amount %
AIR	4.034E-04	1.268E-08	9.901E-07	8.352E-04	99012	99.012
WATER	1.794E-03	5.638E-08	4.404E-06	4.404E-06	880.79	0.881
SOIL	4.764E-03	1.497E-07	1.169E-05	4.871E-06	105	0.105
FISH	1.210E-02	3.803E-07	2.970E-05	2.970E-05	0.0059	5.94E-06
SUSPENDED SEDIMENT	2.977E-02	9.355E-07	7.307E-05	4.871E-05	0.073	7.31E-05
BOTTOM SEDIMENT	9.527E-03	2.994E-07	2.338E-05	9.743E-06	2.34	0.0023
Total					100000	100

f = 3.142E-05 Pa

Level II

	Half-life h	D Values D(reaction)	D(advec'n)	Conc'n mol/m3	Loss Reaction kg/h	Loss Advection kg/h	Removal %
AIR	17	1.64E+09	4.03E+08	2.52E-09	802.28	196.81	99.908
WATER	170	1.46E+06	3.59E+05	1.12E-08	0.714	0.175	0.089
SOIL	550	5.40E+04		2.98E-08	0.026		2.64E-03
FISH				7.56E-08			
SUSPENDED SEDIMENT				1.86E-07			
BOTTOM SEDIMENT	1700	3.88E+02	1.91E+01	5.95E-08	1.89E-04	9.30E-06	1.99E-05
Total		1.65E+09	4.04E+08		803.02	196.98	100
R + A			2.05E+09			1000	

f = 6.246E-06 Pa
Total amount = 19877 kg

Overall residence time = 19.88 h
Reaction time = 24.75 h
Advection time = 100.91 h

Figure 1.7 Fugacity Level I and II calculations for benzene in a generic environment (dimensions defined in Table 1.2)

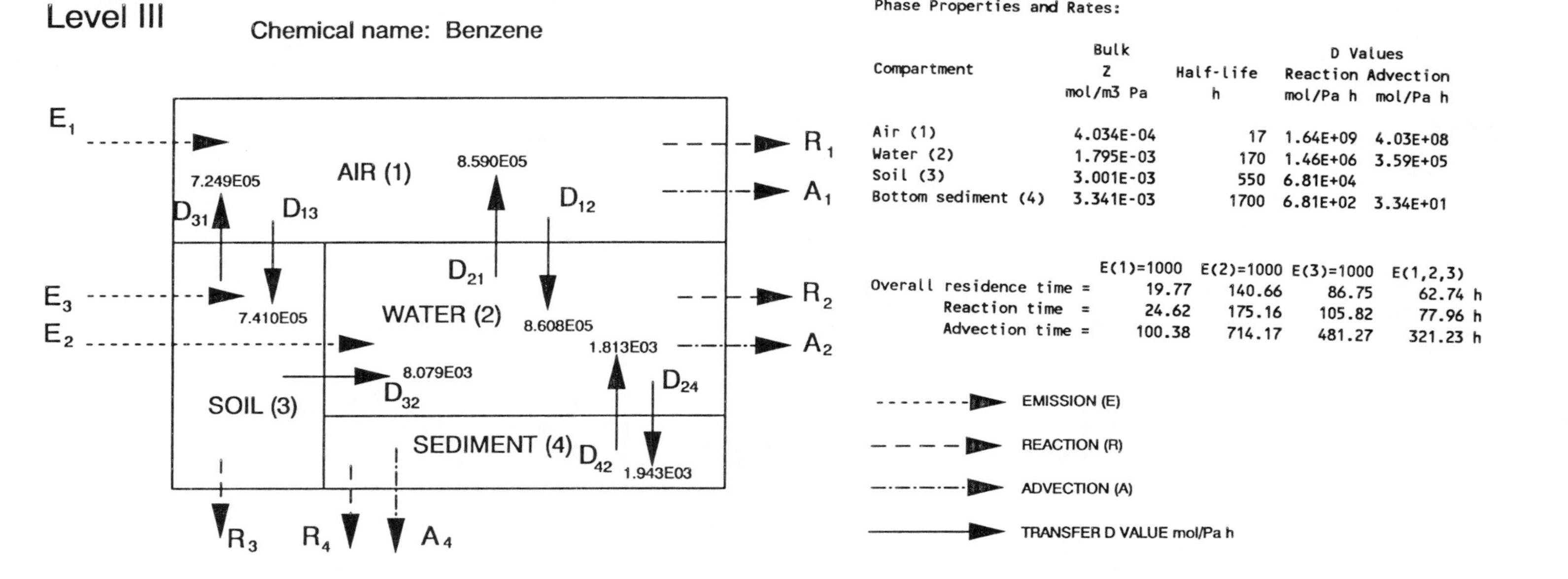

Phase Properties and Rates:

Compartment	Bulk Z mol/m3 Pa	Half-life h	D Values Reaction mol/Pa h	D Values Advection mol/Pa h
Air (1)	4.034E-04	17	1.64E+09	4.03E+08
Water (2)	1.795E-03	170	1.46E+06	3.59E+05
Soil (3)	3.001E-03	550	6.81E+04	
Bottom sediment (4)	3.341E-03	1700	6.81E+02	3.34E+01

	E(1)=1000	E(2)=1000	E(3)=1000	E(1,2,3)
Overall residence time =	19.77	140.66	86.75	62.74 h
Reaction time =	24.62	175.16	105.82	77.96 h
Advection time =	100.38	714.17	481.27	321.23 h

Phase Properties, Composition, Transport and Transformation Rates:

Emission, kg/h			Fugacity, Pa				Concentration, g/m3				Amounts, kg				Total amount, kg
E(1)	E(2)	E(3)	f(1)	f(2)	f(3)	f(4)	C(1)	C(2)	C(3)	C(4)	m(1)	m(2)	m(3)	m(4)	
1000	0	0	6.249E-06	2.023E-06	5.781E-06	1.556E-06	1.969E-07	2.836E-07	1.355E-06	4.059E-07	19692	56.73	24.39	0.203	1.977E+04
0	1000	0	2.002E-06	4.775E-03	1.852E-06	3.671E-03	6.308E-08	6.693E-04	4.341E-07	9.579E-04	6308	133863	7.81	478.96	1.407E+05
0	0	1000	5.676E-06	4.998E-05	1.599E-02	3.842E-05	1.788E-07	7.006E-06	3.748E-03	1.003E-05	17884	1401	67456	5.013	8.675E+04
600	300	100	4.918E-06	1.439E-03	1.603E-03	1.106E-03	1.550E-07	2.017E-04	3.757E-04	2.886E-04	15496	40333	6763	144.31	6.274E+04

Emission, kg/h			Loss, Reaction, kg/h				Loss, Advection, kg/h			Intermedia Rate of Transport, kg/h T12	T21	T13	T31	T32	T24	T42
E(1)	E(2)	E(3)	R(1)	R(2)	R(3)	R(4)	A(1)	A(2)	A(4)	air-water	water-air	air-soil	soil-air	soil-water	water-sed	sed-water
1000	0	0	8.028E+02	2.312E-01	3.07E-02	8.274E-05	1.969E+02	5.673E-02	4.059E-06	4.202E-01	1.358E-01	3.617E-01	3.273E-01	3.648E-03	3.071E-04	2.203E-04
0	1000	0	2.572E+02	5.457E+02	9.85E-03	1.952E-01	6.308E+01	1.339E+02	9.579E-03	1.346E-01	3.204E+02	1.159E-01	1.049E-01	1.169E-03	7.248E-01	5.200E-01
0	0	1000	7.290E+02	5.712E+00	8.50E+01	2.044E-03	1.788E+02	1.401E+00	1.003E-04	3.816E-01	3.353E+00	3.285E-01	9.052E+02	1.009E+01	7.586E-03	5.442E-03
600	300	100	6.317E+02	1.644E+02	8.52E+00	5.883E-02	1.550E+02	1.644E+02	2.886E-03	3.306E-01	9.653E+01	2.846E-01	9.075E+01	1.011E+00	2.184E-01	1.567E-01

Figure 1.8 Fugacity Level III calculation for benzene in a generic environment (dimensions defined in Table 1.2)

In this multimedia discharge scenario the overall residence time is 59 hours, which can be viewed as 60% of the air residence time of 19.7 h, 30% of the water residence time of 140 h and 10% of the overall soil residence time of 53 h. The overall amount in the environment of 59000 kg is thus largely controlled by the discharges to water which account for (0.3 x 133863) or 40,000 kg.

Figure 1.9 shows the distributions of mass and removal process rates for these four scenarios. Clearly when benzene is discharged into a specific medium, most of the chemical is found in that medium. Only in the case of discharges to soil is an appreciable fracition found in another compartment, namely air. This is because benzene evaporates fairly rapidly from soil without being susceptible to reaction or advection.

Finally, it is interesting to note that the fugacity in this final case (in units of mPa) are for the four media 5×10^{-3}, 1.4, 1.6 and 1.1. The soil, sediment and water are fairly close to equilibrium, with the air notably "undersaturated" by a factor of about 200. This is the result of the rapid loss processes from air.

It is believed that these three behavior profiles when combined in the fourth give a comprehensive depiction and explanation of the environmental fate characteristics of the chemical. They show which intermedia transport processes are important and how levels in various media arise from discharges into other media. The same broad fate characteristics as described in the generic environment are believed to be generally applicable to other environments. Certainly this evaluation should, in most cases, identify the key physical-chemical properties, reactions and intermedia transport parameters. With a knowledge of the key parameters, more effort can be devoted to obtaining more accurate site-specific values, and sensitivity analyses can be conducted. In essence, the evaluation translates physical-chemical data into environmental fate information.

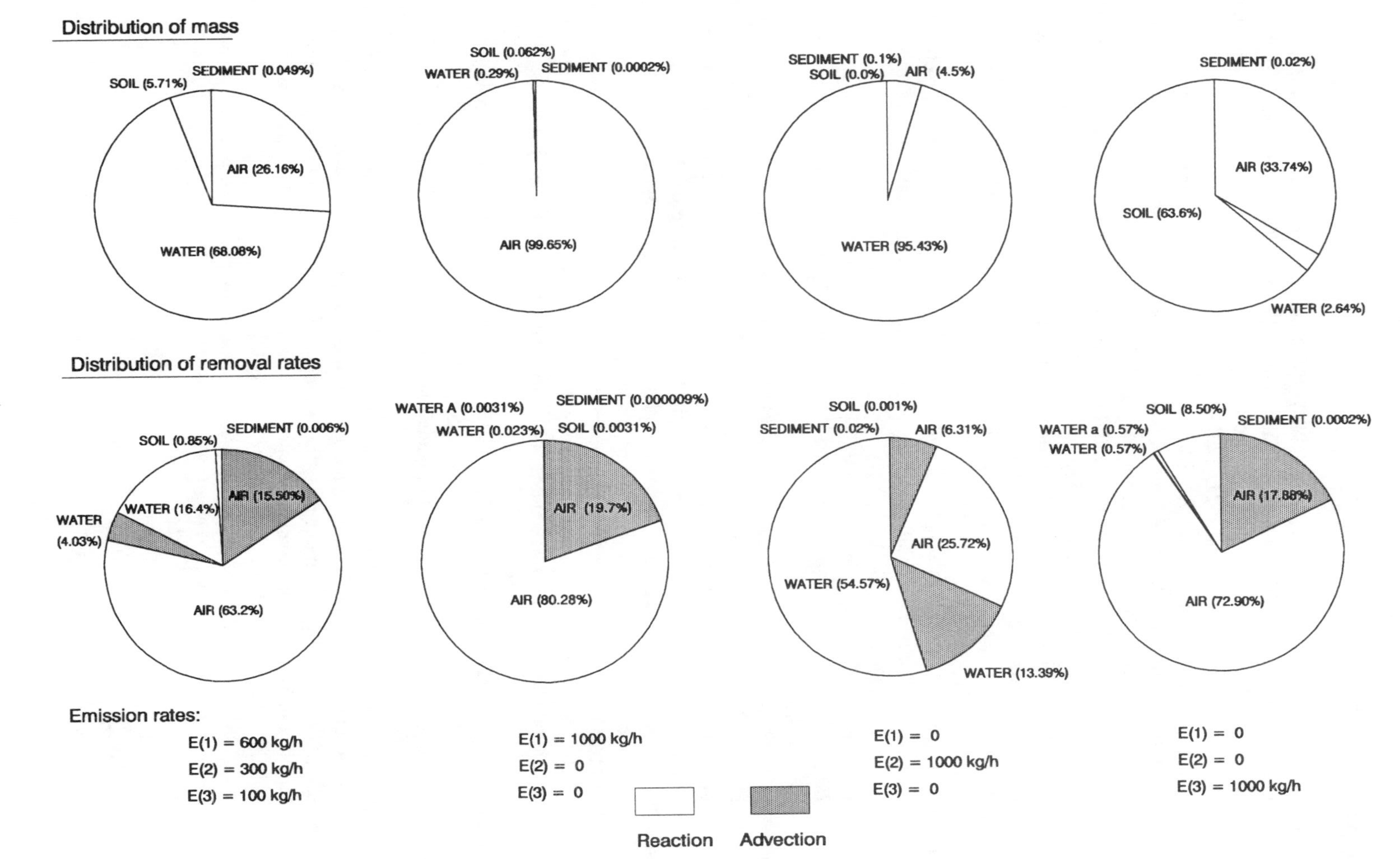

Figure 1.9 Fugacity Level III distributions of benzene for four emission scenarios

1.6 REFERENCES

Abernethy, S., Mackay, D., McCarty, L.S. (1988) "Volume fraction" correlation for narcosis in aquatic organisms: The key role or partitioning. *Environ. Toxicol. Chem.* 7, 469-481.

Abramowitz, R., Yalkowsky, S.H. (1990) Estimation of aqueous solubility and melting point of PCB congeners. *Chemosphere* 21, 1221-1229.

Almgren, M., Grieser, F., Powell, J.R., Thomas, J.K. (1979) A correlation between the solubility of aromatic hydrocarbons in water and micellar solutions, with their normal boiling points. *J. Chem. Eng. Data* 24, 285-287.

Al-Sahhaf, T.A. (1989) Prediction of the solubility of hydrocarbons in water using UNIFAC. *J. Environ. Sci. Health* A24, 49-56.

Ambrose, D. (1981) Reference value of vapor pressure. The vapor pressures of benzene and hexafluorobenzene. *J. Chem. Thermodyn.* 13, 1161-1167.

Amidon, G.L., Williams, N.A. (1982) A solubility equation for non-electrolytes in water. *Intl. J. Pharm.* 11, 249-156.

Amidon, G.L., Anik, S.T. (1981) Application of the surface area approach to the correction and estimation of aqueous solubility and vapor pressure. Alkyl aromatic hydrocarbons. *J. Chem. Eng. Data* 26, 28-33.

Anderson, E., Veith, G.D., Weininger, D. (1987) *SMILES: A Line Notation and Computerized Interpreter for Chemical Structures.* US EPA Environmental Research Brief, EPA/600/M-87/021.

Andren, A.W., Doucette, W.J., Dickhut, R.M. (1987) Methods for estimating solubilities of hydrophobic organic compounds: Environmental modeling efforts. In: *Sources and Fates of Aquatic Pollutants.* Hites, R.A., Eisenreich, S.J., Eds., pp. 3-26, Advances in Chemistry Series 216, American Chemical Society, Washington, D.C.

Andrews, L.J., Keffer, R.M. (1950a) Cation complexes of compounds containing carbon-carbon double bonds. IV. The argentation of aromatic hydrocarbons. *J. Am. Chem. Soc.* 72, 3644-3647.

Andrews, L.J., Keefer, R.M. (1950b) Cation complexes of compounds containing carbon-carbon double bonds. VII. Further studies on the argentation of substituted benzenes. *J. Am. Chem. Soc.* 72, 5034-5037.

Arbuckle, W.B. (1983) Estimating activity coefficients for use in calculating environmental parameters. *Environ. Sci. Technol.* 17, 537-542.

Arbuckle, W.B. (1986) Using UNIFAC to calculate aqueous solubilities. *Environ. Sci. Technol.* 20, 1060-1064.

Ashworth, R.A., Howe, G.B., Mullins, M.E., Roger, T.N. (1988) Air-water partitioning coefficients of organics in dilute aqueous solutions. *J. Hazard. Materials* 18, 25-36.

Balson, E.W. (1947) Studies in vapour pressure measurement. Part III.- An effusion manometer sensitive to $5x10^{-6}$ millimetres of mercury: vapour pressure of D.D.T. and other slightly volatile substances. *Trans. Farad. Soc.* 43, 54-60.

Banerjee, S. (1985) Calculation of water solubility of organic compounds with UNIFAC-derived parameters. *Environ. Sci. Technol.* 19, 369-370.

Banerjee, S., Howard, P.H. (1988) Improved estimation of solubility and partitioning through correction of UNIFAC-derived activity coefficients. *Environ. Sci. Technol.* 22, 839-841.

Banerjee, S., Howard, P.H., Lande, S.S. (1990) General structure-vapor pressure relationships for organics. *Chemosphere* 21, 1173-1180.

Banerjee, S., Yalkowsky, S.H., Valvani, S.C. (1980) Water solubiltiy and octanol/water partition coefficients of organics. Limitations of the solubility-partition coefficient correlation. *Environ. Sci. Technol.* 14, 1227-1229.

Bidleman, T.F. (1984) Estimation of vapor pressures for nonpolar organic compounds by capillary gas chromatography. *Anal. Chem.* 56, 2490-2496.

Bohon, R.L., Claussen, W.F. (1951) The solubility of aromatic hydrocarbons in water. *J. Am. Chem. Soc.* 73, 1571-1576.

Bondi, A. (1964) van der Waals volumes and radii. *J. Phys. Chem.* 68, 441-451.

Booth, H.S., Everson, H.E. (1948) Hydrotropic solublities: solublities in 40 percent sodium xylenesulfonate. *Ind. Eng. Chem.* 40, 1491-1493.

Boublik, T., Fried, V., Hala, E. (1973) *The Vapor Pressure of Pure Substances*, Elsevier, Amsterdam.

Boublik, T., Fried, V., Hala, E. (1984) *The Vapor Pressure of Pure Substances*, 2nd revised ed., Elsevier, Amsterdam.

Bowman, B.T., Sans, W.W. (1983) Determination of octanol-water partitioning coefficient (K_{OW}) of 61 organophosphorus and carbamate insecticides and their relationship to respective water solubility (S) values. *J. Environ. Sci. Health* B18, 667-683.

Bradley, R.S., Cleasby, T.G. (1953) The vapour pressure and lattice energy of some aromatic ring compounds. *J. Chem. Soc.* 1953, 1690-1692.

Briggs, G.G. (1981) Theoretical and experimental relationships between soil adsorption, octanol-water partition coefficients, water solubilities, bioconcentration factors, and the Parachor. *J. Agric. Food Chem.* 29, 1050-1059.

Bruggeman, W.A., van der Steen, J., Hutzinger, O. (1982) Reversed-phase thin-layer chromatography of polynuclear aromatic hydrocarbons and chlorinated biphenyls. Relationship with hydrophobicity as measured by aqueous solubility and octanol-water partition coefficient. *J. Chromatogr.* 238, 335-346.

Budavari, S., Ed. (1989) *The Merck Index. An Encylopedia of Chemicals, Drugs and Biologicals.* 11th ed., Merck & Co. Inc., Rahway, New Jersey.

Burkhard, L.P. (1984) *Physical-Chemical Properties of the Polychlorinated Biphenyls: Measurement, Estimation, and Application to Environmental Systems.* Ph.D. Thesis, University of Wisconsin-Madison, Wisconsin.

Burkhard, L.P., Andren, A.W., Armstrong, D.E. (1985a) Estimation of vapor pressures for polychlorinated biphenyls: A comparison of eleven predictive methods. *Environ. Sci. Technol.* 19, 500-507.

Burkhard, L.P., Armstrong, D.E., Andren, A.W. (1985b) Henry's law constants for polychlorinated biphebyls. *Environ. Sci. Technol.* 590-595.

Burkhard, L.P., Kuehl, D.W., Veith G.D. (1985c) Evaluation of reversed phase liquid chromatograph/mass spectrophotometry for estimation of n-octanol/water partition coefficients of organic chemicals. *Chemosphere* 14, 1551-1560.

Chiou, C.T. (1981) Partition coefficient and water solubility in environmental chemistry. In: *Hazard Assessment of Chemicals Current Developments*. Vol. 1, pp. 117-153. Academic Press, New York.

Chiou, C.T. (1985) Partition coefficients of organic compounds in lipid-water systems and correlations with fish bioconcentration factors. *Environ. Sci. Technol.* 19, 57-62.

Chiou, C.T., Freed, V.H., Schmedding, D.W. (1977) Partition coefficient and bioaccumulation of selected organic chemicals. *Environ. Sci. Technol.* 11, 475-478.

Chiou, C.T., Schmedding, D.W., Manes, M. (1982) Partitioning of organic compounds in octanol-water system. *Environ. Sci. Technol.* 16, 4-10.

Davies, R.P., Dobbs, A.J. (1984) The prediction of bioconcentration in fish. *Water Res.* 18, 1253-1262.

Davis, W.W., Krahl, M.E., Clowes, G.H. (1942) Solubility of carcinogenic and related hydrocarbons in water *J. Am. Chem. Soc.* 64, 108-110.

Davis, W.W., Parke, Jr, T.V. (1942) A nephelometric method for determination of solubilities of extremely low order. *J. Am. Chem. Soc.* 64, 101-107.

Dean, J.D., Ed. (1979) *Lange's Handbook of Chemistry*. 12th ed., McGraw-Hill, New York, N.Y.

Dean, J.D., Ed. (1985) *Lange's Handbook of Chemistry*. 13th ed., McGraw-Hill, New York, N.Y.

Dearden, J.C. (1990) Physico-chemical descriptors. In: *Practical Applications of Quantitative Structure-Activity Relationships (QSAR) in Environmental Chemistry and Toxicology*. Karcher, W. and Devillers, J., Eds., pp. 25-60. Kluwer Academic Publisher, Dordrecht, Netherlands.

De Bruijn, J., Busser, G., Seinen, W., Hermens, J. (1989) Determination of octanol/water partition coefficient for hydrophobic organic chemicals with the "slow-stirring" method. *Environ. Toxicol. Chem.* 8, 499-512.

De Bruijn, J., Hermens, J. (1990) Relationships between octanol/water partition coefficients and total molecular surface area and total molecular volume of hydrophobic organic chemicals. *Quant. Struct.-Act. Relat.* 9, 11-21.

Dobbs, A.J., Grant, C. (1980) Pesticide volatilization rate: a new measurement of the vapor pressure of pentachlorophenol at room temperature. *Pestic. Sci.* 11, 29-32.

Dobbs, A.J., Cull, M.R. (1982) Volatilization of chemical relative loss rates and the estimation of vapor pressures. *Environ. Pollut. (Ser. B)* 3, 289-298.

Doucette, W.J., Andren, A.W. (1987) Correlation of octanol/water partition coefficients and total molecular surface area for highly hydrophobic aromatic compounds. *Environ. Sci. Technol.* 21, 521-524.

Doucette, W.J., Andren, A.W. (1988a) Aqueous solubility of selected biphenyl, furan, and dioxin congeners. *Chemosphere* 17, 243-252.

Doucette, W.J., Andren, A.W. (1988b) Estimation of octanol/water partition coefficients: Evaluation of six methods for highly hydrophobic aromatic hydrocarbons. *Chemosphere* 17, 345-359.

Dunnivant, F.M., Coate, J.T., Elzerman, A.W. (1988) Experimentally determined Henry's law constants for 17 polychlorobiphenyl congeners. *Environ. Sci. Technol.* 22, 448-453.

Fendinger, N.J., Glotfelty, D.E. (1988) A laboratory method for the experimental determination of air-water Henry's law constants for several pesticides. *Environ. Sci. Technol.* 22, 1289-1293.

Fendinger, N.J., Glotfelty, D.E. (1990) Henry's law constants for selected pesticides, PAHs and PCBs. *Environ. Toxicol. Chem.* 9, 731-735.

Foreman, W.T., Bidleman, T.F. (1985) Vapor pressure estimates of individual polychlorinated biphenyls and commercial fluids using gas chromatographic retention data. *J. Chromatogr.* 330, 203-216.

Fredenslund, A., Jones, R.L., Prausnitz, J.M. (1975) Group-contribution estimation of activity coefficients in nonideal liquid mixtures. *AIChE J.* 21, 1086-1099.

Fujita, T., Iwasa, J., Hansch, C. (1964) A new substituent constant, "pi" derived from partition coefficients. *J. Am. Chem. Soc.* 86, 5175-5180.

Garst, J.E. (1984) Accurate, wide-range, automated, high-performance chromatographic method for the estimation of octanol/water partition coefficients. II: Equilibrium in partition coefficient measurements, additivity of substituent constants, and correlation of biological data, *J. Pharm. Sci.* 73, 1623-1629.

Garten, C.T., Trabalka, J.R. (1983) Evaluation of models for predicting terrestrial food chain behavior of xenobiotics. *Environ. Sci. Technol.* 17, 590-595.

Gossett, R. (1987) Measurement of Henry's law constants for C_1 and C_2 chlorinated hydrocarbons. *Environ. Sci. Technol.* 21, 202-208.

Gross, P.M., Saylor, J.H. (1931) The solubilities of certain slightly soluble organic compounds in water. *J. Am. Chem. Soc.* 1931, 1744-1751.

Gückel, W., Rittig, R., Synnatschke, G. (1974) A method for determining the volatility of active ingredients used in plant protection. II. Application to formulated products. *Pestic. Sci.* 5, 393-400.

Gückel, W. Kästel, R., Lawerenz, J., Synnatschke, G. (1982) A method for determining the volatility of active ingredients used in plant protection. Part III: The temperature relationship between vapour pressure and evaporation rate. *Pestic. Sci.* 13, 161-168.

Hafkenscheid, T.L., Tomlinson, E. (1981) Estimation of aqueous solubilities of organic non-electrolytes using liquid chromatographic retention data. *J. Chromatogr.* 218, 409-425.

Hamaker, J.W. Kerlinger, H.O. (1969) Vapor pressures of pesticides. *Adv. Chem. Ser.* 86, 39-54.

Hansch, C., Leo, A. (1979) *Substituent Constants for Correlation Analysis in Chemistry and Biology*. Wiley-Interscience, New York, N.Y.

Hansch, C., Quinlan, J.E., Lawrence, G.L. (1968) The linear-free energy relationship between partition coefficient and aqueous solubility of organic liquids. *J. Org. Chem.* 33, 347-350.

Hawker, D.W. (1989) The relationship between octan-1-ol/water partition coefficient and aqueous solubility in terms of solvatochromic parameters. *Chemosphere* 19, 1586-1593.

Hawker, D.W. (1990a) Vapor pressures and Henry's law constants of polychlorinated biphenyls. *Environ. Sci. Technol.* 23, 1250-1253.

Hawker, D.W. (1990b) Description of fish bioconcentration factors in terms of solvatochromic parameters. *Chemosphere* 20, 267-477.

Hawker, D.W., Connell, D.W. (1988) Octanol-water partition coefficients of polychlorinated biphenyl congeners. *Environ. Sci. Technol.* 22, 382-387.

Hermann, R.B. (1971) Theory of hydrophobic bonding. II. The correlation of hydrocarbon solubility in water with solvent cavity surface area. *J. Phys. Chem.* 76, 2754-2758.

Hinckley, D.A., Bidleman, T.F., Foreman, W.T. (1990) Determination of vapor pressures for nonpolar and semipolar organic compounds from gas chromatographic retention data. *J. Chem. Eng. Data* 35, 232-237.

Hollifield, H.C. (1979) Rapid nephelometric estimate of water solubility of highly insoluble organic chemicals of environmental interest. *Bull. Environ. Contam. Toxicol.* 23, 579-586.

Horvath, A.L. (1982) *Halogenated Hydrocarbons, Solubility - Miscibility with Water.* Marcel Dekker, Inc., New York, N.Y.

Howard, P.H., Ed. (1989) *Handbook of Fate and Exposure Data for Organic Chemicals. Vol. I. Large Production and Priority Pollutants.* Lewis Publishers, Chelsea, Michigan.

Howard, P.H., Ed. (1990) *Handbook of Fate and Exposure Data for Organic Chemicals. Vol. - II - Solvents.* Lewis Publishers, Inc., Chelsea, MI.

Howard, P.H., Boethling, R.S., Jarvis, W.F., Meylan, W.M., Michalenko, E.M. (1991) *Handbook of Environmental Degradation Rates.* Lewis Publishers, Inc., Chelsea, Michigan.

Hussam, A, & Carr, P.W. (1985) A study of a rapid and precise methodology for the measurement of vapor liquid equilibria by headspace gas chromatography. *Anal. Chem.* 57, 793-801.

Isnard, P., Lambert, S. (1988) Estimating bioconcentration factors for octanol-water partition coefficient and aqueous solubility. *Chemosphere* 17, 21-34.

Isnard, P., Lambert, S. (1989) Aqueous solubility/n-octanol-water partition coefficient correlations. *Chemosphere* 18, 1837-1853.

IUPAC Solubility Data Series (1989a) *Vol. 37: Hydrocarbons (C_5 - C_7) with Water and Seawater.* Shaw, D.G., Ed., Pergamon Press, Oxford, England.

IUPAC Solubility Data Series (1989b) *Vol. 38: Hydrocarbons (C_8 -C_{36}) with Water and Seawater.* Shaw, D.G., Ed., Pergamon Press, Oxford, England.

Jury, W.A., Spencer, W.F., Farmer, W.J. (1983) Behavior assessment model for trace organics in soil: I. Model description. *J. Environ. Qual.* 12, 558-566.

Jury, W.A., Farmer, W.J., Spencer, W.F. (1984a) Behavior of assessment model for trace organics in soil: II. Chemical classification and parameter sensitivity. *J. Environ. Qual.* 13, 567-572.

Jury, W.A., Farmer, W.J., Spencer, W.F. (1984b) Behavior assessment model for trace organics in soil: III. Application of screening model. *J. Environ. Qual.* 13, 573-579.

Jury, W.A., Spencer, W.F., Farmer, W.J. (1984) Behavior assessment model for trace organics in soil: IV. Review of experimental evidence. *J. Environ. Qual.* 13, 580-587.

Kabadi, V.N., Danner, R.P. (1979) Nomograph solves for solubilities of hydrocarbons in water. *Hydrocarbon Processing* 68, 245-246.

Kamlet, M.J., Doherty, R.M., Abraham, M.H., Carr, P.W., Doherty, R.F., Raft, R.W. (1987) Linear solvation energy relationships. Important differences between aqueous solublity relationships for aliphatic and aromatic solutes. *J. Phy. Chem.* 91, 1996-

Kamlet, M.J., Doherty, R.M., Carr, P.W., Mackay, D., Abraham, M.H., Taft, R.W. (1988) Linear solvation energy relationships. 44. Parameter estimation rules that allow accurate prediction of octanol/water partition coefficients and other solubility and toxicity properties of polychlorinated biphenyls and polycyclic aromatic hydrocarbons. *Environ. Sci. Technol.* 22, 503-509.

Karcher, W., Devillers, J., Eds., (1990) *Practical Applications of Quantitative-Structure-Activity Relationships (QSAR) in Environmental Chemistry and Toxicology.* Kluwer Academic Publisher, Dordrecht, Netherlands.

Karickhoff, S.W. (1981) Semiempirical estimation of sorption of hydrophobic pollutants on natural sediments and soil. *Chemosphere* 10, 833-846.

Karickhoff, S.W., Brown, D.S., Scott, T.A. (1979) Sorption of hydrophobic pollutants on natural water sediments. *Water Res.* 13, 241-248.

Kenaga, E.E. (1980) Predicted bioconcentration factors and soil sorption coefficients of pesticides and other chemicals. *Ecotox. Environ. Saf.* 4, 26-38.

Kenaga, E.E., Goring, C.A.I. (1980) Relationship between water solubility, soil sorption, octanol-water partitioning, and concentration of chemicals in biota. In: *Aquatic Toxicology.* Eaton, J.G., Parrish, P.R., Hendrick, A.C., Eds., pp. 78-115, Am. Soc. for Testing and Materials, STP 707, Philadelphia.

Kier, L.B., Hall, L.H. (1976) Molar properties and molecular connectivity. In: *Molecular Connectivity in Chemistry and Drug Design.* Medicinal Chem. Vol. 14, pp. 123-167, Academic Press, New York.

Kier, L.B., Hall, L.H. (1986) *Molecular Connectivity in Structure-Activity Analysis.* Wiley, New York.

Kim, Y.-H., Woodrow, J.E., Seiber, J.N. (1984) Evaluation of a gas chromatographic method for calculating vapor pressures with organophosphorus pesticides. *J. Chromotagr.* 314, 37-53.

Könemann, H., van Leeuewen, K. (1980) Toxicokinetics in fish: accumulation of six chlorobenzenes by guppies. *Chemosphere* 9, 3-19.

Könemann, H., Zelle, R., Busser, F. (1979) Determination of log P_{oct} values of chloro-substituted benzenes, toluenes and anilines by high-performance liquid chromatography on ODS-silica. *J. Chromatogr.* 178, 559-565.

Lande, S.S., Banerjee, S. (1981) Predicting aqueous solubility of organic nonelectrolytes from molar volume. *Chemosphere* 10, 751-759.

Leahy, D.E. (1986) Intrinsic molecular volume as a measure of the cavity term in linear solvation energy relationships: octanol-water partition coefficients and aqueous solubilities. *J. Pharm. Sci.* 75, 629-636.

Leo, A., Hansch, C., Elkins, D. (1971) Partition coefficients and their uses. *Chem. Rev.* 71, 525-616.

Lincoff, A.H., Gossett, J.M. (1984) The determination of Henry's law constants for volatile organics by equilibrium partitioning in closed systems. In: *Gas Transfer at Water Surfaces*. Brutsaert, W., Jirka, G.H., Eds., pp. 17-26, D. Reidel Publishing Co., Dordrecht, Holland.

Lyman, W.J., Reehl, W.F., Rosenblatt, D.H. (1982) *Handbook of Chemical Property Estimation Methods*. McGraw-Hill, New York.

Mackay, D. (1979) Finding fugacity feasible. *Environ. Sci. Technol.* 13, 1218-1223.

Mackay, D. (1982) Correlation of bioconcentration factors. *Environ. Sci. Technol.* 16, 274-278.

Mackay, D. (1991) *Multimedia Environmental Models. The Fugacity Approach.* Lewis Publishers, Inc., Chelsea, Michigan.

Mackay, D., Bobra, A.M., Shiu, W.Y., Yalkowsky, S.H. (1980) Relationships between aqueous solubility and octanol-water partition coefficient. *Chemosphere* 9, 701-711.

Mackay, D., Bobra, A.M., Chan, D.W., Shiu, W.Y. (1982) Vapor pressure correlation for low-volatility environmental chemicals. *Environ. Sci. Technol.* 16, 645-649.

Mackay, D., Paterson, S. (1990) Fugacity models. In: *Practical Applications of Quantitative Structure-Activity Relationships (QSAR) in Environmental Chemistry and Toxicology.* Karcher, W., Devillers, J., Eds., pp. 433-460, Kluwer Academic Publishers, Dordrecht, Holland.

Mackay, D., Paterson, S. (1991) Evaluating the multimedia fate of organic chemicals: A Level III fugacity model. *Environ. Sci. Technol.* 25, 427-436.

Mackay, D., Shiu, W.Y. (1977) Aqueous solubility of polynuclear aromatic hydrocarbons. *J. Chem. Eng. Data* 22, 339-402.

Mackay, D., Shiu, W.Y. (1981) A critical review of Henry's law constants for chemicals of environmental interest. *J. Phys. Chem. Ref. Data* 11, 1175-1199.

Mackay, D., Shiu, W.Y., Sutherland, R.P. (1979) Determination of air-water Henry's law constants for hydrophobic pollutants. *Envrion. Sci. Technol.* 13, 333-337.

Mackay, D., Shiu, W.Y., Wolkoff, A.W. (1975) Gas chromatographic determination of low concentration of hydrocarbons in water by vapor phase extration. *ASTM STP 573,* pp. 251-258, American Society for Testing and Materials, Philadelphia, Pa.

Mackay, D., Stiver, W.H. (1991) Predictability and environmental chemistry. In: *Environmental Chemistry of Herbicides*. Vol. II. Grover, R., Lessna, A.J., Eds., pp. 281-297. CRC Press, Boca Raton, FL.

Macknick, A.B., Prausnitz, J.M. (1979) Vapor pressure of high-molecular weight hydrocarbons. *J. Chem. Eng. Data* 24, 175-178.

Mailhot, H., Peters, R.H. (1988) Empirical relationships between the 1-octanol/water partition coefficient and nine physicochemical properties. *Environ. Sci. Technol.* 22, 1479-1488.

May, W.E., Wasik, S.P., Freeman, D.H. (1978a) Determination of the aqueous solubility of polynuclear aromatic hydrocarbons by a coupled-column liquid chromatographic technique. *Anal. Chem.* 50, 175-179.

May, W.E., Wasik, S.P., Freeman, D.H. (1978b) Determination of the solubility behavior of some polycyclic aromatic hydrocarbons in water. *Anal. Chem.* 50, 997-1000.

McAuliffe, C. (1966) Solubility in water of paraffin, cycloparaffin, olefin, acetylene, cycloolefin and aromatic hydrocarbons. *J. Phys. Chem.* 76, 1267-1275.

McDuffie, B. (1981) Estimation of octanol/water partition coefficient for organic pollutants using reversed phase HPLC. *Chemosphere* 10, 73-83.

McGowan, J.C., Mellors, A. (1986) *Molecular Volumes in Chemistry and Biology-Applications including Partitioning and Toxicity.* Ellis Horwood Limited, Chichester, England.

The Merck Index (1989) *An Encyclopedia of Chemicals, Drugs amd Biologicals.* 11th ed., Budavari, S., Ed., Merck and Co., Inc., Rahway, N.J.

Miller, M.M., Ghodbane, S., Wasik, S.P., Tewari, Y.B., Martire, D.E. (1984) Aqueous solubilities, octanol/water partition coefficients and entropies of melting of chlorinated benzenes and biphenyls. *J. Chem. Eng. Data* 29, 184-190.

Miller, M.M., Wasik, S.P., Huang, G.L., Shiu, W.Y., Mackay, D. (1985) Relationships between octanol-water partition coefficient and aqueous solublity. *Environ. Sci. Technol.* 19, 522-529.

Neely, W.B., Branson, D.R., Blau, G.E. (1974) Partition coefficient to measure bioconcentration potential of organic chemicals in fish. *Environ. Sci. Technol.* 8, 1113-1115.

Nirmalakhandan, N.N., Speece, R.E. (1988a) Prediction of aqueous solubility of organic chemicals based on molecular structure. *Environ. Sci. Technol.* 22, 328-338.

Nirmalakhandan, N.N., Speece, R.E. (1988b) QSAR model for predicting Henry's law constant. *Environ. Sci. Technol.* 22, 1349-1357.

Nirmalakhandan, N.N., Speece, R.E. (1989) Prediction of aqueous solubility of organic chemicals based on molecular structure. 2. Application to PNAs, PCBs, PCDDs, etc. *Environ. Sci. Technol.* 23, 708-713.

Oliver, B.G. (1984) The relationship between bioconcentration factor in rainbow trout and physical-chemical properties for some halogenated compounds in: *QSAR in Environmental Toxicology.* Kaiser, K.L.E., Ed., pp. 300-317, D. Reidel Publishing Co., Dordrecht, Holland.

Oliver, B.G., Niimi, A.J. (1988) Trophodynamic analysis of polychlorinated biphenyl congeners and other chlorinated hydrocarbons in the Lake Ontario ecosystem. *Environ. Sci. Technol.* 22, 388-397.

Paterson, S., Mackay, D. (1985) The fugacity concept in environmental modelling, In: *The Handbook of Environmental Chemistry.* Vol. 2/Part C, Hutzinger, O., Ed., pp. 121-140. Springer-Verlag, Heidelberg, Germany.

Pearlman, R.S. (1980) Molecular surface areas and volumes and their use in structure/activity relationships. In: *Physical Chemical Properties of Drugs.* Yalkowsky, S.H., Sinkula, A.A., Valvani, S.C., Eds., Medicinal Research Series, Vol. 10., pp. 321-317, Marcel Dekker, Inc., New York.

Pearlman, R.S. (1986) Molecular surface area and volume: Their calculation and use in predicting solubilities and free energies of desolvation. In: *Partition coefficient, Determination and Estimation.* Dunn III, W.J., Block, J.H., Pearlman R.S., Eds., pp. 3-20, Pergamon Press, New York.

Pearlman, R.S., Yalkowsky, S.H., Banerjee, S. (1984) Water solubilities of polynuclear aromatic and heteroaromatic compounds. *J. Phys. Chem. Ref. Data* 13, 555-562.

Pierotti, C., Deal, C., Derr, E. (1959) Activity coefficient and molecular structure. *Ind. Eng. Chem. Fundam.* 51, 95-101.

Rapaport, R.A., Eisenreich, S.J. (1984) Chromatographic determination of octanol-water partition coefficients (K_{OW}'s) for 58 polychlorinated biphenyl congeners. *Environ. Sci. Technol.* 18, 163-170.

Reid, R.C., Prausnitz, J.M., Polling, B.E. (1987) *The Properties of Gases and Liquids.* 4th ed., McGraw-Hill, New York, N.Y.

Rekker, R.F. (1977) *The Hydrophobic Fragmental Constant.* Elsevier, Amsterdam/New York N.Y.

Sabljic, A. (1984) Predictions of the nature and strength of soil sorption of organic pollutants by molecular topology. *J. Agric. Food Chem.* 32, 243-246.

Sabljic, A. (1987) On the prediction of soil sorption coefficients of organic pollutants from molecular structure: Application of molecular topology model. *Environ. Sci. Technol.* 21, 358-366.

Sabljic, A., Lara, R., Ernst, W. (1989) Modelling association of highly chlorinated biphenyls with marine humic substances. *Chemosphere* 19, 1665-1676.

Sabljic, A., Güsten, H. (1989) Predicting Henry's law constants for polychlorinated biphenyls, *Chemosphere* 19, 1503-1511.

Schwarzenbach, R.P., Westall, J. (1981) Transport of nonpolar compounds from surface water to groundwater. Laboratory sorption studies. *Environ. Sci. Technol.* 11, 1360-1367.

Shiu, W.Y., Mackay, D. (1986) A critical review of aqueous solubilities, vapor pressures, Henry's law constants, and octanol-water partition coefficients of the polychlorinated biphenyls. *J. Phys. Chem. Ref. Data* 15, 911-929.

Shiu, W.Y., Gobas, F.A.P.C., Mackay, D. (1987) Physical-chemical properties of three congeneric series of chlorinated aromatic hydrocarbons. In: *QSAR in Environmental Toxicology -II.* Kaiser, K.L.E., Ed., pp. 347-362. D. Reidel Publishing Co., Dordrecht, Holland.

Shiu, W.Y., Doucette, W., Gobas, F.A.P.C., Mackay, D., Andren, A.W. (1988) Physical-chemical properties of chlorinated dibenzo-p-dioxins. *Environ. Sci. Technol.* 22, 651-658.

Sinke, G.C. (1974) A method for measurement of vapor pressures of organic compounds below 0.1 torr. Naphthalene as reference substance. *J. Chem. Thermodyn.* 6, 311-316.

Sonnefeld, W.J., Zoller, W.H., May, W.E. (1983) Dynamic coupled-column liquid chromatographic determination of ambient temperature vapor pressures of polynuclear aromatic hydrocarbons. *Anal. Chem.* 55, 275-280.

Spencer, W.F., Cliath, M.M. (1969) Vapor density of dieldrin. *Environ. Sci. Technol.* 3, 670-674.

Spencer, W.F., Cliath, M.M. (1970) Vapor density and apparent vapor pressure of lindane (γ-BHC). *J. Agric. Food Chem.* 18, 529-530.

Spencer, W.F., Cliath, M.M. (1972) Volatility of DDT and related compounds. *J. Agric. Food Chem.* 20, 645-649.

Stephenson, R.M., Malanowski, A. (1987) *Handbook of the Thermodynamics of Organic Compounds.* Elsevier, New York.

Stiver, W., Mackay, D. (1989) The linear additivity principle in environmental modelling: Application to chemical behaviour in soil. *Chemosphere* 19, 1187-1198.

Suntio, L.R., Shiu, W.Y., Mackay, D. (1988) Critical review of Henry's law constants for pesticides. *Rev. Environ. Contam. Toxicol.* 103, 1-59.

Swann, R.L., Laskowski, D.A., McCall, P.J., Vander Kuy, K., Dishburger, H.J. (1983) A rapid method for the estimation of the environmental parameters octanol/water partition coefficient, soil sorption constant, water to air ratio, and water solubility. *Residue Rev.* 85, 17-28.

Szabo, G., Prosser, S., Bulman, R.A. (1990) Determination of the adsorption coefficient (K_{OC}) of some aromatics for soil by RP-HPLC on two immobilized humic acid phases. *Chemosphere* 21, 777-788.

Tomlinson, E., Hafkenscheid, T.L. (1986) Aqueous solution and partition coefficient estimation from HPLC data In: *Partition Coefficient, Determination and Estimation.* Dunn III, W.J., Block, J.H., Pearlman, R.S., Eds., pp. 101-141, Pergamon Press, New York.

Tsonopoulos, C., Prausnitz, J.M. (1971) Activity coefficients of aromatic solutes in dilute aqueous solutions. *Ind. Eng. Chem. Fundam.* 10, 593-600.

Tulp, M.T.M., Hutzinger, O. (1978) Some thoughts on the aqueous solubilities and partition coefficients of PCB, and the mathematical correlation between bioaccumulation and physico-chemical properties. *Chemosphere* 7, 849-760.

Veith, G.D., Austin, N.M., Morris, R.T. (1979) A rapid method for estimating log P for organic chemicals. *Water Res.* 13, 43-47.

Veith, G.D., Macek, K.J., Petrocelli, S.R., Caroll, J. (1980) An evaluation of using partition coefficients and water solubilities to estimate bioconcentration factors for organic chemicals in fish. In: *Aquatic Toxicology.* Eaton, J.G., Parrish, P.R., Hendrick, A.C., Eds, pp. 116-129, ASTM ATP 707, Am. Soc. for Testing and Materials, Philadelphia, Pa.

Verschueren, K. (1977) *Handbook of Environmental Data on Organic Chemicals.* Van Nostrand Reinhold, New York, N.Y.

Verschueren, K. (1983) *Handbook of Environmental Data on Organic Chemicals.* Van Nostrand Reinhold, New York, N.Y.

Warne, M., St. J., Connell, D.W., Hawker, D.W. (1990) Prediction of aqueous solubility and the octanol-water partition coefficient for lipophilic organic compounds using molecular descriptors and physicochemical properties. *Chemosphere* 16, 109-116.

Wasik, S.P., Miller, M.M., Tewari, Y.B., May, W.E., Sonnefeld, W.J., DeVoe, H., Zoller, W.H. (1983) Determination of the vapor pressure, aqueous solubility, and octanol/water partition coefficient of hydrophobic substances by coupled generator column/liquid chromatographic mehtods. *Res. Rev.* 85, 29-42.

Weast, R. (1972-73) *Handbook of Chemistry and Physics.* 53th ed., CRC Press, Cleveland, OH.

Weast, R. (1984) *Handbook of Chemistry and Physics.* 64th ed., CRC Press, Boca Raton, FL.

Weil, L., Dure, G., Quentin, K.L. (1974) Solubility in water of insecticide, chlorinated hydrocarbons and polychlorinated biphenyls in view of water pollubion. *Z. Wasser Abwasser Forsch.* 7, 169-175.

Westcott, J.W., Bidleman, T.F. (1982) Determination of polychlorinated biphenyl vapor pressures by capillary gas chromatography. *J. Chromatogr.* 210, 331-336.

Westcott, J.W., Simon, J.J., Bidleman, T.F. (1981) Determination of polychlorinated biphenyl vapor pressures by a semimicro gas saturation method. *Environ. Sci. Technol.* 15, 1375-1378.

Whitehouse, B.G., Cooke, R.C. (1982) Estimating the aqueous solubility of aromatic hydrocarbons by high performance liquid chromatography. *Chemosphere* 11, 689-699.

Winholz, M. Ed. (1983) *The Merck Index, An Encyclopedia of Chemicals, Drugs and Biologicals.* 10th ed., Merck & Co. Inc. Rahway, New Jersey.

Woodburn, K.B., Doucette, W.J., Andren, A.W. (1984) Generator column determination of octanol/water partition coefficients for selected polychlorinated biphenyl congeners. *Environ. Sci. Technol.* 18, 457-459.

Yalkowsky, S.H. (1979) Estimation of entropies of fusion of organic compounds. *Ind. Eng. Chem. Fundam.* 18, 108-111.

Yalkowsky, S.H., Valvani, S.C. (1976) Partition coefficients and surface areas of some alkylbenzenes. *J. Med. Chem.* 19, 727-728.

Yalkowsky, S.H., Valvani, S.C. (1979) Solubility and partitioning. I: Solubility of nonelectrolytes in water. *J. Pharm. Sci.* 69, 912-922.

Yalkowsky, S.H., Orr, R.J., Valvani, S.C. (1979) Solubility and partitioning 3. The solubility of halobenzenes in water. *I&EC Fundam.* 18, 351-353.

Yalkowsky, S.H., Valvani, S.S., Mackay, D. (1983) Estimation of the aqueous solubility of some aromatic compounds. *Res. Rev.* 85, 43-55.

Yoshida, K., Shigeoka, T., Yamauchi, F. (1983) Relationship between molar refraction and n-octanol/water partition coefficient. *Ecotox. Environ. Saf.* 7, 558-565.

Zwolinski, B.J. Wilhoit, R.C. (1971) *Handbook of Vapor Pressures and Heats of Vaporization of Hydrocarbons and Related Compounds.* API-44, TRC Publication No. 101, Texas A&M University, College Station, TX.

2. Monoaromatic Hydrocarbons

2.1 List of Chemicals and Data Compilations

Common Name: Benzene
Synonym: benzol, cyclohexatriene
Chemical Name: benzene
CAS Registry No: 71-43-2
Molecular Formula: C_6H_6
Molecular Weight: 78.11
Melting Point (°C): 5.53
Boiling Point (°C): 80.1
Density (g/cm^3 at 20°C): 0.8765
Molar Volume (cm^3/mol):

96.0 (LeBas method, Reid et al. 1977, 1987; Abernethy & Mackay 1987)
88.7 (20°C, calculated-density, McAuliffe 1966)
89.0 (calculated-density, Klevens 1950; Lande & Banerjee 1981)
89.41 (calculated-density, Chiou et al. 1983; Windholz 1983; Budavari 1989)
89.4 (calculated -density, Wasik et al. 1984)
0.491 (intrinsic volume, V_I/100, Leahy 1986; Kamlet et al. 1987,1988)
76.64 (characteristic molecular volume, McGowan & Mellors 1986)
88.9 (20°C, calculated-density, Stephenson & Malanowski 1987)

Molecular Volume (A^3):

83.95 (Pearlman 1986)

Total Surface Area, TSA (A^2):

109.5 (Yalkowsky & Valvani 1976; Yalkowsky et al. 1979; Whitehouse & Cooke 1982)
255.7 (calculated-with hydration shell, Amidon & Anik 1980,1981)
109.5, 135 (calculated-molecular model, ion mobility, Lande et al. 1985)
255.7, 287 (calculated-molecular model with hydration shell, ion mobility, Lande et al. 1985)
110.01 (Pearlman 1986)
106.8 (calculated- χ , Sabljic 1987a)
109.23 (Doucette & Andren 1988)
110.7 (Warne et al. 1990)

Heat of Fusion, kcal/mol:

2.370 (Tsonopoulos & Prausnitz 1971)

Entropy of Fusion, cal/mol K (e.u.):

8.5 (Tsonopoulos & Prausnitz 1971)

Fugacity Ratio at 25 °C, F: 1.0

Water Solubility (g/m^3 or mg/L at 25°C):

1850 (30°C, shake flask-interferometer, Gross & Saylor 1931)
1786 (shake flask-turbidimetric, Stearns et al. 1947)
1402 (residue-volume method, Booth & Everson 1948)
1740 (shake flask-UV, Andrew & Keefer 1949; quoted, Dreisbach 1955; Chey & Calder 1972; Vesala 1974; API 1985; Shiu et al. 1990)

1860 (shake flask-UV, Klevens 1950; quoted, API 1985; Suntio et al. 1988; Shiu et al. 1990)

1790 (shake flask-UV, Bohon & Claussen 1951; quoted, Chey & Calder 1972; Vesala 1974; API 1985; Suntio et al. 1988; Shiu et al. 1990)

1718 (shake flask-UV, Morrison & Billett 1952)

1755 (shake flask-UV, McDevit & Long 1952; quoted, Suntio et al. 1988; Shiu et al. 1990)

1796 (Hayashi & Sasaki 1956; quoted, Keeley et al. 1988)

1780, 1823 (selected, calculated-molar volume, Lindenburg 1956; quoted, Horvath 1982)

1760 (Brady & Huff 1958)

1740 (shake-UV, Arnold et al. 1958; quoted, Chey & Calder 1972; Vesala 1974; Keeley et al. 1988)

1790 (quoted, Deno & Berkheimer 1960; Tsonopoulos & Prausnitz 1971; Chiou 1981; Lande & Banerjee 1981)

1742 (shake flask-UV, Franks et al. 1963; quoted, Shiu et al. 1990)

1780 (shake flask-GC, McAuliffe 1963; quoted, Shiu et al. 1990)

1778 (calculated-group contribution, Irmann 1965; quoted, Horvath 1982)

1780 (shake flask-GC, McAuliffe 1966; quoted, Hermann 1972; Vesala 1974; Mackay & Shiu 1975; Karickhoff et al. 1979; Hutchinson et al. 1980; Kenaga & Goring 1980; API 1985; Abernethy et al. 1986; Mackay 1988; Suntio et al. 1988; Shiu et al. 1988; 1990)

1718 (calculated, Taha et al. 1966)

2167 (Worley 1967; quoted, IUPAC 1989a)

1797 (quoted, Hansch et al. 1968)

1434 (calculated-K_{OW}, Hansch et al. 1968)

1740 (21°C, extraction by nonpolar resins/elution, Chey & Calder 1972)

1755, 1857 (shake flask-GC, calculated-group contribution, Polak & Lu 1973; quoted, Horvath 1982)

1765 (shake flask-GC, Leinonen & Mackay 1973)

1760 (shake flask-UV, Brown & Wasik 1974)

1906 (shake flask-UV, Vesala 1974)

1769 (shake flask-GC, Mackay et al., 1975)

1780 (shake flask-GC, Mackay & Shiu 1975; quoted, Mackay & Leinonen 1975; Mackay 1981; API 1985; Suntio et al. 1988; Mackay & Shiu 1990)

1789 (quoted, Hine & Moorkerjee 1975; Leahy 1986; Kamlet et al. 1987)

1740 (shake flask-GC, Price 1976; quoted, Suntio et al. 1988; Shiu et al. 1990)

1790, 1270 (selected, calculated-χ, reported as ln S-should be -log S in molar concn., Kier & Hall 1976)

820 (22°C, quoted, Chiou et al. 1977; Gillham & Rao 1990)

1710 (20 °C, quoted, Freed et al. 1977)

1791 (gen.col.-HPLC/UV, May et al. 1978; May 1980; quoted, API 1985; IUPAC 1989a; Howard 1990)

1769 (shake flask-Fluo., Aquan-Yuen et al. 1979; quoted, Suntio et al. 1988; Shiu et al. 1990)
1820-1930 (elution chromatography, Schwarz 1980)
1750 (shake flask-LSC, Banerjee et al. 1980)
1789, 1831 (quoted, calculated-K_{OW} & M.P., Valvani & Yalkowsky 1980)
1790, 2305 (quoted, calculated-K_{OW}, Valvani et al. 1981)
1750 (quoted lit. mean, Amidon & Anik 1981)
1789 (quoted, Eisenreich et al. 1981; Könemann 1981)
1790 (quoted, Lande & Banerjee 1981)
1749, 4006 (quoted, calculated-K_{OW}, Amidon & Williams 1982)
1787 (shake flask-GC, Chiou et al. 1982; 1983)
1789 (quoted, Horvath 1982; Wong et al. 1984)
1786 (quoted lit. average, Whitehouse & Cooke 1982)
1789 (gen. col.-HPLC, Wasik et al. 1983; May et al. 1983)
1632 (calculated-K_{OW}, Yalkowsky et al. 1983)
1809 (calculated-HPLC-k', converted from reported γ_W, Hafkenscheid & Tomlinson 1983a)
1790 (quoted, Miller et al. 1985; Vowles & Mantoura 1987; Eastcott et al. 1988)
1632, 1794 (quoted, calculated-UNIFAC, Banerjee 1985)
1790, 680 (quoted, calculated-TSA, Lande et al. 1985)
1749 (quoted, Brookman et al. 1985; Cline et al. 1991)
1874, 4817 (quoted, calculated-K_{OW} & HPLC-RT, Chin et al. 1986)
1790, 665 (quoted, calculated-characteristic molecular volume, McGowan & Mellors 1986)
897 (calculated-solvatochromic parameters, Leahy 1986)
1852 (calculated-UNIFAC, Banerjee & Howard 1988)
1296 (calculated-V_I, solvatochromic parameters, Kamlet et al. 1987)
1782, 2143 (quoted, calculated-χ, Nirmalakhandan & Speece 1988a)
1695 (shake flask-GC, Keeley et al. 1988)
1650 (shake flask-GC, Coutant & Keigley 1988)
1780, 1766 (quoted, calculated-UNIFAC, Al-Sahhaf 1989)
1820 (quoted, Isnard & Lambert 1988; 1989)
1800 (quoted, Abdul et al. 1990)
1780 (quoted, Olsen & Davis 1990)
1798 (quoted, Warne et al. 1990)

Vapor Pressure (Pa at 25 °C):
12700 (extrapolated, Antoine eqn., Zwolinski & Wilhoit, 1971; quoted, Mackay & Wolkhoff 1973; Mackay & Leinonen 1975; Mackay et al. 1979; Mackay & Shiu 1975; 1981; 1990; Eastcott et al. 1988; Suntio et al. 1988)
12690 (quoted, Hine & Moorkerjee 1975)
12680 (extrapolated, Antoine eqn., Boublik et al. 1973; 1984; quoted, Howard 1990)
12636 (average, bubble cap boilers, Ambrose 1981)

13172 (quoted, Eisenreich et al. 1981; Mackay 1981)
12100 (gas saturation, Politzki et al. 1982)
12690 (extrapolated, Antoine eqn., Dean 1985)
12700 (extrapolated, Antoine eqn., Stephenson & Malanowski 1987)
10133 (quoted, Gillham & Rao 1990)
12690, 3746 (quoted, calculated-UNIFAC, Banerjee et al. 1990)

Henry's Law Constant (Pa m^3/mol):
576 (Taha et al. 1966)
557 (calculated-P/C, Mackay & Leinonen 1975; Mackay 1981; Suntio et al. 1988)
530 (calculated-bond contribution, Hine & Moorkerjee 1975)
555 (calculated as $1/K_{AW}$, C_W/C_A, reported as exptl., Hine & Moorkerjee 1975; Nirmalakhandan & Speece 1988a)
562, 556 (batch stripping, calculated-P/C, Mackay et al. 1979)
552 (shake flask, concn ratio/UV, Green & Frank 1979)
551, 547, 575, 555 (calculated-P/C, Green & Frank 1979)
554 (concentration ratio, Leighton & Calo 1981)
562, 550 (selected exptl., calculated-P/C, Mackay & Shiu 1981; quoted, Howard 1990; Olsen & Davis 1990)
557 (calculated-P/C, Eisenreich et al. 1981)
595 (calculated-C_A/C_W, Matter-Müller et al. 1981)
557 (calculated, Mabey et al. 1982)
556 (calculated-P/C, Mills et al. 1982)
550 (calculated-UNIFAC, Arbuckle 1983)
557 (quoted, Pankow 1986; Pankow & Rosen 1988; Pankow 1990)
740 (20°C, EPICS, Yurteri et al. 1987)
441 (20°C, calculated-P/C, Yurteri et al. 1987)
535 (EPICS, Ashworth et al. 1988)
588 (batch stripping, Ashworth et al. 1988)
552 (calculated-UNIFAC, Ashworth et al. 1988)
586 (concentration ratio, Keeley et al. 1988)
555 (calculated-C_A/C_W, Eastcott et al. 1988)
461 (calculated-QSAR, Nirmalakhandan & Speece 1988b)
453 (calculated-P/C, Munz & Roberts 1989)
557 (calculated-P/C, Mackay & Shiu 1990)
545 (calculated-C_A/C_W, Jury et al. 1990)

Octanol/Water Partition Coefficient, log K_{OW}:
2.13 (shake flask-UV, Fujita et al. 1964; quoted, McCall 1975; Freed et al. 1977; Kenaga & Goring 1980)
2.13 (quoted, Iwasa et al. 1965; Hansch et al. 1968; Hansch et al. 1972; Chiou et al. 1977; 1982; Chiou 1981; 1985; Hitchinson et al. 1980; Tewari et al. 1982b; Brooke et al. 1986)

1.56, 1.65 (shake flask-UV, calculated-M.O. indices, Rogers & Cammarata 1969)
2.15 (calculated-f const., Yalkowsky & Valvani 1976; quoted, Veith et al. 1979; Swann et al. 1983)
2.13, 2.17 (selected, calculated- χ , Kier & Hall 1976)
2.13 (calculated-f const., Rekker 1977; quoted, Harnisch et al. 1983)
2.13, 1.56, 2.15, 2.03, 2.04 (Hansch & Leo 1979; quoted, Suntio et al. 1988)
2.13 (calculated-f constants, Yalkowsky et al. 1979; Könemann 1981; selected, Leegwater 1989)
2.11 (Veith et al. 1979a; quoted, Mackay 1982)
2.39 (HPLC-RT, Veith et al. 1979a; quoted, Suntio et al. 1988)
2.11 (quoted, Karickhoff et al. 1979; Karickhoff 1981; Hudson & Williams 1988; Szabo et al. 1990a,b; Jafvert 1991)
2.14 (Konemann 1979; quoted, Wong et al. 1984)
2.13, 1.90 (quoted, calculated-S, Mackay et al. 1980)
2.12 (shake flask-LSC, Banerjee et al. 1980; quoted, Suntio et al. 1988; Olsen & Davis 1990)
2.13 (calculated-S, Valvani & Yalkowsky 1980)
2.28 (HPLC-k', Hanai et al. 1981; quoted, Suntio et al. 1988)
2.11 (HPLC-RP, McDuffie 1981; quoted, Rapaport & Eisenreich 1984)
2.11, 2.43 (quoted, HPLC-k', McDuffie 1981)
2.01 (quoted, Valvani et al. 1981)
2.11 (quoted, Amidon & Williams 1982)
2.16 (HPLC-k', D'Amboise & Hanai 1982; quoted, Suntio et al. 1988)
2.13 (GC, Wateral et al. 1982)
2.20 (shake flask-HPLC, Hammers et al. 1982; quoted, Suntio et al. 1988)
2.13, 2.18 (quoted, HPLC-k', Miyake & Terada 1982)
1.56-2.15, 2.02 (range, mean; shake flask method, Eadsforth & Moser 1983)
2.01-2.69, 2.38 (range, mean; HPLC method, Eadsforth & Moser 1983)
2.13 (quoted of Hansch & Leo 1979, Harnisch et al. 1983)
2.23, 2.13 (HPLC methods, quoted, Harnish et al. 1983)
2.10 (shake flask-GC, Platford 1983)
2.15, 2.48 (quoted, HPLC-RP, Swann et al. 1983)
2.10 (HPLC-k', Hafkenscheid & Tomlinson 1983b)
2.13 (calculated-f constants, Yalkowsky et al. 1983)
2.13, 2.30, 2.41 (quoted, calculated-molar refraction & ionization potential, molar refraction, Yoshida et al. 1983)
1.99 (selected, Garst 1984)
1.56-2.15, 2.04 (range, HPLC-RV, Garst 1984)
2.25 (HPLC-k', Rapaport & Eisenreich 1984)
2.13 (generator column-GC/ECD, Miller et al. 1984; quoted, Miller et al. 1985; Abernethy et al. 1986; Vowles & Mantoura 1987; Eastcott et al. 1988)
2.13 (Hansch & Leo 1985; quoted, Howard 1990)
2.15 (quoted exptl., Campbell & Luthy 1985)

2.22 (calculated-V_I, solvatochromic parameters, Taft et al. 1985)
2.37 (calculated-V_I, solvatochromic parameters, Leahy 1986)
2.13 (quoted, Burkhard & Kuehl 1986)
2.13, 2.33 (quoted, HPLC-RT, Chin et al. 1986)
2.26 (HPLC-k', De Kock & Lord 1987)
2.01, 1.98 (gen. col.-RPLC, calculated-activity coeff., Schantz & Martire 1987)
2.14, 2.09, 2.50, 2.56, 2.28, 2.52 (calculated-π, f const., HPLC-RT, MW, χ, TSA, Doucette & Andren 1988)
2.13 (quoted, Ryan et al. 1988; Lee et al. 1989)
2.13, 1.91 (quoted, calculated-UNIFAC, Banerjee & Howard 1988)
2.03-2.34 (selected lit. range, Dearden & Bresnen 1988)
2.13 (recommended, Sangster 1989)
2.12 (quoted, Isnard & Lambert 1988, 1989)
2.186 (slow stirring-GC, De Bruijn et al. 1989)
2.13 (quoted as parent compound, De Bruijn et al. 1989)
2.10, 2.43 (quoted, calculated-MO calculation, Bodor et al. 1989)
2.13, 2.142 (quoted, calculated-f constants, Bodor et al. 1989)
2.03-2.34 (quoted from the Medchem Data-base, Dearden 1990)
2.11 (quoted, Szabo et al. 1990a & b)
2.068 (quoted, Warne et al. 1990)

Bioconcentration Factor, log BCF:
0.64 (pacific herring, Korn et al. 1977)
0.54 (eels, Ogata & Miyake 1978; Ogata et al. 1984)
1.10 (fathead minnow, Veith et al. 1980)
1.57 (microorganism, calculated-K_{OW}, Mabey et al. 1982)
1.10, 0.79 (fish-exptl., fish-correlated, Mackay 1982)
1.10 (guppy, calculated- χ, Koch 1983)
1.48, 1.0 (algae, fish, Freitag et al. 1984)
1.48 (algae, Geyer et al. 1984)
0.63 (gold fish, Ogata et al. 1984)
<1.0 (fish, Freitag et al. 1985)
3.23 (activated sludge, Freitag et al. 1985)
1.10 (fish, Isnard & Lambert 1988)
0.54, 0.64, 0.63; 1.38 (selected: eels, pacific herring, gold fish; calculated, Howard 1990)
1.63 (*s. Capricornutum*, Herman et al. 1991)

Sorption Partition Coefficient, log K_{OC}:
1.92 (sediment, sorption isotherm, Karickhoff et al. 1979; quoted, Kenaga & Goring 1980; Lyman 1982; Hodson & Williams 1988)
1.92 (quoted, Kenaga 1980)

1.78 (sediment/soil, quoted, Karickhoff 1981)
1.72 (sediment/soil, estimated from K_{OW} & S, Karickhoff 1981)
2.01 (soil, Schwarzenbach & Westall 1981)
1.81 (sediment, calculated-K_{OW}, Mabey et al. 1982)
1.85, 2.53 (estimated-S, K^{OW}, Lyman 1982)
1.92, 1.95 (quoted, calculated- χ , Sabljic & Protic 1982)
1.26 (soil, Chiou et al. 1983)
1.90 (calculated- χ , Koch 1983)
1.50-2.16 (soil, calculated- χ , Sabljic 1984)
1.99 (Pavlou 1987)
1.49, 1.92, 1.50-2.16 (quoted, Howard 1990)
1.91, 1.87 (quoted, HPLC-k', Szabo et al. 1990a,b)
1.09 (soil, Jury et al. 1990)
1.77 (average, calculated-S, Olsen & Davis 1990)
2.33 (calculated-molecular conductivity index, Olsen & Davis 1990)

Sorption Partition Coefficient, log K_{OM}:
1.04 (untreated soil, Lee et al. 1989)
1.81-2.56 (organic cations treated soil, Lee et al. 1989)
1.26-1.92, 2.09 (observed, calcd.- θ, Olsen & Davis 1990)

Half-Lives in the Environment:

Air: > 5.1 hours, based on a determined rate of disappearance in ambient LA basin air for reaction with OH radicals at 300°K (Doyle et al. 1975); 2.4-24 hours, based on rate of disappearance for the reaction with hydroxy radicals (Darnall et al. 1976); 50.1-501 hours, based on photooxidation half-life in air (Atkinson 1985; Howard et al. 1991).

Surface Water: 4.81 hours, based on evaporation loss at 25°C and 1 m depth of water (Mackay & Leinonen 1975); 120-384 hours, based on unacclimated aerobic biodegradation half-life (Van der Linden 1978; Vaishnav & Babeu 1987; Howard et al. 1991)

Groundwater: estimated half-life from persistence observed in the groundwater of Netherlands, about one year (Zoeteman et al. 1981); 240-17280 hours, based on unacclimated aqueous aerobic biodegradation half-life (Van der Linden 1978; Vaishnav & Babeu 1987; Howard et al. 1991).

Soil: 120-384 hours, based on unacclimated aqueous aerobic biodegradation half-life (Van der Linden 1978; Vaishnav & Babeu 1987; Howard et al. 1991); < 10 days (Ryan et al. 1988); 365 days, assumed first-order biological/chemical degradation in the soil (Jury et al. 1990); disappearance half-life for test soils, < 2.0 days (Anderson et al. 1991).

Environmental Fate Rate Constants or Half-Lives:

Volatilization: $t_{1/2}$ = 4.81 h from water depth of 1 m (calculated, Mackay & Leinonen 1975; Haque et al. 1980); half-lives from marine mesocosm: 23 days at 8-16°C in the spring, 3.1 days at 20 - 22°C in the summer and 13 days at 3-7°C in the winter (Wakeham et al. 1983); estimated half-life from a river of 1 m depth with wind speed 3 m/s and water current of 1 m/s is 2.7 hours at 20°C (Lyman et al. 1982).

Photolysis: atmospheric photolysis half-life: 2808-16152 hours, based on measured photolysis half-lives in deionized water (Hustert et al. 1981; Howard et al. 1991); aqueous photolysis half life: 2808-16152 hours, based on measured photolysis half-lives in deionized water (Hustert et al. 1981; Howard et al. 1991); reaction rate constant in air, 8.64×10^{-4} hour^{-1} and in water, 1.8×10^{-4} hour^{-1} (Mackay et al. 1985).

Oxidation: rate constant for the reaction with OH radicals: determined in an environmental chamber using ambient LA basin air with an initial concentration of 9.0×10^{-10} mole liter^{-1} was $< 2.3 \times 10^{9}$ liter mol^{-1} sec^{-1} (Doyle et al. 1975) and absolute rate constant determined at room temperature using flash photolysis-resonance fluorescence was 1.24×10^{-12} cm^3 molecule^{-1} sec^{-1} (Hansen et al. 1975); rate constant with hydroxyl radicals = 0.85×10^{9} L mol^{-1}s^{-1} with $t_{1/2}$ = 2.4 -24 h (Darnall et al. 1976); photooxidation half-life in water: 8.021×10^{3} -3.21×10^{5} hours, based on measured rate constant for reaction with hydroxy radicals in water (Güesten et al. 1981); reaction with ozone at 300°K having a rate constant of 28 cm^3 mol^{-1} sec^{-1} (Lyman et al. 1982); rate constant k = 0.82×10^{9} M^{-1} sec^{-1}, $t_{1/2}$ = 6.8 days, oxidation by OH radicals in the atmosphere (Mill 1982); room temperature rate constant for the reaction with OH radicals in the gas phase: 1.19×10^{-12} cm^3 sec^{-1} (Atkinson 1985; selected, Howard et al. 1991) and 1.45×10^{-12} cm^3 sec^{-1} (Ohta & Ohyama 1985).

Hydrolysis: k_h = 0, no hydrolyzable functional groups (Mabey et al. 1982).

Biodegradation: Aqueous aerobic biodegradation half-life: 120-384 hours, based on seawater dieaway test data (Van der Linden 1978) and river dieaway data (Vaishnav & Babeu, 1987; Howard et al. 1991); reaction rate constant, k_B = 4.58×10^{-3} hour^{-1} in water (Lee & Ryan 1979; Mackay et al. 1985); k_B = 0.025 day^{-1} and $t_{1/2}$ = 28 days in groundwater, k_B = 0.044 day^{-1} and $t_{1/2}$ = 16 days in Lester River, with nutrient and microbial addition, k_B = 0.082 day^{-1} and $t_{1/2}$ = 8 days in Superior harbor waters (Vaishnav & Babeu 1987); Significant degradation in favorable aerobic environment, k_B = 0.5 day^{-1} (Tabak et al. 1981; Mills et al. 1982); Anaerobic aqueous biodegradation half-life: 2688-17280 hours, based on unacclimated aqueous anaerobic biodegradation screening test data (Horowitz et al. 1982; Howard et al. 1991); Half life in estuarine water is estimated to be 6 days (Lee & Ryan 1976); First order k_B estimated to be 0.12 day^{-1} in river water (Bartholomew & Pfaender 1983; quoted, Battersby 1990); first-order rate constant of 0.2 year^{-1} with a half-life of 110 days (Zoeteman et al 1981; Olsen & Davis 1990).

Bioconcentration, Uptake (k_1) and Elimination (k_2) Constants or Half-Lives: half-life for elimination from eels, 0.5 day (Ogata & Miyake 1978).

Common Name: Toluene
Synonym: methyl benzene, phenylmethane, toluol, methylbenzol, methacide.
Chemical Name: toluene
CAS Registry No: 108-88-3
Molecular Formula: $C_6H_5CH_3$
Molecular Weight: 92.13
Melting Point (°C): -95
Boiling Point (°C): 110.6
Density (g/cm³ at 20°C): 0.8669
Molar Volume (cm³/mol):

106.3 (20 °C, calculated-density, McAuliffe 1966)
106.86 (Windholz 1983; Budavari 1989)
106.3 (25°C, calculated-density, Miller et al. 1985)
106.0 (calculated-density, Klevens 1950; Lande & Banerjee 1981; Chiou 1985)
85.74 (characteristic molecular volume, McGowen & Mellors 1986)
118.0 (LeBas method, Abernethy & Mackay 1987)
106.9 (20°C, calculated-density, Stephenson & Malanowski 1987)
0.591 (intrinsic volume: $V_I/100$, Leahy 1986; Kamlet et al.1987,1988)

Molecular Volume (A³):

99.98 (Pearlman 1986)

Total Surface Area, TSA (A²):

126.5 (Yalkowsky & Valvani 1976, Whitehouse & Cooke 1982)
286.5 (calculated-hydration shell, Amidon & Anik 1981)
128.3 (calculated- χ , Sabljic 1987a)
129.70 (Doucette & Andren 1988)
131.2 (Warne et al. 1990)

Heat of Fusion, kcal/mol:
Fugacity Ratio, F: 1.0

Water Solubility (g/m³ or mg/L at 25°C):

470 (Führer, 1924; quoted, Chiou et al. 1977; Freed et al. 1977)
570 (30°C, shake flask-interferometer, Gross & Saylor 1931; quoted, Vesala 1974)
347 (residue-volume method, Booth & Everson 1948)
530 (shake flask-UV, Andrews & Keffer 1949; quoted, Chey & Calder 1972; Vesala 1974; Shiu et al. 1990)
500 (flask flask-UV, Klevens 1950; quoted, Suntio et al. 1988; Shiu et al. 1990)
627 (shake flask-UV, Bohon & Claussen 1951; quoted, Chey & Calder 1972; Vesala 1974; API 1985; Suntio et al. 1988; Shiu et al. 1990)
546 (shake flask-UV, Morrison & Billett 1952)
550 (Dreisbach 1955; quoted, Chey & Calder 1972; Vesala 1974)
595 (quoted, Deno & Berkheimer 1960)
538 (shake flask-GC, McAuliffe 1963)

515 (shake flask-GC, McAuliffe 1966; quoted, Hermann 1972; Vesala 1974; Mackay & Leinonen 1975; Mackay & Shiu 1975; 1981; 1990; Mackay et al. 1975; Hutchinson et al. 1980; Mackay et al. 1980; Chiou 1981; 1985; Mackay 1981; Chiou et al. 1982; Geyer et al. 1982; API 1985; Garbarnini & Lion 1985; Abernethy et al. 1986; Suntio et al. 1988; Shiu et al. 1988; 1990)
470 (quoted, Hansch et al. 1968)
354 (calculated-K_{OW}, Hansch et al. 1968)
479 (21°C, extraction by nonpolar resins/evolution, Chey & Calder 1972)
573 (shake flask-GC, Polak & Lu 1973; quoted, API 1985)
517 (shake flask-GC, Mackay & Wolkoff 1973)
627 (shake flask-UV, Vesala 1974)
517 (quoted, Hine & Moorkerjee 1975; Kônemann 1981; Kamlet et al.1987)
517 (shake flask-GC, Mackay et al. 1975; Shiu et al. 1990)
520 (shake flask-GC, Mackay & Shiu 1975; quoted, API 1985; Suntio et al. 1988; Shiu et al. 1990)
534.8 (shake flask-GC, Sutton & Calder 1975; quoted, API 1985; Howard 1990)
524, 512 (selected, calcd.- χ , reported as ln S, should be -log S in molar concn., Kier & Hall 1976)
554 (shake flask-GC, Price 1976; quoted, API 1985; Suntio et al. 1988; Shiu et al. 1990)
488 (shake flask-titration, Sada et al. 1975)
534 (shake flask-Fluo., Schwarz 1977; quoted, Shiu et al. 1990)
515 (quoted, Verschueren, 1977 & 1983)
535 (quoted, Callahan et al. 1979; Mabey et al. 1982)
470 (quoted, Neely 1980)
623, 660 (quoted, elution chromatography, Schwarz 1980)
732, 739 (20 °C, exptl.-elution chromatography, shake flask-UV, Schwarz & Miller 1980)
566, 735 (20°C, quoted, average exptl. value, Schwarz & Miller 1980)
1548 (shake flask-LSC, Banerjee et al. 1980)
581 (quoted, Lande & Banerjee 1981)
507 (shake flask-GC, Rossi & Thomas 1981)
542 (quoted lit. mean, Amidon & Anik 1981)
530, 715 (quoted exptl. average, calculated-K_{OW}, Valvani et al. 1981)
585 (gen. col.-HPLC/UV, Teweri et al. 1982b; quoted, Doucette & Andren 1988)
578 (gen. col.-HPLC/UV, Teweri et al. 1982c)
514, 535, 507; 520 (quoted, quoted lit. average, Whitehouse & Cooke 1982)
534 (calculated-HPLC-k', converted from reported γ_W, Hakenscheid & Tomlinson 1983a)
555 (calculated-K_{OW}, Yalkowsky et al. 1983)
580 (gen, col.-HPLC/UV, Wasik et al. 1983)
524 (shake flask-HPLC/UV, Banerjee 1984)
561, 428 (quoted, calculated-UNIFAC, Banerjee 1985)

578 (quoted, Miller et al. 1985; Vowles & Mantoura 1987; Eastcott et al. 1988)
530 (quoted, Brookman et al. 1985; Cline et al. 1991)
529 (quoted, Leahy 1986)
272 (calculated-V_I, solvatochromic parameters, Leahy 1986)
575, 1581 (quoted, calculated-K_{OW} & HPLC-RT, Chin et al. 1986)
515, 265 (quoted, calculated-characteristic volume, McGowen & Mellors 1986)
461 (calculated-V_I, solvatochromic parameters, Kamlet et al. 1987)
707 (calculated- χ , Nirmalakhandan & Speece 1988a)
580 (shake flask-GC, Keeley et al. 1988)
538 (purge and trap-GC, Coutant & Keigley 1988)
525 (quoted, Mackay 1988)
440 (calculated-UNIFAC, Banerjee & Howard 1988)
525 (quoted, Isnard & Lambert 1988, 1989)
515, 530 (quoted, calculated-UNIFAC, Al-Sahhaf 1989)
535 (quoted, Abdul et al. 1990)
545 (quoted, Warne et al. 1990)

Vapor Pressure (Pa at 25 °C):
3800 (interpolated, Antoine eqn., Zwolinski & Wilhoit 1971; quoted, Mackay & Wolkoff 1973; Mackay & Leinonen 1975; Mackay & Shiu 1975; 1981; 1990, Mackay 1981; Eastcott et al. 1988; Suntio et al. 1988)
3792 (quoted, Hine & Moorkerjee 1975)
4000 (quoted, Verschueren, 1977 & 1983)
3826 (quoted, Callahan et al. 1979)
4000 (quoted, Neely 1980)
3749, 3881 (quoted, calculated-B.P., Mackay et al. 1982)
3560 (gas saturation, Politzki et al. 1982)
3786 (Daubert & Danner 1985; quoted, Howard 1990)
3792 (extrapolated, Antoine eqn., Dean 1985)
3800 (extrapolated, Antoine eqn., Stephenson & Malanowski 1987)
3786, 855 (quoted, calculated-UNIFAC, Banerjee et al. 1990)

Henry's Law Constant (Pa m^3/mol):
677 (calculated, Mackay & Leinonen 1975; Mackay 1981)
518 (calculated-bond contribution, Hine & Mookerjee 1975)
682 (calculated as $1/K_{AW}$, C_W/C_A, reported as exptl., Hine & Mookerjee 1975; Nirmalakhandan & Speece 1988b)
673, 675 (batch stripping, calculated-P/C, Mackay et al. 1979; quoted, Suntio et al. 1988)
673 (quoted exptl., Mackay & Shiu 1981)
680 (calculated-P/C, Mackay & Shiu 1981; quoted, Pankow et al. 1984; Pankow 1986; Pankow & Rosen 1988; Pankow 1990; Olsen & Davis 1990)
647 (concentration ratio, Leighton & Calo 1981)

620 (C_A/C_W, Matter-Müller et al. 1981)
675 (calculated, Mabey et al. 1982)
602 (Shen 1982; quoted, Howard 1990)
824 (calculated-UNIFAC, Gmehling et al. 1982)
825 (calculated-UNIFAC, Arbuckle 1983)
647 (EPICS, Garbarnini & Lion 1985)
594 (20°C, EPICS, Yurteri et al. 1987)
519 (20°C, calculated-P/C, Yurteri et al. 1987)
651 (EPICS, Ashworth et al. 1988)
604 (calculated-C_A/C_W, Eastcott et al. 1988)
605 (concentration ratio, Keeley et al. 1988)
637 (calculated-QSAR, Nirmalakhandan & Speece 1988b)
679 (batch stripping, Suntio et al. 1988)
533 (calculated-P/C, Munz & Roberts 1989)
680 (calculated, Mackay & Shiu 1990)
673 (selected, Mackay & Shiu 1990)
694 (calculated-C_A/C_W, Jury et al. 1990)

Octanol/Water Partition Coefficient, log K_{OW}:

2.69 (shake flask-UV, Fujita et al. 1964; quoted, Hansch et al. 1968; Hansch et al. 1972; Chiou et al. 1977; 1982; Chiou 1981; 1985; Hutchinson et al. 1980; McCall 1975; Whitehouse & Cooke 1982; Garbarnini & Lion 1985; Brooke et al. 1986; Hodson & Williams 1988; Suntio et al. 1988; Jafvert 1991)
2.11, 1.83 (shake flask-UV, calculated-M.O. indices, Rogers & Cammarata 1969)
2.69, 2.73, 2.11, 2.80 (quoted, Leo et al. 1971)
2.73, 2.54 (quoted, calculated- χ , Kier & Hall 1976)
2.62 (calculated-f const., Yalkowsky & Valvani 1976)
2.60 (calculated-f const., Rekker 1977; quoted, Harnisch et al. 1983)
2.69, 2.73, 2.71, 2.80 (Hansch & Leo 1979)
2.69 (quoted of Hansch & Leo 1979, Callahan et al. 1979; Harnish et al. 1983; Suntio et al. 1988)
2.69; 2.51 (quoted; calculated-S, Mackay et al. 1980)
2.21 (shake flask-LSC, Banerjee et al. 1980; quoted, Suntio et al. 1988; Olsen & Davis 1990)
2.58, 2.68 (quoted, shake flask-HPLC, Nahum & Horvath 1980)
2.59 (HPLC-k', Hanai et al. 1981)
2.69, 2.97 (quoted, HPLC-k', McDuffie 1981)
2.59 (calculated-f const., Könemann 1981; quoted, Leegwater 1989)
2.58 (quoted, Valvani et al. 1981)
2.79 (calculated-f constants, Mabey et al. 1982)
2.78 (HPLC-k, Hammers et al. 1982)
2.59 (HPLC-k', D'Amboise & Hanai 1982)
2.69, 2.65 (quoted, gen. col.-HPLC/UV, Teweri et al. 1982b,c)

2.69, 2.62 (quoted, HPLC-k', Miyake & Terada 1982)
2.77, 2.65 (quoted of HPLC methods, Harnisch et al. 1983)
2.69 (quoted, Verschueren 1983)
2.63 (calculated-f constants, Yalkowsky et al. 1983)
2.65 (gen. col.-HPLC/UV, Wasik et al. 1983)
2.74 (HPLC-k', Hafkanscheid & Tomlinson 1983b)
2.69, 2.52, 2.76 (quoted, calculated-MR & IP, MR, Yoshida et al. 1983)
2.11-2.80, 2.65 (range, mean; shake flask method, Eadsforth & Moser 1983)
2.51-3.06, 2.88 (range, mean; HPLC method, Eadsforth & Moser 1983)
2.10 (Platford 1979,1983)
2.11-2.80, 2.69 (range, quoted, Garst & Wilson 1984; quoted, Sabljic 1987b)
2.72 (HPLC-RV, Garst & Wilson 1984)
2.89 (HPLC-RT, Rapaport & Eisenreich 1984)
2.58, 2.78, 2.79 (quoted, HPLC/MS, calculated, Burkhard et al. 1985)
2.73 (quoted, Campbell & Luthy 1985)
2.65 (quoted, Miller et al. 1985; Vowles & Mantoura 1987; Szabo et al. 1990a,b)
2.66 (calculated-V_I, solvatochromic parameters, Taft et al. 1985)
2.73 (Hansch & Leo 1985; quoted, Howard 1990)
2.63 (quoted, Burkhard & Kuehl 1986)
2.78 (quoted, Tomlinson & Hafkanscheid 1986)
2.92 (calculated-V_I, solvatochromic parameters, Leahy 1986)
3.00 (HPLC-k', De Kock & Lord 1987)
2.65, 2.64 (gen. col.-RPLC, calculated-activity coeff., Schantz & Martire 1987)
2.69 (quoted, Ryan et al. 1988; Lee et al. 1989)
2.65 (quoted exptl., Doucette & Andren 1988)
2.60, 2.93, 2.81, 2.75, 3.01 (calculated-π, f const., HPLC-RT, MW, TSA, Doucette & Andren 1988)
2.11-2.94 (quoted range, Dearden & Bresnen 1988)
2.73, 2.39 (quoted, calculated-UNIFAC, Banerjee & Howard 1988)
2.73 (recommended, Sangster 1989)
2.63 (quoted, Isnard & Lambert 1988; 1989)
2.786 (slow stirring-GC, De Brujin et al. 1989)
2.74, 2.73 (quoted, calculated-MO calculation, Bodor et al. 1989)
2.73, 2.791 (quoted, calculated-f constants, Bodor et al. 1989)
2.11-2.94 (quoted from the Medchem Data-base, Dearden 1990)
2.65 (quoted, Szabo et al. 1990a,b)
2.687 (quoted, Warne et al. 1990)

Bioconcentration Factor, log BCF:
1.12 (eels, Ogata & Miyake 1978)
0.22 (Manila clam, Nunes & Benville 1979)
0.62 (mussels, Geyer et al. 1982)
2.17 (microorganisms-water, calculated-K_{OW}, Mabey et al. 1982)

0.92 (goldfish, Ogata et al. 1984),
3.28, 2.58, 1.95 (activated sludge, algae, fish, Freitag et al. 1985)
0.89, 0.92 (fish, calculated, correlated, Sabljic 1987b)
1.12, 0.22, 0.72, 2.58, 1.95 (quoted: eels, Manila clam, mussels, algae, fish, Howard 1990)
1.99 (s. capricornutum, Herman et al. 1991)

Sorption Partition Coefficient, log K_{OC}:
2.49 (soil, calculated-K_{OW}, Karickhoff et al. 1979)
2.43 (soil, calculated-K_{OW}, Means et al. 1980)
2.25 (Wilson et al. 1981)
2.39 (soil, Schwarzenbach & Westall 1981)
2.48 (sediment, calcd.-K_{OW}, Mabey et al. 1982)
1.89-2.28 (soil, headspace equilibrium technique, Garbarnini & Lion 1985)
1.93, 2.43, 2.49 (soil: calculated-S, K_{OW}, Garbarnini & Lion 1985)
2.32 (soil, calculated- χ , Sabljic 1987a & b)
1.12-2.85 (soil, calculated-K_{OW}, Sabljic 1987b)
2.58, 1.77, 3.28 (quoted: algae, fish, activated sludge, Halfon & Reggiani 1986)
2.00, 2.18 (quoted, Hodson & Williams 1988)
2.0, 2.18, 2.25 (quoted, Howard 1990)
2.18, 2.26 (quoted, HPLC-k', Szabo et al. 1990a,b)
1.99 (soil, Jury et al. 1990)

Sorption Partition Coefficient, log K_{OM}:
1.29, 1.39 (untreated soil, Lee et al. 1989)
2.16-2.89 (organic cations treated soil, Lee et al. 1989)

Half-Lives in the Environment:

Air: 4.6 hours (Doyle et al. 1975) and 2.4-24 hours (Darnall et al. 1976) based on rate of disappearance for the reaction with hydroxy radicals; 10-104 hours, based on photooxidation half-life in air (Atkinson 1985; Howard et al. 1991).

Surface Water: 5.55 hours, based on evaporative loss at 25°C and 1 m depth of water (calculated, Mackay & Leinonen 1975; Haque et al. 1980); 96-528 hours, based on estimated aqueous aerobic biodegradation half-life (Wakeham et al. 1983; Howard et al. 1991).

Groundwater: estimated half-life from observed persistence in groundwater of Netherlands, 0.3 year (Zoeteman et al. 1981); 168-672 hours, based on unacclimated grab sample data of aerobic soil from groundwater aquifers (Wilson et al. 1983; Swindoll et al. 1987; Howard et al. 1991).

Soil: 96-528 hours, based on estimated aqueous aerobic biodegradation half-life (Wakeham et al. 1983; Howard et al. 1991); < 10 days (Ryan et al. 1988); 5 days assumed first-order biological/chemical degradation in soil (Jury et al. 1990); disappearance half-life from test soils, < 2.0 days (Anderson et al. 1991).

Biota: 10 hours clearance from fish (Neely 1980).

Environmental Fate Rate Constants or Half-Lives:

Volatilization: $t_{1/2}$ = 5.18 hours from water depth of 1 m (Mackay & Leinonen 1975; Haque et al. 1980); half-lives from marine mesocosm: 16 days at 8-16°C in the spring, 1.5 days at 20-22°C in the summer and 13 days at 3-7°C in the winter (Wakeham et al. 1983); estimated evaporation half-life from a river of 1 m depth with wind speed of 3 m/sec and water current of 1 m/s to be 2.9 h at 20°C (Lyman et al. 1982); estimated half-life for evaporation from a river and lake to be 1 and 4 days (Howard 1990).

Photolysis: not environmentally significant or relevant (Mabey et al. 1982).

Oxidation: photooxidation half-life in water: 321-1284 hours, based on measured rate data for hydroxy radicals in aqueous solution (Dorfman & Adams 1973; Howard et al. 1991); calculated rate constant for the reaction with OH radicals in ambient LA basin air at 300°K was 2.5×10^{9} liter $mole^{-1}$ sec^{-1} with an initial concentration of 9.8×10^{-10} mole $liter^{-1}$ (Doyle et al. 1975); absolute room temperature rate constant for the reaction with OH radicals determined by flash photolysis-resonance fluorescence was 5.78×10^{-12} cm^3 $molecule^{-1}$ sec^{-1} (Hansen et al. 1975); rate constant for reaction in gas phase with hydroxyl radicals = 3.6×10^{9} L mol^{-1} s^{-1} and $t_{1/2}$ = 2.4-24 h (Darnall et al. 1976); half-life about 15 hours, probably not an important aquatic fate (Callahan et al. 1979); k (singlet oxygen) $<<$ 360 M^{-1} h^{-1} & k (RO_2) = 144 M^{-1} h^{-1} (Mabey et al. 1982); reaction with ozone at 300°K having a rate constant of 160 cm^3 mol^{-1} sec^{-1} (Lyman et al. 1982); rate constant k = 3.5×10^{9} M^{-1} sec^{-1}, $t_{1/2}$ = 1.6 days, oxidation by OH radicals in the atmosphere (Mill 1982); photooxidation half-life in air: 10-104 hours, based on measured room temperature rate constant of 5.7×10^{-12} cm^3 $molecule^{-1}$ sec^{-1} for the vapor phase reaction with hydroxy radicals in air (Atkinson 1985; Howard et al. 1991); rate constant for the gas phase reaction with OH radicals at 25°C, 6.03×10^{-12} cm^3 $molecule^{-1}$ sec^{-1} (Ohta & Ohyama 1985).

Hydrolysis: not aquatically significant (Callahan et al. 1979); k_h = 0, no hydrolyzable functional groups (Mabey et al. 1982).

Biodegradation: 100% biodegraded after 192 hours at 13°C with an initial concentration of 2.22×10^{-6} l/l (Jamison et al. 1976); half-life in uncontaminated estuarine water estimated to be 90 days and in oil polluted water estimated to be 30 days (Lee 1977); significant degradation in aerobic environment, k_B = 0.5 day^{-1} (Tabak et al. 1981; Mills et al. 1982); aqueous aerobic biodegradation half-life: 96-528 hours, based on an acclimated seawater dieaway test (Wakeham et al. 1983; Howard et al. 1991); anaerobic aqueous biodegradation half-life: 1344-5040 hours, based on anaerobic screening test data and anaerobic sediment grab sample data (Horowitz et al. 1982; Howard et al. 1991); first-order rate constant 0.07 $year^{-1}$ with a half-life of 39 days (Zoeteman et al. 1981; Olsen & Davis 1990).

Biotransformation: $k_b = 1 \times 10^{-7}$ ml cell^{-1} h^{-1} (Mabey et al. 1982).

Bioconcentration, Uptake (k_1) and Elimination (k_2) Rate Constants or Half-Lives: half-life of elimination from eels in seawater estimated to be 1.4 days (Ogata & Miyake 1978).

Common Name: Ethylbenzene
Synonym: phenylethane, ethylbenzol
Chemical Name: ethylbenzene
CAS Registry No: 100-41-4
Molecular Formula: $C_2H_5C_6H_5$
Molecular Weight: 106.2
Melting Point (°C): -95
Boiling Point (°C): 136.2
Density (g/cm^3 at 20°C): 0.867
Molar Volume (cm^3/mol):

122.4 (20°C, calculated-density, McAuliffe 1966; Stephenson & Malanowski 1987)
122 (calculated-density, Lande & Banerjee 1981)
123.1 (Wasik et al. 1984)
123 (calculated-density, Klevens 1950; Chiou 1985)
122.5 (25°C, calculated-density, Miller et al. 1985)
122.5 (calculated-density, Leahy 1986)
0.687 (intrinsic volume: V_I/100, Leahy 1986)
140.4 (LeBas method, Abernethy & Mackay 1987; Eastcott et al. 1988)
0.671 (intrinsic volume: V_I/100, Kamlet et al. 1988)

Molecular Volume (A^3):

116.21 (Pearlman 1986)

Total Surface Area, TSA (A^2):

144.9 (Yalkowsky & Valvani 1976; Whitehouse & Cooke 1982)
315.6 (with hydration shell, Amidon & Anik 1981)
154.66 (Pearlman 1986)
149.2 (calculated- χ , Sabljic 1987a)

Heat of Fusion, kcal/mol:
Fugacity Ratio, F: 1.0

Water Solubility (g/m^3 or mg/L at 25°C):

140 (15°C, volumetric, Führer 1924)
168 (shake flask-UV, Andrews & Keefer 1950; quoted, API 1985)
175 (shake falsk-UV, Klevens 1950)
208 (shake flask-UV, Bohon & Claussen 1952; quoted, Vesala 1974; API 1985)
165 (shake flask-UV, Morrison & Billett 1952)
172 (quoted, Deno & Berkheimer 1960)
159 (shake flask-GC, McAuliffe 1963)
152 (shake flask-GC, McAuliffe 1966; quoted, Hermann 1972; Vesala 1974; Mackay & Wolkoff 1973; Mackay & Leinonen 1975; Mackay & Shiu 1975; 1981; 1990; Hutchinson et al 1980; API 1985; Abernethy et al. 1986)
140 (quoted, Hansch et al. 1968)
112.8 (calculated-K_{OW}, Hansch et al. 1968)
177 (shake flask-GC, Polak & Lu 1973; quoted, API 1985)

180 (GC, Brown & Wasik 1974)
203 (shake flask-UV, Vesala 1974)
161 (shake flask-GC, Sutton & Calder 1975; quoted, API 1985; Howard 1989; Cline et al. 1991)
153.5 (quoted, Hine & Mookerjee 1975; Kamlet et al. 1987)
131 (shake flask-GC, Price 1976; quoted, API 1985)
154, 154 (quoted, calculated- χ , reported as ln S, should be -log S in molar concn., Kier & Hall 1976)
208, 184 (20°C, elution chromatography, shake flask-UV, Schwarz & Miller 1980)
207, 196 (20°C, quoted, average exptl. value, Schwarz & Miller 1980)
164, 217 (quoted exptl. average, calculated-K_{OW}, Valvani et al. 1981)
181 (shake flask-UV, Sanemasa et al. 1981)
159 (quoted lit. mean, Amidon & Anik 1981)
168 (quoted, Lande & Banerjee 1981)
169 (shake flask-UV, Sanemasa et al. 1982)
172 (gen. col.-HPLV/UV, Tewerai et al. 1982a)
187 (gen. col.-HPLC/UV, Teweri et al. 1982c)
143, 266 (quoted, calculated-K_{OW}, Amidon & Williams 1982)
152, 161, 156 (quoted, literature average, Whitehouse & Cooke 1982)
168 (calculated-K_{OW}, Yalkowsky et al. 1983)
152 (20°C, Verschueren, 1983)
166 (calculated-HPLC-k', converted from γ_W, Hakenscheid & Tomlinson 1983a)
187 (gen. col.-HPLC/UV, Wasik et al. 1983)
172 (vapor saturation-UV, Sanemasa et al. 1984)
168, 174 (quoted, calculated-UNIFAC, Banerjee 1985)
135 (quoted, Miller et al. 1985; Vowles & Mantoura 1987; Eastcott et al. 1988)
222, 655 (quoted, calculated-K_{OW} & HPLC-RT, Chin et al. 1986)
192 (gen. col.-HPLC/UV, Owens et al. 1986)
164, 80.6 (quoted, calculated-V_I, solvatochromic parameters, Leahy 1986)
207 (calculated-V_I, solvatochromic parameters, Kamlet et al. 1987)
169 (calculated- χ , Nirmalakhanden & Speece 1988a)
172 (shake flask-purge and trap-GC, Coutant & Keigley 1988)
178 (calculated-UNIFAC, Banerjee & Howard 1988)
60, 57 (quoted, calculated-UNIFAC, Al-Sahhaf 1989 - note: value should be for 1,2,4-trimethylbenzene)
170 (correlated value, Isnard & Lambert 1989)
169 (recommended value, IUPAC 1989b)
157 (quoted, Warne et al. 1990)

Vapor Pressure (Pa at 25°C):

1270 (interpolated, Antoine eqn., Zwolinski & Wilhoit 1971; quoted, Mackay & Wolkoff 1973; Mackay & Leinonen 1975; Mackay & Shiu 1975,1981,1990; Eastcott et al. 1988)

1276 (quoted, Hine & Mookerjee 1975)
1319, 1297 (quoted exptl., calculated-B.P., Mackay et al. 1982)
1266 (extrapolated, Verschueren, 1983)
1270 (Boublik et al. 1984; quoted, Howard 1989)
1270 (quoted, Wasik et al. 1984)
1268 (extrapolated-Antoine eqn., Dean 1985)
1266 (extrapolated-Antoine eqn., Stephenson & Malanowski 1987)
1280, 283 (quoted, calculated-UNIFAC, Banerjee et al. 1990)

Henry's Law Constant (Pa m^3/mol):
757 (calculated-bond contribution, Hine & Mookerjee 1975)
879 (calculated as $1/K_{AW}$, C_W/C_A, reported as exptl., Hine & Mookerjee 1975; Nirmalakhanden & Speece 1988b)
854 (batch stripping, Mackay et al. 1979; quoted, Howard 1989)
884 (calculated-P/C, Mackay et al., 1979)
854, 800 (quoted, exptl, calculated-P/C, Mackay & Shiu 1981,1990; selected, Olsen & Davis 1990)
669 (quoted, Mackay et al. 1982)
800 (quoted, Pankow et al. 1984)
669 (quoted, Pankow 1986; Pankow & Rosen 1988; Pankow 1990)
798 (EPICS, Ashworth et al. 1988)
1001 (calculated-C_A/C_W, Eastcott et al. 1988)
748 (calculated-QSAR, Nirmalakhanden & Speece 1988b)
864, 887 (quoted, calculated-P/C, Mackay & Shiu 1990)

Octanol/Water Partition Coefficient, log K_{OW}:
3.15 (shake flask-UV, Hansch et al. 1968; Hansch et al. 1972)
3.15, 2.98 (selected, calculated- χ , Kier & Hall 1976)
3.13 (calculated, Yalkowsky & Valvani 1976)
3.13 (calculated-f const., Rekker 1977; quoted, Harnisch et al. 1983)
3.15 (Hansch & Leo 1979; selected, Hutchinson et al. 1980; Valvani et al. 1981; Amidon & Williams 1982; Whitehouse & Cooke 1982; Harnish et al. 1983; Burkhard & Kuehl 1986)
3.15, 3.10 (quoted, calculated-S, Mackay et al. 1980)
3.12 (HPLC-k', Hanai et al. 1981)
3.12 (HPLC-k', D'Amboise & Hanai 1982)
3.15, 3.26 (HPLC-k', correlated, Hammers et al. 1982)
3.15 (gen. col.-HPLC/UV, Teweri et al. 1982a)
3.15, 3.13 (quoted, gen. col.-HPLC/UV, Teweri et al. 1982c)
3.15, 3.16 (quoted, HPLC-k', Miyake & Terada 1982)
3.30, 3.17 (quoted from HPLC methods, Harnisch et al. 1983)
3.13 (calculated-f const., Yalkowsky et al. 1983)
3.15 (quoted, Verschueren, 1983)

3.15, 3.24 (quoted, calculated-HPLC-k', Hafkenscheid & Tomlinson 1983b)
3.08, 3.15, 3.11 (quoted, calculated-MR & IP, MR, Yoshida et al. 1983)
3.13 (gen. col.-HPLC/UV, Wasik et al. 1983)
3.13 (quoted, Miller et al. 1985; Vowles & Mantoura 1987; Abernethy et al. 1988; Eastcott et al. 1988; Szabo et al 1990a,b)
3.14 (quoted exptl., Campbell & Luthy 1985)
3.15 (Hansch & Leo 1985; quoted, Howard 1989)
3.43 (calculated-V_I & solvatochromic parameters, Leahy 1986)
3.13, 3.14 (gen. col.-RPLC, calculated-activity coeff., Schantz & Martire 1987)
3.15 (quoted, Ryan et al. 1988; Lee et al. 1989)
3.15; 2.68 (quoted; calcd.-UNIFAC, Banerjee & Howard 1988)
3.13 (quoted, Isnard & Lambert 1989)
3.15 (recommended, Sangster 1989)
3.15, 3.15 (quoted, calculated-MO calculation, Bodor et al. 1989)
3.15, 3.32 (quoted; calculated-f constants, Bodor et al. 1989)
3.141 (quoted, Warne et al. 1990)

Bioconcentration Factor, log BCF:
0.67 (clams, exposed to water-soluble fraction of crude oil, Nunes & Benville 1979; selected, Howard 1989)
2.16 (fish, calculated, Lyman et al. 1982; quoted, Howard 1989)
2.67 (microorganisms-water, calculated-K_{OW}, Mabey et al., 1982)
1.19 (goldfish, Ogata et al. 1984)
1.20, 1.19 (fish: calculated, correlated, Sabljic 1987b)
2.31 (s. capricornutum, Herman et al. 1991)

Sorption Partition Coefficient, log K_{OC}:
3.04 (sediment, calculated-K_{OW}, Mabey et al. 1982)
1.98 (soil, sorption isotherm, Chiou et al. 1983)
2.30 (HPLC-k', Hodson & Williams 1988)
2.21 (silt loam from Chiou et al. 1983, Howard 1989)
2.41, 2.52, 2.47 (quoted, HPLC-k', Szabo et al. 1990a,b)

Sorption Partition Coefficient, log K_{OM}:
1.73-1.97 (untreated soil, Lee et al. 1989)
2.37-3.23 (organic cations treated soil, Lee et al. 1989)

Half-Lives in the Environment:
Air: 0.24-24 hours, based on rate of disappearance for the reaction with hydroxy radicals (Darnall et al. 1976); 8.56-85.6 hours, based on photooxidation half-life in air (Atkinson 1985; Howard et al. 1991).
Surface Water: 5-6 hours (Callahan et al. 1979), based on the estimated evaporative loss of toluene at 25°C and 1 m depth of water (Mackay & Leinonen 1975); 72-

240 hours, based on unacclimated aqueous aerobic biodegradation half-life (Van der Linden 1978; Howard et al. 1991).

Groundwater: 144-5472 hours, based on unacclimated aqueous aerobic biodegradation half-life and seawater dieaway test data (Van der Linden 1978; Howard et al. 1991); estimated half-life from observed persistence in groundwater of the Netherlands, 0.3 year (Zoeteman et al. 1981).

Soil: 72-240 hours, based on unacclimated aqueous aerobic biodegradation half-life (Van der Linden 1978; Howard et al. 1991); < 10 days (Ryan et al. 1988).

Environmental Fate Rate Constants or Half-lives:

Volatilization: half-lives from marine mesocosm: 20 days at 8-16°C in the spring, 2.1 days at 20-22°C in the summer and 13 days at 3-7°C in the winter (Wakeham et al. 1983); estimated half-life of evaporation from a river of 1 m depth with wind speed 3 m/s and water current of 1 m/s to be 3.1 hours at 20°C (Lyman et al. 1982; quoted, Howard 1989).

Photolysis: not environmentally significant or relevant (Mabey et al. 1982).

Oxidation: rate constants for reaction with hydroxyl radicals in gas phase = 4.8×10^9 L mol^{-1} s^{-1} and $t_{1/2}$ = 0.24-24 h (Darnall et al. 1976); half-life in water is about 15 hours, probably not important as aquatic fate (Callahan et al. 1979); k (singlet oxygen) << 360 M^{-1} h^{-1} & k (RO_2 radical) = 720 M^{-1} h^{-1} (Mabey et al. 1982); reaction with ozone at 300°K having a rate constant of 340 cm^3 mol^{-1} s^{-1} (Lyman et al. 1982); photooxidation half-life in air: 8.56-85.6 hours, based on measured room temperature rate constant of 7.5×10^{-12} cm^3 $molecule^{-1}$ sec^{-1} for reaction with hydroxy radicals in air (Atkinson 1985); rate constant for the reaction with OH radicals at 25°C, 6.47×10^{-12} cm^3 $molecule^{-1}$ sec^{-1} (Ohta & Ohyama 1985).

Hydrolysis: not aquatically significant (Callahan et al. 1979); $k_h = 0$, no hydrolyzable functional groups (Mabey et al. 1982); rate constant $k_h = 4.4 \times 10^{-9}$ M^{-1} s^{-1}, $t_{1/2}$ = 1.3 days, oxidation by HO radical in the atmosphere (Mill 1982) .

Biodegradation: aqueous aerobic biodegradation half-life: 72 -240 hours, based on unacclimated aqueous aerobic biodegradation half-life and seawater dieaway test data (Van der Linden 1978; Howard et al. 1991); anaerobic aqueous biodegradation half-life: 4224-5472 hours, based on anaerobic groundwater die-away test data (Wilson et al. 1986; Howard et al. 1991); significant degradation under favorable conditions in an aerobic environment, k_B = 0.5 day^{-1} (Tabak et al. 1981; Mills et al. 1982); 100% biodegraded after 192 hours at 13°C with an initial concentration of 1.36×10^{-6} l/l (Jamison et al. 1976); half-life of degradation by established microorganisms about 2 days depending on body of water and its temperature (Howard 1989); first-order rate constant 0.07 $year^{-1}$ with a half-life of 37 days (Olsen & Davis 1990).

Biotransformation: $k_b = 3 \times 10^{-9}$ ml $cell^{-1}$ h^{-1} (Mabey et al. 1982).

Common Name: o-Xylene
Synonym: 1,2-dimethylbenzene, o-xylol
Chemical Name: o-xylene
CAS Registry No: 95-47-6
Molecular Formula: $C_6H_4(CH_3)_2$
Molecular Weight: 106.2
Melting Point (°C): -25.2
Boiling Point (°C): 144
Density (g/cm^3 at 20°C): 0.8802
Molar Volume (cm^3/mol):

123.47 (calculated-density, Windholz 1983; Budavari 1989)
120.6 (20°C, calculated-density, McAuliffe 1966; Stephenson & Malanowski 1987)
121 (calculated-density, Lande & Banerjee 1981)
121.2 (Wasik et al. 1984)
120.6 (25°C, calculated-density, Miller et al. 1985)
0.671 (intrinsic volume: $V_I/100$, Kamlet et al. 1987,1988)
140.4 (LeBas method, Abernethy & Mackay 1987; Eastcott et al. 1988)

Molecular Volume (A^3):

115.97 (Pearlman 1986)

Total Surface Area, TSA (A^2):

146.8 (Yalkowsky & Valvani 1976)
309.7 (calculated with hydration shell, Amidon & Anik 1981)
153.53 (Pearlman 1986)
146.59 (Doucette & Andren 1988)

Heat of Fusion, kcal/mol:
Entropy of Fusion, cal/mol K (e.u.):

13.2 (Yalkowsky & Valvani 1980)

Fugacity Ratio, F: 1.0

Water Solubility (g/m^3 or mg/L at 25°C):

204 (shake flask-UV, Andrews & Keefer 1949; quoted, Shiu et al. 1990)
202 (quoted, Deno & Berkheimer 1960)
175 (shake flask-GC, McAuliffe 1963)
175 (shake flask-GC, McAuliffe 1966; quoted, Mackay & Leinonen 1975; Mackay & Shiu 1975; 1981; 1990; Mackay 1981; API 1985; Abernethy et al. 1986; Shiu et al. 1990; Cline et al. 1991)
176 (quoted, Hansch et al. 1968)
112.8 (calculated-K_{OW}, Hansch et al. 1968)
176 (shake flask-GC, Hermann 1972)
213 (shake flask-GC, Polak & Lu 1973; quoted, API 1985; Shiu et al. 1990)
176 (quoted, Hine & Mookerjee 1975; Könemann 1981; Kamlet et al. 1987)
170.5 (shake flask-GC, Sutton & Calder 1975; quoted, API 1985; Shiu et al. 1990)

176, 207 (selected, calculated- χ , reported as ln S, should be -log S in molar concn., Kier & Hall 1976)
167 (shake flask-GC, Price 1976; quoted, API 1985)
171 (quoted lit. mean, Amidon & Anik 1981)
193 (quoted, Lande & Banerjee 1981)
185, 352 (quoted exptl. average, calculated-K_{OW}, Valvani et al. 1981)
221 (gen. col.-HPLC/UV, Tewari et al. 1982c; quoted, Doucette & Andren 1988)
221 (gen. col.-HPLC/UV, Wasik et al. 1983; quoted, Shiu et al. 1990)
180 (calculated-K_{OW}, Yalkowsky et al. 1983)
221 (quoted, Miller et al. 1985; Vowles & Mantoura 1987)
189, 655 (quoted, calculated-K_{OW} & HPLC-RT, Chin et al. 1986)
175 (Riddick et al. 1986; quoted, Howard 1990)
207 (calculated-V_I, solvatochromic parameters, Kamlet et al. 1987)
215 (quoted, Eastcott et al. 1988)
223 (calculated- χ , Nirmalakhanden & Speece 1988a)
176 (shake flask-purge & trap-GC, Coutant & Keigley 1988)
175, 170 (quoted, calculated-UNIFAC, Al-Sahhaf 1989)
200 (quoted, Isnard & Lambert 1988, 1989)

Vapor Pressure (Pa at 25 °C):
882 (interpolated, Antoine eqn., Zwolinski & Wilhoit 1971; quoted, Mackay & Shiu 1975,1981,1990; Eastcott et al. 1988)
892 (quoted, Hine & Mookerjee 1975)
933 (extrapolated, Verschueren, 1977& 1983)
871, 908 (quoted exptl., calculated-B.P., Mackay et al. 1982)
882 (quoted, Wasik et al. 1984)
882 (extrapolated-Antoine eqn., Dean 1985)
880 (Riddick et al. 1986; quoted, Howard 1990)
885 (extrapolated-Antoine eqn., Stephenson & Malanowski 1987)

Henry's Law Constant (Pa m^3/mol):
506 (calculated-bond contribution, Hine & Mookerjee 1975)
534 (calculated-P/C, Mackay & Leinonen 1975; Mackay 1981)
542 (calculated as 1/K_{AW}, C_W/C_A, reported as exptl., Hine & Mookerjee 1975; Nirmalakhanden & Speece 1988b)
500 (calculated P/C, Mackay & Shiu 1981; quoted, Olsen & Davis 1990)
647 (concentration ratio, Leighton & Calo 1981)
895 (calculated-UNIFAC, Gmehling et al. 1982)
507 (quoted, Pankow 1986; Pankow & Rosen 1988; Pankow 1990)
594 (20°C, EPICS, Yurteri et al. 1987)
519 (20°C, calculated-P/C, Yurteri et al. 1987)
879 (calculated-QSAR, Nirmalakhanden & Speece 1988b)
493 (EPICS, Ashworth et al. 1988)

436 (calculated-C_A/C_W, Eastcott et al. 1988)
517 (quoted, Howard 1990)
493, 535 (quoted, calculated-P/C, Mackay & Shiu 1990)

Octanol/Water Partition Coefficient, log K_{OW}:
3.15 (calculated-π substituent constant, Hansch et al. 1968)
3.18 (calculated-f const., Yalkowsky & Valvani 1976)
2.77, 2.12 (quoted, Hansch & Leo 1979)
3.12, 3.04 (quoted, calculated-S, Mackay et al. 1980)
2.73 (shake flask-LSC, Banerjee et al. 1980)
3.09 (calculated-f const., Könemann 1981)
2.95 (quoted lit. exptl. value, Valvani et al. 1981)
3.19 (HPLC-k', Hammers et al. 1982)
3.12, 3.13 (quoted, gen. col.-HPLC/UV, Tewari et al. 1982b,c)
3.14 (calculated-f const., Yalkowsky et al. 1983)
2.77 (quoted, Verschueren, 1983)
3.13 (gen. col.-HPLC/UV, Wasik et al. 1983)
3.12, 3.04, 3.11 (quoted, calculated-MR & IP, MR, Yoshida et al. 1983)
2.77-3.12, 3.12 (range, quoted, Garst & Wilson 1984)
3.18 (HPLC-RV, Garst & Wilson 1984)
3.13 (quoted, Miller et al. 1985; Vowles & Mantoura 1987; Eastcott et al. 1988)
3.12 (Hansch & Leo 1985; quoted, Howard 1990)
3.13 (quoted exptl., Doucette & Andren 1988)
3.14, 3.14, 3.06, 3.16, 3.42 (calculated-π, f const., MW, χ, TSA, Doucette & Andren 1988)
3.09 (quoted, Abernethy et al. 1988)
3.12 (recommended, Sangster 1989)
3.16 (quoted, Isnard & Lambert 1988; 1989)
3.13 (quoted, Szabo et al. 1990a,b)

Bioconcentration Factor, log BCF:
1.33 (eels, Ogata & Miyake 1978)
0.79 (clams, Nunes & Benville 1979)
1.15 (goldfish, Ogata et al. 1984)
1.58 (fish, calculated, Sabljic 1987b)
1.15 (fish, correlated, Sabljic 1987b)
1.33 (fish, correlated, Isnard & Lambert 1988)
1.33, 0.79 (quoted, eels, clams, Howard 1990)
2.34 (s. capricornutum, Herman et al. 1991)

Sorption Partition Coefficient, log K_{OC}:
1.68-1.83 (Nathwani & Philip 1977)
2.73 (HPLC-k', Hodson & Williams 1988)
1.68-1.83 (quoted, Howard 1990)

2.34, 2.37 (quoted, calculated-HPLC-k', Szabo et al. 1990a,b)

Half-Lives in the Environment:

Air: 0.24-2.4 hours, based on rate of disappearance for the reaction with hydroxy radicals (Darnall et al. 1976); 4.4-44 hours, based on photooxidation half-life in air (Atkinson 1985; Howard et al. 1991).

Surface Water: 5.18 hours, based on evaporative loss at 25°C and 1 m depth of water (Mackay & Leinonen 1975; Haque et al. 1980); 168-672 hours, based on estimated aqueous aerobic biodegradation half-life (Bridie et al. 1979; Kuhn et al. 1985; Howard et al. 1991); volatilization to be the dominant removal process with a half-life of 1-5 days (Howard 1990).

Groundwater: 336-8640 hours, based on estimated aqueous aerobic and anaerobic biodegradation half-life (Bridie et al. 1979; Kuhn et al. 1985; Wilson et al. 1986; Howard et al. 1991); estimated half-life from observed persistence in groundwater of the Netherlands, 0.3 year (Zoeteman et al. 1981).

Soil: 168-672 hours, based on estimated aqueous aerobic biodegradation half-life (Bridie et al. 1979; Kuhn et al. 1985; Howard et al. 1991).

Environmental Fate Rate Constants or Half-Lives:

Volatilization: $t_{1/2}$ = 5.61 hours from water depth of 1 m (Mackay & Leinonen 1975; Haque et al. 1980); half-life of evaporation from water of 1 m depth with wind speed of 3 m/sec and water current of 1 m/sec estimated to be 3.2 hours (Lyman et al. 1982); half-life of evaporation from a typical river or pond estimated to be 31-125 hours (Howard 1990).

Photolysis:

Oxidation: photooxidation half-life in water: 3.9×10^{5}-2.7×10^{8} hours, based on estimated rate data for alkoxyl radicals in aqueous solution (Hendry et al. 1974); absolute rate constant for the reaction with OH radicals determined by flash photolysis-resonance fluorescence at room temperature was 15.3×10^{-12} cm^3 $molecule^{-1}$ sec^{-1} (Hansen et al. 1975); rate constant for reaction with hydroxyl radicals in gas phase is 8.4×10^{9} L mol^{-1} s^{-1} and $t_{1/2}$ = 0.24-24 h (Darnall et al. 1976); reaction with ozone at 300°K having a rate constant of 950 cm^3 $mole^{-1}$ sec^{-1} (Lyman et al. 1982); rate constant k = 5.9 -12 $\times 10^{-9}$ M^{-1} s^{-1}, $t_{1/2}$ = 0.47-1.0 days, oxidation by HO radical in the atmosphere (Mill 1982); photooxidation half-life in air: 4.4-44 hours, based on a measured room temperature rate constant of 13.4×10^{-12} cm^3 $molecule^{-1}$ sec^{-1} for vapor phase reaction with hydroxy radicals in air (Atkinson 1985); rate constant for the gas phase reaction with OH radicals at 25°C, 12.5×10^{-12} cm^3 $molecule^{-1}$ sec^{-1} (Ohta & Ohyama 1985).

Hydrolysis: k_h = 0, no hydrolyzable functional groups (Mabey et al. 1982).

Biodegradation: 100% biodegraded after 192 hours at 13°C with an initial concentration of 1.62×10^{-6} l/l (Jamison et al. 1976); unacclimated aqueous aerobic biodegradation half-life estimated to be 168-672 hours, based on aqueous screening test data (Bridie et al. 1979; Howard et al. 1991) and soil column study simulating an aerobic river/ground-water infiltration system (Kuhn et al. 1985; Howard et al. 1991); unacclimated aqueous anaerobic biodegradation half-life estimated to be 4320-8640 hours, based on acclimated grab sample data for anaerobic soil from a groundwater aquifer receiving landfill leachate (Wilson et al. 1986) and a soil column study simulating an anaerobic river/groundwater infiltration system (Kuhn et al. 1985; Howard et al. 1991); first-order rate constant 0.06 $year^{-1}$ with a half-life of 32 days (Olsen & Davis 1990).

Bioconcentration, Uptake (k_1) and Elimination (k_2) Rate Constants or Half-Lives: half-life to eliminate from eels in seawater, 2 days (Ogata & Miyake 1978).

Common Name: m-Xylene
Synonym: 1,3-dimethylbenzene, m-xylol
Chemical Name: m-xylene
CAS Registry No: 108-38-3
Molecular Formula: $C_6H_4(CH_3)_2$
Molecular Weight: 106.2
Melting Point (°C): -47.4
Boiling Point (°C): 139.3
Density (g/cm³ at 20°C): 0.8842
Molar Volume (cm³/mol):

121.20 (calculated-density, Windholz 1983; Budavari 1989)
123.2 (25°C, from density, Miller et al. 1985)
0.668 (intrinsic volume: $V_I/100$, Leahy 1986; Kamlet et al. 1987)
0.671 (intrinsic volume: $V_I/100$, Kamlet et al. 1988)
122.9 (20°C, calculated-density, Stephenson & Malanowski 1987)
140.4 (LeBas method, Abernethy & Mackay 1987; Eastcott et al. 1988)

Molecular Volume (A³):

116.0 (Pearlman 1986)

Total Surface Area, TSA (A²):

150.3 (Yalkowsky & Valvani 1976)
317.6 (with hydration shell, Amidon & Anik 1981)
154.08 (Pearlman 1986)
150.15 (Doucette & Andren 1988)
149.8 (calculated- χ , Sabljic 1987a)

Heat of Fusion, kcal/mol:
Entropy of Fusion, cal/mol K (e.U.):

12.4 (Yalkowsky & Valvani 1980)

Fugacity Ratio, F: 1.0

Water Solubility (g/m³ or mg/L at 25°C):

173 (shake flask-UV, Andrew & Keefer 1949; quoted, Shiu et al. 1990)
196 (shake flask-UV, Bohon & Claussen 1951; quoted, Vesala 1974; API 1985)
185 (quoted, Deno & Berkheimer 1960; Hine & Mookerjee 1975)
157 (shake flask-GC, Hermann 1972; quoted, Vesala 1974)
162 (shake flask-GC, Polak & Lu 1973; quoted, API 1985; Abernethy et al. 1986; Shiu et al. 1990)
206 (shake flask-UV, Vesala 1974)
146 (shake flask-GC, Sutton & Calder 1975; quoted, API 1985; Shiu et al. 1990)
185, 202 (quoted, calculated- χ , reported as ln S, should be -log S in molar concn., Kier & Hall 1976)
134 (shake flask-GC, Price 1976; quoted, API 1985; Shiu et al. 1990)
163 (quoted lit. mean, Amidon & Anik 1981)
162, 193 (quoted exptl. average, calculated-K_{OW}, Valvani et al. 1981)

159 (gen. col.-HPLC/UV, Tewari et al. 1982c)
131 (calculated-K_{OW}, Yalkowsky et al. 1983)
160 (gen. col.-HPLC, Wasik et al. 1983; quoted, Shiu et al. 1990)
160 (quoted, Miller et al. 1985; Eastcott et al. 1988)
150 (quoted, Brookman et al. 1985; Cline et al. 1991)
161 (quoted, Leahy 1986; Nirmalakhanden & Speece 1988a)
122 (calculated-V_I, solvatochromic parameters, Leahy 1986)
146 (Riddick et al. 1986; quoted, Howard 1990)
185, 207 (quoted, calculated-V_I, solvatochromic p., Kamlet et al. 1987)
223 (calculated- χ , Nirmalakhanden & Speece 1988a)
158 (quoted, Isnard & Lambert 1989)

Vapor Pressure (Pa at 25 °C):
1100 (extrapolated, Antoine eqn., Zwolinski & Wilhoit 1971; quoted, Mackay & Shiu 1981; Eastcott et al. 1988)
1114 (quoted, Hine & Mookerjee 1975)
1104, 1155 (quoted exptl., calculated-B.P., Mackay et al. 1982)
1160 (extrapolated, Verschueren, 1983)
1166 (extrapolated, Antoine eqn., Boublik et al. 1984)
1106 (extrapolated-Antoine eqn., Dean 1985)
1106 (Riddick et al. 1986; quoted, Howard 1990)
1110 (extrapolated-Antoine eqn., Stephenson & Malanowski 1987)

Henry's Law Constant (Pa m^3/mol):
506 (calculated-bond contribution, Hine & Mookerjee 1975)
637 (calculated as $1/K_{AW}$, C_W/C_A, reported as exptl., Hine & Mookerjee 1975; Nirmalakhanden & Speece 1988b)
778 (NAS 1980; quoted, Howard 1990)
700 (calculated-P/C, Mackay & Shiu 1981)
1115 (calculated-UNIFAC, Gmehling et al. 1982)
709 (quoted, Pankow 1986; Pankow & Rosen 1988; Pankow 1990)
754 (EPICS, Ashworth et al. 1988)
728 (calculated-C_A/C_W, Eastcott et al. 1988)

Octanol/Water Partition Coefficient, log K_{OW}:
3.20 (Hansch et al. 1968)
3.28 (calculated-f const., Yalkowsky & Valvani 1976)
3.18 (gen. col.-HPLC/UV, Wasik et al. 1981; quoted, Jafvert 1991)
3.09 (calculated-f const., Könemann 1981; quoted, Leegwater 1989)
3.20 (quoted, Leo et al. 1971; Hansch & Leo 1979; quoted, Valvani et al. 1981)
3.29 (HPLC-k', Hammers et al. 1982)
3.20, 3.13 (quoted, gen. col.-HPLC/UV, Tewari et al. 1982b,c)
3.20 (quoted, Verschueren, 1983)

3.20 (gen. col.-HPLC/UV, Wasik et al. 1983; quoted, Miller et al. 1985; Eastcott et al. 1988; Szabo et al. 1990a,b)
3.14 (calculated-f const., Yalkowsky et al. 1983)
3.20, 3.04, 3.11 (quoted, calculated-MR & IP, MR, Yoshida et al. 1983)
3.20, 3.28 (quoted, HPLC-RV, Garst & Wilson 1984)
3.37 (HPLC-k', Haky & Young 1984)
3.20 (Hansch & Leo 1985; quoted, Howard 1990)
3.04 (calculated-V_I, solvatochromic parameters, Taft et al. 1985)
3.09 (quoted, Abernethy et al. 1988)
3.18 (quoted exptl., Doucette & Andren 1988)
3.14, 3.14, 3.06, 3.18, 3.50 (calculated-π, f const., MW, χ, TSA, Doucette & Andren 1988)
3.20 (recommended, Sangster 1989)
3.20 (quoted, Isnard & Lambert 1988, 1989)

Bioconcentration Factor, log BCF:
1.37 (eels, Ogata & Miyake 1978)
0.78 (clams, Nunes & Benville 1979)
1.17 (goldfish, Ogata et al. 1984)
1.69, 1.17 (fish: calculated, correlated, Sabljic 1987b)
1.37 (fish, correlated, Isnard & Lambert 1988)
1.37, 0.78, 2.20 (quoted values for eels, clams, calculated, Howard 1990)
2.40 (s. capricornutum, Herman et al. 1991)

Sorption Partition Coefficient, log K_{OC}:
2.26 (observed, Seip et al. 1986; quoted, Sabljic 1987a & b)
2.53 (soil, calculated- χ, Sabljic 1987a & b)
2.04-3.15 (soil, calculated-K_{OW}, Sabljic 1987a & b)
2.22 (Abdul et al. 1987)
2.22 (quoted, Howard 1990)
2.34, 2.37, 2.40 (quoted, HPLC-k', Szabo et al. 1990a,b)

Half-Lives in the Environment:
Air: 0.83 hour, based on rate of disappearance for the reaction with OH radicals in ambient LA basin air at 300°K (Doyle et al. 1975); estimated lifetime is 1.5 hours under photochemical smog conditions in S.E. England (Brice & Derwent 1978) and (Perry et al. 1977); 2.6-26 hours, based on photooxidation half-life in air (Atkinson 1985; Howard et al. 1991).
Surface Water: 168-672 hours, based on estimated aqueous aerobic biodegradation half-life (Bridie et al. 1979; Kuhn et al. 1985; Howard et al. 1991); volatilization appears to be dominant removal process with a half-life of 1-5.5 days (Howard 1990).

Groundwater: 336-8640 hours, based on estimated aqueous aerobic and anaerobic biodegradation half-lives (Bridie et al. 1979; Kuhn et al. 1985; Howard et al. 1991); estimated half-life from observed persistence in groundwater of the Netherlands, 0.3 year (Zoeteman et al. 1981); abiotic hydrolysis or dehydrohalogenation half-life of 377 months (Olsen & Davis 1990).

Soil: 168-672 hours, based on estimated aqueous aerobic biodegradation half-life (Bridie et al. 1979; Kuhn et al. 1985; Howard et al. 1991).

Environmental Fate Rate Constants or Half-Lives:

Volatilization: half-life of evaporation from water of 1 m depth with wind speed of 3 m/sec and water current of 1 m/s estimated to be 3.1 hours (Lyman et al. 1982); half-life for evaporation from a typical river or pond is estimated to be 27-135 hours (Howard 1990).

Photolysis:

Oxidation: photooxidation half-life in water: 4.8×10^6-2.4×10^8 hours, based on estimated rate data for alkoxy radicals in aqueous solution (Hendry et al. 1974); calculated rate constant for the reaction with OH radicals in ambient LA basin air at 300°K was 1.4×10^{10} liter mol^{-1} sec^{-1} with an initial concentration of 2.0×10^{-10} mol $liter^{-1}$ (Doyle et al. 1975); absolute rate constant for the reaction with OH radicals determined by flash photolysis-resonance fluorescence at room temperature was 23.6×10^{-12} cm^3 $molecule^{-1}$ sec^{-1} (Hansen et al. 1975); rate constant for reaction with OH radicals in gas phase is 14.1×10^9 L mol^{-1} s^{-1} and $t_{1/2}$ = 0.24-2.4 h (Darnall et al. 1976); reaction with ozone at 300°K having a rate constant of 780 cm^3 mol^{-1} sec^{-1} (Lyman et al. 1982); k = 5.9-12 $\times 10^9$ M^{-1} s^{-1}, $t_{1/2}$ =0..47-1.0 d (reported for xylenes, oxidation by OH radical in air (Mill 1982); photooxidation half-life in air: 2.6-26 hours, based on measured room temperature rate constant of 23.5×10^{-12} cm^3 $molecule^{-1}$ sec^{-1} for vapor phase reaction with hydroxy radicals in air (Atkinson 1985; Howard et al. 1991); rate constant for the gas phase reaction with OH radicals at 25°C was 22.2×10^{-12} cm^3 $molecule^{-1}$ sec^{-1} (Ohta & Ohyama 1985).

Hydrolysis: k_h = 0, no hydrolyzable functional groups (Mabey et al. 1982).

Biodegradation: 100% biodegraded after 192 hours at 13°C with an initial concentration of 3.28×10^{-6} l/l (Jamison et al. 1976); aqueous aerobic biodegradation half-life: 168-672 hours, based on aqueous screening test data (Bridie et al. 1979; Howard et al. 1991) and soil column study simulating an aerobic river/groundwater infiltration system (Kuhn et al. 1985; Howard et al. 1991); aqueous anaerobic biodegradation half-life: 672-2688 hours, based on unacclimated aqueous aerobic biodegradation half-life (Bridie et al. 1979; Kuhn et al. 1985; Howard et al. 1991); biodegradation half-life of 0.03 days (Olsen & Davis 1990).

Bioconcentration, Uptake (k_1) and Elimination (k_2) Rate Constants or Half-Lives: half-life to eliminate from eels in seawater, 2.6 days (Ogata & Miyake 1978).

Common Name: p-Xylene
Synonym: 1,4-dimethylbenzene, p-xylol
Chemical Name: p-xylene
CAS Registry No: 106-42-3
Molecular Formula: $C_6H_4(CH_3)_2$
Molecular Weight: 106.2
Melting Point (°C): 13.2
Boiling Point (°C): 138
Density (g/cm^3 at 20°C): 0.8611
Molar Volume (cm^3/mol):

123.93 (calculated-density, Windholz 1983, Budavari 1989)
119.8 (25°C, calculated-density, Miller et al. 1985)
99.82 (characteristic molecular volume, McGowen & Mellors 1986)
0.671 (intrinsic volume: $V_I/100$, Kamlet et al. 1987)
123.3 (20°C, calculated-density, Stephenson & Malanowski 1987)
140.4 (LeBas method, Abernethy & Mackay 1987; Eastcott et al. 1988)

Molecular Volume (A^3):

116.0 (Pearlman 1986)

Total Surface Area, TSA (A^2):

150.3 (Yalkowsky & Valvani 1976)
317.6 (with hydration shell, Amidon & Anik 1981)
154.09 (Pearlman 1986)
149.8 (calculated- χ , Sabljic 1987a)
150.15 (Doucette & Andren 1988)

Heat of Fusion, kcal/mol:

4.090 (Tsonopoulos & Prausnitz 1971)

Entropy of Fusion, cal/mol K :

14.2 (Tsonopoulos & Prausnitz 1971; Yalkowsky & Valvani 1980)

Fugacity Ratio, F : 1.0

Water Solubility (g/m^3 or mg/L at 25°C):

200 (shake flask-UV, Andrews & Keefer 1949; quoted, Shiu et al. 1990)
198 (shake flask-UV, Bohon & Claussen 1951; quoted, Deno & Berkheimer 1960; Tsonopoulos & Prausnitz 1971; Shiu et al. 1990)
185 (shake flask-GC, Polak & Lu 1973; quoted, Hutchinson et al. 1980; API 1985; Shiu et al. 1988; 1990)
198 (quoted, Hine & Mookerjee 1975; Könemann 1981; Kamlet et al. 1987)
156 (shake flask-GC, Sutton & Calder 1975; quoted, API 1985; Doucette & Andren 1988; Shiu et al. 1990; Cline et al. 1991)
163 (shake flask-GC, Hermann 1972)
198, 202 (quoted; calculated- χ , reported as ln S, should be -log S in molar concn., Kier & Hall 1976)
157 (shake flask-GC, Price 1976; quoted, API 1985; Shiu et al. 1990)

186 (quoted lit. mean, Amidon & Anik 1981)
180, 217 (quoted, calculated-K_{OW}, Valvani et al. 1981)
214.5 (gen. col.-HPLC/UV, GC/ECD, Tewari et al. 1982c; quoted, Doucette & Andren 1988)
182 (calculated-HPLC-k', converted from reported γ_W, Hafkenscheid & Tomlinson 1983a)
198 (quoted, Verschueren, 1983)
176 (calculated-K_{OW}, Yalkowsky et al. 1983; quoted, Abernethy et al. 1986)
214 (gen. col.-HPLC, Wasik et al. 1983; quoted, Shiu et al. 1990)
215 (gen. col.-GC/ECD, Miller et al. 1985; quoted, Vowles & Mantoura 1987; Eastcott et al. 1988)
180; 104, (quoted; calculated-UNIFAC, Banerjee 1985)
156 (Riddick et al. 1986; quoted, Howard 1990)
185, 78 (quoted, calculated-characteristic volume, McGowen & Mellors 1986)
207 (calculated-V_I, solvatochromic parameters, Kamlet et al. 1987)
223 (calculated- χ , Nirmalakhanden & Speece 1988a)
107 (calculated-UNIFAC, Banerjee & Howard 1988)
182 (correlated, Isnard & Lambert 1988, 1989)
150 (quoted, Abdul et al. 1990)

Vapor Pressure (Pa at 25°C):
1170 (extrapolated, Antoine eqn., Zwolinski & Wilhoit 1971; quoted, Mackay & Shiu 1981; Eastcott et al. 1988)
1176 (quoted, Hine & Mookerjee 1975)
1165, 1206 (quoted exptl., calculated-B.P., Mackay et al. 1982)
1170 (extrapolated, Verschueren 1983)
1170 (extrapolated, Antoine eqn., Boublik et al. 1984)
1167 (extrapolated-Antoine eqn., Dean 1985)
1160 (Riddick et al. 1986; quoted, Howard 1990)
1180 (extrapolated-Antoine eqn., Stephenson & Malanowski 1987)

Henry's Law Constant (Pa m^3/mol):
506 (calulated-bond contribution, Hine & Mookerjee 1975)
637 (calculated as $1/K_{AW}$, C_W/C_A, reported as exptl., Hine & Mookerjee 1975; Nirmalakhanden & Speece 1988b)
778 (quoted, NAS 1980; quoted, Howard 1990)
710 (calculated-P/Cc, Mackay & Shiu 1981)
1185 (calculated-P/C, Gmehling et al. 1982)
754 (EPICS, Ashworth et al. 1988)
762 (batch stripping, Ashworth et al. 1988)
562 (calculated-C_A/C_W, Eastcott et al. 1988)
879 (calculated-QSAR, Nirmalakhanden & Speece 1988b)

Octanol/Water Partition Coefficient, log K_{OW}:

3.15 (Leo et al. 1971)
3.25, 2.90 (selected, calculated- χ , Kier & Hall 1976)
3.28 (calculated-TSA, Yalkowsky & Valvani 1976)
3.20 (quoted, Hansch & Leo 1979)
3.08 (quoted, Hutchinson et al. 1980)
3.10 (HPLC-k', Hanai et al. 1981)
3.15 (quoted, Valvani et al. 1981)
3.28 (HPLC-k', Hammers et al. 1982)
3.15, 3.18 (quoted; gen. col.-HPLC/UV, GC/ECD, Tewari et al. 1982b,c)
3.29 (calculated-HPLC-k', Hafkenscheid & Tomlinson 1983b)
3.15, 3.02, 3.11 (quoted, calculated-MR & IP, MR, Yoshida et al. 1983)
3.15 (quoted, Verschueren, 1983)
3.14 (calculated-f const., Yalkowsky et al. 1983)
3.18 (gen. col.-HPLC, Wasik et al. 1983)
3.15, 3.29 (quoted, HPLC-RV, Garst 1984)
3.18 (quoted, Miller et al. 1985; Vowles & Mantoura 1987; Eastcott et al. 1988)
3.15 (Hansch & Leo 1985; quoted, Howard 1990)
3.14 (quoted, Campbell & Luthy 1985)
3.09 (quoted, Abernethy et al. 1988)
3.20 (quoted exptl., Doucette & Andren 1988)
3.14, 3.14, 3,06, 3.16, 3.50 (calculated-π, f const., MW, χ , TSA, Doucette & Andren 1988)
3.10, 2.85 (quoted, calculated-UNIFAC, Banerjee & Howard 1988)
3.15 (quoted, Isnard & Lambert 1989)
3.15 (recommended, Sangster 1989)

Bioconcentration Factor, log BCF:

1.37 (eels, Ogata & Miyake 1978)
1.17 (goldfish, Ogata et al. 1984)
1.68, 1.17 (fish: calculated, correlated, Sabljic 1987b)
2.22 (fish, calculated, Howard 1990)
1.37 (eels, quoted, Howard 1990)
2.41 (s. capricornutum, Herman et al. 1991)

Sorption Partition Coefficient, log K_{OC}:

2.52 (soil, observed, Schwarzenbach & Westall 1981; quoted, Sabljic 1987a & b; Bahnick & Doucette 1988)
2.53 (soil, calculated- χ , Sabljic 1987a & b)
2.05-3.08 (soil, calculated-K_{OW}, Sabljic 1987a & b)
2.31 (Abdul et al. 1987)
2.65 (calculated- χ , Bahnick & Doucette 1988)
2.31 (quoted, Howard 1990)

Half-Lives in the Environment:

Air: 0.24-2.4 hours, based on rate of disappearance for the reaction with hydroxy radicals (Darnall et al. 1976; Howard et al. 1991); 4.2-42 hours, based on photooxidation half-life (Atkinson 1985; Howard et al. 1991).

Surface Water: 168-672 hours, based on estimated aqueous aerobic biodegradation half-life (Bridie et al. 1979; Kuhn et al. 1985; Howard et al. 1991).

Groundwater: 336-8640 hours, based on estimated aqueous aerobic and anaerobic biodegradation half-lives (Bridie et al. 1979; Kuhn et al. 1985; Howard et al. 1991); estimated half-life from observed persistence in groundwater of the Netherlands, 0.3 year (Zoeteman et al. 1981); abiotic hydrolysis or dehydrohalogenation half-life of 1150 months (Olsen & Davis 1990).

Soil: 168-672 hours, based on estimated aqueous aerobic biodegradation half-life (Bridie et al. 1979; Kuhn et al. 1985; Howard et al. 1991); disappearance half-life from test soils, 2.2 days (Anderson et al. 1991).

Environmental Fate Rate Constants or Half-Lives:

Volatilization: half-life of evaporation from water of 1 m depth with wind speed of 3 m/sec and water current of 1 m/sec estimated to be 3.1 hours (Lyman et al. 1982); estimated half-life of evaporation from a typical river or pond estimated to be 27 to 135 hours, respectively (Howard 1990).

Photolysis:

Oxidation: photooxidation half-life in water: 2.8×10^6-1.4×10^8 hours, based on estimated rate data for alkoxy radicals in aqueous solution (Hendry 1974); absolute rate constant for the reaction with OH radicals determined by flash photolysis-resonance fluorescence at room temperature was 12.2×10^{-12} cm^3 $molecule^{-1}$ sec^{-1} (Hansen et al. 1975); rate constant for reaction with OH radicals in gas phase is 7.45×10^9 L mol^{-1} s^{-1} and $t_{1/2}$ = 0.24-2.4 h (Darnall et al. 1976); reaction with ozone at 300°K having a rate constant of 950 cm^3 $mole^{-1}$ sec^{-1} (Lyman et al. 1982); k = 5.9-12 $x10^9$ M^{-1} sec^{-1}, $t_{1/2}$ = 0.47 - 1.0 days , reported for xylenes as oxidation by OH radicals in atmosphere (Mill 1982); photooxidation half-life in air: 4.2-42 hours, based on measured room temperature rate constant of 14.1×10^{-12} cm^3 $molecule^{-1}$ sec^{-1} for vapor phase reaction with hydroxy radicals in air (Atkinson 1985; Howard et al. 1991); rate constant for the gas phase reaction with OH radicals at 25°C was 12.9×10^{-12} cm^3 $molecule^{-1}$ sec^{-1} (Ohta & Ohyama 1985).

Hydrolysis: k_h = 0, no hydrolyzable functional groups (Mabey et al. 1982).

Biodegradation: 100% biodegraded after 192 hours at 13°C with an initial concentration of 1.03×10^{-6} l/l (Jamison et al. 1976); aqueous aerobic biodegradation half-life: 168-672 hours, based on aqueous screening test data (Bridie et al. 1979) and soil column study simulating an aerobic river/groundwater infiltration system (Kuhn et al. 1985; Howard et al. 1991); anaerobic aqueous biodegradation half-life: 672-2688 hours, based on unacclimated aqueous aerobic biodegradation half life

(Bridie et al. 1979; Kuhn et al. 1985; Howard et al. 1991); biodegradation half-life of 0.03 days (Olsen & Davis 1990).

Bioconcentration, Uptake (k_1) and Elimination (k_2) Rate Constants or Half-Lives: half-life to eliminate from eels in seawater, 2.6 days (Ogata & Miyake 1978).

Common Name: 1,2,3-Trimethylbenzene
Synonym: hemimellitene, hemellitene
Chemical Name: 1,2,3-trimethylbenzene
CAS Registry No: 526-73-8
Molecular Formula: $C_6H_3(CH_3)_3$
Molecular Weight: 120.2
Melting Point (°C): -25.4
Boiling Point (°C): 176.1
Density (g/cm^3 at 20°C): 0.8944
Molar Volume (cm^3/mol):

134.4 (20°C, calculated-density, Stephenson & Malanowski 1987)
162.6 (LeBas method, Eastcott et al. 1988)
0.769 (intrinsic volume: $V_I/100$, Kamlet et al. 1988)

Molecular Volume (A^3):

131.67 (Pearlman 1986)

Total Surface Area, TSA (A^2):

172.50 (Pearlman 1986)
332.4 (with hydration shell, Amidon & Anik 1981)
171.3 (calculated- χ , Sabljic 1987a)
163.49 (Doucette & Andren 1988)

Heat of Fusion, kcal/mol:
Fugacity Ratio, F: 1.0

Water Solubility (g/m^3 or mg/L at 25°C):

75.2 (shake flask-GC, Sutton & Calder 1975; quoted, Amidon & Anik 1981; API 1985; Doucette & Andren 1988; Cline et al. 1991)
65.5 (gen. col.-HPLC/UV, Tewari et al. 1982c; quoted, Miller et al. 1985; Eastcott et al. 1988; Doucette & Andren 1988)
63.1 (calculated-K_{OW}, Yalkowsky et al. 1983)
75.2, 75.5 (quoted, calculated- χ , Nirmalakhanden & Speece 1988a)
66.1 (quoted, Isnard & Lambert 1989)

Vapor Pressure (Pa at 25°C):

198 (extrapolated, Antoine eqn., Zwolinski & Wilhoit 1971; quoted, Mackay & Shiu 1981)
202 (quoted, exptl., Mackay et al. 1982; Eastcott et al. 1988)
217 (calculated-B.P., Mackay et al. 1982)
157 (extrapolated, Antoine eqn., Boublik et al. 1984)
198.4 (extrapolated, Antoine eqn., Dean 1985)
199 (extrapolated, Antoine eqn., Stephenson & Malanowski 1987)

Henry's Law Constant (Pa m^3/mol):

323 (calculated-P/C, Mackay & Shiu 1981)

372 (calculated-K_{AW}, Eastcott et al. 1988)

Octanol/Water Partition Coefficient, log K_{OW}:

3.66 (Hansch & Leo 1979; quoted, Burkhard & Kuehl 1986)

3.66 (HPLC-k', Hammers et al. 1982)

3.58 (calculated-f const., Yalkowsky et al. 1983; quoted, Eastcott et al. 1988)

3.55 (gen. col.-HPLC/UV, Tewari et al. 1982b,1982c; quoted, Miller et al. 1985)

3.55 (quoted exptl., Doucette & Andren 1988)

3.58, 3.58, 3.81, 3.31, 3.57 (calculated-π, f const., HPLC-RT, MW, χ, TSA, Doucette & Andren 1988)

3.55 (quoted, Abernethy et al. 1988)

3.55 (quoted, Isnard & Lambert 1989)

3.60 (recommended, Sangster 1989)

Bioconcentration Factor, log BCF:

Sorption Partition Coefficient, log K_{OC}:

2.80 (observed, Schwarzenbach & Westall 1981; quoted, Sabljic 1987a & b)

2.77 (soil, calculated- χ, Sabljic 1987a & b)

2.46-3.39 (soil, calculated-K_{OW}, Sabljic 1987a &b)

Half-Lives in the Environment:

Air: 0.24-2.4 hours, based on rate of disappearance for the reaction with hydroxy radicals (Darnall et al. 1976; Howard et al. 1991); estimated lifetime under photochemical smog conditions in S.E. England is 1.5 hour. (Brice & Derwent 1978) and (Perry et al. 1977)

Surface Water:

Groundwater:

Soil:

Biota:

Environmental Fate Rate or Half-Lives:

Volatilization:

Photolysis:

Oxidation: absolute rate constant for the reaction with OH radicals determined by flash photolysis-resonance fluorescence at room temperature was 26.4×10^{-12} cm^3 $molecule^{-1}$ sec^{-1} (Hansen et al. 1975); rate constant for reaction with OH radicals in gas phase = 14.9×10^9 L mol^{-1} s^{-1} and $t_{1/2}$ = 0.24-2.4 h (Darnall et al. 1976); rate constant k = 15-30 $\times 10^{-9}$ M^{-1} sec^{-1}, $t_{1/2}$ = 0.2-0.4 day, reported for trimethylbenzenes by OH radicals in the atmosphere (Mill 1982); room temperature rate constant for the gas phase reaction with OH radicals: 31.6×10^{-12}

cm^3 $molecule^{-1}$ sec^{-1} (Atkinson 1985) and $29.6x10^{-12}$ cm^3 $molecule^{-1}$ sec^{-1} (Ohta & Ohyama 1985).

Hydrolysis:

Biotransformation:

Biodegradation:

Bioconcentration:

Common Name: 1,2,4-Trimethylbenzene
Synonym: pseudocumene
Chemical Name: 1,2,4-trimethylbenzene
CAS Registry No: 95-63-6
Molecular Formula: $C_6H_3(CH_3)_3$
Molecular Weight: 120.2
Melting Point (°C): -43.8
Boiling Point (°C): 169.4
Density (g/cm^3 at 20°C): 0.8758
Molar Volume (cm^3/mol):

137.2 (20°C, calculated-density, McAuliffe 1966; Lande & Banerjee 1981)
0.791 (intrinsic volume: $V_I/100$, Kamlet et al. 1987)
137.2 (20°C, calculated-density, Stephenson & Malanowski 1987)
162.6 (LeBas method, Eastcott et al. 1988)
0.768 (intrinsic volume: $V_I/100$, Kamlet et al. 1988)

Molecular Volume (A^3):

132.0 (Pearlman 1986)

Total Surface Area, TSA (A^2):

340.6 (with hydration shell, Amidon & Anik 1981)
175.56 (Pearlman 1986)
163.48 (Doucette & Andren 1988)

Heat of Fusion, kcal/mol:
Fugacity Ratio, F : 1.0

Water Solubility (g/m^3 or mg/L at 25°C):

57 (shake flask-GC, McAuliffe 1966; quoted, Hermann 1972; Mackay et al. 1980; API 1985; Abernethy et al. 1986)
57.5, 31.5 (quoted, calculated-K_{OW}, Hansch et al. 1968)
59.0 (shake flask-GC, Sutton & Calder 1975; quoted, API 1985)
57.5 (quoted, Hine & Mookerjee 1975)
57.5, 79.4 (quoted, calculated- χ , reported as ln S, should be -log S in molar concn., Kier & Hall 1976)
51.9 (shake flask-GC, Price 1976; quoted, API 1985; Eastcott et al. 1988)
616 (quoted, Lande & Banerjee 1981, reported as 1,3,4-TMB)
56.1 (quoted lit. mean, Amidon & Anik 1981)
63.6, 58 (quoted, average, Whitehouse & Cooke 1982)
56.2 (calculated-K_{OW}, Yalkowsky et al. 1983)
66, 282 (quoted, calculated-K_{OW} & PHLC-RT, Chin et al. 1986)
57 (quoted, Nirmalakhanden & Speece 1988a)
77.1 (calculated- χ , Nirmalakhanden & Speece 1988a)
57, 60 (quoted, calculated-UNIFAC, Al-Sahhaf 1989 - mistaken as ethylbenzene)

Vapor Pressure (Pa at 25°C):

271 (extrapolated, Antoine eqn., Zwolinski & Wilhoit 1971; quoted, Mackay & Shiu 1981; Eastcott et al. 1988)

280 (quoted, Hine & Mookerjee 1975)

271, 296 (quoted exptl., calculated-B.P., Mackay et al. 1982)

270 (extrapolated, Antoine eqn., Boublik et al. 1984)

Henry's Law Constant (Pa m^3/mol):

494 (calculated-bond contribution, Hine & Mookerjee 1975)

581 (calculated as $1/K_{AW}$, C_W/C_A, reported as exptl., Hine & Mookerjee 1975; Nirmalakhanden & Speece 1988b)

590 (calculated-P/C, Mackay & Shiu 1981)

644 (calculated-C_A/C_W, Matter-Müller et al. 1981)

1211 (calculated-UNIFAC, Arbuckle 1983)

475 (20°C, EPICS, Yurteri et al. 1987)

385 (20°C, calculated-P/C, Yurteri et al. 1987)

1242 (calculated-QSAR, Nirmalakhanden & Speece 1988b)

627 (calculated-C_A/C_W, Eastcott et al. 1988)

Octanol/Water Partition Coefficient, log K_{OW}:

3.65 (calculated-π substituent constant, Hansch et al. 1968)

3.65, 3.58 (quoted, calculated-S, Mackay et al. 1980)

3.78 (HPLC-k', Hammers et al. 1982)

3.58 (calculated-f const., Yalkowsky et al. 1983; quoted, Eastcott et al. 1988)

3.65, 3.45 (quoted, calculated-MR, Yoshida et al. 1983)

3.63 (quoted exptl., Doucette & Andren 1988)

3.58, 3.58, 3.31, 3.56, 3.82 (calculated-π, f const., MW, χ , TSA, Doucette & Andren 1988)

3.55 (quoted, Abernethy et al. 1988)

3.63 (recommended value, Sangster 1989)

Bioconcentration Factor, log BCF:

Sorption Partition Coefficient, log K_{OC}:

Half-Lives in the Environment:

Air: half-life estimated from the rate of disappearance for the reaction with OH radicals: 0.48 hour (Doyle et al. 1975) and 0.24-2.4 hours (Darnall et al. 1976); 1.6-16 hours, based on photooxidation half-life in air (Atkinson 1985; Howard et al. 1991).

Surface Water: 168-672 hours, based on estimated aqueous aerobic biodegradation half-lives (Kitano 1978; Van der Linden 1978; Tester & Harker 1981; Trzilova & Horska 1988 and Marion & Melaney 1964; Howard et al. 1991).

Groundwater: 336-1344 hours, based on estimated aqueous aerobic biodegradation half-lives (ibid.).
Soil: 168-672 hours, based on estimated aqueous aerobic biodegradation half-lives (ibid.).

Environmental Fate Rate Constants or Half-Lives:
Photolysis: no photolyzable funtional groups (Howard et al. 1991).
Oxidation: calculated rate constant for the reaction with OH radicals in ambient LA basin air at 300°K was 2.0×10^{10} liter mol^{-1} sec^{-1} with an initial concentration of 8.1×10^{-11} mol $liter^{-1}$ (Doyle et al. 1975); absolute rate constant for the reaction with OH radicals determined by flash photolysis-resonance fluorescence at room temperature was 33.5×10^{-12} cm^3 $molecule^{-1}$ sec^{-1} (Hansen et al. 1975); rate constant for reaction with OH radicals in gas phase = 20×10^9 L mol^{-1} s^{-1} and $t_{1/2}$ = 0.24-2.4 h (Darnall et al. 1976); photooxidation half-life in water: 1056-43000 hours, based on measured rate data with hydroxy radicals in aqueous solution (Güesten et al. 1981; Howard et al. 1991); rate constant k = 1.5-30 x 10^{-9} M^{-1} s^{-1}, $t_{1/2}$ = 0.2-0.4 day, reported for trimethylbenzenes by OH radical in the atmosphere (Mill 1982); photooxidation half-life in air: 1.6-16 hours, based on measured room temperature rate constant of 38.4×10^{-12} cm^3 $molecule^{-1}$ sec^{-1} for the vapor phase reaction with hydroxy radicals (Atkinson 1985); rate constant for the gas phase reaction with OH radicals at 25°C, 31.5×10^{-12} cm^3 $molecule^{-1}$ sec^{-1} (Ohta & Ohyama 1985).
Hydrolysis: k_h = 0, no hydrolyzable functional groups (Mabey et al. 1982).
Biodegradation: aqueous aerobic biodegradation half-life: 168-672 hours, based on aqueous screening studies (Marion & Malaney 1964; Kitano 1978; Van der Linden 1978; Tester & Harker 1981; Trzilova & Horska 1988; Howard et al. 1991); anaerobic aqueous biodegradation half-life: 672-2688 hours, based on estimated aqueous aerobic biodegradation half-lives (ibid).

Common Name: 1,3,5-Trimethylbenzene
Synonym: mesitylene
Chemical Name: 1,3,5-trimethylbenzene
CAS Registry No: 108-67-8
Molecular Formula: $C_6H_3(CH_3)_3$
Molecular Weight: 120.2
Melting Point (°C): -44.7
Boiling Point (°C): 164.7
Density (g/cm^3 at 20°C): 0.880
Molar Volume (cm^3/mol):
139.58 (Windholz 1983; Budavari 1989)
139 (calculated-density, Chiou 1985)
113.9 (characteristic molecular volume, McGowen & Mellors 1986)
0.769 (intrinsic volume: $V_I/100$, Leahy 1986; Kamlet 1987)
138.9 (20°C, calculated-density, Stephenson & Malanowski 1987)
162.6 (LeBas method, Eastcott et al. 1988)
Molecular Volume (A^3):
132.03 (Pearlman 1986)
Total Surface Area, TSA (A^2):
348.6 (with hydration shell, Amidon & Anik 1981)
176.11 (Pearlman 1986)
171.3 (calculated- χ , Sabljic 1987a)
Heat of Fusion, kcal/mol:
Fugacity Ratio, F: 1.0

Water Solubility (g/m^3 or mg/L at 25°C):
173 (residue-volume method, Booth & Everson 1948)
97 (shake flask-UV, Andrews & Keffer 1950; quoted, Deno & Berkheimer 1960; Tsonopoulos & Prausnitz 1971; Chiou et al. 1982; Chiou 1985; API 1985)
39.4 (shake flask-UV, Vesala 1974)
48.2 (shake flask-GC, Sutton & Calder 1975; quoted, API 1985; Eastcott et al. 1988)
97.5 (quoted, Mackay & Shiu 1981; Abernethy et al. 1986; Nirmalakhanden & Speece 1988a)
72.8 (quoted lit. mean, Amidon & Anik 1981)
132 (calculated-K_{OW}, Valvani et al. 1981)
69.2 (quoted exptl. average, Valvani et al. 1981)
67.6, 178 (quoted, calculated-K_{OW}, Amidon & Williams 1982)
50.0 (shake flask-UV, Sanemasa et al. 1981)
50 (shake flask-UV, Sanemasa et al. 1982)
49.5 (calculated-HPLC-k', converted from reported γ_W, Hafkenscheid & Tomlinson 1983)
46.4 (calculated-K_{OW}, Yalkowsky et al. 1983)
97.7 (quoted, Chiou & Block 1986)

29.5, 20.4 (quoted, calculated-characteristic volume, McGowen & Mellors 1986)
38.9 (calculated-V_I, solvatochromic parameters, Leahy 1986)
69.2 (quoted, Leahy 1986; Kamlet et al. 1988)
74.1 (calculated-V_I, solvatochromic parameters, Kamlet et al. 1987)
97.7, 77.1 (quoted, calculated- χ , Nirmalakhanden & Speece 1988a)
48.9 (recommended, IUPAC 1989b)
100 (quoted, Isnard & Lambert 1989)

Vapor Pressure (Pa at 25°C):
328 (extrapolated, Antoine eqn., Zwolinski & Wilhoit 1971; quoted, Mackay & Shiu 1981; Eastcott et al. 1988)
322, 366 (quoted exptl., calculated-B.P., Mackay et al. 1982)
323 (extrapolated, Antoine eqn., Boublik et al. 1984)
322.1 (extrapolated-Antoine eqn., Dean 1985)
330 (extrapolated-Antoine eqn., Stephenson & Malanowski 1987)

Henry's Law Constant (Pa m^3/mol):
600 (calculated-P/C, Mackay & Shiu 1981)
1459 (calculated-UNIFAC, Gmehling et al. 1982)
682 (EPICS, Ashworth et al. 1988)
849 (batch stripping, Ashworth et al. 1988)
818 (calculated-C_A/C_W, Eastcott et al. 1988)

Octanol/Water Partition Coefficient, log K_{OW}:
3.42 (Leo et al. 1971; Hansch & Leo 1979; Valvani et al. 1981; Amidon & Williams 1982)
3.58 (calculated, Mackay et al. 1980)
3.78 (HPLC-k', Hammers et al. 1982)
3.42 (quoted, Chiou et al. 1982; Chiou 1985; Chiou & Block 1986)
3.82 (calculated-HPLC-k', Hafkenscheid & Tomlinson 1983)
3.58 (calculated-f const., Yalkowsky et al. 1983)
3.86-3.94, 3.42, 3.42 (range, quoted, HPLC-RV, Garst 1984)
3.44 (calculated-V_I, solvatochromic parameters, Taft et al. 1985)
3.41, 4.15 (quoted, calculated-UNIFAC, Campbell & Luthy 1985)
3.85 (calculated-V_I, solvatochromic parameters, Leahy 1986)
3.42 (quoted, Tomlinson & Hafkanscheid 1986)
4.02 (quoted exptl., Doucette & Andre 1988)
4.28, 4.27, 4.32, 3.34, 3.78 (calculated-π, f const., MW, χ , TSA, Doucette & Andren 1988)
3.55 (quoted, Abernethy et al. 1988)
3.42 (quoted, Isnard & Lambert 1989)
3.42 (recommended, Sangster 1989)

Bioconcentration Factor, log BCF:

Sorption Partition Coefficient, log K_{OC}:

2.82 (soil, observed, Schwarzenbach & Westall 1981; quoted, Sabljic 1987a & b; Bahnick & Doucette 1988)

2.77 (soil, calculated- χ , Sabljic 1987a)

2.75 (soil, calculated- χ , Sabljic 1987b)

2.85 (soil, calculated- χ , Bahnick & Doucette 1988)

Half-Lives in the Environment:

Air: 0.24-2.4 hours, based on rate of disappearance for the reaction with hydroxy radicals (Darnall et al. 1976; Howard et al. 1991); estimated lifetime under photochemical smog conditions in S.E. England is 0.7 hour (Brice & Derwent 1978; Perry et al. 1977 and Darnall et al. 1976); 9.72-97.2 hours, based on estimated photooxidation half-life in air (Atkinson 1987).

Surface Water: 48-192 hours, based on a soil column study in which aerobic groundwater was continuously percolated through quartz sand (Kappeler & Wuhrmann 1978; Howard et al. 1991); 1 day (estimated, Zoeteman et al. 1980).

Groundwater: 96-384 hours, based on a soil column study in which aerobic ground water was continuously percolated through quartz sand (Kappeler & Wuhrmann 1978; Howard et al. 1991).

Soil: 48-192 hours, based on a soil column study in which aerobic groundwater was continuously percolated through quartz sand (Kappeler & Wuhrmann 1978; Howard et al. 1991).

Environmental Fate Rate Constants or Half-Lives:

Volatilization:

Photolysis:

Oxidation: absolute rate constant for the reaction with OH radicals determined by flash photolysis-resonance fluorescence at room temperature was 47.2×10^{-12} cm^3 $molecule^{-1}$ sec^{-1} (Hansen et al. 1975); rate constant for reaction with OH radicals in gas phase is 29.7×10^9L mol^{-1} s^{-1} and $t_{1/2}$ = 0.24-2.4 h (Darnall et al. 1976); photooxidation half-life in water: 3208-1.28×10^5 hours, based on measured rate constant for reaction with hydroxy radical in water (Mill et al. 1980); reaction with ozone at 300°K having a rate constant of 4200 cm^3 $mole^{-1}$ sec^{-1} (Lyman et al. 1982); rate constant k=15-30 x 10^{-9} M^{-1} s^{-1}, $t_{1/2}$ = 0.2-0.4 day, reported for trimethylbenzenes by OH radicals in the atmosphere (Mill 1982); room temperature rate constant for the gas phase reaction with OH radicals: 60.5×10^{-12} cm^3 $molecule^{-1}$ sec^{-1} (Atkinson 1985) and 38.7×10^{-12} cm^3 $molecule^{-1}$ sec^{-1} (Ohta & Ohyama 1985); photooxidation half-life in air: 9.72-97.2 hours, based on estimated rate constant for reaction with hydroxy radical in air (Atkinson 1987; Howard et al. 1991).

Hydrolysis:

Biodegradation: unacclimated aerobic aqueous biodegradation half-life: 48-192 hours, based on a soil column study in which aerobic groundwater was continuously percolated through quartz sand (Kappeler & Wuhrmann 1978; Howard et al. 1991); anaerobic aqueous biodegradation half-life: 192-768 hours, based on unacclimated aqueous aerobic biodegradation half-life (Howard et al. 1991).

Bioconcentration:

Common Name: n-Propylbenzene
Synonym: 1-phenylpropane
Chemical Name: n-propylbenzene
CAS Registry No: 103-65-1
Molecular Formula: $C_6H_5(C_3H_7)$
Molecular Weight: 120.2
Melting Point (°C): -99.6
Boiling Point (°C): 159.2
Density (g/cm³ at 20°C): 0.862
Molar Volume (cm³/mol):
145 (Klevens 1950)
139.4 (calculated-density, Miller et al. 1985)
0.768 (intrinsic volume: $V_I/100$, Kamlet et al. 1987,1988)
139.4 (20°C, Stephenson & Malanowski 1987)
170.0 (LeBas method, Eastcott et al. 1988)
Molecular Volume (A³):
132.49 (Pearlman 1986)
Total Surface Area, TSA (A²):
163.0 (Yalkowsky & Valvani 1976; Whitehouse & Cooke 1982)
347.4 (with hydration shell, Amidon & Anik 1981)
173.36 (Pearlman 1986)
170.1 (Warne et al. 1990)
Heat of Fusion, kcal/mol:
Fugacity Ratio, F: 1.0

Water Solubility (g/m³ or mg/L at 25°C):
60 (15 °C, volumetric, Fuhner 1924; quoted, Chiou et al. 1982; Chiou 1985)
120 (shake flask-turbidimetric, Stearns et al. 1947)
55 (shake flask-UV, Andrews & Keffer 1950; quoted, Deno & Berkheimer 1960; Hutchinson et al. 1980; Mackay et al. 1980; API 1985 or Brookman et al. 1985; Cline et al. 1991)
120 (shake flask-UV, Klevens 1951)
60, 29 (quoted, calculated-K_{OW}, Hansch et al. 1968)
60 (shake flask-GC, Hermann 1972)
54.9 (quoted, Hine & Mookerjee 1975)
70 (GC, Krasnoshchekova 1975)
54.9, 50.1 (quoted, calculated- χ , reported as ln S as should be -log S in molar concn., Kier & Hall 1976)
87.1, 81.3 (quoted, calculated-K_{OW}, Valvani et al. 1981)
51.0 (shake flask-UV, Sanemasa et al. 1981)
55 (quoted lit. mean, Amidon & Anik 1981)
51.9 (gen. col.-HPLC/UV, DeVoe et al. 1981)
55, 57 (quoted, average, Whitehouse & Cooke 1982)

93.3 (calculated-K_{OW} & B.P., Amidon & Williams 1982)
74.1 (calculated-K_{OW}, Yalkowsky et al. 1983)
55.0 (quoted, Mackay & Shiu 1981; Nirmalakhanden & Speece 1988a)
47.1 (gen. col.-HPLC/UV, Tewari et al. 1982a)
52.2 (gen. col.-HPLC/UV, GC/ECD, Tewari et al. 1982c)
59.5 (calculated-HPLC-k', converted from reported γ_W, Hafkenscheid & Tomlinson 1983a)
52.1 (gen. col.-HPLC/UV, Wasik et al. 1983; quoted, Miller et al. 1985; Vowles & Mantorua 1987; Eastcott et al. 1988)
45.2 (shake flask-UV, Sanemasa et al. 1984)
60 (quoted, Dean 1985)
51.7 (gen. col.-HPLC/UV, Owens et al. 1986)
67.6 (calculated-V_I, solvatochromic parameters, Kamlet et al. 1987)
46.1 (calculated- χ , Nirmalakhanden & Speece 1988a)
55.0 (recommended, IUPAC 1989)
60.3 (quoted, Isnard & Lambert 1989)
56.1 (quoted, Warne et al. 1990)

Vapor Pressure (Pa at 25°C):
449 (extrapolated, Antoine eqn., Zwolinski & Wilhoit 1971; quoted, Mackay & Shiu 1981; Eastcott et al. 1988)
457 (quoted, Hine & Mookerjee 1975)
457, 469 (quoted exptl., calculated-B.P., Mackay et al. 1982)
333 (20°C, quoted, Verschueren, 1983)
450 (extrapolated, Antoine eqn., Boublik et al. 1984)
449 (extrapolated-Antoine eqn., Stephenson & Malanowski 1987)

Henry's Law Constant (Pa m^3/mol):
1109 (calculated as $1/K_{AW}$, C_W/C_A, reported as exptl., Hine & Mookerjee 1975)
1159 (calculated-bond contribution, Hine & Mookerjee 1975)
700 (calculated-P/C, Mackay & Shiu 1981)
866 (calculated-UNIFAC, Gmehling et al. 1982)
1094 (EPICS, Ashworth et al. 1988)
942 (calculated-QSAR, Nirmalakhanden & Speece 1988b)
1033 (calculated-C_A/C_W, Eastcott et al. 1988)

Octanol/Water Partition Coefficient, log K_{OW}:
3.68 (shake flask-UV, Iwasa et al. 1965; Hansch et al. 1968; 1972)
3.57, 3.68 (Leo et al. 1971; Hansch & Leo 1979)
3.68, 3.42 (selected, calculated- χ , Kier & Hall 1976)
3.63 (calculated-f const., Yalkowsky & Valvani 1976; selected, Valvani et al. 1981; Amidon & Wiliams 1982)
3.66 (calculated-f const., Rekker 1977; quoted, Harnisch et al. 1983)

3.60 (quoted, Hutchinson et al. 1980)
3.68, 3.59 (quoted, calculated-S, Mackay et al. 1980)
3.62, 3.44 (quoted, shake flask-HPLC, Nahum & Horvath 1980)
3.69 (quoted, Braumann & Grimme 1981)
3.72, 3.68 (quoted, shake flask-HPLC/UV, DeVoe et al. 1981)
3.71 (gen. col.-HPLC/UV, Tewari et al. 1982a)
3.63, 3.90 (HPLC-k', correlated, Hammers et al. 1982)
3.68; 3.69 (quoted, gen. col.-HPLC/GC, Tewari et al. 1982b,c; Wasik et al. 1983)
3.73 (calculated-π, Wasik et al. 1982)
3.89 (calculated-HPLC-k', Hafkenscheid & Tomlinson 1983)
3.63 (quoted from Hansch & Leo 1979, Harnisch et al. 1983)
3.84, 3.69 (quoted of HPLC methods, Harnisch et al. 1983)
3.57, 3.68 (Verschueren, 1983)
3.6 (calculated-f const., Yalkowsky et al. 1983)
3.69 (gen. col.-HPLC/UV, Wasik et al. 1983; quoted, Miller et al. 1985; Vowles & Mantoura 1987; Eastcott et al. 1988; Szabo et al. 1990a,b)
3.68, 3.54, 3.47 (quoted, calculated-MR & IP, MR, Yoshida et al. 1983)
3.70 (quoted, Tomlinson & Hafkenscheid 1986)
3.69, 3.71 (gen. col.-RPLC, calculated-activity coeff., Schantz & Martire 1987)
3.68 (quoted, Isnard & Lambert 1989)
3.69 (recommended, Sangster 1989)
3.63, 3.55 (quoted, calculated-MO calculation, Bodor et al. 1989)
3.72, 3.849 (quoted, calculated-π substituent constant, Bodor et al. 1989)
3.656 (quoted, Warne et al. 1990)
3.69 (quoted, Pussemier et al. 1990)

Bioconcentration Factor, log BCF:

Sorption Partition Coefficient, log K_{OC}:
2.86 (Vowles & Mantoura 1987)
2.86; 2.83, 2.98 (quoted; HPLC-k', Szabo et al. 1990a,b)
2.86 (quoted, Pussemier et al. 1990)

Half-Lives in the Environment:
Air: 2.4-24 hours, based on rate of disappearance for the reaction with hydroxy radicals (Darall et al. 1976; Howard et al. 1991); estimated life time under photochemical smog conditions in S.E. England is 6 hours (Brice & Derwent 1978; Darnall et al. 1976).
Surface water:
Groundwater
Soil:
Sediment:
Biota:

Environmental Fate Rate Constants or Half-Lives:

Volatilization: half-lives from marine mesocosm: 19 days at 8 -16°C in the spring, 1.3 days at 20-22°C in the summer and 13 days at 3-7°C in the winter (Wakeham et al. 1983).

Photolysis:

Oxidation: rate constant for reaction with OH radicals in gas phase = 3.7×10^9 L mol^{-1} s^{-1} and $t_{1/2}$ = 2.4-24 hours (Darnall et al. 1976); rate constant k = 3.5×10^9 M^{-1} s^{-1}, $t_{1/2}$ = 1.6 days, by OH radical in the atmosphere (Mill 1982); room temperature rate constant for the gas phase reaction with OH radicals: 5.7×10^{-12} cm^3 $molecule^{-1}$ sec^{-1} (Atkinson 1985) and 6.58×10^{-12} cm^3 $molecule^{-1}$ sec^{-1} (Ohta & Ohyama 1985).

Hydrolysis:

Biodegradation:

Bioconcentration:

Common Name: Isopropylbenzene
Synonym: cumene, 2-phenylpropane, (1-methylethyl)benzene, cumol
Chemical Name: isopropylbenzene
CAS Registry No: 98-82-8
Molecular Formula: $C_6H_5(C_3H_7)$
Molecular Weight: 120.2
Melting Point (°C): -96.6
Boiling Point (°C): 154.2
Density (g/cm^3 at 20°C): 0.8618
Molar Volume (cm^3/mol):
139.5 (20°C, calculated-density, McAuliffe 1966)
140 (calculated-density, Lande & Banerjee 1981)
140.17 (Windhoz 1983; Budvari 1989)
139.5 (extrapolated, Stephenson & Malanowski 1987)
0.768 (intrinsic volume: V_I/100, Kamlet et al. 1987,1988)
170.0 (LeBas method, Eastcott et al. 1988)
Molecular Volume (A^3):
Total Surface Area, TSA (A^2):
163.4 (Yalkowsky & Valvani 1976; Whitehouse & Cooke 1982)
338.4 (with hydration shell, Amidon & Anik 1981)
Heat of Fusion, kcal/mol:
Fugacity Ratio, F: 1.0

Water Solubility (g/m^3 or mg/L at 25°C):
170 (shake flask-turbidimetric, Stearns et al. 1947)
73 (shake flask-UV, Andrews & Keffer, 1950; quoted, Deno & Berkheimer 1960)
80.4 (batch contacting-UV, Glew & Robertson 1956)
53 (shake flask-GC, McAuliffe 1963)
50 (shake flask-GC, McAuliffe 1966; quoted, Mackay & Wolkoff 1973; Mackay & Leinonen 1975; Hutchinson et al. 1980; Mackay et al. 1980; Mackay & Shiu 1975; 1981; API 1985)
50.1, 58.3 (quoted, calculated-K_{OW}, Hansch et al. 1968)
50 (shake flask-GC, Hermann 1972)
65.3 (shake flask-GC, Sutton & Calder 1975; quoted, API 1985)
50.1 (quoted, Hine & Mookerjee 1975; Kamlet et al. 1987)
50.1, 69.2 (quoted, calculated- χ , reported as ln S should be -log S in molar concn. Kier & Hall 1976)
48.3 (shake flask-GC, Price 1976; quoted, API 1985; Eastcott et al. 1988)
54.9 (quoted, Lande & Banerjee 1981)
74.1, 75.8 (quoted, calculated-K_{OW}, Valvani et el. 1981)
63.6 (quoted lit. mean, Amidon & Anik 1981)
83.1 (calculated-K_{OW} & B.P., Amidon & Williams 1982)
61.5 (shake flask-UV, Sanemasa et al. 1982)

50, 65, 58.1 (quoted, average, Whitehouse & Cooke 1982)
76.7 (calculated-HPLC-k', converted from reported γ_W, Hafkenscheid & Tomlinson 1983)
50 (20°C, quoted, Verschueren, 1983)
50 (quoted, Dean 1985)
67.5 (calculated-V_I, solvatochromic parameters, Kamlet et al. 1987)
59.6 (calculated- χ , Nirmalakhanden & Speece 1988a)
56 (recommended, IUPAC 1989b)
50, 45 (quoted, calculated-UNIFAC, Al-Sahhaf 1989)

Vapor Pressure (Pa at 25°C):
611 (extrapolated, Antoine eqn., Zwolinski & Wilhoit 1971; quoted, Mackay & Wolkoff 1973; Mackay & Leinonen 1975; Mackay & Shiu 1975,1981; Eastcott et al. 1988)
621 (quoted, Hine & Mookerjee 1975)
613, 366 (quoted exptl., calculated-B.P., Mackay et al. 1982)
427 (20°C, quoted, Verschueren, 1983)
610 (extrapolated, Antoine eqn., Boublik et al. 1984)
605 (extrapolated, Antoine eqn., Stephenson & Malanowski 1987)

Henry's Law Constant (Pa m^3/mol):
1492 (calculated as $1/K_{AW}$, C_W/C_A, reported as exptl., Hine & Mookerjee 1975)
1158 (calculated-bond contribution, Hine & Mookerjee 1975)
1480 (calculated-P/C, Mackay & Leinonen 1975)
130 (calculated-P/C, Mackay & Shiu 1981)
942 (calculated-QSAR, Nirmalakhanden & Speece 1988b)
1521 (calculated-C_A/C_W, Eastcott et al. 1988)

Octanol/Water Partition Coefficient, log K_{OW}:
3.43 (calculated-π substituent constant, Hansch et al. 1968)
3.66 (quoted, Leo et al. 1971)
3.66, 3.23 (quoted, calculated- χ , Kier & Hall 1976)
3.64 (calculated-f const., Yalkowsky & Valvani 1976)
3.63 (shake flask-GC, Chiou et al. 1977, 1982)
3.66 (Hansch & Leo 1979; quoted, Valvani et al. 1981; Amidon & Williams 1982; Eastcott et al. 1988)
3.51 (quoted, Hutchinson et al. 1980)
3.66, 3.63 (quoted, calculated-S, Mackay et al. 1980)
3.52 (HPLC-k', Hanai et al. 1981)
3.52 (HPLC-k', D'Amboise & Hanai 1982)
3.66, 3.40 (quoted, HPLC-k', Miyake & Terada 1982)
3.54 (calculated-f const., Yalkowsky et al. 1983)
3.66, 3.52, 3.45 (quoted, calculated-MR & IP, MR, Yoshida et al. 1983)

3.66 (quoted, Hafkenscheid & Tomlinson 1983b)
3.66 (quoted, Tomlinson & Hafkanscheid 1986)
3.50 (quoted, Abernethy et al. 1988)
3.66 (recommended, Sangster 1989)

Bioconcentration Factor, log BCF:
1.55 (goldfish, Ogata et al. 1984)
2.27, 1.55 (fish: calculated, correlated, Sabljic 1987b)

Sorption Partition Coefficient, log K_{OC}:

Half-Lives in the Environment:
Air: 2.4-24 hours, based on rate of disappearance for the reaction with hydroxy radicals (Darnall et al. 1976; Howard et al. 1991); estimated lifetime under photochemical smog conditions in S.E. England is 6 hours (Brice & Derwent 1978) and (Darnall et al. 1976).
Surface Water: 5.79 hours, calculated half-life based on evaporative loss at 25°C and 1 m depth of water (Mackay & Leinonen 1975).
Groundwater:
Soil:
Sediment:
Biota:

Environmental Fate Rate Constants or Half-Lives:
Volatilization: $t_{½}$ = 5.7 h from water depth of 1 m (calculated, Mackay & Leinonen 1975)
Photolysis:
Oxidation: rate constant for the reaction with OH radicals is 3.7×10^9 L mol^{-1} s^{-1} and $t_{½}$ = 2.4-24 h (Darnall et al. 1976); rate constant k = 4.6×10^{-1} M^{-1} sec^{-1}, and $t_{½}$ = 1.2 days, by OH radicals in the atmosphere (Mill 1982); room temperature rate constant for the gas phase reaction with OH radicals: 6.6×10^{-12} cm^3 $molecule^{-1}$ sec^{-1} (Atkinson 1985) and 6.25×10^{-12} cm^3 $molecule^{-1}$ sec^{-1} (Ohta & Ohyama 1985).
Hydrolysis:
Biodegradation:
Bioconcentration:
Hydrolysis:
Biodegradation:
Bioconcentration:

Common Name: 1-Ethyl-2-methylbenzene
Synonym: 2-ethyltoluene
Chemical Name: 1-ethyl-2-methylbenzene, 1-methyl-2-ethylbenzene
CAS Registry No: 611-14-3
Molecular Formula: $C_6H_4CH_3C_2H_5$
Molecular Weight: 120.2
Melting Point (°C): -80.8/-86.6
Boiling Point (°C): 165.2
Density (g/cm^3 at 20°C): 0.8807
Molar Volume (cm^3/mol):

162.6 (LeBas method, Eastcott et al. 1988)
136.5 (calculated-density, Miller et al. 1985)

Molecular Volume (A^3):
Total Surface Area, TSA (A^2):
Heat of Fusion, kcal/mol:
Fugacity Ratio, F: 1.0

Water Solubility (g/m^3 or mg/L at 25°C):

40.0 (estimated from nomograph, Kabadi & Danner 1979; quoted, API 1985 or Brookman et al. 1985; Cline et al. 1991)
93.05 (shake flask-GC, Mackay & Shiu 1981; quoted, Hutchinson et al. 1980)
74.6 (gen. col.-HPLC/UV, Tewari et al. 1982c)
74.6 (quoted, Miller et al. 1985; Eastcott et al. 1988)
93.05 (quoted, Nirmalakhanden & Speece 1988a)
55.8 (calculated- χ , Nirmalakhanden & Speece 1988a)

Vapor Pressure (Pa at 25°C):

330 (extrapolated, Antoine eqn., Zwolinski & Wilhoit 1971; quoted Mackay & Shiu 1981; Eastcott et al. 1988)
330, 358 (quoted exptl., calculated-B.P., Mackay et al. 1982)
328 (extrapolated, Antoine eqn., Boublik et al. 1984)
330 (extrapolated, Antoine eqn., Dean 1985)

Henry's Law Constant (Pa m^3/mol):

427 (calculated-P/C, Mackay & Shiu 1981)
565 (EPICS, Ashworth et al. 1988)
533 (calculated-C_A/C_W, Eastcott et al. 1988)

Octanol/Water Partition Coefficient, log K_{OW}:

3.63 (estimated, Hutchinson et al. 1980)
3.53 (gen. col.-HPLC/UV, Tewari et al. 1982a)
3.53 (quoted, Miller et al. 1985; Eastcott et al. 1988)
3.53 (recommended, Sangster 1989)

Bioconcentration Factor, log BCF:

Sorption Partition Coefficient, log K_{OC}:

Half-Lives in the Environment:

Air: 0.24-2.4 hours, based on rate of disappearance for the reaction with hydroxy radicals (Darnall et al. 1976; Howard et al. 1991).

Surface water: 0.5 day (estimated, Zoeteman et al. 1980).

Groundwater:

Soil:

Sediment:

Biota:

Environmental Fate Rate Constants or Half-Lives:

Photooxidation: rate constant for the reaction with OH radicals in the gas phase is 8.2×10^9 L mol^{-1} s^{-1} and $t_{1/2}$ = 0.24-2.4 hours (Darnall et al. 1976); room temperature rate constant for the gas phase reaction with OH radicals: 12.0×10^{-12} cm^3 molecule^{-1} sec^{-1} (Atkinson 1985) and 12.4×10^{-12} cm^3 molecule^{-1} sec^{-1} (Ohta & Ohyama 1985).

Common Name: 1-Ethyl-4-methylbenzene
Synonym: 4-ethyltoluene
Chemical Name: 1-ethyl-4-methylbenzene, 1-methyl-4-ethylbenzene
CAS Registry No: 622-96-8
Molecular Formula: $C_6H_4CH_3C_2H_5$
Molecular Weight: 120.2
Melting Point (°C): -62.4
Boiling Point (°C): 162
Density (g/cm^3 at 20°C): 0.8614
Molar Volume (cm^3/mol):

162.6 (LeBas method, Eastcott et al. 1988)
139.54 (calculated from density)

Molecular Volume (A^3):
Total Surface Area, TSA (A^2):
Heat of Fusion, kcal/mol:
Fugacity Ratio, F: 1.0

Water Solubility (g/m^3 or mg/L at 25°C):

40.0 (estimated from nomograph, Kabadi & Danner 1979; quoted, API 1985; Brookman et al. 1985)
94.85 (quoted, Hutchinson et al. 1980)
94.85 (shake flask-GC, Mackay & Shiu 1981; quoted, Eastcott et al. 1988)
94.85, 55.8 (quoted, calculated- χ , Nirmalakhanden & Speece 1988a)

Vapor Pressure (Pa at 25°C):

393 (extrapolated, Antoine eqn., Zwolinski & Wilhoit 1971; quoted, Mackay & Shiu 1981; Eastcott et al. 1988)
413, 413 (quoted exptl., calculated-B.P., Mackay et al. 1982)
394 (extrapolated, Antoine eqn., Boublik et al. 1984)
393 (extrapolated, Antoine eqn., Dean 1985)

Henry's Law Constant (Pa m^3/mol):

498 (calculated-P/C, Mackay & Shiu 1981)
498 (calculated-C_A/C_W, Eastcott et al. 1988)

Octanol/Water Partition Coefficient, log K_{OW}:

3.63 (estimated, Hutchinson et al. 1980)
3.63 (recommended, Sangster 1989)

Bioconcentration Factor, log BCF:

Sorption Partition Coefficient, log K_{OC}:

Half-Lives in the Environment:

Air: 0.24-2.4 hours, based on rate of disappearance for the reaction with hydroxy radicals (Darnall et al 1976; Howard et al. 1991).

Water:

Soil:

Sediment:

Biota:

Environmental Fate Rate Constants or Half-Lives:

Oxidation: rate constant for reaction with OH radicals in gas phase is 7.8×10^9 L mol^{-1} s^{-1} and $t_{1/2}$ = 0.24-2.4 hours (Darnall et al. 1976); room temperature rate constant for the gas phase reaction with OH radicals: 11.4×10^{-12} cm^3 $molecule^{-1}$ sec^{-1} (Atkinson 1985) and 12.8×10^{-12} cm^3 $molecule^{-1}$ sec^{-1} (Ohta & Ohyama 1985).

Common Name: 1-Isopropyl-4-methylbenzene
Synonym: p-cymene
Chemical Name: 1-isopropyl-4-methylbenzene
CAS Registry No: 25155-15-1
Molecular Formula: $CH_3C_6H_4C_3H_7$
Molecular Weight: 134.2
Melting Point (°C): -67.9
Boiling Point (°C): 177.1
Density (g/cm³ at 20°C): 0.857
Molar Volume (cm³/mol):

157.3 (Windholz 1983; Budavari 1989)
157.3 (20°C, calculated-density, Stephenson & Malanowski 1987)
184.8 (LeBas method, Eastcott et al. 1988)
156.6 (calculated from density)

Molecular Volume (A³):
Total Surface Area, TSA (A²):

369.3 (with hydration shell, Amidon & Anik 1981)

Heat of Fusion, kcal/mol:
Fugacity Ratio, F: 1.0

Water Solubility (g/m³ or mg/L at 25°C):

34.15 (residue volume method, Booth & Everson 1948; quoted, Eastcott et al. 1988)
31.5 (quoted, Tsonopoulos & Prausnitz 1971; Amidon & Anik 1981)
89 (calculated-B.P. correlation, Almgren et al. 1979; quoted, API 1985; Brookman et al. 1985)
340 (quoted, Dean 1985)
23.35 (shake flask-LSC, Banerjee et al. 1980)
23.35, 17.34 (quoted, calculated-UNIFAC, Banerjee 1985)
17.48 (calculated-UNIFAC, Banerjee & Howard 1988)

Vapor Pressure (Pa at 25°C):

204 (extrapolated, Antoine eqn., Zwolinski & Wilhoit 1971; quoted, Eastcott et al. 1988)
203, 208 (quoted exptl., calculated-B.P., Mackay et al. 1982)
194 (extrapolated, Antoine eqn., Stephenson & Malanowski 1987)

Henry's Law Constant (Pa m³/mol):

800 (calculated-P/C, Mackay & Shiu 1981)

Octanol/Water Partition Coefficient, log K_{OW}:

4.10 (shake flask-LSC, Banerjee et al. 1980)
4.14 (calculated-UNIFAC, Arbuckle 1983)
4.10, 3.45 (quoted, calculated-UNIFAC, Banerjee & Howard 1988)

4.10 (recommended, Sangster 1989)

Bioconcentration Factor, log BCF:

Sorption Partition Coefficient, log K_{OC}:

Half-Lives in the Environment:

Air: tropospheric life time, 1.0-1.4 days (Corchnoy & Atkinson 1990)
Surface water:
Groundwater:
Soil:
Biota:

Environmental Fate Rate Constants or Half-Lives:

Volatilization:
Photolysis:
Photooxidation: gas phase reaction with OH radicals is 1.51×10^{-11} cm^3 $molecule^{-1}$ s^{-1} and tropospheric lifetime is 1.0-1.4 days (Corchnoy & Atkinson 1990).
Hydrolysis:
Biodegradation:
Biotransformation:
Bioconcentration:

Common Name: n-Butylbenzene
Synonym:
Chemical Name: n-butylbenzene
CAS Registry No: 104-51-8
Molecular Formula: $C_6H_5(CH_2)_3$
Molecular Weight: 134.2
Melting Point (°C): -88
Boiling Point (°C): 183
Density (g/cm^3 at 20°C): 0.8601
Molar Volume (cm^3/mol):
156.78 (Windholz 1983; Budavari 1989)
156.1 (calculated-density, Klevens 1950; Miller et al. 1985)
0.868 (intrinsic volume: $V_I/100$, Kamlet et al. 1987)
0.861 (intrinsic volume: $V_I/100$, Kamlet et al. 1988)
184.8 (LeBas method, Eastcott et al. 1988)
150.1 (calculated from density)
Molecular Volume (A^3):
148.78 (Pearlman 1986)
Total Surface Area, TSA (A^2):
379.3 (with hydration shell, Amidon & Anik 1981)
187.8 (calculated-ion mobility, Lande et al. 1985)
379.3, 361.8 (with hydration shell, calculated- molecular model, ion mobility, Lande et al. 1985)
200.08 (Pearlman 1986)
191.0 (calculated- χ , Sabljic 1987a)
Heat of Fusion, kcal/mol:
Fugacity Ratio, F: 1.0

Water Solubility (g/m^3 or mg/L at 25°C):
12.6 (shake flask-UV, Andrews & Keefer 1950; quoted, Hutchinson et al. 1980)
50.5 (shake flask-UV, Klevens 1950)
50.5, 15.4 (quoted, estimated, Deno & Berkheimer 1960)
17.7 (shake flask-GC/ECD, Massaldi & King 1973)
11.8 (shake flask-GC, Sutton & Calder 1975; quoted, Brookman et al. 1985)
15.4 (quoted, Hine & Mookerjee 1975)
13.8, 14.5 (quoted, calculated- χ , reported as ln S as should be -log S in molar concn., Kier & Hall 1976)
12.2 (quoted lit. average, Amidon & Anik 1981)
13.83 (gen. col.-HPLC/UV, GC/ECD, Tewari et al. 1982)
13.8 (gen. col.-HPLC/UV, Wasik et al. 1983)
22.8 (calculated-K_{OW}, Yalkowsky et al. 1983)
12.2, 10.8 (quoted, calculated-TSA, Lande et al. 1985)
13.8 (quoted, Miller et al. 1985; Vowles & Mantoura 1987: Eastcott et al. 1988)

13.76 (gen. col.-HPLC/UV, Owens et al. 1986)
15.0 (recommended, IUPAC 1989b)
13.8 (quoted, Isnard & Lambert 1989)
12.0 (quoted, Warne et al. 1990)

Vapor Pressure (Pa at 25°C):
137 (extrapolated, Antoine eqn., Zwolinski & Wilhoit 1971; quoted, Mackay & Shiu 1981; Eastcott et al. 1988)
144 (quoted, Hine & Mookerjee 1975)
137 (quoted, exptl., Mackay et al. 1982)
158 (calculated-B.P., Mackay et al. 1982)
137 (extrapolated, Antoine eqn., Dean 1985)

Henry's Law Constant (Pa m^3/mol):
1300 (calculated-P/C, Mackay & Shiu 1981)
1333 (calculated-C_A/C_W, Eastcott et al. 1988)

Octanol/Water Partition Coefficient, log K_{OW}:
4.26 (Hansch & Leo 1979; quoted, Harnisch et al. 1983)
4.19 (calculated-f const., Rekker 1977; quoted, Harnisch et al. 1983)
3.86 (quoted, Hutchinson et al. 1980)
4.26, 4.44 (HPLC-k', correlated, Hammers et al. 1982)
4.26, 4.28 (quoted, gen. col.-HPLC/UV, Tewari et al. 1982c)
4.38, 4.21 (quoted of HPLC methods, Harnisch et al. 1983)
3.18 (calculated-f const., Yalkowsky et al. 1983)
4.28 (gen. col.-HPLC/UV, Wasik et al. 1983)
4.28 (quoted, Miller et al. 1985; Vowles & Mantoura 1987; Eastcott et al. 1988; Szabo et al. 1990a,b; Pussemier et al. 1990)
4.29, 4.34 (gen. col.-RPLC, calculated-activity coeff., Schantz & Martire 1987)
4.60 (Klein et al. 1988)
4.28 (quoted, Isnard & Lambert 1989)
4.26 (recommended, Sangster 1989)
4.377 (slow stirring-GC, De Brujin et al. 1989)
4.00 (quoted, Warne et al. 1990)

Bioconcentration Factor, log BCF:

Sorption Partition Coefficient, log K_{OC}:
3.39 (soil, observed, Schwarzenbach & Westall 1981)
3.16 (soil, calculated- χ , Sabljic 1987a & b)
3.40; 3.15, 3.32 (Vowless & Mantoura 1987; seleced; HPLC-k', Szabo et al. 1990a,b)
3.40 (soil, Pussemier et al. 1990)

Half-Lives in the Environment:

Air:

Surface water:

Groundwater:

Soil:

Biota:

Envrionmental Fate Rate Constants or Half-Lives:

Volatilization/Evaporation:

Photolysis:

Oxidation:

Hydrolysis:

Biodegradation:

Bioconcentration:

Common Name: Iso-Butylbenzene
Synonym: 2-methylpropylbenzene, methyl-1-phenylpropane
Chemical Name: iso-butylbenzene
CAS Registry No: 538-93-2
Molecular Formula: $C_6H_5(CH_2)_3$
Molecular Weight: 134.2
Melting Point (°C): -51.4
Boiling Point (°C): 172.8
Density (g/cm^3 at 20°C): 0.8532
Molar Volume (cm^3/mol):

158.1 (Wasik et al. 1984)
184.8 (LeBas method, Eastcott et al. 1988)
157.3 (calculated from density)

Molecular Volume (A^3):
Total Surface Area, TSA (A^2):
Heat of Fusion, kcal/mol:
Fugacity Ratio, F: 1.0

Water Solubility (g/m^3 or mg/L at 25°C):

28.7 (estimated, Deno & Berkheimer 1960)
10.1 (shake flask-GC, Price 1976; quoted, Hutchinson et al. 1980; Mackay & Shiu 1981; API 1985; Eastcott et al. 1988)
28.7 (quoted, Hine & Mookerjee 1975)
17.3 (calculated- χ , Nirmalakhanden & Speece 1988a)

Vapor Pressure (Pa at 25°C):

248 (extrapolated, Antoine eqn., Zwolinski & Wilhoit 1971; quoted, Mackay & Shiu 1981; Eastcott et al. 1988)
251 (quoted, Hine & Mookerjee 1975)
275 (quoted, exptl., Mackay et al. 1982)
252 (calculated-B.P., Mackay et al. 1982)
249 (extrapolated, Antoine eqn., Boublik et al. 1984)
257 (quoted, Wasik et al. 1984)
249 (extrapolated, Antoine eqn., Dean 1985)

Henry's Law Constant (Pa m^3/mol):

1714 (calculated-bond contribution, Hine & Mookerjee 1975)
1159 (calculated as $1/K_{AW}$, C_W/C_A, reported as exptl., Hine & Mookerjee 1975; Nirmalakhanden & Speece 1988b)
3300 (calculated, Mackay & Shiu 1981)
1393 (calculated-QSAR, Nirmalakhanden & Speece 1988b)
3295 (calculated-C_A/C_W, Eastcott et al. 1988)

Octanol/Water Partition Coefficient, log K_{OW}:
4.01 (quoted, Hutchinson et al. 1980)

Bioconcentration Factor, log BCF:

Sorption Partition Coefficient, log K_{OC}:

Half-Lives in the Environment:
Air:
Surface water:
Groundwater:
Soil:
Sediment:

Environmental Fate Rate Constants or Half-Lives:
Volatilization: estimated half-life of evaporation from a river of 1 m depth with wind speed 3 m/sec and water current of 1 m/sec to be 3.2 hours at 20°C (Lyman et al. 1982).

Common Name: *sec*-Butylbenzene
Synonym: 2-phenylbutane, (1-methylpropyl)benzene
Chemical Name: sec-butylbenzene
CAS Registry No: 135-98-8
Molecular Formula: $C_6H_5CH(CH_3)C_2H_5$
Molecular Weight: 134.2
Melting Point (°C): -75
Boiling Point (°C): 174
Density (g/cm^3 at 20°C): 0.8621
Molar Volume (cm^3/mol):
156.44 (Windhoz 1983; Budavari 1989)
156.4 (Wasik et al. 1984)
184.8 (LeBas method, Eastcott et al. 1988)
155.7 (calculated from density)
Molecular Volume (A^3):
Total Surface Area, TSA (A^2):
367.4 (with hydration shell, Amidon & Anik 1981)
176.1 (calculated-ion mobility, Lande et al. 1985)
367.4, 345.5 (calculated- molecular model, ion mobility, with hydration shell, Lande et al. 1985)
Heat of Fusion, kcal/mol:
Fugacity Ratio, F: 1.0

Water Solubility (g/m^3 or mg/L at 25°C):
30.9 (shake flask-UV, Andrews & Keefer 1950)
17.6 (shake flask-GC, Sutton & Calder 1975; quoted, API 1985; Brookman et al. 1985; Eastcott et al. 1988)
28.7 (quoted, Hine & Mookerjee 1975)
28.7, 17.3 (quoted, calculated- χ , reported as ln S as should be -log S in molar concn., Kier & Hall 1976)
24.3 (quoted lit. average, Amidon & Anik 1981)
320 (Dean 1985)
24.4, 24.25 (quoted, calculated-TSA, Lande et al. 1985)
14.9 (calculated- χ , Nirmalakhanden & Speece 1988a)
17.6 (quoted, Abdul et al. 1990)

Vapor Pressure (Pa at 25°C):
241 (extrapolated, Antoine eqn., Zwòlinski & Wilhoit 1971; quoted, Eastcott et al. 1988)
251 (quoted, Hine & Mookerjee 1975)
241 (quoted, exptl., Mackay et al. 1982)
250 (calculated-B.P., Mackay et al. 1982)

147 (20°C, Verschueren, 1983)
250 (quoted, Wasik et al. 1984)
240 (extrapolated, Antoine eqn., Boublik et al. 1984)

Henry's Law Constant (Pa m^3/mol):
1400 (calculated-P/C, Mackay & Shiu 1981)
1838 (calculated-C_A/C_W, Eastcott et al. 1988)

Octanol/Water Partition Coefficient, log K_{OW}:

Bioconcentration Factor, log BCF:

Sorption Partition Coefficient, log K_{OC}:

Half-Lives in the Environment:

Environmental Fate Rate Constants or Half-Lives:

Common Name: *tert*-Butylbenzene
Synonym: (1,1-dimethylethyl)benzene, 2-methyl-2-phenylpropane, trimethylphenylmethane, pseudobutylbenzene
Chemical Name: tert-butylbenzene
CAS Registry No: 98-06-6
Molecular Formula: $C_6H_5(CH_3)_2$
Molecular Weight: 134.2
Melting Point (°C): -58
Boiling Point (°C): 169
Density (g/cm^3 at 20°C): 0.8665
Molar Volume (cm^3/mol):
155.64 (Windhoz 1983; Budavari 1989)
155.6 (Wasik et al. 1984)
156.0 (calculated-density, Chiou & Block 1986)
0.868 (intrinsic volume: $V_I/100$, Kamlet et al. 1987)
184.8 (LeBas method, Eastcott et al. 1988)
154.9 (calculated from density)
Molecular Volume (A^3):
Total Surface Area, TSA (A^2):
176.8 (Yalkowsky & Valvani 1976)
352.7 (with hydration shell, Amidon & Anik 1981)
160.8 (calculated-ion mobility, Lande et al. 1985)
352.7, 323.9 (calculated-molecular model, ion mobility, with hydration shell, Lande et al. 1985)
Heat of Fusion, kcal/mol:
Fugacity Ratio, F: 1.0

Water Solubility (g/m^3 or mg/L at 25°C):
34.0 (shake flask-UV, Andrews & Keefer 1950; quoted, Tsonopoulos & Prausnitz 1971; Chiou et al. 1982; API 1985)
33.7 (quoted, Hine & Mookerjee 1975)
29.5 (shake flask-GC, Sutton & Calder 1975; quoted, API 1985; Eastcott et al. 1988)
33.7, 40.0 (quoted, calculated- χ , reported as ln S as should be -log S in molar concn., Kier & Hall 1976)
31.8 (quoted lit. average, Amidon & Anik 1981)
290 (Dean 1985)
32.4, 93.05 (quoted, calculated-TSA, Lande et al. 1985)
33.7 (quoted, Chiou & Block 1986)
33.9 (correlated, Isnard & Lambert 1989)

Vapor Pressure (Pa at 25°C):

286 (extrapolated, Antoine eqn., Zwolinski & Wilhoit 1971; quoted, Eastcott et al. 1988)
295 (quoted, Hine & Mookerjee 1975)
286 (quoted, exptl., Mackay et al. 1982)
301 (calculated-B.P., Mackay et al. 1982)
200 (20°C, Verschueren, 1983)
294 (quoted, Wasik et al. 1984)
286 (extrapolated, Antoine eqn., Dean 1985)

Henry's Law Constant (Pa m^3/mol):

1200 (calculated-P/C, Mackay & Shiu 1981)
1300 (calculated-C_A/C_W, Eastcott et al. 1988)

Octanol/Water Partition Coefficient, log K_{OW}:

4.11 (Leo et al. 1971; Hansch & Leo 1979; quoted, Chiou et al. 1982; Chiou & Block 1986)
4.11, 4.01 (quoted, calculated-TSA, Yalkowsky & Valvani 1976)
4.11, 4.07 (quoted, shake flask-HPLC, Nahum & Horvath 1980)
4.11 (Verschueren, 1983)
4.11 (recommended, Sangster 1989)
4.11 (quoted, Isnard & Lambert 1989)

Bioconcentration Factor, log BCF:

Sorption Partition Coefficient, log K_{OC}:

Half-Lives in the Environment:

Environmental Fate Rate Constants or Half-Lives:

Oxidation: rate constant for the gas phase reaction with OH radicals at 25°C, 4.58×10^{-12} cm^3 $molecule^{-1}$ sec^{-1} (Ohta & Ohyama 1985).

Common Name: 1,2,3,4-Tetramethylbenzene
Synonym: perhintene, prebnitene
Chemical Name: 1,2,3,4-tetramethylbenzene
CAS Registry No: 488-23-3
Molecular Formula: $C_6H_5CH(CH_3)C_2H_5$
Molecular Weight: 134.2
Melting Point (°C): -6.25
Boiling Point (°C): 205
Density (g/cm^3 at 20°C): 0.9052
Molar Volume (cm^3/mol):

148.3 (20°C, calculated-density, Stephenson & Malanowski 1987)
0.85 (intrinsic volume: $V_I/100$, Kamlet et al. 1987)

Molecular Volume (A^3):

147.07 (Pearlman 1986)

Total Surface Area, TSA (A^2):

188.97 (Pearlman 1986)
180.37 (Doucette & Andren 1988)

Heat of Fusion, kcal/mol:
Fugacity Ratio, F: 1.0

Water Solubility (g/m^3 or mg/L at 25°C):

Vapor Pressure (Pa at 25°C):

45.01 (extrapolated, Antoine eqn., Zwolinski & Wilhoit 1971)
45.01 (extrapolated-Antoine eqn., Dean 1985)
45.02 (extrapolated, Antoine eqn., Stephenson & Malanowski 1987)

Henry's Law Constant (Pa m^3/mol):

Octanol/Water Partition Coefficient, log K_{OW}:

3.84 (gen. col.-HPLC, Wasik et al. 1982)
4.11 (HPLC-k', Hammers et al. 1982)
3.84 (calculated-f const., Yalkowsky et al. 1983)
4.55 (quoted exptl., Doucette & Andren 1988)
4.99, 5.02, 4.85, 4.33, 4.06 (calculated-π, f const., MW, χ, TSA, Doucette & Andren 1988)
4.00 (recommended, Sangster 1989)

Bioconcentration Factor, log BCF:

Sorption Partition Coefficient, log K_{OC}:

Half-Lives in the Environment:

Air:
Water:
Soil:
Sediment:
Biota:

Environmental Fate Rate Constants or Half-Lives:

Volatilization/Evaportaion:
Photolysis:
Photooxidation:
Hydrolysis:
Biodegradation:
Biotransformation:
Bioconcentraion:

Common Name: 1,2,3,5-Tetramethylbenzene
Synonym: isodurene
Chemical Name: 1,2,3,5-tetramethylbenzene
CAS Registry No: 527-53-7
Molecular Formula: $C_6H_2(CH_3)_4$
Molecular Weight: 134.2
Melting Point (°C): -23.68
Boiling Point (°C): 198
Density (g/cm^3 at 20°C): 0.8903
Molar Volume (cm^3/mol):

150.8 (20°C, calculated-density, Stephenson & Malanowski 1987)
0.85 (intrinsic volume: $V_I/100$, Kamlet et al. 1987)

Molecular Volume (A^3):

147.40 (Pearlman 1986)

Total Surface Area, TSA (A^2):

192.04 (Pearlman 1986)
183.94 (Doucette & Andren 1988)

Heat of Fusion, kcal/mol:
Fugacity Ratio, F: 1.0

Water Solubility (g/m^3 or mg/L at 25°C):

Vapor Pressure (Pa at 25°C):

62.22 (extrapolated, Antoine eqn., Zwolinski & Wilhoit 1971)
62.22 (extrapolated, Antoine eqn., Dean 1985)
62.23 (extrapolated, Antoine eqn., Stephenson & Malanowski 1987)

Henry's Law Constant (Pa m^3/mol):

Octanol/Water Partition Coefficient, log K_{OW}:

4.04 (gen. col.-HPLC, Wasik et al. 1982)
4.17 (HLPC-k', Hammers et al. 1982)
4.65 (quoted exptl., Doucette & Andren 1988)
4.99, 5.02, 4.85, 4.06, 4.11 (calculated-π, f const., MW, χ, TSA, Doucette & Andren 1988)
4.10 (recommended, Sangster 1989)

Bioconcentration Factor, log BCF:

Sorption Partition Coefficient, log K_{OC}:

Half-Lives in the Environment:

Environmental Fate Rate Constants or Half-Lives:

Volatilization:

Photolysis:

Oxidation:

Hydrolysis:

Biodegradation:

Bioconcentration:

Common Name: 1,2,4,5-Tetramethylbenzene
Synonym: durene
Chemical Name: 1,2,4,5-tetramethylbenzene
CAS Registry No: 95-93-2
Molecular Formula: $C_6H_2(CH_3)_4$
Molecular Weight: 134.2
Melting Point (°C): 79.2
Boiling Point (°C): 196.8
Density (g/cm³ at 20°C): 0.838
Molar Volume (cm³/mol):
    160.2 (20°C, calculated-density, Stephenson & Malanowski 1987)
    0.90 (intrinsic volume: $V_I/100$, Kamlet et al. 1987)
    0.867 (intrinsic volume: $V_I/100$, Kamlet et al. 1988)
    184.8 (LeBas method, Eastcott et al. 1988)
    128.0 (characteristic molecular volume, McGowen & Mellors 1986)
Molecular Volume (A³):
    147.40 (Pearlman 1986)
Total Surface Area, TSA (A²):
    192.04 (Pearlman 1986)
    192.8 (calculated- χ , Sabljic 1987a)
Heat of Fusion, kcal/mol:
    5.020 (Tsonopoulos & Prausnitz 1971)
Entropy of Fusion, cal/mol K (e.U.):
    14.3 (Tsonopoulos & Prausnitz 1971)
Fugacity Ratio at 25 °C, F: 0.286

Water Solubility (g/m³ or mg/L at 25°C):
    19.5 (Deno & Berkheimer 1960)
    19.5 (quoted, Tsonopoulos & Prausnitz 1971)
    3.48 (shake flask-GC, Price 1976; quoted, Mackay & Shiu 1981; API 1985; Abernethy et al. 1986; Eastcott et al. 1988)
    3.48 (shake flask-GC, Krzyzanowska & Szeliga 1978)
    9.6 (quoted, Hutchinson et al. 1980)
    13.9 (calculated-HPLC-k', converted from reported γ_W, Hafkenscheid & Tomlinson 1983)
    6.13 (calculated-K_{OW}, Yalkowsky et al. 1983)
    3.50 (quoted, Abdul et al. 1990)

Vapor Pressure (Pa at 25°C):
    65.9 (extrapolated, Antoine eqn., Zwolinski & Wilhoit 1971; quoted, Mackay & Shiu 1981; Eastcott et al. 1988)
    65.9 (quoted, exptl., Mackay et al. 1982)
    82.6 (calculated-B.P., Mackay et al. 1982)

65.88 (extrapolated-Antoine eqn., Dean 1985)
65.9 (extrapolated, Antoine eqn., Stephenson & Malanowski 1987)

Henry's Law Constant (Pa m^3/mol):
2540 (calculated-P/C, Mackay & Shiu 1981)
2552 (calculated-C_A/C_W, Eastcott et al. 1988)

Octanol/Water Partition Coefficient, log K_{OW}:
4.0 (Hansch & Leo 1979; quoted, Eastcott et al. 1988)
2.80 (quoted, Hutchinson et al. 1980)
4.11 (calculated-f const., Yalkowsky et al. 1983)
4.24 (HPLC-RV, Garst 1984)
4.00 (quoted, Tomlinson & Hafkanscheid 1986)
4.05 (quoted, Abernethy et al. 1988)
4.10 (recommended, Sangster 1989)

Bioconcentration Factor, log BCF:

Sorption Partition Coefficient, log K_{OC}:
3.12 (soil, observed, Schwarzenbach & Westall 1981)
2.99 (soil, calculated- χ , Sabljic 1987a & b)

Half-Lives in the Environment:
Air:
Surface Water:
Groundwater:
Soil:

Environmental Fate Rate Constants or Half-Lives:
Volatilization:
Photolysis:
Oxidation: rate constant for reaction with ozone at 300°K estimated to be 1.1×10^4 cm^3 $mole^{-1}$ sec^{-1} (Lyman 1982).
Hydrolysis:
Biodegradation:
Bioconcentration:

Common Name: Pentamethylbenzene
Synonym:
Chemical Name: pentamethylbenzene
CAS Registry No: 700-12-9
Molecular Formula: $C_6H(CH_3)_5$
Molecular Weight: 148.25
Melting Point (°C): 54.5
Boiling Point (°C): 231
Density (g/cm^3 at 20°C): 0.917
Molar Volume (cm^3/mol):
207.0 (calculated from LeBas method)
161.7 (calculated from density)
0.965 (intrinsic volume: $V_I/100$, Kamlet et al. 1988)
Molecular Volume (A^3):
162.77 (Pearlman 1986)
Total Surface Area, TSA (A^2):
207.95 (Pearlman 1986)
Heat of Fusion, kcal/mol:
2.95 (Tsonopoulos & Prausnitz 1971)
Entropy of Fusion, cal/mol K (e.u.):
9.01 (Tsonopoulos & Prausnitz 1971)
Fugacity Ratio, F (assuming ΔS_{fusion} = 13.5 e.u.):
0.512

Water Solubility (g/m^3 or mg/L at 25°C):
15.6 (Deno & Berkheimer 1960; quoted, Tsonopoulos & Prausnitz 1971)
15.52 (calculated-K_{OW}, Yalkowsky et al. 1983)

Vapor Pressure (Pa at 25°C):
9.52 (extrapolated, Antoine eqn., Stephenson & Malanowski 1987)

Henry's Law Constant (Pa m^3/mol):

Octanol/Water Partition Coefficient, log K_{OW}:
4.56 (HPLC-k', Hammers et al. 1982)
4.63 (calculated-f const., Yalkowsky et al. 1983)
4.57 (HPLV-RV, Garst 1984)
4.56 (recommended, Sangster 1989)

Bioconcentration Factor, log BCF:

Sorption Partition Coefficient, log K_{OC}:

Half-Lives in the Environment:

Air:
Water:
Soil:
Sediment:
Biota:

Environmental Fate Rate Constants or Half-Lives:

Volatilization:
Photolysis:
Oxidation:
Hydrolysis:
Biodegradation:
Bioconcentration:

Common Name: Pentylbenzene
Synonym: phenylpentane
Chemical Name: pentylbenzene
CAS Registry No: 538-68-1
Molecular Formula: $C_6H_5C_5H_{11}$
Molecular Weight: 148.25
Melting Point (°C): -78.3
Boiling Point (°C): 202.2
Density (g/cm³ at 20°C): 0.8585
Molar Volume (cm³/mol):
207.0 (calculated from LeBas method, Eastcott et al. 1988)
172.7 (calculated from density)
Molecular Volume (A³):
165.06 (Pearlman 1986)
Total Surface Area, TSA (A²):
222.78 (Pearlman 1986)
Heat of Fusion, kcal/mol:
Fugacity Ratio, F: 1.0

Water Solubility (g/m³ or mg/L at 25°C):
10.5 (shake flask-UV, Andrews & Keefer 1950; quoted, Eastcott et al. 1988)
3.84 (gen. col.-HPLC/UV, Tewari et al. 1982c)
3.85 (quoted, Miller et al. 1985)
3.37 (gen. col.-HPLC/UV, Owens et al. 1986)
3.89 (correlated, Isnard & Lambert 1989)

Vapor Pressure (Pa at 25°C):
43.7 (extrapolated, Antoine eqn., Zwolinski & Wilhoit 1971; quoted, Eastcott et al. 1988)
43.7 (quoted, exptl., Mackay et al. 1982)
54.9 (calculated-B.P., Mackay et al. 1982)

Henry's Law Constant (Pa m³/mol):
600 (calculated-P/C, Mackay & Shiu 1981)
617 (calculated-C_A/C_W, Eastcott et al. 1988)

Octanol/Water Partition Coefficient, log K_{OW}:
4.56 (HPLC-k', Hammers et al. 1982)
4.90 (gen. col.-HPLC/UV, Tewari et al. 1982c)
4.90 (quoted, Miller et al. 1985; Eastcott et al. 1988)
4.90 (correlated, Isnard & Lambert 1989)
4.90 (recommended, Sangster 1989)

Bioconcentration Factor, log BCF:

Sorption Partition Coefficient, log K_{OC}:

Half-Lives in the Environment:
Air:
Surface Water:
Groundwater:
Soil:
Environment Fate Rate Constants or Half-Lives:
Volatilization:
Photolysis:
Oxidation: for the reaction with ozone at 300°K having a rate constant of $5.0x10^4$ cm^3 $mole^{-1}$ sec^{-1}. (Lyman 1982).
Hydrolysis:
Biodegradation:

Common Name: n-Hexamethylbenzene
Synonym: mellitene
Chemical Name: n-hexamethylbenzene
CAS Registry No: 89-85-4
Molecular Formula: $C_6(CH_3)_6$
Molecular Weight: 162.28
Melting Point (°C): 166.7
Boiling Point (°C): 265
Density (g/cm^3):
Molar Volume (cm^3/mol):

229.2 (calculated from LeBas method)
1.063 (intrinsic volume: $V_I/100$, Kamlet et al. 1988)

Molecular Volume (A^3):

178.14 (Pearlman 1986)

Total Surface Area, TSA (A^2):

223.85 (Pearlman 1986)
210.61 (Doucette & Andren 1988)

Heat of Fusion, kcal/mol:

4.89 (Tsonopoulos & Prausnitz 1971)

Entropy of Fusion, cal/mol K (e.u.):

11.1 (Tsonopoulos & Prausnitz 1971)

Fugacity Ratio, F (calculated, assuming ΔS_{fusion} = 13.5 e.u.):

0.0402

Water Solubility (g/m^3 or mg/L at 25°C):

0.235 (gen. col.-GC, Doucette & Andren 1988)

Vapor Pressure (Pa at 25°C):

0.155 (extrapolated, Antoine eqn., Stephenson & Malanowski 1987)

Henry's Law Constant (Pa m^3/mol):

Octanol/Water Partition Coefficient, log K_{OW}:

4.61 (gen. col.-HPLC, Wasik et al. 1982)
5.11 (HPLC-k', Hammers et al. 1982)
5.10 (calculated-f const., Yalkowsky et al. 1983)
4.31 (HPLC-RV, Garst & Wilson 1984)
4.60 (slow stirring, Brooke et al. 1986)
4.61 (quoted, Doucette & Andren 1988)
5.16, 5.16, 4.02, 4.73, 4.96 (calculated-π, f const., MW, χ, TSA, Doucette & Andren 1988)
4.75 (recommended, Sangster 1989)

Bioconcentration Factor, log BCF:

Sorption Partition Coefficient, log K_{OC}:

Half-Lives in the Environment:

Air:

Surface Water:

Groundwater:

Soil:

Environmental Fate Rate Constants or Half-Lives:

Volatilization:

Photolysis:

Oxidation: for the reaction with ozone at 300°K having a rate constant of 2.4×10^{5} cm^3 $mole^{-1}$ sec^{-1}. (Lyman et al. 1982)

Common Name: n-Hexylbenzene
Synonym:
Chemical Name: n-hexylbenzene
CAS Registry No: 1077-16-3
Molecular Formula: $C_6H_5(CH_2)_6CH_3$
Molecular Weight: 162.28
Melting Point (°C): -61
Boiling Point (°C): 226
Density (g/cm^3 at 20°C): 0.861
Molar Volume (cm^3/mol):
229.2 (LeBas method, Eastcott et al. 1988)
188.5 (calculated from density)
Molecular Volume (A^3):
181.34 (Pearlman 1986)
Total Surface Area, TSA (A^2):
245.49 (Pearlman 1986)
235.1 (Warne et al. 1990)
Heat of Fusion, kcal/mol:
Fugacity Ratio, F: 1.0

Water Solubility (g/m^3 or mg/L at 25°C):
1.02 (gen. col.-HPLC/UV, Tewari et al. 1982c)
1.02 (quoted, Miller et al. 1985; selected, Eastcott et al. 1988))
0.902 (gen. col.-HPLC/UV, Owens et al. 1986)
1.02 (quoted, Isnard & Lambert 1989)
0.971 (quoted, Warne et al. 1990)

Vapor Pressure (Pa at 25°C):
13.61 (extrapolated, Antoine eqn., Zwolinski & Wilhoit 1971; quoted, Eastcott et al. 1988)

Henry's Law Constant (Pa m^3/mol):
1977 (calculated-C_A/C_W, Eastcott et al. 1988)

Octanol/Water Partition Coefficient, log K_{OW}:
5.25 (calculated-f const., Rekker 1977; quoted, Harnisch et al. 1983)
5.52 (gen. col.-HPLC/UV, Tewari et al. 1982c; quoted, Miller et al. 1985; Eastcott et al. 1988)
5.24 (TLC-RT, Bruggeman et al. 1982)
5.45, 5.25 (quoted of HPLC methods, Harnisch et al. 1983)
5.52 (recommended, Sangster 1989)
5.52 (quoted, Isnard & Lambert 1989)
5.52 (quoted, Warne et al. 1990)

Bioconcentration Factor, log BCF:

Sorption Partition Coefficient, log K_{OC}:

Half-Lives in the Environment:

- Air:
- Water:
- Soil:
- Sediment:
- Biota:

Environmental Fate Rate Constants or Half-Lives:

- Volatilization:
- Photolysis:
- Oxidation:
- Hydrolysis:
- Biodegradation:
- Bioconcentration:

2.2 Summary Tables and QSPR Plots

Table 2.1 Summary of the physical-chemical properties of monoaromatic hydrocarbons

Compound	CAS no.	MW g/mol	mp, °C	bp, °C	Fugacity ratio, F at 25 °C	Density cm^3/mL at 20 °C	V_M (density) cm^3/mol	V_M LeBas cm^3/mol	$V_I/100$ intrinsic (a)	TSA $Å^2$ (b)	TMV $Å^3$ (b)
Benzene	71-43-2	78.11	5.53	80.1	1	0.8765	88.7	96	0.491	110.01	83.95
Toluene	108-88-3	92.13	-95	110.6	1	0.8669	106.3	118.2	0.591	132.05	99.98
Ethylbenzene	100-41-4	106.2	-95	136.2	1	0.867	122.4	140.4	0.671	154.66	116.24
o-Xylene	95-47-6	106.2	-25.2	144	1	0.8802	120.6	140.4	0.671	153.53	115.97
m-Xylene	108-38-3	106.2	-47.9	139	1	0.8842	123.2	140.4	0.671	154.08	116
p-Xylene	106-42-3	106.2	13.2	138	1	0.8611	119.8	140.4	0.671	154.09	116
1,2,3-Trimethyl-	526-73-8	120.2	-25.4	176.1	1	0.8944	134.4	162.6		172.5	131.67
1,2,4-Trimethyl-	95-63-6	120.2	-43.8	169.4	1	0.8758	137.2	162.6	0.768	175.56	132
1,3,5-Trimethyl-	108-67-8	120.2	-44.7	164.7	1	0.8652	139	162.6	0.768	170.11	132.03
n-Propylbenzene	103-65-1	120.2	-101.6	159.2	1	0.862	149.4	170	0.768	177.36	132.49
iso-Propylbenzene	98-82-8	120.2	-96.6	154.2	1	0.8618	139.5	170	0.768		
1-Ethyl-2-methyl-	611-14-3	120.2	-80.8	165.2	1	0.8867	136.5	162.6			
1-Ethyl-3-methyl-	620-14-4	120.2	-95.5	161.5	1	0.8645	139	162.6			
1-Ethyl-4-methyl-	622-96-8	120.2	-62.3	162	1	0.8614	139.54	162.60			
iso-Propyl-4-methyl-	99-87-6	134.22	-67.9	177.1	1	0.857	156.6	184.8			
n-Butylbenzene	104-51-8	134.22	-88	183	1	0.8601	156.1	184.8	0.868	200.08	148.7
iso-Butylbenzene	538-93-2	134.22	-51	170	1	0.8532	157.3	184.8			
sec-Butylbenzene	135-98-8	134.22	-75.5	173	1	0.8621	155.7	184.4			
tert-Butylbenzene	98-06-6	134.22	-57.8	169	1	0.8665	154.9	184.8	0.868		
1,2,3,4-Tetramethyl-	48-23-3	134.22	-6.25	205	1	0.9052	148.3	184.8		188.97	147.07
1,2,3,5-Tetramethyl-	527-53-7	134.22	-23.68	198	1	0.8585	150.7	184.8		192.04	147.4
1,2,4,5-Tetramethyl-	95-93-2	134.22	-79.2	196.8	1	0.838	160.1	184.8		192.04	147.4
n-Pentylbenzene	538-68-1	148.25	-75	205.4	1	0.8585	172.7	207		222.78	165.05
Pentamethyl-	700-12-9	148.25	54.5	231	0.511	0.917	161.7	207		207.95	162.77
n-Hexylbenzene	1077-16-3	162.28	-61	226	1	0.861	188.5	229.2		245.49	181.34
Hexamethyl-	87-85-4	162.28	166.7	265	0.0397	1.063		229.2		207.95	178.14

(a) Kamlet et al. (1987)
(b) Pearlman (1986)

Table 2.2 Summary of physical-chemical properties of monoaromatic hydrocarbons at 25 °C

Selected properties:

Compound	Vapor pressure		Solubility			log K_{OW}	Henry's law const., H
	P^S Pa	P_L Pa	S g/m^3	C^S mol/m^3	C_L mol/m^3		Pa m^3/mol calcd., P/C
Benzene	12700	12700	1780	22.788	22.788	2.13	557
Toluene	3800	3800	515	5.590	5.590	2.69	680
Ethylbenzene	1270	1270	152	1.431	1.431	3.13	887
o-Xylene	1170	1170	220	2.072	2.072	3.15	565
m-Xylene	1100	1100	160	1.507	1.507	3.20	730
p-Xylene	1170	1170	215	2.024	2.024	3.18	578
1,2,3-Trimethyl-	200	200	70	0.582	0.582	3.55	343
1,2,4-Trimethyl-	270	270	57	0.474	0.474	3.60	569
1,3,5-Trimethyl-	325	325	50	0.416	0.416	3.58	781
n-Propylbenzene	450	450	52	0.433	0.433	3.69	1040
iso-Propyl-	610	610	50	0.416	0.416	3.63	1466
1-Ethyl-2-methyl-	330	330	75	0.624	0.624	3.63	529
1-Ethyl-3-methyl-	391	391					
1-Ethyl-4-methyl-	395	395	95	0.790	0.790	3.63	500
iso-Propyl-4-methyl-	204	204	34	0.253	0.253	4.10	805
n-Butylbenzene	137	137	13.8	0.103	0.103	4.26	1332
iso-Butylbenzene	250	250	10.1	0.075	0.075	4.01	3322
sec-Butylbenzene	240	240	17	0.127	0.127		1890
tert-Butylbenzene	286	286	30	0.224	0.224	4.11	1280
1,2,3,4-Tetramethyl-	45	45				3.90	
1,2,3,5-Tetramethyl-	62	62				4.04	
1,2,4,5-Tetramethyl-	66	66	3.48	0.026	0.026	4.10	2546
n-Pentylbenzene	44	44	3.85	0.026	0.026	4.90	1694
Pentamethyl-	9.52	18.63	15.5	0.105	0.205		
n-Hexylbenzene	13.61	13.61	1.02	0.006	0.006	5.52	2165
Hexamethyl-	0.155	3.90	0.235	0.001	0.036	4.61	

Table 2.3 Suggested half-life classes of alkylated monoaromatics in various environmental compartments

Compounds	Air class	Water class	Soil class	Sediment class
Benzene	2	4	5	6
Toluene	2	5	6	7
Ethylbenzene	2	5	6	7
o-Xylene	2	5	6	7
m-Xylene	2	5	6	7
p-Xylene	2	5	6	7
1,2,3-Trimethyl-	2	5	6	7
1,2,4-Trimethyl-	2	5	6	7
1,3,5-Trimethyl-	2	5	6	7
n-Propylbenzene	2	5	6	7
iso-Propylbenzene	2	5	6	7
iso-Propyl-4-methyl-	2	5	6	7
1,2,4,5-Tetramethyl-	2	5	6	7

where,

Class	Mean half-life (hours)	Range (hours)
1	5	< 10
2	17 (~ 1 day)	10-30
3	55 (~ 2 days)	30-100
4	170 (~ 1 week)	100-300
5	550 (~ 3 weeks)	300-1,000
6	1700 (~ 2 months)	1,000-3,000
7	5500 (~ 8 months)	3,000-10,000
8	17000 (~ 2 years)	10,000-30,000
9	55000 (~ 6 years)	> 30,000

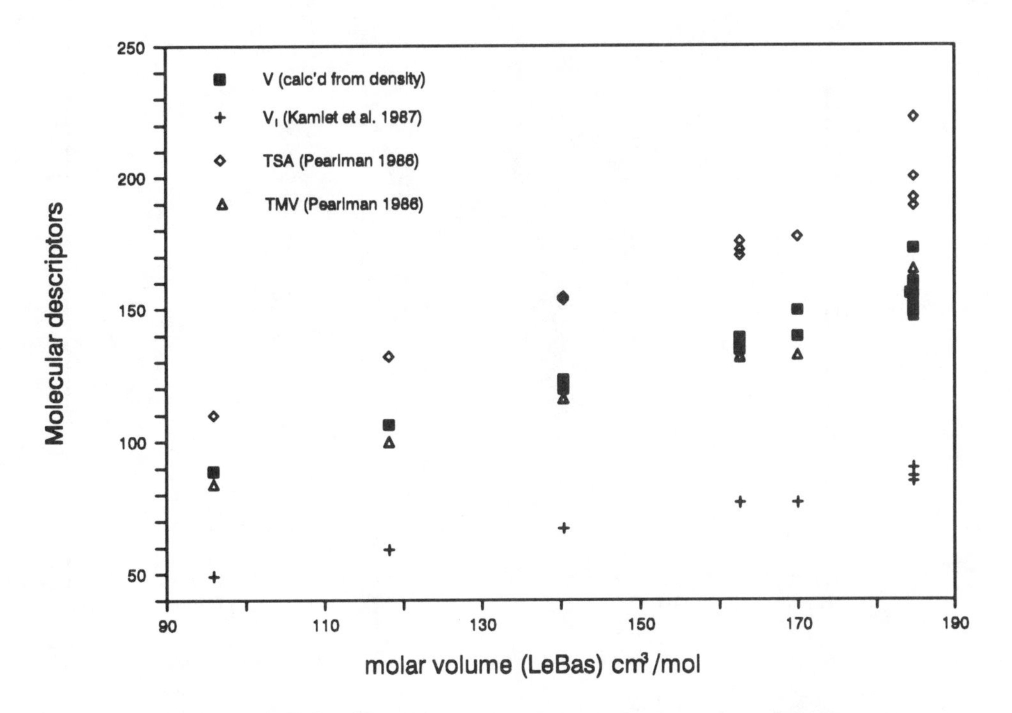

Figure 2.1 Plot of molecular descriptors versus LeBas molar volume

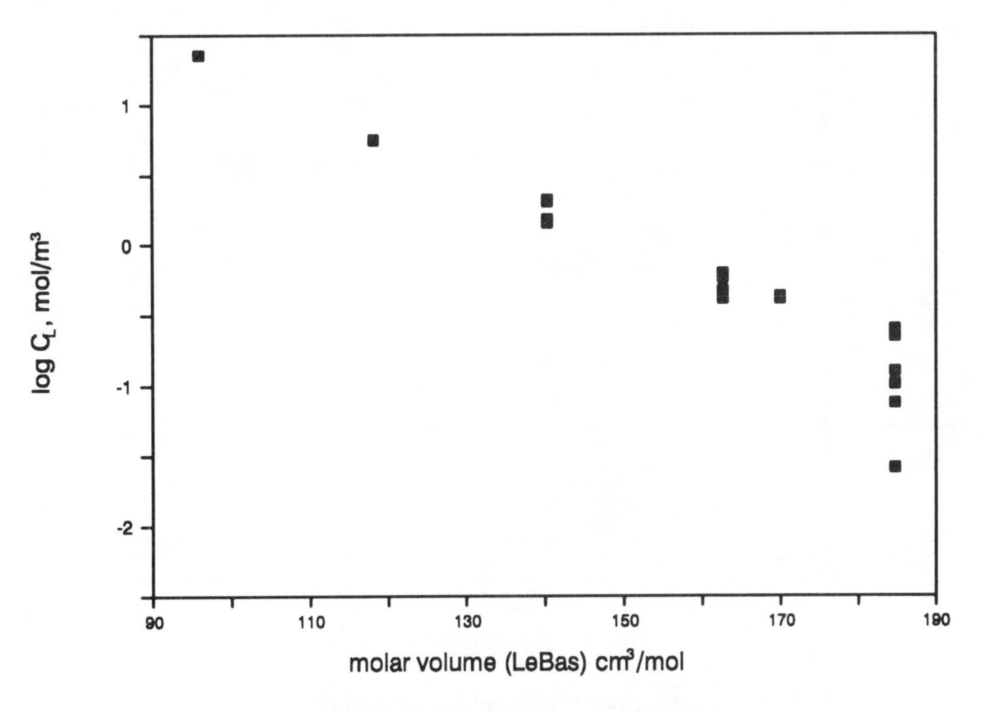

Figure 2.2 Plot of log C_L (liquid solubility) versus molar volume

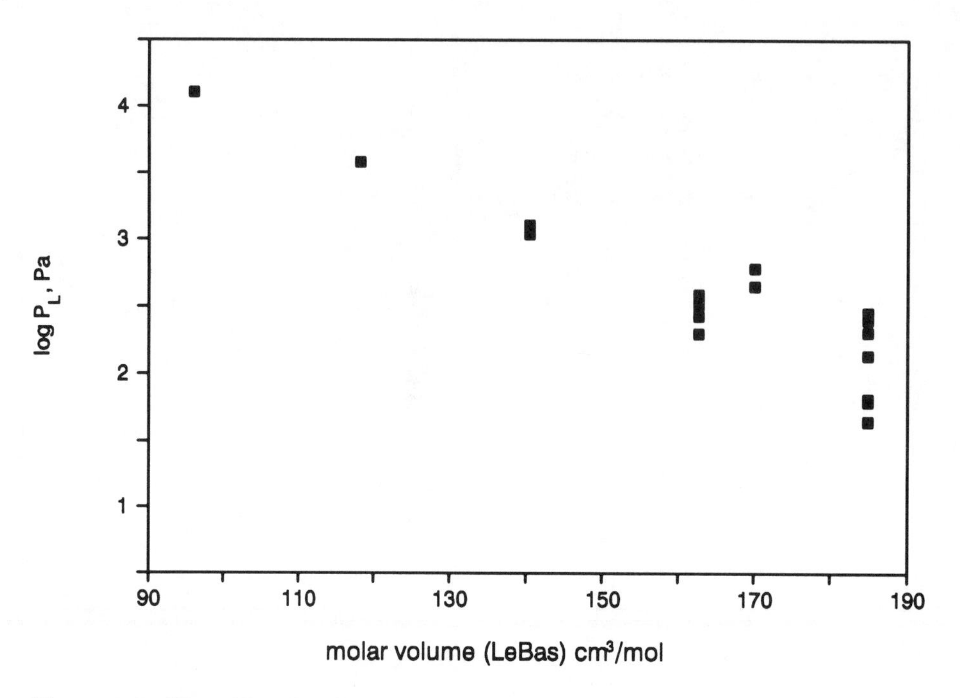

Figure 2.3 Plot of log P_L (liquid vapor pressure) versus molar volume

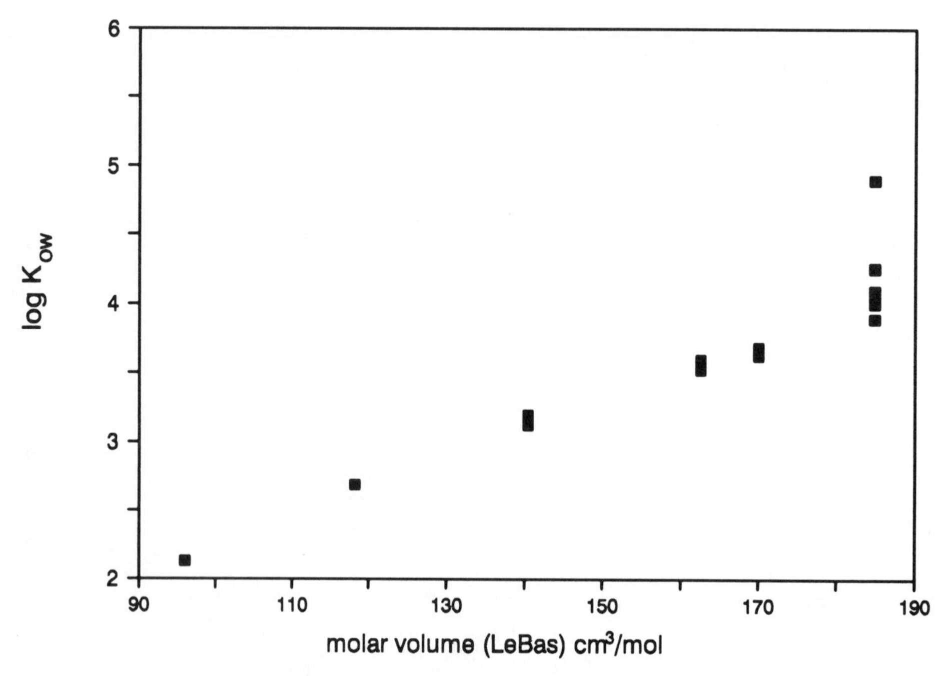

Figure 2.4 Plot of log K_{OW} versus LeBas molar volume

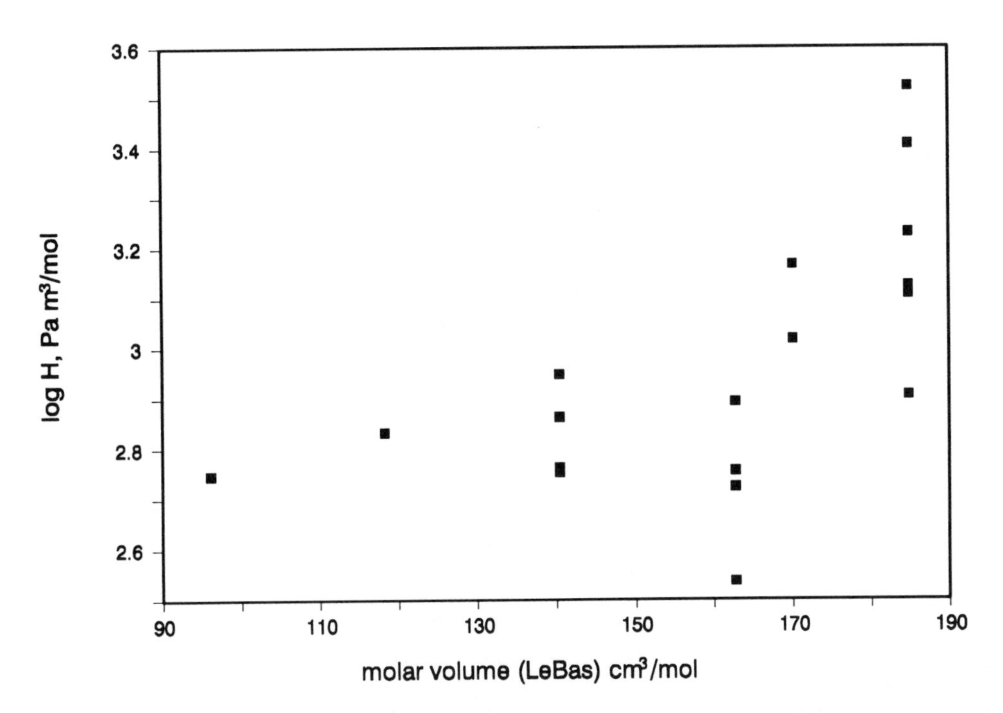

Figure 2.5 Plot of log H (Henry's law constant) versus molar volume

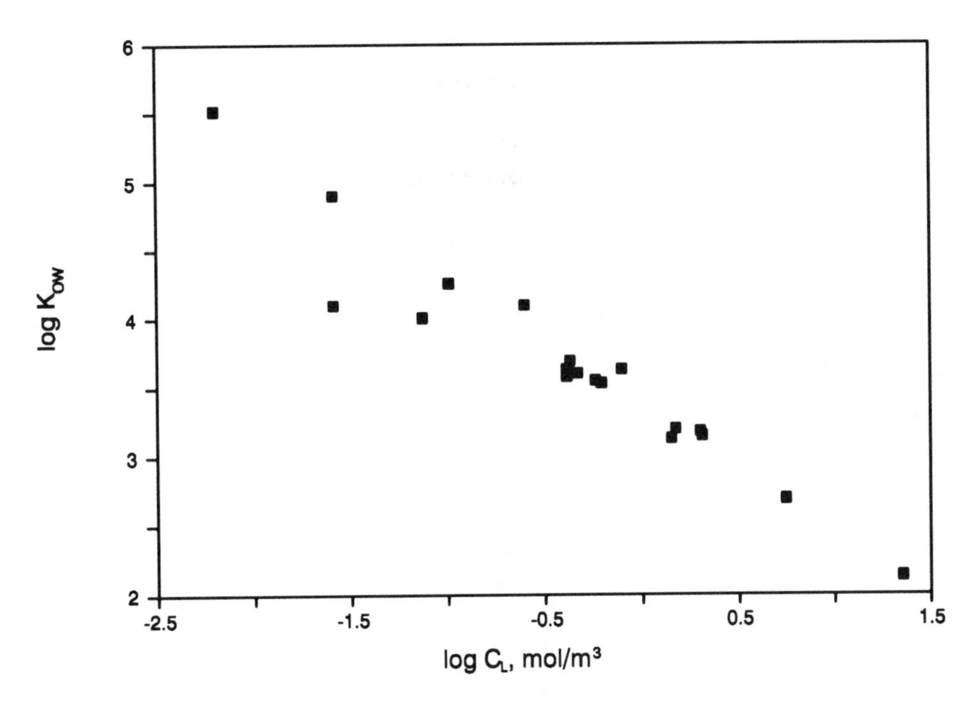

Figure 2.6 Plot of log K_{OW} versus log C_L

2.3 Illustrative Fugacity Calculations: Level I, II, and III

Chemical name: Benzene

Level I

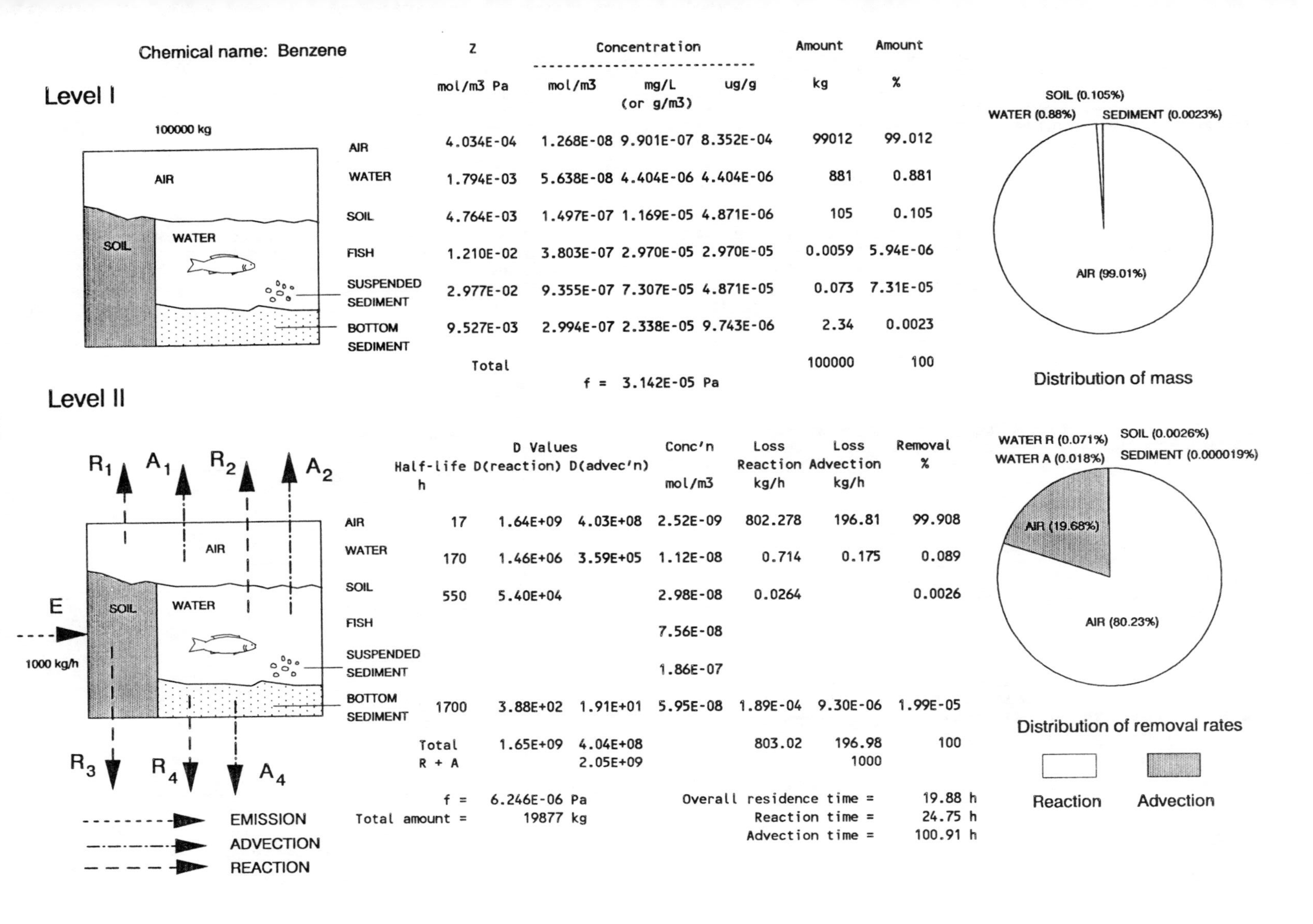

	Z	Concentration			Amount	Amount
	mol/m3 Pa	mol/m3	mg/L (or g/m3)	ug/g	kg	%
AIR	4.034E-04	1.268E-08	9.901E-07	8.352E-04	99012	99.012
WATER	1.794E-03	5.638E-08	4.404E-06	4.404E-06	881	0.881
SOIL	4.764E-03	1.497E-07	1.169E-05	4.871E-06	105	0.105
FISH	1.210E-02	3.803E-07	2.970E-05	2.970E-05	0.0059	5.94E-06
SUSPENDED SEDIMENT	2.977E-02	9.355E-07	7.307E-05	4.871E-05	0.073	7.31E-05
BOTTOM SEDIMENT	9.527E-03	2.994E-07	2.338E-05	9.743E-06	2.34	0.0023
	Total				100000	100

f = 3.142E-05 Pa

Level II

	Half-life	D Values		Conc'n	Loss	Loss	Removal
	h	D(reaction)	D(advec'n)	mol/m3	Reaction kg/h	Advection kg/h	%
AIR	17	1.64E+09	4.03E+08	2.52E-09	802.278	196.81	99.908
WATER	170	1.46E+06	3.59E+05	1.12E-08	0.714	0.175	0.089
SOIL	550	5.40E+04		2.98E-08	0.0264		0.0026
FISH				7.56E-08			
SUSPENDED SEDIMENT				1.86E-07			
BOTTOM SEDIMENT	1700	3.88E+02	1.91E+01	5.95E-08	1.89E-04	9.30E-06	1.99E-05
	Total	1.65E+09	4.04E+08		803.02	196.98	100
	R + A		2.05E+09			1000	

f = 6.246E-06 Pa
Total amount = 19877 kg

Overall residence time = 19.88 h
Reaction time = 24.75 h
Advection time = 100.91 h

Level III

Chemical name: Benzene

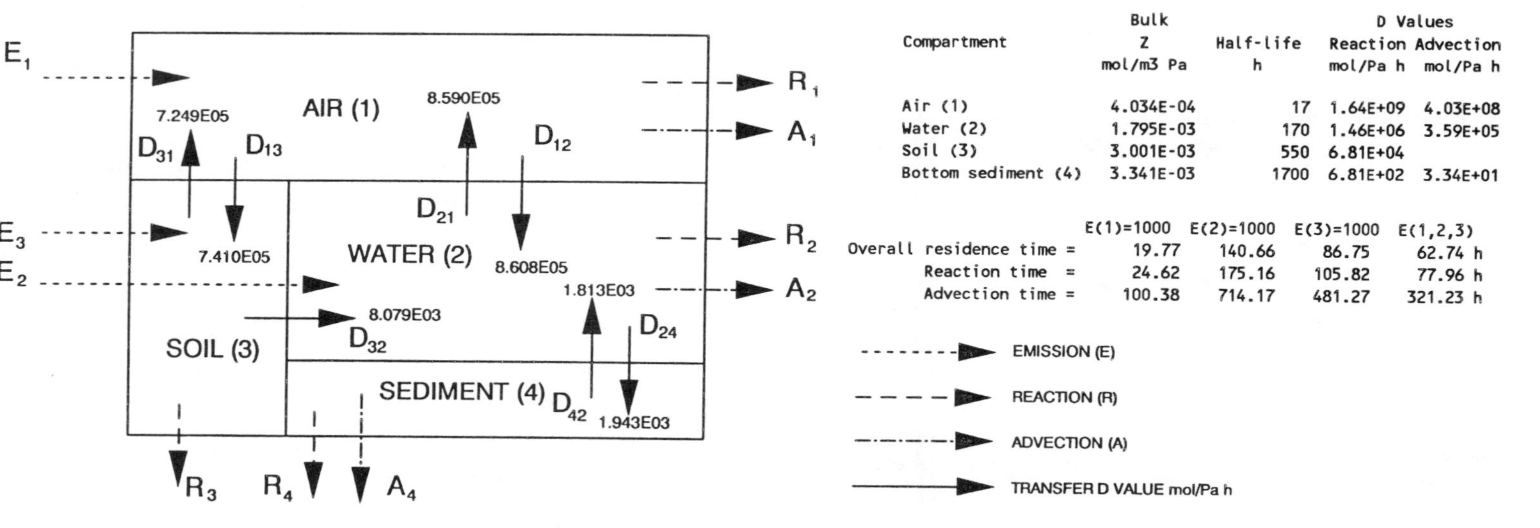

Phase Properties and Rates:

Compartment	Bulk Z mol/m3 Pa	Half-life h	D Values Reaction mol/Pa h	D Values Advection mol/Pa h
Air (1)	4.034E-04	17	1.64E+09	4.03E+08
Water (2)	1.795E-03	170	1.46E+06	3.59E+05
Soil (3)	3.001E-03	550	6.81E+04	
Bottom sediment (4)	3.341E-03	1700	6.81E+02	3.34E+01

	E(1)=1000	E(2)=1000	E(3)=1000	E(1,2,3)
Overall residence time =	19.77	140.66	86.75	62.74 h
Reaction time =	24.62	175.16	105.82	77.96 h
Advection time =	100.38	714.17	481.27	321.23 h

- - - - - - - -▶ EMISSION (E)

— — — —▶ REACTION (R)

—·—·—·—▶ ADVECTION (A)

————————▶ TRANSFER D VALUE mol/Pa h

Phase Properties, Composition, Transport and Transformation Rates:

Emission, kg/h			Fugacity, Pa				Concentration, g/m3				Amounts, kg				Total Amount, kg
E(1)	E(2)	E(3)	f(1)	f(2)	f(3)	f(4)	C(1)	C(2)	C(3)	C(4)	m(1)	m(2)	m(3)	m(4)	
1000	0	0	6.249E-06	2.023E-06	5.781E-06	1.556E-06	1.969E-07	2.836E-07	1.355E-06	4.059E-07	1.969E+04	5.673E+01	2.439E+01	2.030E-01	1.977E+04
0	1000	0	2.002E-06	4.775E-03	1.852E-06	3.671E-03	6.308E-08	6.693E-04	4.341E-07	9.579E-04	6.308E+03	1.339E+05	7.814E+00	4.790E+02	1.407E+05
0	0	1000	5.676E-06	4.998E-05	1.599E-02	3.842E-05	1.788E-07	7.006E-06	3.748E-03	1.003E-05	1.788E+04	1.401E+03	6.746E+04	5.013E+00	8.675E+04
600	300	100	4.918E-06	1.439E-03	1.603E-03	1.106E-03	1.550E-07	2.017E-04	3.757E-04	2.886E-04	1.550E+04	4.033E+04	6.763E+03	1.443E+02	6.274E+04

Emission, kg/h			Loss, Reaction, kg/h				Loss, Advection, kg/h			Intermedia Rate of Transport, kg/h T12	T21	T13	T31	T32	T24	T42
E(1)	E(2)	E(3)	R(1)	R(2)	R(3)	R(4)	A(1)	A(2)	A(4)	air-water	water-air	air-soil	soil-air	soil-water	water-sed	sed-water
1000	0	0	8.028E+02	2.312E-01	3.07E-02	8.274E-05	1.969E+02	5.673E-02	4.059E-06	4.202E-01	1.358E-01	3.617E-01	3.273E-01	3.648E-03	3.071E-04	2.203E-04
0	1000	0	2.572E+02	5.457E+02	9.85E-03	1.952E-01	6.308E+01	1.339E+02	9.579E-03	1.346E-01	3.204E+02	1.159E-01	1.049E-01	1.169E-03	7.248E-01	5.200E-01
0	0	1000	7.290E+02	5.712E+00	8.50E+01	2.044E-03	1.788E+02	1.401E+00	1.003E-04	3.816E-01	3.353E+00	3.285E-01	9.052E+02	1.009E+01	7.586E-03	5.442E-03
600	300	100	6.317E+02	1.644E+02	8.52E+00	5.883E-02	1.550E+02	1.644E+02	2.886E-03	3.306E-01	9.653E+01	2.846E-01	9.075E+01	1.011E+00	2.184E-01	1.567E-01

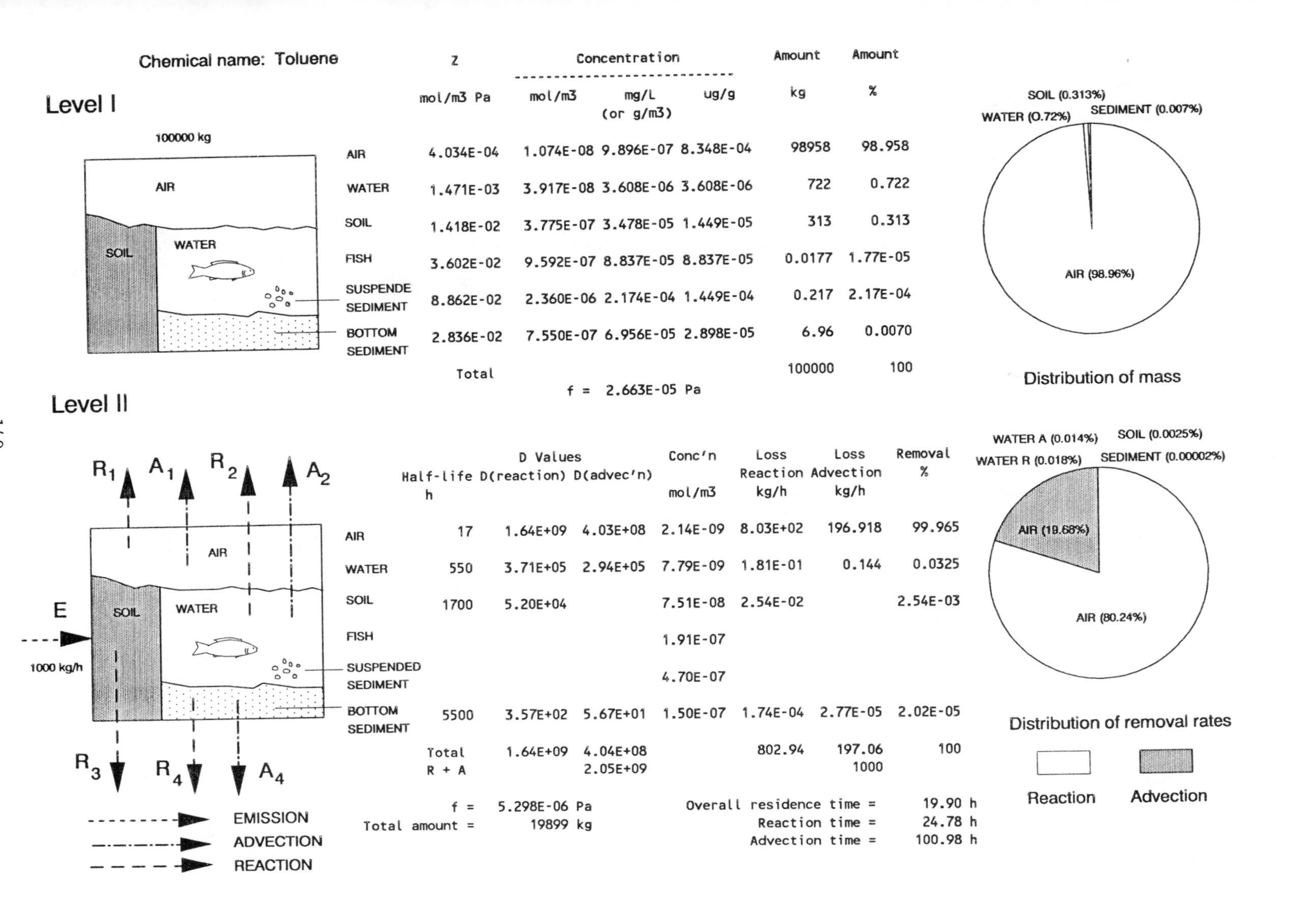

Chemical name: Toluene

Level I

	Z	Concentration			Amount	Amount
	mol/m3 Pa	mol/m3	mg/L (or g/m3)	ug/g	kg	%
AIR	4.034E-04	1.074E-08	9.896E-07	8.348E-04	98958	98.958
WATER	1.471E-03	3.917E-08	3.608E-06	3.608E-06	722	0.722
SOIL	1.418E-02	3.775E-07	3.478E-05	1.449E-05	313	0.313
FISH	3.602E-02	9.592E-07	8.837E-05	8.837E-05	0.0177	1.77E-05
SUSPENDE SEDIMENT	8.862E-02	2.360E-06	2.174E-04	1.449E-04	0.217	2.17E-04
BOTTOM SEDIMENT	2.836E-02	7.550E-07	6.956E-05	2.898E-05	6.96	0.0070
	Total				100000	100

f = 2.663E-05 Pa

Distribution of mass

Level II

	Half-life h	D Values D(reaction)	D(advec'n)	Conc'n mol/m3	Loss Reaction kg/h	Loss Advection kg/h	Removal %
AIR	17	1.64E+09	4.03E+08	2.14E-09	8.03E+02	196.918	99.965
WATER	550	3.71E+05	2.94E+05	7.79E-09	1.81E-01	0.144	0.0325
SOIL	1700	5.20E+04		7.51E-08	2.54E-02		2.54E-03
FISH				1.91E-07			
SUSPENDED SEDIMENT				4.70E-07			
BOTTOM SEDIMENT	5500	3.57E+02	5.67E+01	1.50E-07	1.74E-04	2.77E-05	2.02E-05
	Total R + A	1.64E+09	4.04E+08 2.05E+09		802.94	197.06 1000	100

f = 5.298E-06 Pa
Total amount = 19899 kg

Overall residence time = 19.90 h
Reaction time = 24.78 h
Advection time = 100.98 h

Distribution of removal rates

Level III

Chemical Name: Toluene

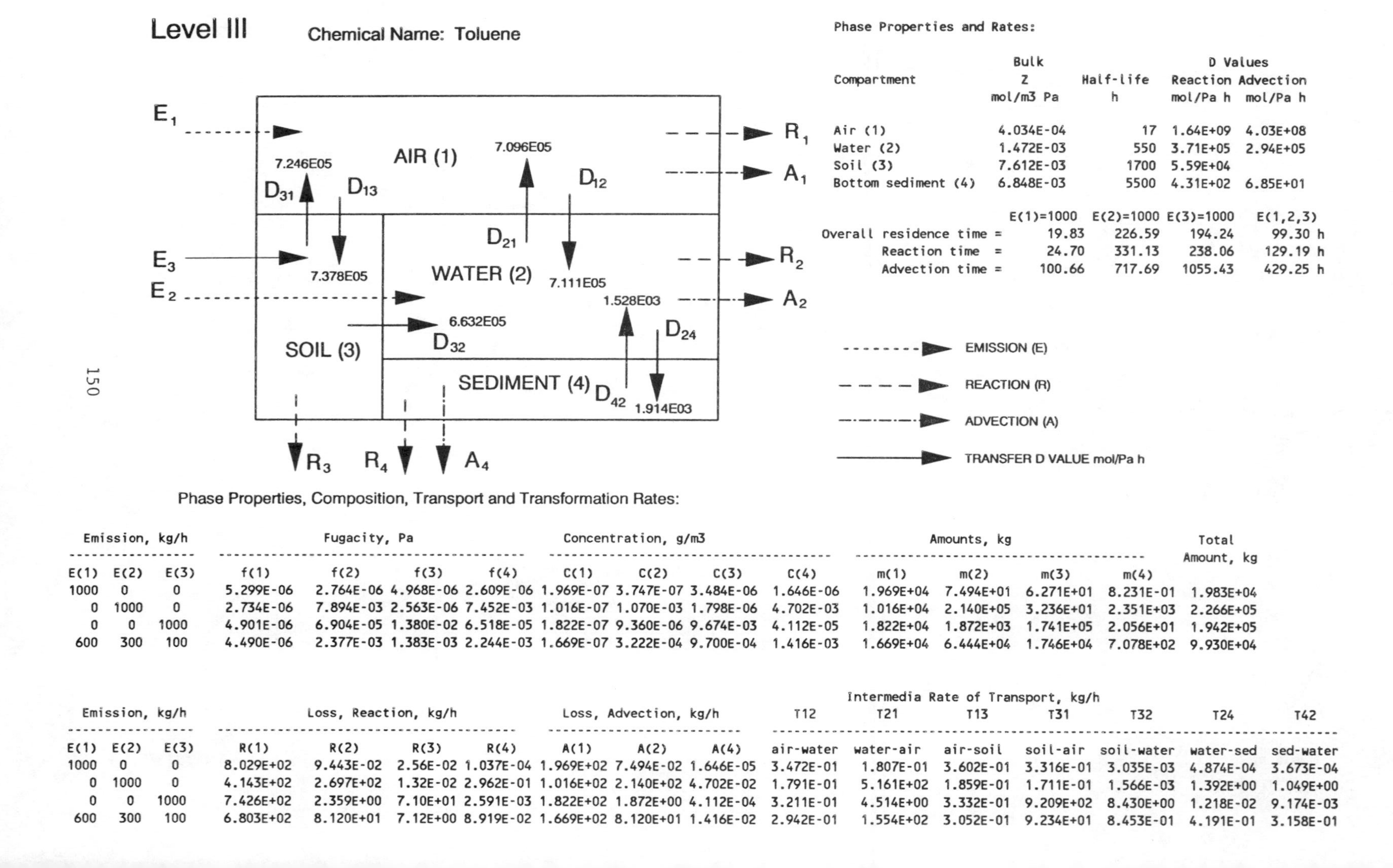

Phase Properties and Rates:

Compartment	Bulk Z mol/m3 Pa	Half-life h	D Values Reaction mol/Pa h	D Values Advection mol/Pa h
Air (1)	4.034E-04	17	1.64E+09	4.03E+08
Water (2)	1.472E-03	550	3.71E+05	2.94E+05
Soil (3)	7.612E-03	1700	5.59E+04	
Bottom sediment (4)	6.848E-03	5500	4.31E+02	6.85E+01

	E(1)=1000	E(2)=1000	E(3)=1000	E(1,2,3)
Overall residence time =	19.83	226.59	194.24	99.30 h
Reaction time =	24.70	331.13	238.06	129.19 h
Advection time =	100.66	717.69	1055.43	429.25 h

Phase Properties, Composition, Transport and Transformation Rates:

Emission, kg/h			Fugacity, Pa				Concentration, g/m3				Amounts, kg				Total Amount, kg
E(1)	E(2)	E(3)	f(1)	f(2)	f(3)	f(4)	C(1)	C(2)	C(3)	C(4)	m(1)	m(2)	m(3)	m(4)	
1000	0	0	5.299E-06	2.764E-06	4.968E-06	2.609E-06	1.969E-07	3.747E-07	3.484E-06	1.646E-06	1.969E+04	7.494E+01	6.271E+01	8.231E-01	1.983E+04
0	1000	0	2.734E-06	7.894E-03	2.563E-06	7.452E-03	1.016E-07	1.070E-03	1.798E-06	4.702E-03	1.016E+04	2.140E+05	3.236E+01	2.351E+03	2.266E+05
0	0	1000	4.901E-06	6.904E-05	1.380E-02	6.518E-05	1.822E-07	9.360E-06	9.674E-03	4.112E-05	1.822E+04	1.872E+03	1.741E+05	2.056E+01	1.942E+05
600	300	100	4.490E-06	2.377E-03	1.383E-03	2.244E-03	1.669E-07	3.222E-04	9.700E-04	1.416E-03	1.669E+04	6.444E+04	1.746E+04	7.078E+02	9.930E+04

Emission, kg/h			Loss, Reaction, kg/h				Loss, Advection, kg/h			Intermedia Rate of Transport, kg/h T12	T21	T13	T31	T32	T24	T42
E(1)	E(2)	E(3)	R(1)	R(2)	R(3)	R(4)	A(1)	A(2)	A(4)	air-water	water-air	air-soil	soil-air	soil-water	water-sed	sed-water
1000	0	0	8.029E+02	9.443E-02	2.56E-02	1.037E-04	1.969E+02	7.494E-02	1.646E-05	3.472E-01	1.807E-01	3.602E-01	3.316E-01	3.035E-03	4.874E-04	3.673E-04
0	1000	0	4.143E+02	2.697E+02	1.32E-02	2.962E-01	1.016E+02	2.140E+02	4.702E-02	1.791E-01	5.161E+02	1.859E-01	1.711E-01	1.566E-03	1.392E+00	1.049E+00
0	0	1000	7.426E+02	2.359E+00	7.10E+01	2.591E-03	1.822E+02	1.872E+00	4.112E-04	3.211E-01	4.514E+00	3.332E-01	9.209E+02	8.430E+00	1.218E-02	9.174E-03
600	300	100	6.803E+02	8.120E+01	7.12E+00	8.919E-02	1.669E+02	8.120E+01	1.416E-02	2.942E-01	1.554E+02	3.052E-01	9.234E+01	8.453E-01	4.191E-01	3.158E-01

Chemical name: Ethylbenzene

Level I

	Z	Concentration			Amount	Amount
	mol/m3 Pa	mol/m3	mg/L (or g/m3)	ug/g	kg	%
AIR	4.034E-04	9.301E-09	9.877E-07	8.332E-04	98774	98.774
WATER	1.127E-03	2.598E-08	2.759E-06	2.759E-06	552	0.552
SOIL	2.992E-02	6.898E-07	7.325E-05	3.052E-05	659	0.659
FISH	7.601E-02	1.752E-06	1.861E-04	1.861E-04	0.0372	3.72E-05
SUSPENDED SEDIMENT	1.870E-01	4.311E-06	4.578E-04	3.052E-04	0.458	4.58E-04
BOTTOM SEDIMENT	5.984E-02	1.380E-06	1.465E-04	6.104E-05	14.65	1.47E-02
	Total				100000	100

f = 2.305E-05 Pa

Level II

	Half-life h	D Values D(reaction)	D(advec'n)	Conc'n mol/m3	Loss Reaction kg/h	Loss Advection kg/h	Removal %
AIR	17	1.64E+09	4.03E+08	1.85E-09	8.03E+02	196.928	99.970
WATER	550	2.84E+05	2.25E+05	5.18E-09	1.39E-01	0.110	0.0249
SOIL	1700	1.10E+05		1.38E-07	5.36E-02		5.36E-03
FISH				3.49E-07			
SUSPENDED SEDIMENT				8.60E-07			
BOTTOM SEDIMENT	5500	7.54E+02	1.20E+02	2.75E-07	3.68E-04	5.84E-05	4.26E-05
	Total	1.64E+09	4.04E+08		802.96	197.04	100
	R + A		2.05E+09			1000	

f = 4.596E-06 Pa
Total amount = 19937 kg

Overall residence time = 19.94 h
Reaction time = 24.83 h
Advection time = 101.18 h

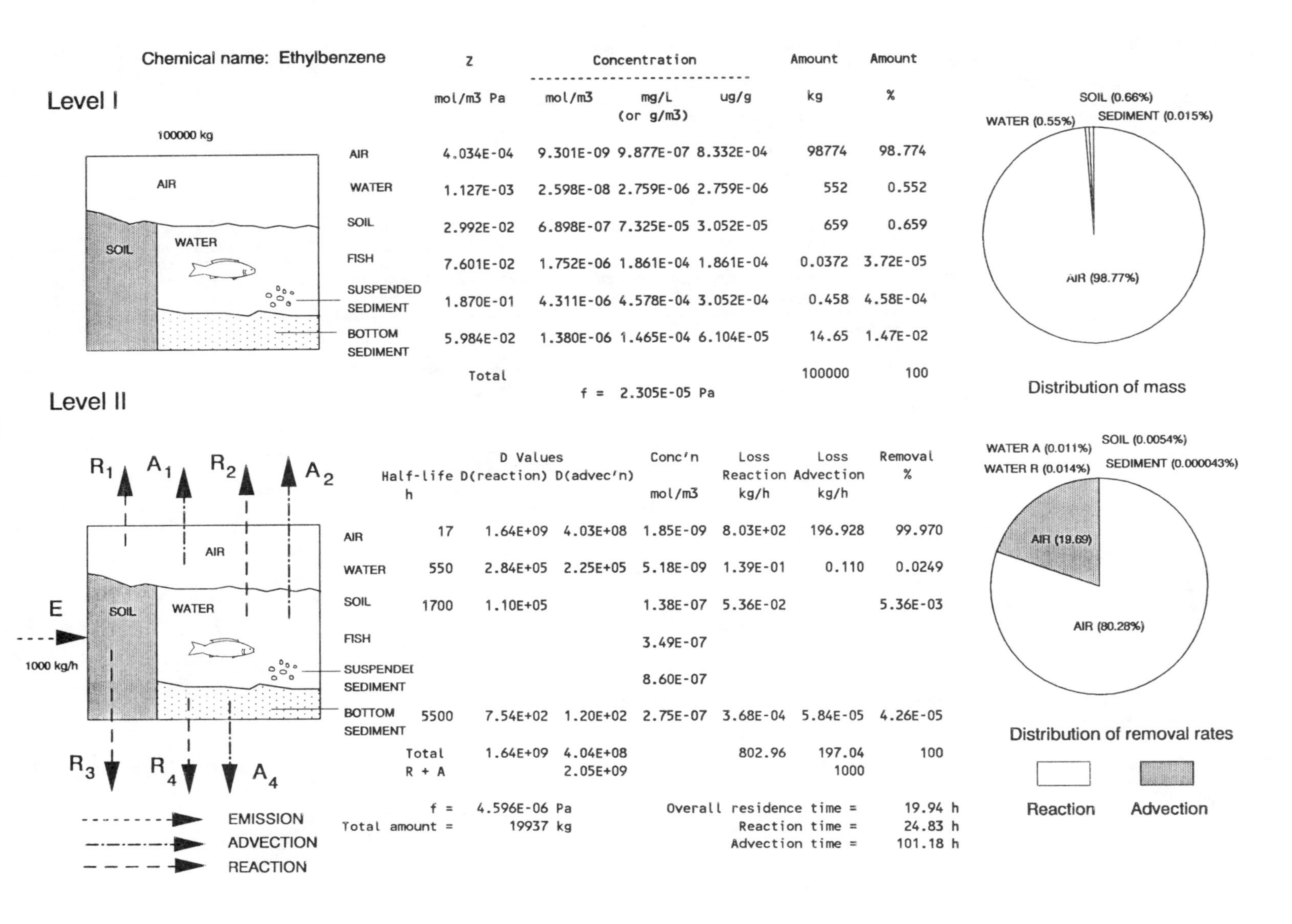

Level III

Chemical Name: Ethylbenzene

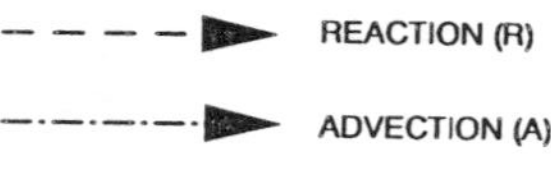
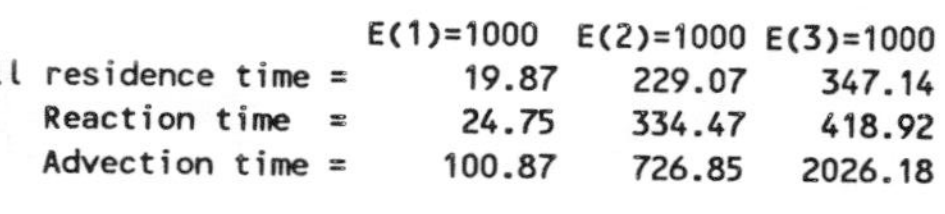
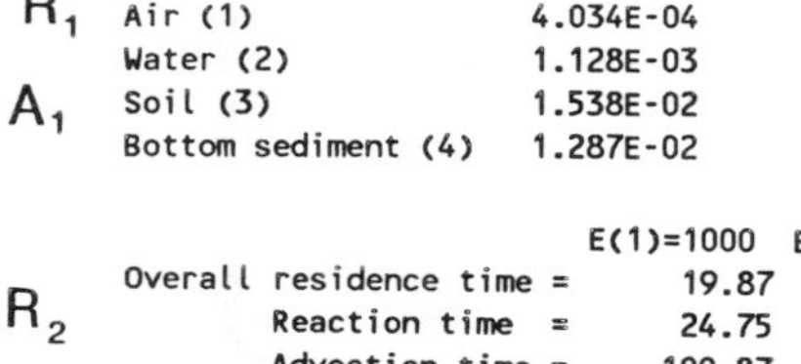

Phase Properties and Rates:

Compartment	Bulk Z mol/m3 Pa	Half-life h	D Values Reaction mol/Pa h	D Values Advection mol/Pa h
Air (1)	4.034E-04	17	1.64E+09	4.03E+08
Water (2)	1.128E-03	550	2.84E+05	2.26E+05
Soil (3)	1.538E-02	1700	1.13E+05	
Bottom sediment (4)	1.287E-02	5500	8.11E+02	1.29E+02

	E(1)=1000	E(2)=1000	E(3)=1000	E(1,2,3)
Overall residence time =	19.87	229.07	347.14	115.36 h
Reaction time =	24.75	334.47	418.92	149.79 h
Advection time =	100.87	726.85	2026.18	501.81 h

Phase Properties, Composirions, Transport and Transformation Rates:

Emission, kg/h			Fugacity, Pa				Concentration, g/m3				Amounts, kg				Total Amount, kg
E(1)	E(2)	E(3)	f(1)	f(2)	f(3)	f(4)	C(1)	C(2)	C(3)	C(4)	m(1)	m(2)	m(3)	m(4)	
1000	0	0	4.597E-06	2.404E-06	4.009E-06	2.268E-06	1.970E-07	2.880E-07	6.548E-06	3.099E-06	1.970E+04	5.760E+01	1.179E+02	1.549E+00	1.987E+04
0	1000	0	2.380E-06	8.894E-03	2.075E-06	8.388E-03	1.020E-07	1.065E-03	3.390E-06	1.146E-02	1.020E+04	2.131E+05	6.101E+01	5.732E+03	2.291E+05
0	0	1000	3.968E-06	5.589E-05	1.118E-02	5.272E-05	1.700E-07	6.695E-06	1.826E-02	7.205E-05	1.700E+04	1.339E+03	3.288E+05	3.602E+01	3.471E+05
600	300	100	3.869E-06	2.675E-03	1.121E-03	2.523E-03	1.658E-07	3.205E-04	1.831E-03	3.448E-03	1.658E+04	6.409E+04	3.297E+04	1.724E+03	1.154E+05

Emission, kg/h			Loss, Reaction, kg/h				Loss, Advection, kg/h			Intermedia Rate of Transport, kg/h T12	T21	T13	T31	T32	T24	T42
E(1)	E(2)	E(3)	R(1)	R(2)	R(3)	R(4)	A(1)	A(2)	A(4)	air-water	water-air	air-soil	soil-air	soil-water	water-sed	sed-water
1000	0	0	8.029E+02	7.257E-02	4.80E-02	1.952E-04	1.970E+02	5.760E-02	3.099E-05	2.682E-01	1.400E-01	3.586E-01	3.084E-01	2.171E-03	5.264E-04	3.002E-04
0	1000	0	4.156E+02	2.685E+02	2.49E-02	7.223E-01	1.020E+02	2.131E+02	1.146E-01	1.388E-01	5.178E+02	1.856E-01	1.596E-01	1.124E-03	1.947E+00	1.111E+00
0	0	1000	6.930E+02	1.687E+00	1.34E+02	4.539E-03	1.700E+02	1.339E+00	7.205E-04	2.315E-01	3.254E+00	3.095E-01	8.602E+02	6.055E+00	1.224E-02	6.979E-03
600	300	100	6.757E+02	8.076E+01	1.34E+01	2.172E-01	1.658E+02	8.076E+01	3.448E-02	2.257E-01	1.557E+02	3.018E-01	8.626E+01	6.072E-01	5.858E-01	3.341E-01

Chemical name: o-Xylene

Level I

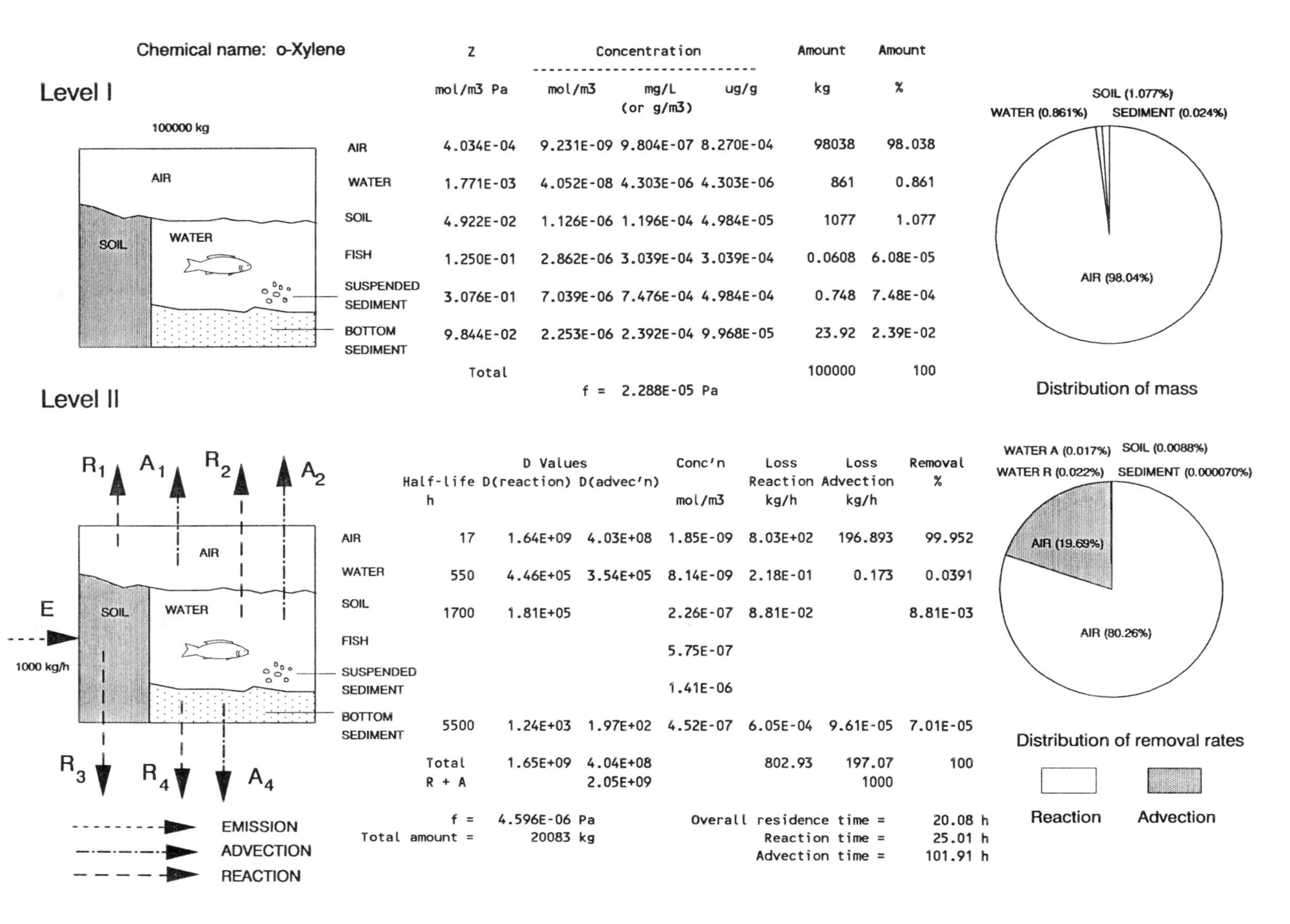

	Z	Concentration			Amount	Amount
	mol/m3 Pa	mol/m3	mg/L (or g/m3)	ug/g	kg	%
AIR	4.034E-04	9.231E-09	9.804E-07	8.270E-04	98038	98.038
WATER	1.771E-03	4.052E-08	4.303E-06	4.303E-06	861	0.861
SOIL	4.922E-02	1.126E-06	1.196E-04	4.984E-05	1077	1.077
FISH	1.250E-01	2.862E-06	3.039E-04	3.039E-04	0.0608	6.08E-05
SUSPENDED SEDIMENT	3.076E-01	7.039E-06	7.476E-04	4.984E-04	0.748	7.48E-04
BOTTOM SEDIMENT	9.844E-02	2.253E-06	2.392E-04	9.968E-05	23.92	2.39E-02
	Total				100000	100

f = 2.288E-05 Pa

Level II

	Half-life h	D Values D(reaction)	D(advec'n)	Conc'n mol/m3	Loss Reaction kg/h	Loss Advection kg/h	Removal %
AIR	17	1.64E+09	4.03E+08	1.85E-09	8.03E+02	196.893	99.952
WATER	550	4.46E+05	3.54E+05	8.14E-09	2.18E-01	0.173	0.0391
SOIL	1700	1.81E+05		2.26E-07	8.81E-02		8.81E-03
FISH				5.75E-07			
SUSPENDED SEDIMENT				1.41E-06			
BOTTOM SEDIMENT	5500	1.24E+03	1.97E+02	4.52E-07	6.05E-04	9.61E-05	7.01E-05
	Total	1.65E+09	4.04E+08		802.93	197.07	100
	R + A		2.05E+09			1000	

f = 4.596E-06 Pa
Total amount = 20083 kg

Overall residence time = 20.08 h
Reaction time = 25.01 h
Advection time = 101.91 h

Level III

Chemical Name: o-Xylene

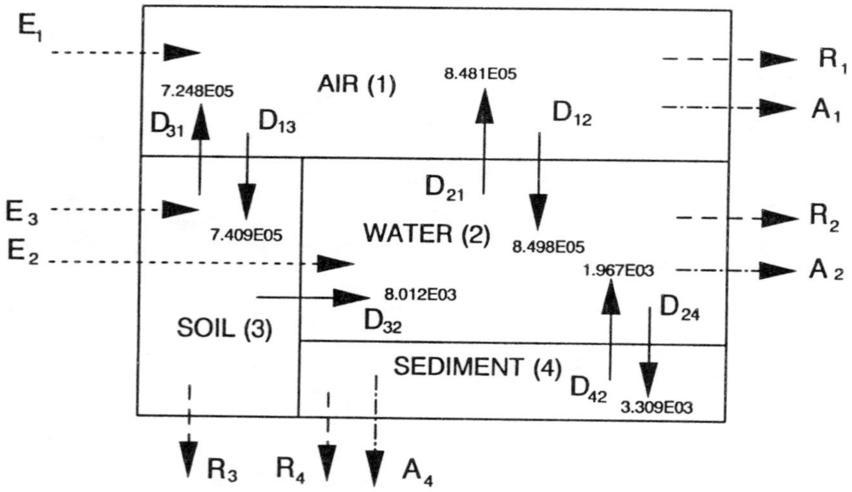

AIR (1) | WATER (2) | SOIL (3) | SEDIMENT (4)

E_1, E_2, E_3; R_1, A_1, R_2, A_2, R_3, R_4, A_4

D_{31} 7.248E05; D_{13} 7.409E05; D_{21} 8.481E05; D_{12} 8.498E05; D_{32} 8.012E03; D_{42} 1.967E03; D_{24} 3.309E03

EMISSION (E)
REACTION (R)
ADVECTION (A)
TRANSFER D VALUE mol/Pa h

Phase Properties and Rates:

Compartment	Bulk Z mol/m3 Pa	Half-life h	D Values Reaction mol/Pa h	D Values Advection mol/Pa h
Air (1)	4.034E-04	17	1.64E+09	4.03E+08
Water (2)	1.772E-03	550	4.47E+05	3.54E+05
Soil (3)	2.522E-02	1700	1.85E+05	
Bottom sediment (4)	2.110E-02	5500	1.33E+03	2.11E+02

	E(1)=1000	E(2)=1000	E(3)=1000	E(1,2,3)
Overall residence time =	19.96	231.03	512.37	132.52 h
Reaction time =	24.86	337.81	608.76	171.86 h
Advection time =	101.33	730.89	3235.90	579.02 h

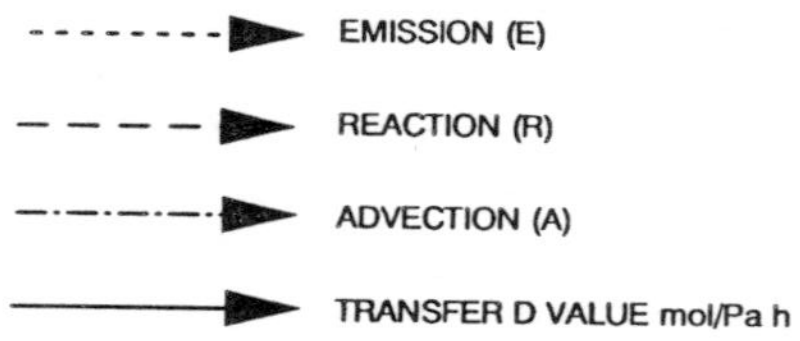

Phase Properties, Compostions, Transport and Transformation Rates:

Emission, kg/h			Fugacity, Pa				Concentration, g/m3				Amounts, kg				Total
E(1)	E(2)	E(3)	f(1)	f(2)	f(3)	f(4)	C(1)	C(2)	C(3)	C(4)	m(1)	m(2)	m(3)	m(4)	Amount, kg
1000	0	0	4.597E-06	2.385E-06	3.710E-06	2.249E-06	1.969E-07	4.488E-07	9.938E-06	5.041E-06	1.969E+04	8.977E+01	1.789E+02	2.521E+00	1.996E+04
0	1000	0	2.362E-06	5.706E-03	1.906E-06	5.382E-03	1.012E-07	1.074E-03	5.106E-06	1.206E-02	1.012E+04	2.148E+05	9.191E+01	6.031E+03	2.310E+05
0	0	1000	3.650E-06	5.166E-05	1.026E-02	4.873E-05	1.564E-07	9.724E-06	2.748E-02	1.092E-04	1.564E+04	1.945E+03	4.947E+05	5.461E+01	5.124E+05
600	300	100	3.832E-06	1.718E-03	1.029E-03	1.621E-03	1.642E-07	3.234E-04	2.756E-03	3.633E-03	1.642E+04	6.468E+04	4.961E+04	1.816E+03	1.325E+05

Emission, kg/h			Loss, Reaction, kg/h				Loss, Advection, kg/h			Intermedia Rate of Transport, kg/h T12	T21	T13	T31	T32	T24	T42
E(1)	E(2)	E(3)	R(1)	R(2)	R(3)	R(4)	A(1)	A(2)	A(4)	air-water	water-air	air-soil	soil-air	soil-water	water-sed	sed-water
1000	0	0	8.028E+02	1.131E-01	7.29E-02	3.176E-04	1.969E+02	8.977E-02	5.041E-05	4.149E-01	2.148E-01	3.617E-01	2.856E-01	3.157E-03	8.379E-04	4.700E-04
0	1000	0	4.125E+02	2.706E+02	3.75E-02	7.599E-01	1.012E+02	2.148E+02	1.206E-01	2.132E-01	5.139E+02	1.858E-01	1.467E-01	1.622E-03	2.005E+00	1.124E+00
0	0	1000	6.375E+02	2.450E+00	2.02E+02	6.880E-03	1.564E+02	1.945E+00	1.092E-03	3.295E-01	4.653E+00	2.872E-01	7.899E+02	8.731E+00	1.815E-02	1.018E-02
600	300	100	6.692E+02	8.150E+01	2.02E+01	2.289E-01	1.642E+02	8.150E+01	3.633E-02	3.458E-01	1.548E+02	3.015E-01	7.920E+01	8.755E-01	6.038E-01	3.386E-01

Chemical name: m-Xylene

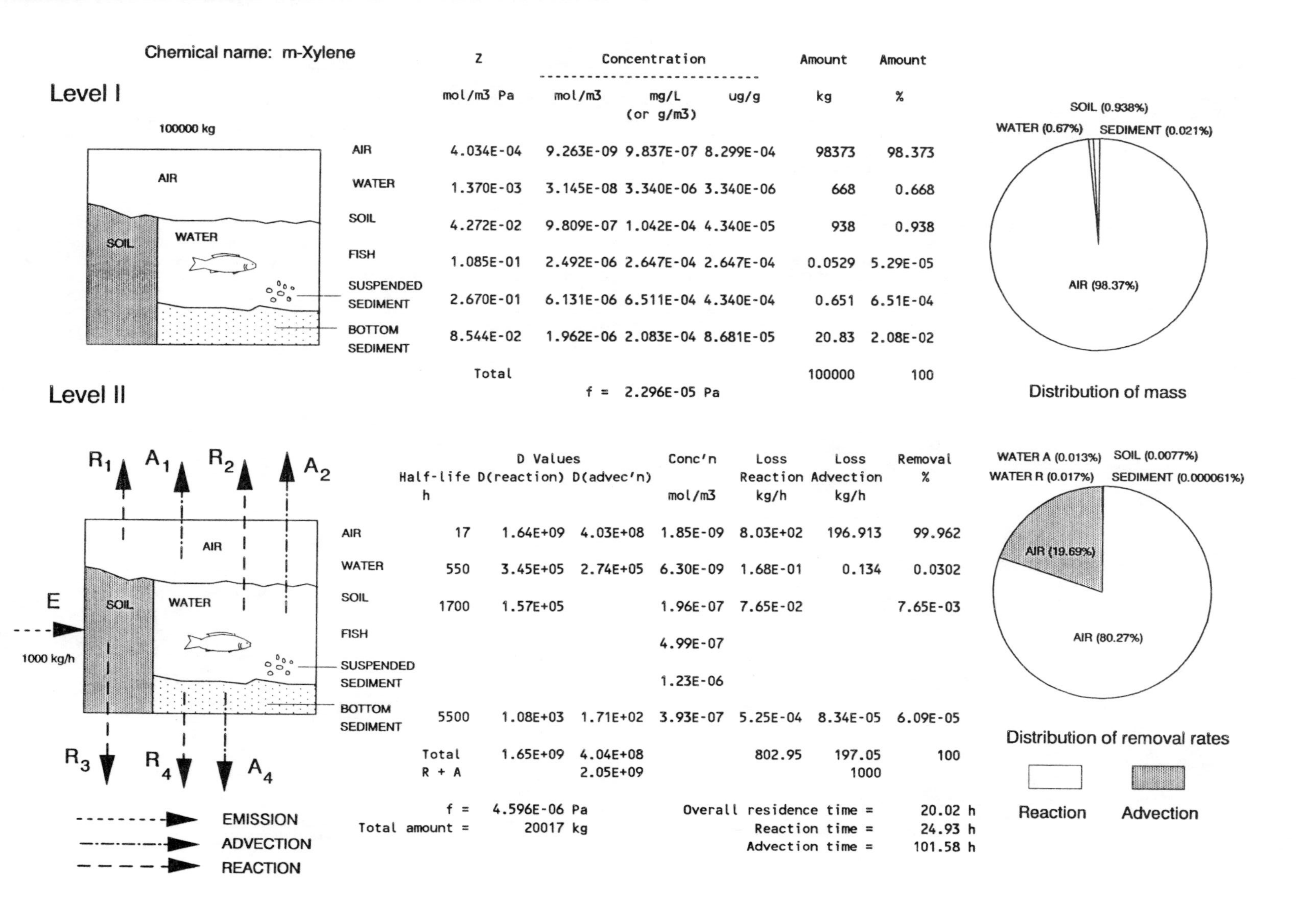

Level I

	Z	Concentration			Amount	Amount
	mol/m3 Pa	mol/m3	mg/L (or g/m3)	ug/g	kg	%
AIR	4.034E-04	9.263E-09	9.837E-07	8.299E-04	98373	98.373
WATER	1.370E-03	3.145E-08	3.340E-06	3.340E-06	668	0.668
SOIL	4.272E-02	9.809E-07	1.042E-04	4.340E-05	938	0.938
FISH	1.085E-01	2.492E-06	2.647E-04	2.647E-04	0.0529	5.29E-05
SUSPENDED SEDIMENT	2.670E-01	6.131E-06	6.511E-04	4.340E-04	0.651	6.51E-04
BOTTOM SEDIMENT	8.544E-02	1.962E-06	2.083E-04	8.681E-05	20.83	2.08E-02
	Total				100000	100

f = 2.296E-05 Pa

Level II

	Half-life h	D Values D(reaction)	D(advec'n)	Conc'n mol/m3	Loss Reaction kg/h	Loss Advection kg/h	Removal %
AIR	17	1.64E+09	4.03E+08	1.85E-09	8.03E+02	196.913	99.962
WATER	550	3.45E+05	2.74E+05	6.30E-09	1.68E-01	0.134	0.0302
SOIL	1700	1.57E+05		1.96E-07	7.65E-02		7.65E-03
FISH				4.99E-07			
SUSPENDED SEDIMENT				1.23E-06			
BOTTOM SEDIMENT	5500	1.08E+03	1.71E+02	3.93E-07	5.25E-04	8.34E-05	6.09E-05
	Total	1.65E+09	4.04E+08		802.95	197.05	100
	R + A		2.05E+09			1000	

f = 4.596E-06 Pa
Total amount = 20017 kg

Overall residence time = 20.02 h
Reaction time = 24.93 h
Advection time = 101.58 h

Level III

Chemical name: m-Xylene

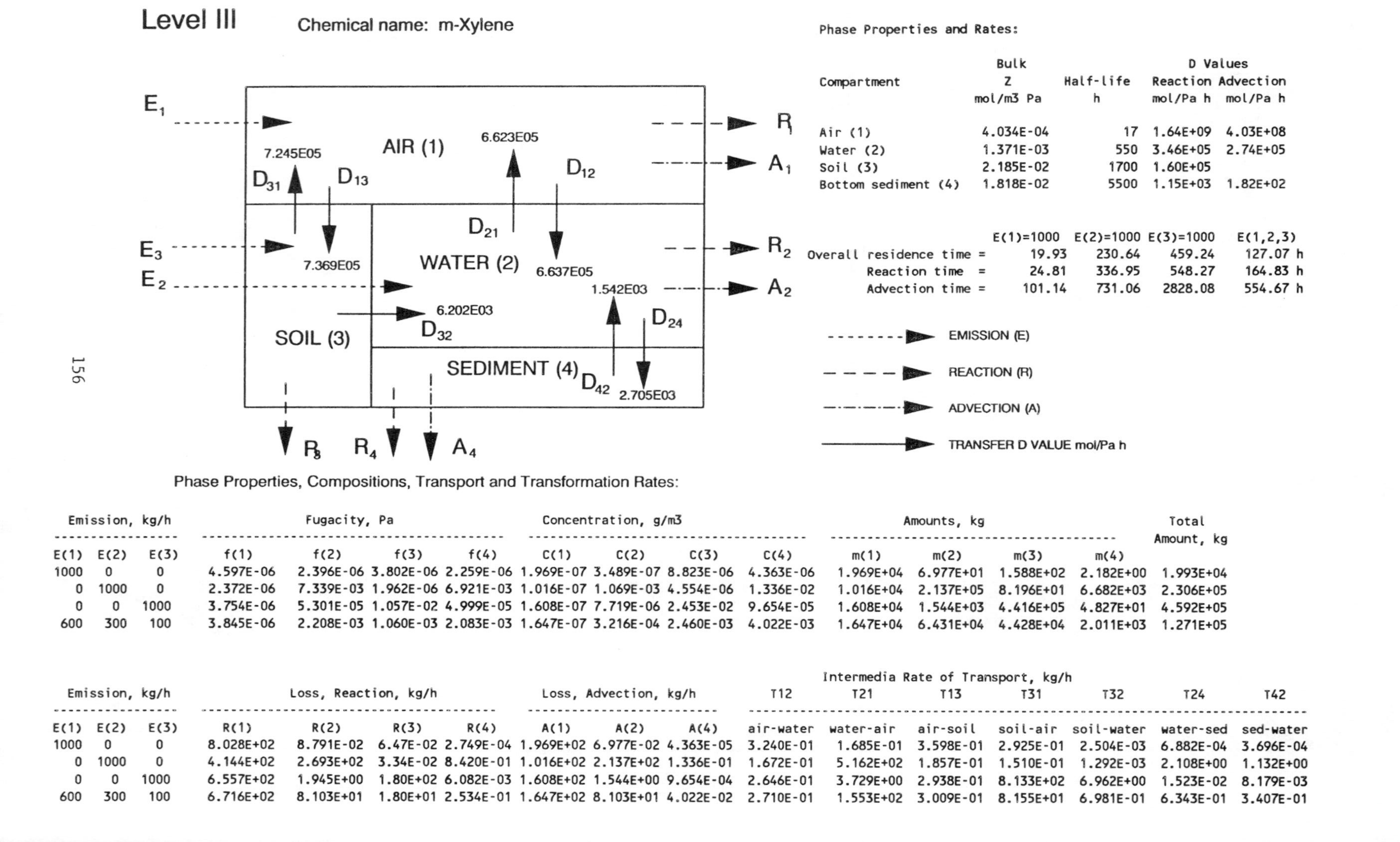

Phase Properties and Rates:

Compartment	Bulk Z mol/m3 Pa	Half-life h	D Values Reaction mol/Pa h	D Values Advection mol/Pa h
Air (1)	4.034E-04	17	1.64E+09	4.03E+08
Water (2)	1.371E-03	550	3.46E+05	2.74E+05
Soil (3)	2.185E-02	1700	1.60E+05	
Bottom sediment (4)	1.818E-02	5500	1.15E+03	1.82E+02

	E(1)=1000	E(2)=1000	E(3)=1000	E(1,2,3)
Overall residence time =	19.93	230.64	459.24	127.07 h
Reaction time =	24.81	336.95	548.27	164.83 h
Advection time =	101.14	731.06	2828.08	554.67 h

Phase Properties, Compositions, Transport and Transformation Rates:

Emission, kg/h			Fugacity, Pa				Concentration, g/m3				Amounts, kg				Total
E(1)	E(2)	E(3)	f(1)	f(2)	f(3)	f(4)	C(1)	C(2)	C(3)	C(4)	m(1)	m(2)	m(3)	m(4)	Amount, kg
1000	0	0	4.597E-06	2.396E-06	3.802E-06	2.259E-06	1.969E-07	3.489E-07	8.823E-06	4.363E-06	1.969E+04	6.977E+01	1.588E+02	2.182E+00	1.993E+04
0	1000	0	2.372E-06	7.339E-03	1.962E-06	6.921E-03	1.016E-07	1.069E-03	4.554E-06	1.336E-02	1.016E+04	2.137E+05	8.196E+01	6.682E+03	2.306E+05
0	0	1000	3.754E-06	5.301E-05	1.057E-02	4.999E-05	1.608E-07	7.719E-06	2.453E-02	9.654E-05	1.608E+04	1.544E+03	4.416E+05	4.827E+01	4.592E+05
600	300	100	3.845E-06	2.208E-03	1.060E-03	2.083E-03	1.647E-07	3.216E-04	2.460E-03	4.022E-03	1.647E+04	6.431E+04	4.428E+04	2.011E+03	1.271E+05

Emission, kg/h			Loss, Reaction, kg/h				Loss, Advection, kg/h			Intermedia Rate of Transport, kg/h T12	T21	T13	T31	T32	T24	T42
E(1)	E(2)	E(3)	R(1)	R(2)	R(3)	R(4)	A(1)	A(2)	A(4)	air-water	water-air	air-soil	soil-air	soil-water	water-sed	sed-water
1000	0	0	8.028E+02	8.791E-02	6.47E-02	2.749E-04	1.969E+02	6.977E-02	4.363E-05	3.240E-01	1.685E-01	3.598E-01	2.925E-01	2.504E-03	6.882E-04	3.696E-04
0	1000	0	4.144E+02	2.693E+02	3.34E-02	8.420E-01	1.016E+02	2.137E+02	1.336E-01	1.672E-01	5.162E+02	1.857E-01	1.510E-01	1.292E-03	2.108E+00	1.132E+00
0	0	1000	6.557E+02	1.945E+00	1.80E+02	6.082E-03	1.608E+02	1.544E+00	9.654E-04	2.646E-01	3.729E+00	2.938E-01	8.133E+02	6.962E+00	1.523E-02	8.179E-03
600	300	100	6.716E+02	8.103E+01	1.80E+01	2.534E-01	1.647E+02	8.103E+01	4.022E-02	2.710E-01	1.553E+02	3.009E-01	8.155E+01	6.981E-01	6.343E-01	3.407E-01

Chemical name: p-Xylene

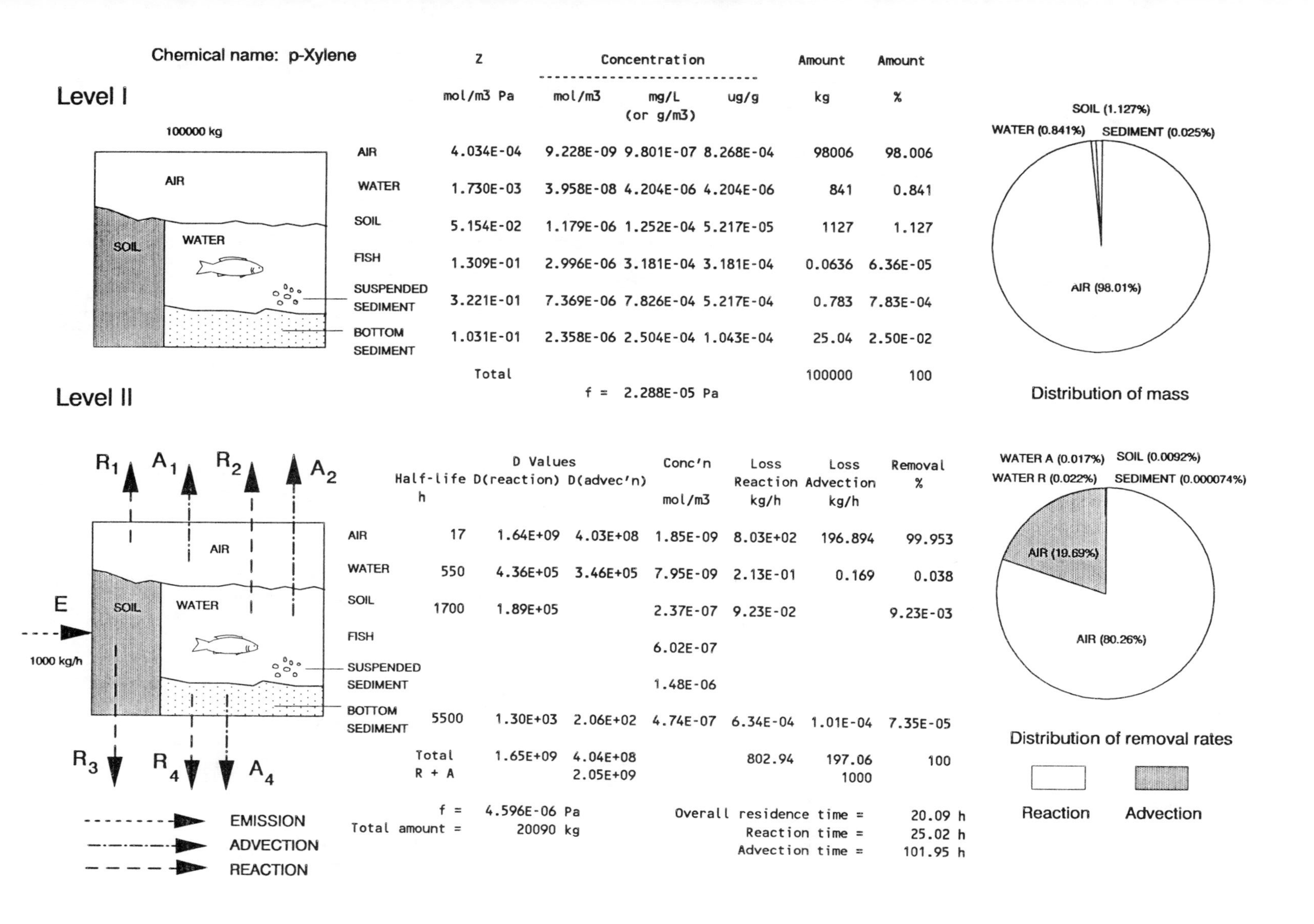

Level I

	Z mol/m3 Pa	Concentration mol/m3	mg/L (or g/m3)	ug/g	Amount kg	Amount %
AIR	4.034E-04	9.228E-09	9.801E-07	8.268E-04	98006	98.006
WATER	1.730E-03	3.958E-08	4.204E-06	4.204E-06	841	0.841
SOIL	5.154E-02	1.179E-06	1.252E-04	5.217E-05	1127	1.127
FISH	1.309E-01	2.996E-06	3.181E-04	3.181E-04	0.0636	6.36E-05
SUSPENDED SEDIMENT	3.221E-01	7.369E-06	7.826E-04	5.217E-04	0.783	7.83E-04
BOTTOM SEDIMENT	1.031E-01	2.358E-06	2.504E-04	1.043E-04	25.04	2.50E-02
Total					100000	100

f = 2.288E-05 Pa

Distribution of mass

Level II

	Half-life h	D Values D(reaction)	D(advec'n)	Conc'n mol/m3	Loss Reaction kg/h	Loss Advection kg/h	Removal %
AIR	17	1.64E+09	4.03E+08	1.85E-09	8.03E+02	196.894	99.953
WATER	550	4.36E+05	3.46E+05	7.95E-09	2.13E-01	0.169	0.038
SOIL	1700	1.89E+05		2.37E-07	9.23E-02		9.23E-03
FISH				6.02E-07			
SUSPENDED SEDIMENT				1.48E-06			
BOTTOM SEDIMENT	5500	1.30E+03	2.06E+02	4.74E-07	6.34E-04	1.01E-04	7.35E-05
Total		1.65E+09	4.04E+08		802.94	197.06	100
R + A			2.05E+09			1000	

f = 4.596E-06 Pa
Total amount = 20090 kg

Overall residence time = 20.09 h
Reaction time = 25.02 h
Advection time = 101.95 h

Distribution of removal rates

Level III

Chemical name: p-Xylene

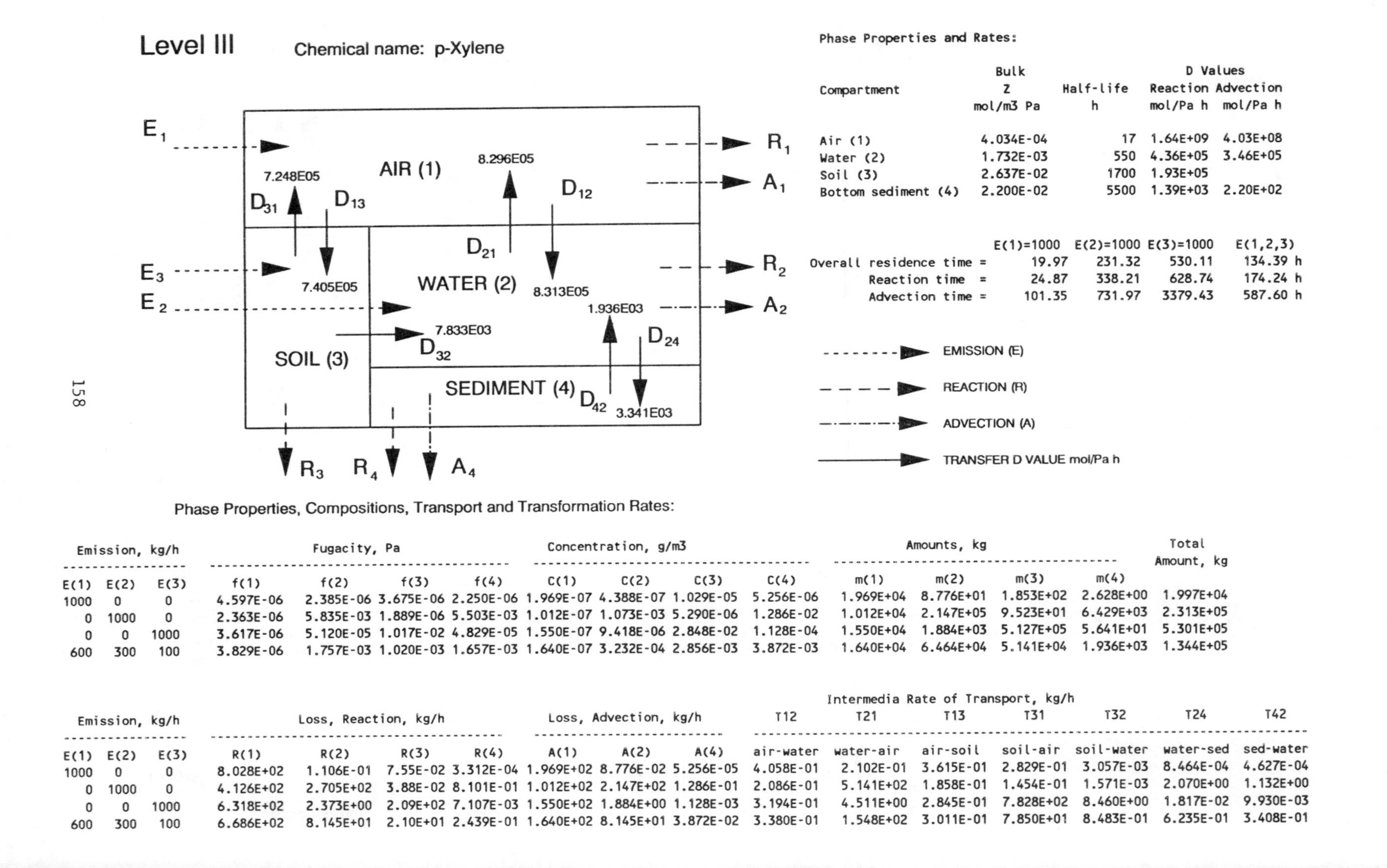

Phase Properties and Rates:

Compartment	Bulk Z mol/m3 Pa	Half-life h	D Values Reaction mol/Pa h	D Values Advection mol/Pa h
Air (1)	4.034E-04	17	1.64E+09	4.03E+08
Water (2)	1.732E-03	550	4.36E+05	3.46E+05
Soil (3)	2.637E-02	1700	1.93E+05	
Bottom sediment (4)	2.200E-02	5500	1.39E+03	2.20E+02

	E(1)=1000	E(2)=1000	E(3)=1000	E(1,2,3)
Overall residence time =	19.97	231.32	530.11	134.39 h
Reaction time =	24.87	338.21	628.74	174.24 h
Advection time =	101.35	731.97	3379.43	587.60 h

Phase Properties, Compositions, Transport and Transformation Rates:

Emission, kg/h			Fugacity, Pa				Concentration, g/m3				Amounts, kg				Total Amount, kg
E(1)	E(2)	E(3)	f(1)	f(2)	f(3)	f(4)	C(1)	C(2)	C(3)	C(4)	m(1)	m(2)	m(3)	m(4)	
1000	0	0	4.597E-06	2.385E-06	3.675E-06	2.250E-06	1.969E-07	4.388E-07	1.029E-05	5.256E-06	1.969E+04	8.776E+01	1.853E+02	2.628E+00	1.997E+04
0	1000	0	2.363E-06	5.835E-03	1.889E-06	5.503E-03	1.012E-07	1.073E-03	5.290E-06	1.286E-02	1.012E+04	2.147E+05	9.523E+01	6.429E+03	2.313E+05
0	0	1000	3.617E-06	5.120E-05	1.017E-02	4.829E-05	1.550E-07	9.418E-06	2.848E-02	1.128E-04	1.550E+04	1.884E+03	5.127E+05	5.641E+01	5.301E+05
600	300	100	3.829E-06	1.757E-03	1.020E-03	1.657E-03	1.640E-07	3.232E-04	2.856E-03	3.872E-03	1.640E+04	6.464E+04	5.141E+04	1.936E+03	1.344E+05

Emission, kg/h			Loss, Reaction, kg/h				Loss, Advection, kg/h			Intermedia Rate of Transport, kg/h T12	T21	T13	T31	T32	T24	T42
E(1)	E(2)	E(3)	R(1)	R(2)	R(3)	R(4)	A(1)	A(2)	A(4)	air-water	water-air	air-soil	soil-air	soil-water	water-sed	sed-water
1000	0	0	8.028E+02	1.106E-01	7.55E-02	3.312E-04	1.969E+02	8.776E-02	5.256E-05	4.058E-01	2.102E-01	3.615E-01	2.829E-01	3.057E-03	8.464E-04	4.627E-04
0	1000	0	4.126E+02	2.705E+02	3.88E-02	8.101E-01	1.012E+02	2.147E+02	1.286E-01	2.086E-01	5.141E+02	1.858E-01	1.454E-01	1.571E-03	2.070E+00	1.132E+00
0	0	1000	6.318E+02	2.373E+00	2.09E+02	7.107E-03	1.550E+02	1.884E+00	1.128E-03	3.194E-01	4.511E+00	2.845E-01	7.828E+02	8.460E+00	1.817E-02	9.930E-03
600	300	100	6.686E+02	8.145E+01	2.10E+01	2.439E-01	1.640E+02	8.145E+01	3.872E-02	3.380E-01	1.548E+02	3.011E-01	7.850E+01	8.483E-01	6.235E-01	3.408E-01

Chemical name: n-Propylbenzene

Level I

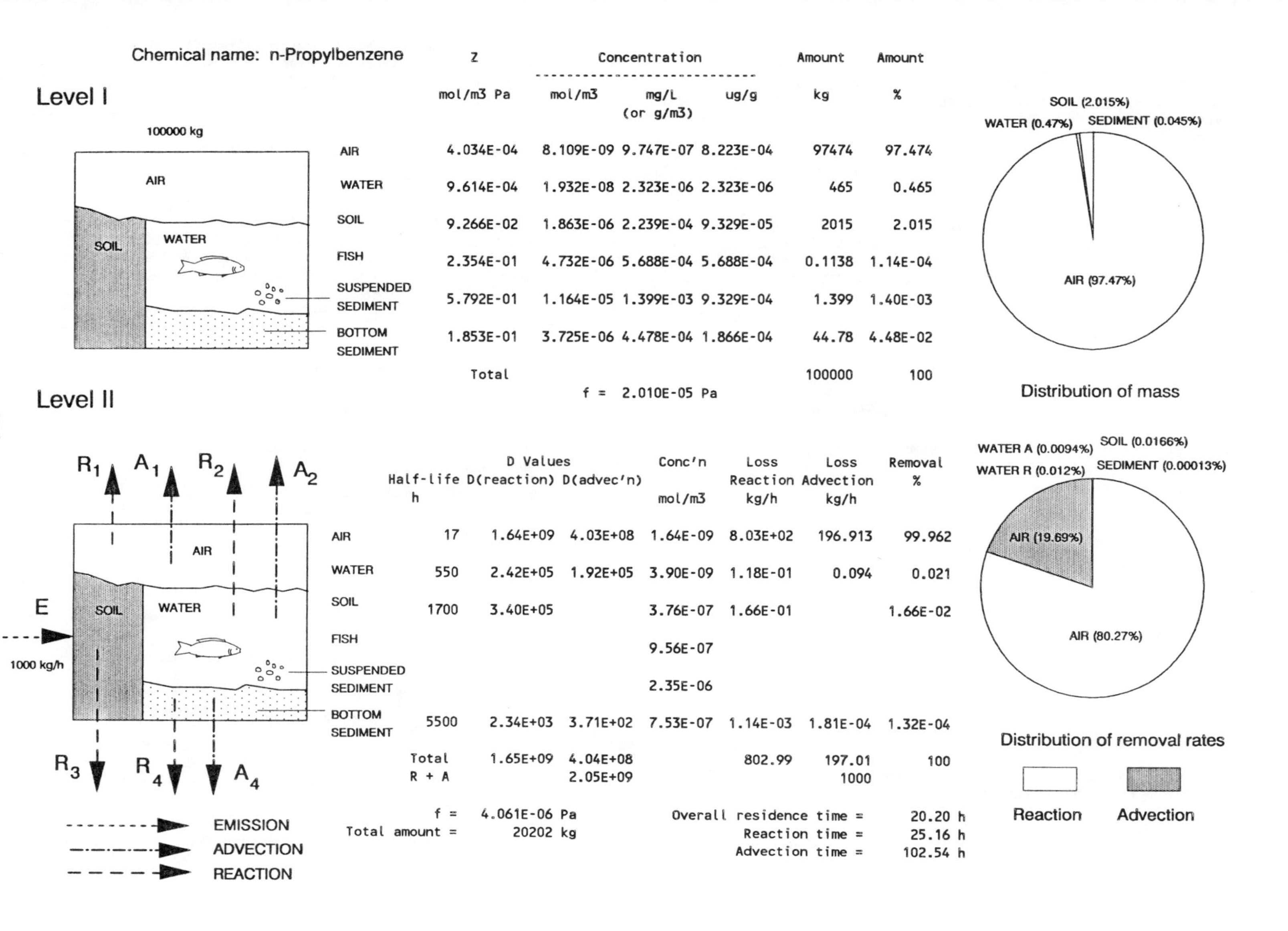

	Z	Concentration			Amount	Amount
	mol/m3 Pa	mol/m3	mg/L (or g/m3)	ug/g	kg	%
AIR	4.034E-04	8.109E-09	9.747E-07	8.223E-04	97474	97.474
WATER	9.614E-04	1.932E-08	2.323E-06	2.323E-06	465	0.465
SOIL	9.266E-02	1.863E-06	2.239E-04	9.329E-05	2015	2.015
FISH	2.354E-01	4.732E-06	5.688E-04	5.688E-04	0.1138	1.14E-04
SUSPENDED SEDIMENT	5.792E-01	1.164E-05	1.399E-03	9.329E-04	1.399	1.40E-03
BOTTOM SEDIMENT	1.853E-01	3.725E-06	4.478E-04	1.866E-04	44.78	4.48E-02
	Total				100000	100

f = 2.010E-05 Pa

Level II

	Half-life h	D Values D(reaction)	D(advec'n)	Conc'n mol/m3	Loss Reaction kg/h	Loss Advection kg/h	Removal %
AIR	17	1.64E+09	4.03E+08	1.64E-09	8.03E+02	196.913	99.962
WATER	550	2.42E+05	1.92E+05	3.90E-09	1.18E-01	0.094	0.021
SOIL	1700	3.40E+05		3.76E-07	1.66E-01		1.66E-02
FISH				9.56E-07			
SUSPENDED SEDIMENT				2.35E-06			
BOTTOM SEDIMENT	5500	2.34E+03	3.71E+02	7.53E-07	1.14E-03	1.81E-04	1.32E-04
	Total R + A	1.65E+09	4.04E+08 2.05E+09		802.99	197.01 1000	100

f = 4.061E-06 Pa
Total amount = 20202 kg

Overall residence time = 20.20 h
Reaction time = 25.16 h
Advection time = 102.54 h

Level III

Chemical name: n-Propylbenzene

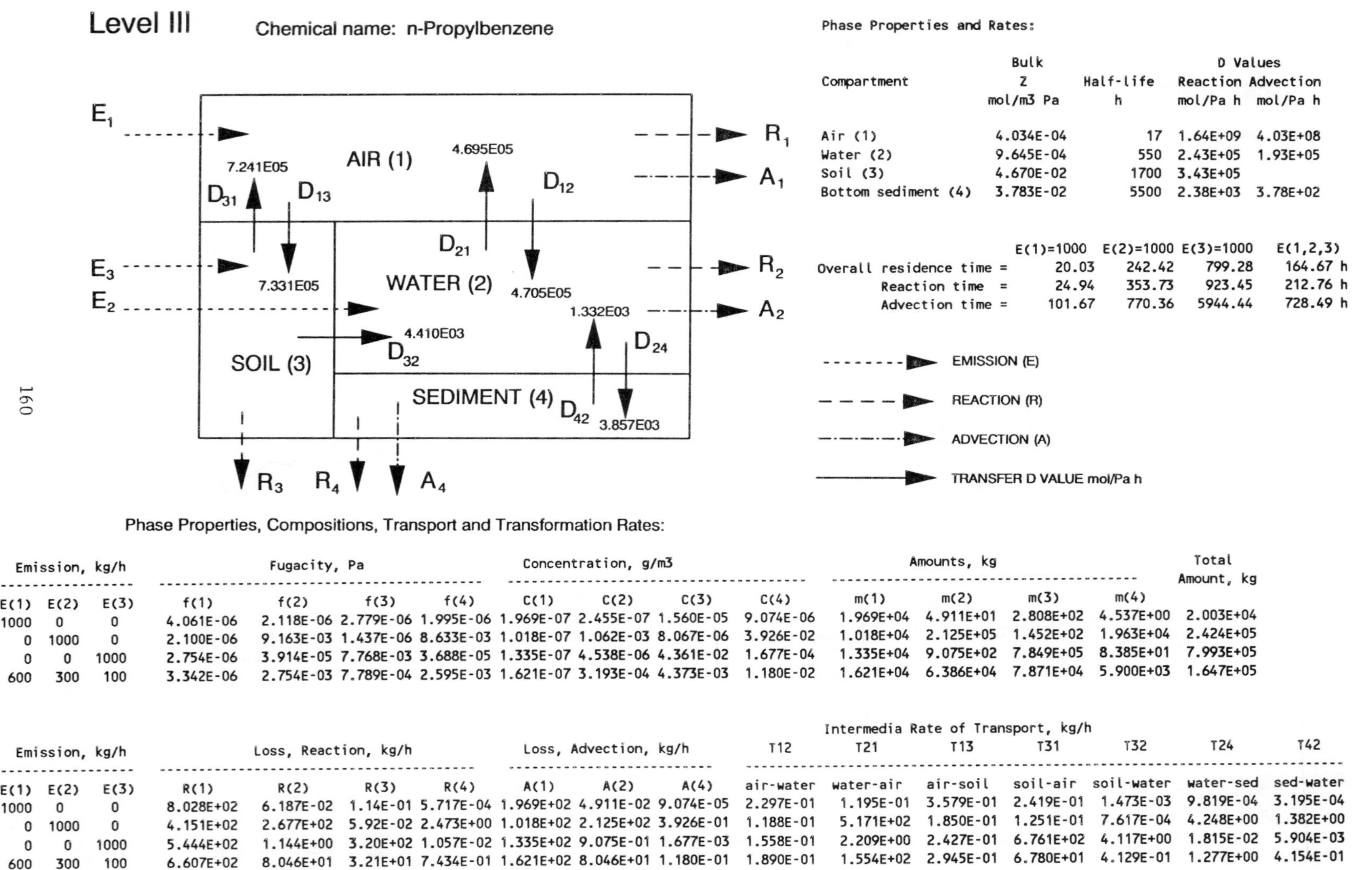

Phase Properties and Rates:

Compartment	Bulk Z mol/m3 Pa	Half-life h	D Values Reaction mol/Pa h	D Values Advection mol/Pa h
Air (1)	4.034E-04	17	1.64E+09	4.03E+08
Water (2)	9.645E-04	550	2.43E+05	1.93E+05
Soil (3)	4.670E-02	1700	3.43E+05	
Bottom sediment (4)	3.783E-02	5500	2.38E+03	3.78E+02

	E(1)=1000	E(2)=1000	E(3)=1000	E(1,2,3)
Overall residence time =	20.03	242.42	799.28	164.67 h
Reaction time =	24.94	353.73	923.45	212.76 h
Advection time =	101.67	770.36	5944.44	728.49 h

Phase Properties, Compositions, Transport and Transformation Rates:

Emission, kg/h			Fugacity, Pa				Concentration, g/m3				Amounts, kg				Total Amount, kg
E(1)	E(2)	E(3)	f(1)	f(2)	f(3)	f(4)	C(1)	C(2)	C(3)	C(4)	m(1)	m(2)	m(3)	m(4)	
1000	0	0	4.061E-06	2.118E-06	2.779E-06	1.995E-06	1.969E-07	2.455E-07	1.560E-05	9.074E-06	1.969E+04	4.911E+01	2.808E+02	4.537E+00	2.003E+04
0	1000	0	2.100E-06	9.163E-03	1.437E-06	8.633E-03	1.018E-07	1.062E-03	8.067E-06	3.926E-02	1.018E+04	2.125E+05	1.452E+02	1.963E+04	2.424E+05
0	0	1000	2.754E-06	3.914E-05	7.768E-03	3.688E-05	1.335E-07	4.538E-06	4.361E-02	1.677E-04	1.335E+04	9.075E+02	7.849E+05	8.385E+01	7.993E+05
600	300	100	3.342E-06	2.754E-03	7.789E-04	2.595E-03	1.621E-07	3.193E-04	4.373E-03	1.180E-02	1.621E+04	6.386E+04	7.871E+04	5.900E+03	1.647E+05

Emission, kg/h			Loss, Reaction, kg/h				Loss, Advection, kg/h			Intermedia Rate of Transport, kg/h T12	T21	T13	T31	T32	T24	T42
E(1)	E(2)	E(3)	R(1)	R(2)	R(3)	R(4)	A(1)	A(2)	A(4)	air-water	water-air	air-soil	soil-air	soil-water	water-sed	sed-water
1000	0	0	8.028E+02	6.187E-02	1.14E-01	5.717E-04	1.969E+02	4.911E-02	9.074E-05	2.297E-01	1.195E-01	3.579E-01	2.419E-01	1.473E-03	9.819E-04	3.195E-04
0	1000	0	4.151E+02	2.677E+02	5.92E-02	2.473E+00	1.018E+02	2.125E+02	3.926E-01	1.188E-01	5.171E+02	1.850E-01	1.251E-01	7.617E-04	4.248E+00	1.382E+00
0	0	1000	5.444E+02	1.144E+00	3.20E+02	1.057E-02	1.335E+02	9.075E-01	1.677E-03	1.558E-01	2.209E+00	2.427E-01	6.761E+02	4.117E+00	1.815E-02	5.904E-03
600	300	100	6.607E+02	8.046E+01	3.21E+01	7.434E-01	1.621E+02	8.046E+01	1.180E-01	1.890E-01	1.554E+02	2.945E-01	6.780E+01	4.129E-01	1.277E+00	4.154E-01

Chemical name: iso-Propylbenzene

Level I

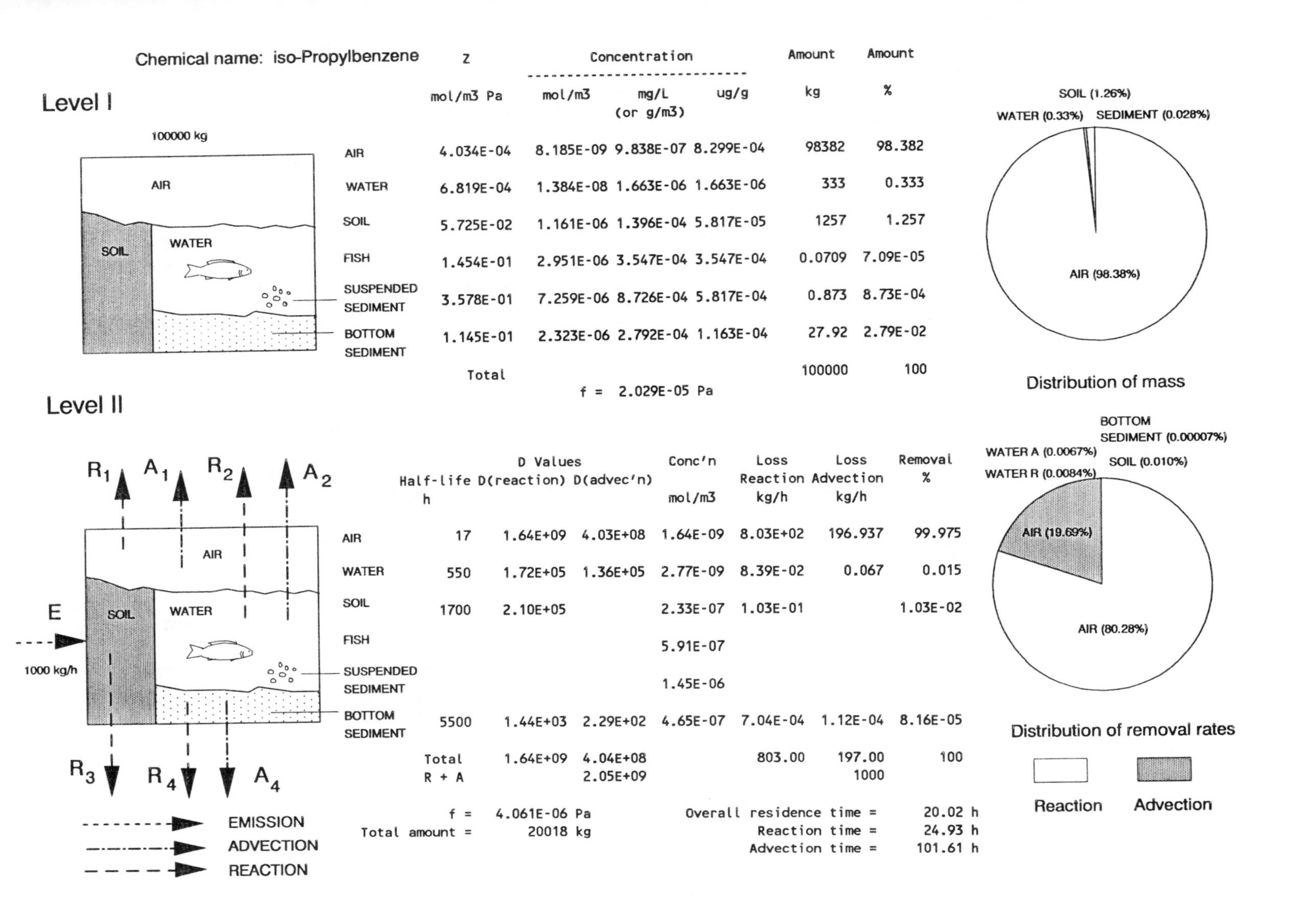

	Z	Concentration			Amount	Amount
	mol/m3 Pa	mol/m3	mg/L (or g/m3)	ug/g	kg	%
AIR	4.034E-04	8.185E-09	9.838E-07	8.299E-04	98382	98.382
WATER	6.819E-04	1.384E-08	1.663E-06	1.663E-06	333	0.333
SOIL	5.725E-02	1.161E-06	1.396E-04	5.817E-05	1257	1.257
FISH	1.454E-01	2.951E-06	3.547E-04	3.547E-04	0.0709	7.09E-05
SUSPENDED SEDIMENT	3.578E-01	7.259E-06	8.726E-04	5.817E-04	0.873	8.73E-04
BOTTOM SEDIMENT	1.145E-01	2.323E-06	2.792E-04	1.163E-04	27.92	2.79E-02
	Total				100000	100

f = 2.029E-05 Pa

Level II

	Half-life h	D Values D(reaction)	D(advec'n)	Conc'n mol/m3	Loss Reaction kg/h	Loss Advection kg/h	Removal %
AIR	17	1.64E+09	4.03E+08	1.64E-09	8.03E+02	196.937	99.975
WATER	550	1.72E+05	1.36E+05	2.77E-09	8.39E-02	0.067	0.015
SOIL	1700	2.10E+05		2.33E-07	1.03E-01		1.03E-02
FISH				5.91E-07			
SUSPENDED SEDIMENT				1.45E-06			
BOTTOM SEDIMENT	5500	1.44E+03	2.29E+02	4.65E-07	7.04E-04	1.12E-04	8.16E-05
	Total	1.64E+09	4.04E+08		803.00	197.00	100
	R + A		2.05E+09			1000	

f = 4.061E-06 Pa
Total amount = 20018 kg

Overall residence time = 20.02 h
Reaction time = 24.93 h
Advection time = 101.61 h

Level III

Chemical name: iso-Propylbenzene

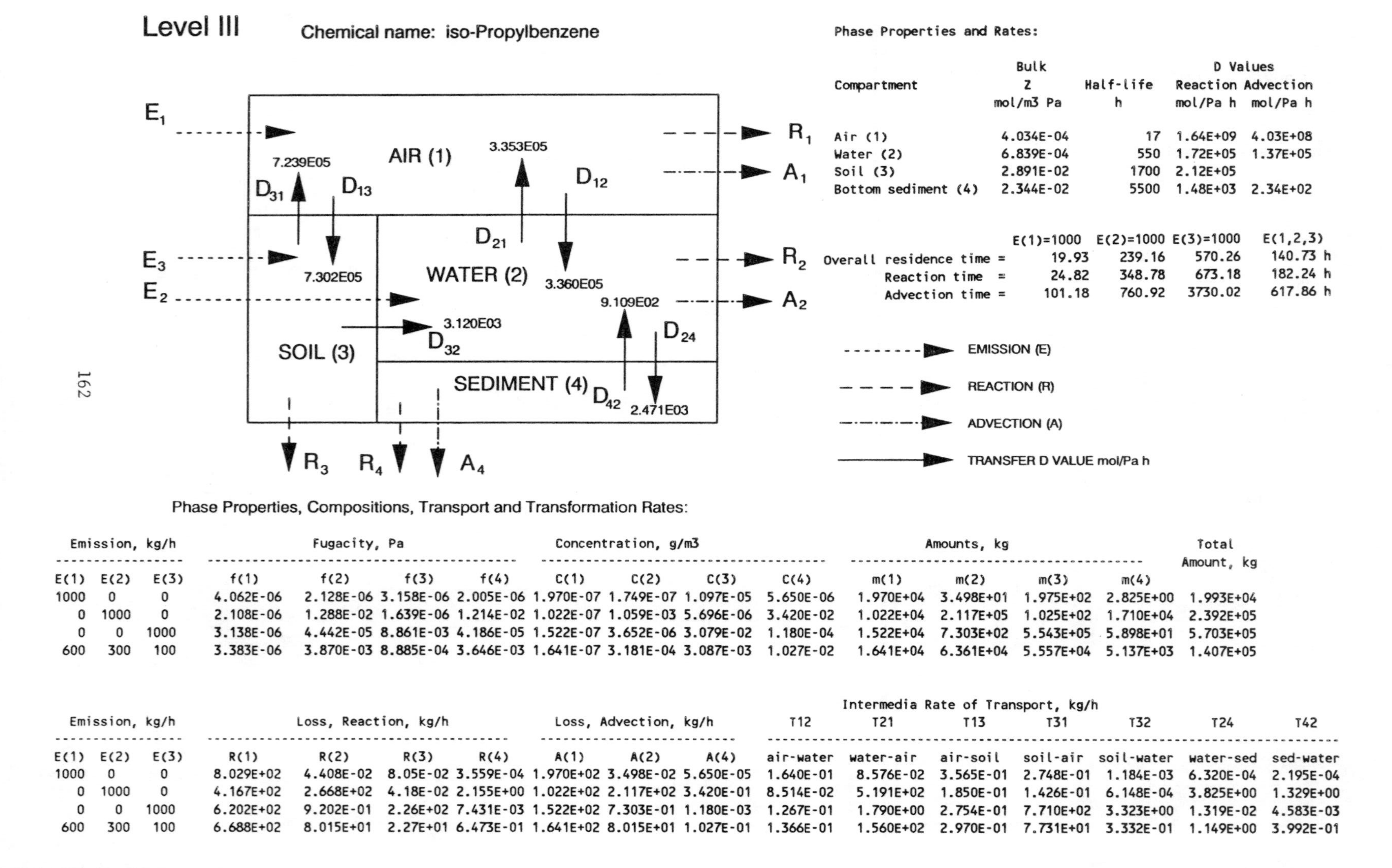

Phase Properties and Rates:

Compartment	Bulk Z mol/m3 Pa	Half-life h	D Values Reaction mol/Pa h	D Values Advection mol/Pa h
Air (1)	4.034E-04	17	1.64E+09	4.03E+08
Water (2)	6.839E-04	550	1.72E+05	1.37E+05
Soil (3)	2.891E-02	1700	2.12E+05	
Bottom sediment (4)	2.344E-02	5500	1.48E+03	2.34E+02

	E(1)=1000	E(2)=1000	E(3)=1000	E(1,2,3)
Overall residence time =	19.93	239.16	570.26	140.73 h
Reaction time =	24.82	348.78	673.18	182.24 h
Advection time =	101.18	760.92	3730.02	617.86 h

Phase Properties, Compositions, Transport and Transformation Rates:

Emission, kg/h			Fugacity, Pa				Concentration, g/m3				Amounts, kg				Total
E(1)	E(2)	E(3)	f(1)	f(2)	f(3)	f(4)	C(1)	C(2)	C(3)	C(4)	m(1)	m(2)	m(3)	m(4)	Amount, kg
1000	0	0	4.062E-06	2.128E-06	3.158E-06	2.005E-06	1.970E-07	1.749E-07	1.097E-05	5.650E-06	1.970E+04	3.498E+01	1.975E+02	2.825E+00	1.993E+04
0	1000	0	2.108E-06	1.288E-02	1.639E-06	1.214E-02	1.022E-07	1.059E-03	5.696E-06	3.420E-02	1.022E+04	2.117E+05	1.025E+02	1.710E+04	2.392E+05
0	0	1000	3.138E-06	4.442E-05	8.861E-03	4.186E-05	1.522E-07	3.652E-06	3.079E-02	1.180E-04	1.522E+04	7.303E+02	5.543E+05	5.898E+01	5.703E+05
600	300	100	3.383E-06	3.870E-03	8.885E-04	3.646E-03	1.641E-07	3.181E-04	3.087E-03	1.027E-02	1.641E+04	6.361E+04	5.557E+04	5.137E+03	1.407E+05

Emission, kg/h			Loss, Reaction, kg/h				Loss, Advection, kg/h			Intermedia Rate of Transport, kg/h T12	T21	T13	T31	T32	T24	T42
E(1)	E(2)	E(3)	R(1)	R(2)	R(3)	R(4)	A(1)	A(2)	A(4)	air-water	water-air	air-soil	soil-air	soil-water	water-sed	sed-water
1000	0	0	8.029E+02	4.408E-02	8.05E-02	3.559E-04	1.970E+02	3.498E-02	5.650E-05	1.640E-01	8.576E-02	3.565E-01	2.748E-01	1.184E-03	6.320E-04	2.195E-04
0	1000	0	4.167E+02	2.668E+02	4.18E-02	2.155E+00	1.022E+02	2.117E+02	3.420E-01	8.514E-02	5.191E+02	1.850E-01	1.426E-01	6.148E-04	3.825E+00	1.329E+00
0	0	1000	6.202E+02	9.202E-01	2.26E+02	7.431E-03	1.522E+02	7.303E-01	1.180E-03	1.267E-01	1.790E+00	2.754E-01	7.710E+02	3.323E+00	1.319E-02	4.583E-03
600	300	100	6.688E+02	8.015E+01	2.27E+01	6.473E-01	1.641E+02	8.015E+01	1.027E-01	1.366E-01	1.560E+02	2.970E-01	7.731E+01	3.332E-01	1.149E+00	3.992E-01

Chemical name: 1,2,3-Trimethylbenzene

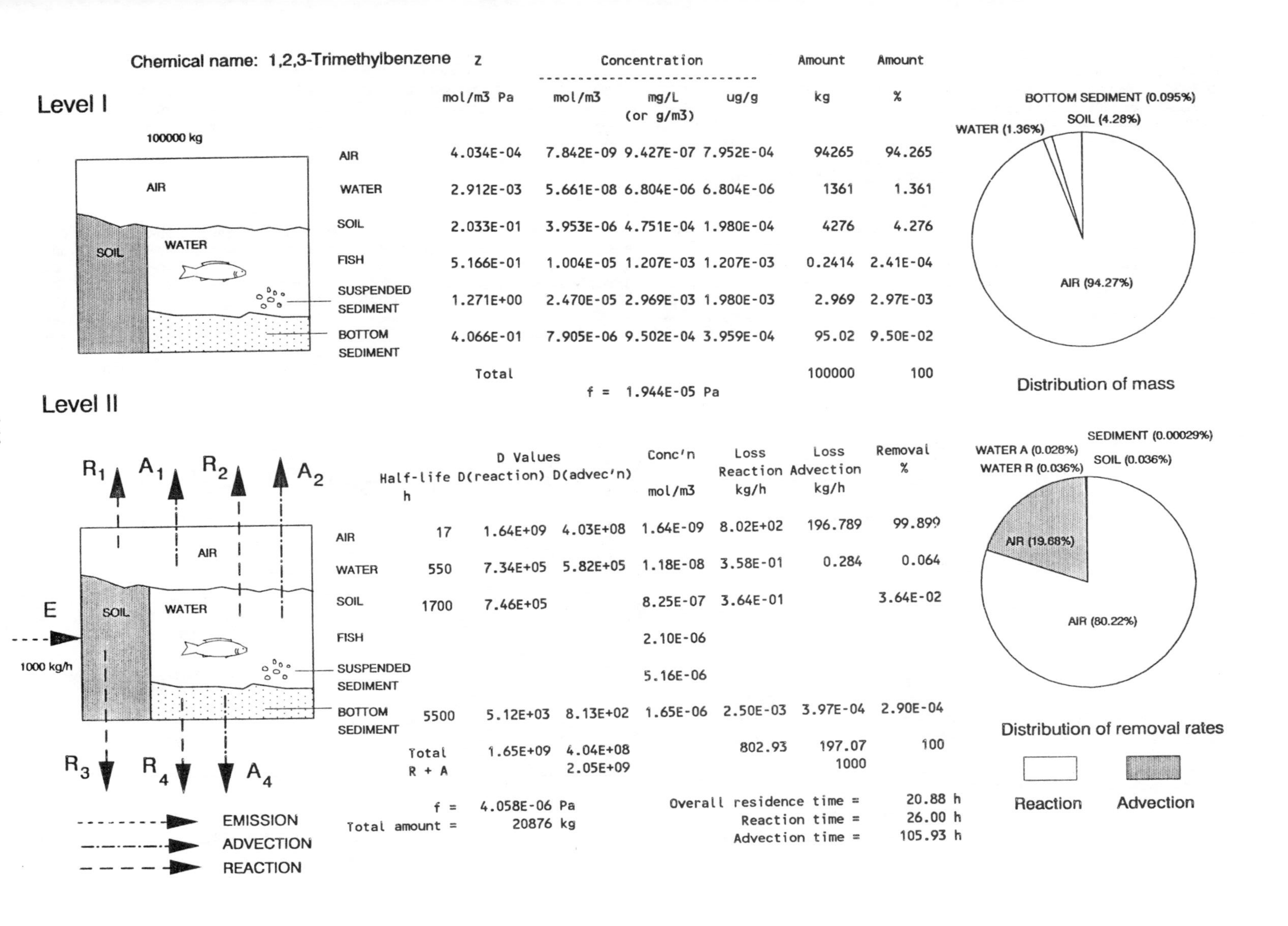

Level I

	Z mol/m3 Pa	Concentration mol/m3	mg/L (or g/m3)	ug/g	Amount kg	Amount %
AIR	4.034E-04	7.842E-09	9.427E-07	7.952E-04	94265	94.265
WATER	2.912E-03	5.661E-08	6.804E-06	6.804E-06	1361	1.361
SOIL	2.033E-01	3.953E-06	4.751E-04	1.980E-04	4276	4.276
FISH	5.166E-01	1.004E-05	1.207E-03	1.207E-03	0.2414	2.41E-04
SUSPENDED SEDIMENT	1.271E+00	2.470E-05	2.969E-03	1.980E-03	2.969	2.97E-03
BOTTOM SEDIMENT	4.066E-01	7.905E-06	9.502E-04	3.959E-04	95.02	9.50E-02
Total					100000	100

f = 1.944E-05 Pa

Level II

	Half-life h	D Values D(reaction)	D(advec'n)	Conc'n mol/m3	Loss Reaction kg/h	Loss Advection kg/h	Removal %
AIR	17	1.64E+09	4.03E+08	1.64E-09	8.02E+02	196.789	99.899
WATER	550	7.34E+05	5.82E+05	1.18E-08	3.58E-01	0.284	0.064
SOIL	1700	7.46E+05		8.25E-07	3.64E-01		3.64E-02
FISH				2.10E-06			
SUSPENDED SEDIMENT				5.16E-06			
BOTTOM SEDIMENT	5500	5.12E+03	8.13E+02	1.65E-06	2.50E-03	3.97E-04	2.90E-04
Total		1.65E+09	4.04E+08		802.93	197.07	100
R + A			2.05E+09			1000	

f = 4.058E-06 Pa
Total amount = 20876 kg

Overall residence time = 20.88 h
Reaction time = 26.00 h
Advection time = 105.93 h

Level III

Chemical name: 1,2,3-Trimethylbenzene

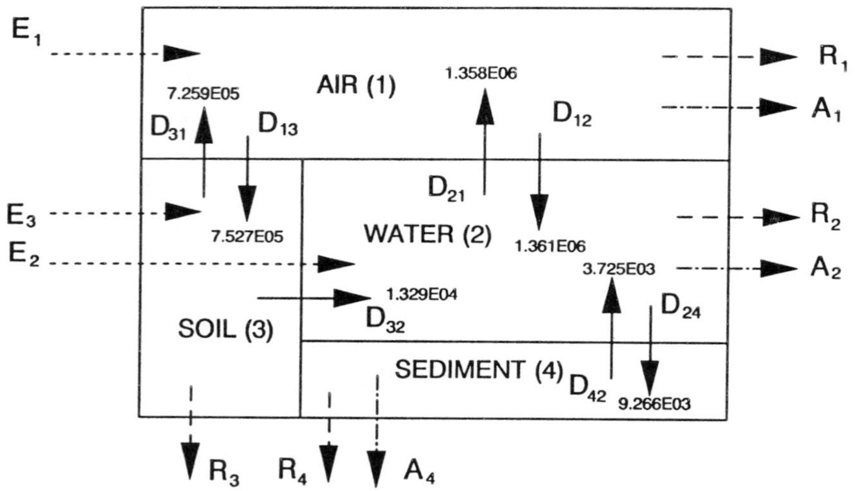

E1 E3 E2
AIR (1) WATER (2) SOIL (3) SEDIMENT (4)
D31 7.259E05 D13 7.527E05
D21 1.358E06 D12 1.361E06
D32 1.329E04
D42 3.725E03 D24 9.266E03
R1 A1 R2 A2 R3 R4 A4

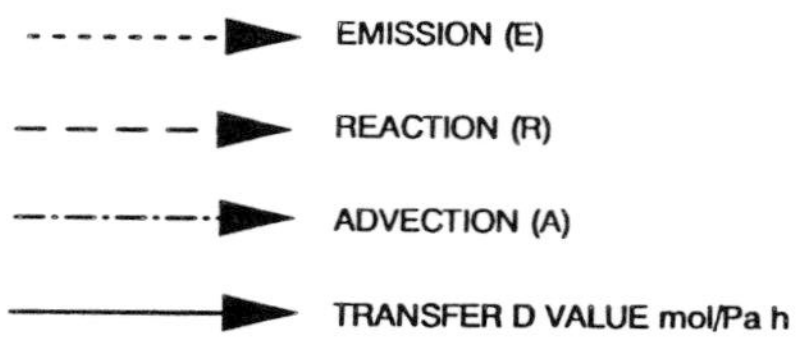

EMISSION (E)
REACTION (R)
ADVECTION (A)
TRANSFER D VALUE mol/Pa h

Phase Properties and Rates:

Compartment	Bulk Z mol/m3 Pa	Half-life h	D Values Reaction mol/Pa h	D Values Advection mol/Pa h
Air (1)	4.034E-04	17	1.64E+09	4.03E+08
Water (2)	2.919E-03	550	7.36E+05	5.84E+05
Soil (3)	1.026E-01	1700	7.53E+05	
Bottom sediment (4)	8.366E-02	5500	5.27E+03	8.37E+02

	E(1)=1000	E(2)=1000	E(3)=1000	E(1,2,3)
Overall residence time =	20.30	242.54	1249.95	209.94 h
Reaction time =	25.28	355.42	1386.79	270.32 h
Advection time =	103.02	763.69	12667.27	939.88 h

Phase Properties, Compositions, Transport and Transformation Rates:

Emission, kg/h			Fugacity, Pa				Concentration, g/m3				Amounts, kg				Total
E(1)	E(2)	E(3)	f(1)	f(2)	f(3)	f(4)	C(1)	C(2)	C(3)	C(4)	m(1)	m(2)	m(3)	m(4)	Amount, kg
1000	0	0	4.060E-06	2.070E-06	2.048E-06	1.950E-06	1.969E-07	7.261E-07	2.526E-05	1.961E-05	1.969E+04	1.452E+02	4.548E+02	9.807E+00	2.030E+04
0	1000	0	2.055E-06	3.102E-03	1.037E-06	2.923E-03	9.965E-08	1.088E-03	1.279E-05	2.940E-02	9.965E+03	2.177E+05	2.302E+02	1.470E+04	2.425E+05
0	0	1000	1.993E-06	2.861E-05	5.577E-03	2.696E-05	9.666E-08	1.004E-05	6.879E-02	2.711E-04	9.666E+03	2.007E+03	1.238E+06	1.356E+02	1.250E+06
600	300	100	3.252E-06	9.347E-04	5.592E-04	8.808E-04	1.577E-07	3.279E-04	6.898E-03	8.858E-03	1.577E+04	6.558E+04	1.242E+05	4.429E+03	2.099E+05

Emission, kg/h			Loss, Reaction, kg/h				Loss, Advection, kg/h			Intermedia Rate of Transport, kg/h T12	T21	T13	T31	T32	T24	T42
E(1)	E(2)	E(3)	R(1)	R(2)	R(3)	R(4)	A(1)	A(2)	A(4)	air-water	water-air	air-soil	soil-air	soil-water	water-sed	sed-water
1000	0	0	8.026E+02	1.830E-01	1.85E-01	1.236E-03	1.969E+02	1.452E-01	1.961E-04	6.642E-01	3.378E-01	3.674E-01	1.787E-01	3.271E-03	2.305E-03	8.733E-04
0	1000	0	4.062E+02	2.742E+02	9.38E-02	1.852E+00	9.965E+01	2.177E+02	2.940E-01	3.362E-01	5.063E+02	1.859E-01	9.045E-02	1.656E-03	3.455E+00	1.309E+00
0	0	1000	3.941E+02	2.529E+00	5.05E+02	1.708E-02	9.666E+01	2.007E+00	2.711E-03	3.261E-01	4.669E+00	1.804E-01	4.866E+02	8.906E+00	3.186E-02	1.207E-02
600	300	100	6.428E+02	8.263E+01	5.06E+01	5.580E-01	1.577E+02	8.263E+01	8.858E-02	5.320E-01	1.526E+02	2.942E-01	4.879E+01	8.930E-01	1.041E+00	3.944E-01

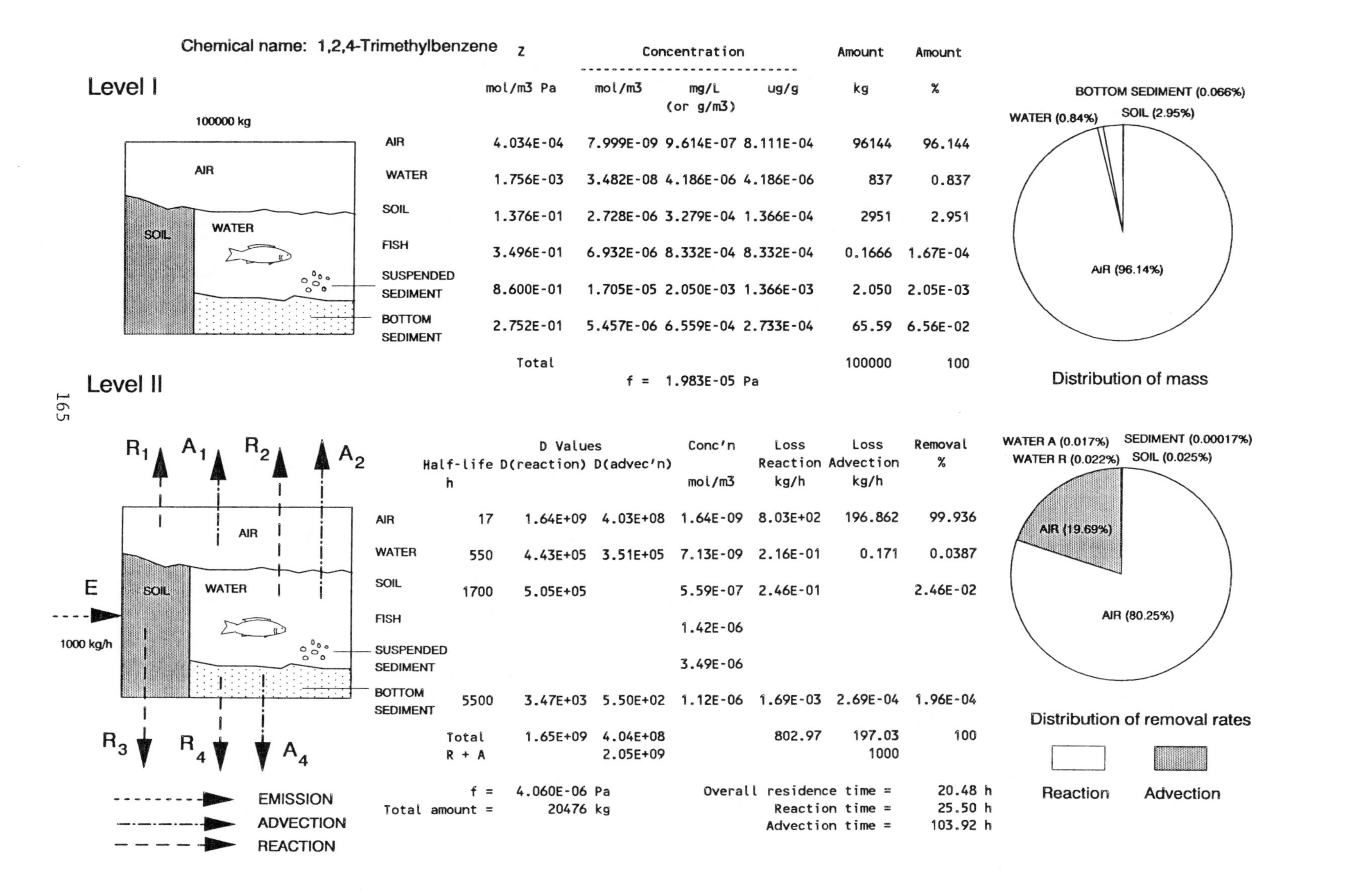

Chemical name: 1,2,4-Trimethylbenzene

Level I

100000 kg

	Z	Concentration			Amount	Amount
	mol/m3 Pa	mol/m3	mg/L (or g/m3)	ug/g	kg	%
AIR	4.034E-04	7.999E-09	9.614E-07	8.111E-04	96144	96.144
WATER	1.756E-03	3.482E-08	4.186E-06	4.186E-06	837	0.837
SOIL	1.376E-01	2.728E-06	3.279E-04	1.366E-04	2951	2.951
FISH	3.496E-01	6.932E-06	8.332E-04	8.332E-04	0.1666	1.67E-04
SUSPENDED SEDIMENT	8.600E-01	1.705E-05	2.050E-03	1.366E-03	2.050	2.05E-03
BOTTOM SEDIMENT	2.752E-01	5.457E-06	6.559E-04	2.733E-04	65.59	6.56E-02
	Total				100000	100

f = 1.983E-05 Pa

Distribution of mass

Level II

	Half-life h	D Values D(reaction)	D(advec'n)	Conc'n mol/m3	Loss Reaction kg/h	Loss Advection kg/h	Removal %
AIR	17	1.64E+09	4.03E+08	1.64E-09	8.03E+02	196.862	99.936
WATER	550	4.43E+05	3.51E+05	7.13E-09	2.16E-01	0.171	0.0387
SOIL	1700	5.05E+05		5.59E-07	2.46E-01		2.46E-02
FISH				1.42E-06			
SUSPENDED SEDIMENT				3.49E-06			
BOTTOM SEDIMENT	5500	3.47E+03	5.50E+02	1.12E-06	1.69E-03	2.69E-04	1.96E-04
	Total	1.65E+09	4.04E+08		802.97	197.03	100
	R + A		2.05E+09			1000	

f = 4.060E-06 Pa
Total amount = 20476 kg

Overall residence time = 20.48 h
Reaction time = 25.50 h
Advection time = 103.92 h

Distribution of removal rates

Level III

Chemical name: 1,2,4-Trimethylbenzene

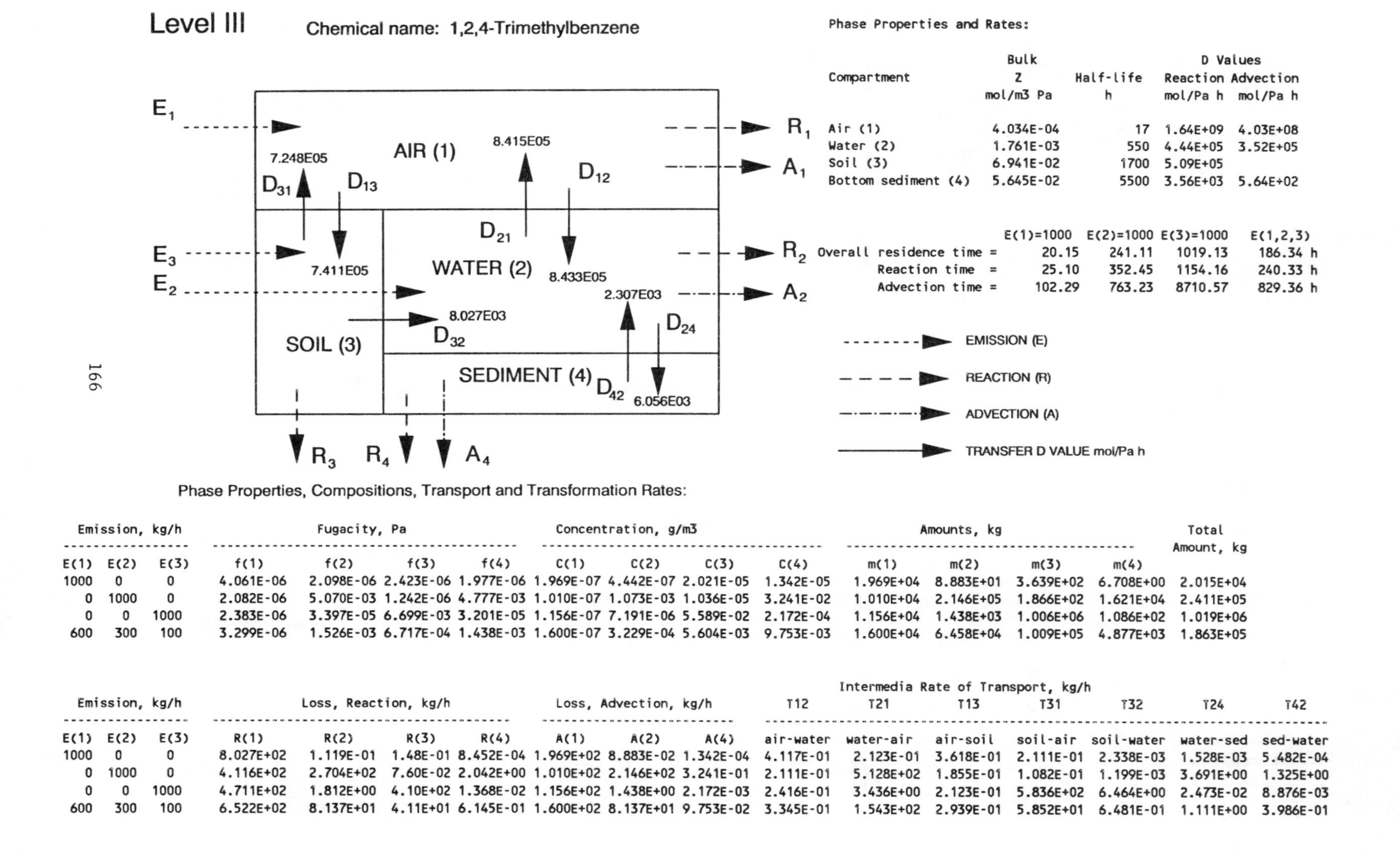

Phase Properties and Rates:

Compartment	Bulk Z mol/m3 Pa	Half-life h	D Values Reaction mol/Pa h	D Values Advection mol/Pa h
Air (1)	4.034E-04	17	1.64E+09	4.03E+08
Water (2)	1.761E-03	550	4.44E+05	3.52E+05
Soil (3)	6.941E-02	1700	5.09E+05	
Bottom sediment (4)	5.645E-02	5500	3.56E+03	5.64E+02

	E(1)=1000	E(2)=1000	E(3)=1000	E(1,2,3)
Overall residence time =	20.15	241.11	1019.13	186.34 h
Reaction time =	25.10	352.45	1154.16	240.33 h
Advection time =	102.29	763.23	8710.57	829.36 h

Phase Properties, Compositions, Transport and Transformation Rates:

Emission, kg/h			Fugacity, Pa				Concentration, g/m3				Amounts, kg				Total Amount, kg
E(1)	E(2)	E(3)	f(1)	f(2)	f(3)	f(4)	C(1)	C(2)	C(3)	C(4)	m(1)	m(2)	m(3)	m(4)	
1000	0	0	4.061E-06	2.098E-06	2.423E-06	1.977E-06	1.969E-07	4.442E-07	2.021E-05	1.342E-05	1.969E+04	8.883E+01	3.639E+02	6.708E+00	2.015E+04
0	1000	0	2.082E-06	5.070E-03	1.242E-06	4.777E-03	1.010E-07	1.073E-03	1.036E-05	3.241E-02	1.010E+04	2.146E+05	1.866E+02	1.621E+04	2.411E+05
0	0	1000	2.383E-06	3.397E-05	6.699E-03	3.201E-05	1.156E-07	7.191E-06	5.589E-02	2.172E-04	1.156E+04	1.438E+03	1.006E+06	1.086E+02	1.019E+06
600	300	100	3.299E-06	1.526E-03	6.717E-04	1.438E-03	1.600E-07	3.229E-04	5.604E-03	9.753E-03	1.600E+04	6.458E+04	1.009E+05	4.877E+03	1.863E+05

Emission, kg/h			Loss, Reaction, kg/h				Loss, Advection, kg/h			Intermedia Rate of Transport, kg/h T12	T21	T13	T31	T32	T24	T42
E(1)	E(2)	E(3)	R(1)	R(2)	R(3)	R(4)	A(1)	A(2)	A(4)	air-water	water-air	air-soil	soil-air	soil-water	water-sed	sed-water
1000	0	0	8.027E+02	1.119E-01	1.48E-01	8.452E-04	1.969E+02	8.883E-02	1.342E-04	4.117E-01	2.123E-01	3.618E-01	2.111E-01	2.338E-03	1.528E-03	5.482E-04
0	1000	0	4.116E+02	2.704E+02	7.60E-02	2.042E+00	1.010E+02	2.146E+02	3.241E-01	2.111E-01	5.128E+02	1.855E-01	1.082E-01	1.199E-03	3.691E+00	1.325E+00
0	0	1000	4.711E+02	1.812E+00	4.10E+02	1.368E-02	1.156E+02	1.438E+00	2.172E-03	2.416E-01	3.436E+00	2.123E-01	5.836E+02	6.464E+00	2.473E-02	8.876E-03
600	300	100	6.522E+02	8.137E+01	4.11E+01	6.145E-01	1.600E+02	8.137E+01	9.753E-02	3.345E-01	1.543E+02	2.939E-01	5.852E+01	6.481E-01	1.111E+00	3.986E-01

Chemical name: 1,3,5-Trimethylbenzene

Level I

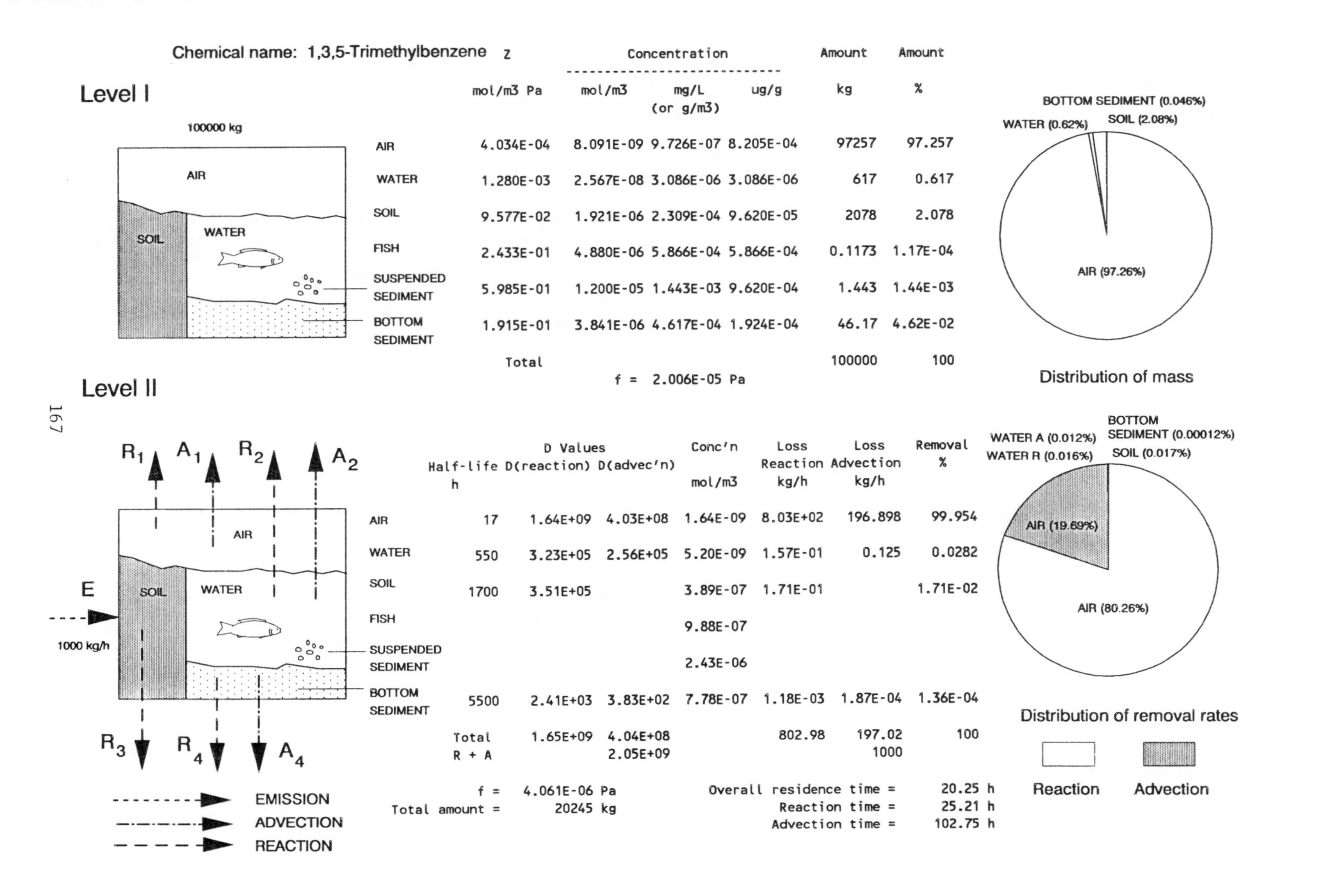

	Z	Concentration			Amount	Amount
	mol/m3 Pa	mol/m3	mg/L (or g/m3)	ug/g	kg	%
AIR	4.034E-04	8.091E-09	9.726E-07	8.205E-04	97257	97.257
WATER	1.280E-03	2.567E-08	3.086E-06	3.086E-06	617	0.617
SOIL	9.577E-02	1.921E-06	2.309E-04	9.620E-05	2078	2.078
FISH	2.433E-01	4.880E-06	5.866E-04	5.866E-04	0.1173	1.17E-04
SUSPENDED SEDIMENT	5.985E-01	1.200E-05	1.443E-03	9.620E-04	1.443	1.44E-03
BOTTOM SEDIMENT	1.915E-01	3.841E-06	4.617E-04	1.924E-04	46.17	4.62E-02
	Total				100000	100

f = 2.006E-05 Pa

Level II

	Half-life h	D Values D(reaction)	D Values D(advec'n)	Conc'n mol/m3	Loss Reaction kg/h	Loss Advection kg/h	Removal %
AIR	17	1.64E+09	4.03E+08	1.64E-09	8.03E+02	196.898	99.954
WATER	550	3.23E+05	2.56E+05	5.20E-09	1.57E-01	0.125	0.0282
SOIL	1700	3.51E+05		3.89E-07	1.71E-01		1.71E-02
FISH				9.88E-07			
SUSPENDED SEDIMENT				2.43E-06			
BOTTOM SEDIMENT	5500	2.41E+03	3.83E+02	7.78E-07	1.18E-03	1.87E-04	1.36E-04
	Total	1.65E+09	4.04E+08		802.98	197.02	100
	R + A		2.05E+09			1000	

f = 4.061E-06 Pa
Total amount = 20245 kg

Overall residence time = 20.25 h
Reaction time = 25.21 h
Advection time = 102.75 h

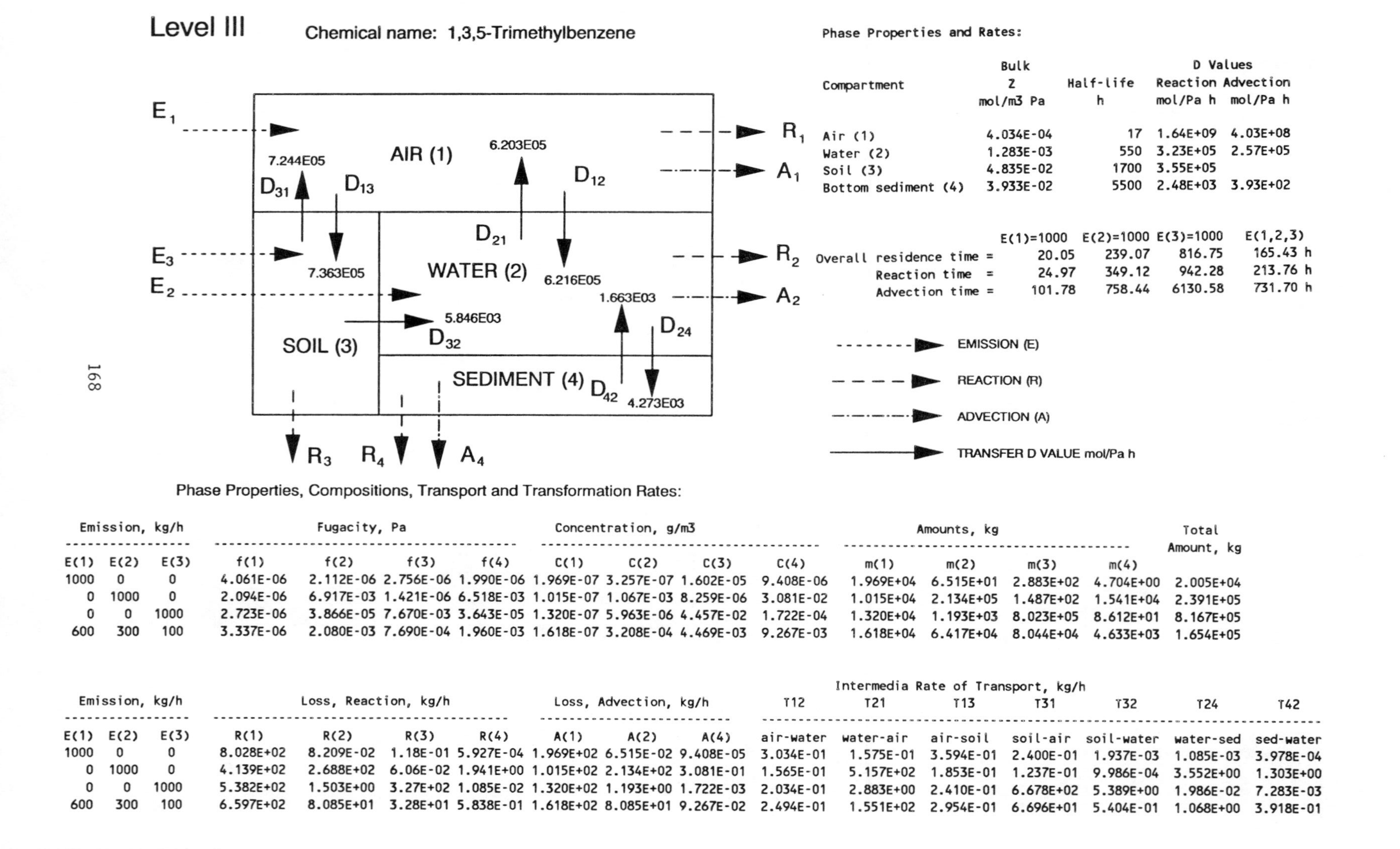

Phase Properties and Rates:

Compartment	Bulk Z mol/m3 Pa	Half-life h	D Values Reaction mol/Pa h	D Values Advection mol/Pa h
Air (1)	4.034E-04	17	1.64E+09	4.03E+08
Water (2)	1.283E-03	550	3.23E+05	2.57E+05
Soil (3)	4.835E-02	1700	3.55E+05	
Bottom sediment (4)	3.933E-02	5500	2.48E+03	3.93E+02

	E(1)=1000	E(2)=1000	E(3)=1000	E(1,2,3)
Overall residence time =	20.05	239.07	816.75	165.43 h
Reaction time =	24.97	349.12	942.28	213.76 h
Advection time =	101.78	758.44	6130.58	731.70 h

Phase Properties, Compositions, Transport and Transformation Rates:

Emission, kg/h			Fugacity, Pa				Concentration, g/m3				Amounts, kg				Total
E(1)	E(2)	E(3)	f(1)	f(2)	f(3)	f(4)	C(1)	C(2)	C(3)	C(4)	m(1)	m(2)	m(3)	m(4)	Amount, kg
1000	0	0	4.061E-06	2.112E-06	2.756E-06	1.990E-06	1.969E-07	3.257E-07	1.602E-05	9.408E-06	1.969E+04	6.515E+01	2.883E+02	4.704E+00	2.005E+04
0	1000	0	2.094E-06	6.917E-03	1.421E-06	6.518E-03	1.015E-07	1.067E-03	8.259E-06	3.081E-02	1.015E+04	2.134E+05	1.487E+02	1.541E+04	2.391E+05
0	0	1000	2.723E-06	3.866E-05	7.670E-03	3.643E-05	1.320E-07	5.963E-06	4.457E-02	1.722E-04	1.320E+04	1.193E+03	8.023E+05	8.612E+01	8.167E+05
600	300	100	3.337E-06	2.080E-03	7.690E-04	1.960E-03	1.618E-07	3.208E-04	4.469E-03	9.267E-03	1.618E+04	6.417E+04	8.044E+04	4.633E+03	1.654E+05

Emission, kg/h			Loss, Reaction, kg/h				Loss, Advection, kg/h			Intermedia Rate of Transport, kg/h						
										T12	T21	T13	T31	T32	T24	T42
E(1)	E(2)	E(3)	R(1)	R(2)	R(3)	R(4)	A(1)	A(2)	A(4)	air-water	water-air	air-soil	soil-air	soil-water	water-sed	sed-water
1000	0	0	8.028E+02	8.209E-02	1.18E-01	5.927E-04	1.969E+02	6.515E-02	9.408E-05	3.034E-01	1.575E-01	3.594E-01	2.400E-01	1.937E-03	1.085E-03	3.978E-04
0	1000	0	4.139E+02	2.688E+02	6.06E-02	1.941E+00	1.015E+02	2.134E+02	3.081E-01	1.565E-01	5.157E+02	1.853E-01	1.237E-01	9.986E-04	3.552E+00	1.303E+00
0	0	1000	5.382E+02	1.503E+00	3.27E+02	1.085E-02	1.320E+02	1.193E+00	1.722E-03	2.034E-01	2.883E+00	2.410E-01	6.678E+02	5.389E+00	1.986E-02	7.283E-03
600	300	100	6.597E+02	8.085E+01	3.28E+01	5.838E-01	1.618E+02	8.085E+01	9.267E-02	2.494E-01	1.551E+02	2.954E-01	6.696E+01	5.404E-01	1.068E+00	3.918E-01

Chemical name: iso-Propyl-4-methylbenzene

Level I

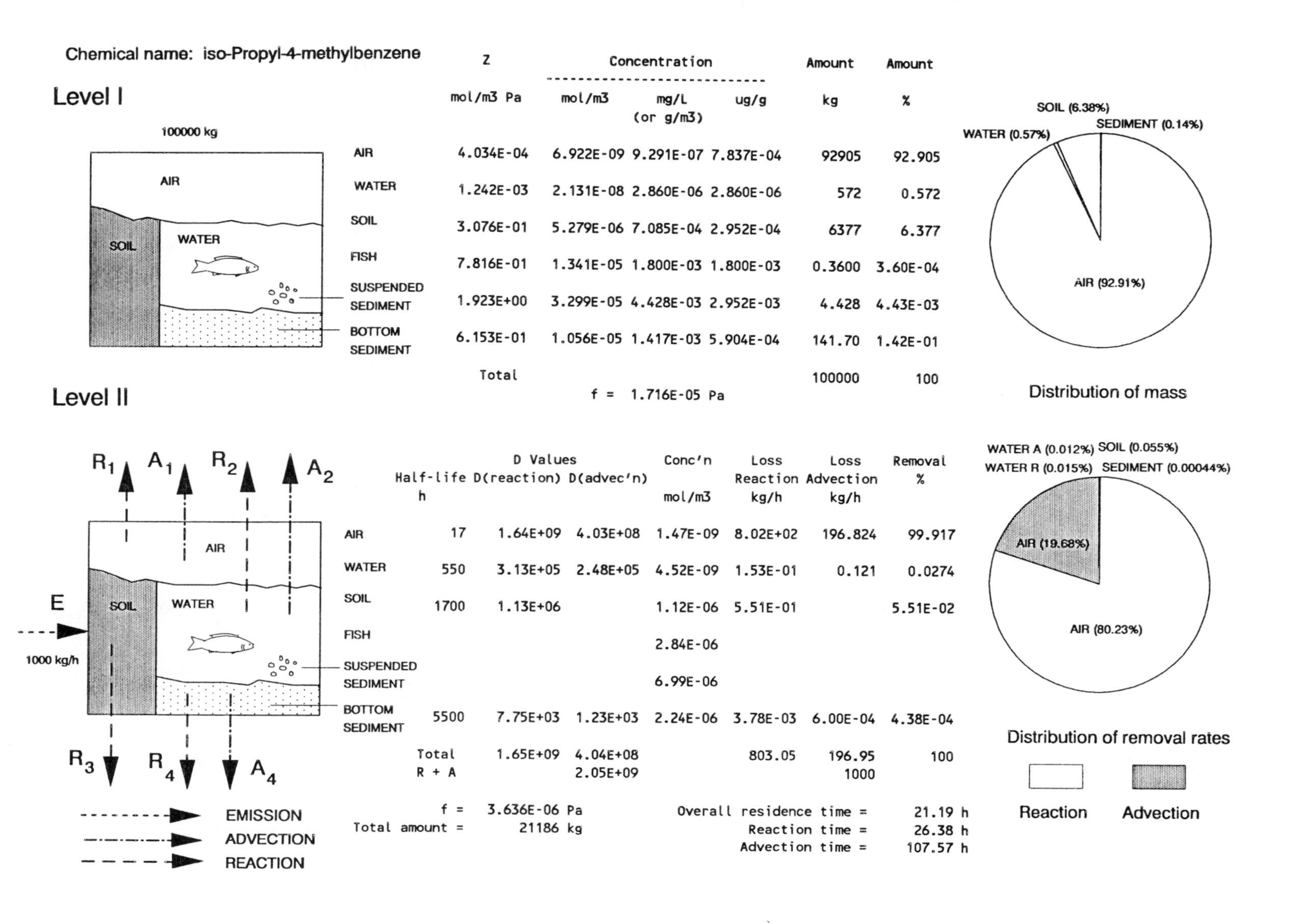

	Z	Concentration			Amount	Amount
	mol/m3 Pa	mol/m3	mg/L (or g/m3)	ug/g	kg	%
AIR	4.034E-04	6.922E-09	9.291E-07	7.837E-04	92905	92.905
WATER	1.242E-03	2.131E-08	2.860E-06	2.860E-06	572	0.572
SOIL	3.076E-01	5.279E-06	7.085E-04	2.952E-04	6377	6.377
FISH	7.816E-01	1.341E-05	1.800E-03	1.800E-03	0.3600	3.60E-04
SUSPENDED SEDIMENT	1.923E+00	3.299E-05	4.428E-03	2.952E-03	4.428	4.43E-03
BOTTOM SEDIMENT	6.153E-01	1.056E-05	1.417E-03	5.904E-04	141.70	1.42E-01
	Total				100000	100

f = 1.716E-05 Pa

Level II

	Half-life h	D Values D(reaction)	D(advec'n)	Conc'n mol/m3	Loss Reaction kg/h	Loss Advection kg/h	Removal %
AIR	17	1.64E+09	4.03E+08	1.47E-09	8.02E+02	196.824	99.917
WATER	550	3.13E+05	2.48E+05	4.52E-09	1.53E-01	0.121	0.0274
SOIL	1700	1.13E+06		1.12E-06	5.51E-01		5.51E-02
FISH				2.84E-06			
SUSPENDED SEDIMENT				6.99E-06			
BOTTOM SEDIMENT	5500	7.75E+03	1.23E+03	2.24E-06	3.78E-03	6.00E-04	4.38E-04
	Total	1.65E+09	4.04E+08		803.05	196.95	100
	R + A		2.05E+09			1000	

f = 3.636E-06 Pa
Total amount = 21186 kg

Overall residence time = 21.19 h
Reaction time = 26.38 h
Advection time = 107.57 h

Level III

Chemical Name: iso-Propyl-4-methylbenzene

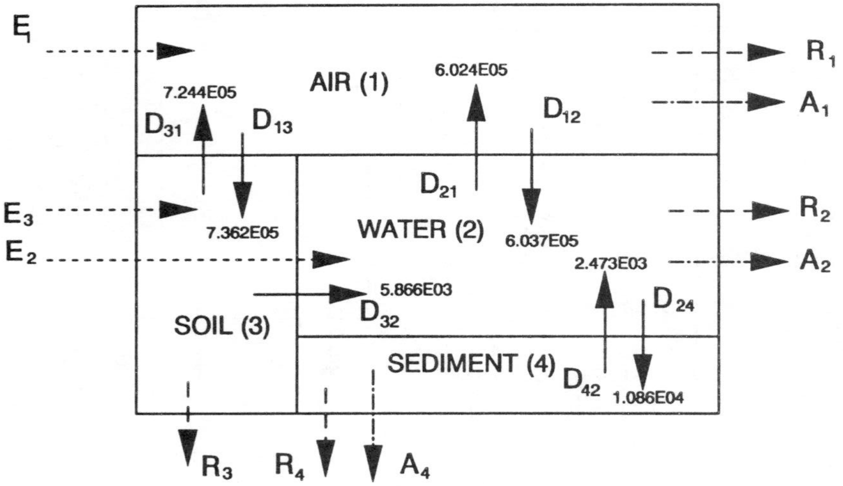

E1, E3, E2, AIR (1), WATER (2), SOIL (3), SEDIMENT (4), R1, A1, R2, A2, R3, R4, A4

D31 7.244E05, D13 7.362E05, D21 6.024E05, D12 6.037E05, D32 5.866E03, D42 2.473E03, D24 1.086E04

EMISSION (E) ; REACTION (R) ; ADVECTION (A) ; TRANSFER D VALUE mol/Pa h

Phase Properties and Rates:

Compartment	Bulk Z mol/m3 Pa	Half-life h	D Values Reaction mol/Pa h	D Values Advection mol/Pa h
Air (1)	4.034E-04	17	1.64E+09	4.03E+08
Water (2)	1.252E-03	550	3.16E+05	2.50E+05
Soil (3)	1.543E-01	1700	1.13E+06	
Bottom sediment (4)	1.241E-01	5500	7.82E+03	1.24E+03

	E(1)=1000	E(2)=1000	E(3)=1000	E(1,2,3)
Overall residence time =	20.31	272.83	1500.04	244.03 h
Reaction time =	25.29	398.06	1626.24	313.00 h
Advection time =	103.08	867.18	19329.73	1107.58 h

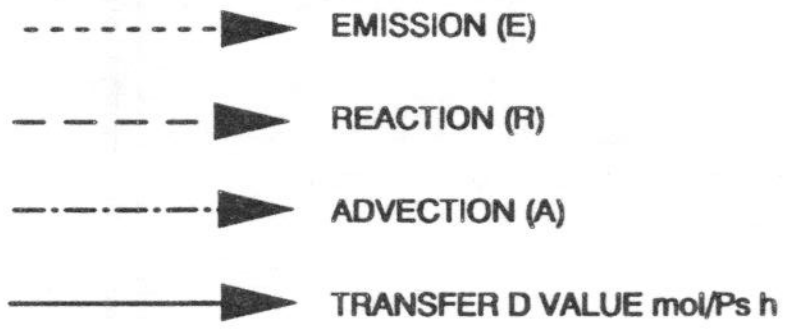

Phase Properties, Compositions, Transport and Transformation Rates:

Emission, kg/h			Fugacity, Pa				Concentration, g/m3				Amounts, kg				Total
E(1)	E(2)	E(3)	f(1)	f(2)	f(3)	f(4)	C(1)	C(2)	C(3)	C(4)	m(1)	m(2)	m(3)	m(4)	Amount, kg
1000	0	0	3.637E-06	1.873E-06	1.438E-06	1.764E-06	1.969E-07	3.148E-07	2.977E-05	2.936E-05	1.969E+04	6.295E+01	5.359E+02	1.468E+01	2.031E+04
0	1000	0	1.862E-06	6.332E-03	7.359E-07	5.963E-03	1.008E-07	1.064E-03	1.524E-05	9.928E-02	1.008E+04	2.128E+05	2.743E+02	4.964E+04	2.728E+05
0	0	1000	1.421E-06	2.066E-05	4.002E-03	1.946E-05	7.690E-08	3.473E-06	8.286E-02	3.240E-04	7.690E+03	6.946E+02	1.491E+06	1.620E+02	1.500E+06
600	300	100	2.883E-06	1.903E-03	4.012E-04	1.792E-03	1.561E-07	3.198E-04	8.308E-03	2.983E-02	1.561E+04	6.396E+04	1.496E+05	1.492E+04	2.440E+05

Emission, kg/h			Loss, Reaction, kg/h				Loss, Advection, kg/h			Intermedia Rate of Transport, kg/h T12	T21	T13	T31	T32	T24	T42
E(1)	E(2)	E(3)	R(1)	R(2)	R(3)	R(4)	A(1)	A(2)	A(4)	air-water	water-air	air-soil	soil-air	soil-water	water-sed	sed-water
1000	0	0	8.027E+02	7.932E-02	2.18E-01	1.850E-03	1.969E+02	6.295E-02	2.936E-04	2.947E-01	1.514E-01	3.593E-01	1.398E-01	1.132E-03	2.729E-03	5.852E-04
0	1000	0	4.109E+02	2.682E+02	1.12E-01	6.255E+00	1.008E+02	2.128E+02	9.928E-01	1.508E-01	5.119E+02	1.839E-01	7.153E-02	5.792E-04	9.226E+00	1.979E+00
0	0	1000	3.135E+02	8.752E-01	6.08E+02	2.041E-02	7.690E+01	6.946E-01	3.240E-03	1.151E-01	1.671E+00	1.403E-01	3.890E+02	3.150E+00	3.011E-02	6.457E-03
600	300	100	6.362E+02	8.059E+01	6.10E+01	1.880E+00	1.561E+02	8.059E+01	2.983E-01	2.336E-01	1.538E+02	2.848E-01	3.900E+01	3.158E-01	2.772E+00	5.946E-01

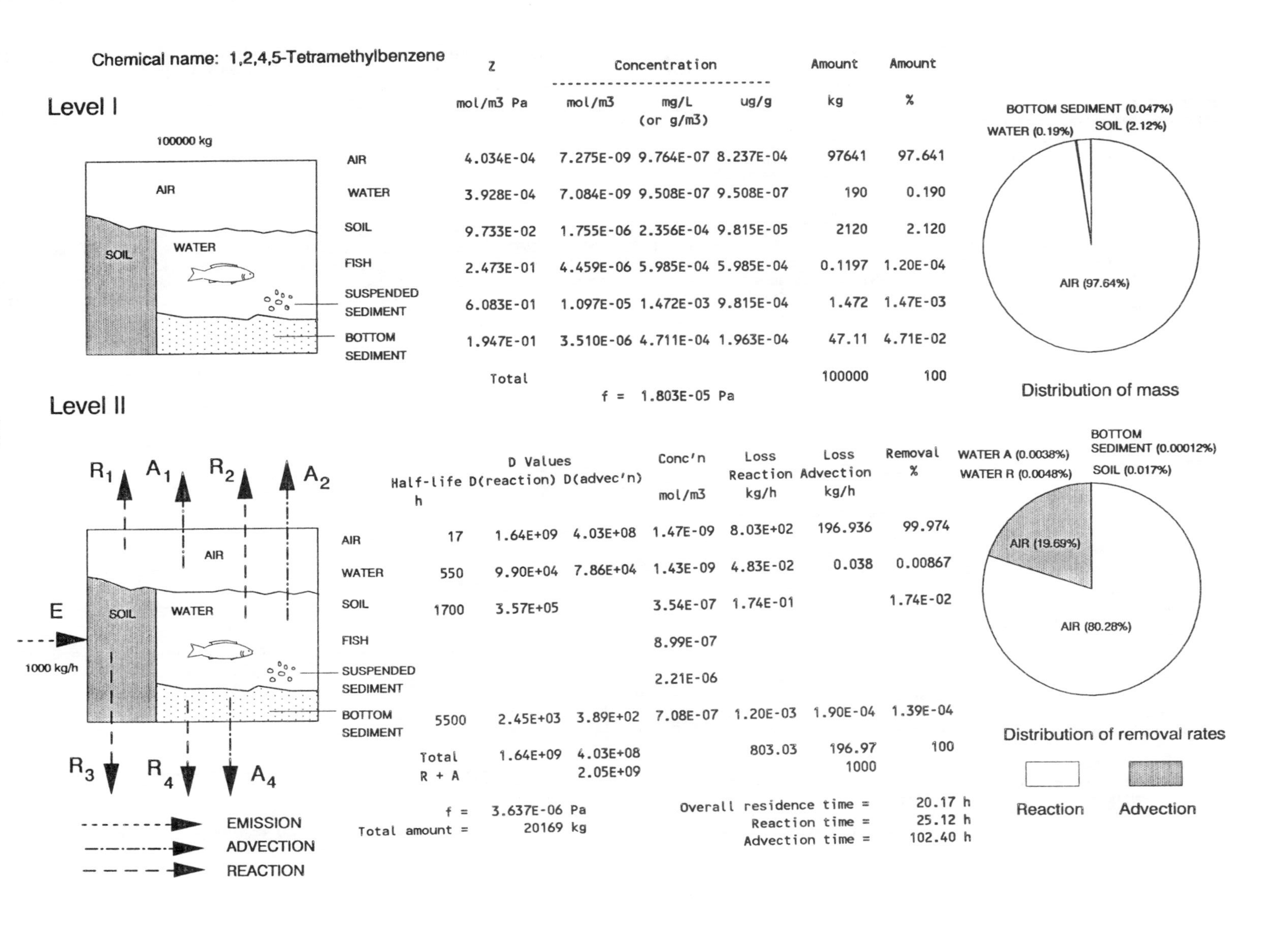

Chemical name: 1,2,4,5-Tetramethylbenzene

Level I

	Z	Concentration			Amount	Amount
	mol/m3 Pa	mol/m3	mg/L (or g/m3)	ug/g	kg	%
AIR	4.034E-04	7.275E-09	9.764E-07	8.237E-04	97641	97.641
WATER	3.928E-04	7.084E-09	9.508E-07	9.508E-07	190	0.190
SOIL	9.733E-02	1.755E-06	2.356E-04	9.815E-05	2120	2.120
FISH	2.473E-01	4.459E-06	5.985E-04	5.985E-04	0.1197	1.20E-04
SUSPENDED SEDIMENT	6.083E-01	1.097E-05	1.472E-03	9.815E-04	1.472	1.47E-03
BOTTOM SEDIMENT	1.947E-01	3.510E-06	4.711E-04	1.963E-04	47.11	4.71E-02
	Total				100000	100

f = 1.803E-05 Pa

Level II

	Half-life h	D Values D(reaction)	D(advec'n)	Conc'n mol/m3	Loss Reaction kg/h	Loss Advection kg/h	Removal %
AIR	17	1.64E+09	4.03E+08	1.47E-09	8.03E+02	196.936	99.974
WATER	550	9.90E+04	7.86E+04	1.43E-09	4.83E-02	0.038	0.00867
SOIL	1700	3.57E+05		3.54E-07	1.74E-01		1.74E-02
FISH				8.99E-07			
SUSPENDED SEDIMENT				2.21E-06			
BOTTOM SEDIMENT	5500	2.45E+03	3.89E+02	7.08E-07	1.20E-03	1.90E-04	1.39E-04
	Total R + A	1.64E+09	4.03E+08 2.05E+09		803.03	196.97 1000	100

f = 3.637E-06 Pa
Total amount = 20169 kg

Overall residence time = 20.17 h
Reaction time = 25.12 h
Advection time = 102.40 h

Level III

Chemical name: 1,2,4,5-Tetramethylbenzene

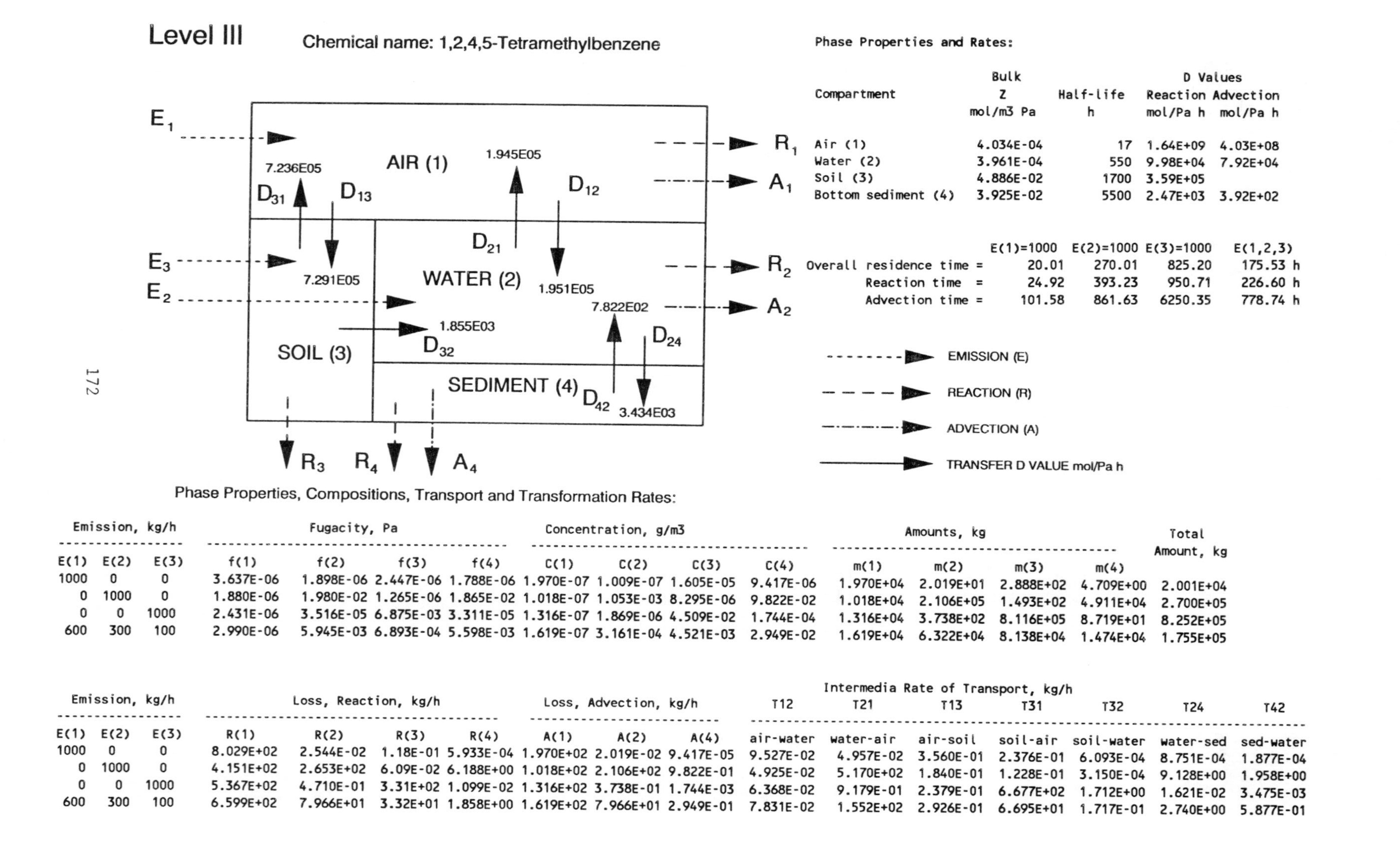

Phase Properties and Rates:

Compartment	Bulk Z mol/m3 Pa	Half-life h	D Values Reaction mol/Pa h	D Values Advection mol/Pa h
Air (1)	4.034E-04	17	1.64E+09	4.03E+08
Water (2)	3.961E-04	550	9.98E+04	7.92E+04
Soil (3)	4.886E-02	1700	3.59E+05	
Bottom sediment (4)	3.925E-02	5500	2.47E+03	3.92E+02

	E(1)=1000	E(2)=1000	E(3)=1000	E(1,2,3)
Overall residence time =	20.01	270.01	825.20	175.53 h
Reaction time =	24.92	393.23	950.71	226.60 h
Advection time =	101.58	861.63	6250.35	778.74 h

Phase Properties, Compositions, Transport and Transformation Rates:

Emission, kg/h			Fugacity, Pa				Concentration, g/m3				Amounts, kg				Total Amount, kg
E(1)	E(2)	E(3)	f(1)	f(2)	f(3)	f(4)	C(1)	C(2)	C(3)	C(4)	m(1)	m(2)	m(3)	m(4)	
1000	0	0	3.637E-06	1.898E-06	2.447E-06	1.788E-06	1.970E-07	1.009E-07	1.605E-05	9.417E-06	1.970E+04	2.019E+01	2.888E+02	4.709E+00	2.001E+04
0	1000	0	1.880E-06	1.980E-02	1.265E-06	1.865E-02	1.018E-07	1.053E-03	8.295E-06	9.822E-02	1.018E+04	2.106E+05	1.493E+02	4.911E+04	2.700E+05
0	0	1000	2.431E-06	3.516E-05	6.875E-03	3.311E-05	1.316E-07	1.869E-06	4.509E-02	1.744E-04	1.316E+04	3.738E+02	8.116E+05	8.719E+01	8.252E+05
600	300	100	2.990E-06	5.945E-03	6.893E-04	5.598E-03	1.619E-07	3.161E-04	4.521E-03	2.949E-02	1.619E+04	6.322E+04	8.138E+04	1.474E+04	1.755E+05

Emission, kg/h			Loss, Reaction, kg/h				Loss, Advection, kg/h			Intermedia Rate of Transport, kg/h						
										T12	T21	T13	T31	T32	T24	T42
E(1)	E(2)	E(3)	R(1)	R(2)	R(3)	R(4)	A(1)	A(2)	A(4)	air-water	water-air	air-soil	soil-air	soil-water	water-sed	sed-water
1000	0	0	8.029E+02	2.544E-02	1.18E-01	5.933E-04	1.970E+02	2.019E-02	9.417E-05	9.527E-02	4.957E-02	3.560E-01	2.376E-01	6.093E-04	8.751E-04	1.877E-04
0	1000	0	4.151E+02	2.653E+02	6.09E-02	6.188E+00	1.018E+02	2.106E+02	9.822E-01	4.925E-02	5.170E+02	1.840E-01	1.228E-01	3.150E-04	9.128E+00	1.958E+00
0	0	1000	5.367E+02	4.710E-01	3.31E+02	1.099E-02	1.316E+02	3.738E-01	1.744E-03	6.368E-02	9.179E-01	2.379E-01	6.677E+02	1.712E+00	1.621E-02	3.475E-03
600	300	100	6.599E+02	7.966E+01	3.32E+01	1.858E+00	1.619E+02	7.966E+01	2.949E-01	7.831E-02	1.552E+02	2.926E-01	6.695E+01	1.717E-01	2.740E+00	5.877E-01

2.4 COMMENTARY ON THE PHYSICAL-CHEMICAL PROPERTIES AND ENVIRONMENTAL FATE

QSPR Plots

A discussion of the QSPR plots of the monoaromatic chemicals has been presented earlier in Chapter 1. There is clearly a strong, consistent relationship between molar volume and the properties, and between molar volume and other common descriptors.

Addition of a methyl group generally increases molar volume by about 22 to 25 cm^3/mol which coincidentally is about the volume of a chlorine molecule. It is thus useful for "rule of thumb" purposes to note the effect of a 25 cm^3/mol increase. Such an increase causes (approximately):

(i) A decrease in log solubility by 0.62 units (a factor of 4.2);
(ii) A decrease in log vapor pressure by 0.55 units (a factor of 3.5);
(iii) An increase in log Henry's law constant of 0.07 units (a factor of 1.2);
(iv) An increase in log K_{OW} of 0.5 units (a factor of 3.2).

The slope of the log solubility versus log K_{OW} line is thus about 0.5/0.62 or 0.8. The series is generally "well-behaved" and the accuracy of the reported values is probably high, and certainly sufficient for estimating environmental fate.

Evaluative Calculations

The Level I calculations suggest that this group of chemicals will partition predominantly into the atmosphere. The more alkylated compounds have higher affinities for soil and sediment because of their larger K_{OW} values, but at equilibrium these solid phases account for only a few percent of the total sorptive capacity.

The Level II calculations also reflect this distribution and suggest that atmospheric reaction and advection will account for most removal, with about 20% advection and 80% reaction. Removal processes in other media are negligible, i.e., 0.02% or less, partly because of the slower reaction rate (longer half-lives) but mainly because so little of the chemical resides in these media.

The Level III calculations show a strong dependence of fate on how the chemical is discharged.

If discharged to air, there is an approximate 20 h residence time for all chemicals, with negligible transfer to other media.

If discharged to water, the overall residence time is longer, ranging from 100 to 300 h with the dominant removal processes being advection and reaction from water, and evaporation (with subsequent advective and reaction loss from the atmosphere).

If discharged to soil the residence times are more variable as a result of the differences in sorption to the soil causing differences in evaporation rate. Reaction in soil and evaporation are the most significant mechanisms. Residence times range from 80 to 1500 h depending on affinity for the soil. The strongly sorbed more alkylated chemicals tend to remain longer in the soil and are thus more subject to degrading reactions in that medium.

In summary, this series of chemicals shows broadly similar partitioning characteristics, but with environmental fate strongly dependent on the medium of discharge. The key processes are advective and reaction loss from air, evaporation from water and soil, and reaction in water and soil.

2.5 REFERENCES

Abdul, A.S., Gibson, T.L., Rai, D.N. (1987) Statistical correlations for predicting the partition coefficient for nonpolar organic contaminants between aquifer organic carbon and water. *Hazard Waste Hazard Material* 4(3), 211.

Abdul, A.S., Gibson, T.L., Rai, D.N. (1990) Use of humic acid solution to remove organic contaminants from hydrogeologic systems. *Environ. Sci. Technol.* 24(3), 328-333.

Abernethy, S., Bobra, A.M., Shiu, W.Y., Wells. P.G., Mackay, D. (1986) Acute lethal toxicity of hydrocarbon to two planktonic crustaceans: the key role of organism-water partitioning. *Aquat. Toxicol.* 8, 163-174.

Abernethy, S., Mackay, D. (1987) A discussion of correlations for narcosis in aquatic species. In: *QSAR in Environmental Toxicology - II*, Kaiser, K.L.E., Ed., pp.1-16, D. Reidel Publ. Co., Dordrecht, Holland.

Abernethy, S., Mackay, D., McCarty, L.S. (1988) "Volume fraction" correlation for narcosis in aquatic organisms: the key role of partitioning. *Environ. Toxicol. Chem.* 7, 469-481.

Almgren, M., Grieser, F., Powell, J.R., Thomas, J.K. (1979) A correlation between the solubility of aromatic hydrocarbons in water and micellar solutions, with their normal boiling points. *J. Chem. Eng. Data* 24, 285-287.

Al-Sahhaf, T.A. (1989) Prediction of the solubility of hydrocarbons in water using UNIFAC. *J. Environ. Sci. Health* A24, 49-56.

Ambrose, D. (1981) Reference value of vapour pressure. The vapour pressures of benzene and hexafluorobenzene. *J. Chem. Thermodyn.* 13, 1161-1167.

Amidon, G.L., Anik, S.T. (1980) Hydrophobicity of polycyclic aromatic compound, thermodynamic partitioning analysis. *J. Phys. Chem.* 84, 970-974.

Amidon, G.L., Anik, S.T. (1981) Application of the surface area approach to the correlation and estimation of aqueous solubility and vapor pressure. Alkyl aromatic hydrocarbons. *J. Chem. Eng. Data* 26, 28-33. (supplementary material)

Amidon, G.L., Williams, N.A. (1982) A solubility equation for non-electrolytes in water. *Intl. J. Pharm.* 11, 249-256.

Anderson, T.A., Beauchamp, J.J., Walton, B.T. (1991) Organic chemicals in the environment. *J. Environ. Qual.* 20, 420-424.

Andrews L.J., Keefer, R.M. (1949) Cation complexed of compounds containing carbon-carbon double bonds. IV. The argentation of aromatic hydrocarbons. *J. Am. Chem. Soc.* 71, 3644-3647.

Andrews, L.J., Keefer, R.M. (1950) Cation complexes of compounds containing carbon-carbon double bonds. VII. Further studies on the argentation of substituted benzenes. *J. Am. Chem. Soc.* 72, 5034-5037.

API (1985) *Literature survey: hydrocarbon solubilities and attenuation mechanisms*. Health and Environmental Sci. Dept. API Publication No. 4414, Am. Petroleum Institute, Washington, D.C.

Aquan-Yuen, M., Mackay, D., Shiu, W.Y. (1979) Solubility of hexane, phenanthrene, chlorobenzene, and p-dichlorobenzene in aqueous electrolyte solutions. *J. Chem. Eng. Data* 24, 30-34.

Arbuckle, W.B. (1983) Estimating activity coefficients for use in calculating environmental parameters. *Environ. Sci. Technol.* 17, 537-542.

Arnold D.S., Plank, C.A., Erickson, E.E., Pike, F.P. (1958) Solubility of benzene in water. *Chem. Eng. Data Ser.* 3, 253.

Ashworth, R.A., Howe, G.B., Mullins, M.E., Rogers, T.N. (1988) Air-water partitioning coefficients of organics in dilute aqueous solutions. *J. Hazard. Materials* 18, 25-36.

Atkinson, R. (1985) Kinetics and mechanisms of the gas phase reaction of hydroxy radicals with organic compounds under atmospheric conditions. *Chem. Rev.* 85, 69-201.

Atkinson, R. (1987) Structure-activity relationship for the estimation of the rate constants for the gas phase reactions of OH radicals with organic compounds. *Int. J. Chem. Kinetics* 19, 799-828.

Bahnick, D.A., Doucette, W.J. (1988) Use of molecular indices to estimate soil sorption coefficient for organic chemicals. *Chemosphere* 17, 1703-1715.

Banerjee, S. (1984) Solubility of organic mixture in water. *Environ. Sci. Technol.* 18, 587-591.

Banerjee S. (1985) Calculation of water solubility of organic compounds with UNIFAC-derived parameters. *Environ. Sci. Technol.* 19, 369-370.

Banerjee, S., Howard, P.H. (1988) Improved estimation of solubility and partitioning through correction of UNIFAC-derived activity coefficients. *Environ. Sci. Technol.* 22, 839-841.

Banerjee, S., Howard, P.H., Lande, S.S. (1990) General structure vapor pressure relationship for organics. *Chemosphere* 21, 1173-1180.

Banerjee, S., Yalkowsky, S.H., Valvani, S.C. (1980) Water solubility and octanol/water partition coefficient of organics. Limitations of solubility-partition coefficient correlation. *Environ. Sci. Technol.* 14, 1227-1229.

Bartholomew, G.W., Pfaender, F.K. (1983) Influence of spatial and temporal variations on organic pollutant biodegradation rates in an estuarine environment. *Appl. Environ. Microbiol.* 45, 103-109.

Battersby, N.S. (1990) A review of biodegradation kinetics in the aquatic environment. *Chemosphere* 21, 1243-1284.

Bodor, N., Gabanyi, Z., Wong, C.-K. (1989) A new method for the estimation of partition coefficient. *J. Am. Chem. Soc.* 111, 3783-3786.

Bohon, R.L., Claussen, W.F. (1951) The solubility of aromatic hydrocarbons in water. *J. Am. Chem. Soc.* 72, 1571-1576.

Booth, H.S., Everson, H.E. (1948) Hydrotropic solubilities: solubilities in 40 per cent sodium xylenesulfonate. *Ind. Eng. Chem.* 40(8), 1491-1493.

Boublik, T., Fried, V., Hala, E. (1973) *The Vapour Pressure of Pure Substances.* Elsevier, Amsterdam.

Boublik, T., Fried, V., Hala, E. (1984) *The Vapour Pressures of Pure Substances.* (second revised edition), Elsevier, Amsterdam.

Brady & Huff (1958) Vapor pressure of benzene over aqueous detergent solutions. *J. Phys. Chem.* 62, 644-649.

Braumann, T., Grimme, L.H. (1981) Determination of hydrophobic parameters for pyriazinone herbicides by liquid-liquid partition and reversed-phase high performance liquid chromatography. *J. Chromatogr.* 206, 7-15.

Brice, K.A., Derwent, R.G. (1978) Emissions inventory for hydrocarbons in United Kingdom. *Atom. Environ.* 12, 2045-2054.

Bridie, A.L., Wolff, C.J.M., Winter, M. (1979) BOD and COD of some petrochemicals. *Water Res.* 13, 627-630.

Brooke, D.N., Dobbs, A.J., Williams, N. (1986) Octanol/water partition coefficients P: measurement, estimation, and interpretation particularly for chemicals with P > 10^6. *Ecotoxicol. Environ. Saf.* 11, 251-260.

Brookman, G.T., Flanagan, M., Kebe, J.O. (1985) *Literature Survey: Hydrocarbon solubilities and attenuation mechanisms.* Prepared for Environmental Affairs Dept. of American Petroleum Institute. as API Publication No. 4414, August, 1985, Washington D.C.

Brown, R.L., Wasik, S.P. (1974) A method of measuring the solubilities of hydrocarbons in aqueous solutions. *J. Res. Nat. Bur. Std.* 78A, 453-460.

Bruggeman, W.A, Van Der Steen, J., Hutzinger, O. (1982). Reversed-phase thin-layer chromatography of polynuclear aromatic hydrocarbons and chlorinated biphenyls. Relationship with hydrophobicity as measured by aqueous solubility and octanol-water partition coefficient. *J. Chromatogr.* 238, 335-346.

Budavari, S., Ed. (1989) *The Merck Index. An Encyclopedia of Chemicals, Drugs and Biologicals.* 11th edition, Merck & Co. Inc., Rahway, New Jersey.

Burkhard, L.P., Kuehl, D.W. (1986) n-octanol/water partition coefficients by reversed phase liquid chromatograph/mass-spectrophotometry for eight tetra-chlorinated planar molecules. *Chemosphere* 15(2), 163-167.

Burkhard, L.P., Kuehl, D.W., Veith, G.D. (1985) Evaluation of reversed-phase liquid chromatography/mass spectrometry for estimation of n-octanol/water partition coefficients organic chemicals. *Chemosphere* 14(10), 1551-1560.

Callahan, M.A., Slimak, M.W., Gabel, N.W., May, I.P., Fowler, C.F., Freed, J.R., Jennings, P., Durfee, R.L., Whitmore, F.C., Maestri, B., Mabey, W.R., Holt, B.R., Gould, C. (1979). *Water-Related Environmental fate of 129 Priority Pollutants*, Vol. I, EPA Report No. 440/4-79-029a. Versar, Inc., Springfield, Virginia.

Campbell, J.R., Luthy, R.G. (1985) Prediction of aromatic solute partition coefficient using the UNIFAC group contribution model. *Environ. Sci. Technol.* 19, 980-985.

Chey, W., Calder, G.V. (1972) Method for determining solubility of slightly soluble organic compounds. *J. Chem. Eng. Data* 17(2), 199-200.

Chin, Y.P., Weber, Jr., W.J., Voice, T.C. (1986) Determination of partition coefficient and aqueous solubilities by reversed phase chromatography - II. Evaluation of partioning and solubility models. *Water Res.* 20, 1443-1350.

Chiou, C.T. (1981) Partition coefficient and water solubility in environmental chemistry. In: *Hazard Assessment of Chemicals, Current Developments*, vol. 1, pp. 117-153. Academic Press, New York.

Chiou, C.T. (1985) Partition coefficients of organic compounds in lipid-water system and correlations with fish bioconcentration factors. *Environ. Sci. Technol.* 19, 57-62.

Chiou, C.T., Block, J.B. (1986) Parameters affecting the partition coefficient of organic compounds in solvents-water and lipid water systems. In: *Partition Coefficient, Determination and Estimation.* Dunn, III, W.J., Block J.H., Pearlman, R.S., Eds., pp. 37-60. Pergamon Press, New York.

Chiou, C., Freed, D., Schmedding, D., Kohnert, R. (1977) Partition coefficient and bioaccumulation of selected organic chemicals. *Environ. Sci. Technol.* 11(5),475-478.

Chiou, C.T., Porter, P.E., Schmedding, D.W. (1983) Partition equilibria of nonionic organic compounds between soil organic matter and water. *Environ. Sci. Technol.* 17, 227-231.

Chiou, C.T., Schmedding, D.W., Manes, M. (1982) Partitioning of organic compounds in octanol-water systems. *Environ. Sci. Technol.* 16, 4-10.

Cline, P.V., Delfino, J.J., Rao, P.S.C. (1991) Partitioning of aromatic constituents into water from gasoline and other complex solvent mixtures. *Environ. Sci. Technol.* 25, 914-920.

Corchnoy, S.B., Atkinson, R. (1990) Kinetics of the gas-phase reactions of OH and NO_3 with 2-carene, 1,8-cineole, p-cymene, and terpinolene. *Environ. Sci. Technol.* 24, 1497-1502.

Coutant, R.W., Keigley, G.W. (1988) An alternative method for gas chromatographic determination of volatile organic compounds in water. *Anal. Chem.* 60, 2436-2537.

D'Amboise, M., Hanai, T. (1982) Hydrophobicity and retention in reverse phase liquid chromatography. *J. Liq. Chromatogr.* 5, 229-244.

Darnall, K.R., Loyld, A.C., Winer, A.M., Pitts, J.N. (1976) Reactivity scale for atmospheric hydrocarbons based on reaction with hydroxy radicals. *Environ. Sci. Technol.* 10, 692-696.

Daubert, T.E., Danner, R.P. (1985) *Data Compilation Tables of Properties of Pure Compounds.* Am. Institute of Chem. Engineers. pp 450.

Dean, J.D., Ed. (1979). *Lange's Handbook of Chemistry.* 12th ed., McGraw-Hill, Inc., New York.

Dean, J.D., Ed. (1985) *Lange's Handbook of Chemistry.* 13th ed. McGraw-Hill, Inc., New York.

Dearden, J.C. (1990) Physico-chemical descriptors. In: *Practical Application of Quantitative Structure-Activity Relationships (QSAR) in Environmental Chemistry and Toxicology. pp.25-29. W. Karcher & J. Devillers Eds. ECSC, EEC, EAEC, Brussels and Luxembourg.*

Dearden, J.C., Bresnen, G.M. (1988) The measurement of partition coefficients. *Quant. Struct.-Act. Relat.* 7, 133-144.

De Bruijn, J., Busser, F., Seinen, W., Hermens, J. (1989) Determination of octanol/water partition coefficient for hydrophobic organic chemicals with the "slow-stirring" method. *Environ. Toxicol. Chem.* 8, 499-512.

De Kock, A.C., Lord, D.A. (1987) A simple procedure for determining octanol-water partition coefficients using reversed phase high performance liquid chromatography (RPHPLC). *Chemosphere* 16(1), 133-142.

Deno, N.C., Berkheimer, H.E. (1960) Phase equilibria molecular transport thermodynamics: activity coefficients as a function of structure and media. *J. Chem. Eng. Data* 5, 1-5.

DeVoe, H., Miller, M.M., Wasik, S.P. (1981) Generator columns and high pressure liquid chromatography for determining aqueous solubilities and octanol-water partition coefficients of hydrophobic substances. *J. Res. Natl. Bur. Std.* 86, 361-.

Dorfman, L.M., Adams, G.E. (1973) *Reactivity of the hydroxy radical in aqueous solution.* NSRD-NDB-46. NTIS COM-73-50623. National Bureau Standards, pp. 51,. Washington, DC.

Doucette, W.J., Andren, A.W. (1988) Estimation of octanol/water partition coefficients: Evaluation of six methods for highly hydrophobic aromatic hydrocarbons. *Chemosphere* 17, 345-359.

Doyle, G.J., Lloyd, A.C., Darnall, K.R., Winer, A.M., Pitts, J.N. Jr. (1975) Gas phase kinetic study of relative rates of reaction of selected aromatic compounds with hydroxy radicals in an environmental chamber. *Environ. Sci. Technol.* 9(3), 237-241.

Dreisbach, R.R. (1955) Physical Properties of Chemical Compounds. *Adv. Chem. Ser.* 15, 134.

Eadsforth, C.V., Moser, P. (1983) Assessments of reversed phase chromatographic methods for determining partition coefficients. *Chemosphere* 12, 1459-1475.

Eastcott, L., Shiu, W.Y., Mackay, D. (1988) Environmentally relevant physical-chemical properties of hydrocarbons: A review of data and development of simple correlations. *Oil & Chem. Pollut.* 4, 191-216.

Eisenreich, S.J., Looney, B.B., Thornton, J,D. (1981) Airborne organic contaminants in the Great Lakes ecosystem. *Environ. Sci. Technol.* 15, 30-38.

Franks, F., Gent, M., Johnson, H.H. (1963) The solubility of benzene in water. *J. Chem. Soc.* 2716-2723.

Freed, V.H., Chiou, C.T., Hague, R. (1977) Chemodynamics: Transport and behavior of chemicals in the environment - A problem in environmental health. *Environ. Health Perspect.* 20, 55.

Freitag, D., Lay, J.P. Korte, F. (1984) Environmental hazard profile-test results as related to structures and translation into the environment. In: *QSAR in Environmental Toxicology*, Kaiser, K.L.E., Ed., D. Reidel Publ. Co., Dordrecht, Netherlands.

Freitag, D., Ballhorn, L., Geyer, H., Korte, F. (1985) Environmental hazard profile of organic chemicals. An experimental method for the assessment of the behaviour of organic chemicals in the ecosphere by means of simple laboratory tests with ^{14}C labelled chemicals. *Chemosphere* 14, 1589-1616.

Fühner, H. (1924) Die Wasserlöslichkeit in homologen Reihen. *Chem. Ber.* 57, 510-515.

Fujita, T., Iwasa, Hansch, C. (1964). A new substituent constant, "pi" derived from partition coefficients. *J. Am. Chem. Soc.* 86, 5175-5180.

Garbarnini, D.R., Lion, L.W. (1985) Evaluation of sorptive partitioning of nonionic pollutants in closed systems by headspace analysis. *Environ. Sci. Technol.* 19, 1122-1128.

Garst, J.E. (1984) Accurate, wide range, automated, high performance liquid chromatographic method for the estimation of octanol/water partition coefficients. II: Equilibration in partition coefficient measurements, additivity of substituent-constants and correlation of biological data. *J. Pharm. Sci.* 73(11), 1623-1629.

Garst, J.E., Wilson, W.C. (1984) Accurate, wide range, automated, high performance liquid chromatographic method for the estimation of octanol/water partition coefficients. II: Effects of chromatographic method and procedure variables on accuracy and reproducibility of the method. *J. Pharm. Sci.* 73(11), 1616-1623.

Geyer, H., Sheenhan, P., Kotzias, D., Freitag, D., Korte, F. (1982) Prediction of ecotoxicological behaviour of chemicals: relationship between physico-chemical properties and bioaccumulation of organic chemicals in the mussel *mytilus edulis*. *Chemosphere* 11, 1121-1134.

Geyer, H.J., Politzki, G., Freitag, D. (1984) Prediction of ecotoxicological behaviour of chemicals: relationship between n-octanol/water partition coefficient and bioaccumulation of organic chemicals by alga chlorella. *Chemosphere* 13(2), 269-284.

Gillham, R.W., Rao, P.S.C. (1990) Transport, distribution and fate of volatile organic compounds in groundwater. In: *Significance and Treatment of Volatile Organic Compounds in Water Supplies*. Ram, N.M., Christman, R.F., Cantor, K.P., Eds., pp. 141-181. Lewis Publishers, Inc., Chelsea, Michigan.

Glew, D.N., Roberson, R.E. (1956) The spectrophotometric determination of the solubility of cumene in water by a kinetic method. *J. Phys. Chem.* 60, 332.

Gmehling, J., Rasmussen, P., Fredenslund, A. (1982) Vapor-liquid equilibria by UNIFAC group contribution. Revision and extension 2. *Ind. Eng. Chem. Proc. Des. Dev.* 21, 118-127.

Green, W.J., Frank, H.S. (1979) The state of dissolved benzene in aqueous solution. *J. Soln. Chem.* 8(3), 187-196.

Gross, P.M., Saylor, J.H. (1931) The solubilities of certain slightly soluble organic compounds in water. *J. Am. Chem. Soc.* 53, 1744-1751.

Güesten, H., Filby, W.G., Schoop, S. (1981) Prediction of hydroxy radical reaction rates with organic compounds in the gas phase. *Atmos. Environ.* 15, 1763-1765.

Halfon, E., Reggiani, M.G. (1986) On ranking chemicals for environmental hazard. *Environ. Sci. Technol.* 20, 1173-1179.

Haky, J.E., Young, A.M. (1984) Evaluation of a simple HPLC correlation method for the estimation of the octanol-water partition coefficients of organic compounds. *J. Liq. Chromatogr.* 7, 675-689.

Hafkenscheid, T.L., Tomlinson, E. (1983a) Isocratic chromatographic retention data for estimating aqueous solubilities of acidic, basic and neutral drugs. *Intl. J. Pharm.* 16, 1-21.

Hafkenscheid, T.L., Tomlinson, E. (1983b) Correlations between alkane/water and octan-1-ol/water distribution coefficients and isocratic reversed-phase liquid chromatographic capacity factor of acids, bases and neutrals. *Intl. J. Pharm.* 16, 225-240.

Hammers, W.E., Meurs, G.J., De Ligny, C.L. (1982) Correlations between liquid chromatographic capacity ratio data on Lichrosorb RP-18 and partition coefficients in the octanol-water system. *J. Chromatog.* 247, 1-13.

Hanai, T., Tran, C., Hubert, J. (1981) An approach to the prediction of retention times in liquid chromatography. *J. HRC & CC* 4, 454-460.

Hansch, C., Leo, A. (1979) *Substituent Constants for Correlation Analysis in Chemistry and Biology.* Wiley, New York.

Hansch, C., Leo, A. (1985) Medichem Project. Pomona College, Claremont, California.

Hansch, C., Leo, A., Nickaitani, D. (1972) On the additive - constitutive character of partition coefficients. *J. Org. Chem.* 37, 3090-3092.

Hansch, C., Quinlan, J.E., Lawrance, G.L. (1968) The linear free-energy relationship between partition coefficients and the aqueous solubility of organic liquids. *J. Am. Chem. Soc.* 33, 345-350.

Hansen, D.A., Atkinson, R., Pitts, J.N. Jr. (1975) Rate constants for the reaction of OH radicals with a series of aromatic hydrocarbons. *J. Phys. Chem.* 79(17), 1763-1766.

Haque, R., Falco, J., Cohen, S., Riordan, C. (1980) Role of transport and fate studies in the exposure, assessment and screening of toxic chemicals. In: *Dynamics, Exposure and Hazard Assessment of Toxic Chemicals.* Haque, R., Ed., pp. 47-67, Ann Arbor Sci. Publ., Ann Arbor, Michigan.

Harnisch, M., Möckel, H.J., Schulze, G. (1983) Relationship between log P_{OW} shake-flask values and capacity factors derived from reversed-phase HPLC for n-alkylbenzenes and some OECD reference substances. *J. Chromatogr.* 282, 315-332.

Hayashi, M., Sasaki, T. (1956) Measurements of solubilities of sparingly soluble liquids in water and aqueous detergent solutions using nonionic surfactant. *Bull. Chem. Soc, Jpn.* 29, 857.

Hendry, D.G., Mill, T., Piszkiewicz, L., Howard, J.A., Eigenman, H.K. (1974) A critical review of H-atom transfer in the liquid phase: chlorine atom, alkyl, trichloromethyl, alkoxy and alkylperoxy radicals. *J. Phys. Chem. Ref. Data* 3, 944-978.

Herman, D.C., Mayfield, C.I., Innis, W.E. (1991) The relationship between toxicity and bioconcentration of volatile aromatic hydrocarbons by the *alga selenastrum capricornutum. Chemosphere* 22(7), 665-676.

Hermann, R.B. (1972) Theory of hydrophobic bonding. II. The correlation of hydrocarbon solubility in water with solvent cavity surface area. *J. Phys. Chem.* 76, 2754-2758.

Hine, J., Mookerjee, P.K. (1975) The intrinsic hydrophilic character of organic compounds. Correlations in terms of structural contributions. *J. Org. Chem.* 40(3), 292-298.

Hodson, J., Williams, N.A. (1988) The estimation of the adsorption coefficient (K_{OC}) for soils by high performance liquid chromatography. *Chemosphere* 17, 67-77.

Horowitz, A., Shelton, D.R., Cornell, C.P., Tiedje, J.M. (1982) Anaerobic degradation of aromatic compounds in sediment and digested sludge. *Dev. Ind. Microbiol.* 23, 435-444.

Horvath, A.L. (1982) *Halogenated Hydrocarbons, Solubility-Miscibility with Water.* Marcel Dekker, Inc., New York, N.Y.

Howard, P.H., Ed. (1989) *Handbook of Fate and Exposure Data for Organic Chemicals. Vol.I - Large Production and Priority Pollutants.* Lewis Publishers, Chelsea, Michigan.

Howard, P.H., Ed., (1990) *Handbook of Fate and Exposure Data for Organic Chemicals. Vol. II - Solvents.* Lewis Publ., Inc., Chelsea, Michigan.

Howard, P.H., Boethling, R.S., Jarvis, W.F., Meylan, W.M., Michalenko, E.M., Eds. (1991) *Handbook of Environmental Degradation Rates.* Lewis Publishers, Inc., Chelsea, Michigan.

Hustert, K., Mansour, M., Korte, F. (1981) The EPA Test- a method to determine the photochemical degradation of organic compounds in aqueous systems. *Chemosphere* 10, 995-998.

Hutchinson, T.C., Hellebust, J.A., Tam, D., Mackay, D., Mascarenhas, R.A., Shiu, W.Y. (1980) The correlation of the toxicity to algae of hydrocarbons and halogenated hydrocarbons with their physical-chemical properties. In: *Hydrocarbons and Halogenated Hydrocarbons in the Aquatic Envrionment.* Afghan, B.K., Mackay, D., Eds., pp. 577-586. Plenum Press, New York.

Irmann, F. (1965) Eine einfache korrelation zwischen wasserlöslichkeit und struktur vor kohlenwasserstoffen und hologen kohlen wasserstoffen. *Chem. Eng. Tech.* 37(8), 789-798.

Isnard, P., Lambert, S. (1988) Estimating bioconcentration factors from octanol-water partition coeffient and aqueous solubility. *Chemosphere* 17, 21-34.

Isnard, P., Lambert, S. (1989) Aqueous solubility/n-octanol water partition coefficient correlations. *Chemosphere* 18, 1837-1853.

IUPAC Solubility Data Series (1989a) Vol. 37: Hydrocarbons (C_5-C_7) with Water and Seawater. Shaw, D.G., Ed., Pergamon Press, Oxford, England.

IUPAC Solubility Data Series (1989b) *Vol. 38: Hydrocarbons (C_8-C_{36}) with Water and Seawater.* Shaw, D.G., Ed., Pergamon Press, Oxford, England.

Iwasa, J., Fujita, T., Hansch, C. (1965) Substituent constants for aliphatic functions obtained from partition coefficients. *J. Med. Chem.* 8, 150-153.

Jafvert, C.T. (1991) Sediment- and saturated-soil-associated reactions involving an anionic surfactant (dodecylsulfate). 2. Partition of PAH compounds amoung phases. *Environ. Sci. Technol.* 25, 1039-1045.

Jamison, V.W., Raymond, R.L., Hudson, J.O. (1976) Biodegradation of high octane gasoline. Proceedings of The Third International Biodegradation Symposium. J.M. Sharpley, A.M. Kaplan Eds., P.187-196.

Jury, W.A., Russo, D., Streile, G., El Abd, H. (1990) Evaluation of volatilization by organic chemicals residing below the soil surface. *Water Resources Res.* 26, 13-26.

Kabadi, V.N., Danner, R.P. (1979) Nomograph solves for solubilties of hydrocarbons in water. *Hydrocarbon Processing* 58, 245-246.

Kamlet, M.J., Doherty, R.M., Abraham, M.H., Carr, P.W., Doherty, R.F., Taft, R.W. (1987) Linear solvation energy relationships. Important differences between aqueous solubility relationships for aliphatic and aromatic solutes. *J. Phys. Chem.* 91, 1996-

Kamlet, M.J., Doherty, R.M., Carr, P.W., Mackay, D., Abraham, M.H., Taft, R.W. (1988) Linear solvation energy relationships. 44. Parameter estimation rules that allow accurate prediction of octanol/water partition coefficients and other solubility and toxicity properties of polychlorinated biphenyls and polycyclic aromatic hydrocarbons. *Environ. Sci. Technol.* 22(5), 503-509.

Kappeler, T., Wuhrmann, K. (1978) Microbial degradation of water soluble fraction of gas oil. *Water Res.* 12, 327-333.

Karickhoff, S.W. (1981) Semiempirical estimation of sorption of hydrophobic pollutants on natural sediments and soils. *Chemosphere* 10, 833-846.

Karickhoff, S.W., Brown, D.S., Scott, T.A. (1979) Sorption of hydrophobic pollutants on natural water sediments. *Water Res.* 13, 241-248.

Keeley, D.F., Hoffpauir, M.A., Meriweather, J.R. (1988) Solubility of aromatic hydrocarbons in water and sodium chloride solutions of different ionic strengths: Benzene and Toluene. *J. Chem. Eng. Data* 33, 87-89.

Kenaga, E.E. (1980) Predicted bioconcentration factors and soil sorption coefficients of pesticides and other chemicals. *Ecotoxicol. Environ. Safety* 4, 26-38.

Kenaga, E.E., Goring, C.A.I. (1980) Relationship between water solubility, soil sorption, octanol-water partitioning and concentration of chemicals in biota. In: *Aquatic Toxicology.* Eaton, J.G., Parrish, P.R., Hendricks, A.C. Eds., Amer. Soc. for Testing and Materials, STP 707, pp. 78-115.

Kier, L.B., Hall, L.H. (1976) Molar properties and molecular connectivity. In: *Molecular Connectivity in Chemistry and Drug Design.* Medicinal Chem. Vol.14, pp. 123-167. Academic Press, New York.

Kier, L.B., Hall, L.H. (1986) *Molecular Connectivity in Structure-Activity Analysis.* Wiley, New York.

Kitano, M. (1978) Biodegradation and bioaccumulation of chemical substances. OECD Tokyo Meeting. Reference Book TSU-No.3.

Klein, W., Ködel, W., Weiß, M., Poremski, H.J. (1988) Updating the OECD test guideline 107 "partition coefficient N-octanol/water" : OECD laboratory intercomparison test on HPLC method. *Chemosphere* 17(2), 361-386.

Klevens, H.B. (1950) Solubilization of polycyclic hydrocarbons. *J. Phys. Colloid Chem.* 54, 283-298.

Koch, R. (1983) Molecular connectivity index for assessing ecotoxicological behaviour of organic compounds. *Toxicol. Environ. Chem.* 6, 87-96.

Könemann, W.H. (1979) *Quantitative Structure Activity Relationship for Kinetics and Toxicity of Aquatic Pollutants and their Mixtures in Fish.* University Utrecht, Netherlands.

Könemann, W.H. (1981) Quantitative structure-activity relationships in fish toxicity studies. Part 1: Relationship for 50 industrial pollutants. *Toxicology* 19, 209-221.

Korn, S. et al. (1977) The uptake, distribution and depuration of carbon-14 labelled benzene and carbon-14 labelled toluene in pacific herring. *Fish Bull. Natl. Marine Fish Ser.* 75, 633-636.

Krasnoshchekova, R.Ya. (1975) Solubility of alkylbenzenes in fresh and salt waters. *Vodnye. Resursy.* 2, 170-173.

Krzyzanowska, T., Szeliga, J. (1978) Determination of the solubility of individual hydrocarbons. *Nafta (Katowice)* 34(12), 413-417.

Kuhn, E.P., Coldberg, P.J., Schnoor, J.L., Waner, O., Zehnder, A.J.B. Schwarzenbach, R.P. (1985) Microbial transformation of substituted benzenes during infiltration of river water to ground water: laboratory column studies. *Environ. Sci. Technol.* 19, 961-968.

Lande, S.S., Banerjee, S. (1981) Predicting aqueous solubility of organic nonelectrolytes from molar volume. *Chemosphere* 10, 751-759.

Lande, S.S., Hagen, D.F., Seaver, A.E. (1985) Computation of total molecular surface area from gas phase ion mobility data and its correlation with aqueous solubilities of hydrocarbons. *Environ. Toxicol. Chem.* 4, 325-334.

Leahy, D.E. (1986) Intrinsic molecular volume as a measure of the cavity term in linear solvation energy relationships: octanol-water partition coefficients and aqueous solubilities. *J. Pharm. Sci.* 75, 629-636.

Lee, R.F. (1977) Oil Spill Conference, Am. Petroleum Institute pp.611-616.

Lee, J.F., Crum, J.R., Boyd, S.A. (1989) Enhanced retention of organic contaminants by soils exchanged with organic cations. *Environ. Sci. Technol.* 23, 1365-1372.

Lee, R.F., Ryan, C. (1976) Biodegradation of petroleum hydrocarbons by marine microbes. In: *Proceedings of the third International Biodegradation Symposium,* PP.119-125.

Lee, R.F., Ryan, C. (1979) Microbial degradation of organochlorine compounds in estuarine waters and sediments. In: *Procceedings of the Workshop of Microbial Degradation of Pollutants in Marine Environments.* EPA-600/9-79-012. Washington, D.C.

Leegwater, D.C. (1989) QSAR-analysis of acute toxicity of industrial pollutants to the guppy using molecular connectivity indices. *Aqua. Toxicol.* 15, 157-168.

Leighton, D.T., Calo, J.M. (1981) Distribution coefficients of chlorinated hydrocarbons in dilute air-water systems for groundwater contamination applications. *J. Chem. Eng. Data* 26, 381-385.

Leinonen, P.J., Mackay, D. (1973) The multicomponent solubility of hydrocarbons in water. *Can. J. Chem. Eng.* 51, 230-233.

Leo, A. (1985) Medchem. Proj. Issue No.26, Pomona College, Claremont, CA.

Leo, A., Hansch, C., Elkins, D. (1971) Partition coefficients and their uses. *Chemical Reviews* 71, 525-616.

Lindberg, A.B. (1956) Physicochime des solutions. - Sur une relation simple entre le volume moléculaire et la solubilité dans l'eau des hydrocarbures et dérivé halogénés. *C.R. Acad. Sci.* 243, 2057-2060.

LOGP and Related Computerized Data Base, Pomona College Med-Chem. Project, Pomona College, Claremont, CA.. Technical Data Base (TDS) Inc.

Lyman, W.J. (1982) Adsorption coefficients for soil and sediments. chapter 4, In: *Handbook of Chemical Property Estimation Methods,* W.J. Lyman, W.F. Reehl, D.H. Rosenblatt, Eds., McGraw-Hill, New York.

Lyman, W.J., Reehl, W.F., Rosenblatt, D.H. (1982) *Handbook of Chemical Property Estimation Methods,* McGraw-Hill, New York.

Mabey, W., Smith, , J.H., Podoll, R.T., Johnson, H.L., Mill, T., Chou, T.W., Gates, J., Waight-Partridge, I., Vanderberg, D. (1982) *Aquatic Fate Process for Organic Priority Pollutants.* EPA Report No. 440/4-81-014.

Mackay, D. (1981) Environmental and laboratory rates of volatilization of toxic chemicals from water. In: *Hazardous Assessment of Chemicals, Current Development.* Volume 1, Academic Press.

Mackay, D. (1982) Correlation of bioconcentration factors. *Environ. Sci. Technol.* 16, 274-278.

Mackay, D. (1988) The chemistry and modelling of soil contamination with petroleum. In: *Soils Contaminated by Petroleum: Environmental and Public Health Effects.* Calabrese, E.J., Kostecki, P.T., Editors, John Wiley & Sons, New York.

Mackay, D., Bobra, A.M., Chan, D.W., Shiu, W.Y. (1982) Vapor pressure correlation for low-volatility environmental chemicals. *Environ. Sci. Technol.* 16, 645-649.

Mackay, D., Bobra, A., Shiu, W.Y., Yalkowsky, S.H. (1980) Relationships between aqueous solubility and octanol-water partition coefficient. *Chemosphere* 9, 701-711.

Mackay, D., Leinonen, P.J. (1975) Rate of evaporation of low-solubility contaminants from water to atmosphere. *Environ. Sci. Technol.* 7, 1178-1180.

Mackay, D., Paterson, S., Chung, B., Neely, W.B. (1985) Evaluation of the environmental behavior of chemicals with a level III fugacity model. *Chemosphere* 14(3/4), 335-374.

Mackay, D., Shiu, W.Y. (1975) The aqueous solubility and air-water exchange characteristics of hydrocarbons under environmental conditions. In: *Chemistry and Physics of Aqueous Gas Solutions.* Adams, W.A., Greer, G., Desnoyers, J.E., Atkinson, G., Kell, K.B., Oldham, K.B., Walkey, J., Eds., pp. 93-110, Electrochem. Soc. , Inc., Princeton, N.J.

Mackay, D., Shiu, W.Y. (1981) A critical review of Henry's law constants for chemicals of environmental interest. *J. Phys. Chem. Ref. Data* 10, 1175-1199.

Mackay, D., Shiu, W.Y. (1990) Physical-chemical properties and fate of volatile organic compounds: an application of the fugacity approach. In: *Significance and Treatment of Volatile Organic Compounds in Water Supplies.* Ram, N.M., Christman, R.F., Cantor, K.P., Eds., pp.183-203, Lewis Publishers, Inc., Chelsea, Michigan.

Mackay, D., Shiu, W.Y., Sutherland, R.P. (1979) Determination of air-water Henry's law constants for hydrophobic pollutants. *Environ. Sci. Technol.* 13, 333-337.

Mackay, D., Shiu, W.Y., Wolkoff, A.W. (1975) Gas chromatographic determination of low concentrations of hydrocarbons in water by vapor phase extraction. *Water Quality Parameters. ASTM STP* 573, pp. 251-258, Am. Soc. Testing and Materials, Philadelphia.

Mackay, D., Wolkoff, A.W. (1973) Rate of evaporation of low-solubility contaminants from water bodies to atmosphere. *Environ. Sci. Technol.* 7, 611-614.

Marion, C.V., Malaney, G.W. (1964) Ability of activated sludge microorganisms to oxidize aromatic organic compounds. In: *Proc. Ind. Waste Conf., Eng. Bull.,* Purdue Univ., Eng. Ext. Ser., pp. 297-308.

Matter-Müller, C., Gujer, W., Giger, W. (1981) Transfer of volatile substances from water to the atmosphere. *Water Res.* 15, 1271-1279.

Massaldi, H.A., King, C.J. (1973) Simple technique to determine solubilities of sparingly soluble organics: solubility and activity coefficients of d-limonene, butylbenzene, and n-hexyl acetate in water and sucrose solutions. *J. Chem. Eng. Data* 18, 393-397.

May, W.E. (1980) The solubility behaviour of polycyclic aromatic hydrocarbons in aqueous systems. *Petroleum Mar. Environ.: Adv. Chem. Ser. 185*, chapter 7, Am. Chem. Soc., Washington DC.

May, W.E., Wasik, S.P., Freeman, D.H. (1978) Determining of the solubility behavior of some polycyclic aromatic hydrocarbons in water. *Anal. Chem.* 50(7), 997-1000.

May, W.E., Wasik, S.P., Miller, M.M., Tewari Y.B., Brown-Thomas, J.M., Goldberg, R.N. (1983) Solution thermodynamics of some slightly soluble hydrocarbons in water. *J. Chem. Eng. Data* 28, 197-200.

McAuliffe, C. (1963) Solubility in water of C_1 - C_9 hydrocarbons. *Nature (London)* 200, 1092-1093.

McAuliffe, C. (1966) Solubility in water of paraffin, cycloparaffin, olefin, acetylene, cycloolefin and aromatic hydrocarbons. *J. Phys. Chem.* 76, 1267-1275.

McCall, J.M. (1975) Liquid-liquid partition coefficient by high-pressure liquid chromatography. *J. Med. Chem.* 18, 549-552.

McDevit, W.F., Long, F.A. (1952) The activity coefficient of benzene in aqueous salt solutions. *J. Am. Chem. Soc. 74, 1773-1777.*

McDuffie, B. (1981) Estimation of octanol/water partition coefficients for organic pollutants using reversed phase HPLC. *Chemosphere* 10, 73-83.

McGowan, J.C., Mellors, A. (1986) *Molecular Volumes in Chemistry and Biology-applications including Partitioning and Toxicity.* Ellis Horwood Limited, Chichester, England.

Means, J.C., Wood, S.C., Hassett, J.J., Banwart, W.L. (1980) Sorption of polynuclear aromatic hydrocarbons by sediments and soils. *Environ. Sci. Technol.* 14, 1524-1631.

The Merck Index. An Encyclopedia of Chemicals, Drugs and Biologicals (1983). Windholz, M., Editor, Merck and Co., Inc., Rahway, N.J., U.S.A., 10th ed.

The Merck Index. An Encyclopedia of Chemicals, Drugs and Biologicals. (1989). Budavari, S., Editor, Merck & Co., Inc., Rahway, N.J., 11th ed.

Mill, T. (1982) Hydrolysis and oxidation processes in the environment. *Environ. Toxicol. Chem.* 1, 135-141.

Mill, T., Hendry, D.G., Richardson, H. (1980) Free radical oxidants in natural waters. *Science* 207, 886-887.

Miller, M.M., Ghodbane, S., Wasik, S.P., Tewari, Y.B., Martire, D.E. (1984) Aqueous solubilities, octanol/water partition coefficients and entropies of melting of chlorinated benzenes and biphenyls. *J. Chem. Eng. Data* 29, 184-190.

Miller, M.M., Wasik, S.P., Huang, G.L., Shiu, W.Y., Mackay, D. (1985) Relationships between octanol-water partition coefficient and aqueous solubility. *Environ. Sci. Technol.* 19, 522-529.

Mills, W.B., Dean, J.D., Porcella, D.B., Gherini, S.A., Hudson, R.J.M., Frick, W.E., Rupp, G.L., Bowie, G.L. (1982) *Water Quality Assessment: A Screening Procedure for Toxic and Conventional Pollutants.* Part 1, U.S. EPA, EPA-600/6-82-004a.

Miyake, K., Terada, H. (1982) Determination of partition coefficients of very hydrophobic compounds by high performance liquid chromatography on glyceryl-coated controlled-pore glass. *J. Chromatogr.* 240, 9-20.

Morrison, T.J., Billett, F. (1952) The salting out of non-electrolytes. Part II. The effect of variation in non-electrolyte. *J. Chem. Soc.* 3819-3822.

Munz, C., Robert, P.V. (1989) Gas- and liquid-phase mass transfer resistances of organic compounds during mechanical surface aeration. *Water Res.* 23, 589-601.

Nahum, A., Horvath, C. (1980) Evaluation of octanol-water partition coefficients by using high-performance liquid chromatography. *J. Chromatogr.* 192, 315-322.

NAS (1980) *The Alkyl Benzenes.* page I-1 to II-51, National Academy of Science, Washington, D.C.

Nathwani, J.S., Philip, C.R. (1977) Absorption-desorption of selected hydrocarbons in crude oils on soils. *Chemosphere* 6, 157-162.

Neely, W.B. (1980) A method for selecting the most appropriate environmental experiments on a new chemical. In: *Dynamics, Exposure and Hazard Assessment of Toxic Chemicals.* Haque, R., Ed., pp. 287-298, Ann Arbor Sci. Publ., Ann Arbor, Michigan.

Nirmalakhandan, N.N., Speece, R.E. (1988a) Prediction of aqueous solubility of organic chemicals based on molecular structure. *Environ. Sci. Technol.* 22, 328-338.

Nirmalakhandan, N.N., Speece, R.E. (1988b) QSAR model for predicting Henry's law constant. *Environ. Sci. Technol.* 22, 1349-1357.

Nunes, P., Benville, P.E., Jr. (1979) Uptake and depuration of petroleum hydrocarbons in the Manila clams, Tapes semidecussata Reeve. *Bull Environ. Contam. Toxicol.* 21, 719-724.

Ogata, M., Miyake, Y. (1978) Disappearance of aromatic hydrocarbons and organic sulfur compounds from fish flesh reared in crude oil suspension. *Water Res.* 12, 1041-1044.

Ogata, M., Fujisawa, K., Ogino, Y., Mano, E. (1984) Partition coefficients as a measure of bioconcentration potential of crude oil compounds in fish and shellfish. *Bull. Environ. Contam. Toxicol.* 33, 561-567.

Ohta, T., Ohyama, T. (1985) *A set of rate constants for the reactions of hydroxy radicals with aromatic hydrocarbons. Bull. Chem. Soc. Jpn.* 58, 3029.

Olsen, R.L., Davis, A. (1990) Predicting the fate and transport of organic compounds in groundwater. *Hazard. Mat. Control* 3, 40-64.

Owens, J.W., Wasik, S.P., DeVoe, H. (1986) Aqueous solubilities and ethalpies of solution of n-alkybenzenes. *J. Chem. Eng. Data* 31, 47-51.

Pankow, J.F. (1986) Magnitude of artifacts caused by bubbles and head-space in the determination of volatile compounds in water. *Anal. Chem.* 58, 1822-1826.

Pankow, J.F. (1990) Minimization of volatilization losses during sampling and analysis of volatile organic compounds in water. In: *Significance, Treatment of Volatile Organic Compounds in Water Supplies.* Ram, N.M., Christman, R.F., Cantor, K.P., Eds., pp. 73-86, Lewis Publishers, Inc., Chelsea, Michigan.

Pankow, J.F., Rosen, M.E. (1988) Determination of volatile compounds in water by purging directly to a capillary column with whole column cryotrapping. *Environ. Sci. Technol.* 22, 398-405.

Pankow, J.F., Isabelle, L.M., Asher, W.E. (1984) Trace organic compounds in rain. 1. Sample design and analysis by adsorption/thermal desorption (ATD). *Environ. Sci. Technol.* 18, 310-318.

Pavlou, S.P. (1987) The use of the equilibrium partitioning approach in determining safe levels of contaminants in marine sediments. In: *Fate and Effects of Sediment-Bound Chemicals in Aqueous Systems.* K.L. Dickson, A.W. Maki, W.A. Brungs, Eds., pp.395-412, Pergamon Press, New York.

Pearlman, R.S. (1986) Molecular surface area and volume: their calculation and use in predicting solubilities and free energies of desolvation. In: *Partition Coefficient, Determination and Estimation.* Dunn, III, W.J., Block, J.H., Pearlman, R.S., Eds., pp. 3-20. Pergamon Press, New York.

Perry, R.A., Atkinson, R., Pitts, J.N. (1977) Kinetics and mechanisms of the gas phase reaction of the hydroxy radicals with aromatic hydrocarbons over temperature range 296-473°K. *J. Phys. Chem.* 81, 296-304.

Platford, R.F. (1979) Glyceryl trioleate-water partition coefficients for three simple organic compounds. *Bull. Environ. Contam. Toxicol.* 21, 68.

Platford, R.F. (1983) The octanol-water partitioning of some hydrophobic and hydrophilic compounds. *Chemosphere* 12(7/8), 1107-1111.

Polak, J., Lu, B.C.Y. (1973) Mutual solubilities of hydrocarbons and water at 0° and 25°C. *Can. J. Chem.* 51, 4018-4023.

Politzki, G.R., Bieniek, D., Lahaniatis, E.S., Sheunert, I., Klein, W., Korte, F. (1982) Determination of vapour pressures of nine organic chemicals on silicagel. *Chemosphere* 11, 1217-1229.

Prausnitz, J.M. (1969) *Molecular Thermodynamic of Fluid-Phase Equilibria.* Prentice Hall, Englewood Cliffs, N.J.

Price, L.C. (1976) Aqueous solubility of petroleum as applied to its origin and primary migration. *Am. Assoc. Petrol. Geol. Bull.* 60, 213-244.

Pussemier, L., Szabo, G., Bulman, R.A. (1990) Prediction of the soil adsorption coefficient K_{OC} for aromatic pollutants. *Chemosphere* 21, 1199-1212.

Radding, S.B., Liu, D.H., Johnson, H.L., Mill, T. (1977) *Review of the Environmental Fate of Selected Chemicals.* U.S. Environmental Protection Agency Report No. EPA-560/5-77-003.

Rappaport, R.A., Eisenreich, S. (1984) Chromatographic determination of octanol-water partition coefficients (K_{OW}'s) for 58 PCB congeners. *Environ. Sci. Technol.* 18, 163-170.

Reid, R.C., Prausnitz, J.M., Sherwood, T.K. (1977) *The Properties of Gases and Liquids,* 3rd ed. McGraw Hill, New York.

Reid, R.C., Prausnitz, J.M., Poling, B.E. (1987) *The Properties of Gases and Liquids.* 4th ed., McGraw-Hill, New York.

Rekker, R.F. (1977) *The Hydrophobic Fragmental Constants. Its Derivation and Application, a Means of Characterizing Membrane Systems.* Elsevier Sci. Publ. Co., Oxford, England.

Riddick, J.A. et al. (1986) *Organic Solvents.* Wiley Interscience, Ney York.

Rogers, K.S., Cammarata, A. (1969) Superdelocalizability and charge density. A correlation with partition coefficients. *J. Med. Chem.* 12, 692-693.

Rossi, S.S., Thomas, W.H. (1981) Solubility behavior of three aromatic hydrocarbons in distilled water and natural seawater. *Environ. Sci. Technol.* 15, 715-716.

Ryan, J.A., Bell, R.M., Davidson, J.M., O'Connor, G.A. (1988) Plant uptake of non-ionic organic chemicals from soils. *Chemosphere* 17, 2299-2323.

Sabljic, A. (1984) Predictions of the nature and strength of soil sorption of organic pollutants by molecular topology. *J. Agric. Food Chem.* 32, 243-246.

Sabljic, A. (1987a) On the prediction of soil sorption coefficients of organic pollutants from molecular structure: Application of molecular topology model. *Environ. Sci. Technol.* 27, 358-366.

Sabljic, A. (1987b) Nonempirical modeling of environmental distribution and toxicity of major organic pollutants. In: *QSAR in Environmental Toxicology - II.* Kaiser, K.L.E., Ed., pp. 309-332, D. Reidel Publ. Co., Dordrecht, Netherlands.

Sabljic, A., Protic, M. (1982) Relationship between molecular connectivity indices and soil sorption coefficients of polycyclic aromatic hydrocarbons. *Bull. Environ. Contam. Toxicol.* 28, 162-165.

Sada, E., Kito, S., Ito, Y. (1975) Solubility of toluene in aqueous salt solutions. *J. Chem. Eng. Data* 20(4), 373-375.

Sanemasa, I., Araki, M., Deguchi, T., Nagai, H. (1981) Solubilities of benzene and alkylbenzenes in water. Methods for obtaining aqueous solutions saturated with vapors in equilibrium with organic liquids. *Chem. Lett* 2, 255-258.

Sanemasa, I., Araki, M., Deguchi, T., Nagai, H. (1982) *Bull. Chem. Soc. Jpn.* 53, 1054-1062.

Sanemasa, I., Arakawa, S., Araki, M., Deguchi, T. (1984) The effects of salts on the solubility of benzene, toluene, ethylbenzene and propylbenzene in water. *Bull. Chem. Soc. Jpn.* 57, 1359-1544.

Sangster, J. (1989) Octanol-water partition coefficients of simple organic compounds. *J. Phys. Chem. Ref. Data* 18, 1111-1230.

Schantz, M.M., Martire, D.E. (1987) Determination of hydrocarbon-water partition coefficients from chromatographic data and based on solution thermodynamics and theory. *J. Chromatogr.* 391, 35- .

Schwarz, F.P. (1977) Determination of temperature dependence of solubilities of polycyclic aromatic hydrocarbons in aqueous solutions by a fluorescence method. *J. Chem. Eng. Data* 22, 273-277.

Schwarz, F.P. (1980) Measurement of the solubilities of slightly soluble organic liquids in water by elution chromatography. *Anal. Chem.* 52, 10-15.

Schwarz, F.P., Miller, J. (1980) Measurement of the solubilities of slightly soluble organic liquids in water by elution chromatography. *Anal. Chem.* 52, 2161-2164.

Schwarzenbach, R.P., Westall, J. (1981) Transport of nonpolar compounds from surface water to groundwater. Laboratory sorption studies. *Environ. Sci. Technol.* 11, 1360-1367.

Seip, H.M., Alstad, J., Carlberg, G.E., Martinsen, K., Skaane, P. (1986) Measurement of mobility of organic compounds in soils. *Sci. Total Environ.* 50, 87-101.

Shen, T.T. (1982) Estimation of organic compound emissions from water lagoons. *J. Pollut. Control Assoc.* 32, 79-82.

Shiu, W.Y., Maijanen, A., Ng, A.L.Y., Mackay, D. (1988) Preparation of aqueous solutions of sparingly soluble organic substances: II. Multicomponent systems - hydrocarbon mixtures and petroleum products. *Environ. Toxicol. Chem.* 7, 125-137.

Shiu, W.Y., Ma, K.C., Mackay, D., Seiber, J.N., Wauchope, R.D.(1990) Solubilities of Pesticides in Water. Part I: Environmental Physical Chemistry and Part II: Data Compilation. *Reviews of Environ. Contam. Toxicol.* 116, 1-187.

Smith, J.H., Mabey, W.R., Bahonos, N., Holt, B.R., Lee, S.S., Chou, T.W., Bomberger, D.C., Mill, T. (1978) *Environmental Pathways of Selected Chemicals in Freshwater Systems: Part II. Laboratory Studies.* Interagency Energy-environment Research and Development Program Report. EPA-600/7-78-074. Environmental Research Laboratory Office of Research and Development. U.S. Environmental Protection Agency, Athens, Georgia 30605, p. 304.

Stearns, R.S., Oppenheimer, H., Simon, E., Harkins, W.D. (1947) Solubilization by solutions of long chain colloidal electrolytes. *J. Chem. Phys.* 15, 496-507.

Stephen, H., Stephen, Y. (1963) *Solubilities of Inorganic and Organic Compounds.* Vol. 1 & 2, Pergamon Press, Oxford.

Stephenson, R.M., Malanowski, S. (1987) *Handbook of the Thermodynamic of Organic Compounds.* Elsevier Science publishing Co. Inc., New York, N.Y.

Suntio, L.R., Shiu, W.Y., Mackay, D. (1988) A review of the nature and properties of chemicals present in pulp mill effluents. *Chemosphere* 17, 1249-1290.

Sutton. C., Calder, J.A. (1975) Solubility of alkybenzenes in distilled water and seawater at 25 °C. *J. Chem Eng. Data* 20, 320-322.

Swann, R.L., Laskowski, D.A., McCall, P.J., Vender Kuy, K., Dishburger, J.J. (1983) A rapid method for the estimation of the environmental parameters octanol/water partition coefficient, soil sorption constant, water to air ratio, and water solubility. *Res. Rev.* 85, 17-28.

Swindoll, C.M., Aelion, C.M., Pfaender, F.K. (1987) Inorganic and organic amendment effects of biodegradation of organic pollutants by ground water microorganisms. Am. Soc. Microbiol. Abst., 87th Annual Meeting, pp.298, Atlanta, GA.

Szabo, G., Prosser, S.L., Bulman, R.A. (1990a) Prediction of the adsorption coefficient (K_{OC}) for soil by a chemically immobilized humic acid column using RP-HPLC. *Chemosphere* 21, 729-740.

Szabo, G., Prosser, S.L., Bulman, R.A. (1990b) Determination of the adsorption coefficient (K_{OC}) of some aromatics for soil by RP-HPLC on two immobilized humic acid phases. *Chemosphere* 21, 777-788.

Tabak, H.H., Quave, S.A., Moshni, C.I., Barth, E.F. (1981) Biodegradability studies with organic priority pollutant compounds. *J. Water Pollut. Control Fed.* 53, 1503-1518.

Taft, R.W., Abrahm, M.H., Famini, G.R., Doherty, R.M., Abboud, J.L., Kamlet, M.J. (1985) Solubility properties in polymers and biological media. 5: An analysis of physicochemical properties which influence octanol-water partition coefficients of aliphatic and aromatic solutes. *J. Pharmaceutical Sci.* 74(8), 807-814.

Taha, A.A., Grisby, R.D., Johnson, J.R., Christian, S.D., Affsprung, H.E. (1966) Monometric apparatus for vapor and solution studies. *J. Chem. Education* 43, 432.

Tester, D.J., Harker, R.J. (1981) Ground water pollution investigations in the Great Ouse Basin. *Water Pollut. Control* 80, 614-631.

Tewari, Y.B., Martire, D.E., Wasik, S.P., Miller, M.M. (1982a) Aqueous solubilities and octanol-water partition coefficients of binary liquid mixtures of organic compounds at 25 °C. *J. Solution Chem.* 11, 435-445.

Tewari, Y.B., Miller, M.M., Wasik, S.P. (1982b) Calculation of aqueous solubilities of organic compounds. *NBS J. Res.* 87, 155-158.

Tewari, Y.B., Miller, M.M., Wasik, S.P., Martire, D.E. (1982c) Aqueous solubility and octanol/water partition coefficient of organic compounds at 25.0 °C. *J. Chem. Eng. Data* 27, 451-454.

Tomlinson, E., Hafkenscheid, T.L. (1986) Aqueous solubility and partition coefficient estimation from HPLC data. In: *Partition Coefficient, Determination and Estimation.* Dunn, III, W.J., Block, J.H., Pearlman, R.S., Eds., pp. 101-141. Pergamon Press, New York.

Trzilova, B., Horska, E. (1988) Biodegradation of amines and alkanes in aquatic environment. *Biologia (Bratislava)* 43, 209-218.

Tsonopoulos, C., Prausnitz, J.M. (1971) Activity coefficients of aromatic solutes in dilute aqueous solutions. *I & EC Fundum.* 10, 593-600.

Vaishnav, D.D., Babeu, L. (1987) Comparison of occurrence and rates of chemical biodegradation in natural waters. *Bull. Environ. Contam. Toxicol.* 39, 237-244.

Valvani, S.C., Yalkowsky, S.H. (1980) Solubility and partitioning in drug design. In: *Physical Chemical Properties of Drugs.* Med. Res. Ser. vol 10, Yalkowsky, S.H., Sinkula, A.A., Valvani, S.C., Eds., pp. 201-229, Marcel Dekker Inc., New York.

Valvani, S.C., Yalkowsky, S.H., Roseman, T.J. (1981) Solubility and partitioning. IV. Aqueous solubility and octanol-water partition coefficient of liquid electrolytes. *J. Pharm. Sci.* 70, 502-507.

Van der Linden, A.C. (1978) Degradation of oil in the marine environment. *Dev. Biodegrad. Hydrocarbons* 1, 165-200.

Veith, G.D., Morris, R.T. (1978) *A Rapid Method for Estimating log P for Organic Chemicals.* EPA-600/3-78-049. U.S. Environmental Protection Agency, Ecological Research Series.

Veith, G.D., Austin, N.M., Morris, R.T. (1979) A rapid method for estimating log P for organic chemicals. *Water Res.* 13, 43-47.

Veith, G.D., Macek, K.J., Petrocelli, S.R., Caroll, J. (1980) An evaluation of using partition coefficients and water solubilities to estimate bioconcentration factors for organic chemicals in fish. In: *Aquatic Toxicology.* J.G. Eaton, P.R. Parrish, A.C. Hendricks, Eds., ASTM STP 707, Am. Soc. for Testing and Materials, pp.116-129.

Verschueren, K. (1977) *Handbook of Environmental Data on Organic Chemicals.* Van Nostrand Reinhold, New York.

Verschueren, K. (1983) *Handbook of Environmental Data on Organic Chemicals,* 2nd edition, Van Nostrand Reinhold, New York.

Vesala, A. (1974) Thermodynamics of transfer nonelectrolytes from light and heavy water. I. Linear free energy correlations of free energy of transfer with solubility and heat of melting of nonelectrolyte. *Acta Chem. Scand.* 28A(8), 839-845.

Vowles, P.D., Mantoura, R.F.C. (1987) Sediment-water partition coefficients and HPLC retention factors of aromatic hydrocarbons. *Chemosphere* 16, 109-116.

Wakeham, S.G., Davis, A.C., Karas, J.L. (1983) Microcosm experiments to determine the fate and persistence of volatile organic compounds in coastal seawater. *Environ. Sci. Technol.* 17, 611-617.

Warne, M.St.J., Connell, D.W., Hawker, D.W., Schuurmann, G. (1990) Prediction of aqueous solubility and the octanol-water partition coefficient for lipophilic organic compounds using molecular descriptors and physicochemical properties. *Chemosphere* 21, 877-888.

Wasik, S.P., Miller, M.M., Tewari, Y.B., May, W.E., Sonnefeld, W.J., DeVoe, H., Zoller, W.H. (1983) Determination of the vapor pressure, aqueous solubility, and octanol/water partition coefficient of hydrophobic substances by coupled generator column/liquid chromatographic methods. *Res. Rev.* 85, 29-42.

Wasik, S.P., Schwarz, F.P., Tewari, Y.B., Miller, M.M., Purnell, J.H. (1984) A head-space method for measuring activity coefficients, partition coefficients and solubilities of hydrocarbons in saline solutions. *J. Res. Natl. Bur. Std.* 89, 273-277.

Wasik, S.P., Tewari, Y.B., Miller, M.M., Martire, D.E. (1981) Octanol/water partition coefficients and aqueous solubilities of organic compounds. PB82-141797, U.S. EPA, Washington, D.C.

Wasik, S.P., Tewari, Y.B., Miller, M.M. (1982) Measurements of octanol/water partition coefficient by chromatographic method. J. Res. Natl. Bur. Std. 87, 311-315.

Wateral, H., Tanaka, M., Suzaki, N. (1982) Determination of partition coefficient of halobenzenes in heptane/water and 1-octanol/water systems and comparison with the scaled particle calculation. *Anal. Chem.* 54, 702-705.

Watts, C.D., Moore, K. (1987) Fate and transport of organic compounds in river. In: *Organic Micropollutants in the Aquatic Environment.* Angelletti, G., Bjorseth, A., Eds., Proceedings of the 5th European Symposium, Rome, Italy, pp.154-169, Kluwer Academic Publ.

Weast, R.C., Ed. (1972-73). *Handbook of Chemistry and Physics,* 53th ed. CRC Press, Cleveland.

Weast, R.C. (1983-84). *Handbook of Chemistry and Physics,* 64th ed., CRC Press, Florida.

Whitehouse, B.G., Cooke, R.C. (1982) Estimating the aqueous solubility of aromatic hydrocarbons by high performance liquid chromatography. *Chemosphere* 11, 689-699.

Wilson, B.H., Smith, G.B., Rees, J.F. (1986) Biotransformations of selected alkylbenzenes and halogenated aliphatic hydrocarbons in methanogenic aquifer material: A Microcosm Study. *Environ. Sci. Technol.* 20, 997-1002.

Wilson, J.T., Enfield, C.G., Dunlap, W.J., Crosby, R.L., Foster, D.A., Baskin, L.B. (1981) Transport and fate of selected organic pollutants in a sandy soil. *J. Environ. Qual.* 10, 501-506.

Wilson, J.T., McNabb, J.F., Balkwill, D.L., Ghiorse, W.C. (1983) Enumeration and characterization of bacteria indigenous to a shallow water-table aquifer. *Ground Water* 21, 134-142.

Wilson, J.T., McNabb, J.F., Wilson, R.H., Noonan, M.J. (1983) Biotransformation of selected organic pollutants in ground water. *Dev. Ind. Microbiol.* 24, 225-233.

Windholz, M., Ed. (1983) *The Merck Index, An Encyclopedia of Chemicals, Drugs and Biologicals.* 10th edition, Merck & Co., Inc., Rahway, New Jersey.

Wong, P.T.S., Chau, Y.K., Rhamey, J.S., Docker, M. (1984) Relationship between water solubility of chlorobenzenes and their effects on a fresh water green algae. *Chemosphere* 13(9), 991-996.

Worley, J.D. (1967) Benzene as a solute in water. *Can. J. Chem.* 45, 2465-2467.

Yalkowsky, S.H., Valvani, S.C. (1976) Partition coefficients and surface areas of some alkylbenzenes. *J. Med. Chem.* 19, 727-728.

Yalkowsky, S.H. (1979) Estimation of entropies of fusion of organic compounds. *Ind. Eng. Chem. Fundam.* 18, 108-111.

Yalkowsky, S.H., Orr, R.J., Valvani, S.C. (1979) Solubility and partitioning. 3.The solubility of halobenzenes in water. *I & EC Fundamentals* 18, 351-353.

Yalkowsky, S.H., Valvani, S.C. (1980) Solubility and Partitioning I: Solubility of nonelectrolytes in water. *J. Pharm. Sci.* 69(8), 912-922.

Yalkowsky, S.H., Valvani, S.C., Mackay, D. (1983) Estimation of the aqueous solubility of some aromatic compounds. *Res. Rev.* 85, 43-55.

Yoshida, K., Shigeoka, T., Yamauchi, F. (1983) Relationship between molar refraction and n-octanol/water partition coefficient. *Ecotox. Environ. Saf.* 7, 558-565.

Yurteri, C. Ryan, D.F., Callow, J.J., Gurol, J.J. (1987) The effect of chemical composition of water on Henry's law constant. *J. WPCF* 59, 950-956.

Zeyer, J., Kuhn, E.P., Schwazenbach, R.P. (1986) Rapid microbial mineralization of toluene and 1,3-dimethylbenzene in absence of molecular oxygen. *Appl. Environ. Microbiol.* 52, 944-947.

Zwolinski, B.J., Wilhoit, R.C. (1971) *Handbook of Vapor Pressures and Heats of Vaporization of Hydrocarbons and Related Compounds.* API-44 TRC Publication No.101, Texas A & M University, Evans Press, Fort Worth, Texas.

Zoeteman, B.C.J., Harmsen, K., Linders, J.B.H. (1980) Persistent organic pollutants in river water and ground water of the Netherlands. *Chemosphere* 9, 231-249.

Zoeteman, B.C.J., De Greef, E., Brinkmann, F.J.J. (1981) Persistency of organic contaminants in groundwater. Lessons from soil pollution incidents in the Netherlands. *Sci. Total Environ.* 21, 187-202.

3. Chlorobenzenes

3.1 List of Chemicals and Data Compilations

Common Name: Chlorobenzene
Synonym: monochlorobenzene, benzene chloride, phenyl chloride
Chemical Name: chlorobenzene
CAS Registry No: 108-90-7
Molecular Formula: C_6H_5Cl
Molecular Weight: 112.56
Melting Point (°C):
-45.2 (Doolittle 1935)
-45.0 (Verschueren 1977,1983)
-45.6 (Mackay et al. 1980,1982b; Mackay & Shiu 1981; Kishii et al. 1987)
-45.8 (Miller et al. 1985)
Boiling Point (°C):
130 (Doolittle 1935)
125.7 (Mackay et al. 1982a)
131.7 (Sato & Nakajima 1979; Miller et al. 1984)
Density (g/cm³ at 20°C):
1.107
1.1058 (Weast 1972-73)
Molar Volume (cm³/mol):
101.8 (20°C, calculated-density, Wing & Johnston 1957)
101.74 (20°C, calculated-density, Weast 1972-73; Horvath 1982; Windholz 1983; Budavari 1989)
116.9 (LeBas method, Miller et al. 1985; Shiu et al., 1987)
102 (calculated-density, Lande & Banerjee 1981; Stephenson & Malanowski 1987)
102 (Chiou et al. 1983; Chiou 1985)
0.581 (intrinsic volume: $V_I/100$, Leahy 1986; Kamlet et al. 1988; Hawker 1989)
117 (Abernethy & Mackay 1987)
112 (Abernethy et al. 1988)
58 (intrinsic, Abernethy et al. 1988)
115 (Valsaraj 1988)
102.42 (calculated, Govers et al. 1990)
Molecular Volume (A³):
119.7 (De Bruijn & Hermens 1990)
Total Surface Area, TSA (A²):
127.1 (Yalkowsky et al. 1979)
127.9 (Kishii et al. 1987)
123.9 (Sabljic 1987a)
126.73 (Doucette & Andren 1988)
127 (Valsaraj 1988)
158.4 (De Bruijn & Hermens 1990)
Heat of Fusion, kcal/mol:
Entropy of Fusion, cal/mol K (e.u.):
Fugacity Ratio, F: 1.0

Water Solubility (g/m^3 or mg/L at 25 °C):

488 (30°C, shake flask-interferometer, Gross & Saylor 1931; quoted, Vesala 1974)
551 (Landolt-Börnstein 1951)
488 (Seidell 1941; quoted, Hansch et al. 1968; McKim et al. 1985)
< 200 (residue-volume method, Booth & Everson 1948)
500 (shake flask-UV, Andrews & Keefer 1950; quoted, Tsonopoulos & Prausnitz 1971)
501 (Dreisbach 1955; quoted, Chey & Calder 1972; Vesala 1974)
285 (calculated-K_{OW}, Hansch et al. 1968)
534 (21°C, extraction by nonpolar resins/evolution, Chey & Calder 1972)
557 (Chey & Calder 1972; quoted, Vasala 1974)
463 (shake flask-UV, Vasala 1974)
100 (Stephen & Stephen 1963; Lu & Metcalf 1975)
409 (quoted, Hine & Mookerjee 1975)
448 (30°C, quoted, Chiou et al. 1977; Freed et al 1977; Kenaga & Goring 1980)
503 (quoted, Verschueren 1977; Könemann 1981)
472 (shake flask-GC, Aquan-Yuen et al. 1979)
472 (shake flask-GC, Mackay et al. 1979,1980,1982b; quoted, Suntio et al. 1988b; Howard 1989)
503 (shake flask-UV, Yalkowsky et al. 1979; quoted, Miller et al. 1984)
539 (calculated from K_{OW}, Yalkowsky et al. 1979; Yalkowsky & Valvani 1980; Valvani & Yalkowsky 1980; Valvani et al. 1981)
500 (quoted, Callahan et al. 1979; Neely 1984; Neuhauser et al. 1985)
448 (quoted, Kenaga 1980)
420, 480, 450 (20°C; elution chromatography, UV adsorption, average, Schwarz & Miller 1980)
502 (quoted, Lande & Banerjee 1981)
491 (quoted value, Mackay & Shiu 1981)
473 (quoted, Eisenreich et al. 1981)
490 (quoted, Neely 1982; McKim et al. 1985)
488 (quoted, Mabey et al. 1982; Yoshida et al. 1983b)
295 (gen. col.-HPLC/UV, Tewari et al. 1982)
491 (quoted, Chiou et al. 1982; 1983; Chiou 1985)
498 (quoted and recommended, Horvath 1982; Wong et al. 1984)
499 (generator column-HPLC/UV, Wasik et al. 1983)
500 (quoted, Calamari et al. 1983)
327 (calculated-UNIFAC, Arbuckle 1983)
508 (calculated-HPLC-k', converted from reported γ_W, Hafkenscheid & Tomlinson 1983a)
503 (calculated from K_{OW}, Yalkowsky et al. 1983)
500 (20°C, quoted, Verschueren 1983)
295 (gen. col.-GC, Miller et al. 1984; 1985; quoted, Doucette & Andren 1988)
495 (recommended, IUPAC 1985)

490 (30°C, quoted, Dean 1985)
502 (shake flask-HPLC, Banerjee 1984)
502, 326 (quoted, calculated-UNIFAC, Banerjee 1985)
503, 1353 (quoted, calculated-K_{OW} & HPLC-RT, Chin et al. 1986)
364 (calculated-V_I, sovatochromic parameters, Leahy 1986)
348 (Lo et al. 1986)
484 (quoted, Abernethy et al. 1986; Suntio et al. 1988b)
428 (Sanemasa et al. 1987)
336 (calculated-UNIFAC, Banerjee & Howard 1988)
465 (calculated- χ , Nirmalakahandan & Speece 1988a)
472 (quoted, Valsaraj 1988)
408.7 (quoted, subcooled liquid, Hawker 1989)
470 (quoted, Figueroa & Simmons 1991)

Vapor Pressure (Pa at 25 °C):
1580 (extrapolated, Antoine eqn., Weast, 1972-73)
1596, 1610 (extrapolated, Antoine eqn., Boublik et al. 1973; 1984)
1613 (quoted, Hine & Mookerjee 1975)
1573 (quoted, Verschueren 1977,1983)
1173 (quoted, Callahan et al. 1979; Neuhauser et al. 1985)
1621 (Mackay 1981)
1585 (quoted, Mackay & Shiu 1981)
1196 (quoted, reported as 11.8 atm, Eisenreich et al. 1981)
1580, 1560 (quoted exptl., calculated-B.P., Mackay et al. 1982a)
1573 (quoted, Neely 1982; Yoshida et al. 1983b)
1560 (quoted, Mabey et al. 1982)
1580 (calculated-HLC, Arbuckle 1983)
1580 (selected, Bobra et al. 1985; Suntio et al. 1988b)
1427 (20°C, quoted, Chiou & Shoup 1985)
1586 (Daubert & Danner 1985; quoted, Howard 1989)
1596 (extrapolated, Antoine eqn., Dean 1985)
1600 (extrapolated, Antoine eqn., Stephenson & Malanowski 1987)
1570 (quoted, Valsaraj 1988)
1600, 765 (quoted, calculated-UNIFAC, Banerjee et al. 1990)

Henry's Law Constant (Pa m^3/mol):
451 (calculated as $1/K_{AW}$, C_W/C_A, reported as exptl., Hine & Mookerjee 1975)
441 (calculated-bond contribution, Hine & Mookerjee 1975)
382 (batch stripping, Mackay et al. 1979; quoted, Mackay & Shiu 1981; Suntio et al. 1988b)
379 (calculated, Mackay et al. 1979)
330 (concentration ratio, Leighton & Calo 1981)
314 (batch stripping, Mackay & Shiu 1981; quoted, Pankow et al. 1984; Suntio et al.

1988b; Olsen & Davis 1990)

350 (quoted average of Mackay & Shiu 1981, Howard 1989)

375 (quoted, Eisenreich et al. 1981)

363 (quoted, Mabey et al. 1982)

545 (calculated-UNIFAC, Arbuckle 1983)

263 (calculated-P/C, Calamari et al. 1983)

372 (calculated-P/C, Yoshida et al. 1983b)

367 (selected, Bobra et al., 1985)

319 (20°C, EPICS, Yurteri et al. 1987)

273 (20°C, calculated-P/C, Yurteri et al. 1987)

365 (EPICS, Ashworth et al. 1988)

346 (20°C, EPICS, Ashworth et al. 1988)

365 (quoted, Pankow et al. 1984; Pankow 1986; Pankow & Rosen, 1988; Pankow 1990)

237 (calculated-QSAR, Nirmalakhandan & Speece 1988b)

451 (observed, Nirmalakhandan & Speece 1988b)

367.7 (calculated-P/C, Suntio et al. 1988b)

374.8 (quoted, Valsaraj 1988)

377, 382 (calculated-P/C, lit. exptl., Mackay & Shiu 1990)

Octanol/Water Partition Coefficient, log K_{OW}:

2.84 (Fujita et al. 1964; Hansch et al. 1968; Leo et al. 1971; quoted, Chiou et al. 1977; 1982; 1983; Chiou & Schmedding 1981; Chiou 1985; Callahan et al. 1979; Kenaga & Goring 1980; Mackay et al. 1980,1982b; Ellgehausen et al. 1981; Kaiser 1983; Hodson et al. 1984; Freitag et al. 1985; Neuhauser et al. 1985; Dobbs et al. 1989)

2.83 (calculated-f const., Yalkowsky et al. 1979; Yalkowsky & Valvani 1980; Valvani & Yalkowsky 1980; Valvani et al. 1981; quoted, Miller 1984; Suntio et al. 1988b)

2.18 (^{14}C, LSC, Lu & Metcalf 1975; quoted, Suntio et al. 1988b)

2.84, 2.46, 2.18 (Hansch & Leo 1979; quoted, McKim et al. 1985; Suntio et al. 1988b)

2.84, 2.81 (calculated-f constants, Rekker 1977; quoted, Leegwater 1989)

2.79 (HPLC-RT, Veith et al. 1979b)

2.84, 2.63 (quoted, calculated-S, Mackay et al. 1980,1982b)

2.84 (HPLC-k', Könemann et al. 1979; quoted, Wong et al. 1984; Figueroa & Simmons 1991)

2.81 (calculated-f const., Könemann et al. 1979; quoted, Könemann 1981; Leegwater 1989)

2.80 (HPLC-k', Hanai et al. 1981)

2.81 (HPLC-k', D'amboise & Hanai 1982)

2.79 (quoted, Mackay 1982a)

2.80 (quoted, Neely 1982; Wateral et al.1982)

2.83 (shake flask-HPLC, Hammers et al. 1982)
2.13, 2.18 (quoted, HPLC-k', Miyake & Terada 1982)
2.49 (quoted, Calamari et al. 1983)
2.80 (calculated-HPLC-k', Hafkenscheid & Tomlinson 1983a)
2.83 (calculated-f const., Yalkowsky et al. 1983)
2.84; 2.92 (quoted, calculated-chlorine substituted number, Kaiser 1983)
2.98 (gen. column-HPLC/UV, Wasik et al. 1983)
2.84; 2.65, 2.78 (quoted, calculated-molar refraction & ionization potential, molar refraction, Yoshida et al. 1983b)
2.83 (quoted, Yoshida et al. 1983a)
2.98 (gen. col.-GC/ECD, Miller et al. 1984; 1985; quoted, Sarna et al. 1984; Kamlet et al. 1988; Doucette & Andren 1988; Suntio et al. 1988b; Hawker et al. 1989b)
2.81-2.84 (HPLC-RV, Garst & Wilson 1984; Garst 1984)
2.84 (quoted, Garst & Wilson 1984; Garst 1984)
2.49 (HPLC-k', Haky & Young 1984)
2.80 (selected, Bobra et al. 1985; Abernethy & Mackay 1987; Suntio et al. 1988b)
2.84 (Hansch & Leo 1985; quoted, Howard 1989)
2.69 (calculated-V_I & solvatochromic parameters, Taft et al. 1985)
2.79 (calculated-V_I & solvatochromic parameters, Leahy 1986)
3.00 (HPLC-k', De Kock & Lord 1987)
2.98 (quoted exptl., Doucette & Andren 1988)
2.84, 3.07, 3.18, 2.82, 2.94 (calculated-π, f const., MW, χ, TSA, Doucette & Andren 1988)
2.84 (quoted, Ryan et al. 1988)
2.18-2.84 (quoted lit. range, Dearden & Bresnen 1988)
2.84, 2.51 (quoted, calculated-UNIFAC, Banerjee & Howard 1988)
2.80 (calculated-V_I & solvatochromic parameters, Kamlet et al. 1988)
2.84 (recommended, Sangster 1989)
2.99 (quoted, Isnard & Lambert 1989)
2.898 (slow stirring, De Bruijn et al. 1989; De Bruijn & Hermens 1990)
2.89 (calculated-f const., De Bruijn et al. 1989)
2.98 (quoted, Hawker 1989)
2.18-2.84 (quoted from the Medchem Data-base, Dearden 1990)

Bioconcentration Factor, log BCF:
2.65 (fathead minnow, Veith et al. 1979b; 1980; quoted, Mackay 1982; Suntio et al. 1988b; Howard 1989)
1.30 (Kenaga 1980a)
1.08 (fish, flowing water, Kenaga & Goring 1980; quoted, Yoshida et al. 1983b; Suntio et al. 1988b)
2.21 (microorganisms-water, Mabey et al. 1982)
2.47 (fish, correlated, Mackay 1982)
2.65 (fathead minnow, quoted, Bysshe 1982)

1.08 (quoted, Bysshe 1982)
1.70 (guppy, calculated- χ , Koch 1983)
1.41 (calculated-K_{OW}, Calamari et al. 1983)
1.70 (algae, Freitag et al. 1984,1985; Halfon & Raggiani 1986; quoted, Suntio et al. 1988b)
1.88 (fish, Freitag et al. 1984; Halfon & Raggiani 1986; quoted, Suntio et al. 1988b)
3.23 (activated sludge, Freitag et al. 1984; Halfon & Raggiani 1986)
1.85 (fish, Freitag et al. 1985)
3.23 (activated sludge, Fritag et al. 1985)
2.65 (correlated, Isnard & Lambert 1988)
1.93 (fish, calculated, Figueroa & Simmons 1991)

Sorption Partition Coefficient, log K_{OC}:
2.52 (sediment, Mabey et al. 1982)
2.18 (Kenaga 1980a)
2.59 (field data, Robert et al. 1979; Schwarzenbach & Westall 1981; quoted, Voice & Weber 1985)
2.44, 2.50 (calculated from K_{OW}, Schwarzenbach & Westall 1981)
2.73 (soil, calculated-K_{OW}, Calamari et al. 1983)
2.68 (log K_{OM}, soil organic matter, sorption isotherm, Chiou et al.1983)
2.10 (calculated- χ , Koch 1983)
2.92 (calculated-K_{OW}, Yoshida et al. 1983b)
2.44 (calculated- χ , Bahnick & Doucette 1988)
1.92-2.59 (soil, Howard 1989)
Sorption Partition Coefficient, log K_{OM}:
2.10, 2.32 (quoted, calculated- χ , Sabljic 1984)

Half-Lives in the Environment:
Air: 72.9-729 hours, based on photooxidation half-life in air (Howard et al. 1991).
Surface Water: 1632-3600 hours, based on unacclimated aerobic river dieaway tests (Hungspreugs et al. 1984; Lee & Ryan 1976); 75 days for an estuarine river with near natural conditions at 22°C (Lee & Ryan 1976); 0.3 days (estimated, Zoeteman et al. 1980).
Groundwater: 3264-7200 hours, based on estimated aqueous aerobic biodegradation half-life (Howard et al. 1991).
Sediment: 75 days (Lee & Ryan 1976,1979).
Soil: disappearance half-life from testing soils, 2.1 days (Anderson et al. 1991); 1632-3600 hours, based on estimated aqueous aerobic biodegradation half-life (Howard et al. 1991).

Environmental Fate Rate Constants or Half-Lives:
Volatilization: estimated half-life from a flowing stream is 1-12 hours (Cadena et al. 1984; selected, Howard 1989); estimated half-lives from marine mesocosm: 21

days in the spring at 8-16°C, 4.6 days in the summer at 20-22°C and 13 days in the winter at 3-7°C; estimated half-lives from soil: 0.3 days of 1 cm depth and 12.6 days for 10 cm depth (Wakeham et al. 1983).

Photolysis: not environmentally significant or relevant (Mabey et al. 1982); estimated half-life for photolysis by sunlight in surface water at 40°N in the summer is 170 years with a rate constant of 1.1 day^{-1} (Dulin et al. 1986); the half-life for a measured rate constant of $9.4x10^{-13}$ cm^3 $molecule^{-1}$ sec^{-1} at room temperature for reaction with $5.0x10^5$ molecule cm^{-3} hydroxy radicals estimated to be 17 days (Atkinson 1987).

Oxidation: photooxidation half-life in water is 1553-62106 hours, based on a measured rate for hydroxy radicals in aqueous solution (Dorfman & Adams, 1973); k (singlet oxygen) $<<$ 360 M^{-1} h^{-1} and k (RO_2 radical) $<<$1 M^{-1} h^{-1} (Mabey et al. 1982); reaction rate constant in air, $2.88x10^{-3}$ $hour^{-1}$ (Yoshida et al. 1983b; selected, Mackay et al. 1985); photooxidation half-life in air is 72.9-729 hours, based on a measured rate constant of $0.94x10^{-12}$ cm^3 $molecule^{-1}$ sec^{-1} for the vapor phase reaction with hydroxy radicals in air (Atkinson 1985; Atkinson et al. 1985).

Hydrolysis: not environmentally significant (Mabey et al. 1982); first order half-life > 879 years, based on rate constant k_h < 0.9 M^{-1} h^{-1} extrapolated to pH 7 at 25°C from 1% disappearance after 16 days at 85°C and pH 9.7 (Ellington et al. 1988).

Biodegradation: aqueous aerobic half-life is 1632-3600 hours, based on unacclimated aerobic river dieaway tests (Hungspreugs et al. 1984; Lee & Ryan 1976,1979); aqueous anaerobic half-life is 6528-14400 hours, based on estimated aqueous aerobic biodegradation half-life; significant degradation on anaerobic environment, k_B = 0.5 day^{-1} (Tabak et al. 1981; Mills et al. 1982); degradation rate constants: in air, $1.88x10^{-4}$ $hour^{-1}$ and $3.83x10^{-4}$ $hour^{-1}$ in sediments (Lee & Ryan 1979; selected, Mackay et al. 1985); 0.24 L/day in air, no degradation in water and on the ground (Neely 1982); first-order rate constant k_B = 0.07-0.3 day^{-1} in river water; 0.04-0.2 day^{-1} in estuary water; and 0.01 day^{-1} in marine water (Bartholomew & Pfaender 1983; selected, Battersby 1990); first-order rate constant 0.07 $year^{-1}$ with a half-life of 37 days (Olsen & Davis 1990).

Biotransformation: k_b = $3x10^{-9}$ ml $cell^{-1}$ h^{-1} (estimated, Mabey et al. 1982).

Common Name: 1,2-Dichlorobenzene
Synonym: o-dichlorobenzene, dowtherm E
Chemical Name: 1,2-dichlorobenzene
CAS Registry No: 95-50-1
Molecular Formula: $C_6H_4Cl_2$
Molecular Weight: 147.01
Melting Point (°C):
-17.6 (Doolittle 1935)
-17 (Pirsch 1956; Mackay & Shiu 1981; Miller et al. 1985; Kishii et al. 1987; Howard 1989)
-16.7 (Verschueren 1977,1983)
25 (Mackay et al. 1982b; Yalkowsky et al. 1983)
Boiling Point (°C):
177.0 (Doolittle 1935)
179.6 (Mellan 1970)
179.0 (Verschueren 1977,1983)
180.5 (Sato & Nakajima 1979; Mackay & Shiu 1981; Mackay et al. 1982b; Miller et al. 1984)
Density (g/cm^3 at 20°C):
1.306
1.3048 (Weast 1972-73; Horvath 1982)
Molar Volume (cm^3/mol):
112.7 (Weast 1972-73)
113 (calculated-density, Lande & Banerjee 1981)
112.82 (calculated-density, Windholz 1983; Budavari 1989)
137.8 (LeBas method, Miller et al. 1985; Shiu et al. 1987)
113 (Chiou et al. 1983; Chiou 1985; Stephenson & Malanowski 1987)
138 (Abernethy & Mackay 1987)
123 (Abernethy et al. 1988)
0.671 (intrinsic volume: $V_I/100$, Kamlet et al. 1988; Hawker et al. 1989b; Hawker 1990)
67 (intrinsic, Abernethy et al. 1988)
113.36 (calculated, Govers et al. 1990)
Molecular Volume (A^3):
150.0 (De Bruijn & Hermens 1990)
Total Surface Area, TSA (A^2):
142.7 (Yalkowsky et al. 1979; Mackay et al. 1982b)
143.3 (Kishii et al. 1987)
141.0 (Sabljic 1987a)
142.22 (Doucette & Andren 1988)
150.0 (De Bruijn & Hermens 1990)
Heat of Fusion, kcal/mol:
3.09 (Weast 1972-73)

Entropy of Fusion, cal/mole K (e.u.):
12.06 (Pirsch 1956)
12.1 (Yalkowsky & Valvani 1980)

Fugacity Ratio, F:
1.0 (Suntio et al. 1988b)

Water Solubility (g/m^3 or mg/L at 25 °C):
145 (volumetric, Klemenc & Low 1930; Seidell 1941; quoted, Suntio et al. 1988b)
< 260 (residue-volume method, Booth & Everson 1948)
92.7 (Landolt-Börnstein 1951)
145, 51.7 (quoted, calculated-K_{OW}, Hansch et al. 1968)
145 (quoted, Tsonopoulos & Prausnitz 1971)
79.0 (radiolabeled ^{14}C, Metcalf et al. 1975)
144 (quoted, Hine & Mookerjee 1975)
148 (20°C, shake flask-GC/ECD, Chiou & Freed 1977; Chiou et al. 1979)
79 (quoted, Chiou et al. 1977)
145 (quoted, Callahan et al. 1979; Neely 1984; Neuhauser et al. 1985)
92.8 (shake flask-UV, Yalkowsky et al. 1979; quoted, Miller et al. 1984; Suntio et al. 1988b)
125 (calculated-K_{OW}, Yalkowsky et al. 1979; Yalkowsky & Valvani 1980; Valvani & Yalkowsky 1980)
154 (shake flask-LSC/^{14}C, Veith et al. 1980)
131 (calculated, Veith et al. 1980)
128 (20°C, elution chromatography, Schwarz & Miller 1980)
124 (20°C, UV, Schwarz & Miller 1980)
145 (quoted, Haque et al. 1980; Mackay et al. 1982b)
155.8 (shake flask-LSC, Banerjee et al. 1980; Banerjee 1985; quoted, Suntio et al. 1988b)
99.1 (shake flask-GC, Könemann 1981)
109 (Valvani et al. 1981; selected, Amidon & Williams 1982)
378 (calculated-M.P., Amidon & Williams 1982)
92.6 (recommended, Horvath 1982; quoted, Wong et al. 1984)
154 (shake flask-GC, Chiou et al. 1982, 1983; Chiou 1981,1985; Chiou & Schmedding 1981; quoted, Suntio et al. 1988b)
125 (quoted average, Yalkowsky et al. 1983)
48.7 (calculated-UNIFAC, Arbuckle 1983)
145 (quoted, Verschueren 1977, 1983; Calamari et al. 1983)
92.3 (gen. col.-GC, Miller et al. 1984, 1985; quoted, Doucette & Andren 1988; Suntio et al. 1988b)
137 (shake flask-HPLC, Banerjee 1984)
147 (recommended, IUPAC 1985)
100 (Dean 1985)
92.8, 534 (quoted, calculated-K_{OW} & HPLC-RT, Chin et al. 1986)

145 (quoted, Abernethy et al. 1986)
169 (Lo et al. 1986)
156 (quoted, Riddick et al. 1986; Howard 1989)
96.4 (calculated- χ , Nirmalakhandan & Speece 1988a)
145 (correlated, Isnard & Lambert 1988; 1989)
118.8 (selected, Suntio et al. 1988b)
143.7 (quoted, Hawker 1989)
85.1 (quoted, Figueroa & Simmons 1991)

Vapor Pressure (Pa at 25°C):
137 (20°C, Stull 1947)
208 (20°C, McDonald et al. 1959)
196 (extrapolated, Antoine eqn., Weast 1972-73; quoted, Mackay & Shiu 1981; Suntio et al. 1988b; Howard 1989)
174 (extrapolated, Antoine eqn., Boublik et al. 1973; 1984)
237 (quoted, Hine & Mookerjee 1975)
200 (quoted, Verschueren 1977,1983; Suntio et al. 1988b)
133 (quoted, Callahan et al. 1979; Neuhauser et al. 1985)
133 (20°C, Haque et al. 1980)
188 (gas saturation, Grayson & Fosbracy 1982)
196, 177 (quoted exptl., calculated-B.P., Mackay et al. 1982a&b)
197 (extrapolated, Antoine eqn., Dean 1985)
198 (extrapolated, Antoine eqn., Stephenson & Malanowski 1987)
196 (selected, Suntio et al. 1988b)

Henry's Law Constant (Pa m^3/mol):
248 (calculated as $1/K_{AW}$, C_W/C_A, reported as exptl., Hine & Mookerjee 1975; selected exptl., Nirmalakhadan & Speece 1988a)
375 (calculated-bond contribution, Hine & Mookerjee 1975)
193 (batch stripping, Mackay & Shiu 1981; quoted, Pankow et al. 1984; Pankow & Rosen 1988; Pankow 1986,1990)
190 (calculated-P/C, Mackay & Shiu 1981)
195.5 (calculated, Mabey et al. 1982)
132 (calculated-P/C, Calamari et al. 1983)
245 (selected, Bobra et al. 1985)
122 (20°C, batch stripping, Oliver 1985; quoted, Howard 1989)
159 (EPICS, Ashworth et al. 1988)
170 (20°C, EPICS, Ashworth et al., 1988)
248, 119 (observed, calculated-QSAR, Nirmalakhandan & Speece 1988b)
244.2 (calculated-P/C, Suntio et al. 1988b)

Octanol/Water Partition Coefficient, log K_{OW}:

3.55 (Hansch et al. 1968)
3.38 (Leo et al. 1971; Hansch & Leo 1979; quoted, Chiou & Schmedding 1981; Kaiser 1983)
3.38, 3.57 (quoted, calculated-f constants, Rekker 1977; quoted, Harnisch et al. 1983)
3.38 (quoted, Chiou & Freed 1977; Chiou 1981; 1985; Chiou & Schmedding 1981; Chiou et al. 1982; 1983)
3.38 (quoted, Callahan et al. 1979; Neuhauser et al. 1985)
3.59 (calculated-f const., Yalkowsky et al. 1979; Yalkowsky & Valvani 1980; Valvani et al. 1980; quoted, Miller et al. 1984; Suntio et al. 1988b)
3.55, 3.39 (shake flask-GC, HPLC-k', Könemann et al. 1979; quoted, Kaiser et al. 1984; Wong et al. 1984; Figueroa & Simmons 1991)
3.53 (calculated-f const., Könemann et al. 1979; quoted, Könemann 1981; Leegwater 1989)
3.40 (concentration ratio, Veith et al. 1980)
3.75 (HPLC-RT, Veith et al. 1980)
3.55 (calculated-π, Veith et al. 1980)
3.40 (shake flask-LSC, Banerjee et al. 1980; quoted, Oliver 1987a; Suntio et al. 1988b)
3.38 (quoted, Ellgehausen et al. 1981)
3.34 (shake flask-HPLC, Hammers et al. 1982)
3.56 (calculated, Mabey et al. 1982)
3.38 (Verschueren 1983)
2.97 (quoted, Calamari et al. 1983)
3.34, 3.20 (quoted of HPLC methods, Harnisch et al. 1983)
3.61 (quoted od OECD/EEC shake-flask method, Harnisch et al. 1983)
3.38 (gen. column-HPLC/UV, Wasik et al. 1983)
3.59 (calculated-f const., Yalkowsky et al. 1983)
3.40 (shake flask-GC, Wateral et al. 1982; quoted, Suntio et al. 1988b)
3.90, 3.67 (calculated-UNIFAC, Arbuckle 1983)
3.55 (quoted, Wong et al. 1984)
3.38 (gen. col.-GC, Miller et al. 1984, 1985; quoted, Suntio et al. 1988b; Hawker 1989)
3.56 (HPLC-RV, Garst 1984)
3.40 (quoted, Garst 1984)
3.38 (quoted, Sarna et al. 1984)
3.40 (selected, Bobra et al. 1985; Abernethy & Mackay 1987)
3.40 (quoted, Oliver & Niimi 1983; 1985)
3.38 (Hansch & Leo 1985; quoted, Howard 1989)
3.38 (quoted, Ryan et al. 1988)
3.38 (quoted exptl., Doucette & Andren 1988)
3.57, 3.59, 3.77, 3.35, 3.36 (calculated-π, f const., MW, χ, TSA, Doucette & Andren 1988)
3.38, 3.28 (quoted, calculated-V_I, solvatochromic parameters, Kamlet et al. 1988)

3.433 (slow stirring, De Bruijn et al. 1989; De Bruijn & Hermens 1990)
3.65 (calculated-f const., De Bruijn et al. 1989)
3.39 (quoted, Isnard & Lambert 1988; 1989)
3.49 (shake flask-GC, Pereira et al. 1988)

Bioconcentration Factor, log BCF:
1.95 (bluegill sunfish, Veith et al. 1979b; 1980; quoted, Suntio et al. 1988b)
2.19-2.48 (fish, calculated, Veith et al. 1980)
1.95 (bluegill sunfish, whole body, flow system, Barrows et al. 1980)
2.86 (microorganisms-water, Mabey et al. 1982)
1.95 (bluegill sunfish, quoted, Bysshe 1982)
2.43-2.75 (rainbow trout, Oliver & Niimi 1983)
3.51-3.80 (rainbow trout, liqid-based, Oliver & Niimi 1983; quoted, Suntio et al. 1988b)
1.82 (calculated-K_{OW}, Calamari et al. 1983)
2.43-2.75 (fish, Oliver 1984)
2.47 (fish, calculated, Garst 1984)
1.95 (correlated-flow through method, bluegill sunfish, Davis & Dobbs 1984)
1.60 (fish-normalized, Tadokoro & Tomita 1987)
1.89 (fish, calculated- χ , Sabljic 1987b)
1.95 (fish, Correlated, Isnard & Lambert 1988; 1989)
3.94, 4.46, 3.79, 3.82 (field data: Atlantic croaker, blue crabs, spotted sea trout, blue catfish, lipid-based, Pereira et al. 1988)
2.47 (fish, calculated, Figueroa & Simmons 1991)

Sorption Partition Coefficient, log K_{OC}:
2.26 (soil, Chiou et al. 1979)
3.0, 2.96, 2.62 (calculated-K_{OW}, C_L, C_S, Karickhoff 1981)
3.23 (calculated, Mabey et al. 1982)
2.43-3.57 (calculated- χ , K_{OW}, Lyman et al. 1982)
2.99 (soil, calculated-K_{OW}, Calamari et al. 1983)
4.5 (field data, Oliver & Niimi 1985)
4.1 (Niagara River organic matter, Oliver & Niimi 1985)
3.0 (calculated from K_{OW}, Oliver & Niimi 1985)
2.59 (Appalachee soil, Stauffer & MacIntyre 1986)
2.45-3.51 (aquafer materials, Stauffer et al. 1989)

Sorption Coefficient, log K_{OM}:
2.54 (soil, Chiou et al. 1979; quoted, Howard 1989)
2.50 (soil, Chiou et al. 1983; quoted, Howard 1989)
2.27 (soil-org. matter, sorption isotherm, Chiou et al. 1983)
2.26, 2.54 (quoted, calculated- χ , Sabljic 1984)

Half-Lives in the Environment:

Air: 152.8-1528 hours, based on the photooxidation half-life in air.

Surface water: 672-4320 hours, based on estimated unacclimated aqueous aerobic biodegration half-life (Haider et al. 1981); 1.2-37 days estimated from field data, 0.3-3 days for river, 3-30 days for lakes (Zoeteman et al. 1980).

Groundwater: 1344-8640 hours, based on unacclimated aqueous aerobic biodegradation half-life (Haider et al. 1981); 30-300 days, estimated from persistence in water (Zoeteman et al. 1980); estimated half-life from observed persistence in groundwater of the Netherlands, one year (Zoeteman et al. 1981).

Soil: 672-4320 hours, based on unacclimated aerobic screening test data (Canton et al. 1985).

Biota: half-life in fish , < 1 day (Veith et al. 1980); half-life in worms at 8°C, < 5 days (Oliver 1987a); half-life in bluegill sunfish, < 1 day. (Barrows et al. 1980).

Environmental Fate Rate Constants or Half-Lives:

Volatilization/Evaporation: experimental evaporation rate into air, 1.18×10^{-6} g cm^{-2} sec^{-1} (Chiou et al. 1980); estimated half-life for a model river of 1 m depth with a flow rate of 1 m/sec and wind velocity of 3 m/sec at 20°C is 4.4 hours (Lyman et al. 1982).

Photolysis: not environmentally significant (Mabey et al. 1982).

Oxidation: k (singlet oxygen) << 360 M^{-1} h^{-1} & k (RO_2) << 1 M^{-1} h^{-1} (Mabey et al. 1982); photooxidation half-life in air is 24 days based on a measured room temperature rate constant of 0.42×10^{-12} cm^3 $molecule^{-1}$ sec^{-1} for the vapor phase reaction with hydroxy radicals in air (Atkinson 1985).

Hydrolysis: not environmentally significant (Mabey et al. 1982).

Biodegradation: aqueous aerobic half-life is 672-4320 hours, based on unacclimated soil grab sample data, (Haider et al. 1981) and aerobic screening test data (Canton et al. 1985); aqueous anaerobic half-life is 2880-17280 hours, based on estimated unacclimated aqueous aerobic biodegradation half-life (Haider et al. 1981); Significant degradation in an aerobic environment, k_B = 0.05 d^{-1} (Tabak et al. 1981; Mills et al. 1982); in a continuous flow of activated sludge system, virtually 100% (78% biodegradation & 22% stripping) was observed (Kincannon et al. 1983; selected, Howard 1989).

Biotransformation: k_b = 1×10^{-10} ml $cell^{-1}$ h^{-1}. (estimated, Mabey et al. 1982).

Common Name: 1,3-Dichlorobenzene
Synonym: m-dichlorobenzene
Chemical Name: 1,3-dichlorobenzene
CAS Registry No: 541-73-1
Molecular Formula: $C_6H_4Cl_2$
Molecular Weight: 147.01
Melting Point (°C):
-24.0 (Pirsch 1956)
-24.76 (Verschueren 1977,1983; Mackay & Shiu 1981)
-24.9 (Miller et al. 1985)
-24.4 (Kishii et al. 1987)
Boiling Point (°C):
173 (Sato & Nakajima 1979; Mackay & Shiu 1981; Miller et al. 1984)
Density (g/cm^3 at 20°C):
1.2884 (Weast 1972-73; Horvath 1982)
Molar Volume (cm^3/mol)
114 (calculated-density, Weast 1972-73; Lande & Banerjee 1981; Horvath 1982)
114 (Chiou et al. 1983; Chiou 1985; Stephenson & Malanowski 1987)
114.60 (calculated-density, Windholz 1983; Budavari 1989)
137.8 (LeBas method, Miller et al. 1985; Shiu et al. 1987)
138 (Abernethy & Mackay 1987)
124, 67 (from density, intrinsic, Abernethy et al. 1988)
0.671 (intrinsic volume: V_I/100, Kamlet et al. 1988; Hawker 1989b)
133 (selected, Valsaraj 1988)
114.8 (calculated, Govers et al. 1990)
Molecular Volume (A^3):
153.7 (De Bruijn & Hermens 1990)
Total Surface Area, TSA (A^2):
144.7 (Yalkowsky et al. 1979)
144.7 (Mackay et al. 1982a; selected, Valsaraj 1988; Valsaraj & Thibodeaux 1989)
141.0 (calculated- χ , Sabljic 1987a)
144.24 (Doucette & Andren 1988)
168.8 (De Bruijn & Hermens 1990)
Heat of Fusion, kcal/mol: 3.021 (Weast 1972-73)

Entropy of Fusion, cal/mol K (e.u.):
12.1 (Pirsch 1956)
12.3 (Yalkowsky 1979)
12.2 (Yalkowsky & Valvani 1980)
Fugacity Ratio, F:
1.0 (Suntio et al. 1988b)

Water Solubility (g/m^3 or mg/L at 25 °C):

123 (volumetric, Klemenc & Löw 1930; quoted, Tsonopoulos & Prausnitz 1971; Suntio et al. 1988b)

123.5 (Ginnings et al. 1939; quoted, Vesala 1974)

124 (Landolt-Börnstein 1951)

123, 51.1 (quoted, calculated-K_{OW}, Hansch et al. 1968)

102.9 (shake flask-UV, Vesala 1974; quoted, Suntio et al. 1988b)

122 (quoted, Hine & Mookerjee 1975)

123 (Verschueren 1977; 1983)

119.5 (shake flask-UV, Yalkowsky et al. 1979; quoted, Miller et al. 1984)

125 (calculated-K_{OW}, Yalkowsky et al. 1979; Yalkowsky & Valvani 1980; Valvani & Yalkowsky 1980)

131, 119 (LSC-^{14}C, calculated-K_{OW}, Veith et al. 1980)

123, 144, 149 (quoted, 23.5 °C, elution chromatography, Schwarz 1980)

111, 89, 113, 101 (20 °C, quoted, UV, elution chromatography, average exptl. value, Schwarz & Miller 1980)

133.5 (shake flask-LSC, Banerjee et al. 1980; quoted, Suntio et al. 1988b)

68.6 (shake flask-GC, Könemann 1981)

123 (Mackay et al. 1982b; quoted, Valsaraj 1988; Valsaraj & Thibodeaux 1989)

134 (shake-flask-GC/ECD, Chiou & Schmedding 1981; Chiou et al. 1982; 1983; Chiou 1985; quoted, Suntio et al. 1988b)

124 (reommended, Horvath 1982; quoted, Wong et al. 1984)

125 (quoted lit. average, Yalkowsky et al. 1983)

124.5 (gen. col.-GC/ECD, Miller et al. 1984; 1985; quoted, Doucette & Andren 1988; Suntio et al. 1988b; Hawker 1989)

143 (shake flask-HPLC, Bnaerjee 1984; quoted, Suntio et al. 1988b)

143, 47.2 (quoted, calculated-UNIFAC, Banerjee 1985)

106 (recommended, IUPAC 1985)

122 (selected, Bobra et al. 1985)

110 (Dean 1985)

111 (20°C, Riddick et al. 1986; quoted, Howard 1989)

144 (quoted, Kamlet et al. 1987)

131 (calculated-solvatochromic parameters, Kamlet et al. 1987)

120 (selected, Suntio et al. 1988b)

48.1 (calculated-UNIFAC, Banerjee & Howard 1988)

123 (correlated, Isnard & Lambert 1988; 1989)

150 (quoted, Lee et al. 1989)

98.6 (calculated- χ , Nirmalakhandan & Speece 1988a)

109.65 (quoted, Figueroa & Simmons 1991)

Vapor Pressure (Pa at 25°C):

307 (extrapolated, Antoine eqn., Weast 1972-73; quoted, Mackay & Shiu 1981; Howard 1989)
389 (quoted, Hine & Mookerjee 1975)
303.9 (quoted, Mabey et al. 1982)
307, 250 (quoted, exptl. calculated-B.P., Mackay et al. 1982a)
307 (selected, Bobra et al. 1985; Suntio et al. 1988b)
266 (extrapolated, Antoine eqn., Boublik et al. 1984)
265 (extrapolated, Antoine eqn., Dean 1985)
243 (extrapolated, Antoine eqn., Stephenson & Malanowski 1987)
287, 160 (quoted, calculated-UNIFAC, Banerjee et al. 1990)

Henry's Law Constant (Pa m^3/mol):

472 (calculated as $1/K_{AW}$, C_W/C_A, reported as exptl., Hine & Mookerjee 1975; selected exptl., Nirmalakhandan & Speece 1988a)
375 (calculated-bond contribution, Hine & Mookerjee 1975)
365 (calculated, Mackay & Shiu 1981; quoted, Pankow et al. 1984; Pankow & Rosen 1988; Pankow 1986,1990)
366 (25°C, calculated, Mabey et al. 1982)
370 (selected, Bobra et al. 1985)
182 (20°C, batch stripping, Oliver 1985; quoted, Howard 1989)
289 (EPICS, Ashworth et al. 1988)
298 (20°C, EPICS, Ashworth et al., 1988)
472, 119 (observed, calculated-QASR, Nirmalakhadan & Speece, 1988b)
376.1 (calculated-P/C, Suntio et al. 1988b)
365 (quoted, Valsaraj 1988)

Octanol/Water Partition Coefficient, log K_{OW}:

3.55 (Hansch et al. 1968)
3.38 (Leo et al. 1971; Hansch & Leo 1979; quoted, Chiou & Freed 1977; Chiou & Schmedding 1981; Chiou et al. 1982; 1983; Chiou 1985; Veith et al. 1983; Suntio et al. 1988b)
3.38; 3.57, 3.55 (quoted, calculated-f constants, Rekker 1977)
3.59 (calculated-f const., Yalkowsky et al. 1979; Yalkowsky & Valvani 1980; Valvani & Yalkowsky 1980; quoted, Miller et al. 1984; Suntio et al. 1988b)
3.60 (shake flask-GC, Könemann et al. 1979; quoted, Kaiser et al. 1984; Suntio et al. 1988b; Figueroa & Simmons 1991)
3.62 (HPLC-k', Könemann et al. 1979)
3.53 (calculated-f const., Könemann et al. 1979; quoted, Könemann 1981; Leegwater 1989)
3.44 (shake flask-LSC, Banerjee et al. 1980; quoted, Oliver 1987a; Suntio et al. 1988b)
3.44 (shake flask-LSC, Veith et al. 1980)
3.95, 3.55 (HPLC-RT, calculated-f constant, Veith et al. 1980)

3.48 (Chiou 1981; quoted, Valsaraj & Thibodeaux 1989)
3.53 (shake flask-GC, Wateral et al. 1982)
3.56 (calculated, Mabey et al. 1982)
3.46 (shake flask-HPLC, Hammers et al. 1982)
3.38-3.62, 3.52 (range, mean, shake flask method, Eadsforth & Moser 1983)
3.62-3.95, 3.73 (range, mean, HPLC method, Eadsforth & Moser 1983)
3.48 (gen. column-HPLC/UV, Wasik et al. 1983)
3.59 (calculated-f const., Yalkowsky et al. 1983)
3.38 (quoted, Verschueren 1983)
3.44 (quoted, Oliver & Niimi 1983)
3.60 (quoted, Kaiser 1983)
3.48 (gen. col.-GC/ECD, Miller et al. 1984; 1985; quoted, Suntio et al. 1988b; Hawker & Connell 1989; Hawker 1989)
3.40 (quoted, Oliver & Charlton 1984)
3.57 (HPLC-RV, Garst 1984)
3.40 (quoted, Garst 1984)
3.48 (Sarna et al. 1984; quoted, Kamlet et al. 1988)
3.40 (selected, Bobra et al. 1985; Abernethy & Mackay 1987; Suntio et al. 1988b)
3.60 (Hansch & Leo 1985; quoted, Howard 1989)
3.48 (quoted exptl., Doucette & Andren 1988)
3.60, 3.13 (quoted, calculated-UNIFAC, Banerjee & Howard 1988)
3.55 (quoted, Ryan et al. 1988)
3.67, 3.59, 3.77, 3.34, 3.36 (calculated-π, f const., MW, χ , TSA, Doucette & Andren 1988)
3.35 (calculated-V_I, solvatochromic parameters, Kamlet et al. 1988)
3.50 (shake flask-GC, Pereira et al. 1988)
3.433 (slow stirring, De Bruijn et al. 1989; De Bruijn & Hermens 1990)
3.65 (calculated-f const., De Bruijn et al. 1989)
3.48 (quoted, Isnard & Lambert 1988; 1989)

Bioconcentration Factor, log BCF:
1.82 (bluegill sunfish, Veith et al. 1979b; 1980; quoted, Suntio et al. 1988b)
2.22-2.68 (bluegill sunfish, calculated, Veith et al. 1980)
1.82 (bluegill sunfish, whole body, flow system, Barrows et al. 1980)
2.86 (microorganisms-water, Mabey et al. 1982)
1.82 (bluegill sunfish, quoted, Bysshe 1982)
2.62-2.87 (rainbow trout, Oliver & Niimi 1983)
3.70-4.02 (rainbow trout, lipid base, Oliver & Niimi 1983; quoted, Suntio et al. 1988b)
2.62-2.87 (fish, Oliver 1984)
2.48 (fish-calculated, Garst 1984)
1.82 (bluegill sunfish, quoted, Davis & Dobbs 1984)
3.70-4.02 (quoted, rainbow trout, Chiou 1985)
1.99 (fathead minnow, flowing water, Carlson & Kosian 1987)

2.01 (fish, calculated- χ , Sabljic 1987b)
3.60, 3.86, 3.25, 3.40 (field data-lipid based: Atlantic croakers, blue crabs, spotted sea trout, blue catfish, Pereira et al. 1988)
2.62 (fish-correlated, Isnard & Lambert 1989)
2.51 (fish, calculated, Figueroa & Simmons 1991)

Sorption Partition Coefficient, log K_{OC}:
3.23 (calculated, Mabey et al. 1982)
2.47-3.39 (estimated from solubility & K_{OW}, Lyman et al. 1982)
4.60 (field data, Oliver & Charlton 1984)
4.60 (Niagara River-organic matter, Oliver & Charlton 1984)
3.00 (calculated from K_{OW}, Oliver & Charlton 1984)
2.65 (calculated-m. connectivity, Bahnick & Doucette, 1988)
2.14 (soil, Lee et al. 1989)

Sorption Partition Coefficient, log K_{OM}:
2.23 (soil organic matter, sorption isotherm, Chiou et al. 1982)
2.47 (soil, Chiou et al. 1983)
2.23, 2.53 (selected, calculated- χ , Sabljic 1984)
3.88 (micelle-water, Valsaraj & Thibodeaux 1989)

Half-Lives in the Environment:
Air: 200.6-2006 hours, based on photooxidation half-life in air (Atkinson 1985).
Surface Water: 672-4320 hours, based on aqueous aerobic biodegradation half-life (Haider et al. 1981);
0.9-50 days, estimated (Zoeteman et al. 1980).
Groundwater: 1334-8640 hours, based on aqueous aerobic biodegradation half-life (Haider et al. 1981); estimated half-life from observed persistence in groundwater of the Netherlands, one year (Zoeteman et al. 1981).
Soil: 672-4320 hours, based on unacclimated aerobic screening test data (Canton et al. 1985) and aerobic soil grab sample data (Haider et al. 1981).
Biota: half-life in bluegill sunfish, < 1 day (Veith et al. 1980; Barrows et al. 1980); in worms at 8°C, < 5 days (Oliver 1987a).

Environmental Fate Rate Constants or Half-Lives:
Volatilization: estimated half-life from a model river of 1 m depth with a current of 1 m/sec and wind velocity of 3 m/sec at 20°C is 4.1 hours (Lyman et al. 1982).
Photolysis: not environmentally significant or relevant (Mabey et al. 1982).
Oxidation: k (singlet oxygen) << 360 M^{-1} h^{-1} and k (RO_2 radical) <<1 M^{-1} h^{-1} (Mabey et al. 1982);
photooxidation half-life is 200.6-2006 hours, based on a measured room temperature rate constant of 0.72×10^{-12} cm^3 $molecule^{-1}$ sec^{-1} for the vapor phase reaction with hydroxy radicals in air (Atkinson 1985).

Hydrolysis: not environmentally significant (Mabey et al. 1982); first-order half-life > 879 years, based on $k_h < 0.9\ M^{-1}\ h^{-1}$ extrapolated to pH 7 at 25°C from 1% disappearance after 16 days at 85°C and pH 9.7 (Ellington et al. 1988).

Biodegradation: aqueous aerobic half-life is 672-4320 hours, estimated from unacclimated soil grab sample data (Haider et al. 1981; Marinucci & Bartha 1979); aqueous anaerobic half-life is 2688-17280 hours, based on estimated unacclimated aqueous aerobic biodegradation half-life (Howard et al. 1991); significant degradation in an aerobic environment. $k_B = 0.05\ d^{-1}$ (Tabak et al. 1981; Mills et al. 1982); in a continuous flow activated sludge system nearly 100% removed by an apparent combination of biodegradation and stripping (Kincannon et al. 1983; selected, Howard 1989).

Biotransformation: $k_b = 1 \times 10^{-10}\ ml\ cell^{-1}\ h^{-1}$ (estimated, Mabey et al. 1982).

Common Name: 1,4-Dichlorobenzene
Synonym: p-dichlorobenzene, paradichlorobenzene
Chemical Name: 1,4-dichlorobenzene
CAS Registry No: 106-46-7
Molecular Formula: $C_6H_4Cl_2$
Molecular Weight: 147.01
Melting Point (°C):
53.0 (Pirsch 1956; Verschueren 1977,1983)
53.1 (Mackay et al. 1980; Mackay & Shiu 1981; Kishii et al. 1987; Howard 1989)
25.0 (Yalkowsky et al. 1983)
53.5 (Miller et al. 1984)
52.9 (Miller et al. 1985)
Boiling Point (°C):
173.4 (Verschueren 1977,1983)
174.0 (Mackay & Shiu 1981; Miller et al. 1984)
Density (g/cm^3 at 20°C):
1.241
1.2457 (Weast 1972-73; Horvath 1982)
Molar Volume (cm^3/mol):
118 (calculated-density, Weast 1972-73; Lande & Banerjee 1981; Horvath 1982)
137.8 (LeBas method, Miller et al. 1985; Shiu et al. 1987; Suntio et al. 1988a)
138 (Abernethy & Mackay 1987)
124, 67 (from density, intrinsic, Abernethy et al. 1988)
0.671 (intrinsic volume: $V_I/100$, Kamlet et al. 1988; Hawker 1989, 1990)
Molecular Volume (A^3):
153.7 (De Bruijn & Hermens 1990)
Total Surface Area, TSA (A^2):
144.7 (Yalkowsky et al. 1979)
144.9 (Kishii et al. 1987)
141.0 (Sabljic 1987a)
144.24 (Doucette & Andren 1988)
171.3 (De Bruijn & Hermens 1990)
Heat of Fusion, kcal/mol:
4.274 (Weast 1972-73)
4.34, 4.47 (Wauchope & Getzen 1972)
4.35 (Dean, 1985)
4.35 (Tsonopoulos & Prausnitz 1971)
4.54 (Miller et al. 1984)
Entropy of Fusion, cal/mol K (e.u.):
13.5 (Pirsch 1956)
13.3 (Tsonopoulos & Prausnitz 1971)
13.1 (Weast 1976-77; Amidon & Williams 1982)
13.1 (Yalkowsky 1979)

13.4 (Yalkowsky & Valvani 1980)
13.9 (Miller et al. 1984)

Fugacity Ratio, F:

0.53 (Mackay et al. 1980; Suntio et al. 1988b)
0.47 (20°C, Suntio et al. 1988a)

Water Solubility (g/m^3 or mg/L at 25 °C):

79.1 (volumetric, Klemenc & Löw 1931; Seidell 1941; quoted, Suntio et al. 1988b)
77 (30 °C, shake flask-interferometer, Gross & Saylor 1931; quoted, Vesala 1974)
< 500 (residue-volume method, Booth & Everson 1948)
76 (shake flask-UV, Andrew & Keefer 1950; quoted, Suntio et al. 1988b)
89.8 (Landold-Börnstein 1951)
80 (Irmann 1965; quoted, Schwarzenbach et al. 1979)
71.3 (quoted, Tsonopoulos & Prausnitz 1971)
83.1 (shake flask-UV, Wauchope & Getzen 1972; quoted, Vesala 1974; Suntio et al. 1988b)
85.5 (shake flask-UV, Vesala 1974)
77.2 (quoted, Hine & Mookerjee 1975)
56.9 (20°C, shake flask-GC/ECD, Chiou & Freed 1977)
79.0 (quoted, Chiou et al. 1977; Freed et al. 1977; Haque et al. 1980; Kenaga & Goring 1980)
87.2 (shake flask-GC, Aquan-Yuen et al. 1979; quoted, Suntio et al. 1988b)
90.6 (shake flask-UV, Yalkowsky et al. 1979; quoted, Mackay & Shiu 1981; Miller et al. 1984; Suntio et al. 1988b)
68.8 (calculated-K_{OW}, Yalkowsky et al. 1979; Yalkowsky & Valvani 1980; Valvani & Yalkowsky 1980)
73.7, 140 (shake flask-LSC, calculated-K_{OW}, Veith et al. 1980)
73.8 (shake flask-LSC, Banerjee et al. 1980; quoted, Suntio et al. 1988b; Howard 1989)
79 (quoted, Kenaga 1980a; Kenaga & Goring; Mackay 1980; Neely 1980)
48.7 (shake flask-GC, Könemann 1981)
73, 137 (shake flask-GC, subcooled liquid, Chiou et al. 1982)
189 (calculated-melting point, Amidon & Williams 1982)
79 (quoted, Mabey et al. 1982)
90.0 (recommended, Horvath 1982; quoted, Wong et al 1984; Oliver 1987b)
25 (calculated-UNIFAC, Arbuckle 1983)
79 (quoted, Calamari et al. 1983)
175 (calculated-HPLC-k', converted from reported γ_W, Hafkenscheid & Tomlinson 1983a)
86.6 (quoted lit. average, Yalkowsky et al. 1983)
30.9 (gen. col.-GC, Miller et al. 1984,1985; quoted, Doucette & Andren 1988; Suntio et al. 1988b; Hawker 1989)
65.3 (shake flask-HPLC, Banerjee 1984; quoted, Suntio et al. 1988b)
82.9 (recommended, IUPAC 1985)

77.5 (selected, Bobra et al. 1985)
100 (Dean 1985)
76.32, 47.2 (quoted, calculated-UNIFAC, Banerjee 1985)
70.0 (quoted, Mackay et al. 1985)
83.5 (quoted, Abernethy et al. 1986; Suntio et al. 1988b)
87.0 (quoted, Riddick et al. 1986; Howard 1989)
30.9, 60-70 (quoted, calculated-K_{OW}, Anliker & Moser 1987)
98.6 (calculated- χ , Nirmalakhandan & Speece 1988a)
148.5 (selected, Suntio et al. 1988b)
69.2 (correlated, Isnard & Lambert 1989)
79.4 (quoted, Figueroa & Simmons 1991)

Vapor Pressure (Pa at 25 °C):
90.2 (solid vapour pressure, extrapolated, Antoine eqn., Weast 1972-73; quoted, Mackay & Shiu 1981; Suntio et al. 1988b)
235 (quoted, Hine & Mookerjee 1975)
53.3 (quoted, Haque et al. 1980)
133 (quoted, Neely 1980)
90, 125.6 (quoted exptl., calculated-B.P., Mackay et al. 1982a)
134 (calculated-HLC, Arbuckle 1983)
128 (extrapolated, Antoine eqn., Boublik et al. 1984)
243 (extrapolated, Antoine eqn., Dean 1985)
86.7 (20 °C, gas saturation, Chiou & Shoup 1985)
47.5 (quoted, Mackay et al. 1985)
235 (quoted, Riddick et al. 1986; Howard 1989)
140 (extrapolated, Antoine eqn., Stephenson & Malanowski 1987)
266 (selected, Suntio et al. 1988a)
173 (selected, Suntio et al. 1988b)

Henry's Law Constant (Pa m^3/mol):
451 (calculated as $1/K_{AW}$, C_W/C_A, reported as exptl., Hine & Mookerjee 1975; selected exptl., Nirmalakhandan & Speece 1988a)
375 (calculated-bond contribution, Hine & Mookerjee 1975)
157.3 (Callahan et al. 1979)
744 (Matter 1979; Schwarzenbach et al. 1979)
330 (direct concentration ratio, Leighton & Calo 1981)
162 (Mackay 1981)
240 (batch stripping, Mackay & Shiu 1981; quoted, Suntio et al. 1988b)
160 (selected value, Mackay & Shiu 1981; quoted, Pankow et al. 1984)
347 (calculated-C_A/C_W, Matter-Müller et al. 1981)
314 (25°C, calculated, Mabey et al. 1982; quoted, Pankow 1986,1990; Pankow & Rosen 1988)
465 (calculated-UNIFAC, Arbuckle 1983)

243 (calculated-P/C, Calamari et al. 1983)
171 (selected, Bobra et al. 1985; Suntio et al. 1988b)
152 (20°C, batch stripping, Oliver 1985; quoted, Howard 1989)
365 (EPICS, Ashworth et al. 1988)
262.4 (20°C, EPICS, Ashworth et al. 1988)
197, 119 (observed, calculated-QSAR, Nirmalakhandan & Speece 1988b)
262.5 (20°C, calculated-P/C, Suntio et al. 1988a)

Octanol/Water Partition Coefficient, log K_{OW}:

3.39 (Leo et al. 1971; Hansch & Leo 1979; quoted, Kenaga & Goring 1980; Mackay et al. 1980; Chiou & Schmedding 1981; Freitag et al. 1985)
3.39; 3.57, 3.55 (quoted; calculated-f constants, Rekker 1977)
3.38 (Hansch & Leo 1979; quoted, Suntio et al. 1988b)
3.39 (quoted, Chiou et al. 1977, 1982; Chiou & Freed 1977; Freed et al. 1977; Chiou 1981; 1985; Schwarzenbach et al. 1979; Schwarzenbach & Westall 1981; Brooke et al. 1986)
3.59 (calculated-f const., Yalkowsky et al. 1979; Yalkowsky & Valvani 1980; Valvani & Yalkowsky 1980; Yalkowsky et al. 1983; quoted, Miller et al. 1984; Suntio et al. 1988b)
3.62, 3.39 (shake flask-GC, HPLC-k', Könemann et al. 1979; quoted, Figueroa & Simmons 1991)
3.53 (calculated-f const., Könemann et al. 1979; Könemann 1981; quoted, Opperhuizen 1986; Suntio et al. 1988b; Leegwater 1989)
3.24 (calculated, Mackay et al. 1980)
3.37, 3.78, 3.55 (shake flask-LSC, HPLC-RT, calculated-f constant, Veith et al. 1980)
3.37 (shake flask-LSC, Banerjee et al. 1980; quoted, Oliver 1987a & b; Suntio et al. 1988b)
3.38, 3.24 (quoted, calculated-S, Mackay et al. 1980)
3.46 (HPLC-k', Hammers et al. 1982)
3.56 (calculated, Mabey et al. 1982)
3.53 (quoted, Mackay 1982)
3.52 (shake flask-GC, Wateral et al. 1982; quoted, Suntio et al. 1988b)
3.38, 3.43 (quoted, HPLC-k', Miyake & Terada 1982)
3.67, 3.9 (calculated-UNIFAC, Arbuckle 1983)
3.37 (quoted, Oliver & Niimi 1983; Veith et al. 1983)
3.17 (quoted, Calamari et al. 1983)
3.62 (quoted, Kaiser 1983; Kaiser et al. 1984)
3.39 (quoted, Hafkenscheid & Tomlinson 1983b)
3.37 (gen. column-HPLC/UV, Wasik et al. 1983)
3.38, 3.14 (quoted, calculated-molar refraction, Yoshida et al. 1983a)
3.38 (gen. col.-GC/ECD, Miller et al. 1984; 1985; quoted, Suntio et al. 1988b; Clark et al. 1990)
3.37, 3.37 (quoted, HPLC-RV, Garst 1984; Garst & Wilson 1984)

3.40 (quoted, Oliver & Charlton 1984; Oliver & Niimi 1985; Oliver 1987a,b)
3.38 (Sarna et al. 1984)
3.40 (selected, Bobra et al. 1985; Abernethy & Mackay 1987; Suntio et al. 1988b)
3.52 (Hansch & Leo 1985; quoted, Howard 1989)
3.37 (quoted, Hawker & Connell 1985, 1988; Hawker 1990)
3.42 (quoted, Mackay et al. 1985)
3.38 (quoted, Gobas et al. 1987, 1989)
3.38 (quoted exptl., Doucette & Andren 1988)
3.55 (quoted, Ryan et al. 1988)
3.57, 3.59, 3.77, 3.34, 3.36 (calculated-π, f const., MW, χ , TSA, Doucette & Andren 1988)
3.38, 3.48 (selected, calculated-V_I, solvatochromic parameters, Kamlet et al. 1988)
3.48 (quoted, Isnard & Lambert 1989; Hawker 1989)
3.444 (slow stirring-GC, De Bruijn et al. 1989)
3.65 (calculated-f const., De Bruijn et al. 1989)
3.37 (quoted, Thomann 1989)
3.37 (quoted, Hawker 1990)

Bioconcentration Factor, log BCF:
2.33 (rainbow trout, Neely et al. 1974; quoted, Mackay 1982b; Suntio et al. 1988b)
2.33 (fish, flowing water, Kenaga & Goring 1980; quoted, Kenaga 1980a)
1.78 (bluegill sunfish, Veith et al. 1979b,1980; quoted, Suntio et al. 1988b)
2.16-2.51 (fish, calculated, Veith et al. 1980)
-0.347 (activated sludge, dry weight, Schwarzenbach et al. 1979)
1.78 (bluegill sunfish, whole body, flow system, Barrows et al. 1980)
3.26 (guppy, lipid basis, Könemann & van Leeuwen 1980; quoted, Suntio et al. 1988b)
2.21 (fish, correlated, Mackay 1982)
2.86 (microorganisms-water, Mabey et al. 1982)
1.78 (bluegill sunfish, quoted, Bysshe 1982)
3.64-3.96 (rainbow trout, lipid basis, Oliver & Niimi 1983; quoted, Suntio et al. 1988b)
2.57-2.86 (rainbow trout, whole body, Oliver & Niimi 1983)
1.78 (fish, quoted, Dobbs & Williams 1983; Davies & Dobbs 1984; Anliker & Moser 1987)
1.99 (fish, calculated-K_{OW}, Calamari et al. 1983)
3.40 (15°C, guppy, Banerjee et al. 1984a)
2.83-2.86 (fish, Oliver 1984)
2.00, 1.70, 2.75 (algae, fish, activated sludge, Freitag et al. 1984; quoted, Suntio et al. 1988b)
2.33 (fish, calculated, Garst 1984)
2.0, 1.70, 2.75 (algae, fish, activated sludge, Freitag et al. 1985)
2.71-2.95 (rainbow trout, Oliver & Niimi 1985)
2.04 (fathead minnow, flowing water, Carlson & Kosian 1987)
5.3-5.6, 5.5; 3.0 (Niagara River plume, range, mean; calculated-K_{OW}, Oliver 1987b)

1.99 (guppy, calculated-K_{OW}, Gobas et al. 1987)
3.25 (guppy-lipid phase, calculated-K_{OW}, Gobas et al. 1987)
3.91, 4.53, 4.09, 3.51 (field data-lipid based: Atlantic creakers, blue crabs, spotted sea trout, blue catfish, Pereira et al. 1988)
3.64 (guppy-lipid phase, calculated-K_{OW}, Gobas et al. 1989)
1.98 (fish, calculated-C_B/C_W or k_1/k_2, Connell & Hawker 1988)
1.98 (fish, calculated-solvatochromic parameters, Hawker 1990)
2.47, 3.56 (flagfish: whole fish, lipid, Smith et al. 1990)
2.52 (fish, calculated, Figueroa & Simmons 1991)

Sorption Partition Coefficient, log K_{OC}:
2.59 (Kenaga 1980a)
2.78-3.26 (organic carbon, Schwarzenbach & Westall 1981)
3.23 (calculated, Mabey et al. 1982)
2.61-3.18 (calculated from solubility & K_{OW}, Lyman et al. 1982)
3.10 (soil, calculated-K_{OW}, Calamari et al. 1983)
4.80 (field data, Oliver & Charlton 1984)
5.0 (Niagara River organic matter, Oliver & Charlton 1984)
3.0 (calculated from K_{OW}, Oliver & Charlton 1984)
5.3-5.6, 5.5, 3.0 (Niagara River plume, range, average, calculated-K_{OW}, Oliver 1987b)
2.65 (calculated- χ , Bahnick & Doucette 1988)
2.92 (calculated-polymaleic acid, Chin & Weber 1989)
2.91, 2.43 (Aldrich humic acid, organic polymers, Chin et al. 1990)

Sorption Coefficient, log K_{OM}:
2.20 (soil org. matter, sorption isotherm, Chiou et al. 1983)
2.40, 2.53 (quoted, calculated- χ , Sabljic 1984)

Half-Lives in the Environment:
Air: 200.6-2006 hours, based on photooxidation in air (Atkinson 1985).
Surface Water: 672-4320 hours, based on aqueous aerobic biodegradation half-life (Haider et al. 1981); 1.1-26 days, estimated (Zoeteman et al. 1980).
Groundwater: 1344-8640 hours, based on aqueous aerobic biodegradation half-life (Haider et al. 1981); estimated half-life from observed persistence in groundwater of the Netherlands, one year (Zoeteman et al. 1981).
Soil: 672-4320 hours, based on unacclimated aerobic screening test data (Canton et al. 1985) and aerobic soil grab sample data (Haider et al. 1981).
Biota: half-life in fish tissues < 1 day (Veith et al. 1980); half-life in bluegill sunfish < 1 day (Barrows et al. 1980); 16 hours, clearance from fish (Neely 1980); half-life in worms at 8°C, < 5 days (Oliver 1987a); 0.70 days, 0.59 days clearance from flagfish: whole fish, fish lipid (Smith et al 1990).

Environmental Fate Rate Constants or Half-Lives:

Volatilization: half-life from a model river of 1 m depth and with a current of 1 m/s and wind velocity of 3 m/s at 20°C is 4.3 hours (Lyman et al. 1982); half-lives from marine mesocosm: 18 days in the spring at 8-16°C, 10 days in the summer at 20-22°C and 31 days in the winter at 3-7°C (Wakeham et al. 1983).

Photolysis: not environmentally significant or relevant (Mabey et al. 1982). Direct photolysis unimportant (Zepp & Cline 1977; Zepp 1978).

Oxidation: k (singlet oxygen) << 360 M^{-1} h^{-1} and k (RO_2 radicals) << 1 M^{-1} h^{-1} (Mabey et al. 1982); photooxidation half-life is 200.6-2006 hours (31 days), based on measured rate for the vapor phase reaction with hydroxy radicals in air with rate constant of 3.2×10^{-13} cm^3 $molecule^{-1}$ sec^{-1} at room temperature (Atkinson 1985); reaction rate constant in air, k = 9.63×10^{-3} $hour^{-1}$ (Ware & West 1977; selected, Mackay et al. 1985).

Hydrolysis: not environmentally significant (Mabey et al. 1982); first order half-life > 879 years, based on k_h < 0.9 M^{-1} h^{-1} extrapolated to pH 7 at 25°C from 1% disappearance after 16 days at 85°C and pH 9.7 (Ellington et al. 1988).

Biodegradation: degradation in water, $k_B = 1.09 \times 10^{-2}$ $hour^{-1}$ (Verschueren 1977; selected, Mackay et al. 1985); aqueous aerobic half-life is 672-4320 hours, based on unacclimated aerobic screening test data (Canton et al. 1985) and aerobic soil grab sample data (Haider et al. 1981); significant degradation in aerobic environment, k_B = 0.05 day^{-1} Tabak et al. 1981; Mills et al. 1982); aqueous anaerobic biodegradation half-life is 2688-17280 hours, based on estimated unacclimated aqueous aerobic biodegradation half-life (Howard et al. 1991).

Biotransformation: $k_b = 1 \times 10^{-10}$ ml $cell^{-1}$ h^{-1} (estimated, Mabey et al. 1982).

Bioconcentration, Uptake (k_1) and Elimination (k_2) Rate Constants:

- k_1: 5.67 $hour^{-1}$ (trout muscle, Neely et al. 1974; Neely 1979)
- k_1: 7.44 $hour^{-1}$ (10°C, yellow perch, Neely et al. 1974; Neely 1979)
- k_1: 17.0 $hour^{-1}$ (25°C, yellow perch, Neely et al. 1974; Neely 1979)
- k_2: 0.0264 $hour^{-1}$ (trout muscle, Neely et al. 1974)
- k_1: 1800 day^{-1} (guppy, Könemann & van Leeuwen 1980)
- k_2: 1.00 day^{-1} (guppy, Könemann & van Leeuwen 1980; quoted, Clark et al. 1990)
- k_1: 4.1 $hour^{-1}$ (guppy, quoted from Könemann & van Leeuwen 1980, Hawker & Connell 1985)
- k_1: 5.7 $hour^{-1}$ (rainbow trout, quoted from Neely et al. 1974, Hawker & Connell 1985)
- $1/k_2$: 38 hour (trout, quoted, Hawker & Connell 1985, 1988a & b; Connell & Hawker 1988)
- $1/k_2$: 24 hour (guppy, quoted, Hawker & Connell 1985, 1988; Connell & Hawker 1988)
- k_1: 97 day^{-1} (fish, quoted, Opperhuizen 1986)
- $\log k_2$: 0, -0.2 day^{-1} (fish, calculated-K_{OW}, Thomann 1989)
- $\log k_1$: 1.99 day^{-1} (guppy, Gobas et al. 1989)

log k_2: 0.0 (guppy, Gobas et al. 1989)

log k_1: 1.99 day^{-1} (fish, quoted, Connell & Hawker 1988)

k_1: 291 d^{-1}, 4230 d^{-1} (American flagfish: whole fish, fish lipid, Smith et al. 1990)

k_2: 0.98 d^{-1}, 1.18 d^{-1} (American flagfish: whole fish, fish lipid, Smith et al. 1990)

k_2: 0.98 d^{-1}, 1.46 d^{-1} (American flagfish: bioconcentration exptl., toxicity data, Smith et al. 1990)

Common Name: 1,2,3-Trichlorobenzene
Synonym: vic-trichlorobenzene
Chemical Name: 1,2,3-trichlorobenzene
CAS Registry No: 81-61-6
Molecular Formula: $C_6H_3Cl_3$
Molecular Weight: 181.45
Melting Point (°C):
52.0 (Verschueren 1977,1983)
53.0 (Mackay & Shiu 1981)
51.0 (Yalkowsky et al. 1983)
52.6 (Miller et al. 1984,1985; Kishii et al. 1987)
Boiling Point (°C):
221
219 (Verschueren 1977,1983)
218 (Mackay & Shiu 1981; Miller et al. 1984)
Density (g/cm^3 at 20°C):
1.69
1.4533 (40°C, Weast 1972-73; Horvath 1982)
Molar Volume (cm^3/mol):
124.9 (calculated-density, Weast 1972-73; Horvath 1982)
158.7 (LeBas method, Miller et al. 1985; Shiu et al. 1987)
125 (calculated-density, Chiou 1985)
159 (Abernethy & Mackay 1987)
135, 76 (from density, intrinsic, Abernethy et al. 1988)
0.761 (intrinsic volume: $V_I/100$, Kamlet et al. 1988; Hawker 1989b,1990)
Molecular Volume (A^3):
163.3 (De Bruijn & Hermens 1990)
Total Surface Area, TSA (A^2):
158.3 (Yalkowsky et al. 1979)
158.7 (Kishii et al. 1987)
158.1 (calculated- χ , Sabljic 1987a)
157.7 (Doucette & Andren 1988)
179.2 (De Bruijn & Hermens 1990)
Heat of Fusion, kcal/mol:
4.15 (Tsonopoulos & Prausnitz 1971)
4.30 (Miller et al. 1984)
Entropy of Fusion, cal/mol K (e.u.):
12.7 (Tsonopoulos & Prausnitz 1971)
13.6 (Miller et al. 1984)
Fugacity Ratio, F (assuming ΔS_{fusion} = 13.5 e.u.):
0.53 (25°C, Suntio et al. 1988b)

Water Solubility (g/m^3 or mg/L at 25°C):

25.10 (quoted, Tsonopoulos & Prausnitz 1971)
31.5 (shake flask-UV, Yalkowsky et al. 1979,1983; quoted, Hodson et al. 1984; Miller et al. 1984; Doucette & Andren 1988; Suntio et al. 1988b; Paya-Perez et al. 1991)
18.1 (calculated-K_{OW}, Yalkowsky et al. 1979; Yalkowsky & Valvani 1980; Valvani & Yalkowsky 1980)
16.6 (shake flask -GC, Mackay & Shiu 1981; quoted, Calamari et al. 1983; Suntio et al. 1988b)
12.0 (shake flask-GC, Könemann 1981)
31.6 (recommended, Horvath 1982; quoted, Wong et al. 1984)
12.27 (gen. col.-GC/ECD, Miller et al. 1984; 1985; quoted, Doucette & Andren 1988; Suntio et al. 1988b)
18.0 (shake flask-HPLC, Banerjee 1984; quoted, Suntio et al. 1988b)
16.3 (23°C, shake flask-GC, Chiou 1985)
21.0 (selected, Bobra et al. 1985; Suntio et al. 1988b)
16.7 (quoted, Abernethy et al. 1986)
31.5, 158 (quoted, calculated-K_{OW} & HPLC-RT, Chin et al. 1986)
18.0 (shake flask-GC, Chiou et al. 1986; Chiou et al. 1991)
12.0 (quoted, Isnard & Lambert 1988, 1989)
20.5 (calculated- χ , Nirmalakhandan & Speece 1988a)
40.6 (quoted, subcooled liquid, Hawker 1989)
30.9 (quoted, Figueroa & Simmons 1991)

Vapor Pressure (Pa at 25°C):

52.4 (extrapolated subcooled liquid, Antoine eqn., Weast 1972-73; quoted, Suntio et al. 1988b)
28 (selected, Mackay & Shiu 1981)
28.0, 15.8 (quoted exptl., calculated-B.P., Mackay et al. 1982a)
28 (selected, Bobra et al. 1985)
17.7 (extrapolated, Antoine eqn., Stephenson & Malanowski 1987)
66.5 (selected, subcooled liquid, Suntio et al. 1988b)
28, 18.7 (quoted, calculated-UNIFAC, Banerjee et al. 1990)

Henry's Law Constant (Pa m^3/mol):

127 (batch stripping, Mackay & Shiu, 1981; quoted, Suntio et al. 1988b)
306, 161 (calculated-P/C, Mackay & Shiu 1981)
101 (calculated-P/C, Calamari et al. 1983)
239 (calculated-P/C, Bobra et al. 1985)
90.2 (20°C, batch stripping, Oliver 1985)
124, 55.5 (observed, calculated-QSAR, Nirmalakahandan & Speece 1988b)
234 (calculated-P/C, Suntio et al. 1988b)

Octanol/Water Partition Coefficient, log K_{OW}:

4.27 (Leo et al. 1971; Hansch & Leo, 1979; quoted, Suntio et al. 1988b)
4.27 (calculated-f const., Yalkowsky et al. 1979; Yalkowsky & Valvani 1980; Valvani & Yalkowsky 1980; Yalkowsky et al. 1983; quoted, Miller et al. 1984)
4.11, 3.99 (shake flask-GC, HPLC-k', Könneman et al. 1979; quoted, Figueroa & Simmons 1991)
4.20 (calculated-f const., Könemann et al. 1979; Könemann 1981; quoted, Suntio et al. 1988b; Leegwater 1989)
4.02 (HPLC-k', McDuffie 1981; quoted, Suntio et al. 1988b)
3.97 (shake flask-GC, Wateral et al. 1982; quoted, Suntio et al. 1988b)
3.96 (HPLC-k', Hammers et al. 1982)
3.75 (quoted, Calamari et al. 1983)
4.11 (quoted, Oliver & Niimi 1983; Hawker & Connell 1985)
4.26 (quoted, Hodson et al. 1984)
4.04 (generator column-GC, Miller et al. 1984; 1985; quoted, Suntio et al. 1988b; Clark et al. 1990; Paya-Perez et al. 1991)
4.10 (quoted, Oliver & Charlton 1984; Oliver 1987a)
4.04 (Sarna et al. 1984)
4.11 (quoted, Kaiser 1983; Kaiser et al. 1984; Wong et al. 1984)
4.14 (shake flask-GC, Chiou 1985)
4.10 (selected, Bobra et al. 1985; Abernethy & Mackay 1987)
4.04 (quoted, Gobas et al. 1987; 1989; Hawker 1989)
4.10 (selected, Suntio et al. 1988b)
4.04 (selected exptl., Docette & Andren 1988)
4.28, 4.27, 4.32, 3.85, 3.69 (calculated-π, f const., MW, χ , TSA, Doucette & Andren 1988)
4.04, 3.79 (quoted, calculated-V_I, solvatochromic parameters, Kamlet et al. 1988)
4.14 (shake flask-GC, Pereira et al. 1988)
4.11 (selected, Thomann 1989)
4.11 (quoted, Connell & Hawker 1988; Hawker 1990)
4.40 (calculated-f const., De Bruijn et al. 1989)
4.139 (slow stirring-GC, De Bruijn et al. 1989)
4.27 (correlated, Isnard & Lambert 1988, 1989)
3.85-4.30, 4.0 (range, average: round robin work, shake flask or HPLC-k', Kishi & Hashimoto 1989)

Bioconcentration Factor, log BCF:

4.11 (guppy, lipid basis, Könneman & van Leeuwen 1980; quoted, Chiou 1985; Suntio et al. 1988b)
3.08-3.42 (rainbow trout, whole fish, Oliver & Niimi 1983)
4.15-4.47 (rainbow trout, lipid basis, Oliver & Niimi 1983; quoted, Chiou 1985; Suntio et al. 1988b)
2.49 (calculated-K_{OW}, Calamari et al. 1983)

3.08-3.42 (fish, Oliver 1984)
2.85 (quoted, rainbow trout, hatching, 3.2 % lipid, wet wt basis, Geyer et al. 1985)
4.35 (quoted, rainbow trout, hatching, 3.2 % lipid, lipid basis, Geyer et al. 1985)
2.85 (quoted, guppy, 5.4 % lipid, wet wt basis, Geyer et al. 1985)
4.11 (quoted, guppy, 5.4 % lipid, lipid basis, Geyer et al. 1985)
2.84 (guppy, calculated, Gobas et al. 1987)
4.11 (guppy-lipid phase, calculated, Gobas et al. 1987, 1989)
4.25 (guppy-lipid phase, calculated, Gobas et al. 1989)
2.85 (fish, calculated-C_B/C_W or k_1/k_2, Connell & Hawker 1988; Hawker 1990)
3.28 (guppy, Van Hoogan & Opperhuizen 1988)
4.76, 4.90, 3.54, 4.68 (field data-lipid based: Atlantic croakers, blue crabs, spotted sea trout, blue catfish, Pereira et al. 1988)
2.89 (fish, calculated, Figueroa & Simmons 1991)

Sorption Partition Coefficient, log K_{OC}:
3.37 (calculated-K_{OW}, Schwarzenbach & Westall 1981)
3.42 (soil, calculated-K_{OW}, Calamari et al. 1983)
4.70 (field data, Oliver & Charlton 1984)
4.10 (Niagara River-organic matter, Oliver & Charlton 1984)
3.70 (calculated-K_{OW}, Oliver & Charlton 1984)
3.0, 2.0 (soil, river humic acid, GC/ECD, Chiou et al. 1986)
2.3, 2.0 (soil, river fulvic acid, GC/ECD, Chiou et al. 1986)
2.77 (soil, calculated- χ , Sabljic 1987b)
3.18-3.43 (soil, batch-equilibration, Kishi et al. 1990)
3.91 (soil, Paya-Perez et al. 1991)

Half-Lives in the Environment:
Air:
Surface Water: 1.9-30 days, estimated (Zoeteman et al. 1980).
Groundwater:
Sediment:
Soil:
Biota: in worms at 8°C, < 5 days (Oliver 1987a).

Environment Fate Rate Constants or Half-Lives:
Volatilization:
Photolysis:
Oxidation:
Biodegradation:
Biotransformation:
Bioconcentration, Uptake (k_1) and Elimination (k_2) Rate Constants or Half-Lives:
k_1: 8300 day^{-1} (guppy, exptl., Könemann & van Leeuwen 1980)
k_2: 0.45 day^{-1} (guppy, exptl., Könemann & van Leeuwen 1980; selected,

Clark et al. 1990)

k_1: 18.7 hour^{-1} (guppy, quoted, Hawker & Connell 1985)

$1/k_2$: 53 hour (guppy, quoted, Hawker & Connell 1985; 1988)

k_2: -0.34 day^{-1} (fish, calculated-K_{OW}, Thomann 1989)

log k_1: 2.50 day^{-1} (guppy, Gobas et al. 1989)

log k_2: -0.35 day^{-1} (guppy, Gobas et al. 1989)

log k_1: 2.65 day (fish, quoted, Connell & Hawker 1988)

log k_2: 0.34 day (fish, quoted, Connell & Hawker 1988)

k_1: 780 mL/g.day (guppy, Van Hoogan & Opperhuizen 1988)

k_2: 0.42 day^{-1} (guppy, Van Hoogan & Opperhuizen 1988)

Common Name: 1,2,4-Trichlorobenzene
Synonym: unsym-trichlorobenzene
Chemical Name: 1,2,4-trichlorobenzene
CAS Registry No: 120-82-1
Molecular Formula: $C_6H_3Cl_3$
Molecular Weight: 181.45
Melting Point (°C):
17.0 (Verschueren 1977,1983; Mackay & Shiu 1981; Miller et al. 1985; Kishii et al. 1987)
18.0 (Schmidt-Bleek et al. 1982)
25.0 (Yalkowsky et al. 1983)
Boiling Point (°C):
213.0 (Verschueren 1977,1983)
213.5 (Mackay & Shiu 1981; Miller et al. 1984)
214.0 (Schmidt-Bleek et al. 1982)
Density (g/cm³ at 20°C):
1.5707
1.4542 (Weast 1972-73; Horvath 1982)
Molar Volume (cm³/mol):
124.8 (calculated-density, Weast 1972-73; Horvath 1982)
125 (calculated-density, Lande & Banerjee 1981)
158.7 (LeBas method, Miller et al. 1985; Shiu et al. 1987)
125 (Chiou et al. 1983; Chiou 1985)
159 (Abernethy & Mackay 1987)
135, 76 (from density, intrinsic, Abernethy et al. 1988)
0.761 (intrinsic volume: $V_I/100$, Kamlet et al. 1988; Hawker 1989b)
151 (selected, Valsaraj 1988)
125.55 (calculated, Govers et al. 1990)
Molecular Volume (A³):
167 (De Bruijn & Hermens 1990)
Total Surface Area, TSA (A²):
160.2 (Yalkowsky et al. 1979)
160.3 (Kishii et al. 1987)
158.1 (calculated- χ , Sabljic 1987a)
159.72 (Doucette & Andren 1988)
160 (selected, Valsaraj 1988)
181.7 (De Bruijn & Hermens 1990)
Heat of Fusion, kcal/mol:
3.70 (Tsonopoulos & Prausnitz 1971)
Entropy of Fusion, cal/mol K (e.u.):
12.8 (Tsonopoulos & Prausnitz 1971)
Fugacity Ratio, F: 1.0 (Suntio et al. 1988b)

Water Solubility (g/m^3 or mg/L at 25 °C):

25.03 (Irmann 1965; quoted, Tsonopoulos & Prausnitz 1971; Suntio et al. 1988b)
34.7 (shake flask-UV, Yalkowsky et al. 1979; quoted, Hodson et al. 1984; Miller et al. 1984; Doucette & Andren 1988; Suntio et al. 1988b; Paya-Perez et al. 1991)
30.0 (quoted, Callahan et al. 1979; Neuhauser et al. 1985)
30.0 (quoted, Kenaga & Goring 1980; Kenaga 1980a; Yoshida et al. 1983b)
33.0 (calculated-K_{OW}, Yalkowsky et al. 1979; Yalkowsky & Valvani 1980; Valvani & Yalkowsky 1980)
19.4 (shake flask-GC, Könemann 1981)
30.0 (recommended, Hovarth 1982; quoted, Wong et al. 1984; Nirmalakhandan & Speece 1988a; Oliver 1987b)
48.8 (20°C, shake flask-GC, Chiou & Schmedding 1981; Chiou et al. 1982; 1983; selected, Howard 1989)
34.6 (quoted lit. average, Yalkowsky et al. 1983)
30.0 (quoted, Calamari et al. 1983)
46.1 (gen. col.-GC/ECD, Miller et al. 1984; 1985; quoted, Doucette & Andren 1988; Suntio et al. 1988b)
31.3 (shake flask-HPLC, Banerjee 1984; quoted, Suntio et al. 1988b)
30.9 (recommended, IUPAC 1985)
34.6 (shake flask-GC, Chiou 1985)
40.0 (selected, Bobra et al. 1985; Suntio et al. 1988b)
25.0 (quoted, Mackay et al. 1985)
46, 29-45 (quoted, calculated-K_{OW}, Anliker & Moser 1987)
48.8 (quoted, Elzerman & Coates 1987)
20.9 (calculated- χ , Nirmalakhandan & Speece 1988a)
30.2 (quoted, Isnard & Lambert 1988, 1989)
30.0 (quoted, Valsaraj 1988)
20.0 (quoted, Lee et al. 1989)
45.6 (quoted, Hawker 1989)
19.5 (quoted, Figueroa & Simmons 1991)

Vapor Pressure (Pa at 25°C):

38.6 (Dreisbach 1955)
60.6 (extrapolated subcooled liquid, Antoine eqn., Weast 1972-73; quoted, Mackay & Shiu 1981)
56.0 (quoted, Callahan et al. 1979; Neuhauser et al. 1985)
38.6 (Mabey et al. 1982)
60.6, 37.3 (quoted, calculated-B.P., Mackay et al. 1982a)
26.4 (gas saturation, Politzki et al. 1982)
66.7 (quoted, Yoshida et al. 1983b)
60.8 (quoted, Mackay et al. 1985)
38.7 (extrapolated, Antoine eqn., Stephenson & Malanowski 1987)
61.0 (selected, Bobra et al. 1985; Suntio et al. 1988b)

38.5 (quoted, Valsaraj 1988)
38.7 (quoted, Howard 1989)

Henry's Law Constant (Pa m^3/mol):
144 (calculated-P/C, Lyman et al. 1982; quoted, Howard 1989)
233 (Mabey et al., 1982)
435 (calculated-P/C, Calamari et al. 1983)
396 (calculated-P/C, Yoshida et al. 1983b)
334 (calculated-P/C, Pankow et al. 1984)
122 (20°C, batch stripping, Oliver 1985)
277 (calculated-P/C, Bobra et al. 1985)
367 (calculated-P/C, Suntio et al. 1988b)
195 (EPICS, Ashworth et al. 1988)
235 (quoted, Valsaraj 1988)

Octanol/Water Partition Coefficient, log K_{OW}:
4.05 (Leo et al. 1971; quoted, Schwarzenbach & Westall 1981; Freitag et al. 1985)
4.26 (quoted, Callahan et al. 1979; Neuhauser et al. 1985)
4.02 (Hansch & Leo 1979; quoted, Suntio et al. 1988b; Lee et al. 1989)
4.27 (calculated-f const., Yalkowsky et al. 1979; Yalkowsky & Valvani 1980; Valvani & Yalkowsky 1980; quoted, Miller et al. 1984; Suntio et al. 1988b)
3.93, 4.12 (shake flask-GC, HPLC-k', Könemann et al. 1979; quoted, Figueroa & Simmons 1991)
4.20 (calculated-f const., Könemann et al. 1979; Könemann 1981; quoted, Suntio et al. 1988b; Leegwater 1989)
4.18 (quoted, Kenaga & Goring 1980; Yoshida et al. 1983b)
4.27 (calculated-f constant, Ellgehausen et al. 1981)
4.02 (20°C, shake flask-GC, Chiou & Schmedding 1981; Chiou et al. 1982; Chiou 1985; selected, Oliver & Niimi 1985; Oliver 1987b & c)
3.97 (shake flask-GC, Wateral et al. 1982; quoted, Suntio et al. 1988b)
4.28 (calculated, Mabey et al. 1982)
3.96 (HPLC-k', Hammers et al. 1982)
4.23 (selected, Mackay 1982)
3.93-4.18, 4.09 (range, mean, shake flask method, Eadsforth & Moser 1983)
4.12-4.32, 4.21 (range, mean, HPLC method, Eadsforth & Moser 1983)
4.28 (calculated-f const., Veith et al. 1983)
4.27 (calculated-f const., Yalkowsky et al. 1983)
3.51 (calculated-molar refraction, Yoshida et al. 1983a)
4.18 (selected, Yoshida et al. 1983a & b)
4.02 (selected, Oliver & Niimi 1983; Oliver 1987b)
3.93 (selected, Kaiser 1983; Kaiser et al. 1984; Wong et al. 1984; Oliver 1987a)
3.55 (selected, Calamari et al. 1983)
3.93, 4.02, 4.12 (selected, Geyer et al. 1984)

4.26 (selected, Hodson et al. 1984)
3.98 (gen. col.-GC/ECD, Miller et al. 1984; 1985; quoted, Suntio et al. 1988b; Paya-Perez et al. 1991)
4.00 (selected, Oliver & Charlton 1984)
3.98 (Sarna et al. 1984)
4.10 (selected, Bobra et al. 1985; Abernethy & Mackay 1987; Suntio et al. 1988)
4.02 (Hansch & Leo 1985; quoted, Howard 1989)
4.04 (quoted, Mackay et al. 1985)
4.18 (quoted, Brooke et al. 1986)
4.10 (quoted, Elzerman & Coates 1987)
4.22 (HPLC-k', De Kock & Lord 1987)
3.98 (quoted exptl., Doucette & Andren 1988)
4.28, 4.27, 4.32, 3.34, 3.78 (calculated-π, f const., MW, χ , TSA, Doucette & Andren 1988)
3.98, 3.92 (selected, calculated-V_I, solvatochromic parameters, Kamlet et al. 1988)
4.26 (quoted, Ryan et al. 1988)
4.04 (quoted, Isnard & Lambert 1988, 1989)
4.26 (quoted, Hodson et al. 1988)
4.02 (shake flask-GC, Pereira et al. 1988)
4.05, 4.40 (slow stirring-GC, calculated-f const., De Bruijn et al. 1989)
4.00, 4.2 (selected, Thomann 1989)

Bioconcentration Factor, log BCF:
3.32 (fathead minnow, 32 day exposure, Veith et al. 1979b; quoted, Suntio et al. 1988b)
3.45 (fathead minnow, Veith et al. 1979b)
3.37 (green sunfish, Veith et al. 1979b; quoted, Suntio et al. 1988b)
2.95 (rainbow trout, Veith et al. 1979b; quoted, Suntio et al. 1988b)
2.69 (fish, flowing water, Kenaga & Goring 1980; Kenaga 1980a; quoted, Yoshida et al. 1983b; Suntio et al. 1988b)
1.96 (calculated from water solubility, Kenaga 1980a)
2.26 (bluegill sunfish, whole body, flow system, Barrows et al. 1980)
2.91 (fish, correlated, Mackay 1982)
3.52 (microorganism-water, Mabey et al. 1982)
3.45 (fathead minnow, selected, Bysshe 1982)
3.11-3.51 (rainbow trout, Oliver & Niimi 1983)
4.19-4.56 (rainbow trout, lipid base, Oliver & Niimi 1983; quoted, Chiou 1985; Suntio et al. 1988b)
2.32 (calculated-K_{OW}, Calamari et al. 1983)
2.80 (fish, quoted, Dobbs & Williams 1983)
3.26-3.61 (fish, Oliver 1984, 1985)
2.40, 2.90 (algae: exptl, calculated, Geyer et al. 1984)
3.15 (activated sludge, Freitag et al. 1984; Halfon & Reggiani 1986)

2.40, 2.69, 3.15 (algae, fish, sludge, Klein et al. 1984)
3.11 (rainbow trout, laboratory data, Oliver & Niimi 1985)
3.08 (rainbow trout, field data, Oliver & Niimi 1985)
3.36-3.57 (rainbow trout, Oliver & Niimi 1985)
4.19-4.56 (rainbow trout, lipid base, Oliver & Niimi 1985)
2.09 (quoted, rainbow trout, 1.8 % lipid, wet wt basis, Geyer et al. 1985)
3.84 (quoted, rainbow trout, 1.8 % lipid, lipid basis, Geyer et al. 1985)
2.28, 2.40, 2.34, 2.66 (quoted, carp, 2.2 % lipid, wet wt basis, Geyer et al. 1985)
3.94, 3.96, 4.0, 4.32 (quoted, carp, 2.2 % lipid, lipid basis, Geyer et al. 1985)
2.54 (quoted, rainbow trout, hatching, 3.2 % lipid, wet wt basis, Geyer et al. 1985)
4.04 (quoted, rainbow trout, hatching, 3.2 % lipid, lipid basis, Geyer et al. 1985)
2.66, 2.73 (quoted, carp, 4.4 % lipid, wet wt basis, Geyer et al. 1985)
4.02, 4.09 (quoted, carp, 4.4 % lipid, lipid basis, Geyer et al. 1985)
2.96 (quoted, golden ide, 5.0 % lipid, wet wt basis, Geyer et al. 1985)
4.26 (quoted, golden ide, 5.0 % lipid, lipid basis, Geyer et al. 1985)
2.86, 2.94 (quoted, zebra fish, 5.2 % lipid, wet wt basis, Geyer et al. 1985)
4.15, 4.19 (quoted, zebra fish, 5.2 % lipid, lipid basis, Geyer et al. 1985)
2.83, 3.12 (quoted, tilapia, 5.2 % lipid, wet wt basis, Geyer et al. 1985)
4.12, 4.22 (quoted, tilapia, 5.2 % lipid, lipid basis, Geyer et al. 1985)
2.98, 3.12 (quoted, bluegill sunfish, 5.7 % lipid, wet wt bisis, Geyer et al. 1985)
4.23, 4.36 (quoted, bluegill sunfish, 5.7 % lipid, lipid basis, Geyer et al. 1985)
3.13, 3.14 (quoted, guppy, 5.8 % lipid, wet wt basis, Geyer et al. 1985)
4.37, 4.38 (quoted, guppy, 5.8 % lipid, lipid basis, Geyer et al. 1985)
3.11, 3.20 (quoted, rainbow trout, 7.7 %, wet wt basis, Geyer et al. 1985)
4.23, 4.38 (quoted, rainbow trout, 7.7 %, lipid basis, Geyer et al. 1985)
2.96, 3.03 (quoted, guppy, 8.2 % lipid, wet wt basis, Geyer et al. 1985)
4.05, 4.23 (quoted, guppy, 8.2 % lipid, lipid basis, Geyer et al. 1985)
3.11 (quoted, rainbow trout, 8.3 % lipid, wet wt basis, Geyer et al. 1985)
4.19 (quoted, rainbow trout, 8.3 % lipid, lipid basis, Geyer et al. 1985)
3.32 (quoted, fathead minnow, 10.5 % lipid, wet wt basis, Geyer et al. 1985)
4.30 (quoted, fathead minnow, 10.5 % lipid, wet wt basis, Geyer et al. 1985)
2.69 (fish, Freitag et al. 1985; Halfon & Reggiani 1986)
2.40 (algae, Freitag et al. 1985; Halfon & Reggiani 1986)
3.15 (activated sludge, Freitag et al. 1985)
2.15 (fish, quoted, Hawker & Connell 1986; Suntio et al. 1988b)
2.61 (fathead minnow, Carlson & Kosian 1987)
2.09-3.32 (fish, selected, Anliker & Moser 1987)
2.72 (fish, calculated- χ , Sabljic 1987b)
3.30 (fish, correlated, Sabljic 1987b)
4.76, 4.90, 3.54, 4.68 (field data-lipid base: Atlantic croakers, blue crabs, spotted sea trout, blue catfish, Pereira et al. 1988)
4.18 (fish, calculated-K_{OW}, Thomann 1989)
3.31, 4.25 (American flagfish, whole fish, fish lipid, Smith et al. 1990)

2.76 (fish, calculated, Figueroa & Simmons 1991)

Sorption Partition Coefficient, log K_{OC}:

3.96 (Mabey et al. 1982)

3.31 (soil, calculated-K_{OW}, Calamari et al. 1983)

3.65 (calculated-K_{OW}, Yoshida et al. 1983b)

4.70 (field data, Oliver & Charlton 1984)

4.40 (Niagara River-organic matter, Oliver & Charlton 1984)

3.60 (calculated from K_{OW}, Oliver & Charlton 1984)

3.09, 3.16 (soil, Banerjee et al. 1985)

3.69 (sediment, calculated-K_{OW}, Elzerman & Coates 1987)

4.3-5.1, 5.1 (suspended sediment, average, Oliver 1987c)

5.1 (algae > 50 μm, Oliver 1987c)

4.8-5.3, 5.0; 3.6 (Niagara River plume, range, mean; calculated-K_{OW}, Oliver 1987b)

4.02 (soil, Paya-Perez et al. 1991)

Sorption Partition Coefficient, log K_{OM}:

2.70 (soil organic matter, sorption isotherm, Chiou et al. 1983)

2.70, 2.76 (selected, calculated- χ , Sabljic 1984)

2.89, 2.97 (untreated Marlette soil, Lee et al. 1989)

2.87-3.97 (organic cations treated Marlette soil, Lee et al. 1989)

Half-Lives in the Environment:

Air: 128.4-1284 hours, based on the photooxidation half-life in air.

Surface Water: 672-4320 hours, based on unacclimated aqueous aerobic biodegradation half-life (Howard et al. 1991); 2.1-28 days estimated, 0.3-3 days for river water, 3-30 days for lakes estimated from persistence (Zoeteman et al. 1980).

Groundwater: 1344-8640 hours, based on unacclimated aqueous aerobic biodegradation half-life (Howard et al. 1991); 30-300 days estimated from persistence (Zoeteman et al. 1980).

Soil: 672-4320 hours, based on unacclimated aerobic soil grab sample data (Haider et al. 1981; Marinucci & Bartha 1979); < 10 days (Ryan et al. 1988).

Biota: half-life in worms at 8° C, < 5 days (Oliver 1987a); half-life in bluegill sunfish, > 1 < 3 day (Barrows et al. 1980); 1.21 days, 1.20 days clearance from American flagfish: whole fish, fish lipid (Smith et al. 1990).

Environmental Fate Rate Constants or Half-Lives:

Volatilization: half-lives from marine mesocosm: 22 days at 8-16°C in the spring, 11 days at 20-22°C in the summer, and 10 days at 3-7°C in the winter (Wakeham et al. 1983).

Photolysis: not environmentally relevant (Mabey et al. 1982); with a half life of 450 years for surface water photolysis at 40° north latitude in the summer (Dulin et al. 1986).

Oxidation: rate constant in air, 7.22×10^{-3} $hour^{-1}$ (Simmons et al. 1976; selected, Mackay et al. 1985); k(singlet oxygen) << 360 M^{-1} h^{-1} (Mabey et al. 1982); k (RO_2 radical) << 1 M^{-1} h^{-1} (Mabey et al. 1982); photooxidation half-life in air: 128.4-1284 hours, based on a measured room temperature rate constant of 5.32×10^{-13} cm^3 $molecule^{-1}$ sec^{-1} for vapor phase reaction with hydroxy radicals in air (Atkinson et al. 1985).

Hydrolysis: not environmentally significant (Mabey et al. 1982); first-order rate constant $k = 2.3 \times 10^{-5}$ h^{-1} ($t_{1/2}$ = 29784 h) at pH 7 & 25°C (Ellington et al. 1988).

Biodegradation: degradation rate constant in water, $k_B = 1.92 \times 10^{-2}$ $hour^{-1}$ (Simmons et al. 1976; selected, Mackay et al. 1985); aqueous aerobic half-life estimated from unacclimated soil grab sample data is 672-4320 hours (Haider et al. 1981; Marinucci & Bartha 1979); aqueous anaerobic half-life is 2688-17280 hours based on estimated unacclimated aqueous aerobic biodegradation half-life (Howard et al. 1991); significant degradation in an aerobic environment, k_B = 0.05 d^{-1} (Tabak et al. 1981; Mills 1982); First-order biodegradation rate constant, k_B (day^{-1}) in river water 0.029, estuarine water 0.026, and marine water 0.012 (Bartholomew & Pfaender 1983; Battersby 1990).

Biotransformation: $k_b = 1 \times 10^{-10}$ ml $cell^{-1}$ h^{-1} (estimated, Mabey et al. 1982).

Bioconcentration, Uptake (k_1) and Elimination (k_2) Rate Constants:

k_1: 8300 day^{-1} (guppy, exptl., Könemann & Van Leeuwen 1980)
k2: 0.45 day^{-1} (guppy, exptl., Könemann & Van Leeuwen 1980)
k1: 18.7 $hour^{-1}$ (guppy, selected, Hawker & Connell 1985)
1/k2: 53.0 $hour^{-1}$ (guppy, selected, Hawker & Connell 1985)
log k_1: 2.65 day^{-1} (guppy, selected, Connell & Hawker 1988)
log $1/k_2$: 0.34 day^{-1} (guppy, selected, Connell & Hawker 1988)
log k_2: -1.85 day^{-1} (fish, calculated-K_{OW}, fast-biphasic, Thomann 1989)
log k_2: 0.23 day^{-1} (fish, calculated-K_{OW}, slow-biphasic, Thomann 1989)
k_1: 1158 day^{-1}, 10140 day^{-1} (American flagfish: whole fish, fish lipid, Smith et al. 1990)
k_2: 0.57 day^{-1}, 0.57 day^{-1} (American flagfish: whole fish, fish lipid, Smith et al. 1990)
k_2: 0.57 day^{-1}, 0.46 day^{-1} (American flagfish: bioconcentration data, toxicity data, Smith et al. 1990)

Common Name: 1,3,5-Trichlorobenzene
Synonym: sym-trichlorobenzene
Chemical Name: 1,3,5-trichlorobenzene
CAS Registry No: 108-70-3
Molecular Formula: $C_6H_3Cl_3$
Molecular Weight: 181.45
Melting Point (°C):
63.0 (Verschueren 1977,1983; Mackay & Shiu 1981)
64.0 (Yalkowsky et al. 1983; Miller et al. 1985)
63.1 (Miller et al. 1984)
63.4 (Kishii et al. 1987)
63-64 (Howard 1989)
Boiling Point (°C):
208.5 (Verschueren 1977,1983)
208.0 (Mackay & Shiu 1981; Miller et al. 1984)
Density (g/cm^3 at 20°C):
1.3865 (64°C, Weast 1972-73; Horvath 1982)
Molar Volume (cm^3/mol):
130.9 (calculated-density, Weast 1972-73; Horvath 1982)
158.7 (LeBas method, Miller et al. 1985; Shiu et al. 1987)
125 (calculated-density,liquid molar volume, Chiou 1985)
159 (Abernethy & Mackay 1987)
135, 76 (from density, intrinsic, Abernethy et al. 1988)
0.761 (intrinsic volume: $V_I/100$, Kamlet et al. 1988; Hawker et al. 1989b; Hawker 1990)
Molecular Volume (A^3):
170 (De Bruijn & Hermens 1990)
Total Surface Area, TSA (A^2):
162.2 (Yalkowsky et al. 1979)
162.9 (Kishii et al. 1987)
158.1 (calculated- χ , Sabljic 1987a)
161.75 (Doucette & Andren 1988)
181.7 (De Bruijn & Hermens 1990)
Heat of Fusion, kcal/mol:
4.69 (Tsonopoulos & Prausnitz 1971)
4.49 (Miller et al. 1984)
Entropy of Fusion, cal/mol k (e.u.):
13.9 (Tsonopoulos & Prausnitz 1971)
13.4 (Miller et al. 1984)
Fugacity Ratio, F (assuming ΔS_{fusion} = 13.5 e.u.):
0.413 (25°C, Suntio et al. 1988b)

Water Solubility (g/m^3 or mg/L at 25°C):

25.03 (Tsonopoulos & Prausnitz 1971; quoted, Suntio et al. 1988b)
6.59 (shake flask-UV, Yalkowsky et al. 1979; quoted, Hodson et al. 1984; Miller et al. 1984; Doucette & Andren 1988; Paya-Perez et al. 1991)
14.4 (calculated-K_{OW}, Yalkowsky et al. 1979; Yalkowsky & Valvani 1980; Valvani & Yalkowsky 1980)
5.87 (shake flask-GC, Könemann 1981)
6.61 (recommended, Hovarth 1982; quoted, Wong et al. 1984; Nirmalakhandan & Speece 1988a)
6.59 (quoted lit. average, Yalkowsky et al. 1983; quoted, Suntio et al. 1988b)
4.12 (gen. col.-GC/ECD, Miller et al. 1984,1985; quoted, Doucette & Andren 1988)
6.01 (shake flask-HPLC, Banerjee 1984; quoted, Suntio et al. 1988b; Howard 1989)
10.6 (shake flask-GC, Chiou 1985)
6.09, 6.55 (quoted, calculated-UNIFAC, Banerjee 1985)
6.53 (recommended, IUPAC 1985)
5.3 (selected, Suntio et al. 1988b)
4.40 (calculated-UNIFAC, Banerjee & Howard 1988)
20.9 (calculated- χ , Nirmalakhandan & Speece 1988a)
5.25 (quoted, Isnard & Lambert 1988, 1989)
13.8 (quoted, subcooled liquid, Hawker 1989)
6.03 (quoted, Figueroa & Simmons 1991)

Vapor Pressure (Pa at 25°C):

77 (extrapolated subcooled liquid, Antoine eqn., Weast 1972-73; quoted, Suntio et al. 1988b)
77 (selected, Mackay & Shiu 1981; Howard 1989)
77, 20.4 (selected exptl., calculated-B.P., Mackay et al. 1982a)
32 (selected, Bobra et al. 1985)
24.4 (extrapolated, solid, Antoine eqn., Stephenson & Malanowski 1987)
76.0 (selected, Suntio et al. 1988b)

Henry's Law Constant (Pa m^3/mol):

1100 (calculated-P/C, Bobra et al. 1985)
192.5 (20°C, batch stripping, Oliver 1985; quoted, Howard 1989)
367 (calculated-P/C, Suntio et al. 1988b)

Octanol/Water Partition Coefficient, log K_{OW}:

4.02 (Leo et al. 1971; Hansch & Leo 1979)
4.27 (calculated-f const., Yalkowsky et al. 1979; Yalkowsky & Valvani 1980; Valvani et al. 1980; quoted, Miller et al. 1984)
4.15, 4.49 (shake flask-GC, HPLC-k', Könemann et al. 1979; quoted, Figueroa & Simmons 1991)
4.20 (calculated-f const., Könemann et al. 1979; Könemann 1981; quoted, Opperhuizen

1986; Suntio et al. 1988b; Leegwater 1989)
4.17 (shake flask-GC, Wateral et al. 1982; quoted, Suntio et al. 1988b)
4.17 (HPLC-k', Hammers et al. 1982)
4.27 (calculated-f const., Yalkowsky et al. 1983)
4.15 (quoted, Oliver & Niimi 1983)
4.15 (quoted, Kaiser 1983; Kaiser et al. 1984; Wong et al. 1984)
4.02 (gen. col.-GC/ECD, Miller et al. 1984; 1985; quoted, Suntio et al. 1988b; Hawker & Connell 1989; Hawker 1989; Clark et al. 1990; Paya-Perez et al. 1991)
4.18 (HPLC-RV, Garst 1984)
4.26 (quoted, Hodson et al. 1984)
4.20 (quoted, Oliver & Charlton 1984; Oliver 1987a)
4.02 (Sarna et al. 1984)
4.31 (shake flask-GC, Chiou 1985; quoted, Suntio et al. 1988b)
4.49 (Hansch & Leo 1985; quoted, Howard 1989)
4.10 (selected, Bobra et al. 1985; Abernethy & Mackay 1987; Suntio et al. 1988b)
4.15 (selected, Hawker & Connell 1985; Connell & Hawker 1988; Hawker 1990)
4.02 (selected, Gobas et al. 1987; 1989)
4.49, 3.75 (quoted, calculated-UNIFAC, Banerjee & Howard 1988)
4.02 (quoted exptl., Doucette & Andren 1988)
4.28, 4.27, 4.32, 3.34, 3.78 (calculated-π, f const., MW, χ , TSA, Doucette & Andren 1988)
4.02, 3.98 (quoted, calculated-V_I, solvatochromatic parameters, Kamlet et al. 1988)
4.31 (shake flask-GC, Pereira et al. 1988)
4.26 (quoted, Hodson et al. 1988)
4.08 (quoted, Isnard & Lambert 1988, 1989)
4.189, 4.40 (slow stirring-GC, calculated-f const., De Bruijn et al. 1989)
4.15 (selected, Thomann 1989)
4.15 (quoted, Hawker 1990)

Bioconcentration Factor, log BCF:
4.15 (guppy, lipid basis, Könemann & van Leeuwen 1980; quoted, Suntio et al. 1988b)
3.11-3.51 (rainbow trout, Oliver & Niimi 1983)
4.34-4.67 (rainbow trout, lipid basis, Oliver & Niimi 1983; quoted, Chiou 1985; Suntio et al. 1988b)
3.18 (calculated from solubility & K_{OW}, Lyman et al. 1982)
3.26-3.61 (fish, Oliver 1984)
2.88 (quoted, guppy, female, 5.4 % lipid, wet wt basis, Geyer et al. 1985)
4.15 (quoted, guppy, female, 5.4 % lipid, lipid basis, Geyer et al. 1985)
2.88 (guppy, calculated-K_{OW}, Gobas et al. 1987)
2.39-2.55 (fish, normalized, Tadokoro & Tomita 1987)
4.14 (guppy-lipid phase, calculated-K_{OW}, Gobas et al. 1987, 1989)
4.23 (guppy-lipid phase, calculated-K_{OW}, Gobas et al. 1989)

2.88 (fish, calculated-C_B/C_W or k_1/k_2, Connell & Hawker 1988; Hawker 1990)
4.40, 4.45, 3.51, 4.22 (field data-lipid base: Atlantic croakers, blue crabs, spotted sea trout, blue catfish, Pereira et al. 1988)
2.92 (fish, calculated, Figueroa & Simmons 1991)

Sorption Partition Coefficient, log K_{OC}:
3.82 (calculated from solubility & K_{OW}, Lyman et al. 1982)
5.10 (field data, Oliver & Charlton 1984)
4.20 (Niagara River-organic matter, Oliver & Charlton 1984)
3.80 (calculated from K_{OW}, Oliver & Charlton 1984)
2.85 (observed, Seip et al. 1986)
2.75 (soil, calculated- χ , Sabljic 1987a & b)
2.85 (calculated- χ , Bahnick & Doucette 1988)
4.13 (soil, Paya-Perez et al. 1991)

Half-Lives in the Environment:
Air:
Surface Water:
Groundwater: 18 days, estimated (Zoeteman et al. 1980).
Sediment:
Soil:
Biota: half-life in worms at 8°C, < 5 days (Oliver 1987a).

Environmental Fate Rate Constants or Half-Lives:
Volalitilization: half-life from a model river of 1 m depth with water current 1 m/sec and wind velocity 3 m/sec at 20°C is 4.5 hours (Lyman et al. 1982).
Photolysis: may be susceptible to direct photolysis by sunlight (Howard 1989); half-life for sunlight photolysis estimated to be 450 years at 40°N in the summer (Dulin et al. 1986).
Oxidation: photooxidation half-life in air is 6.17 months, based on a measured rate for the vapor phase reaction with photochemically produced hydroxy radicals in air (Atkinson et al. 1985).
Hydrolysis: will not hydrolyze under normal environmental conditions (Howard 1989).
Biodegradation: resistant to biodegradation (Tabak et al. 1964; Howard 1989).
Biotransformation:
Bioconcentration Uptake (k_1) and Elimination (k_2) Rate Constants:
k_1: 8000 day^{-1} (guppy, Könemann & van Leeuwen 1980)
k_2: 0.40 day^{-1} (guppy, Könemann & van Leeuwen 1980; quoted, Clark et al. 1990)
$1/k_2$: 60 hour (guppy, quoted, Hawker & Connell 1985)
k_1: 18.0 hour^{-1} (guppy, quoted, Hawker & Connell 1985)
k_1: 430 day^{-1} (fish, quoted, Opperhuizen 1986)
log k_1: 2.63 day^{-1} (fish, quoted, Connell & Hawker 1988)

log k_2: 0.440 day^{-1} (fish, quoted, Connell & Hawker 1988)
log k_1, 2.48 day^{-1} (guppy, Gobas et al. 1989)
log k_2, -0.40 day^{-1} (guppy, Gobas et al. 1989)
log k_2, -0.40 day^{-1} (fish, calculated-K_{OW}, Thomann 1989)

Common Name: 1,2,3,4-Tetrachlorobenzene
Synonym:
Chemical Name: 1,2,3,4-tetrachlorobenzene
CAS Registry No: 634-66-2
Molecular Formula: $C_6H_2Cl_4$
Molecular Weight: 215.9
Melting Point (°C):
    47.5 (Mackay & Shiu 1981; Kishii et al. 1987)
    48.0 (Yalkowsky et al. 1983)
    47.0 (Miller et al. 1984,1985)
Boiling Point (°C):
    254 (Mackay & Shiu 1981; Miller et al. 1984)
Density (g/cm³):
Molar Volume (cm³/mol):
    179.6 (LeBas method, Miller et al. 1985; Shiu et al. 1987)
    142 (calculated-density, liquid molar volume, Chiou 1985)
    180 (Abernethy & Mackay 1987)
    180, 85 (calculated from density, intrinsic, Abernethy et al. 1988)
    0.851 (intrinsic volume: $V_I/100$, Kamlet et al. 1988; Hawker 1989)
Molecular Volume (A³):
    176.6 (De Brujin & Hermens 1990)
Total Surface Area, TSA (A²):
    173.5 (Gross & Saylor 1931)
    173.8 (Yalkowsky et al. 1979)
    175.2 (calculated- χ , Sabljic 1987a)
    173.18 (Doucette & Andren 1988)
    189.6 (De Brujin & Hermens 1990)
Heat of Fusion, kcal/mol:
    4.06 (Miller et al. 1984)
Entropy of Fusion, cal/mol K (e.u.):
    12.7 (Miller et al. 1984)
Fugacity Ratio, F (assuming ΔS_{fusion} =13.5 e.u.):
    0.608 (25°C, Suntio et al. 1988b)

Water Solubility (g/m³ or mg/L at 25°C):
    4.31 (shake flask-UV, Yalkowsky et al. 1979; quoted Miller et al. 1984; Doucette & Andren 1988; Suntio et al. 1988b)
    4.11 (calculated-K_{OW}, Yalkowsky et al. 1979; Yalkowsky & Valvani 1980; Valvani & Yalkowsky 1980; quoted, Paya-Perez et al. 1991)
    3.42 (shake flask-GC, Könemann 1981)
    4.32 (recommended, Horvath 1982; quoted, Wong et al. 1984; Nirmalakhandan & Speece 1988a; Oliver 1987b)
    3.50 (22°C, Verschueren 1983)

1.06 (quoted lit average, Yalkowsky et al. 1983)
12.2 (gen. col.-GC/ECD, Miller et al. 1984,1985; quoted, Doucette & Andren 1988; Suntio et al. 1988b)
7.18 (23 °C, shake flask-GC, Chiou 1985)
5.92 (shake flask-HPLC, Banerjee et al. 1984; quoted, Suntio et al. 1988b)
4.33 (recommended, IUPAC 1985)
7.8 (selected, Bobra et al. 1985; Suntio et al. 1988b; Ma et al. 1990)
4.35 (calculated- χ , Nirmalakhandan & Speece 1988a)
12.2 (gen. col.-GC, Doucette & Andren, 1988)
0.13 (quoted, Isnard & Lambert 1988, 1989)
7.8 (selected, Suntio et al. 1988b; quoted, Ballschmitter & Wittlinger 1991)
12.7 (quoted, subcooled liquid, Hawker 1989)
4.07 (quoted, Figueroa & Simmons 1991)

Vapor Pressure (Pa at 25°C):
8.76 (extrapolated subcooled liquid, Antoine eqn., Weast 1972-73; quoted, Mackay & Shiu 1981; Suntio et al. 1988b)
5.21 (calculated, Antoine eqn., solid, Weast 1972-73; quoted, Mackay & Shiu 1981)
5.21, 3.04 (quoted exptl., calculated-B.P., Mackay et al. 1982a)
3.41 (subcooled liquid, GC-RT, Bildleman 1984)
5.20 (selected, Bobra et al. 1985)
2.44 (subcooled liq., Antoine eqn., Stephenson & Malanowski 1987)
5.60 (selected, subcooled liquid, Suntio et al. 1988b; quoted, Ballschmitter & Wittlinger 1991)
8.0 (selected, subcooled liquid, Hinckley et al. 1990)
5.6 (selected, Ma et al. 1990)

Henry's Law Constant (Pa m^3/mol):
261 (calculated- P/C, Mackay & Shiu 1981)
1100 (calculated-P/C, Bobra et al. 1985)
70 (20°C, batch stripping, Oliver 1985)
1100 (calculated-P/C, Suntio et al. 1988b; quoted, Ballschmiter & Wittlinger 1991)

Octanol/Water Partition Coefficient, log K_{OW}:
4.72 (Leo et al. 1971; quoted Schwarzenbach & Westall)
5.05 (calculated-f const., Yalkowsky et al. 1979; Yalkowsky & Valvani 1980; Valvani & Yalkowsky 1980; quoted, Miller et al. 1984; Suntio et al. 1988b)
4.46, 4.92 (shake flask-GC, HPLC-k', Könemann et al. 1979; quoted, Figueroa & Simmons 1991)
4.94 (calculated-f const., Könemann et al. 1979; Könemann 1981; quoted, Suntio et al. 1988b; Leegwater 1989)
4.68 (reported as tetrachlorobenzene, Briggs 1981)
4.37 (shake flask-GC, Wateral et al. 1982; quoted, Suntio et al. 1988b)

4.94 (TLC-RT, Bruggeman et al. 1982)
4.75 (shake flask-GC, Bruggeman et al. 1982)
4.41 (HPLC-k', Hammers et al. 1982)
4.46 (quoted, Kaiser 1983; Kaiser et al. 1984; Oliver & Niimi 1983; Oliver 1987a & c)
5.02 (calculated-f constants, Yalkowsky et al. 1983)
4.55 (generator column-GC, Miller et al. 1984; quoted, Suntio et al. 1988b; Paya-Perez et al. 1991)
4.60 (shake flask-GC, Chiou 1985; quoted, Suntio et al. 1988b)
4.80 (quoted, Oliver & Niimi 1985)
4.50 (selected, Bobra et al. 1985; Abernethy & Mackay 1987; Suntio et al. 1988b; quoted, Ballschmiter & Wittlinger 1991)
5.8, 4.65 (quoted, HPLC-RV/MS, Burkhard & Kuehl 1986)
4.55 (quoted exptl., Doucette & Andren 1988)
4.99, 5.02, 4.85, 4.33, 4.06 (calculated-π, f const., MW, χ, TSA, Doucette & Andren 1988)
4.55, 4.32 (quoted, calculated-V_I, solvatochromic parameters, Kamlet et al. 1988)
4.46 (quoted, Hodson & Williams 1988)
4.60 (shake flask-GC, Pereira et al. 1988)
4.635, 5.16 (slow stirring-GC, calcd.-f const., De Bruijn et al. 1989)
4.65 (quoted, Isnard & Lambert 1988, 1989)
4.5 (selected, Oliver 1987a & b; Thomann 1989; Ma et al. 1990)

Bioconcentration Factor, log BCF:
4.86 (guppy, lipid basis, Könemann & van Leeuwen 1980; quoted, Suntio et al. 1988b)
3.25 (Briggs 1981)
3.72-4.08 (rainbow trout, Oliver & Niimi 1983)
4.80-5.13 (rainbow trout, lipid basis, Oliver & Niimi 1983; quoted, Chiou 1985; Suntio et al. 1988b)
3.72-4.08 (fish, Oliver 1984)
3.70 (15°C, rainbow trout, Banerjee et al. 1984)
3.32 (fish, calculated, Garst 1984)
3.89 (rainbow trout, Oliver & Niimi 1985)
3.80-3.91 (rainbow trout, Oliver & Niimi 1985)
3.30 (fish, correlated, Sabljic 1987b)
3.7 (calculated- χ, Sabljic 1987b)
3.38 (fathead minnow, Carlson & Kosian 1987)
4.9-5.4, 5.1; 4.1 (Niagara River plume, range, mean; calculated-K_{OW}, Oliver 1987b)
3.80 (correlated, Isnard & Lambert 1988)
3.36 (guppy, Van Hoogan & Opperhuizen 1988)
4.98 (fish, calculated-K_{OW}, Thomann 1989)
5.46, 5.70, 4.68, 5.30 (field data-lipid base: Atlantic croakers, blue crabs, spotted sea trout, blue catfish, Pereira et al. 1988)

3.50 (selected, Ma et al. 1990)
3.15 (fish, calculated, Figueroa & Simmons 1991)

Sorption Partition Coefficient, log K_{OC}:
3.83 (calculated from K_{OW}, Schwarzenbach & Westall 1981; Sabljic 1987b)
5.00 (field data, Oliver & Charlton 1984)
4.10 (calculated from K_{OW}, Oliver & Charlton 1984)
3.00 (soil, calculated- χ , Sabljic 1987a & b)
4.1-6.0, 5.2 (suspended sediment, average, Oliver 1987c)
3.48-3.91 (soil, batch-equilibration, Kishi et al. 1990)
4.28 (soil, Paya-Perez et al. 1991)

Sorption Partition Coefficient, log K_{OM}:
3.25 (shake flask-GC, soil org. matter, Briggs 1981)
4.90 (Niagara River-organic matter, Oliver & Charlton 1984)
4.9-5.4, 5.1, 4.1 (Niagara-River plume, range, average, calculated-K_{OW}, Oliver 1987b)
4.5 (algae > 50 μm, Oliver 1987c)

Half-Lives in the Environment:
Air:
Surface Water:
Groundwater:
Sediment:
Soil:
Biota: half-life in worms at 8°C, < 5 days (Oliver 1987a).

Environmental Fate Rate Constants or Half-Lives:
Volatilization:
Photolysis:
Hydrolysis:
Oxidation:
Biodegradation:
Biotransformation:
Bioconcentration, Uptake (k_1) and Elimination (k_2) Rate Constants:
k_1: 140 hour^{-1} (rainbow trout, 15°C, Banerjee et al. 1984)
k_2: 0.021 hour^{-1} (rainbow trout, 15°C, Banerjee et al. 1984)
k_1: 670 mL/g.d (guppy, Van Hoogan & Opperhuizen 1988)
k_2: 0.29 day^{-1} (guppy, Van Hoogan & Opperhuizen 1988)

Common Name: 1,2,3,5-Tetrachlorobenzene
Synonym:
Chemical Name: 1,2,3,5-tetrachlorobenzene
CAS Registry No: 634-90-2
Molecular Formula: $C_6H_2Cl_4$
Molecular Weight: 215.9
Melting Point (°C):
 54.5 (Mackay & Shiu 1981; Kishii et al. 1987)
 51.0 (Yalkowsky et al. 1983)
 50.7 (Miller et al. 1984,1985)
Boiling Point (°C):
 246 (Miller et al. 1984)
Density (g/cm^3):
Molar Volume (cm^3/mol):
 179.6 (LeBas method, Miller et al. 1985; Shiu et al. 1987)
 180 (Abernethy & Mackay 1987)
 180, 85 (calculated-density, intrinsic, Abernethy et al. 1988)
 0.851 (intrinsic volume: $V_I/100$, Kamlet et al. 1988; Hawker 1989, 1990)
Molecular Volume (A^3):
 180.3 (De Bruijn & Hermens 1990)
Total Surface Area, TSA (A^2):
 175.8 (Yalkowsky et al. 1979)
 176.1 (Kishii et al. 1987)
 175.2 (calculated- χ , Sabljic 1987a)
 175.2 (Doucette & Andren 1988)
 192.1 (De Bruijn & Hermens 1990)
Heat of Fusion, kcal/mol:
 4.54 (Miller et al. 1984)
Entropy of Fusion, cal/mol K (e.u.):
 14.0 (Miller et al. 1984)
Fugacity Ratio, F (assuming Δ_{fusion} = 13.5 e.u.):
 0.556 (25°C, Suntio et al. 1988b)

Water Solubility (g/m^3 or mg/L at 25°C):
 3.50 (shake flask-UV, Yalkowsky et al. 1979; quoted, Miller et al. 1984; Doucette & Andren 1988; Suntio et al. 1988b; Paya-Perez et al. 1991)
 3.58 (calculated-K_{OW}, Yalkowsky et al. 1979; Yalkowsky & Valvani 1980; Valvani & Yalkowsky 1980)
 4.02 (shake flask-LSC, Banerjee et al. 1980; quoted, Suntio et al. 1988b)
 4.11, 17.1 (LSC-^{14}C, calculated-K_{OW}, Veith et al. 1980)
 2.48 (shake flask-GC, Könemann 1981)
 3.51 (recommended, Horvath 1982; quoted, Wong et al. 1984)
 2.40 (22°C, Verschueren 1983)

0.541 (calculated-UNIFAC, Arbuckle 1983)
3.67 (quoted lit. average, Yalkowsky et al. 1983)
5.10 (shake flask-HPLC, Banerjee 1984; quoted, Suntio et al. 1988b)
2.89 (gen.-col.-GC/ECD, Miller et al. 1984; 1985; quoted, Doucette & Andren 1988; Suntio et al. 1988b; Hawker 1989)
3.23 (shake flask-GC, Chiou 1985)
3.46 (recommended, IUPAC 1985)
5.61, 2.88 (quoted, calculated-UNIFAC, Banerjee 1985)
3.60 (selected, Bobra et al. 1985; Suntio et al. 1988b; Ma et al. 1990)
3.45 (quoted, Abernethy et al. 1986)
2.89 (quoted, Doucette & Andren 1988)
0.931 (calculated-UNIFAC, Banerjee & Howard 1988)
4.35 (calculated-molecular connectivity, Nirmalakhandan & Speece 1988a)
4.32 (generator column-GC, Doucette & Andren 1988)
6.67 (quoted, subcooled liquid, Hawker 1989)
4.65 (correlated, Isnard & Lambert, 1988, 1989)
3.02 (quoted, Figueroa & Simmons 1991)

Vapor Pressure (Pa at 25°C):
18.6 (extrapolated subcooled liquid, Antoine eqn., Weast 1972-73; quoted, Mackay & Shiu 1981; Suntio et al. 1988b)
9.8 (calculated from extrapolated vapor pressure with a fugacity ratio correction, Mackay & Shiu 1981; quoted, Bobra et al. 1985; Ma et al. 1990)
9.8, 3.88 (selected exptl., calculated-B.P., Mackay et al. 1982a)
17.2 (extrapolated subcooled liquid, Antoine eqn., Stephenson & Malanowski 1987)
9.56 (converted to solid pressure, Stephenson & Malanowski 1987)
10.8 (selected subcooled liquid, Suntio et al. 1988b)

Henry's Law Constant (Pa m^3/mol):
159 (batch stripping, Mackay & Shiu 1981; quoted, Suntio et al. 1988b)
593 (calculated-P/C, Mackay & Shiu 1981)
160, 41 (observed, calculated-QSAR, Nirmalakahandan & Speece 1988b)
590 (calculated-P/C, Bobra et al. 1985; Suntio et al. 1988b)

Octanol/Water Partition Coefficient, log K_{OW}:
5.05 (calculated-f const., Yalkowsky et al. 1979; Yalkowsky & Valvani 1980; Valvani et al. 1980; quoted, Miller et al. 1984; Suntio et al. 1988b)
4.50, 4.92 (shake flask-GC, HPLC-k', Könemann et al. 1979; quoted, Figueroa & Simmons 1991)
4.94 (calculated-f const., Könemann et al. 1979; Könemann 1981; quoted, Opperhuizen 1986; Suntio et al. 1988b; Leegwater 1989)
4.46 (HPLC-RT, Veith et al. 1979b)
4.46, 5.0, 4.97 (shake flask-LSC, HPLC-RT, calculated-f constants, Veith et al. 1980)

4.52 (shake flask-LSC, Banerjee et al. 1980)
4.56 (shake flask-GC, Wateral et al. 1982; quoted, Suntio et al. 1988b)
4.53 (HPLC-k', Hammers et al. 1982)
4.46 (selected, Mackay 1982)
4.50 (quoted, Kaiser 1983; Kaiser et al. 1984; Wong et al. 1984)
5.02 (calculated-f const., Yalkowsky et al. 1983)
4.46, 3.88 (quoted, calculated-molar refraction, Yoshida et al. 1983a)
5.18, 5.56 (calculated-UNIFAC, Arbuckle 1983)
4.51 (gen.-col.-GC/ECD, Miller et al. 1984; 1985; quoted, Suntio et al. 1988b; Hawker 1989; Clark et al. 1990; Paya-Perez et al. 1991)
4.61-4.73 (HPLC-RV, Garst 1984)
4.46 (selected, Garst 1984)
4.65 (Sarna et al. 1984)
4.59 (shake flask-GC, Chiou 1985; quoted, Suntio et al. 1988b)
4.65 (selected, Gobas et al. 1987)
4.50 (selected, Bobra et al. 1985; Abernethy & Mackay 1987; Suntio et al. 1988b; Ma et al. 1990)
4.92, 4.36 (quoted, calculated-UNIFAC, Banerjee & Howard 1988)
4.48 (quoted, Hawker & Connell 1985; Connell & Hawker 1988; Hawker 1990)
4.65 (quoted exptl., Doucette & Andren 1988)
4.99, 5.02, 4.65, 4.06, 4.11 (calculated-π, f const., MW, χ , TSA, Doucette & Andren 1988)
4.56, 4.32 (quoted, calculated-V_I, salvatochromic parameters, Kamlet et al. 1988)
4.50 (quoted, Hodson et al. 1988)
4.59 (shake flask-GC, Pereira et al. 1988)
4.658, 5.16 (slow stirring-GC, calculated-f constants, De Bruijn et al. 1989)
4.65 (correlated, Isnard & Lambert 1988, 1989)
4.48 (selected, Thomann 1989)

Bioconcentration Factor, log BCF:
3.26 (fathead minnow, Veith et al. 1979b; quoted, Suntio et al. 1988b)
4.86 (guppy-lipid basis, Könemann et al. 1979; quoted, Suntio et al. 1988b)
4.15, 4.86 (guppy-lipid basis, Könemann & van Leeuwen 1980)
3.26 (bluegill sunfish, whole body, flow system, Barrows et al. 1980)
3.14 (fish, correlated, Mackay 1982)
4.80-5.13 (rainbow trout, lipid base, Oliver & Niimi 1983; quoted, Chiou 1985)
3.46 (22°C, bluegill sunfish, Banerjee et al. 1984)
3.5 (calculated- χ , Sabljic 1987b)
3.34 (fish, calculated, Sabljic 1987b)
3.59 (guppy, calculated, Gobas et al. 1987)
4.85 (guppy-lipid phase, calcd., Gobas et al. 1987)
3.59 (fish, calculated-C_A/C_W or k_1/k_2, Connell & Hawker 1988; Hawker 1990)
4.73, 4.86 (guppy-lipid phase, calculated-K_{OW}, Gobas et al. 1989)

3.26 (quoted, Isnard & Lambert 1988)
5.05, 5.20, 4.27, 4.90 (field data-lipid base: Atlantic croakers, blue crabs, spotted sea trout, blue catfish, Pereira et al. 1988)
4.86 (guppy, Gobas et al. 1989)
3.19 (fish, calculated, Figueroa & Simmons 1991)

Sorption Partition Coefficient, log K_{OC}:
3.25 (Briggs 1981)
3.20 (Koch 1983)
3.08 (calculated- χ , Bahnick & Doucette 1988)
4.25 (soil, Paya-Perez et al. 1991)
Sorption Partition Coefficient, log K_{OM}:
3.20, 2.98 (quoted, calculated- χ , Sabljic 1984)

Half-Lives in the Environment:
Air:
Surface Water:
Groundwater:
Sediment:
Soil:
Biota: half-life in fish 2-4 days (Veith et al. 1980); half-life in bluegill sunfish, > 2 and < 4 days (Barrows et al. 1980).

Environmental Fate Rate Constants or Half-Lives:
Volatilization:
Photolysis:
Oxidation:
Hydrolysis:
Biodegradation:
Biotransformation:
Bioconcentration, Uptake (k_1) and Elimination (k_2) Rate Constants:
k_1: 15000 day^{-1} (guppy, Könemann & van Leeuwen 1980)
k_2: 0.26 day^{-1} (guppy, Könemann & van Leeuwen 1980; selected, Clark et al. 1990)
k_1: 74 $hour^{-1}$ (bluegill sunfish, Banerjee et al. 1984)
k_2: 0.026 $hour^{-1}$ (bluegill sunfish, Banerjee et al. 1984)
$1/k_2$: 92.0 hour (guppy, quoted, Hawker & Connell 1985; 1988)
k_1: 33.8 $hour^{-1}$ (guppy, Hawker & Connell 1985)
k_1: 810 day^{-1} (fish quoted, Opperhuizen 1986)
log k_1: 2.91 day^{-1} (guppy, quoted, Connell & Hawker 1988)
log $1/k_2$:-0.42 day^{-1} (guppy, quoted, Connell & Hawker 1988)
log k_1: 3.00 day^{-1} (guppy, Gobas et al. 1989)
log k_2: -0.59 day^{-1} (guppy, Gobas et al. 1989)

log k_2: -0.58 day^{-1} (fish, calculated-K_{OW}, Thomann 1989)

Common Name: 1,2,4,5-Tetrachlorobenzene
Synonym:
Chemical Name: 1,2,4,5-tetrachlorobenzene
CAS Registry No: 95-94-3
Molecular Formula: $C_6H_2Cl_4$
Molecular Weight: 215.9
Melting Point (°C):
140 (Mackay & Shiu 1981; Kishii et al. 1987)
139 (Yalkowsky et al. 1983; Miller et al. 1984,1985)
Boiling Point (°C):
243 (Mackay & Shiu 1981; Miller et al. 1984)
Density (g/cm³ at 20°C):
1.858 (22°C, Weast 1972-1973; Horvath 1982)
Molar Volume (cm³/mol):
116.2 (calculated-density, Weast 1972-73; Horvath 1982)
179.6 (LeBas method, Miller et al. 1985; Shiu et al. 1987)
116 (calculated-density, Lande & Banerjee 1981; Mailhot 1987)
142 (liquid molar volume, Chiou 1985)
180 (Abernethy & Mackay 1987)
0.851 (intrinsic volume: $V_I/100$, Kamlet et al. 1988; Hawker 1989)
94 (intrinsic, Abernethy et al. 1988)
Molecular Volume (A³):
180.3 (De Bruijn & Hermens 1990)
Total Surface Area, TSA (A²):
175.8 (Yalkowsky et al. 1979)
175.7 (Kishii et al. 1987)
175.2 (Sabljic 1987a)
175.2 (Doucette & Andren 1988)
192.1 (De Bruijn & Hermens 1990)
Heat of Fusion, kcal/mol:
5.76 (Miller et al. 1984)
Entropy of Fusion, cal/mol K (e.u.):
14.0 (Miller et al. 1984)
Fugacity Ratio, F (assuming ΔS_{fusion} = 13.5 e.u.):
0.073 (25°C, Suntio et al. 1988b)

Water Solubility (g/m³ or mg/L at 25 °C):
0.595 (shake flask-UV, Yalkowsky et al. 1979; quoted, Miller et al. 1984; Doucette & Andren 1988; Suntio et al. 1988b)
0.542 (calculated-K_{OW}, Yalkowsky et al. 1979; Yalkowsky & Valvani 1980; Valvani & Yalkowsky 1980)
0.63 (quoted, Geyer et al. 1980)
6.0 (Kenaga & Goring 1980; Kenaga 1980a)

0.29 (shake flask-GC, Könemann 1981)
0.596 (recommended, Horvath 1982; quoted, Wong et al. 1984)
0.3 (22°C, quoted, Verschueren 1983)
0.595 (calculated-K_{OW}, Yalkowsky et al. 1983)
2.35 (gen. col.-GC/ECD, Miller et al. 1984, 1985; quoted, Doucette & Andren 1988; Suntio et al. 1988b; Hawker 1989)
0.465 (shake flask-HPLC, Banerjee 1984; quoted, Suntio et al. 1988b)
0.606 (recommended, IUPAC 1985)
1.27 (selected, Bobra et al. 1985; Suntio et al. 1988b)
1.80 (quoted, Mailhot 1987)
4.35 (calculated- χ , Nirmalakhandan & Speece 1988a)
2.34 (correlated, Isnard & Lambert, 1988, 1989)
21.6 (quoted, subcooled liquid, Hawker 1989)
0.70 (selected, Ma et al. 1990)
0.603 (quoted, Figueroa & Simmons 1991)

Vapor Pressure (Pa at 25°C):
10.1 (extrapolated subcooled liquid, Antoine eqn., Weast 1972-73; quoted, Mackay & Shiu 1981; Suntio et al. 1988b)
0.72 (selected, Mackay & Shiu 1981; Bobra et al. 1985; Ma et al. 1990)
0.72, 0.64 (quoted exptl., calculated-B.P., Mackay et al. 1982a)
0.2 (evaporation rate, Dobbs & Cull 1982)
2.98, 0.22 (subcooled liquid, solid, Stephenson & Malanowski 1987)
9.86 (selected subcooled liquid, Suntio et al. 1988b)
0.72, 0.615 (quoted, calculated-UNIFAC, Banerjee et al. 1990)

Henry's Law Constant (Pa m^3/mol):
261 (calculated-P/C, Mackay & Shiu 1981)
101 (20°C, batch stripping, Oliver 1985)
122 (calculated-P/C, Bobra et al. 1985; Suntio et al. 1988b)

Octanol/Water Partition Coefficient, log K_{OW}:
4.72 (Leo et al. 1971; quoted, Schwarzenbach & Westall 1981)
5.05 (calculated-f const., Yalkowsky et al. 1979; quoted, Miller et al. 1984; Suntio et al. 1988b)
4.82, 4.56 (shake flask-GC, HPLC-k', Könemann et al. 1979)
4.67 (quoted, Kenaga & Goring 1980)
4.94 (calculated-f const., Könemann et al. 1979; Könemann 1981; quoted, Suntio et al. 1988b; Leegwater 1989)
4.46 (shake flask-GC, Wateral et al. 1982; quoted, Suntio et al. 1988b)
4.52 (HPLC-k', Hammers et al. 1982)
4.52 (quoted, Oliver & Niimi 1983; Oliver 1987a)
4.50 (quoted, Kaiser 1983; Kaiser et al. 1984)

5.02 (quoted lit. average, Yalkowsky et al. 1983)
4.82 (quoted, Wong et al. 1984)
4.51 (Sarna et al. 1984)
4.51 (gen. col.-GC/ECD, Miller et al. 1984,1985; quoted, Suntio et al. 1988b; Hawker & Connell 1989; Hawker 1989)
4.70 (shake flask-GC, Chiou 1985)
4.50 (selected, Bobra et al. 1985; Suntio et al. 1988b)
4.60 (HPLC-k', Mailhot 1987)
4.50 (quoted, Oliver 1987a)
4.51 (quoted exptl., Doucette & Andren 1988)
4.99, 5.02, 4.85, 4.49, 4.11 (calculated-π, f const., MW, χ , TSA, Doucette & Andren 1988)
4.51, 4.44 (quoted, calculated-V_I, solvatochromic parameters, Kamlet et al. 1988)
4.604, 5.16 (slow stirring-GC, calculated-f const., De Bruijn et al. 1989)
4.67 (quoted, Isnard & Lambert 1988, 1989)
4.70 (shake flask-GC, Pereira et al. 1988)
4.52 (selected, Thomann 1989; Figueroa & Simmons 1991)

Bioconcentration Factor, log BCF:
0.20 (rats, adipose tissue, Geyer et al. 1980)
3.65 (fish, flowing water, Kenaga & Goring 1980; Kenaga 1980a; quoted, Suntio et al. 1988b)
2.35 (calculated from water solubility, Kenaga 1980a)
3.72-4.11 (rainbow trout, Oliver & Niimi 1983)
4.80-5.13 (rainbow trout, lipid base, Oliver & Niimi 1983; quoted, Chiou 1985; Suntio et al. 1988b)
3.72-4.11 (fish, Oliver 1984)
2.80 (Tadokoro & Tomita 1987)
3.89 (green algae, Mailhot 1987)
3.65 (correlated, Isnard & Lambert 1988)
5.05, 5.20, 4.27, 4.90 (field data-lipid base: Atlantic croakers, blue crabs, spotted sea trout, blue catfish, Pereira et al. 1988)
4.80 (fish, calculated-K_{OW}, Thomann 1989)
2.76 (picea omorika, Reischl et al. 1989)
3.61, 4.70 (American flagfish: whole fish, fish lipid, Smith et al. 1990)
3.20 (fish, calculated, Figueroa & Simmons 1991)
4.5 (selected, Ma et al. 1990)

Sorption Partition Coefficient, log K_{OC}:
3.20 (Kenaga 1980a)
3.25 (shake flask-GC, soil-organic matter, Briggs 1981)
3.86 (calculated-K_{OW}, Schwarzenbach & Westall 1981; Sabljic 1987b)
5.10 (field data, Oliver & Charlton 1984)

4.70 (Niagara River-organic matter, Oliver & Charlton 1984)
4.10 (calculated-K_{OW}, Oliver & Charlton 1984)
2.99 (soil, calculated- χ , Sabljic 1987a & b)

Half-Lives in the Environment:

Air:
Surface Water:
Groundwater:
Sediment:
Soil:
Biota: half-life in worms at 8°C, < 5 days (Oliver 1987a); half-life in picea omorika, 33 days (Reischl et al. 1989); 1.72 days clearance from American flagfish (Smith et al. 1990).

Environmental Fate Rate Constants or Half-Lives:

Volatilization:
Photolysis:
Oxidation:
Hydrolysis:
Biodegradation:
Biotransformation:
Bioconcentration, Uptake (k_1) and Elimination (k_2) Rate Constants:

k_1: 1630 day^{-1}, 171000 day^{-1} (American flagfish: whole fish, fish lipid, Smith et al. 1990)
k_2: 0.4 day^{-1}, 0.34 day^{-1} (American flagfish: whole fish, fish lipid, Smith et al. 1990)
k_2: 0.4 day^{-1}, 2.35 day^{-1} (American flagfish: bioconcentration data, toxicity data, Smith et al. 1990)

Common Name: Pentachlorobenzene
Synonym:
Chemical Name: Pentachlorobenzene
CAS Registry No: 608-93-5
Molecular Formula: C_6HCl_5
Molecular Weight: 250.3
Melting Point (°C):
86 (Mackay & Shiu 1981; Yalkowsky et al. 1983; Kishii et al. 1987)
84.5 (Miller et al. 1984,1985)
Boiling Point (°C):
277 (Mackay & Shiu 1981; Miller et al. 1984)
Density (g/cm³ at 20°C):
1.609
1.8342 (16.5°C, Weast 1972-73; Horvath 1982)
Molar Volume (cm³/mol):
136.5 (calculated-density, Weast 1972-73; Horvath 1982)
136 (calculated-density, Lande & Banerjee 1981)
200.5 (LeBas method, Miller et al. 1985; Shiu et al. 1987)
166 (liquid molar volume, Chiou 1985)
200 (Abernethy & Mackay 1987)
0.941 (intrinsic volume: V_I/100, Kamlet et al. 1988; Hawker 1989b,1990)
94 (intrinsic, Abernethy et al. 1988)
Molecular Volume (A³):
189.9 (De Bruijn & Hermens 1990)
Total Surface Area, TSA (A²):
189.4 (Yalkowsky et al. 1979)
188.9 (Kishii et al. 1987)
192.3 (Sabljic 1987a)
188.66 (Doucette & Andren 1988)
200.0 (De Bruijn & Hermens 1990)
Heat of Fusion, kcal/mol:
4.92 (Miller et al. 1984)
Entropy of Fusion, cal/mol K (e.u.):
13.8 (Miller et al. 1984)

Fugacity Ratio, F (assuming ΔS_{fusion} = 13.5 e.u.):
0.258 (25°C)
0.230 (Suntio et al. 1988b)

Water Solubility (g/m³ or mg/L at 25°C):
0.56 (shake flask-UV, Yalkowsky et al. 1979; quoted, Miller et al. 1984; Doucette & Andren 1988; Suntio et al. 1988b; Paya-Perez et al. 1991)
0.379 (calculated-K_{OW}, Yalkowsky et al. 1979; Yalkowsky & Valvani 1980; Valvani

& Yalkowsky 1980)

1.33 (shake flask-LSC, Banerjee et al. 1980; quoted, Suntio et al. 1988b)

0.135 (Kenaga & Goring 1980; Kenaga 1980a)

1.34, 6.58 (shake flask-LSC-^{14}C, calculated-K_{OW}, Veith et al. 1980)

0.24 (shake flask-GC, Könemann 1981; quoted, Bruggeman et al. 1984)

0.562 (recommended, Horvath 1982; quoted, Wong et al. 1984; Oliver 1987b)

0.24 (quoted, Verschueren 1983)

3.51 (calculated-UNIFAC, Arbuckle, 1983)

0.87 (quoted lit. average, Yalkowsky et al. 1983)

1.32 (quoted, Garten & Trabalka 1983)

0.831 (gen. col.-GC/ECD, Miller et al. 1984,1985; quoted, Doucette & Andren 1988; Suntio et al. 1988b; Hawker 1989)

0.24 (quoted, Bruggeman et al. 1984)

0.385 (23 °C, shake flask-GC, Chiou 1985)

0.87, 0.12 (selected,calculated-UNIFAC, Banerjee 1985)

0.180 (gen. col.-GC/ECD, Opperhuizen et al. 1985)

0.552 (recommended, IUPAC 1985)

0.65 (selected, Bobra et al. 1985; Suntio et al. 1988b; quoted, Ballschmiter & Wittlinger 1991)

0.55 (quoted, Abernethy et al. 1986)

0.56, 3.46 (quoted, calculated-K_{OW} & HPLC-RT, Chin et al. 1986)

0.831 (quoted, Isnard & Lambert 1988, 1989)

3.15 (quoted, subcooled liquid, Hawker 1989)

0.135 (quoted, Figueroa & Simmons 1991)

Vapor Pressure (Pa at 25°C):

0.889 (extrapolated subcooled liquid, Antoine eqn., Weast 1972-73; quoted, Mackay & Shiu 1981; Suntio et al. 1988b)

0.219 (calculated, Weast 1972-73; quoted, Mackay & Shiu 1981)

0.22 (selected, Bobra et al. 1985)

1.08, 0.28 (subcooled liquid, solid, Antoine eqn., Stephenson & Malanowski 1987)

0.88 (selected subcooled liquid, Suntio et al. 1988b; quoted, Ballschmiter & Wittlinger 1991)

Henry's Law Constant (Pa m^3/mol):

97.7 (calculated-P/C, Mackay & Shiu 1981)

85 (calculated-P/C, Bobra et al. 1985; Suntio et al. 1988b; quoted, Ballschmiter & Wittlinger 1991)

71.9 (20 °C, batch stripping, Oliver 1985)

Octanol/Water Partition Coefficient, log K_{OW}:

5.79 (calculated-f const., Yalkowsky et al. 1979; Yalkowsky & Valvani 1980; Valvani & Yalkowsky 1980; quoted, Miller et al. 1984; Suntio et al. 1988b)

4.88, 5.52 (shake flask-GC, HPLC-k', Könemann et al. 1979; quoted, Figueroa & Simmons 1991)
5.69 (calculated-f const., Könemann et al. 1979; Könemann 1981; quoted, Opperhuizen 1986; Suntio et al. 1988b; Leegwater 1989)
4.94 (shake flask-LSC, Banerjee et al. 1980; quoted, Oliver 1987a,b & c; Suntio et al. 1988b)
4.94, 5.29 (shake flask-LSC, HPLC-RT, Veith et al. 1980)
5.19 (quoted, Kenaga & Goring 1980; Brooke et al. 1986)
5.17 (shake flask-GC, Wateral et al. 1982; quoted, Suntio et al. 1988b)
5.06 (HPLC-k', Hammers et al. 1982)
5.69 (HPLC-RT, Bruggeman et al. 1982; quoted, Bruggeman et al. 1984)
5.19 (quoted, Mackay 1982)
5.94, 6.45 (calculated-UNIFAC, Arbuckle 1983)
5.71 (calculated-f const., Veith et al. 1983)
5.77 (calculated-f const., Yalkowsky et al. 1983)
4.94 (quoted, Garten & Trabalka 1983; Oliver & Niimi 1983)
4.88 (quoted, Kaiser 1983; Kaiser et al. 1984)
4.94, 4.24 (quoted, calcd-molar refraction, Yoshida et al. 1983a)
5.03 (gen. col.-GC/ECD, Miller et al. 1984; 1985; quoted, Suntio et al. 1988b; Clark et al. 1990; Paya-Perez et al. 1991)
4.88 (quoted, Geyer et al. 1984; Freitag et al. 1985)
5.17 (shake flask-HPLC, Banerjee 1984; quoted, Suntio et al. 1988b)
4.94, 5.11-5.21 (quoted, HPLC-RV, Garst 1984)
5.03 (Sarna et al. 1984)
4.90 (quoted, Oliver & Charlton, 1984; Oliver 1987a)
5.70 (quoted, Bruggeman et al. 1984)
5.20 (shake flask-GC, Chiou 1985; quoted, Suntio et al. 1988b)
5.00 (selected, Bobra et al. 1985; Abernethy & Mackay 1987; Suntio et al. 1988b; quoted, Ballschmiter & Wittlinger 1991)
5.46 (Opperhuizen et al. 1985)
5.17, 4.97 (quoted, calculated-K_{OW} & HPLC-RT, Chin et al. 1986)
6.12 (HPLC-k', De Kock & Lord 1987)
5.03 (quoted, Gobas et al. 1987; 1989)
4.94, 5.46 (quoted, Hawker & Connell 1985; Connell & Hawker 1988; Hawker 1990)
5.03 (quoted exptl., Doucette & Andren 1988)
5.71, 5.77, 5.47, 5.34, 4.86, 4.43 (calculated-π, f const., HPLC-RT, MW, χ, TSA, Doucette & Andren 1988)
5.03, 4.84 (quoted, calculated-V_I, solvatochromic parameters, Kamlet et al. 1988)
5.20 (shake flask-GC, Pereira et al. 1988)
5.19 (quoted, Hodson et al. 1988)
5.17, 4.97 (quoted, calculated-UNIFAC, Banerjee & Howard 1988)
5.183, 5.92 (slow stirring-GC, calculated-f constants, De Bruijn et al. 1989)
5.19 (quoted, Isnard & Lambert 1988, 1989)

4.94 (selected, Thomann 1989)

Bioconcentration Factor, log BCF:

3.89 (trout muscle, Neely et al. 1974)
3.70 (fish, flowing water, Kenaga & Goring 1980; Kenaga 1980a; quoted, Suntio et al. 1988b)
5.40 (guppy, lipid content, Könemann et al. 1979)
3.53 (bluegill sunfish, Veith et al. 1980; quoted, Suntio et al. 1988b)
3.50-3.80 (fish, calculated, Veith et al. 1980)
3.53 (bluegill sunfish, whole body, flow system, Barrows 1980)
5.41 (guppy, lipid basis, Könemann & van Leeuwen 1980; quoted, Suntio et al. 1988b)
3.87 (fish, correlated, Mackay 1982)
1.79 (poultry, Garten & Trabalka 1983)
3.84 (fish, flowing water, Garten & Trabalka 1983)
3.50 (guppy, calculated- χ , Koch 1983)
4.11-4.30 (rainbow trout, Oliver & Niimi 1983)
5.19-5.36 (rainbow trout, lipid basis, Oliver & Niimi 1983; quoted, Suntio et al. 1988b)
4.11-4.30 (rainbow trout, Oliver 1984)
3.71 (22°C, bluegill sunfish, 1.5 % lipid, Banerjee et al. 1984)
3.65 (15°C, rainbow trout, 1.8 % lipid, Banerjee et al. 1984)
3.86 (15°C, guppy, 2.8 % lipid, Banerjee et al. 1984)
4.41 (guppy, Bruggeman et al. 1984; quoted, Suntio et al. 1988b)
3.60 (algae, Geyer et al. 1984)
3.49 (algae, calculated, Geyer et al. 1984)
3.65 (algae, Freitag et al. 1984; Halfon & Reggiani 1986; quoted, Suntio et al. 1988b)
3.48 (fish, Freitag et al. 1984; Halfon & Reggiani 1986; quoted, Suntio et al. 1988b)
3.69 (fish-calculated, Garst 1984)
3.48, 3.65, 4.16 (fish, algae, activated sludge, Freitag et al. 1985)
4.23 (guppy, k_1/k_2, Opperhuizen et al. 1985)
4.38, 4.14 (guppy, calculated, Gobas et al. 1987)
5.40, 5.41 (guppy, lipid-phase, calculated, Gobas et al. 1987)
3.92 (fathead minnow, Carlson & Kosian 1987)
2.83, 2.65 (human fat, Geyer et al. 1987)
4.28, 4.30 (worms, fish, Oliver 1987a)
3.36-3.42 (fish, Tadokoro & Tomita 1987)
5.0, 5.46 (guppy, lipid-phase, calculated-K_{OW}, Gobas et al. 1989)
3.87 (picea omorika, Reischl et al. 1989)
5.33 (fish, calculated-K_{OW}, Thomann 1989)
4.14, 2.96, 4.23 (fish, Connell & Hawker 1988; Hawker 1990)
3.67 (guppy, Van Hoogan & Opperhuizen 1988)
5.93, 6.12, 4.96, 5.57 (field data-lipid base: Atlantic croakers, blue crabs, spotted sea trout, blue catfish, Pereira et al. 1988)
3.48 (fish, calculated, Figueroa & Simmons 1991)

Sorption Partition Coefficient, log K_{OC}:

4.11 (Kenaga 1980a)
5.30 (field data, Oliver & Charlton 1984)
5.40 (Niagara River-organic matter, Oliver & Charlton 1984)
4.50 (calculated-K_{OW}, Oliver & Charlton 1984)
4.60 (bottom sediment, Karickhoff & Morris 1985a)
4.9-6.2, 5.8 (suspended sediment, average, Oliver 1987c)
5.7 (algae > 50 μm, Oliver 1987c)
5.5-5.9, 5.7, 4.6 (Niagara River plume, range, mean; calculated-K_{OW}, Oliver 1987b)
4.49 (soil, Paya-Perez et al. 1991)

Sorption Partition Coefficient, log K_{OM}:

3.50, 3.22 (quoted, calculated- χ , Sabljic 1984)

Half-Lives in the Environment:

Air:
Surface Water:
Groundwater:
Sediment:
Soil:
Biota: half-life in fish > 7 days (Veith et al. 1980); half-life in bluegill sunfish > 7 days (Barrows et al. 1980); half-life in guppy, 4.6 days (Bruggeman et al. 1984); half-life in guppy, 8.9 days (Opperhuizen et al. 1985); half-life in worms at 8°C, < 5 days (Oliver 1987a); half-life in picea omorika, 27 days (Reischl et al. 1989).

Environmental Fate Rate Constants or Half-Lives:

Volatilization:
Photolysis:
Oxidation:
Hydrolysis:
Biodegradation:
Biotransformation:
Bioconcentration, Uptake (k_1) and Elimination (k_2) Rate Constants:

k_1: 18.76 $hour^{-1}$ (trout, Neely et al. 1974)
k_2: 0.00238 $hour^{-1}$ (trout, Neely et al. 1974)
k_1: 22000 day^{-1} (guppy, Könemann & van Leeuwen 1980)
k_2: 0.40 day^{-1} (guppy, Könemann & van Leeuwen 1980; selected, Clark et al. 1990)
k^1: 130 day^{-1} (guppy, Bruggeman et al. 1984)
k_2: 0.15 day^{-1} (guppy, Bruggeman et al. 1984)
k_1: 110 $hour^{-1}$ (22°C, bluegill sunfish, Banerjee et al. 1984)
k_2: 0.021 $hour^{-1}$ (22°C, bluegill sunfish, Banerjee et al. 1984)
k_1: 170 $hour^{-1}$ (15°C, rainbow trout, Banerjee et al. 1984)

k_2: 0.036 hour^{-1} (15°C, rainbow trout, Banerjee et al. 1984)
k_1: 98 hour^{-1} (15°C, guppy, Banerjee et al. 1984)
k_2: 0.014 hour^{-1} (15°C, guppy, Banerjee et al. 1984)
k_1: 49.5 hour^{-1} (guppy, quoted, Hawker & Connell 1985)
k_1: 5.4 hour^{-1} (guppy, quoted, Hawker & Connell 1985)
$1/k_2$: 220 hour (guppy, quoted, Hawker & Connell 1985)
$1/k_2$: 160 hour (guppy, quoted, Hawker & Connell 1985)
k_1: 1400 day^{-1} (guppy, Opperhuizen et al. 1985)
k_2: 0.078 day^{-1} (guppy, Opperhuizen et al. 1985; quoted, Clark et al. 1990)
k_1: 1200 day^{-1} (fish, quoted, Opperhuizen 1986)
k_2: 0.00309, 0.00402 day^{-1} (rainbow trout, calculated-fish mean body weight, Barber et al. 1988)
log k_1: 2.11 day^{-1} (fish, quoted, Connell & Hawker 1988)
log k_1: 3.08 day^{-1} (fish, quoted, Connell & Hawker 1988)
log $1/k_2$: 2.11 day^{-1} (fish, quoted, Connell & Hawker 1988)
log $1/k_2$:-0.96 day^{-1} (fish, quoted, Connell & Hawker 1988)
k_1: 710 mL/g.d (guppy, Van Hoogan & Opperhuizen 1988)
k_2: 0.15 day^{-1} (guppy, Van Hoogan & Opperhuizen 1988)
log k^1: 3.24 day^{-1} (guppy, Gobas et al. 1989a)
log k_2: -0.96 day^{-1} (guppy, Gobas et al. 1989a)
k_s: 0.049 hour^{-1} (uptake of mayfly-sediment model II, Gobas et al. 1989b)
k_t: 0.027 hour^{-1} (depuration of mayfly-sediment model II, Gobas et al. 1989b)
log k_2: -0.82, day^{-1} (fish, calculated-K_{OW}, Thomann 1989)
log k_2: -0.96 day^{-1} (fish, calculated-K_{OW}, Thomann 1989)
$1/k_2$: 9.12, 12.9 day (guppy, Clark et al. 1990)

Sediment Exchange Rate Constant:
0.083-0.31 day^{-1} (natural sediment, Karickhoff & Morris 1985b).

Common Name: Hexachlorobenzene
Synonym: HCB, perchlorobenzene, anticarie, Bunt-cure, Bunt-no-more, Julin's carbon chloride
Chemical Name: hexachlorobenzene
CAS Registry No: 118-74-1
Molecular Formula: C_6Cl_6
Molecular Weight: 284.79
Melting Point (°C):
227.0 (Verschueren 1977,1983)
230.0 (Mackay & Shiu 1981; Yalkowsky et al. 1983; Kishii et al. 1987)
228.0 (Schmidt-Bleek et al. 1982)
227.9 (Miller et al. 1984)
228.3 (Miller et al. 1985)
Boiling Point (°C):
322 (Verschueren 1977,1983; Mackay & Shiu 1981; Miller et al. 1984)
318 (Schmidt-Bleek et al. 1982)
Density (g/cm^3 at 20°C):
2.044
1.5691 (23.6°C, Weast 1972-73; Horvath 1982)
Molar Volume (cm^3/mol):
181.5 (calculated-density, Weast 1972-73; Horvath 1982)
182 (calculated-density, Lande & Banerjee 1981; Mailhot 1987)
221.4 (LeBas method, Miller et al. 1985; Shiu et al. 1987)
1.031 (intrinsic volume: $V_I/100$, Kamlet et al. 1988; Hawker 1989, 1990)
Molecular Volume (A^3):
199.5 (De Bruijn & Hermens 1990)
Total Surface Area, TSA (A^2):
203.0 (Yalkowsky et al. 1979)
202.2 (Kishii et al. 1987)
209.4 (Sabljic 1987)
202.12 (planar, Doucette & Andren 1988)
207.9 (De Bruijn & Hermens 1990)
Heat of Fusion, kcal/mol:
6.87 (Tsonopoulo & Prausnitz 1971)
5.354 (Miller et al. 1984)
Entropy of Fusion, cal/mol K (e.u.):
13.7 (Tsonopoulos & Prausnitz 1971)
10.7 (Miller et al. 1984)
Fugacity Ratio, F (assuming ΔS_{fusion} = 13.5 e.u.):
0.009 (25°C, Miller et al. 1985)
0.0075 (20°C, Suntio et al. 1988a)
0.0094 (25°C, Suntio et al. 1988b)

Water Solubility (g/m^3 or mg/L at 25 °C):

0.005 (generator column-GC/ECD, Weil et al. 1974; quoted, Korte, et al. 1978; Kilzer et al. 1979; Chiou et al. 1982; Pereira et al. 1988; Suntio et al. 1988b; Shiu et al. 1990)

0.006 (shake flask-LSC/^{14}C, Lu & Metcalf 1975; quoted, Shiu et al. 1990)

0.11 (shake flask-nephelometric spectrophotofluorometry, Hollifield 1979)

0.006 (quoted, Callahan et al. 1979; Niimi & Cho 1980)

0.005 (shake flask-UV, Yalkowsky et al. 1979; quoted, Miller et al. 1984; Suntio et al. 1988b; Shiu et al. 1990)

0.004 (quoted, Geyer et al. 1980)

0.035 (quoted, Kenaga & Goring 1980; Kenaga 1980a; Shiu et al. 1990)

0.0034 (calculated-K_{OW}, Yalkowsky et al. 1979; Yalkowsky & Valvani 1980; Valvani & Yalkowsky 1980)

0.0035 (quoted, Neely 1980)

0.036 (quoted, Briggs 1981)

0.005 (quoted, Geyer et al. 1981)

0.0039 (shake flask-GC, Könemann 1981)

0.0054 (gen. col.-GC/ECD, Hashimoto et al. 1982; quoted, McKim et al. 1985)

0.0012-0.014 (shake flask-GC/ECD, Hashimoto et al. 1982)

0.005 (recommended, Horvath 1982; quoted, Wong et al. 1984; Oliver 1987b)

0.0084 (quoted lit. average, Yalkowsky et al. 1983)

0.035 (quoted, Calamari et al. 1983)

0.0066 (quoted, Yoshida et al. 1983b)

0.047 (gen. col.-GC/ECD, Miller et al. 1984; 1985; quoted, Doucette & Andren 1988; Suntio et al. 1988b; Shiu et al. 1990; Mackay & Paterson 1991)

0.0084, 0.0162 (quoted, calculated-UNIFAC, Banerjee 1985)

0.005 (recommended, IUPAC 1985)

0.005 (selected, Bobra et al. 1985; Suntio et al. 1988a & b; quoted, Ballschmiter & Wittlinger 1991)

0.006 (selected, Mackay et al. 1985)

0.0047 (selected, Abernethy et al. 1986)

0.00495, 0.0146 (quoted, calculated-K_{OW} & HPLC-RT, Chin et al. 1986)

0.005 (quoted, Eadie & Robbins 1987)

0.0124 (quoted, Mailhot 1987)

0.005-0.05, 0.006-0.2 (quoted exptl., calculated-K_{OW}, Anliker & Moser 1987)

0.006 (quoted, Isnard & Lambert 1988, 1989)

0.860 (quoted, subcooled liquid, Hawker 1989)

0.0024, 0.00537 (quoted, calculated-UNIFAC, Banerjee et al. 1990)

0.0063 (quoted, Figueroa & Simmons 1991)

Vapor Pressure (Pa at 25°C):

0.00028 (Sears & Hopke 1949; quoted, Suntio et al. 1988b)

0.0015 (quoted, Callahan et al. 1979; Neuhauser et al. 1985)

0.0026 (selected, OECD 1979; quoted, Suntio et al. 1988b)
0.00145 (20°C, Kiltzer et al. 1979)
0.0023 (gas saturation, Farmer et al. 1980; quoted, Suntio et al. 1988b)
0.0013 (quoted, Neely 1980)
0.000453; 0.000167 (Klein et al. 1981; quoted, Suntio et al. 1988b)
0.0019 (OECD 1981; quoted, Dobbs et al. 1984)
0.0015 (Mackay & Shiu 1981; quoted, Mackay & Paterson 1991)
0.00046 (evaporation rate, Dobbs & Cull 1982; quoted, Suntio et al. 1988b)
0.00121 (extrapolated, Antoine eqn., Gückel et al. 1982)
0.0006 (20°C, evaporation rate & gravimetric, Gückel et al. 1982)
0.00147 (quoted, Yoshida et al. 1983b)
0.303; 0.159; 0.121 (subcooled liquid, quoted; GC-RT, Bidleman 1984)
0.0031 (selected, Mackay et al. 1985)
0.00147, 0.187 (20°C, selected, solid, subcooled liquid, Bidleman & Foreman 1987)
0.13 (selected, Suntio et al. 1988a)
0.245 (selected, Suntio et al. 1988b; quoted, Ballschmiter & Wittlinger 1991)
0.303, 0.127 (subcooled liquid, quoted, Hinckley et al. 1990)

Henry's Law Constant (Pa m^3/mol):
68.2 (20°C, Callahan et al. 1979)
5.07 (calculated-P/C, Mackay & Shiu 1981; quoted, Pankow et al. 1984)
131.3 (batch stripping, Atlas et al. 1982; quoted, Suntio et al. 1988b)
68.9 (20°C, calculated, Mabey et al. 1982)
12.16 (calculated-P/C, Calamari et al. 1983)
62 (calculated-P/C, Yoshida et al. 1983b)
139 (calculated-P/C, Bobra et al. 1985; Suntio et al. 1988b; quoted, Ballschmiter & Wittlinger 1991)
48.6 (20°C, batch stripping, Oliver 1985)
133, 115.9 (observed, calculated-QSAR, Nirmalakahandan & Speece 1988b)
7.12 (20°C, calculated-P/C, Suntio et al. 1988a)

Octanol/Water Partition Coefficient, log K_{OW}:
6.18 (Neely et al. 1974; quoted, McKim et al. 1985)
4.13 (radioisotope tracer-^{14}C, Lu & Metcalf 1975)
6.51 (calculated-f const., Rekker 1977; quoted, Harnisch et al. 1983)
6.18 (quoted, Callahan et al. 1979; Neuhauser et al. 1985)
4.13 (Hansch & Leo 1979; quoted, Harnisch et al. 1983; Suntio et al. 1988b)
5.0, 6.27 (shake flask-GC, HPLC-k', Könemann et al. 1979; quoted, Figueroa & Simmons 1991)
6.44 (calculated-f constant, Könemann et al. 1979; Könemann 1981; quoted, Opperhuizen 1986)
5.23 (HPLC-RT, Veith et al. 1979b)
6.18, 7.42 (HPLC-RT, calculated, Veith et al. 1979a; quoted, Suntio et al. 1988b)

6.53 (calculated-f const., Yalkowsky et al. 1979,1982,1983; Yalkowsky & Valvani 1980; Valvani & Yalkowsky 1980; quoted, Miller et al. 1984)
5.23 (quoted, Kenaga & Goring 1980; Yoshida et al. 1983b)
5.44 (quoted, Briggs 1981)
6.22 (HPLC-RT, McDuffie 1981)
5.50 (shake flask-GC, Chiou et al. 1982; Chiou 1985; quoted, Oliver 1987a,b & c; Suntio et al. 1988b)
5.23 (quoted, Mackay 1982; Freitag et al. 1985)
5.66 (HPLC-RT, Hammers et al. 1982)
5.40 (shake flask-GC, Wateral et al. 1982; quoted, Suntio et al. 1988b)
6.13-6.27, 5.66 (range, mean, shake flask method, Eadsforth & Moser 1983)
6.27-6.48, 6.38 (range, mean, HPLC method, Eadsforth & Moser 1983)
5.46, 5.26 (quoted of HPLC methods, Harnisch et al. 1983)
5.47 (quoted of OECD/EEC shake-flask method, Harnisch et al. 1983)
5.50 (quoted, Oliver & Niimi 1983; Oliver 1987)
5.0, 5.19 (quoted, calculated, Kaiser 1983; Kaiser et al. 1984)
5.89 (quoted, Calamari et al. 1983)
6.42 (calculated-f const., Veith et al. 1983)
5.23, 4.61 (quoted, calculated-molar refraction, Yoshida et al. 1983a)
5.47 (generator column-GC/ECD, Miller et al. 1984; 1985; quoted, Suntio et al. 1988b; Mackay & Paterson 1991)
5.50 (quoted, Oliver & Charlton 1984)
5.75 (quoted, Garst 1984)
5.70-5.79 (HPLC-RV, Garst & Wilson 1984; Garst 1984)
5.20, 5.23, 5.44, 5.50, 5.55 (reported lit., Geyer et al. 1984)
5.47 (Sarna et al. 1984)
5.50 (selected, Bobra et al. 1985; Suntio et al. 1988b; quoted, Ballschmiter & Wittlinger 1991)
5.47, 6.86, 6.42 (quoted, HPLC/MS, calculated-π, Burkhard et al. 1985)
5.50 (quoted, Hawker & Connell 1985; Connell & Hawker 1988; Hawker 1990)
5.61 (selected, Mackay et al. 1985)
5.75 (selected OECD value, Brooke et al. 1986)
5.6, 5.9 (HPLC-RV, Brooke et al. 1986)
6.51, 6.18 (quoted, calculated-K_{OW} & HPLC-RT, Chin et al. 1986)
5.50 (quoted, Geyer et al. 1987)
6.92 (HPLC-k', De kock & Lord 1987)
5.64 (HPLC-k', Mailhot 1987)
5.45 (quoted, Gobas et al. 1987,1989; Travis & Arms 1988)
5.47 (quoted exptl., Doucette & Andren 1988)
6.42, 6.55, 6.22, 5.34, 4.86, 4.75 (calculated-π, f const., HPLC-RT, MW, χ , TSA, Doucette & Andren 1988)
5.47, 5.37 (quoted, calculated-V_I, solvatochromic parameters, Kamlet et al. 1988)
6.18 (quoted, Ryan et al. 1988)

5.50 (shake flask-GC, Pereira et al. 1988)
6.0 (selected, Suntio et al. 1988a)
5.31, 6.58 (quoted, calculated-UNIFAC, Banerjee & Howard 1988)
6.68 (calculated-f const., De Bruijn et al. 1989)
5.73 (slow stirring-GC, De Bruijn et al. 1989; De Bruijn & Hermens 1990)
5.66 (correlated, Isnard & Lambert 1988, 1989)
5.50 (selected, Thomann 1989)

Bioconcentration Factor, log BCF:
3.89 (rainbow trout, calculated-k_1/k_2, Neely et al. 1974)
3.09 (fish, Körte et al. 1978)
4.27, 3.73, 4.34 (fathead minnow, rainbow trout, green sunfish, Veith et al. 1979b)
5.46 (guppy, lipid basis, Könemann & van Leeuwen 1980; quoted, Chiou 1985)
4.27 (fish, Ciam et al. 1980)
1.20 (rats, adipose tissue, Geyer et al. 1980)
3.93, 2.46 (fish, flowing water, static water, Kenaga & Goring 1980; Kenaga 1980a)
3.61, 2.45 (calculated from water solubility, K_{OC}, Kenaga 1980a)
4.39, 4.2 (algae, calculated, Geyer et al. 1981)
3.91 (fish, correlated, Mackay 1982)
4.27, 3.89 (fathead minnow, rainbow trout, quoted, Bysshe 1982)
4.27 (fish, quoted, Dobbs & Williams 1983)
4.60 (guppy, calculated- χ , Koch 1983)
4.08-4.30 (rainbow trout, Oliver & Niimi 1983)
5.16-5.37 (rainbow trout, lipid basis, Oliver & Niimi 1983; quoted, Chiou 1985)
4.31 (calculated-K_{OW}, Calamari et al. 1983)
3.93 (quoted, Yoshida et al. 1983b)
4.39, 3.36, 4.54 (algae, fish, activated sludge, Klein et al. 1984)
4.39, 3.83 (algae: exptl., calculated, Geyer et al. 1984)
4.27 (fathead minnow, 25°C, calculated, Davis & Dobbs 1984; Anliker & Moser 1987)
4.34 (green sunfish, 15°C, calculated, Davis & Dobbs 1984)
3.74 (rainbow trout, 15°C, calculated, Davis & Dobbs 1984)
4.39, 3.36, 4.54 (algae, fish, sludge, Klein et al. 1984)
4.54 (activated sludge, Freitag et al. 1984; Halfon & Reggiani 1986)
4.39, 3.41, 4.54 (algae, fish, activated sludge, Freitag et al. 1985)
3.05 (fish, quoted, Hawker & Connell 1986)
2.62-2.97 (human fat, lipid basis, Geyer et al. 1987)
2.44-2.79 (human fat, wet weight, Geyer et al. 1987)
4.41 (algae, Mailhot 1987)
4.34 (fathead minnow, Carlson & Kosian 1987)
4.38, 4.30 (worms, fish, Oliver 1987a)
3.48 (fish-normalized, Tadokoro & Tomita 1987)
4.19 (guppy, calculated, Gobas et al. 1987)
5.46 (guppy-lipid phase, calculated-K_{OW}, Gobas et al. 1987, 1989)

6.42, 6.71, 5.96, 5.98 (field data-lipid base: Atlantic croakers, blue crabs, spotted sea trout, blue catfish, Pereira et al. 1988)
-1.35 (beef, reported as biotransfer factor log B_b, Travis & Arms 1988)
-2.07 (milk, reported as biotransfer factor log B_m, Travis & Arms 1988)
-0.32 (vegetable, reported as biotransfer factor log B_v, Travis & Arms 1988)
5.30 (guppy-lipid phase, calculated-K_{OW}, Gobas et al. 1989)
3.90, 4.19 (fish, quoted, Connell & Hawker 1988; Hawker 1990)
5.30 (guppy, correlated, Gobas et al. 1989)
3.53 (picea omorika, Reischl et al. 1989)
3.57 (fish, calculated, Figueroa & Simmons 1991)

Sorption Partition Coefficient, log K_{OC}:
3.59 (Kenaga & Goring 1980; Kenaga 1980a; quoted, Lyman 1982; Yoshida et al. 1983b)
4.45 (Kenaga 1980a)
4.44, 4.21, 3.59 (estimated-S, K_{OW}, BCF, Lyman 1982)
6.08 (calculated, Mabey et al. 1982)
3.59 (quoted, Bysshe 1982; Lyman et al. 1982)
2.56 (Speyer soil < 2.00 mm, Freundlich isotherm, Rippen et al. 1982)
2.70 (Alfisol, Freundlich isotherm, Rippen et al. 1982)
4.58 (calculated-K_{OW}, Calamari et al. 1983)
5.90 (field data, Oliver & Charlton 1984)
4.90 (bottom sediment, Karickhoff & Morris 1985a)
5.10 (calculated-K_{OW}, Oliver & Charlton 1984)
5.2-6.7, 6.1 (suspended sediment, average, Oliver 1987c)
5.80 (algae > 50 μm, Oliver 1987c)
6.0-6.5, 6.3; 5.1 (Niagara River plume, range, mean; calculated-K_{OW}, Oliver 1987b)
4.77 (HPLC-k', Hodson & Williams 1988)

Sorption Partition Coefficient, log K_{OM}:
4.25 (shake flask-GC, soil-organic matter, Briggs 1981)
5.50 (Niagara River-organic matter, Oliver & Charlton 1984)

Half-Lives in the Environment:
Air: 3753-37530 hours, based on estimated photooxidation half-life (Atkinson 1987).
Surface Water: 23256-50136 hours, based on estimated unacclimated aqueous aerobic biodegradation half-life (Beck & Hansen 1974); 1.4-50 days estimated, 0.3-3 days for river water and 30-300 days for lakes, estimated from persistence (Zoeteman et al. 1980).
Groundwater: 46512-100272 hours, based on unacclimated aqueous aerobic biodegradation half-life (Beck & Hansen 1974); 30-300 days, estimated from persistence (Zoeteman et al. 1980).
Soil: 23256-50136 hours, based on unacclimated aerobic soil grab sample data (Beck & Hansen, 1974); > 50 days (Ryan et al. 1988); disappearance half-life from

testing soils, 11.3 days (Anderson et al. 1991).

Biota: half-life in rainbow trout, >224 days (Niimi & Cho 1981); in subadult rainbow trout-calculated, 210 days at 4°C, 80 days at 12°C, 70 days at 18°C (Niimi & Palazzo 1985); in worms at 8°C, 27 days (Oliver 1987a); picea omorika, 30 days (Reischl et al. 1989).

163 hours, clearance from fish (Neely 1980)

Environmental Fate Rate Constants or Half-Lives:

Volatilization/Evaporation: 3.45×10^{-10} mol/m^{2}h (Gückel et al. 1982).

Photolysis:

Oxidation: rate constant in air, 1.44×10^{-2} hour^{-1} (Brown et al. 1975; quoted, Mackay et al. 1985); photooxidation half-life in air: 3753-37530 hours, based on estimated rate constant for the vapor phase reaction with hydroxy radicals in air (Atkinson 1987).

Hydrolysis: not expected to be important, based on $k_h = 0$ was observed after 13 days at pH 3, 7, 11 and 85°C (Ellington et al. 1987).

Biodegradation: aqueous aerobic biodegradation half-life: 23256-50136 hours, based on unacclimated aerobic soil grab sample data (Beck & Hansen 1974); anaerobic aqueous biodegradation half-life: 93024-200544 hours, based on estimated unacclimated aqueous aerobic biodegradation half-life (Beck & Hansen 1974); degradation rate constant in soil, 1.9×10^{-5} hour^{-1} (Beck & Hansen 1974; selected, Mackay et al. 1985; Mackay & Paterson 1991); not significant in an aerobic environment, and no significant degradation rate (Tabak et al. 1981; Mills et al. 1982).

Bioconcentration Uptake (k_1) and Elimination (k_2) Rate Constants:

k_1: 18.76 hour^{-1} (trout muscle, Neely et al. 1974)
k_2: 0.00238 hour^{-1} (trout muscle, Neely et al. 1974)
k_1: 10000 day^{-1} (guppy, Könemann & van Leeuwen 1980)
k_1: 22.5 hour^{-1} (guppy, quoted, Hawker & Connell 1985)
k_1: 18.8 hour^{-1} (trout, quoted, Hawker & Connell 1985)
k_1: 540 day^{-1} (fish, quoted, Opperhuizen 1986)
k_2: 0.00510, 0.00818, 0.00640, 0.0047 day^{-1} (rainbow trout, calculated-fish mean body weight, Barber et al. 1988)
$1/k_2$: 420 hour (trout, quoted, Hawker & Connell 1985)
log k_1: 2.73 day^{-1} (fish, quoted, Connell & Hawker 1988)
log k_1: 2.65 day^{-1} (fish, quoted, Connell & Hawker 1988)
log $1/k_2$: 1.24 day^{-1} (fish, quoted, Connell & Hawker 1988)
log k_2: -1.24 day^{-1} (fish, calculated-K_{OW}, Thomann 1989)
k_s: 0.049 hour^{-1} (uptake of mayfly-sediment model II, Gobas et al. 1989b)
k_t: 0.023 hour^{-1} (depuration of mayfly-sediment model II, Gobas et al. 1989b)

Sediment Exchange Rate Constant:

0.026-1.2 day^{-1} (natural sediment, Karickhoff & Morris 1985b).

Sediment Burial Constant:

4.6×10^{-6} $hour^{-1}$ (Di Toro et al. 1981; quoted, Mackay et al. 1985)

Stratospheric Diffusion Rate Constant:

1.7×10^{-6} $hour^{-1}$ (Mackay et al. 1985)

3.2 Summary Tables and QSPR Plots

Table 3.1 Summary of physical-chemical properties of chlorobenzenes at 25 °C

Compound	CAS no.	Formula	MW g/mol	mp, °C	bp, °C	Density g/cm³ 20 °C	Fugacity ratio, F at 25 °C
Chlorobenzene	108-90-7	C_6H_5Cl	112.6	-45.6	132.2	1.1058	1
1,2-Dichloro-	95-50-1	$C_6H_4Cl_2$	147.01	-17	180.5	1.3048	1
1,3-Dichloro-	541-73-1	$C_6H_4Cl_2$	147.01	-24.9	173.5	1.2884	1
1,4-Dichloro-	106-46-7	$C_6H_4Cl_2$	147.01	53.1	174.6	1.2475	0.53
1,2,3-Trichloro-	87-61-6	$C_6H_3Cl_3$	181.45	53	218		0.53
1,2,4-Trichloro-	120-82-1	$C_3H_3Cl_3$	181.45	16.95	213.5	1.4542	1
1,3,5-Trichloro-	108-70-3	$C_6H_3Cl_3$	181.45	64	208		0.41
1,2,3,4-Tetrachloro-	634-66-2	$C_6H_2Cl_4$	215.9	47.5	254		0.6
1,2,3,5-Tetrachloro-	634-90-2	$C_6H_2Cl_4$	215.9	54.5	246		0.51
1,2,4,5-Tetrachloro-	95-94-3	$C_6H_2Cl_4$	215.9	140	243		0.075
Pentachlorobenzene	608-93-5	C_6HCl_5	250.3	86	277	1.8342	0.23
Hexachlorobenzene	118-74-1	C_6Cl_6	284.8	230	322	1.5691	0.0094

	Molar volume, cm³/mol			Total surface area, Å²			TMV, Å³
Compound	V_M from ρ (a)	V_M LeBas	$V_I/100$ intrinsic (b)	TSA (c)	TSA (d)	TSA (e)	(e)
Chlorobenzene	102	117	0.581	127.1	126.73	158.4	136.7
1,2-Dichloro-	113	138	0.671	142.7	142.22	168.8	150.0
1,3-Dichloro-	113	138	0.671	144.7	144.24	171.3	153.7
1,4-Dichloro-	118	138	0.671	144.7	144.24	171.3	153.7
1,2,3-Trichloro-	125	159	0.761	158.3	157.7	179.2	163.3
1,2,4-Trichloro-	125	159	0.761	160.2	159.72	181.7	167.0
1,3,5-Trichloro-	125	159	0.761	162.2	161.75	184.2	170.0
1,2,3,4-Tetrachloro-	142	180	0.851	173.8	173.18	189.6	176.6
1,2,3,5-Tetrachloro-	142	180	0.851	175.8	175.2	192.1	180.3
1,2,4,5-Tetrachloro-	142	180	0.851	175.8	175.2	192.1	180.3
Pentachlorobenzene	166	200	0.941	189.4	188.66	200	189.9
Hexachlorobenzene	186	221.4	1.031	203	202.12	207.9	199.5

(a) Chiou et al. (1982); (b) Kamlet et al. (1988); (c) Yalkowsky and Valvani (1976); (d) Doucette and Andren (1988); (e) De Bruijn and Hermens (1990)

Table 3.2 Summary of physical-chemical properties of chlorobenzenes at 25 °C

Selected properties:

Compound	Vapor pressure		Solubility			log K_{OW}	Henry's const., H
	P^S, Pa	P_L, Pa	S, g/m^3	C^S mol/m^3	C_L mol/m^3		Pa m^3/mol calcd., P/C
Chlorobenzene	1580	1580	484	4.2984	4.2984	2.80	368
1,2-Dichloro-	196	196	118	0.8027	0.8027	3.40	244
1,3-Dichloro-	307	307	120	0.8163	0.8163	3.40	376
1,4-Dichloro-	90.2	170.19	83	0.5646	1.0653	3.40	160
1,2,3-Trichloro-	28	52.83	21	0.1157	0.2184	4.10	242
1,2,4-Trichloro-	61	61	40	0.2204	0.2204	4.10	277
1,3,5-Trichloro-	32	78.05	5.3	0.0292	0.0712	4.10	1096
1,2,3,4-Tetrachloro-	5.2	8.67	7.8	0.0361	0.0602	4.50	144
1,2,3,5-Tetrachloro-	9.8	19.22	3.6	0.0167	0.0327	4.50	588
1,2,4,5-Tetrachloro-	0.72	9.6	1.27	0.00588	0.0784	4.50	122
Pentachlorobenzene	0.22	0.9565	0.65	0.00260	0.0113	5.00	85
Hexachlorobenzene	0.0023	0.2447	0.005	0.0000176	0.00187	5.50	131

Table 3.3 Suggested half-life classes of chlorobenzenes in various environmental compartments

Compounds	Air class	Water class	Soil class	Sediment class
Chlorobenzene	4	6	7	8
1,2-Dichloro-	5	6	7	8
1,3-Dichloro-	5	6	7	8
1,4-Dichloro-	5	6	7	8
1,2,3-Trichloro-	5	6	7	8
1,2,4-Trichloro-	5	6	7	8
1,3,5-Trichloro-	5	6	7	8
1,2,3,4-Tetrachloro-	6	7	7	8
1,2,3,5-Tetrachloro-	6	7	7	8
1,2,4,5-Tetrachloro-	6	7	7	8
Pentachlorobenzene	7	8	8	8
Hexachlorobenzene	8	9	9	9

where,

Class	Mean half-life (hours)	Range (hours)
1	5	< 10
2	17 (~ 1 day)	10-30
3	55 (~ 2 days)	30-100
4	170 (~ 1 week)	100-300
5	550 (~ 3 weeks)	300-1,000
6	1700 (~ 2 months)	1,000-3,000
7	5500 (~ 8 months)	3,000-10,000
8	17000 (~ 2 years)	10,000-30,000
9	55000 (~ 6 years)	> 30,000

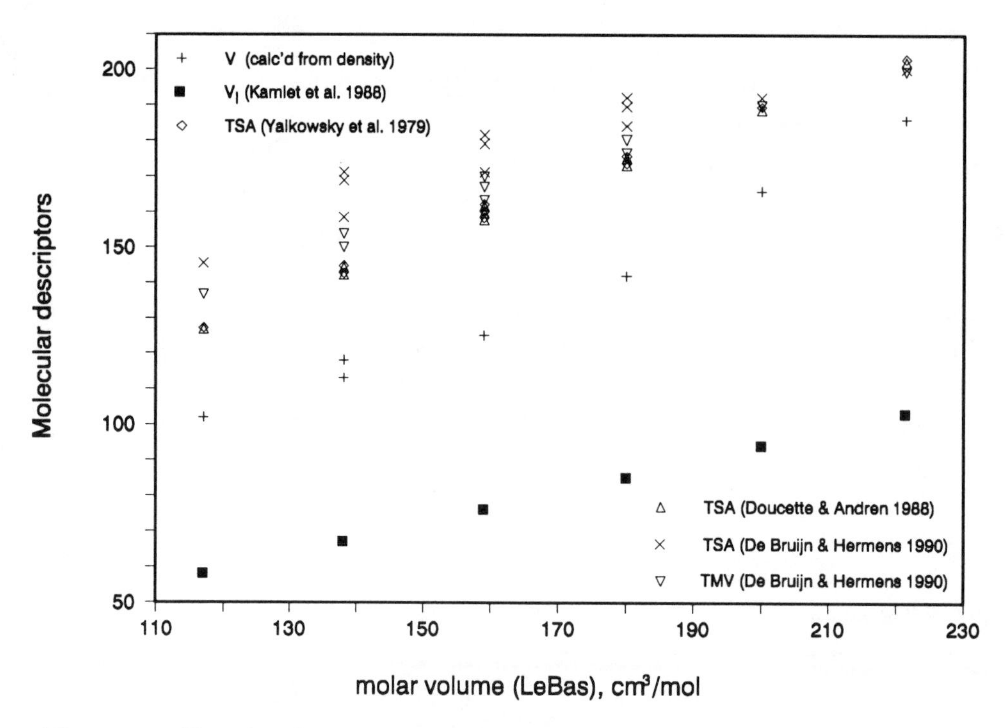

Figure 3.1 Plot of molecular descriptors versus LeBas molar volume

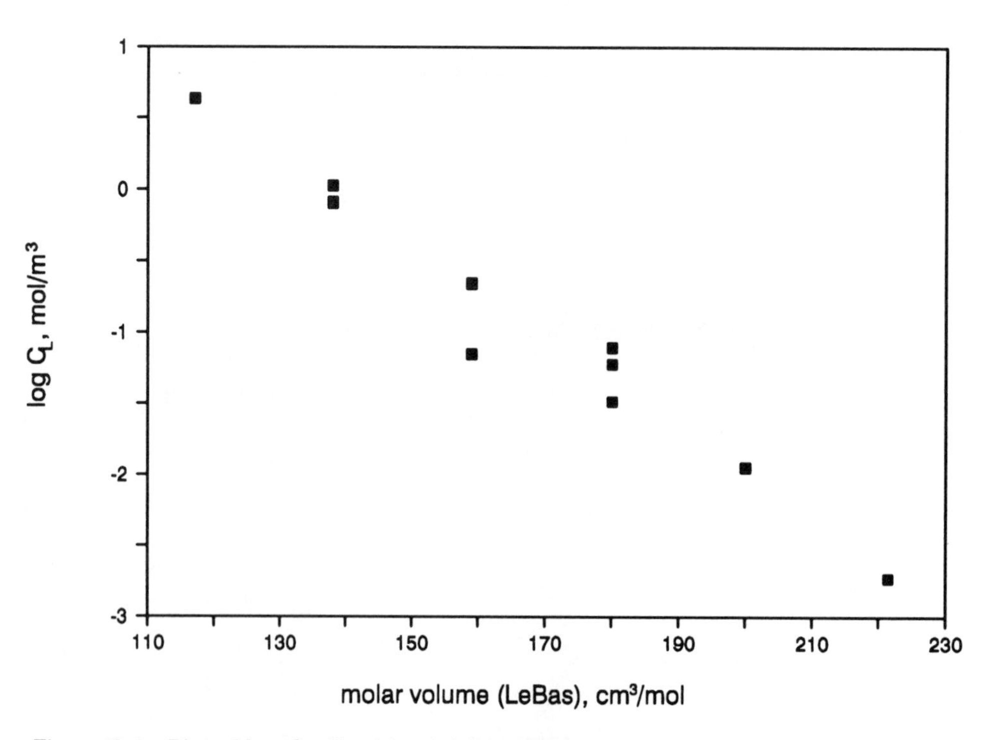

Figure 3.2 Plot of log C_L (liquid solubility) versus molar volume

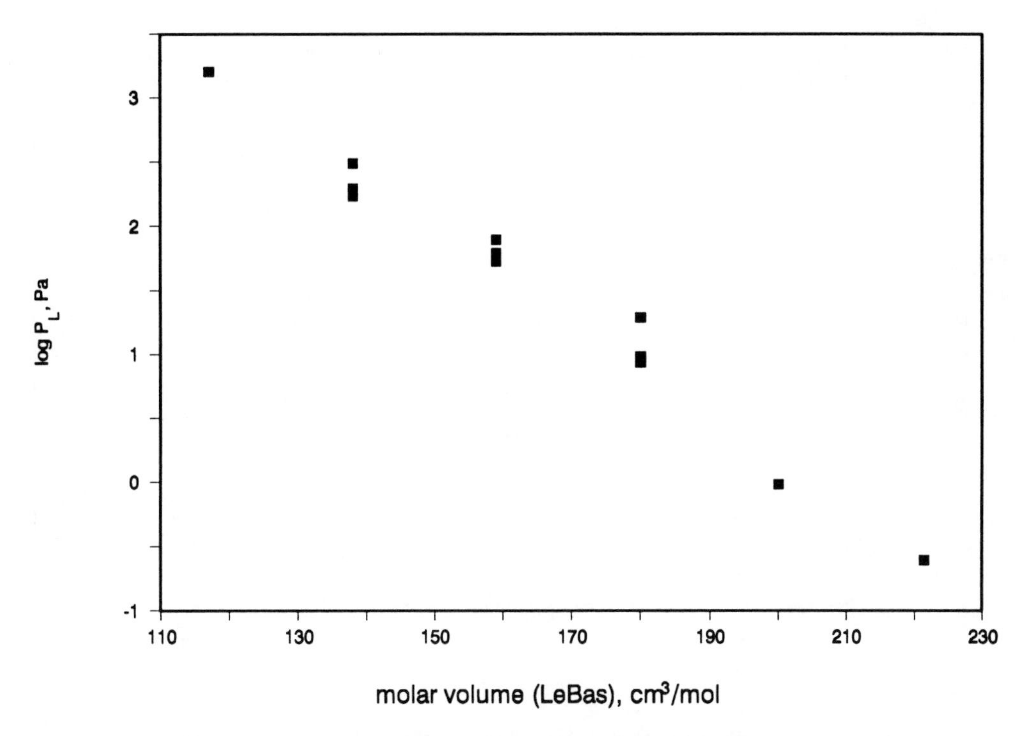

Figure 3.3 Plot of P_L (liquid vapor pressure) versus molar volume

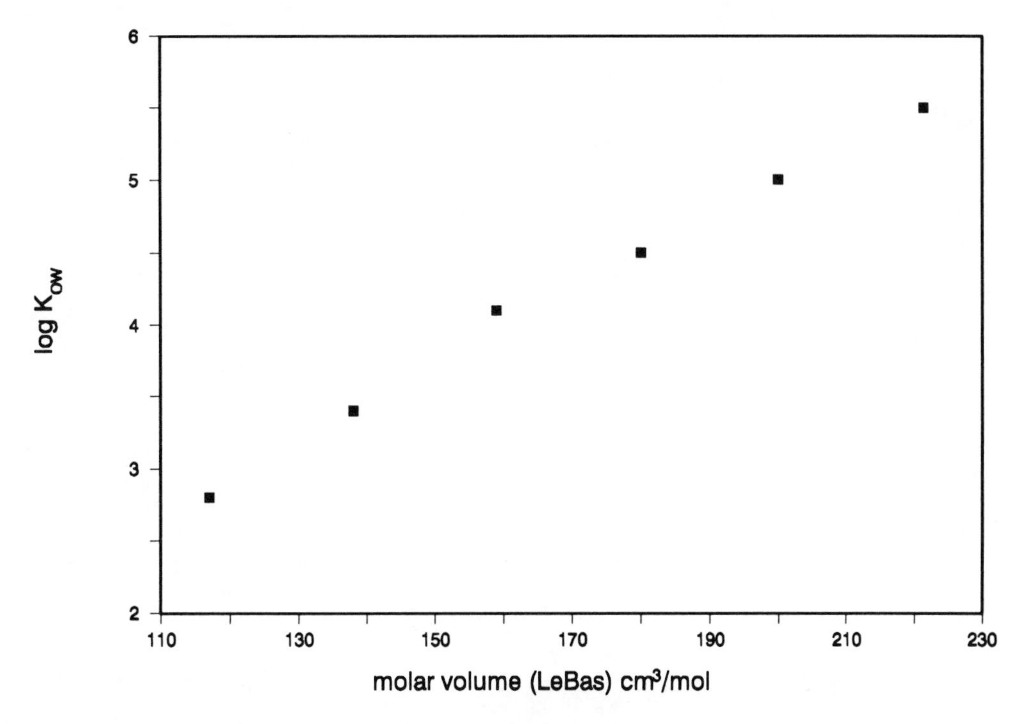

Figure 3.4 Plot of log K_{OW} versus LeBas molar volume

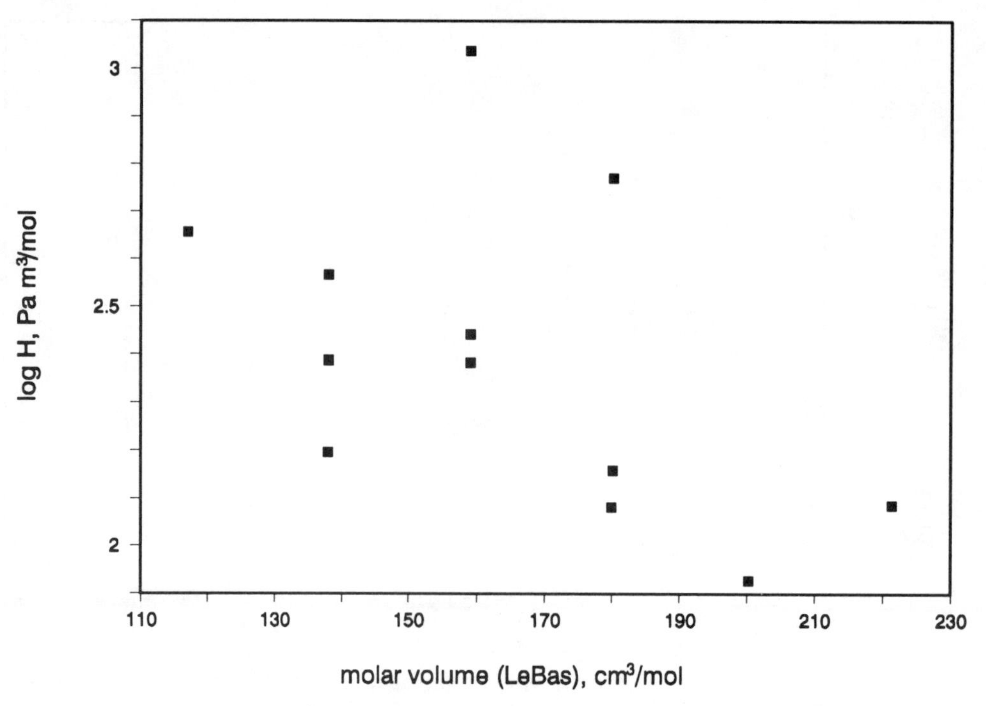

Figure 3.5 Plot of log H (Henry's law constant) versus molar volume

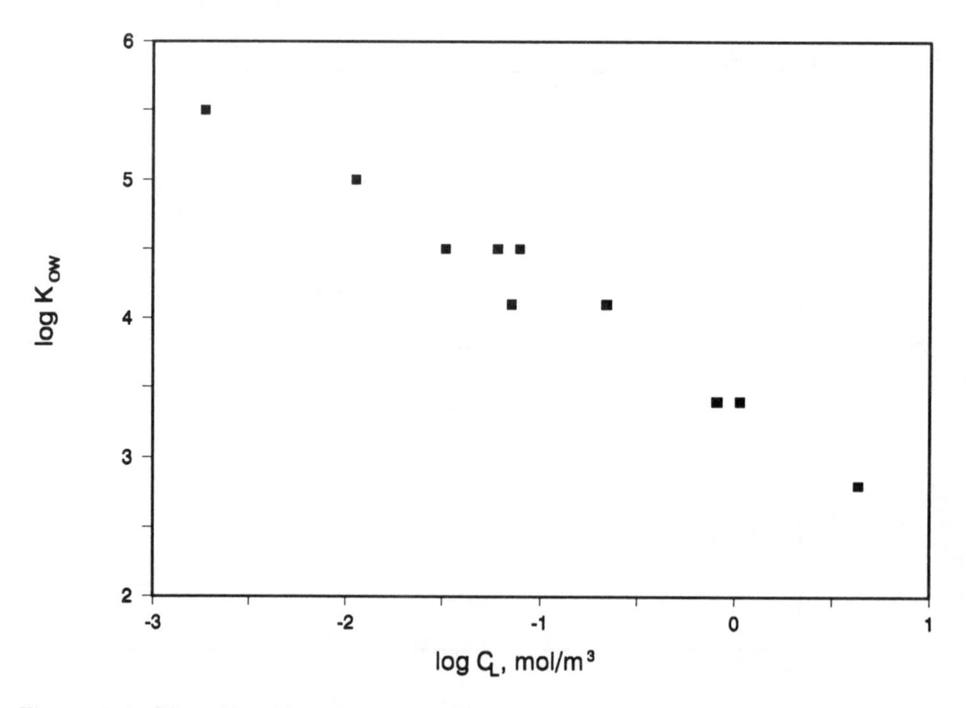

Figure 3.6 Plot of log K_{OW} versus log C_L

3.3 Illustrative Fugacity Calculations: Level I, II, III

Chemical name: Chlorobenzene

Level I

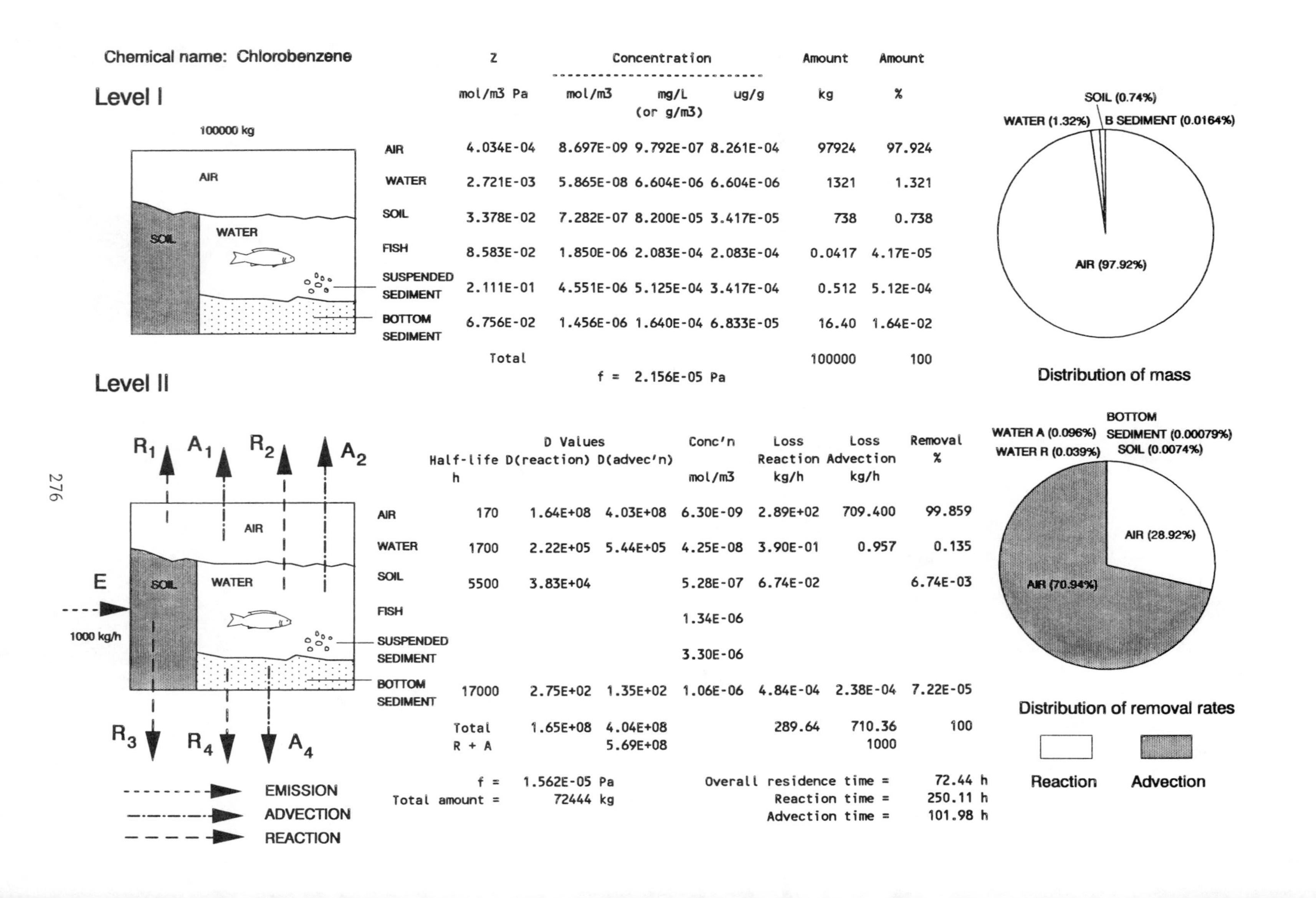

	Z	Concentration			Amount	Amount
	mol/m3 Pa	mol/m3	mg/L (or g/m3)	ug/g	kg	%
AIR	4.034E-04	8.697E-09	9.792E-07	8.261E-04	97924	97.924
WATER	2.721E-03	5.865E-08	6.604E-06	6.604E-06	1321	1.321
SOIL	3.378E-02	7.282E-07	8.200E-05	3.417E-05	738	0.738
FISH	8.583E-02	1.850E-06	2.083E-04	2.083E-04	0.0417	4.17E-05
SUSPENDED SEDIMENT	2.111E-01	4.551E-06	5.125E-04	3.417E-04	0.512	5.12E-04
BOTTOM SEDIMENT	6.756E-02	1.456E-06	1.640E-04	6.833E-05	16.40	1.64E-02
	Total				100000	100

f = 2.156E-05 Pa

Level II

	Half-life h	D Values D(reaction)	D(advec'n)	Conc'n mol/m3	Loss Reaction kg/h	Loss Advection kg/h	Removal %
AIR	170	1.64E+08	4.03E+08	6.30E-09	2.89E+02	709.400	99.859
WATER	1700	2.22E+05	5.44E+05	4.25E-08	3.90E-01	0.957	0.135
SOIL	5500	3.83E+04		5.28E-07	6.74E-02		6.74E-03
FISH				1.34E-06			
SUSPENDED SEDIMENT				3.30E-06			
BOTTOM SEDIMENT	17000	2.75E+02	1.35E+02	1.06E-06	4.84E-04	2.38E-04	7.22E-05
	Total	1.65E+08	4.04E+08		289.64	710.36	100
	R + A		5.69E+08			1000	

f = 1.562E-05 Pa
Total amount = 72444 kg

Overall residence time = 72.44 h
Reaction time = 250.11 h
Advection time = 101.98 h

Level III

Chemical name: Chlorobenzene

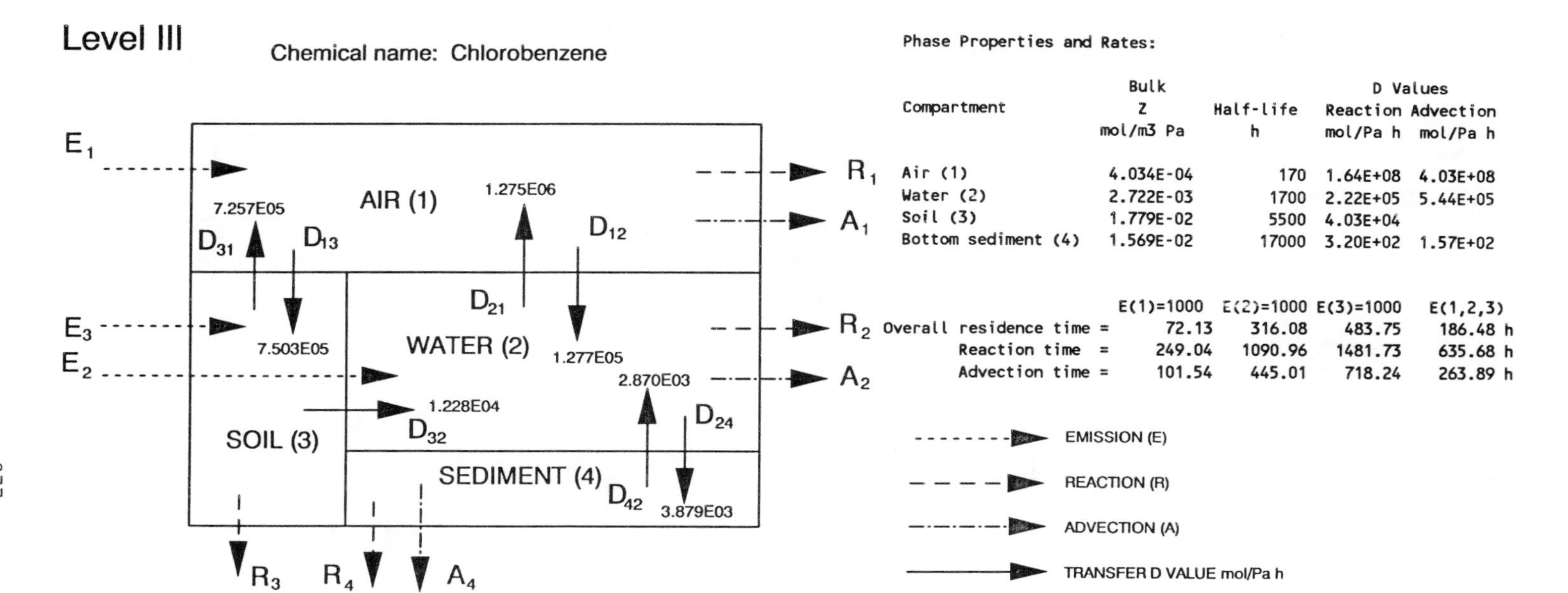

Phase Properties and Rates:

Compartment	Bulk Z mol/m3 Pa	Half-life h	D Values Reaction mol/Pa h	D Values Advection mol/Pa h
Air (1)	4.034E-04	170	1.64E+08	4.03E+08
Water (2)	2.722E-03	1700	2.22E+05	5.44E+05
Soil (3)	1.779E-02	5500	4.03E+04	
Bottom sediment (4)	1.569E-02	17000	3.20E+02	1.57E+02

	E(1)=1000	E(2)=1000	E(3)=1000	E(1,2,3)
Overall residence time =	72.13	316.08	483.75	186.48 h
Reaction time =	249.04	1090.96	1481.73	635.68 h
Advection time =	101.54	445.01	718.24	263.89 h

Phase Properties, Compositions, Transport and Transformation Rates:

Emission, kg/h			Fugacity, Pa				Concentration, g/m3				Amounts, kg				Total Amount, kg
E(1)	E(2)	E(3)	f(1)	f(2)	f(3)	f(4)	C(1)	C(2)	C(3)	C(4)	m(1)	m(2)	m(3)	m(4)	
1000	0	0	1.563E-05	9.870E-06	1.507E-05	1.118E-05	7.098E-07	3.025E-06	3.018E-05	1.976E-05	7.098E+04	6.049E+02	5.432E+02	9.879E+00	7.213E+04
0	1000	0	9.758E-06	4.357E-03	9.407E-06	4.938E-03	4.431E-07	1.335E-03	1.884E-05	8.723E-03	4.431E+04	2.671E+05	3.391E+02	4.361E+03	3.161E+05
0	0	1000	1.473E-05	7.793E-05	1.143E-02	8.832E-05	6.687E-07	2.388E-05	2.289E-02	1.560E-04	6.687E+04	4.777E+03	4.120E+05	7.801E+01	4.838E+05
600	300	100	1.378E-05	1.321E-03	1.155E-03	1.497E-03	6.257E-07	4.048E-04	2.313E-03	2.644E-03	6.257E+04	8.096E+04	4.163E+04	1.322E+03	1.865E+05

Emission, kg/h			Loss, Reaction, kg/h				Loss, Advection, kg/h			Intermedia Rate of Transport, kg/h T12	T21	T13	T31	T32	T24	T42
E(1)	E(2)	E(3)	R(1)	R(2)	R(3)	R(4)	A(1)	A(2)	A(4)	air-water	water-air	air-soil	soil-air	soil-water	water-sed	sed-water
1000	0	0	2.893E+02	2.466E-01	6.84E-02	4.027E-04	7.098E+02	6.049E-01	1.976E-04	2.248E+00	1.416E+00	1.320E+00	1.231E+00	2.082E-02	4.197E-03	3.596E-03
0	1000	0	1.806E+02	1.089E+02	4.27E-02	1.778E-01	4.431E+02	2.671E+02	8.723E-02	1.403E+00	6.252E+02	8.241E-01	7.684E-01	1.300E-02	1.853E+00	1.588E+00
0	0	1000	2.726E+02	1.947E+00	5.19E+01	3.180E-03	6.687E+02	4.777E+00	1.560E-03	2.118E+00	1.118E+01	1.244E+00	9.335E+02	1.579E+01	3.314E-02	2.840E-02
600	300	100	2.550E+02	3.300E+01	5.25E+00	5.390E-02	6.257E+02	3.300E+01	2.644E-02	1.981E+00	1.895E+02	1.164E+00	9.432E+01	1.596E+00	5.617E-01	4.813E-01

Chemical name: 1,2-Dichlorobenzene

Level I

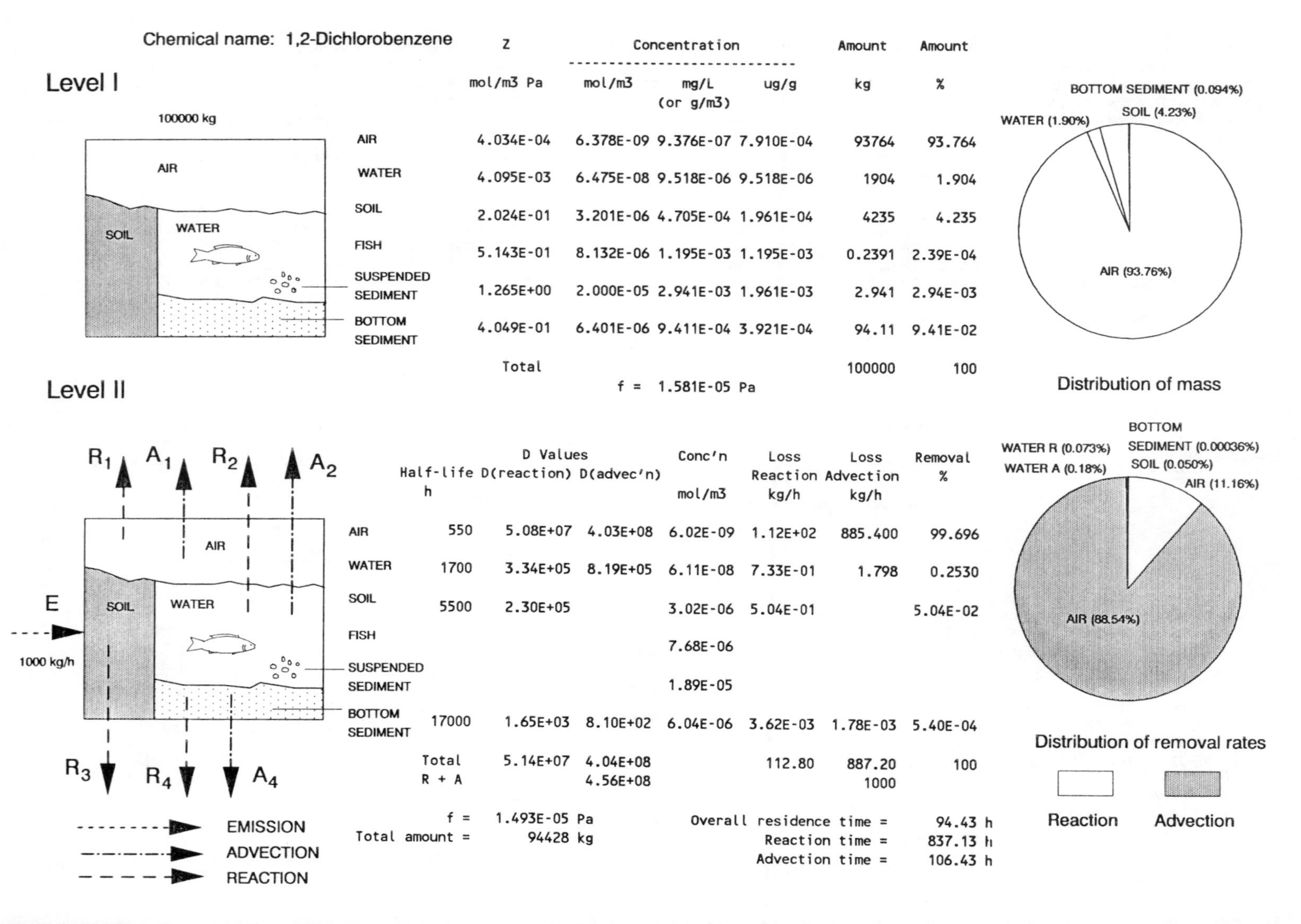

	Z	Concentration			Amount	Amount
	mol/m3 Pa	mol/m3	mg/L (or g/m3)	ug/g	kg	%
AIR	4.034E-04	6.378E-09	9.376E-07	7.910E-04	93764	93.764
WATER	4.095E-03	6.475E-08	9.518E-06	9.518E-06	1904	1.904
SOIL	2.024E-01	3.201E-06	4.705E-04	1.961E-04	4235	4.235
FISH	5.143E-01	8.132E-06	1.195E-03	1.195E-03	0.2391	2.39E-04
SUSPENDED SEDIMENT	1.265E+00	2.000E-05	2.941E-03	1.961E-03	2.941	2.94E-03
BOTTOM SEDIMENT	4.049E-01	6.401E-06	9.411E-04	3.921E-04	94.11	9.41E-02
	Total				100000	100

f = 1.581E-05 Pa

Level II

	Half-life h	D Values D(reaction)	D(advec'n)	Conc'n mol/m3	Loss Reaction kg/h	Loss Advection kg/h	Removal %
AIR	550	5.08E+07	4.03E+08	6.02E-09	1.12E+02	885.400	99.696
WATER	1700	3.34E+05	8.19E+05	6.11E-08	7.33E-01	1.798	0.2530
SOIL	5500	2.30E+05		3.02E-06	5.04E-01		5.04E-02
FISH				7.68E-06			
SUSPENDED SEDIMENT				1.89E-05			
BOTTOM SEDIMENT	17000	1.65E+03	8.10E+02	6.04E-06	3.62E-03	1.78E-03	5.40E-04
	Total R + A	5.14E+07	4.04E+08 4.56E+08		112.80	887.20 1000	100

f = 1.493E-05 Pa
Total amount = 94428 kg

Overall residence time = 94.43 h
Reaction time = 837.13 h
Advection time = 106.43 h

Level III

Chemical name: 1,2-Dichlorobenzene

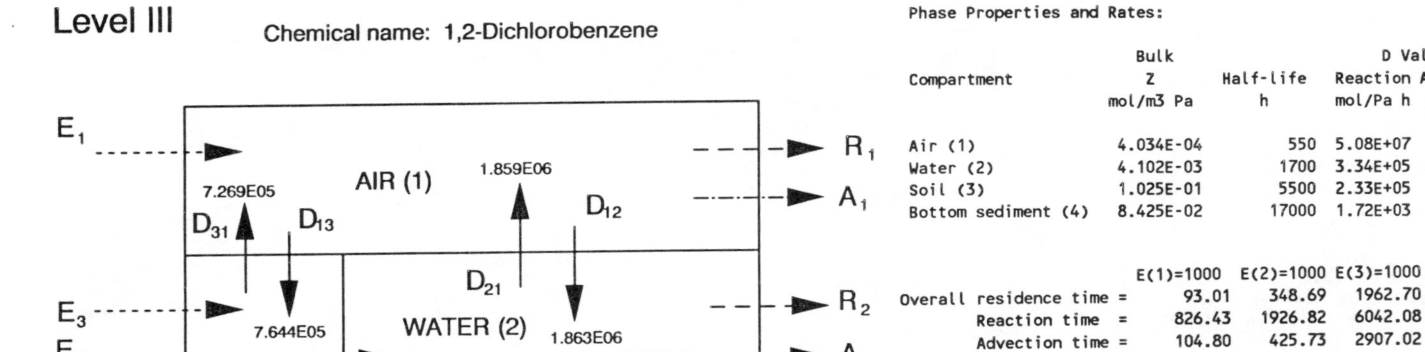

Phase Properties and Rates:

Compartment	Bulk Z mol/m3 Pa	Half-life h	D Values Reaction mol/Pa h	D Values Advection mol/Pa h
Air (1)	4.034E-04	550	5.08E+07	4.03E+08
Water (2)	4.102E-03	1700	3.34E+05	8.20E+05
Soil (3)	1.025E-01	5500	2.33E+05	
Bottom sediment (4)	8.425E-02	17000	1.72E+03	8.43E+02

	E(1)=1000	E(2)=1000	E(3)=1000	E(1,2,3)
Overall residence time =	93.01	348.69	1962.70	356.68 h
Reaction time =	826.43	1926.82	6042.08	2311.65 h
Advection time =	104.80	425.73	2907.02	421.76 h

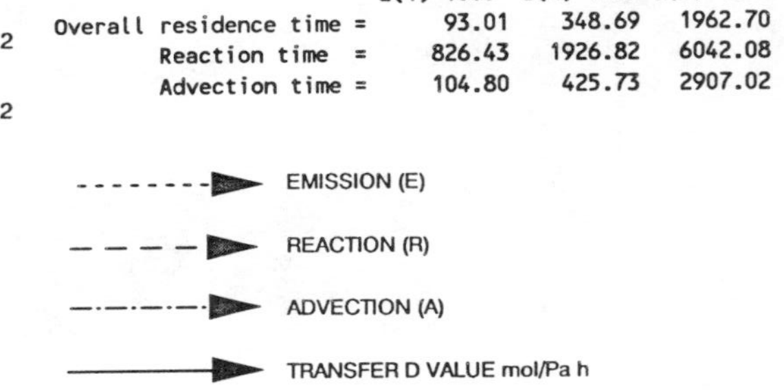

Phase Properties, Compositions, Transport and Transformation Rates:

Emission, kg/h			Fugacity, Pa				Concentration, g/m3				Amounts, kg				Total Amount, kg
E(1)	E(2)	E(3)	f(1)	f(2)	f(3)	f(4)	C(1)	C(2)	C(3)	C(4)	m(1)	m(2)	m(3)	m(4)	
1000	0	0	1.495E-05	9.300E-06	1.168E-05	1.298E-05	8.863E-07	5.608E-06	1.761E-04	1.608E-04	8.863E+04	1.122E+03	3.169E+03	8.041E+01	9.301E+04
0	1000	0	9.207E-06	2.260E-03	7.196E-06	3.155E-03	5.461E-07	1.363E-03	1.085E-04	3.908E-02	5.461E+04	2.726E+05	1.952E+03	1.954E+04	3.487E+05
0	0	1000	1.128E-05	4.992E-05	6.964E-03	6.969E-05	6.691E-07	3.010E-05	1.050E-01	8.632E-04	6.691E+04	6.021E+03	1.889E+06	4.316E+02	1.963E+06
600	300	100	1.286E-05	6.886E-04	7.055E-04	9.614E-04	7.625E-07	4.153E-04	1.063E-02	1.191E-02	7.625E+04	8.305E+04	1.914E+05	5.954E+03	3.567E+05

Emission, kg/h			Loss, Reaction, kg/h				Loss, Advection, kg/h			Intermedia Rate of Transport, kg/h T12	T21	T13	T31	T32	T24	T42
E(1)	E(2)	E(3)	R(1)	R(2)	R(3)	R(4)	A(1)	A(2)	A(4)	air-water	water-air	air-soil	soil-air	soil-water	water-sed	sed-water
1000	0	0	1.117E+02	4.572E-01	3.99E-01	3.278E-03	8.863E+02	1.122E+00	1.608E-03	4.093E+00	2.541E+00	1.680E+00	1.248E+00	3.196E-02	1.425E-02	9.362E-03
0	1000	0	6.880E+01	1.111E+02	2.46E-01	7.966E-01	5.461E+02	2.726E+02	3.908E-01	2.522E+00	6.176E+02	1.035E+00	7.690E-01	1.969E-02	3.463E+00	2.275E+00
0	0	1000	8.431E+01	2.454E+00	2.38E+02	1.759E-02	6.691E+02	6.021E+00	8.632E-03	3.090E+00	1.364E+01	1.268E+00	7.442E+02	1.905E+01	7.648E-02	5.025E-02
600	300	100	9.608E+01	3.386E+01	2.41E+01	2.427E-01	7.625E+02	3.386E+01	1.191E-01	3.522E+00	1.882E+02	1.445E+00	7.540E+01	1.930E+00	1.055E+00	6.932E-01

Chemical name: 1,3-Dichlorobenzene

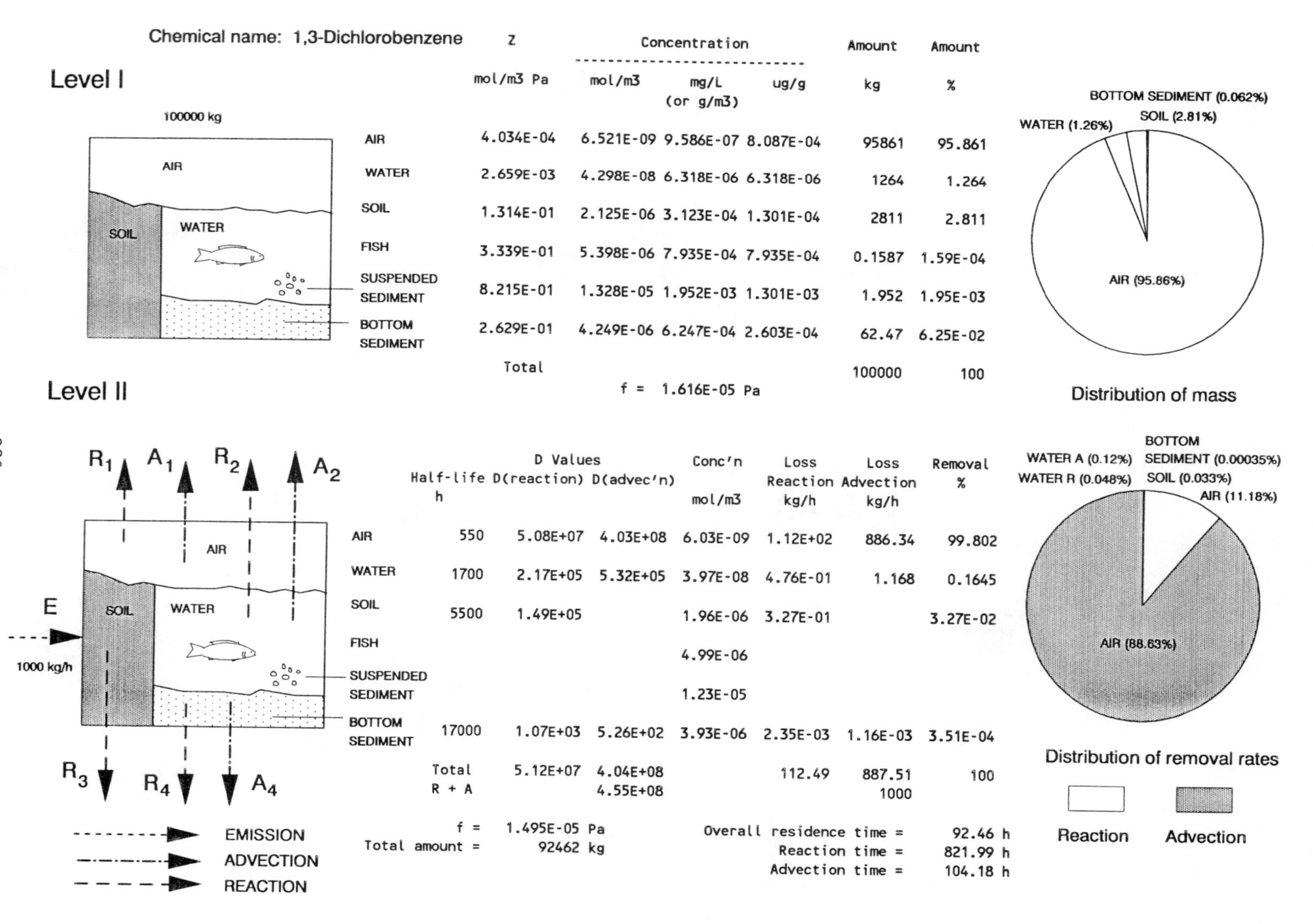

Level I

	Z	Concentration			Amount	Amount
	mol/m3 Pa	mol/m3	mg/L (or g/m3)	ug/g	kg	%
AIR	4.034E-04	6.521E-09	9.586E-07	8.087E-04	95861	95.861
WATER	2.659E-03	4.298E-08	6.318E-06	6.318E-06	1264	1.264
SOIL	1.314E-01	2.125E-06	3.123E-04	1.301E-04	2811	2.811
FISH	3.339E-01	5.398E-06	7.935E-04	7.935E-04	0.1587	1.59E-04
SUSPENDED SEDIMENT	8.215E-01	1.328E-05	1.952E-03	1.301E-03	1.952	1.95E-03
BOTTOM SEDIMENT	2.629E-01	4.249E-06	6.247E-04	2.603E-04	62.47	6.25E-02
	Total				100000	100

f = 1.616E-05 Pa

Distribution of mass

Level II

	Half-life h	D Values D(reaction)	D(advec'n)	Conc'n mol/m3	Loss Reaction kg/h	Loss Advection kg/h	Removal %
AIR	550	5.08E+07	4.03E+08	6.03E-09	1.12E+02	886.34	99.802
WATER	1700	2.17E+05	5.32E+05	3.97E-08	4.76E-01	1.168	0.1645
SOIL	5500	1.49E+05		1.96E-06	3.27E-01		3.27E-02
FISH				4.99E-06			
SUSPENDED SEDIMENT				1.23E-05			
BOTTOM SEDIMENT	17000	1.07E+03	5.26E+02	3.93E-06	2.35E-03	1.16E-03	3.51E-04
	Total	5.12E+07	4.04E+08		112.49	887.51	100
	R + A		4.55E+08			1000	

f = 1.495E-05 Pa
Total amount = 92462 kg

Overall residence time = 92.46 h
Reaction time = 821.99 h
Advection time = 104.18 h

Distribution of removal rates

Level III

Chemical name: 1,3-Dichlorobenzene

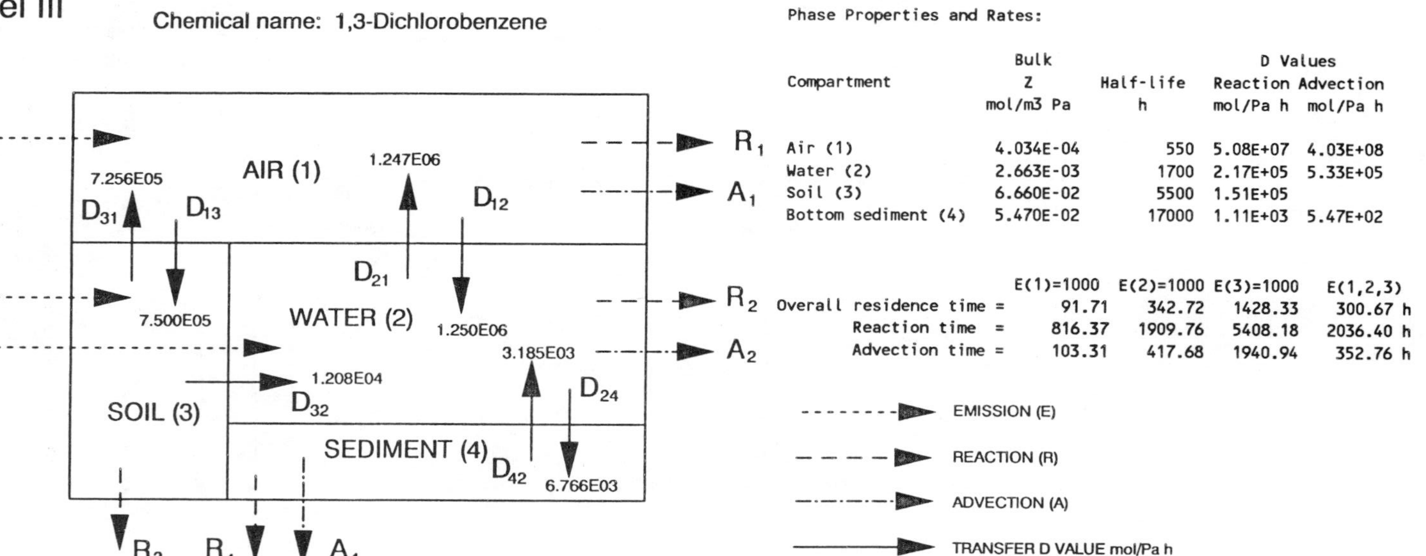

Phase Properties and Rates:

Compartment	Bulk Z mol/m3 Pa	Half-life h	D Values Reaction mol/Pa h	D Values Advection mol/Pa h
Air (1)	4.034E-04	550	5.08E+07	4.03E+08
Water (2)	2.663E-03	1700	2.17E+05	5.33E+05
Soil (3)	6.660E-02	5500	1.51E+05	
Bottom sediment (4)	5.470E-02	17000	1.11E+03	5.47E+02

	E(1)=1000	E(2)=1000	E(3)=1000	E(1,2,3)
Overall residence time =	91.71	342.72	1428.33	300.67 h
Reaction time =	816.37	1909.76	5408.18	2036.40 h
Advection time =	103.31	417.68	1940.94	352.76 h

Phase Properties, Compositions, Transport and Transformation Rates:

Emission, kg/h			Fugacity, Pa				Concentration, g/m3				Amounts, kg				Total
E(1)	E(2)	E(3)	f(1)	f(2)	f(3)	f(4)	C(1)	C(2)	C(3)	C(4)	m(1)	m(2)	m(3)	m(4)	Amount, kg
1000	0	0	1.495E-05	9.426E-06	1.262E-05	1.316E-05	8.869E-07	3.690E-06	1.236E-04	1.058E-04	8.869E+04	7.381E+02	2.224E+03	5.291E+01	9.171E+04
0	1000	0	9.329E-06	3.408E-03	7.873E-06	4.758E-03	5.533E-07	1.334E-03	7.708E-05	3.826E-02	5.533E+04	2.669E+05	1.387E+03	1.913E+04	3.427E+05
0	0	1000	1.234E-05	5.403E-05	7.664E-03	7.543E-05	7.317E-07	2.115E-05	7.503E-02	6.066E-04	7.317E+04	4.231E+03	1.351E+06	3.033E+02	1.428E+06
600	300	100	1.301E-05	1.033E-03	7.763E-04	1.443E-03	7.713E-07	4.046E-04	7.601E-03	1.160E-02	7.713E+04	8.093E+04	1.368E+05	5.802E+03	3.007E+05

Emission, kg/h			Loss, Reaction, kg/h				Loss, Advection, kg/h			Intermedia Rate of Transport, kg/h T12	T21	T13	T31	T32	T24	T42
E(1)	E(2)	E(3)	R(1)	R(2)	R(3)	R(4)	A(1)	A(2)	A(4)	air-water	water-air	air-soil	soil-air	soil-water	water-sed	sed-water
1000	0	0	1.118E+02	3.009E-01	2.80E-01	2.157E-03	8.869E+02	7.381E-01	1.058E-03	2.748E+00	1.728E+00	1.649E+00	1.346E+00	2.242E-02	9.376E-03	6.161E-03
0	1000	0	6.971E+01	1.088E+02	1.75E-01	7.799E-01	5.533E+02	2.669E+02	3.826E-01	1.714E+00	6.249E+02	1.029E+00	8.398E-01	1.398E-02	3.390E+00	2.228E+00
0	0	1000	9.219E+01	1.725E+00	1.70E+02	1.236E-02	7.317E+02	4.231E+00	6.066E-03	2.267E+00	9.907E+00	1.360E+00	8.176E+02	1.361E+01	5.375E-02	3.532E-02
600	300	100	9.718E+01	3.299E+01	1.72E+01	2.365E-01	7.713E+02	3.299E+01	1.160E-01	2.390E+00	1.895E+02	1.434E+00	8.282E+01	1.379E+00	1.028E+00	6.755E-01

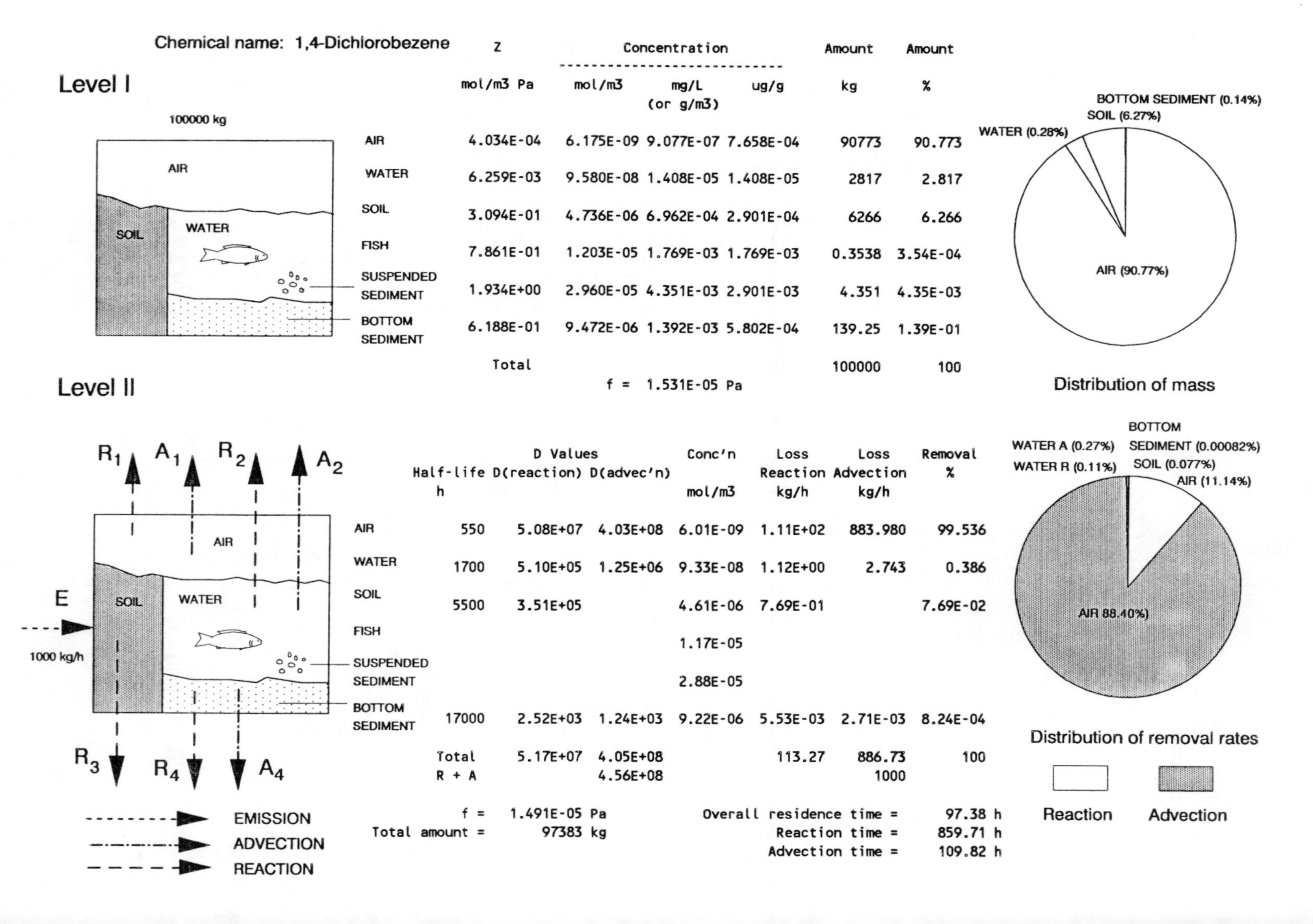

Chemical name: 1,4-Dichlorobezene

Level I

	Z	Concentration			Amount	Amount
	mol/m3 Pa	mol/m3	mg/L (or g/m3)	ug/g	kg	%
AIR	4.034E-04	6.175E-09	9.077E-07	7.658E-04	90773	90.773
WATER	6.259E-03	9.580E-08	1.408E-05	1.408E-05	2817	2.817
SOIL	3.094E-01	4.736E-06	6.962E-04	2.901E-04	6266	6.266
FISH	7.861E-01	1.203E-05	1.769E-03	1.769E-03	0.3538	3.54E-04
SUSPENDED SEDIMENT	1.934E+00	2.960E-05	4.351E-03	2.901E-03	4.351	4.35E-03
BOTTOM SEDIMENT	6.188E-01	9.472E-06	1.392E-03	5.802E-04	139.25	1.39E-01
	Total				100000	100

f = 1.531E-05 Pa

Level II

	Half-life h	D Values D(reaction)	D(advec'n)	Conc'n mol/m3	Loss Reaction kg/h	Loss Advection kg/h	Removal %
AIR	550	5.08E+07	4.03E+08	6.01E-09	1.11E+02	883.980	99.536
WATER	1700	5.10E+05	1.25E+06	9.33E-08	1.12E+00	2.743	0.386
SOIL	5500	3.51E+05		4.61E-06	7.69E-01		7.69E-02
FISH				1.17E-05			
SUSPENDED SEDIMENT				2.88E-05			
BOTTOM SEDIMENT	17000	2.52E+03	1.24E+03	9.22E-06	5.53E-03	2.71E-03	8.24E-04
	Total	5.17E+07	4.05E+08		113.27	886.73	100
	R + A		4.56E+08			1000	

f = 1.491E-05 Pa
Total amount = 97383 kg

Overall residence time = 97.38 h
Reaction time = 859.71 h
Advection time = 109.82 h

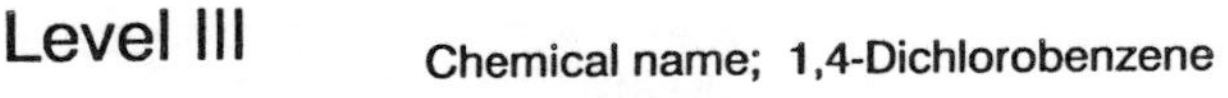

Chemical name; 1,4-Dichlorobenzene

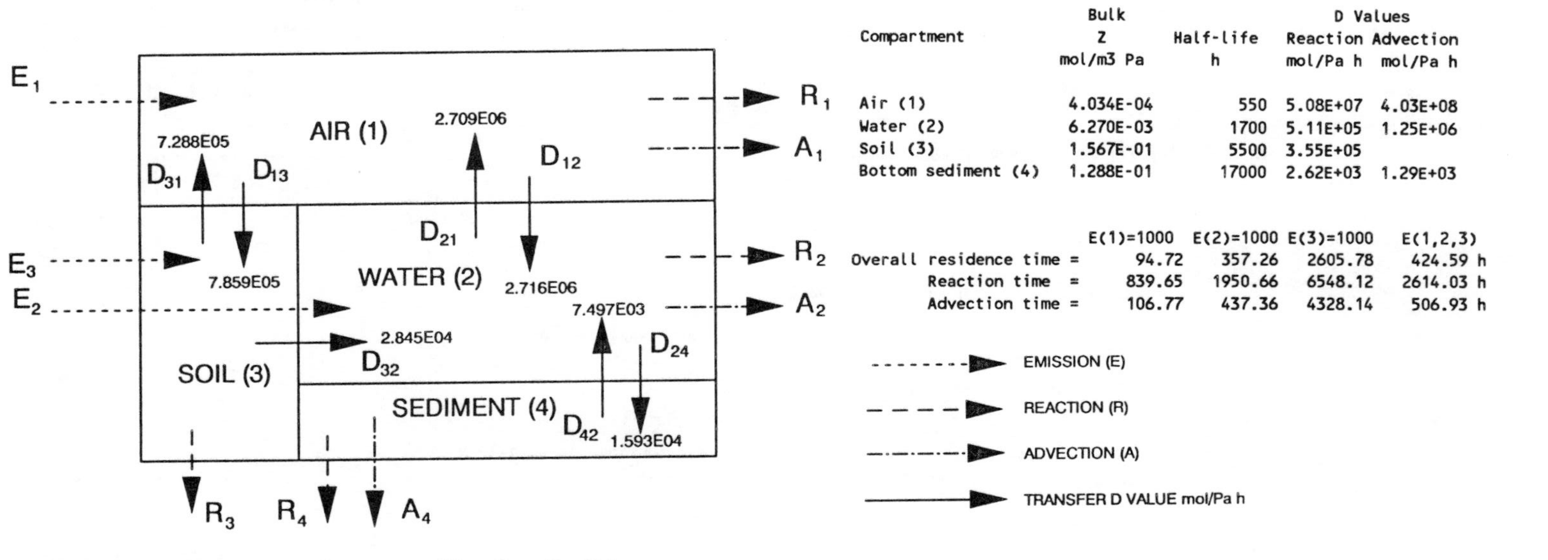

Phase Properties and Rates:

Compartment	Bulk Z mol/m3 Pa	Half-life h	D Values Reaction mol/Pa h	D Values Advection mol/Pa h
Air (1)	4.034E-04	550	5.08E+07	4.03E+08
Water (2)	6.270E-03	1700	5.11E+05	1.25E+06
Soil (3)	1.567E-01	5500	3.55E+05	
Bottom sediment (4)	1.288E-01	17000	2.62E+03	1.29E+03

	E(1)=1000	E(2)=1000	E(3)=1000	E(1,2,3)
Overall residence time =	94.72	357.26	2605.78	424.59 h
Reaction time =	839.65	1950.66	6548.12	2614.03 h
Advection time =	106.77	437.36	4328.14	506.93 h

Phase Properties, Compositions, Transport and Transformation Rates:

Emission, kg/h			Fugacity, Pa				Concentration, g/m3				Amounts, kg				Total Amount, kg
E(1)	E(2)	E(3)	f(1)	f(2)	f(3)	f(4)	C(1)	C(2)	C(3)	C(4)	m(1)	m(2)	m(3)	m(4)	
1000	0	0	1.493E-05	9.118E-06	1.055E-05	1.273E-05	8.855E-07	8.404E-06	2.429E-04	2.410E-04	8.855E+04	1.681E+03	4.373E+03	1.205E+02	9.472E+04
0	1000	0	9.030E-06	1.524E-03	6.379E-06	2.128E-03	5.355E-07	1.405E-03	1.469E-04	4.028E-02	5.355E+04	2.809E+05	2.644E+03	2.014E+04	3.573E+05
0	0	1000	1.001E-05	4.493E-05	6.121E-03	6.273E-05	5.938E-07	4.142E-05	1.410E-01	1.188E-03	5.938E+04	8.283E+03	2.538E+06	5.938E+02	2.606E+06
600	300	100	1.267E-05	4.671E-04	6.203E-04	6.522E-04	7.513E-07	4.306E-04	1.429E-02	1.235E-02	7.513E+04	8.611E+04	2.572E+05	6.173E+03	4.246E+05

Emission, kg/h			Loss, Reaction, kg/h				Loss, Advection, kg/h			Intermedia Rate of Transport, kg/h T12	T21	T13	T31	T32	T24	T42
E(1)	E(2)	E(3)	R(1)	R(2)	R(3)	R(4)	A(1)	A(2)	A(4)	air-water	water-air	air-soil	soil-air	soil-water	water-sed	sed-water
1000	0	0	1.116E+02	6.852E-01	5.51E-01	4.912E-03	8.855E+02	1.681E+00	2.410E-03	5.961E+00	3.632E+00	1.725E+00	1.130E+00	4.411E-02	2.135E-02	1.403E-02
0	1000	0	6.748E+01	1.145E+02	3.33E-01	8.209E-01	5.355E+02	2.809E+02	4.028E-01	3.605E+00	6.070E+02	1.043E+00	6.835E-01	2.667E-02	3.569E+00	2.345E+00
0	0	1000	7.481E+01	3.377E+00	3.20E+02	2.421E-02	5.938E+02	8.283E+00	1.188E-02	3.997E+00	1.790E+01	1.157E+00	6.558E+02	2.560E+01	1.052E-01	6.914E-02
600	300	100	9.467E+01	3.510E+01	3.24E+01	2.517E-01	7.513E+02	3.510E+01	1.235E-01	5.058E+00	1.861E+02	1.464E+00	6.647E+01	2.594E+00	1.094E+00	7.188E-01

Chemical name: 1,2,3-Trichlorobenzene z

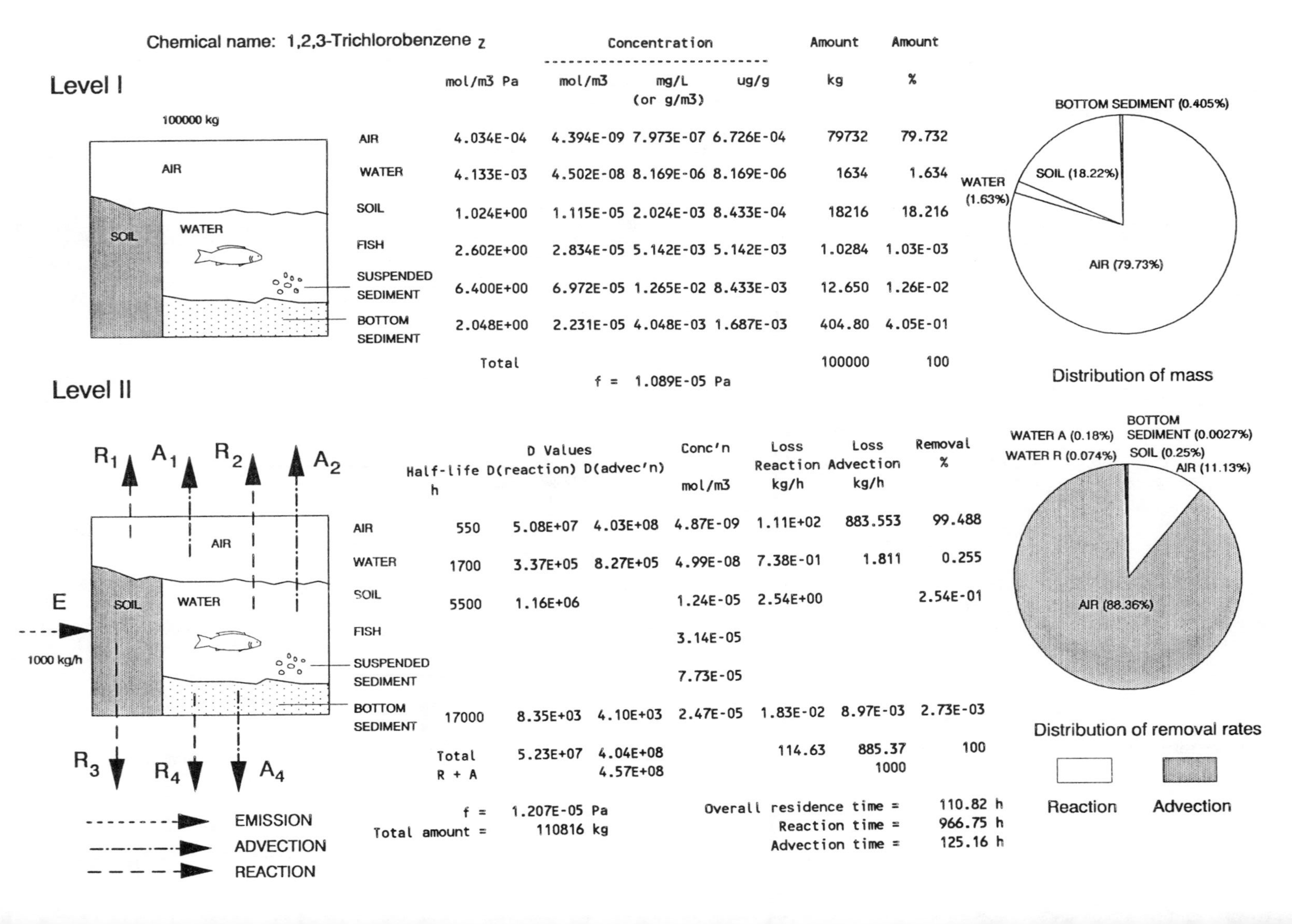

Level I

	mol/m3 Pa	Concentration mol/m3	mg/L (or g/m3)	ug/g	Amount kg	Amount %
AIR	4.034E-04	4.394E-09	7.973E-07	6.726E-04	79732	79.732
WATER	4.133E-03	4.502E-08	8.169E-06	8.169E-06	1634	1.634
SOIL	1.024E+00	1.115E-05	2.024E-03	8.433E-04	18216	18.216
FISH	2.602E+00	2.834E-05	5.142E-03	5.142E-03	1.0284	1.03E-03
SUSPENDED SEDIMENT	6.400E+00	6.972E-05	1.265E-02	8.433E-03	12.650	1.26E-02
BOTTOM SEDIMENT	2.048E+00	2.231E-05	4.048E-03	1.687E-03	404.80	4.05E-01
	Total				100000	100

f = 1.089E-05 Pa

Level II

	Half-life h	D Values D(reaction)	D(advec'n)	Conc'n mol/m3	Loss Reaction kg/h	Loss Advection kg/h	Removal %
AIR	550	5.08E+07	4.03E+08	4.87E-09	1.11E+02	883.553	99.488
WATER	1700	3.37E+05	8.27E+05	4.99E-08	7.38E-01	1.811	0.255
SOIL	5500	1.16E+06		1.24E-05	2.54E+00		2.54E-01
FISH				3.14E-05			
SUSPENDED SEDIMENT				7.73E-05			
BOTTOM SEDIMENT	17000	8.35E+03	4.10E+03	2.47E-05	1.83E-02	8.97E-03	2.73E-03
	Total	5.23E+07	4.04E+08		114.63	885.37	100
	R + A		4.57E+08			1000	

f = 1.207E-05 Pa
Total amount = 110816 kg

Overall residence time = 110.82 h
Reaction time = 966.75 h
Advection time = 125.16 h

Level III

Chemical name: 1,2,3-Trichlorobenzene

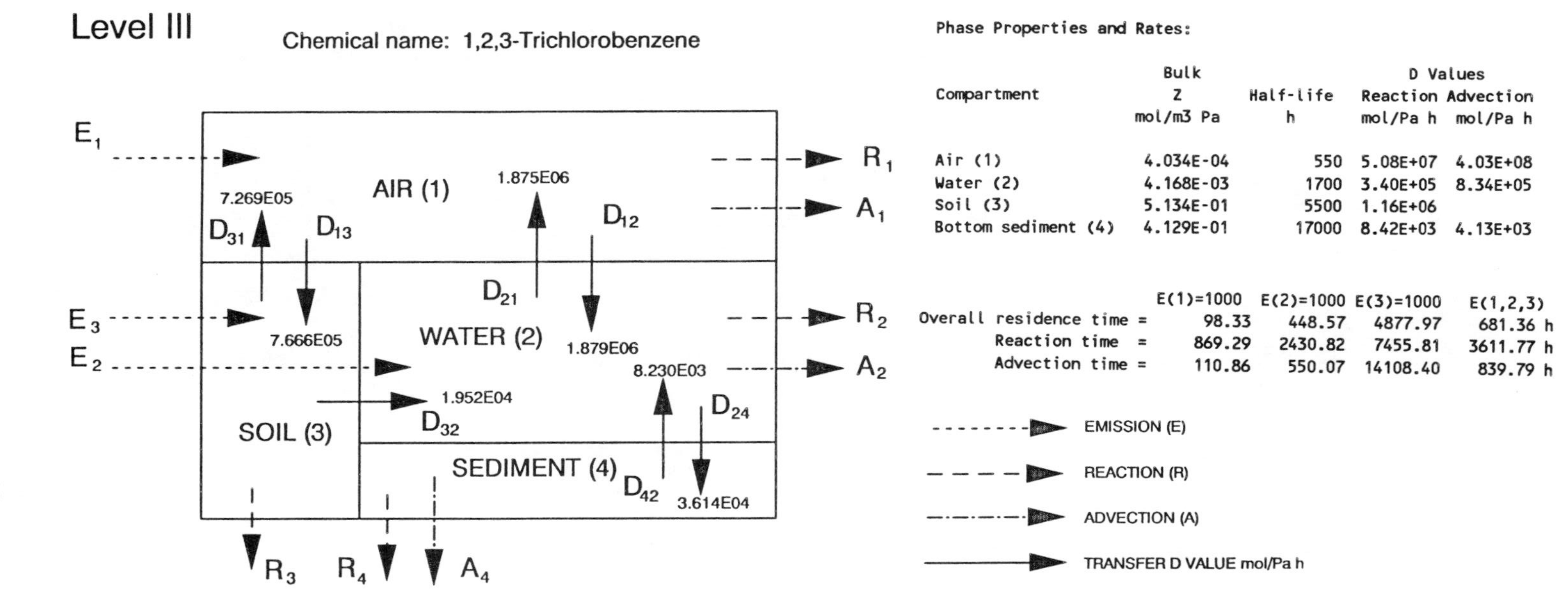

Phase Properties and Rates:

Compartment	Bulk Z mol/m3 Pa	Half-life h	D Values Reaction mol/Pa h	D Values Advection mol/Pa h
Air (1)	4.034E-04	550	5.08E+07	4.03E+08
Water (2)	4.168E-03	1700	3.40E+05	8.34E+05
Soil (3)	5.134E-01	5500	1.16E+06	
Bottom sediment (4)	4.129E-01	17000	8.42E+03	4.13E+03

	E(1)=1000	E(2)=1000	E(3)=1000	E(1,2,3)
Overall residence time =	98.33	448.57	4877.97	681.36 h
Reaction time =	869.29	2430.82	7455.81	3611.77 h
Advection time =	110.86	550.07	14108.40	839.79 h

Phase Properties, Compositions, Transport and Transformation Rates:

Emission, kg/h			Fugacity, Pa				Concentration, g/m3				Amounts, kg				Total Amount, kg
E(1)	E(2)	E(3)	f(1)	f(2)	f(3)	f(4)	C(1)	C(2)	C(3)	C(4)	m(1)	m(2)	m(3)	m(4)	
1000	0	0	1.210E-05	7.437E-06	4.855E-06	1.294E-05	8.858E-07	5.625E-06	4.522E-04	9.693E-04	8.858E+04	1.125E+03	8.140E+03	4.846E+02	9.833E+04
0	1000	0	7.389E-06	1.800E-03	2.965E-06	3.130E-03	5.409E-07	1.361E-03	2.761E-04	2.346E-01	5.409E+04	2.722E+05	4.971E+03	1.173E+05	4.486E+05
0	0	1000	4.679E-06	2.122E-05	2.886E-03	3.690E-05	3.425E-07	1.605E-05	2.688E-01	2.765E-03	3.425E+04	3.209E+03	4.839E+06	1.383E+03	4.878E+06
600	300	100	9.945E-06	5.465E-04	2.924E-04	9.506E-04	7.280E-07	4.133E-04	2.724E-02	7.122E-02	7.280E+04	8.266E+04	4.903E+05	3.561E+04	6.814E+05

Emission, kg/h			Loss, Reaction, kg/h				Loss, Advection, kg/h			Intermedia Rate of Transport, kg/h T12	T21	T13	T31	T32	T24	T42
E(1)	E(2)	E(3)	R(1)	R(2)	R(3)	R(4)	A(1)	A(2)	A(4)	air-water	water-air	air-soil	soil-air	soil-water	water-sed	sed-water
1000	0	0	1.116E+02	4.586E-01	1.03E+00	1.976E-02	8.858E+02	1.125E+00	9.693E-03	4.126E+00	2.530E+00	1.683E+00	6.404E-01	1.720E-02	4.877E-02	1.932E-02
0	1000	0	6.815E+01	1.110E+02	6.26E-01	4.781E+00	5.409E+02	2.722E+02	2.346E+00	2.519E+00	6.122E+02	1.028E+00	3.910E-01	1.050E-02	1.180E+01	4.675E+00
0	0	1000	4.316E+01	1.308E+00	6.10E+02	5.636E-02	3.425E+02	3.209E+00	2.765E-02	1.595E+00	7.217E+00	6.509E-01	3.807E+02	1.022E+01	1.391E-01	5.511E-02
600	300	100	9.172E+01	3.370E+01	6.18E+01	1.452E+00	7.280E+02	3.370E+01	7.122E-01	3.391E+00	1.859E+02	1.383E+00	3.857E+01	1.036E+00	3.583E+00	1.419E+00

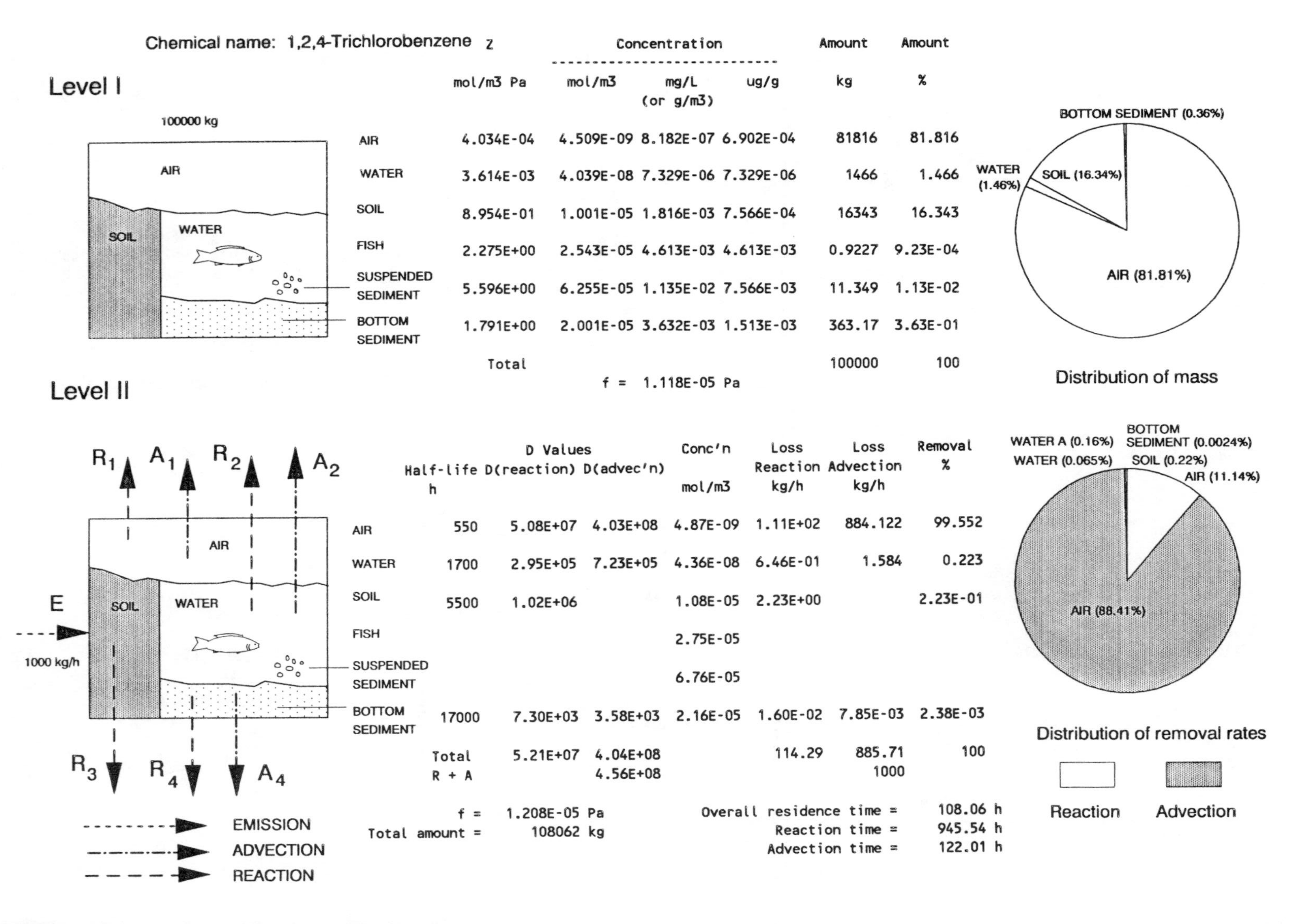

Chemical name: 1,2,4-Trichlorobenzene

Level I

	Z	Concentration			Amount	Amount
	mol/m3 Pa	mol/m3	mg/L (or g/m3)	ug/g	kg	%
AIR	4.034E-04	4.509E-09	8.182E-07	6.902E-04	81816	81.816
WATER	3.614E-03	4.039E-08	7.329E-06	7.329E-06	1466	1.466
SOIL	8.954E-01	1.001E-05	1.816E-03	7.566E-04	16343	16.343
FISH	2.275E+00	2.543E-05	4.613E-03	4.613E-03	0.9227	9.23E-04
SUSPENDED SEDIMENT	5.596E+00	6.255E-05	1.135E-02	7.566E-03	11.349	1.13E-02
BOTTOM SEDIMENT	1.791E+00	2.001E-05	3.632E-03	1.513E-03	363.17	3.63E-01
	Total				100000	100

f = 1.118E-05 Pa

Level II

	Half-life h	D Values D(reaction)	D(advec'n)	Conc'n mol/m3	Loss Reaction kg/h	Loss Advection kg/h	Removal %
AIR	550	5.08E+07	4.03E+08	4.87E-09	1.11E+02	884.122	99.552
WATER	1700	2.95E+05	7.23E+05	4.36E-08	6.46E-01	1.584	0.223
SOIL	5500	1.02E+06		1.08E-05	2.23E+00		2.23E-01
FISH				2.75E-05			
SUSPENDED SEDIMENT				6.76E-05			
BOTTOM SEDIMENT	17000	7.30E+03	3.58E+03	2.16E-05	1.60E-02	7.85E-03	2.38E-03
	Total	5.21E+07	4.04E+08		114.29	885.71	100
	R + A		4.56E+08			1000	

f = 1.208E-05 Pa
Total amount = 108062 kg

Overall residence time = 108.06 h
Reaction time = 945.54 h
Advection time = 122.01 h

Level III

Chemical name: 1,2,4-Trichlorobenzene

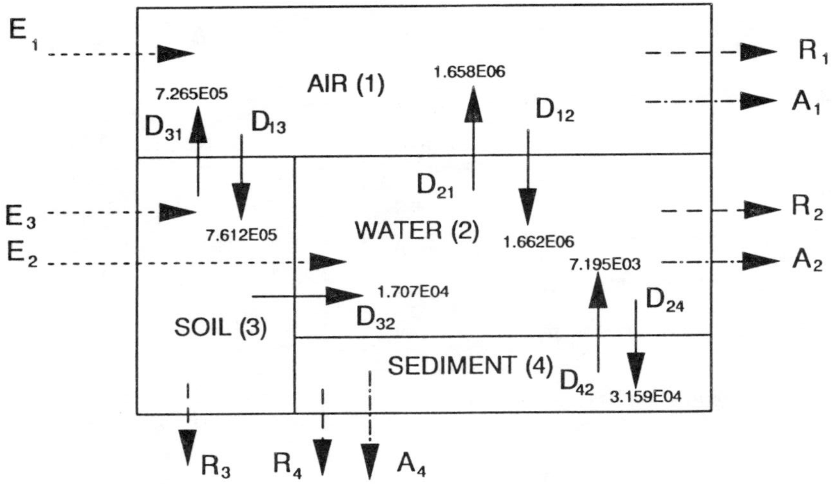

E1, E2, E3; AIR (1); WATER (2); SOIL (3); SEDIMENT (4); R1, A1, R2, A2, R3, R4, A4

D12 1.658E06; D21 1.662E06; D13 7.612E05; D31 7.265E05; D32 1.707E04; D24 3.159E04; D42 7.195E03

EMISSION (E)
REACTION (R)
ADVECTION (A)
TRANSFER D VALUE mol/Pa h

Phase Properties and Rates:

Compartment	Bulk Z mol/m3 Pa	Half-life h	D Values Reaction mol/Pa h	D Values Advection mol/Pa h
Air (1)	4.034E-04	550	5.08E+07	4.03E+08
Water (2)	3.644E-03	1700	2.97E+05	7.29E+05
Soil (3)	4.488E-01	5500	1.02E+06	
Bottom sediment (4)	3.610E-01	17000	7.36E+03	3.61E+03

	E(1)=1000	E(2)=1000	E(3)=1000	E(1,2,3)
Overall residence time =	97.68	445.66	4631.07	655.41 h
Reaction time =	864.27	2422.51	7394.46	3530.73 h
Advection time =	110.13	546.13	12392.08	804.81 h

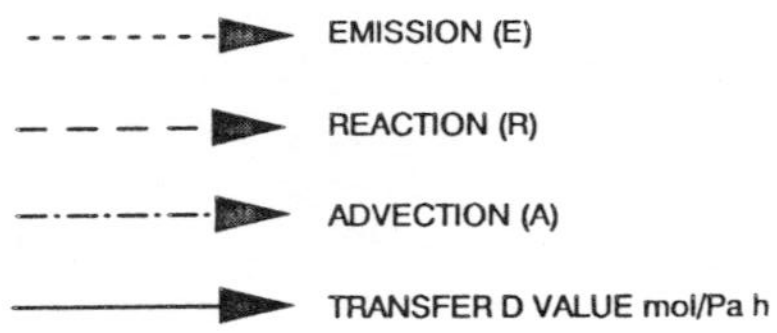

Phase Properties, Compositions, Transport and Transformation Rates:

Emission, kg/h			Fugacity, Pa				Concentration, g/m3				Amounts, kg				Total Amount, kg
E(1)	E(2)	E(3)	f(1)	f(2)	f(3)	f(4)	C(1)	C(2)	C(3)	C(4)	m(1)	m(2)	m(3)	m(4)	
1000	0	0	1.210E-05	7.475E-06	5.230E-06	1.300E-05	8.860E-07	4.943E-06	4.259E-04	8.517E-04	8.860E+04	9.886E+02	7.667E+03	4.259E+02	9.768E+04
0	1000	0	7.425E-06	2.043E-03	3.208E-06	3.554E-03	5.435E-07	1.351E-03	2.613E-04	2.328E-01	5.435E+04	2.702E+05	4.703E+03	1.164E+05	4.457E+05
0	0	1000	5.064E-06	2.288E-05	3.131E-03	3.980E-05	3.707E-07	1.513E-05	2.550E-01	2.607E-03	3.707E+04	3.026E+03	4.590E+06	1.304E+03	4.631E+06
600	300	100	9.996E-06	6.197E-04	3.172E-04	1.078E-03	7.317E-07	4.098E-04	2.583E-02	7.061E-02	7.317E+04	8.196E+04	4.650E+05	3.531E+04	6.554E+05

Emission, kg/h			Loss, Reaction, kg/h				Loss, Advection, kg/h			Intermediate Rate of Transport, kg/h T12	T21	T13	T31	T32	T24	T42
E(1)	E(2)	E(3)	R(1)	R(2)	R(3)	R(4)	A(1)	A(2)	A(4)	air-water	water-air	air-soil	soil-air	soil-water	water-sed	sed-water
1000	0	0	1.116E+02	4.030E-01	9.66E-01	1.736E-02	8.860E+02	9.886E-01	8.517E-03	3.651E+00	2.249E+00	1.672E+00	6.894E-01	1.620E-02	4.285E-02	1.697E-02
0	1000	0	6.848E+01	1.101E+02	5.93E-01	4.745E+00	5.435E+02	2.702E+02	2.328E+00	2.239E+00	6.148E+02	1.025E+00	4.229E-01	9.936E-03	1.171E+01	4.640E+00
0	0	1000	4.670E+01	1.233E+00	5.78E+02	5.314E-02	3.707E+02	3.026E+00	2.607E-02	1.527E+00	6.885E+00	6.993E-01	4.127E+02	9.696E+00	1.312E-01	5.196E-02
600	300	100	9.219E+01	3.341E+01	5.86E+01	1.439E+00	7.317E+02	3.341E+01	7.061E-01	3.015E+00	1.865E+02	1.381E+00	4.181E+01	9.823E-01	3.553E+00	1.407E+00

Chemical name: 1,3,5-Trichlorobenzene

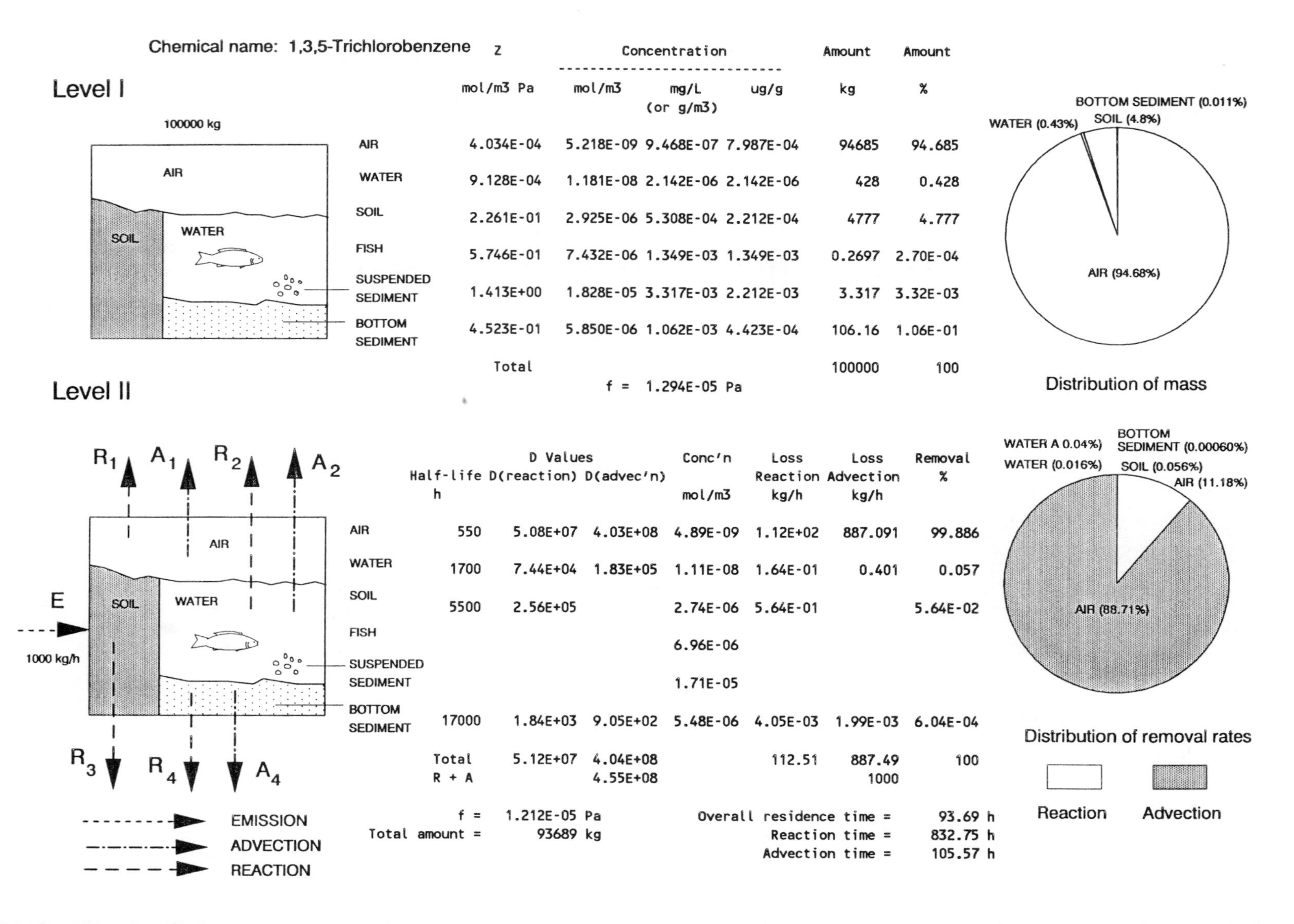

Level I

	Z	Concentration			Amount	Amount
	mol/m3 Pa	mol/m3	mg/L (or g/m3)	ug/g	kg	%
AIR	4.034E-04	5.218E-09	9.468E-07	7.987E-04	94685	94.685
WATER	9.128E-04	1.181E-08	2.142E-06	2.142E-06	428	0.428
SOIL	2.261E-01	2.925E-06	5.308E-04	2.212E-04	4777	4.777
FISH	5.746E-01	7.432E-06	1.349E-03	1.349E-03	0.2697	2.70E-04
SUSPENDED SEDIMENT	1.413E+00	1.828E-05	3.317E-03	2.212E-03	3.317	3.32E-03
BOTTOM SEDIMENT	4.523E-01	5.850E-06	1.062E-03	4.423E-04	106.16	1.06E-01
	Total				100000	100

f = 1.294E-05 Pa

Level II

	Half-life h	D Values D(reaction)	D(advec'n)	Conc'n mol/m3	Loss Reaction kg/h	Loss Advection kg/h	Removal %
AIR	550	5.08E+07	4.03E+08	4.89E-09	1.12E+02	887.091	99.886
WATER	1700	7.44E+04	1.83E+05	1.11E-08	1.64E-01	0.401	0.057
SOIL	5500	2.56E+05		2.74E-06	5.64E-01		5.64E-02
FISH				6.96E-06			
SUSPENDED SEDIMENT				1.71E-05			
BOTTOM SEDIMENT	17000	1.84E+03	9.05E+02	5.48E-06	4.05E-03	1.99E-03	6.04E-04
	Total	5.12E+07	4.04E+08		112.51	887.49	100
	R + A		4.55E+08			1000	

f = 1.212E-05 Pa
Total amount = 93689 kg

Overall residence time = 93.69 h
Reaction time = 832.75 h
Advection time = 105.57 h

Level III

Chemical name: 1,3,5-Trichlorobenzene

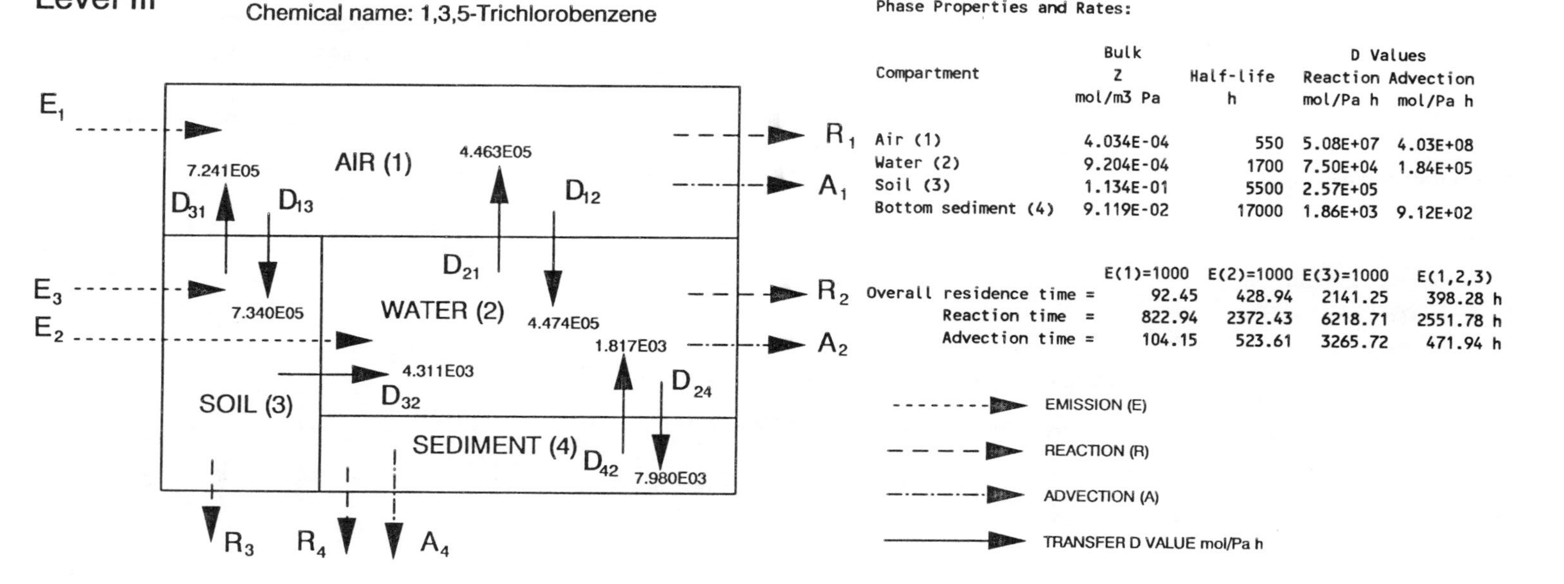

Phase Properties and Rates:

Compartment	Bulk Z mol/m3 Pa	Half-life h	D Values Reaction mol/Pa h	D Values Advection mol/Pa h
Air (1)	4.034E-04	550	5.08E+07	4.03E+08
Water (2)	9.204E-04	1700	7.50E+04	1.84E+05
Soil (3)	1.134E-01	5500	2.57E+05	
Bottom sediment (4)	9.119E-02	17000	1.86E+03	9.12E+02

	E(1)=1000	E(2)=1000	E(3)=1000	E(1,2,3)
Overall residence time =	92.45	428.94	2141.25	398.28 h
Reaction time =	822.94	2372.43	6218.71	2551.78 h
Advection time =	104.15	523.61	3265.72	471.94 h

EMISSION (E)
REACTION (R)
ADVECTION (A)
TRANSFER D VALUE mol/Pa h

Phase Properties, Compositions, Transport and Transformation Rates:

Emission, kg/h			Fugacity, Pa				Concentration, g/m3				Amounts, kg				Total
E(1)	E(2)	E(3)	f(1)	f(2)	f(3)	f(4)	C(1)	C(2)	C(3)	C(4)	m(1)	m(2)	m(3)	m(4)	Amount, kg
1000	0	0	1.212E-05	7.691E-06	9.027E-06	1.338E-05	8.874E-07	1.285E-06	1.858E-04	2.214E-04	8.874E+04	2.569E+02	3.344E+03	1.107E+02	9.245E+04
0	1000	0	7.618E-06	7.764E-03	5.673E-06	1.350E-02	5.576E-07	1.297E-03	1.168E-04	2.235E-01	5.576E+04	2.593E+05	2.102E+03	1.117E+05	4.289E+05
0	0	1000	8.939E-06	3.961E-05	5.598E-03	6.890E-05	6.543E-07	6.615E-06	1.152E-01	1.140E-03	6.543E+04	1.323E+03	2.074E+06	5.700E+02	2.141E+06
600	300	100	1.045E-05	2.338E-03	5.669E-04	4.066E-03	7.652E-07	3.905E-04	1.167E-02	6.728E-02	7.652E+04	7.809E+04	2.100E+05	3.364E+04	3.983E+05

Emission, kg/h			Loss, Reaction, kg/h				Loss, Advection, kg/h			Intermedia Rate of Transport, kg/h T12	T21	T13	T31	T32	T24	T42
E(1)	E(2)	E(3)	R(1)	R(2)	R(3)	R(4)	A(1)	A(2)	A(4)	air-water	water-air	air-soil	soil-air	soil-water	water-sed	sed-water
1000	0	0	1.118E+02	1.047E-01	4.21E-01	4.512E-03	8.874E+02	2.569E-01	2.214E-03	9.841E-01	6.228E-01	1.615E+00	1.186E+00	7.062E-03	1.114E-02	4.411E-03
0	1000	0	7.026E+01	1.057E+02	2.65E-01	4.555E+00	5.576E+02	2.593E+02	2.235E+00	6.184E-01	6.288E+02	1.015E+00	7.453E-01	4.437E-03	1.124E+01	4.453E+00
0	0	1000	8.245E+01	5.394E-01	2.61E+02	2.324E-02	6.543E+02	1.323E+00	1.140E-02	7.257E-01	3.208E+00	1.190E+00	7.355E+02	4.379E+00	5.735E-02	2.272E-02
600	300	100	9.641E+01	3.183E+01	2.65E+01	1.371E+00	7.652E+02	3.183E+01	6.728E-01	8.486E-01	1.893E+02	1.392E+00	7.448E+01	4.435E-01	3.385E+00	1.341E+00

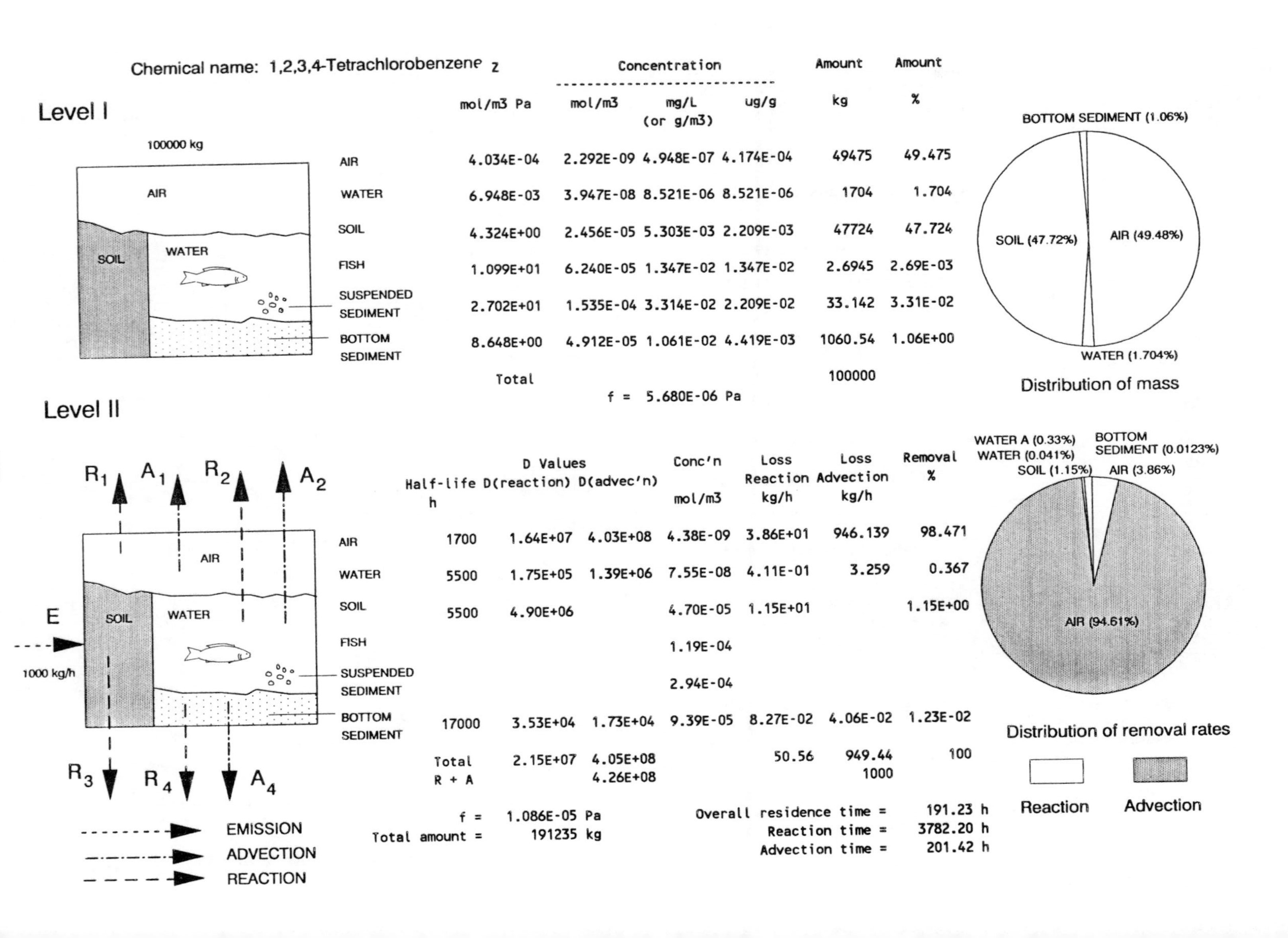

Level I

	mol/m3 Pa	Concentration mol/m3	mg/L (or g/m3)	ug/g	Amount kg	Amount %
AIR	4.034E-04	2.292E-09	4.948E-07	4.174E-04	49475	49.475
WATER	6.948E-03	3.947E-08	8.521E-06	8.521E-06	1704	1.704
SOIL	4.324E+00	2.456E-05	5.303E-03	2.209E-03	47724	47.724
FISH	1.099E+01	6.240E-05	1.347E-02	1.347E-02	2.6945	2.69E-03
SUSPENDED SEDIMENT	2.702E+01	1.535E-04	3.314E-02	2.209E-02	33.142	3.31E-02
BOTTOM SEDIMENT	8.648E+00	4.912E-05	1.061E-02	4.419E-03	1060.54	1.06E+00
Total					100000	

f = 5.680E-06 Pa

Level II

	Half-life h	D Values D(reaction)	D Values D(advec'n)	Conc'n mol/m3	Loss Reaction kg/h	Loss Advection kg/h	Removal %
AIR	1700	1.64E+07	4.03E+08	4.38E-09	3.86E+01	946.139	98.471
WATER	5500	1.75E+05	1.39E+06	7.55E-08	4.11E-01	3.259	0.367
SOIL	5500	4.90E+06		4.70E-05	1.15E+01		1.15E+00
FISH				1.19E-04			
SUSPENDED SEDIMENT				2.94E-04			
BOTTOM SEDIMENT	17000	3.53E+04	1.73E+04	9.39E-05	8.27E-02	4.06E-02	1.23E-02
Total		2.15E+07	4.05E+08		50.56	949.44	100
R + A			4.26E+08			1000	

f = 1.086E-05 Pa
Total amount = 191235 kg

Overall residence time = 191.23 h
Reaction time = 3782.20 h
Advection time = 201.42 h

Level III

Chemical name: 1,2,3,4-Tetrachlorobenzene

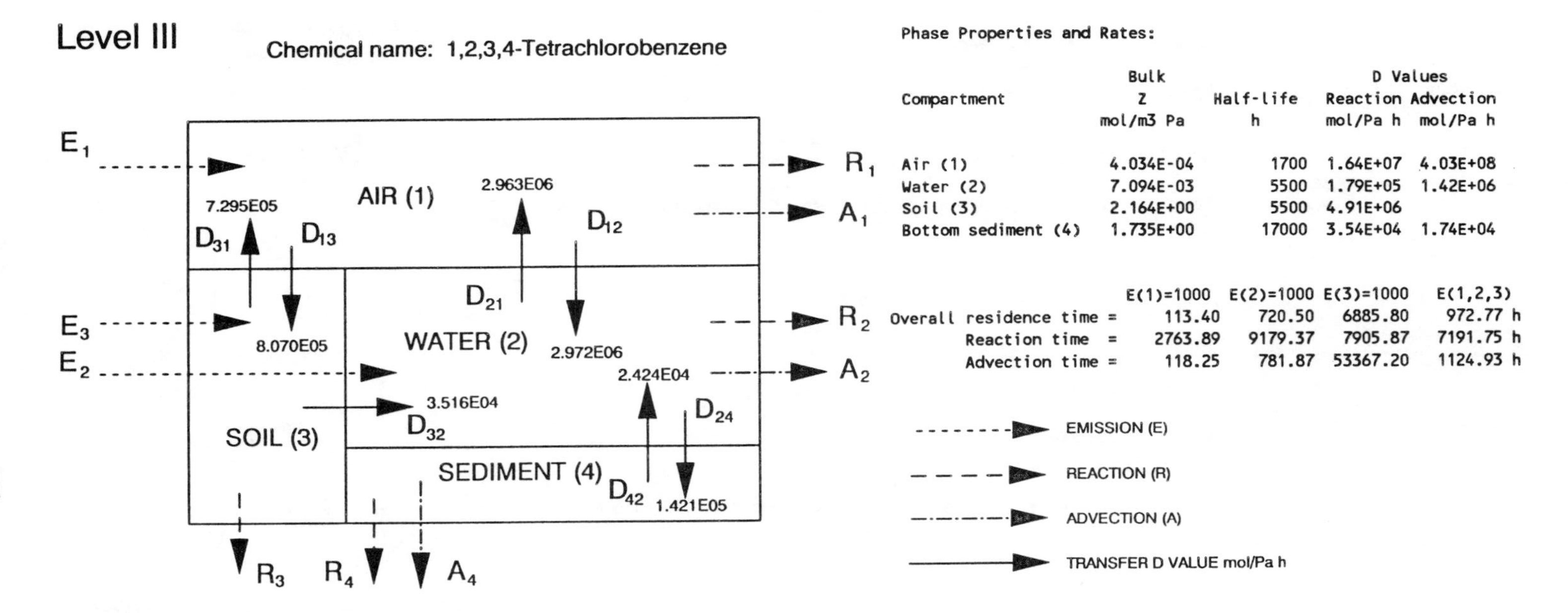

Phase Properties and Rates:

Compartment	Bulk Z mol/m3 Pa	Half-life h	D Values Reaction mol/Pa h	D Values Advection mol/Pa h
Air (1)	4.034E-04	1700	1.64E+07	4.03E+08
Water (2)	7.094E-03	5500	1.79E+05	1.42E+06
Soil (3)	2.164E+00	5500	4.91E+06	
Bottom sediment (4)	1.735E+00	17000	3.54E+04	1.74E+04

	E(1)=1000	E(2)=1000	E(3)=1000	E(1,2,3)
Overall residence time =	113.40	720.50	6885.80	972.77 h
Reaction time =	2763.89	9179.37	7905.87	7191.75 h
Advection time =	118.25	781.87	53367.20	1124.93 h

Phase Properties, Compositions, Transport and Transformation Rates:

Emission, kg/h			Fugacity, Pa				Concentration, g/m3				Amounts, kg				Total Amount, kg
E(1)	E(2)	E(3)	f(1)	f(2)	f(3)	f(4)	C(1)	C(2)	C(3)	C(4)	m(1)	m(2)	m(3)	m(4)	
1000	0	0	1.098E-05	7.020E-06	1.563E-06	1.296E-05	9.568E-07	1.075E-05	7.302E-04	4.855E-03	9.568E+04	2.150E+03	1.314E+04	2.427E+03	1.134E+05
0	1000	0	6.988E-06	9.988E-04	9.942E-07	1.844E-03	6.087E-07	1.530E-03	4.645E-04	6.907E-01	6.087E+04	3.059E+05	8.361E+03	3.453E+05	7.205E+05
0	0	1000	1.456E-06	7.093E-06	8.167E-04	1.309E-05	1.268E-07	1.086E-05	3.816E-01	4.905E-03	1.268E+04	2.173E+03	6.868E+06	2.452E+03	6.886E+06
600	300	100	8.833E-06	3.046E-04	8.291E-05	5.622E-04	7.693E-07	4.664E-04	3.874E-02	2.106E-01	7.693E+04	9.329E+04	6.972E+05	1.053E+05	9.728E+05

Emission, kg/h			Loss, Reaction, kg/h				Loss, Advection, kg/h			Intermedia Rate of Transport, kg/h T12	T21	T13	T31	T32	T24	T42
E(1)	E(2)	E(3)	R(1)	R(2)	R(3)	R(4)	A(1)	A(2)	A(4)	air-water	water-air	air-soil	soil-air	soil-water	water-sed	sed-water
1000	0	0	3.900E+01	2.710E-01	1.66E+00	9.895E-02	9.568E+02	2.150E+00	4.855E-02	7.049E+00	4.492E+00	1.914E+00	2.461E-01	1.186E-02	2.153E-01	6.783E-02
0	1000	0	2.481E+01	3.855E+01	1.05E+00	1.408E+01	6.087E+02	3.059E+02	6.907E+00	4.484E+00	6.390E+02	1.218E+00	1.566E-01	7.546E-03	3.063E+01	9.650E+00
0	0	1000	5.169E+00	2.737E-01	8.65E+02	9.997E-02	1.268E+02	2.173E+00	4.905E-02	9.342E-01	4.538E+00	2.537E-01	1.286E+02	6.199E+00	2.175E-01	6.853E-02
600	300	100	3.136E+01	1.175E+01	8.79E+01	4.293E+00	7.693E+02	1.175E+01	2.106E+00	5.668E+00	1.949E+02	1.539E+00	1.306E+01	6.293E-01	9.341E+00	2.943E+00

Chemical name: 1,2,3,5-Tetrachlorobenzene

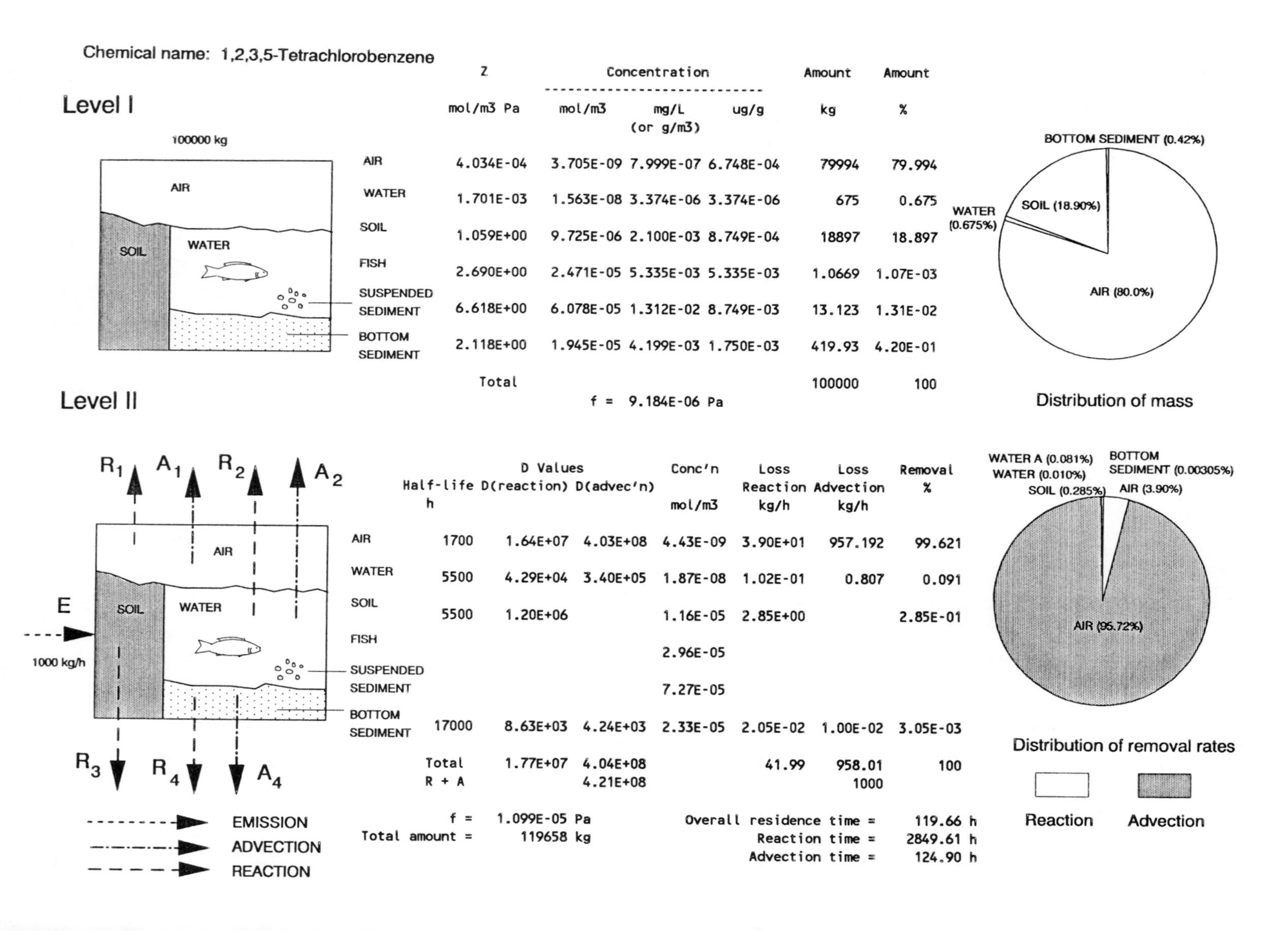

Level I

	Z	Concentration			Amount	Amount
	mol/m3 Pa	mol/m3	mg/L (or g/m3)	ug/g	kg	%
AIR	4.034E-04	3.705E-09	7.999E-07	6.748E-04	79994	79.994
WATER	1.701E-03	1.563E-08	3.374E-06	3.374E-06	675	0.675
SOIL	1.059E+00	9.725E-06	2.100E-03	8.749E-04	18897	18.897
FISH	2.690E+00	2.471E-05	5.335E-03	5.335E-03	1.0669	1.07E-03
SUSPENDED SEDIMENT	6.618E+00	6.078E-05	1.312E-02	8.749E-03	13.123	1.31E-02
BOTTOM SEDIMENT	2.118E+00	1.945E-05	4.199E-03	1.750E-03	419.93	4.20E-01
	Total				100000	100

f = 9.184E-06 Pa

Level II

	Half-life h	D Values D(reaction)	D Values D(advec'n)	Conc'n mol/m3	Loss Reaction kg/h	Loss Advection kg/h	Removal %
AIR	1700	1.64E+07	4.03E+08	4.43E-09	3.90E+01	957.192	99.621
WATER	5500	4.29E+04	3.40E+05	1.87E-08	1.02E-01	0.807	0.091
SOIL	5500	1.20E+06		1.16E-05	2.85E+00		2.85E-01
FISH				2.96E-05			
SUSPENDED SEDIMENT				7.27E-05			
BOTTOM SEDIMENT	17000	8.63E+03	4.24E+03	2.33E-05	2.05E-02	1.00E-02	3.05E-03
	Total	1.77E+07	4.04E+08		41.99	958.01	100
	R + A		4.21E+08			1000	

f = 1.099E-05 Pa
Total amount = 119658 kg

Overall residence time = 119.66 h
Reaction time = 2849.61 h
Advection time = 124.90 h

Level III

Chemical name: 1,2,3,5-Tetrachlorobenzene

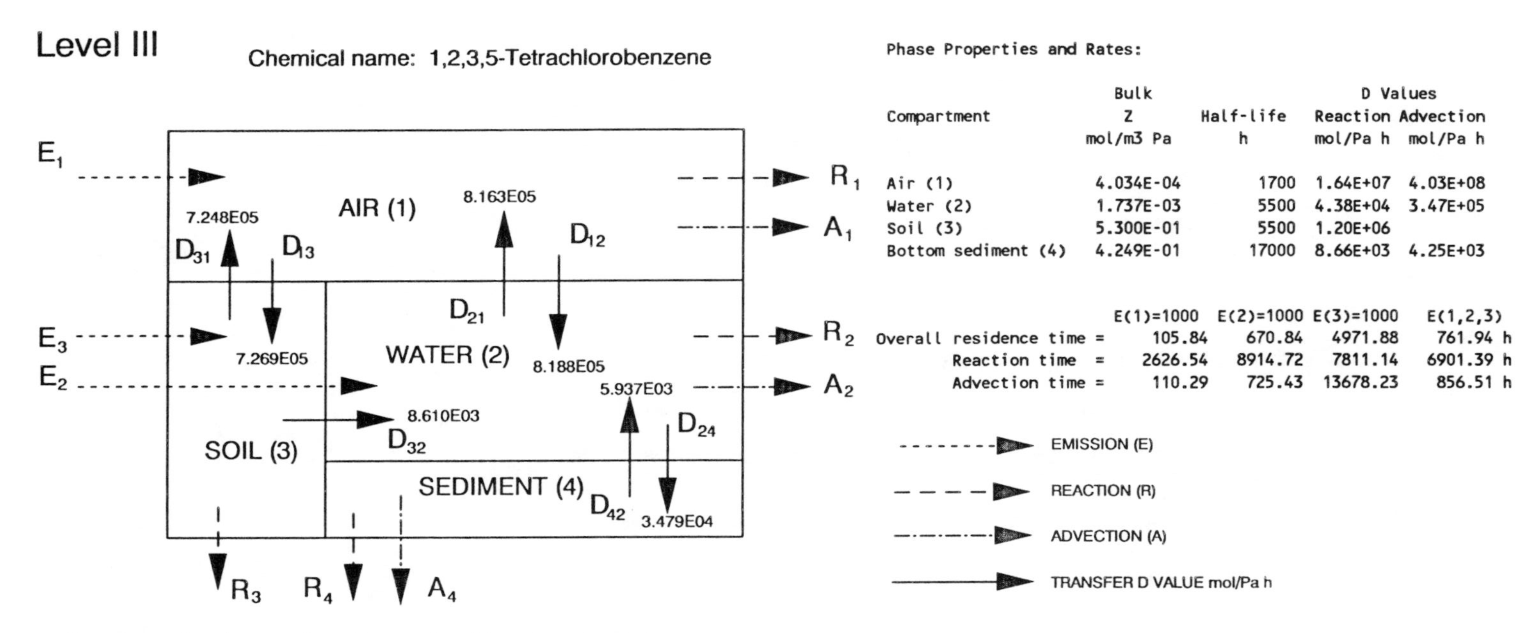

Phase Properties and Rates:

Compartment	Bulk Z mol/m3 Pa	Half-life h	D Values Reaction mol/Pa h	D Values Advection mol/Pa h
Air (1)	4.034E-04	1700	1.64E+07	4.03E+08
Water (2)	1.737E-03	5500	4.38E+04	3.47E+05
Soil (3)	5.300E-01	5500	1.20E+06	
Bottom sediment (4)	4.249E-01	17000	8.66E+03	4.25E+03

	E(1)=1000	E(2)=1000	E(3)=1000	E(1,2,3)
Overall residence time =	105.84	670.84	4971.88	761.94 h
Reaction time =	2626.54	8914.72	7811.14	6901.39 h
Advection time =	110.29	725.43	13678.23	856.51 h

Phase Properties, Compositions, Transport and Transformation Rates:

Emission, kg/h			Fugacity, Pa				Concentration, g/m3				Amounts, kg				Total Amount, kg
E(1)	E(2)	E(3)	f(1)	f(2)	f(3)	f(4)	C(1)	C(2)	C(3)	C(4)	m(1)	m(2)	m(3)	m(4)	
1000	0	0	1.101E-05	7.352E-06	4.250E-06	1.357E-05	9.591E-07	2.758E-06	4.863E-04	1.245E-03	9.591E+04	5.515E+02	8.753E+03	6.225E+02	1.058E+05
0	1000	0	7.300E-06	3.766E-03	2.817E-06	6.953E-03	6.358E-07	1.413E-03	3.224E-04	6.378E-01	6.358E+04	2.825E+05	5.803E+03	3.189E+05	6.708E+05
0	0	1000	4.156E-06	1.951E-05	2.395E-03	3.601E-05	3.620E-07	7.316E-06	2.740E-01	3.304E-03	3.620E+04	1.463E+03	4.933E+06	1.652E+03	4.972E+06
600	300	100	9.213E-06	1.136E-03	2.429E-04	2.098E-03	8.024E-07	4.262E-04	2.779E-02	1.924E-01	8.024E+04	8.524E+04	5.002E+05	9.621E+04	7.619E+05

Emission, kg/h			Loss, Reaction, kg/h				Loss, Advection, kg/h			Intermedia Rate of Transport, kg/h T12	T21	T13	T31	T32	T24	T42
E(1)	E(2)	E(3)	R(1)	R(2)	R(3)	R(4)	A(1)	A(2)	A(4)	air-water	water-air	air-soil	soil-air	soil-water	water-sed	sed-water
1000	0	0	3.910E+01	6.949E-02	1.10E+00	2.538E-02	9.591E+02	5.515E-01	1.245E-02	1.947E+00	1.296E+00	1.776E+00	6.650E-01	7.899E-03	5.522E-02	1.740E-02
0	1000	0	2.592E+01	3.560E+01	7.31E-01	1.300E+01	6.358E+02	2.825E+02	6.378E+00	1.290E+00	6.638E+02	1.177E+00	4.408E-01	5.237E-03	2.829E+01	8.912E+00
0	0	1000	1.476E+01	1.844E-01	6.22E+02	6.733E-02	3.620E+02	1.463E+00	3.304E-02	7.347E-01	3.438E+00	6.702E-01	3.747E+02	4.451E+00	1.465E-01	4.616E-02
600	300	100	3.271E+01	1.074E+01	6.30E+01	3.922E+00	8.024E+02	1.074E+01	1.924E+00	1.629E+00	2.003E+02	1.486E+00	3.800E+01	4.514E-01	8.535E+00	2.689E+00

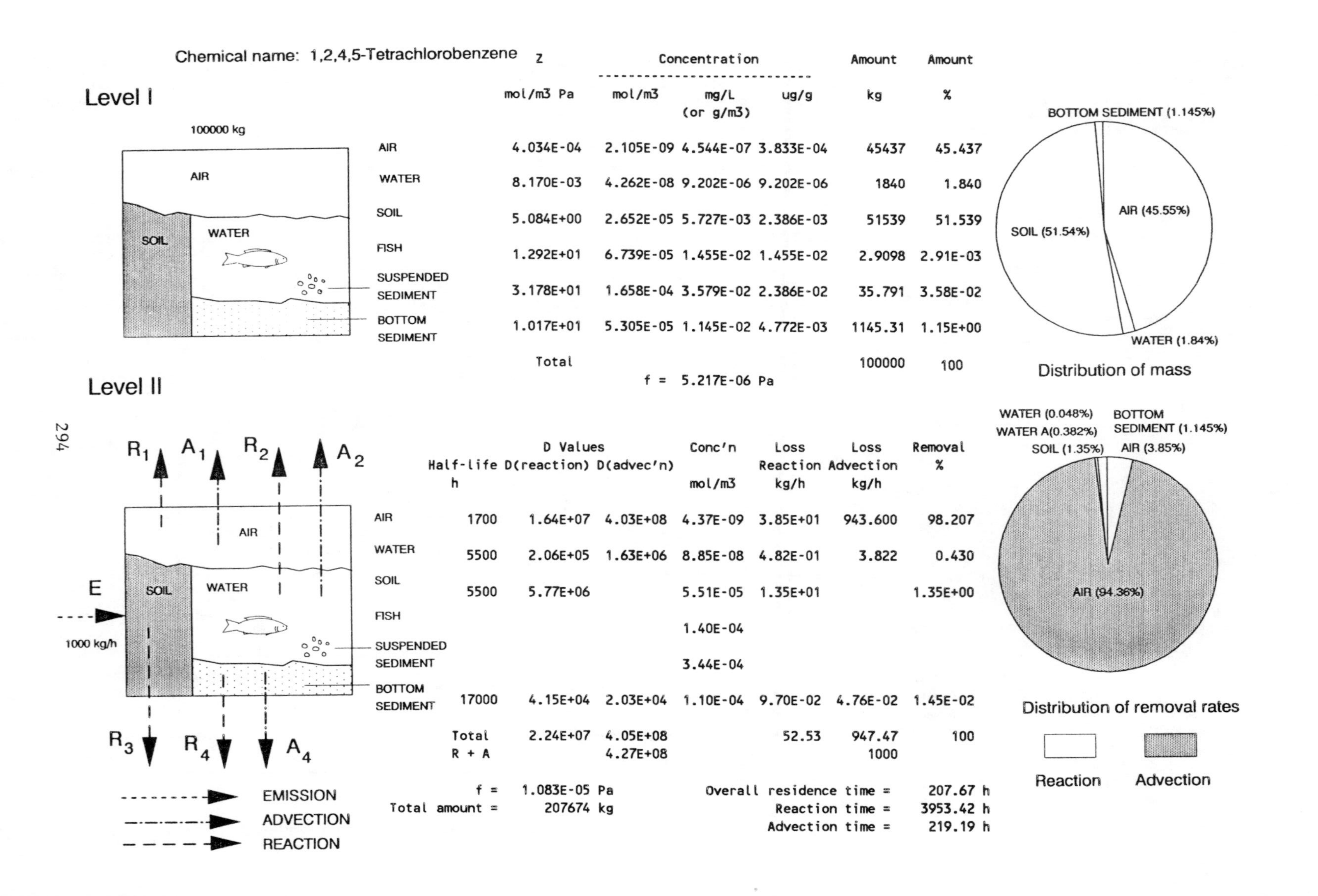

Chemical name: 1,2,4,5-Tetrachlorobenzene

Level I

	Z mol/m3 Pa	Concentration mol/m3	mg/L (or g/m3)	ug/g	Amount kg	Amount %
AIR	4.034E-04	2.105E-09	4.544E-07	3.833E-04	45437	45.437
WATER	8.170E-03	4.262E-08	9.202E-06	9.202E-06	1840	1.840
SOIL	5.084E+00	2.652E-05	5.727E-03	2.386E-03	51539	51.539
FISH	1.292E+01	6.739E-05	1.455E-02	1.455E-02	2.9098	2.91E-03
SUSPENDED SEDIMENT	3.178E+01	1.658E-04	3.579E-02	2.386E-02	35.791	3.58E-02
BOTTOM SEDIMENT	1.017E+01	5.305E-05	1.145E-02	4.772E-03	1145.31	1.15E+00
Total					100000	100

f = 5.217E-06 Pa

Level II

	Half-life h	D Values D(reaction)	D(advec'n)	Conc'n mol/m3	Loss Reaction kg/h	Loss Advection kg/h	Removal %
AIR	1700	1.64E+07	4.03E+08	4.37E-09	3.85E+01	943.600	98.207
WATER	5500	2.06E+05	1.63E+06	8.85E-08	4.82E-01	3.822	0.430
SOIL	5500	5.77E+06		5.51E-05	1.35E+01		1.35E+00
FISH				1.40E-04			
SUSPENDED SEDIMENT				3.44E-04			
BOTTOM SEDIMENT	17000	4.15E+04	2.03E+04	1.10E-04	9.70E-02	4.76E-02	1.45E-02
Total		2.24E+07	4.05E+08		52.53	947.47	100
R + A			4.27E+08			1000	

f = 1.083E-05 Pa
Total amount = 207674 kg

Overall residence time = 207.67 h
Reaction time = 3953.42 h
Advection time = 219.19 h

Level III

Chemical name: 1,2,4,5-Tetrachlorobenzene

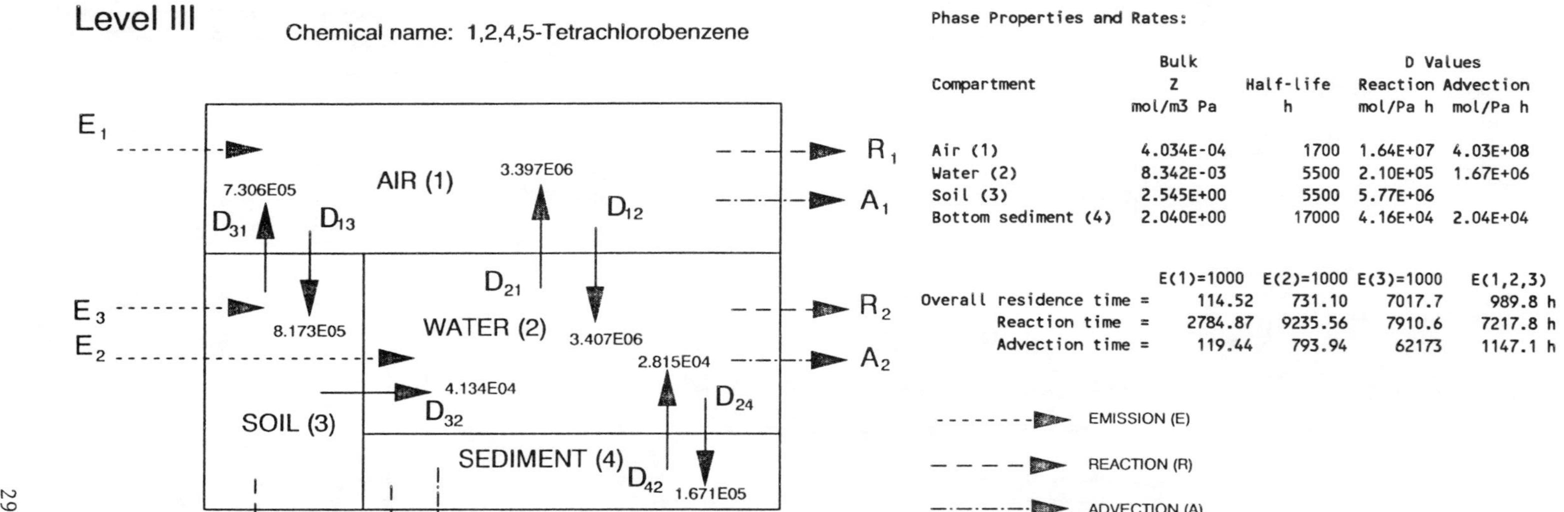

Phase Properties and Rates:

Compartment	Bulk Z mol/m3 Pa	Half-life h	D Values Reaction mol/Pa h	D Values Advection mol/Pa h
Air (1)	4.034E-04	1700	1.64E+07	4.03E+08
Water (2)	8.342E-03	5500	2.10E+05	1.67E+06
Soil (3)	2.545E+00	5500	5.77E+06	
Bottom sediment (4)	2.040E+00	17000	4.16E+04	2.04E+04

	E(1)=1000	E(2)=1000	E(3)=1000	E(1,2,3)
Overall residence time =	114.52	731.10	7017.7	989.8 h
Reaction time =	2784.87	9235.56	7910.6	7217.8 h
Advection time =	119.44	793.94	62173	1147.1 h

Phase Properties, Compositions, Transport and Transformation Rates:

Emission, kg/h			Fugacity, Pa				Concentration, g/m3				Amounts, kg				Total
E(1)	E(2)	E(3)	f(1)	f(2)	f(3)	f(4)	C(1)	C(2)	C(3)	C(4)	m(1)	m(2)	m(3)	m(4)	Amount, kg
1000	0	0	1.098E-05	6.950E-06	1.371E-06	1.283E-05	9.563E-07	1.252E-05	7.535E-04	5.652E-03	9.563E+04	2.503E+03	1.356E+04	2.826E+03	1.145E+05
0	1000	0	6.920E-06	8.637E-04	8.643E-07	1.594E-03	6.027E-07	1.556E-03	4.749E-04	7.023E-01	6.027E+04	3.111E+05	8.548E+03	3.512E+05	7.311E+05
0	0	1000	1.270E-06	6.233E-06	7.080E-04	1.151E-05	1.106E-07	1.123E-05	3.890E-01	5.068E-03	1.106E+04	2.245E+03	7.002E+06	2.534E+03	7.018E+06
600	300	100	8.791E-06	2.639E-04	7.188E-05	4.872E-04	7.657E-07	4.753E-04	3.949E-02	2.146E-01	7.657E+04	9.506E+04	7.109E+05	1.073E+05	9.898E+05

Emission, kg/h			Loss, Reaction, kg/h				Loss, Advection, kg/h			Intermedia Rate of Transport, kg/h T12	T21	T13	T31	T32	T24	T42
E(1)	E(2)	E(3)	R(1)	R(2)	R(3)	R(4)	A(1)	A(2)	A(4)	air-water	water-air	air-soil	soil-air	soil-water	water-sed	sed-water
1000	0	0	3.898E+01	3.154E-01	1.71E+00	1.152E-01	9.563E+02	2.503E+00	5.652E-02	8.075E+00	5.097E+00	1.937E+00	2.163E-01	1.224E-02	2.507E-01	7.896E-02
0	1000	0	2.457E+01	3.920E+01	1.08E+00	1.432E+01	6.027E+02	3.111E+02	7.023E+00	5.090E+00	6.335E+02	1.221E+00	1.363E-01	7.715E-03	3.115E+01	9.813E+00
0	0	1000	4.508E+00	2.829E-01	8.82E+02	1.033E-01	1.106E+02	2.245E+00	5.068E-02	9.338E-01	4.571E+00	2.240E-01	1.117E+02	6.319E+00	2.248E-01	7.082E-02
600	300	100	3.121E+01	1.198E+01	8.96E+01	4.374E+00	7.657E+02	1.198E+01	2.146E+00	6.466E+00	1.936E+02	1.551E+00	1.134E+01	6.416E-01	9.519E+00	2.998E+00

Chemical name: Pentachlorobenzene

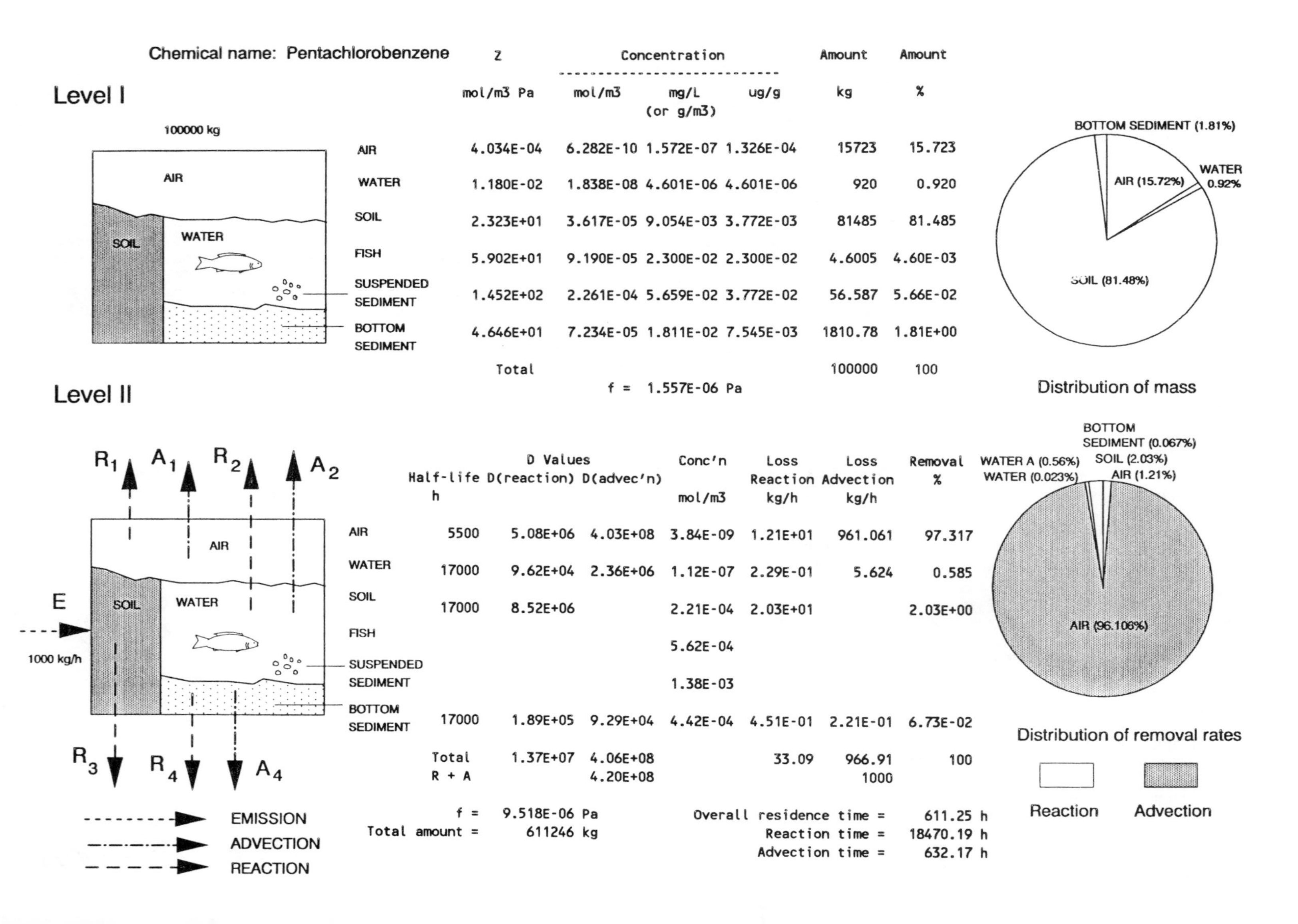

Level I

	Z mol/m3 Pa	Concentration mol/m3	Concentration mg/L (or g/m3)	Concentration ug/g	Amount kg	Amount %
AIR	4.034E-04	6.282E-10	1.572E-07	1.326E-04	15723	15.723
WATER	1.180E-02	1.838E-08	4.601E-06	4.601E-06	920	0.920
SOIL	2.323E+01	3.617E-05	9.054E-03	3.772E-03	81485	81.485
FISH	5.902E+01	9.190E-05	2.300E-02	2.300E-02	4.6005	4.60E-03
SUSPENDED SEDIMENT	1.452E+02	2.261E-04	5.659E-02	3.772E-02	56.587	5.66E-02
BOTTOM SEDIMENT	4.646E+01	7.234E-05	1.811E-02	7.545E-03	1810.78	1.81E+00
	Total				100000	100

f = 1.557E-06 Pa

Level II

	Half-life h	D Values D(reaction)	D Values D(advec'n)	Conc'n mol/m3	Loss Reaction kg/h	Loss Advection kg/h	Removal %
AIR	5500	5.08E+06	4.03E+08	3.84E-09	1.21E+01	961.061	97.317
WATER	17000	9.62E+04	2.36E+06	1.12E-07	2.29E-01	5.624	0.585
SOIL	17000	8.52E+06		2.21E-04	2.03E+01		2.03E+00
FISH				5.62E-04			
SUSPENDED SEDIMENT				1.38E-03			
BOTTOM SEDIMENT	17000	1.89E+05	9.29E+04	4.42E-04	4.51E-01	2.21E-01	6.73E-02
	Total R + A	1.37E+07	4.06E+08 4.20E+08		33.09	966.91 1000	100

f = 9.518E-06 Pa
Total amount = 611246 kg

Overall residence time = 611.25 h
Reaction time = 18470.19 h
Advection time = 632.17 h

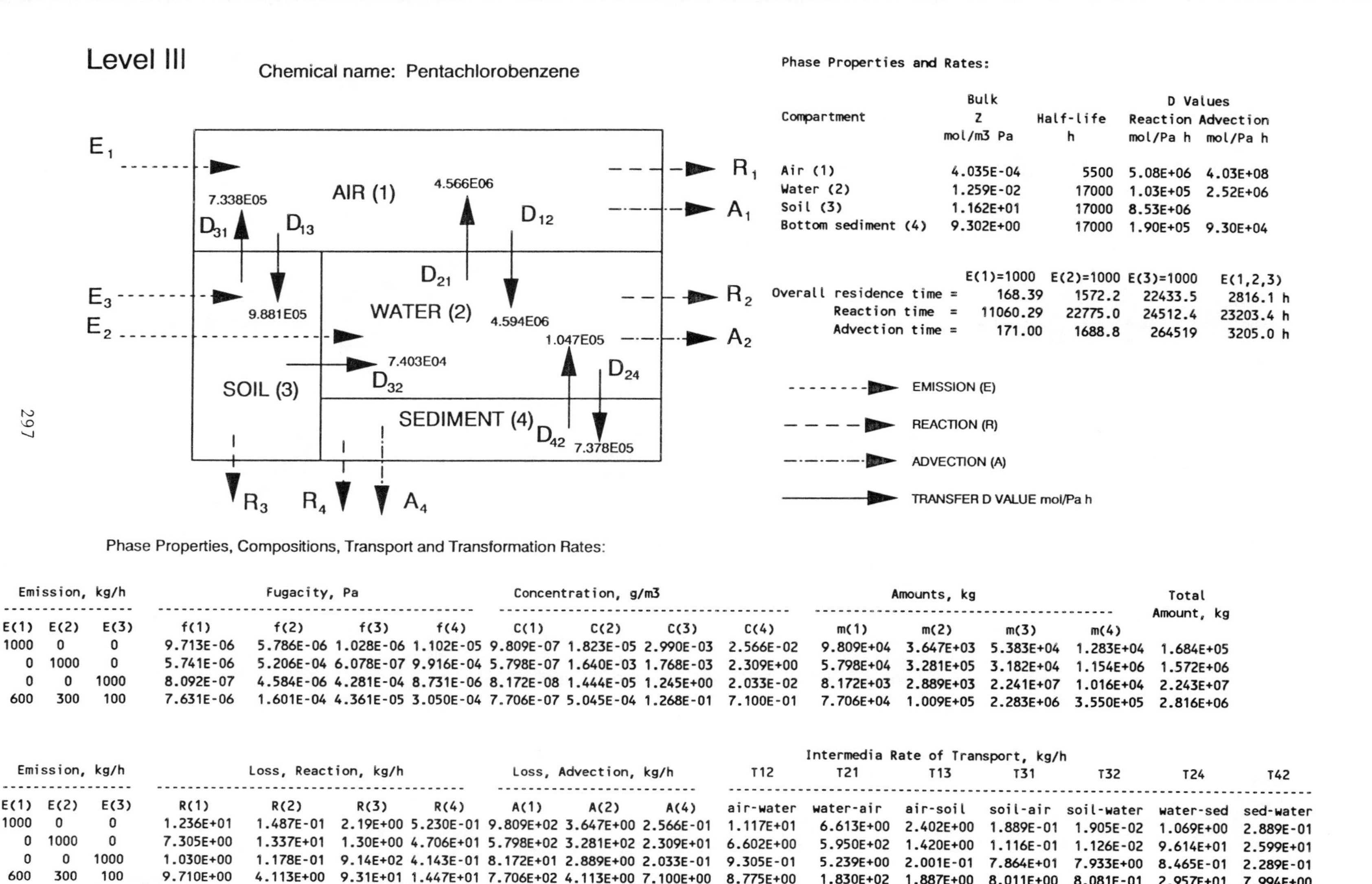

Phase Properties and Rates:

Compartment	Bulk Z mol/m3 Pa	Half-life h	D Values Reaction mol/Pa h	Advection mol/Pa h
Air (1)	4.035E-04	5500	5.08E+06	4.03E+08
Water (2)	1.259E-02	17000	1.03E+05	2.52E+06
Soil (3)	1.162E+01	17000	8.53E+06	
Bottom sediment (4)	9.302E+00	17000	1.90E+05	9.30E+04

	E(1)=1000	E(2)=1000	E(3)=1000	E(1,2,3)
Overall residence time =	168.39	1572.2	22433.5	2816.1 h
Reaction time =	11060.29	22775.0	24512.4	23203.4 h
Advection time =	171.00	1688.8	264519	3205.0 h

Phase Properties, Compositions, Transport and Transformation Rates:

Emission, kg/h			Fugacity, Pa				Concentration, g/m3				Amounts, kg				Total
E(1)	E(2)	E(3)	f(1)	f(2)	f(3)	f(4)	C(1)	C(2)	C(3)	C(4)	m(1)	m(2)	m(3)	m(4)	Amount, kg
1000	0	0	9.713E-06	5.786E-06	1.028E-06	1.102E-05	9.809E-07	1.823E-05	2.990E-03	2.566E-02	9.809E+04	3.647E+03	5.383E+04	1.283E+04	1.684E+05
0	1000	0	5.741E-06	5.206E-04	6.078E-07	9.916E-04	5.798E-07	1.640E-03	1.768E-03	2.309E+00	5.798E+04	3.281E+05	3.182E+04	1.154E+06	1.572E+06
0	0	1000	8.092E-07	4.584E-06	4.281E-04	8.731E-06	8.172E-08	1.444E-05	1.245E+00	2.033E-02	8.172E+03	2.889E+03	2.241E+07	1.016E+04	2.243E+07
600	300	100	7.631E-06	1.601E-04	4.361E-05	3.050E-04	7.706E-07	5.045E-04	1.268E-01	7.100E-01	7.706E+04	1.009E+05	2.283E+06	3.550E+05	2.816E+06

Emission, kg/h			Loss, Reaction, kg/h				Loss, Advection, kg/h			Intermedia Rate of Transport, kg/h T12	T21	T13	T31	T32	T24	T42
E(1)	E(2)	E(3)	R(1)	R(2)	R(3)	R(4)	A(1)	A(2)	A(4)	air-water	water-air	air-soil	soil-air	soil-water	water-sed	sed-water
1000	0	0	1.236E+01	1.487E-01	2.19E+00	5.230E-01	9.809E+02	3.647E+00	2.566E-01	1.117E+01	6.613E+00	2.402E+00	1.889E-01	1.905E-02	1.069E+00	2.889E-01
0	1000	0	7.305E+00	1.337E+01	1.30E+00	4.706E+01	5.798E+02	3.281E+02	2.309E+01	6.602E+00	5.950E+02	1.420E+00	1.116E-01	1.126E-02	9.614E+01	2.599E+01
0	0	1000	1.030E+00	1.178E-01	9.14E+02	4.143E-01	8.172E+01	2.889E+00	2.033E-01	9.305E-01	5.239E+00	2.001E-01	7.864E+01	7.933E+00	8.465E-01	2.289E-01
600	300	100	9.710E+00	4.113E+00	9.31E+01	1.447E+01	7.706E+02	4.113E+00	7.100E+00	8.775E+00	1.830E+02	1.887E+00	8.011E+00	8.081E-01	2.957E+01	7.994E+00

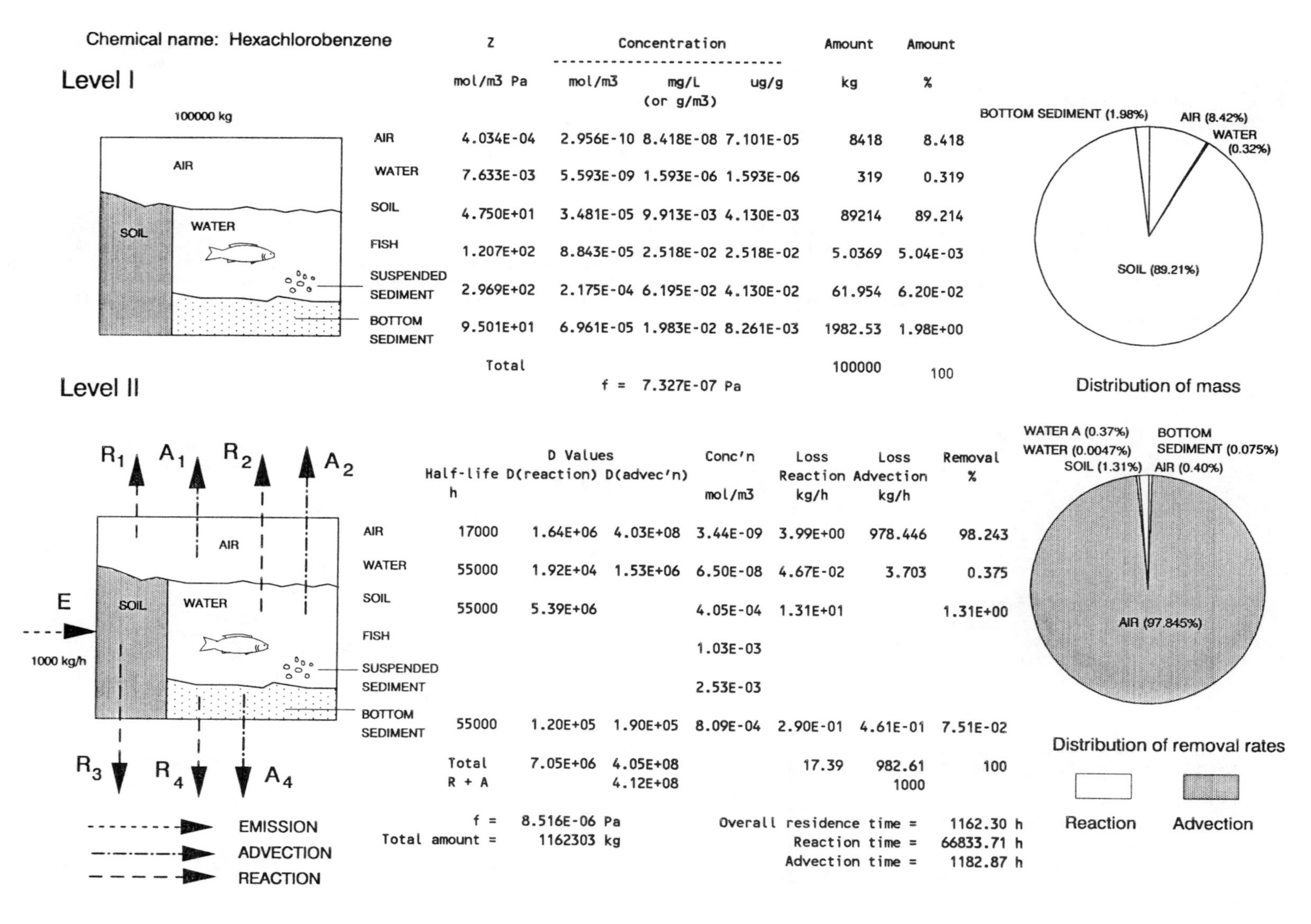

Chemical name: Hexachlorobenzene

Level I

	Z	Concentration			Amount	Amount
	mol/m3 Pa	mol/m3	mg/L (or g/m3)	ug/g	kg	%
AIR	4.034E-04	2.956E-10	8.418E-08	7.101E-05	8418	8.418
WATER	7.633E-03	5.593E-09	1.593E-06	1.593E-06	319	0.319
SOIL	4.750E+01	3.481E-05	9.913E-03	4.130E-03	89214	89.214
FISH	1.207E+02	8.843E-05	2.518E-02	2.518E-02	5.0369	5.04E-03
SUSPENDED SEDIMENT	2.969E+02	2.175E-04	6.195E-02	4.130E-02	61.954	6.20E-02
BOTTOM SEDIMENT	9.501E+01	6.961E-05	1.983E-02	8.261E-03	1982.53	1.98E+00
	Total				100000	100

f = 7.327E-07 Pa

Level II

	Half-life h	D Values D(reaction)	D(advec'n)	Conc'n mol/m3	Loss Reaction kg/h	Loss Advection kg/h	Removal %
AIR	17000	1.64E+06	4.03E+08	3.44E-09	3.99E+00	978.446	98.243
WATER	55000	1.92E+04	1.53E+06	6.50E-08	4.67E-02	3.703	0.375
SOIL	55000	5.39E+06		4.05E-04	1.31E+01		1.31E+00
FISH				1.03E-03			
SUSPENDED SEDIMENT				2.53E-03			
BOTTOM SEDIMENT	55000	1.20E+05	1.90E+05	8.09E-04	2.90E-01	4.61E-01	7.51E-02
	Total R + A	7.05E+06	4.05E+08 4.12E+08		17.39	982.61 1000	100

f = 8.516E-06 Pa
Total amount = 1162303 kg

Overall residence time = 1162.30 h
Reaction time = 66833.71 h
Advection time = 1182.87 h

Level III

Chemical name: Hexachlorobenzene

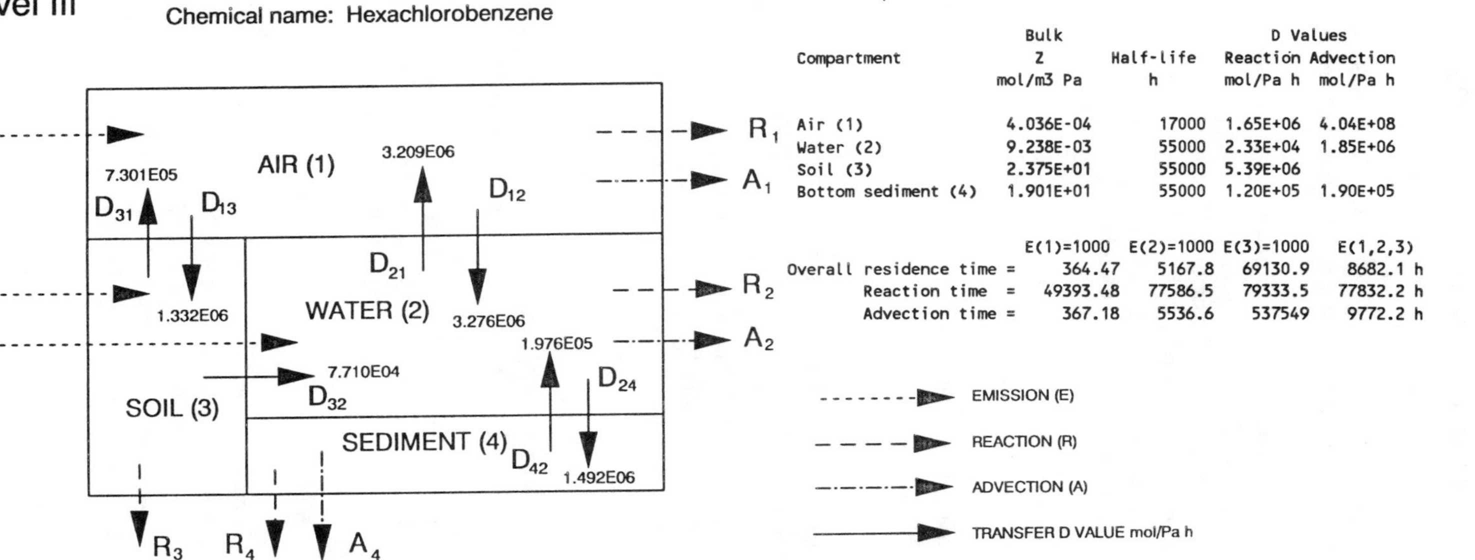

Phase Properties and Rates:

Compartment	Bulk Z mol/m3 Pa	Half-life h	D Values Reaction mol/Pa h	D Values Advection mol/Pa h
Air (1)	4.036E-04	17000	1.65E+06	4.04E+08
Water (2)	9.238E-03	55000	2.33E+04	1.85E+06
Soil (3)	2.375E+01	55000	5.39E+06	
Bottom sediment (4)	1.901E+01	55000	1.20E+05	1.90E+05

	E(1)=1000	E(2)=1000	E(3)=1000	E(1,2,3)
Overall residence time =	364.47	5167.8	69130.9	8682.1 h
Reaction time =	49393.48	77586.5	79333.5	77832.2 h
Advection time =	367.18	5536.6	537549	9772.2 h

Phase Properties, Compositions, Transport and Transformation Rates:

Emission, kg/h			Fugacity, Pa				Concentration, g/m3				Amounts, kg				Total Amount, kg
E(1)	E(2)	E(3)	f(1)	f(2)	f(3)	f(4)	C(1)	C(2)	C(3)	C(4)	m(1)	m(2)	m(3)	m(4)	
1000	0	0	8.607E-06	4.730E-06	1.851E-06	1.391E-05	9.894E-07	1.245E-05	1.252E-02	7.529E-02	9.894E+04	2.489E+03	2.254E+05	3.765E+04	3.645E+05
0	1000	0	4.611E-06	5.886E-04	9.915E-07	1.731E-03	5.300E-07	1.549E-03	6.708E-03	9.369E+00	5.300E+04	3.097E+05	1.207E+05	4.684E+06	5.168E+06
0	0	1000	1.072E-06	7.884E-06	5.670E-04	2.318E-05	1.232E-07	2.074E-05	3.836E+00	1.255E-01	1.232E+04	4.148E+03	6.905E+07	6.274E+04	6.913E+07
600	300	100	6.655E-06	1.802E-04	5.811E-05	5.299E-04	7.649E-07	4.741E-04	3.931E-01	2.868E+00	7.649E+04	9.483E+04	7.077E+06	1.434E+06	8.682E+06

Emission, kg/h			Loss, Reaction, kg/h				Loss, Advection, kg/h			Intermedia Rate of Transport, kg/h T12	T21	T13	T31	T32	T24	T42
E(1)	E(2)	E(3)	R(1)	R(2)	R(3)	R(4)	A(1)	A(2)	A(4)	air-water	water-air	air-soil	soil-air	soil-water	water-sed	sed-water
1000	0	0	4.033E+00	3.136E-02	2.84E+00	4.744E-01	9.894E+02	2.489E+00	7.529E-01	8.031E+00	4.324E+00	3.265E+00	3.849E-01	4.064E-02	2.010E+00	7.829E-01
0	1000	0	2.160E+00	3.903E+00	1.52E+00	5.902E+01	5.300E+02	3.097E+02	9.369E+01	4.302E+00	5.380E+02	1.749E+00	2.062E-01	2.177E-02	2.501E+02	9.742E+01
0	0	1000	5.022E-01	5.227E-02	8.70E+02	7.905E-01	1.232E+02	4.148E+00	1.255E+00	1.000E+00	7.206E+00	4.066E-01	1.179E+02	1.245E+01	3.350E+00	1.305E+00
600	300	100	3.118E+00	1.195E+00	8.92E+01	1.807E+01	7.649E+02	1.195E+00	2.868E+01	6.209E+00	1.647E+02	2.525E+00	1.208E+01	1.276E+00	7.658E+01	2.983E+01

3.4 COMMENTARY ON THE PHYSICAL-CHEMICAL PROPERTIES AND ENVIRONMENTAL FATE

QSPR Plots

The QSPR plots of the chlorobenzenes given in Figures 3.1 to 3.6 show consistently linear relationships. The various commonly used molecular descriptors are shown in Figure 3.1.

An increase of 20 cm^3/mol (approximately one chlorine atom) causes (approximately):
(i) a decrease in log solubility by 0.65 units (a factor of 4.5);
(ii) a decrease in log vapor pressure by 0.72 units (a factor of 5.2);
(iii) an increase in log K_{OW} by 0.53 units (a factor of 3.4);
(iv) a decrease in log Henry's law constant by 0.16 units (a factor of 1.4).
The slope of the log K_{OW} versus log C_L line in Figure 3.6 is thus 0.53/0.65 or 0.81.

It must be noted that the plotted solubilities and vapor pressures of the higher chlorobenzenes are those of the subcooled liquids, not the actual solid phase values.

It is concluded that the accuracy of the selected data is adequate for most environmental purposes.

Evaluative Calculations

The Level I calculations show that as the chlorine number increases, there is a corresponding increase in partitioning into soil and sediment. Whereas about 1 % of chlorobenzene (MCB) is sorbed to these solid phases, this increases to nearly 90% for hexachlorobenzene (HCB). This is caused by the marked increase in K_{OW} noted above. The water compartment accounts for only about 1%. Notable is the increase in concentration in fish from 0.2 ng/g for MCB to 25 ng/g for HCB. This is consistent with the observed bioaccumulative properties of the higher chlorobenzenes.

The Level II calculations show that atmospheric advection is the primary removal process, suggesting that substances such as HCB will tend to be distributed widely by atmospheric transport. Reaction rates are slow, with reaction residence times ranging from 250 to 3000h.

The Level III calculations suggest that when discharged to air these chemicals tend to be primarily advected from the air. For the higher CBs, appreciable quantities accumulate in soil from which reaction or transport is slow. For example, at steady-state, 62% of HCB is found in soil, but the rate of transfer to or from soil is very slow. Essentially the soil acts as a "buffer." The overall residence time ranges from 70 to 400 h.

When discharged to water, the residence time increases to 300 to 5200 h with appreciable loss by evaporation and advection. The lower CBs show considerable loss by reaction but for the higher CBs, transfer to sediment becomes more important. For HCB, sediment contamination becomes very significant with concentrations reaching 9.4 g/m^3 and 90% of the HCB residing in the bottom sediment, which acts as a low reactivity "buffer." Clearly, any HCB discharged to water will have a profound effect on aquatic systems, especially benthic systems.

When discharged to soil, residence times range from 500 to 70000 h (8 years). Most chemical either reacts in the soil or evaporates. The evaporation rate decreases as chlorine number increases, so that discharge of 1000 kg/h of HCB to soil would result in the presence of nearly 70 million kg of HCB in soil with a corresponding residence time of 70000 h and a high concentration of 3.8 g/m^3.

In general, as chlorination increases, these chemicals become of much greater concern because they tend to be more persistent and accumulate to high concentrations in soil, sediments and their resident biota. The primary removal is by advection, but this merely relocates the problem, it does not solve it. Discharges to water and soil are likely to be of particular local concern because of the tendency to accumulate caused by slow intermedia transport and slow reaction.

It is interesting to examine the amounts (kg) of MCB and HCB which reside in the environment if there is discharge of 1 kg/h to the three receiving media (i.e. 1/1000th of the rates in the figures)

	MCB	HCB
Discharge to air	72	364
Discharge to water	316	5168
Discharge to soil	519	69130

These amounts control the exposure experienced by all organisms, including humans. Clearly there are profound differences (factor of 100) between the exposures which will result from discharge of 1 kg of MCB and 1 kg of HCB. Likewise, large differences in exposure result when the chemicals are discharged variously to air, water or soil.

3.5 REFERENCES

Abernethy, S., Bobra, A.M., Shiu, W.Y., Wells, P.G., Mackay, D. (1986) Acute lethal toxicity of hydrocarbons and chlorinated hydrocarbons to two planktonic crustaceans: The key role of organisms-water partitioning. *Aquatic Toxicol.* 8, 163-174.

Abernethy, S., Mackay, D. (1987) A discussion of correlations for narcosis in aquatic species. In: *QSAR in Environmental Toxicology* II, Kaiser, K.L.E., Ed., pp.1-16, D. Reidel Publ. Co., Dordrecht, Holland.

Abernethy, S., Mackay, D., McCarty, L.S. (1988) "Volume fraction" correlation for narcosis in aquatic organisms: the key role of partitioning. *Environ. Toxicol. Chem.* 7, 469-481.

Amidon, G.L., Williams, N.A. (1982) An solubility equation for non-electrolytes in water. *Int'l. J. Pharm.* 11, 249-256.

Anderson, T.A., Beauchamp, J.J., Walton, B.T. (1991) Organic chemicals in the environment. *J. Environ. Qual.* 20, 420-424.

Andrews, L.J., Keefer, R.M. (1950) Cation complexes of compounds containing of carbon-carbon double bonds. VI The argentation of substituted benzenes. *J. Am. Chem. Soc.* 72, 3110-3116.

Anliker, R., Moser, P. (1987) The limits of bioaccumulation of organic pigments in fish: Their relation to the partition coefficient and the solubility in water and octanol. *Ecotoxicol. Environ. Saf.* 13, 43-52.

Aquan-Yuen, M., Mackay, D., Shiu, W.Y. (1979) Solubility of hexane, phenanthrene, chlorobenzene, and p-dichlorobenzene in aqueous electrolyte solutions. *J. Chem. Eng. Data* 24, 30-34.

Arbuckle, W.B. (1983) Estimating activity coefficients for use in calculating environmental parameters. *Environ. Sci. Technol.* 17, 537-542.

Ashworth, R.A., Howe, G.B., Mullins, M.E., Rogers, T.N. (1988) Air-water partitioning coefficients of organics in dilute aqueous solutions. *J. Hazard. Materials* 18, 25-36.

Atkinson, R. (1985) Kinetics and mechanisms of the gas phase reactions of hydroxy radicals with organic compounds under atmospheric conditions. *Chem. Rev.* 85, 69-201.

Atkinson, R. (1987a) Estimation of OH radical reaction rate constants and atmospheric life times for polychlorobiphenyls, dibenzo-p-dioxins and dibenzofurans. *Environ. Sci. Technol.* 21, 305-307.

Atkinson, R. (1987b) A structure-activity relationship for the estimation of rate constants for the gas phase reaction of OH radicals with organic compounds. *Int'l. J. Chem. Kinetics* 19, 799-828.

Atkinson, R., Aschmann, S.M., Winner, A. M., Jr., Pitts, J.N. (1985) Atmosphere gas phase loss process of chlorobenzene, benzotrifluoride, and 4-chlorobenzotrifluoride and generalization of predictive technique for atmospheric life times of aromatic compounds. *Archiv. Environ. Contam. Toxicol.* 14, 417-425.

Atlas, E., Foster, R., Giam, C.S. (1982) Air-sea exchange of high molecular weight organic pollutants: laboratory studies. *Environ. Sci. Technol.* 16, 283-286.

Bahnick, D.A., Doucette, W.J. (1988) Use of molecular indices to estimate soil sorption coefficient for organic chemicals. *Chemosphere* 17, 1703-1715.

Ballschmiter, K., Wittlinger, R. (1991) Interhemisphere exchange of hexachlorohexanes, hexachlorobenzene,polychlorobiphenylsand1,1,1-trichloro-2,2-bis(p-chlorophenyl)ethane in the lower troposphere. *Environ. Sci. Technol.* 25(6), 1103-1111.

Banerjee, S. (1984) Solubility of organic mixture in water. *Environ. Sci. Technol.* 18, 587-591.

Banerjee, S. (1985) Calculation of water solubility of organic compounds with UNIFAC-derived parameters. *Environ. Sci. Technol.* 19, 369-370.

Banerjee, S., Howard, P.H. (1988) Improved estimation of solubility and partitioning through correction of UNIFAC-derived activity coefficients. *Environ. Sci. Technol.* 22, 839-841.

Banerjee, S., Howard, P.H., Lande, S.S. (1990) General structure vapor pressure relationship for organics. *Chemosphere* 21, 1173-1180.

Banerjee, S., Howard, P.H., Rosenberg, A.M., Dombrowski, A.E., Solla. H., Tullis, D.L. (1984) Development of a general kinetic model for biodegradation and its application to chlorophenols and related compounds. *Environ. Sci. Technol.* 18, 416-422.

Banerjee, P., Piwoni, M.D., Ebeid, K. (1985) Sorption of organic contaminants to a low carbon subsurface core. *Chemosphere* 14, 1057-1067.

Banerjee, S., Sugatt, R.H., O'Grady, D.P.(1984) A simple method for determining bioconcentration parameters of hydrophobic compounds. *Environ. Sci. Technol.* 18, 79-81.

Banerjee, S., Yalkowsky, S.H., Valvani, S.C. (1980) Water solubility and octanol/water partition coefficients of organics. Limitations of the solubility-partition coefficient correlation. *Environ. Sci. Technol.* 14, 1227-1229.

Barber, M.G., Suarez, L.A., Lassiter, R.R. (1988) Modelling bioconcentration of nonpolar organic pollutants by fish. *Environ. Toxicol. Chem.* 7, 545-558.

Barrows M.E., Petrocelli, S.R., Macek, K.J. (1980) Bioconcentration and elimination of selected water pollutants by bluegill sunfish (*Lepomis macrochirus*). In: *Dynamic, Exposure, Hazard Assessment Toxic Chemicals.* Haque, R. Ed., pp. 379-392, Ann Arbor Science Publisher Inc., Ann Arbor, Michigan.

Bartholomew, G.W., Pfaender, F.K. (1983) Influence of spatial and temporal variations on organic pollutant biodegradation rates in an estuarine environment. *Appl. Environ. Microbiol.* 45, 103-109.

Battersby, N.S. (1990) A review of biodegradation kinetics in the aquatic environment. *Chemosphere* 21, 1243-1284.

Baughman, G.L., Paris, D.F. (1981) Microbial bioconcentration of organic pollutants from aquatic systems - a critical review. *CRC Critical Reviews in Microbiology*, pp. 205-228.

Beck, J., Hansen, K.E. (1974) The degradation of quintozene, pentachlorobenzene, hexachlorobenzene and pentachloroaniline in soil. *Pestic. Sci.* 5, 41-8.

Bidleman, T.F. (1984) Estimation of vapor pressures for nonpolar organic compounds by capillary gas chromatography. *Anal. Chem.* 56, 2490-2496.

Bidleman, T.F., Foreman, W.T. (1987) Vapor-particle partitioning of semivolatile organic compounds. In: *Sources and Fates of Aquatic Pollutants.* Hite, R.A., Eisenreich, S.J., Eds., Advances in Chemistry Series 216, American Chemical Society, Washington, D.C.

Bobra, A.M., Shiu, W.Y., Mackay, D. (1985) Quantitative structure-activity relationships for the acute toxicity of chlorobenzenes to *Daphnia Magna. Environ. Toxical. Chem.* 4, 297-305.

Booth, H.S., Everson, H.E. (1948) Hydrotropic solubilities: solubilities in 40 per cent sodium xylenesulfonate. *Ind. Eng. Chem.* 40(8), 1491-1493.

Boublik, T., Fried, V., Hala, E. (1973) *The Vapour Pressure of Pure Substances.* Elsevier, Amsterdam.

Boublik T. et al. (1984) *Dynamics, Exposure, Hazard Assessment Toxic Chemicals.* p.379-392.

Boublik, T., Fried, V., Hala, E. (1984) *The Vapour Pressures of Pure Substances.* (second revised edition), Elsevier, Amsterdam.

Briggs, G.G. (1981) Theoretical and experimental relationships between soil adsorption, octanol-water partition coefficients, water solubilities, bioconcentration factors, and the Parachor. *J. Agric. Food Chem.* 29, 1050-1059.

Brooke, D.N., Dobbs, A.J., Williams, N. (1986) Octanol/water partition coefficients (P): Measurement, estimation, and interpretation, particularly for chemicals with P > 10^5. *Ecotoxicol. Environ. Safety* 11, 251-260.

Brown, S., Chan, F., Jones, J., Liu, D., McCalab, K., Mill, T., Supios, K., Schendel, D. (1975) *Research program on hazard priority ranking of manufactured chemicals: Phase II. Final Report: Chemicals 1-19 & 21-40.* Stanford Research Institute, Menlo Park, California.

Bruggeman, W.A., Van Der Steen, J., Hutzinger, O. (1982) Reversed-phase thin-layer chromatography of polynuclear aromatic hydrocarbons and chlorinated biphenyls. Relationship with hydrophobicity as measured by aqueous solubility and octanol-water partition coefficient. *J. Chromatogr.* 238, 335-346.

Bruggeman, W.A., Opperhuizen, A. Wizbeuga, A., Hutzinger, O. (1984) Bioaccumulation of super-lipophilic chemicals in fish. *Toxicol. Environ. Chem.* 7, 173-189.

Budavari, S., Ed. (1989) *The Merck Index. An Encyclopedia of Chemicals, Drugs and Biologicals.* 11th Edition, Merck & Co. Inc., Rahway, New Jersey.

Burkhard, L.P., Kuehl, D.W., Veith, G.D. (1985) Evaluation of reversed phase LC/MS for estimation of n-octanol/water partition coefficients of organic chemicals. *Chemosphere* 14, 1551-1560.

Burkhard, L.P., Kuehl, D.W. (1986) n-Octanol/water partition coefficients by reversed phase liquid chromatography/mass spectrometry for eight tetrachlorinated planar molecules. *Chemosphere* 15, 163-167.

Bysshe, S.E. (1982) Bioconcentration factor in aquatic organisms. In: *Handbook of Chemical Property Estimation Methods.* Lyman, W.J., Reehl, W.F., Rosenblatt, D.H. Editors, Chapter 5, Ann Arbor Sci., Ann Arbor, Michigan.

Cadena, F. (1984) Removal of volatile organic pollutants from rapid streams. *J. Water Pollut. Control Fed.* 460-463.

Calamari, D., Galassi, S., Sette, F., Vighi, M. (1983) Toxicity of selected chlorobenzenes to aquatic organisms. *Chemosphere* 12, 253-262.

Callahan, M.A.,Slimak, M.W., Gabel, N.W., May, I.P., Fowler, C.F., Freed, J.R., Jennings, P., Durfee, R.L., Whitmore, F.C., Maestri, B., Mabey, W.R., Holt, B.R., Gould, C. (1979) *Water Related Environmental Fate of 129 Priority Pollutants.* EPA-440-4-79-029a,b. Versar, Inc., Springfield, Virginia.

Canton, J.H., Sloof, W., Kool, H.J., Struys, J., Pouw, T.J.M., Wegman, R.C.C., Piet, G.J. (1985) Toxicity biodegradability and accumulation of a number of chlorine/nitrogen containing compounds for classification and establishing water quality criteria. *Regul. Toxicol. Pharmocol.* 5, 123- 31.

Carlson, A.R., Kosian, P.A. (1987) Toxicity of chlorinated benzenes to fathead minnows *(pimephales promelas)*. *Arch. Environ. Contam. Toxicol.* 16, 129-135.

Chey, W., Calder, G.V. (1972) Method for determining solubility of slightly soluble organic compounds. *J. Chem. Eng. Data* 17, 199.

Chin, Y.P., Weber, Jr., W.J., Voice, T.C. (1986) Determination of partition coefficient and aqueous solubilities by reversed phase chromatography-II. Evaluation of partitioning and solubility models. *Water Res.* 20, 1443-1450.

Chin, Y.-P., Weber, Jr., W.J. (1989) Estimating the effects of dispersed organic polymers on the sorption of contaminants by natural solids. 1. A predictive thermodynamic humic substance-organic solute interaction model. *Environ. Sci. Technol.* 23, 976-984.

Chin, P.-Y., Weber, Jr., W.J., Eadie, B.J. (1990) Estimating the effects of dispersed organic polymers on the sorption of contaminants by natural solids. 2. Sorption in the presence of humic and other natural macromolecules. *Environ. Sci. Technol.* 24, 837-842.

Chiou, C.T. (1981) Partition coefficient and water solubility in environmental chemistry. In: *Hazard Assessment of Chemicals. Current Development,* Vol.1, 117-153. Academic Press, Inc., New York.

Chiou, C.T. (1985) Partition coefficients of organic compounds in lipid-water systems and correlations with fish bioconcentration factors. *Environ. Sci. Technol.* 19, 57-62.

Chiou, C.T., Freed, V.H. (1977) *Chemodynamic studies on bench mark industrial chemicals.* NSF/RA-770286, National Science Foundation, Washington D.C.

Chiou, C.T., Schmedding, D.W. (1981) Measurement and interrelation of octanol-water partition coefficient and water solubility of organic chemicals. In: *Test Protocols for Environmental Fate and Movement of Toxicants. J. Assoc. Anal. Chem.* 28-42, Arlington, Virginia.

Chiou, C.T., Freed, V.H., Schmedding, D.W., Kohnert, R.L. (1977) Partition coefficient and bioaccumulation of selected organic chemicals. *Environ. Sci. Technol.* 11, 475-478.

Chiou, C.T., Kile, D.E., Rutherford, D.W. (1991) The natural oil in commercial linear alkylbenzenesulfonate and its effect on organic solute solubility in water. *Environ. Sci. Tecnol.* 25(4), 660-665.

Chiou, C.T., Malcolm, R.L., Brinton, T.I., Kile, D.E. (1986) Water solubility enhancement of some organic pollutants and pesticides by dissolved humic and fulvic acids. *Environ. Sci. Technol.* 20, 502-508.

Chiou, C.T., Peters, L.J., Freed, V.H. (1979) A physical concept of soil-water equilibria for nonionic organic compounds. *Science* 206, 831-832.

Chiou, C.T., Freed, V.H., Peters, L.J., Kohnert, H.L. (1980) Evaporation of solutes from water. *Environ. Int'l.* 3, 231-236.

Chiou, C.T., Porter, P.E., Schmedding, D.W. (1983) Partition equilibria of nonionic organic compounds between soil organic matter and water. *Environ. Sci. Technol.* 17, 227-231.

Chiou, C.T., Schmedding, D.W., Manes, M. (1982) Partitioning of organic compounds in octanol-water system. *Environ. Sci. Technol.* 16, 4-10.

Chiou, C.T., Shoup, T.D. (1985) Soil sorption of organic vapors and effects of humidity on sorptive mechanism and capacity. *Environ. Sci. Technol.* 19, 1196-1200.

Ciam, C.S., Murray, H.E., Ray, L.E., Kira, S. (1980) Bioaccumulation of hexachlorobenzene in killifish. *Bull. Environ. Contam. Toxic.* 25, 891-897.

Clark, K.E., Gobas, F.A.P.C., Mackay, D. (1990) Model of organic chemical uptake and clearance by fish from food and water. *Environ. Sci. Technol.* 24(8), 1203-1213.

Connell, D.W., Hawker, D.W. (1988) Use of polynomial expressions to describe the bioconcentration of hydrophobic chemicals by fish. *Ecotoxicol. Envrion. Safety* 16, 242-257.

D'Amboise, M.,Hanai, T. (1982) Hydrophobicity and retention in reversed phase liquid chromatography. *J. Liq. Chromatogr.* 5, 229-244.

Darnall, K.R., Loyld, A.C., Winer, A.M., Pitts, J.N. (1976) Reactivity scale for atomospheric hydrocarbons based on reaction with hydroxy radical. *Environ. Sci. Technol.* 10, 692-696.

Daubert, T.E., Danner, R.P. (1985) *Data Compilation Tables of Properties of Pure Compounds.* Am. Inst. of Chem. Engineers, pp.450.

Davies, R.P., Dobbs, A.J. (1984) The prediction of bioconcentration in fish. *Water Res.* 18 (10), 1253-1262.

De Bruijn, J., Busser, F., Seinen, W., Hermens, J. (1989) Determination of octanol/water partition coefficients for hydrophobic organic chemicals with the "slowing-stirring" method. *Environ. Toxicol. Chem.* 8, 499-512.

De Bruijn, J., Hermens, J. (1990) Relationships between octanol/water partition coefficients and total molecular surface area and total molecular volume of hydrophobic organic chemicals. *Quant. Struct.-Act. Relat.* 9, 11-21.

De Kock, A.C., Lord, D.A. (1987) A simple procedure for determining octanol-water partition coefficients using reversed phase high performance liquid chromatography (RPHPLC). *Chemosphere* 16(1), 133-142.

Dean, J.D., Ed. (1979) *Lange's Handbook of Chemistry.* 12th ed., McGraw-Hill, Inc., New York.

Dean, J.D., Ed. (1985) *Lange's Handbook of Chemistry.* 13th ed. McGraw-Hill, Inc., New York.

Dearden, J.C. (1990) Physico-chemical descriptors. In: *Practical Applications of Quantitative Structure-Activity Relationships (QSAR) in Environmental Chemistry and Toxicology.* pp.25-29. W. Karcher & J. Devillers, Eds. ECSC, EEC, EAEC, Brussels and Luxembourg.

Dearden, J.C., Bresnen, G.M. (1988) The measurement of partition coefficients. *Quant. Strut. Act. Relat.* 7, 133-144.

Deno, N.C., Berkheimer, H.E. (1960) Phase equilibria molecular transport thermodynamics: activity coefficients as a function of structure and media. *J. Chem. Eng. Data* 5, 1-5.

DiToro, D.M., O'Connor, D.J., Thomann, R.V., St. John, J.P. (1981) *Analysis of Fate of Chemicals in Receiving Waters.* Phase I, Hydroqual Inc., prepared for the Chemical Manufacturers Association, Washington, D.C., May, 1981.

Dobbs, A.J., Cull, M.R. (1982) Volatilization of chemicals-relative loss rates and the estimation of vapor pressures. *Environ. Pollut.* (series B) 3, 289-298.

Dobbs, A.J., Hart, G.F., Parsons, A.H. (1984) The determination of vapour pressures from relative volatilization rates. *Chemosphere* 13(5/6), 687-692.

Dobbs, A.J., Williams, N. (1983) Fat solubility-A property of environmental relevance? *Chemosphere* 12(1), 97-104.

Dobbs, R.A., Wang, L., Govind, R. (1989) Sorption of toxic organic compounds on wastewater solids: correlation with fundamental properties. *Environ. Sci. Technol.* 23, 1092-1097.

Doolittle, A.K. (1935) Lacquer solvents in commercial use. *Ind. Eng. Chem.* 27, 1169-1179.

Dorfman, L.M., Adams, G.E. (1973) *Reactivity of the hydroxy radical in aqueous solution.* NSRD-NDB-46. NTIS COM-73-50623. Washington, DC. National Bureau of Standards. pp. 51.

Doucette, W.J., Andren, A.W. (1988). Estimation of octanol/water partition coefficients: Evaluation of six methods for highly hydorphobic aromatic hydrocarbons. *Chemosphere* 17:345-359.

Dreisbach, R.R. (1955) *Physical Properties of Chemical Compounds.* Am. Chem. Soc. Adv. Chem. Ser. 15, Washington DC.

Dulin, D., Drossman, H., Mill, T. (1986) Products and quantum yields for photolysis of chloroaromatics in water. *Environ. Sci. Technol.* 20, 72-77.

Eadie, B.J., Robbins, J.A. (1987) The role of particulate matter in the movement of contaminants in the Great Lakes. In: *Sources and Fate of Aquatic Pollutants.* R.A. Hite, S.J. Eisenreich, Eds., pp.320-364. Advances in Chemistry Series 216, Am. Chem. Soc. Washington DC.

Eadsforth, C.V., Moser, P.(1983) Assessment of reversed phase chromatographic methods for determining partition coefficients. *Chemosphere* 12, 1459-1475.

Eisenreich, S.J., Looney, B.B., Thornton, J.D. (1981) Airborne organic contaminants in the Great Lakes ecosystem. *Environ. Sci. Technol.* 15, 30-38.

Ellgehausen, H., D'Hondt, C., Fuerer, R. (1981) Reversed-phase chromatography as a general method for determining octanol/water partition coefficients. *Pestic. Sci.* 12, 219.

Ellington, J.J., Stancil, F.E., Payne, W.D. (1987) *Measurements of hydrolysis rate constants for evaluation of hazardous waste land disposal. Vol. 1. data on 32 chemicals*. U.S. EPA-600/3-86-043. NTIS PB87-140349/GAR.

Ellington, J.J., Stancil, F.E., Peyne, W.D., Trusty, C.D. (1988) *Measurements of hydrolysis rate constants for evaluation of hazadous waste land disposal: vol. 3, data on 70 chemicals* (preprint). EPA/600/s3-88/028. NTIS PB 88-234042/AS.

Elzerman, A.W., Coates, J.T. (1987) Hydrophobic organic compounds in sediments: equilibria and kinetics of sorption. In: *Sources and Fate of Aquatic Pollutants*. R.A. Hite, S.J. Eisenreich, Eds., pp.263-317. Advances in Chemistry Series 216, Am. Chem. Soc. Washington DC.

Farmer, W.J., Yang, M.S., Spencer, W.F. (1980) Hexachlorobenzene: its vapor pressure and vapor phase diffusion in soil. *Soil Sci. Soc. Am. J.* 44, 676-680.

Figueroa, I. del C., Simmons, M.S. (1991) Structure-activity relationships of chlorobenzenes using DNA measurement as a toxicity parameter in algae. *Environ. Toxicol. Chem.* 10, 323-329.

Freed, V.H., Chiou, C.T., Haque, R. (1977) Chemodynamics: transport and behavior of chemicals in the environment - a problem in environmental health. *Environ. Health Perspect.* 20, 55-70.

Freitag, D., Lay, J.P. Korte, F. (1984) Environmental hazard profile-test results as related to structures and translation into the environment. In: *QSAR in Environmental Toxicology*. Kaiser, K.L.E., Ed., D. Reidel Publ. Co., Dordrecht, Netherlands.

Freitag, D., Ballhorn, L., Geyer, H., Korte, F. (1985) Environmental hazard profile of organic chemicals. An experimental method for the assessment of the behaviour of chemicals in the ecosphere by simple laboratory tests with C-14 labelled chemicals. *Chemosphere* 14, 1589-1616.

Fujita, T., Iwasa, J., Hansch, C. (1964) A new substituent constant, "pi" derived from partition coefficients. *J. Am. Chem. Soc.* 86, 5175-5180.

Garst, J.E. (1984) Accurate, wide-range, automated, high-performance liquid chromatographic method for the estimation of octanol/water partition coefficients. II: Equilibrium in partition coefficient measurements, additivity of substituent constants, and correlation of biological data. *J. Pharm. Sci.* 73, 1623-1629.

Garst, J.E., Wilson, W.C. (1984) Accurate, wide-range, automated, high-performance chromatographic method for the estimation of octanol/water partition coefficients. I: Effect of chromatographic conditions and procedure variables on accuracy and reproducibility of the method. *J. Pharm. Sci.* 73, 1616-1622.

Garten, Jr., C.T., Trabalka, J.R. (1983) Evaluation of models for predicting terrestrial food chain behavior of xenobiotics. *Environ. Sci. Technol.* 17, 590-595.

Geyer, H., Kraus, A.G., Klein, W., Richter, E., Korte, F. (1980) Relationship between water solubility and bioaccumulation potential of organic chemicals in rats. *Chemosphere* 9, 277-291.

Geyer, H., Visvanathan, R., Freitag, D., Korte, F. (1981) Relationship between water solubility of organic chemicals and their bioaccumulation by the *Alga Chlorella*. *Chemosphere* 10, 1307-1313.

Geyer, H., Politzki, Freitag, D. (1984) Prediction of ecotoxicological behaviour of chemicals: relationship between n-octanol/water partition coefficient and bioaccumulation of organic chemicals by *Alga Chlorella*. *Chemosphere* 13, 269-184.

Geyer, H., Scheunert, I., Korte, F. (1985) Relationship between the lipid content of fish and their bioconcentration potential of 1,2,4-trichlorobenzene. *Chemosphere* 14, 545-555.

Geyer, H. J., Scheunert, I., Korte, F. (1987) Correlation between the bioconcentration potential of organic environmental chemicals in humans and their n-octanol/water partition coefficients. *Chemosphere* 16(1), 239-252.

Gibson, D.T. (1976) Microbial degradation of carcinogenic hydrocarbons and related compounds. In: *Proceedings of Symposium on Sources, Effects and Sinks of Hydrocarbons in the Aquatic Environment.* pp.224-238, American Institute of Biological Sciences, Washington, D.C., August 1976.

Ginnings, P.M., Hering, E., Coltrane, D.J. (1939) *J. Am. Chem. Soc.* 61, 807.

Gobas, F.A.P.C., Shiu, W.Y., Mackay, D. (1987) Factors determining partitioning of hydrophobic organic chemicals in aquatic organisms. In: *QSAR in Environmental Toxicology* II. Kaiser, K.L.E., Ed., D. Reidel Publ. Co., Dordrecht, Holland.

Gobas, F.A.P.C., Clark, K., Shiu, W.Y., Mackay, D. (1989a) Bioconcentration of polybrominated benzenes and biphenyls and related superhydrophobic chemicals in fish: role of bioavailability and elimination into the feces. *Environ. Toxicol. Chem.* 8, 231-245.

Gobas, F.A.P.C., Bedard, D.C., Ciborowski, J.J.H. (1989b) Bioconcentration of chlorinated hydrocarbons by the mayfly *(Hexgenia Limbata)* in Lake St. Clair. *J. Great Lakes Res.* 15(4), 581-588.

Govers, H.A.J., Luijk, R. Evers, E.H.G. (1990) Calculation of heat of vaporization, molar volume and solubility parameters of polychlorodibenzo-p-dioxins. *Chemosphere* 30(3/4), 287-294.

Grayson, B.T., Fosbracy, L.A. (1982) Determination of the vapour pressure of pesticides. *Pest. Sci.* 13(3), 269-278.

Gross, P.M., Saylor, J.H. (1931) The solubilities of certain slightly soluble organic compounds in water. *J. Am. Chem. Soc.* 53, 1744-1751.

Gückel, W., Kstel, R., Lewerenz, J., Synnatschke, G. (1982) A method for determining the volatility of active ingredients used in plant protection. Part III: The temperature relationship between vapor pressure and evaporation rate. *Pest. Sci.* 13, 161-168.

Güesten, H., Filby, W.G., Schoop, S. (1981) Prediction of hydroxy radical reaction rates with organic compounds in the gas phase. *Atmos. Environ.* 15, 1763-1765.

Hafkenscheid, T.L., Tomlinson, E. (1983a) Isocratic chromatographic retention data for estimating aqueous solubilities of acidic, basic and neutral drugs. *Intl. J. Pharm.* 16, 1-21.

Hafkenscheid, T.L., Tomlinson, E. (1983b) Correlations between alkane/water and octan-1-ol/water distribution coefficients and isocratic reversed phase liquid chromatographic capacity factor of acids, bases and neutrals. *Intl. J. Pharm.* 16, 225-240.

Haider, K., Jagnow, G., Kohnen, R., Lim, S.U. (1981) Degradation of chlorinated benzenes, phenols and cyclohexane derivatives by benzene and phenol utilizing soil bacteria under aerobic conditions. In: *Decomposition of Toxic and Non-toxic Organic Compounds in Soil.* Overcash, V.R., Ed., pp. 207-223, Ann Arbor Sci. Publ., Ann Arbor, Michigan.

Haky, J.E., Young, A.M. (1984) Evaluation of a simple HPLC correlation method for the estimation of the octanol-water partition coefficients of organic compounds. *J. Liq. Chromatogr.* 7, 675-689.

Halfon, E., Reggiani, M.G. (1986) On ranking chemicals for environmental hazard. *Environ. Sci. Technol.* 20, 1173-1179.

Hambrick, G.A., Delaune, R.D., Patrick, W.H., Jr. (1980) Effects of estuarine sediment, pH and oxidation-reduction potential on microbial hydrocarbon degradation. *Appl. Environ. Microbiol.* 40, 365-9.

Hammers, W.E., Meurs, G.J., De Ligny, C.L. (1982) Correlations betweeen liquid chromatographic capacity ratio data on Lichrosorb RP-18 and partition coefficients in the octanol-water system. *J. Chromatogr.* 247, 1-13.

Hanai, T., Tran, C., Hrbert, J. (1981) An approach to the prediction of retention times in liquid chromatography. *J. High Resolution Chromatography & Chromatography Communication* (J. HRC & CC) 4, 454-460.

Hansch, C., Leo, A. (1979) *Substituent Constants for Correlation Analysis in Chemistry and Biology.* Wiley, New York.

Hansch, C., Leo, A. (1985) *Medchem Project Issue No. 26.* Pomona College, Claremont, California.

Hansch, C., Quinlan, J.E., Lawrence, G.L (1968) The linear free-energy relationship between partition coefficients and the aqueous solubility of organic liquids. *J. Org. Chem.* 33, 347-350.

Haque, R., Falco, J., Cohen, S., Riordan, C. (1980) Role of transport and fate studies in exposure, assessment and screening of toxic chemicals. In: *Dynamics, Exposure and Hazard Assessment of Toxic Chemicals.* Haque, R., Ed., pp. 47-67. Ann Arbor Sci. Publishing, Ann Arbor, Michigan.

Harnisch, M., Möckel, H.J., Schulze, G. (1983) Relationship between log P_{OW} shake-flask values and capacity factors derived from reversed HPLC for n-alkylbenzenes and some OECD reference substances. *J. Chromatogr.* 282, 315-332.

Hashimoto, Y., Tokura, K., Ozaki, K., Strachan, W.M.J. (1982) A comparison of water solubility by the flask and micro-column methods. *Chemosphere* 11 (10), 991-1001.

Hawker, D. (1989) The relationship between octan-1-ol/water partition coefficient and aqueous solubility in terms of solvatochromic parameters. *Chemosphere* 19(10/11), 1585-1593.

Hawker, D. (1990) Description of fish bioconcentration factors in terms of solvatochromic parameters. *Chemosphere* 20, 467-477.

Hawker, D.W., Connell, D.W. (1985) Relationships between partition coefficient uptake rate constant, clearance rate constant and time to equilibration for bioaccumulation. *Chemosphere* 14, 1205-1219.

Hawker, D.W., Connell, D.W. (1986) Bioconcentration of lipophilic compounds by some aquatic organisms. *Ecotox. Environ. Safety* 11, 184-197.

Hawker, D.W., Connell, D.W. (1988a) Octanol/water partition coefficient of polychlorinated biphenyl congeners. *Environ. Sci. Technol.* 22, 382-387.

Hawker, D.W., Connell, D.W. (1988b) Influence of partition coefficient of lipophilic compounds on bioconcentration kinetics with fish. *Water Res.* 22, 701-707.

Hawker, D.W., Connell, D.W. (1989) A simple water/octanol partition system for bioconcentration investigations. *Environ. Sci. Technol.* 23, 961-965.

Hinckley, D.A., Bidleman, T.F., Foreman, W.T. (1990) Determination of vapor pressures for nonpolar and semipolar organic compounds from gas chromatographic retention data. *J.Chem. Eng. Data* 35, 232-237.

Hine, J., Mookerjee, P.K. (1975) The intrinsic hydrophilic character of organic compounds. Correlations in terms of structural contributions. *J. Org. Chem.* 40, 292-298.

Hodson, P.V., Dixon, D.G., Kaiser, K.L.E. (1984) Measurement of median lethal dose as a rapid indication of contaminant toxicity to fish. *Environ. Toxicol. Chem.* 3, 243-254.

Hodson, P.V., Dixon, D.G., Kaiser, K.L.E. (1988) Estimating the acute toxicity of waterborne chemicals in trout from measurements of median lethal dose and the octanol-water partition coefficient. *Environ. Toxicol. Chem.* 7, 443-454.

Hodson, J., Williams, N.A. (1988) The estimation of the adsorption coefficient (K_{oc}) for soils by high performance liquid chromatography. *Chemosphere* 17, 67-77.

Hollifield, H.C. (1979) Rapid nephelometric estimate of water solubility of highly insoluble organic chemicals of environmental interests. *Bull. Environ. Contam. Toxicol.* 23, 579-586.

Horvath, A.L. (1982) *Halogenated Hydrocarbons, Solubility-Miscibility with Water.* Marcel Dekker, Inc., New York, N.Y.

Howard, P.H., Ed. (1989) *Handbook of Fate and Exposure Data for Organic Chemicals.* Vol. I - Large Production and Priority Pollutants. Lewis Publishers, Chelsea, Michigan.

Howard, P.H., Boethling, R.S., Jarvis, W.F., Meylan, W.M., Michalenko, E.M., Eds. (1991) *Handbook of Environmental Degradation Rates.* Lewis Publ., Inc., Chelsea, Michigan.

Hungspreugs, M., Silpapat, S., Tonapong, C., Lee, R.F., Windom, H.L., Tenore, K.R. (1984) Heavy metals and polycyclic hydrocarbon compounds in benthic organisms of the upper Gulf of Thailand. *Mar. Pollut. Bull.* 15, 213-218.

Irmann, F. (1965) Eine einfache korrelation zwischen wasserlöslichkeit und struktur vor kohlenwasserstoffen und hologen kohlen wasserstoffen. *Chem.-Ing.-Tech.* 37(8), 789-798.

Isnard, P., Lambert, S. (1988) Estimating bioconcentration factors from octanol-water partition coefficient and aqueous solubility. *Chemosphere* 17, 21-34.

Isnard, P., Lambert, S. (1989) Aqueous solubility/n-octanol water partition coefficient correlations. *Chemosphere* 18, 1837-1853.

IUPAC Solubility Data Series (1985) Vol. 20: *Halogenated Benzenes, Toluenes and Phenols with Water.* Horvath, A.L., Getzen, F.W. Eds., Pergamon Press, Oxford, England.

Jaber, H.M., Smith, J.H., Cwirla, A.N. (1982) *Evaluation of gas saturation methods to measure vapor pressure.* (EPA Contract No. 68-01-5117) SRI International, Menlo Park, CA.

Johnson, H. (1982) In: *Aquatic Fate Process Data for Organic Priority Pollutants.* Mabey, W.R., Ed., *EPA Final Report on Contract* No. 68-01-3867; U.S. Government Printing Office, Washington, D.C.

Kaiser, K.L.E. (1983) A non-linear function for the approximation of octanol/water partition coefficients of aromatic compounds with multiple chlorine substitution. *Chemosphere* 12, 1159-1167.

Kaiser, K.L.E., Dixon, D.G., Hodson, P.V. (1984) QSAR studies on chlorophenols, chlorobenzenes and para-substituted phenols. In: *QSAR in Environmental Toxicology.* Kaiser, K.L.E., Ed., pp.189-206, D. Reidel Publ. Co., Dordrecht, Netherlands.

Kamlet, M.J., Doherty, R.M., Abraham, M.H., Carr, P.W., Doherty, R.F., Taft, R.W. (1987) Linear solvation energy relationships. 42. Important differences between aqueous solubility relationships for aliphatic and aromatic solutes. *J. Phys. Chem.* 91, 1996.

Kamlet, M.J., Doherty, R.M., Carr, P.W., Mackay, D., Abraham, M.H., Taft, R.W. (1988) Linear solvation energy relationship. 44. Parameter estimation rules that allow accurate prediction of octanol/water partition coefficients and other solubility and toxicity properties of polychlorinated biphenyls and polycyclic aromatic hydrocarbons. *Environ. Sci. Technol.* 22, 503-509.

Karickhoff, S.W. (1981) Semiempirical estimation of sorption of hydrophobic pollutants on natural sediments and soils. *Chemosphere* 10, 833-846.

Karickhoff, S.W., Morris, K.R. (1985a) Impact of tubificid oligochaetes on pollutant transport in bottom sediments. *Environ. Sci. Technol.* 19(1), 51-56.

Karickhoff, S.W., Morris, K.R. (1985b) Sorption dynamics of hydrophobic pollutants in sediment suspensions. *Environ. Toxicol. Chem.* 4, 469-479.

Kenaga, E.E. (1980a) Predicted bioconcentration factors and soil sorption coefficients of pesticides and other chemicals. *Ecotoxicol. Environ. Safety* 4, 26-38.

Kenaga, E.E. (1980b) Correlation of bioconcentration factors of chemicals in aquatic and terrestrial organisms with their physical and chemical properties. *Environ. Sci. Technol.* 14, 553-556.

Kenaga, E.E., Goring, C.A.I. (1980) In: *Aquatic Toxicology.* Eaton, J.G., Parrish, P.R., Hendricks, A.C. Eds., Am. Soc. for Testing and Materials, STP 707, pp. 78-115.

Kilzer, L., Scheunert, I., Geyer, H., Klein, W., Korte, F. (1979) Laboratory screening of the volatilization rates of organic chemicals from water and soil. *Chemosphere* 10, 751-761.

Kincannon, D.F. et al. (1983) Removal mechanisms for toxic priority pollutants. *J. Water Pollut. Control Fed.* 55, 157-163.

Kishi, H., Hashimoto, Y. (1989) Evaluation of the procedures for the measurement of water solubility and n-octanol/water partition coefficient of chemicals. *Chemosphere* 18, 1749-1759.

Kishi, H., Kogure, N., Hashimoto, Y. (1990) Contribution of soil constituents in adsorption coefficient of aromatic compounds, halogenated alicyclic and aromatic compounds in soil. *Chemosphere* 12(7), 867-876.

Kishii, H., Nakamura, M., Hashimoto, Y. (1987) Prediction of solubility of aromatic compounds in water by using total molecular surface area. *Nippon Kagaku Kaishi* 8, 1615-1622.

Klein, W. Geyer, H., Freitag, D., Rohleder, H. (1984) Sensitivity of schemes for ecotoxicological hazard banking of chemicals. *Chemosphere* 13(1), pp 203-211.

Klein, A.W., Harnish, M., Porenski, H.J., Schmidt-Bleek, F. (1981) OECD chemicals testing programme physico-chemical tests. *Chemosphere* 10, 153-207.

Klemenc, A., Löw, M. (1930) Die löslichkeit in wasser und ihr zusammenhang der drei dichlorbenzole. Eine methode zur bestimmung derlöslichkeit sehr wenig löslicher und zuglich sehr flüchtiger stöffe. *Rec. Trav. Chim. Pays-Bas.* 49(4), 629-640.

Koch, R. (1983) Molecular connectivity index for assessing ecotoxicological behaviour of organic compounds. *Toxicol. Environ. Chem.* 6, 87-96.

Könemann, W.H. (1979) *Quantitative Structure Activity Relationship for Kinetics and Toxicity of Aquatic Pollutants and Their Mixtures in Fish.* Ph.D. Thesis, University Utrecht, Netherlands.

Könemann, H. (1981) Quantitative structure-activity relationships in fish toxicity studies. Part 1: Relationship for 50 industrial pollutants. *Toxicology* 19, 209-221.

Könemann, H., van Leeuwen, K. (1980) Toxicokinetics in fish: accumulation and elimination of six chlorobenzenes by guppies. *Chemosphere* 9, 3-19.

Könemann, H., Zelle, R., Busser, F. (1979) Determination of log P_{oct} values of chloro-substituted benzenes, toluenes and anilines by high-performance liquid chromatography on ODS-sillca. *J. Chromatogr.* 178, 559-565.

Körte, F., Freitag, D., Geyer, H., Klein, W., Kraus, A.G., Lahaniatis, E. (1978) Ecotoxicologic profile analysis - a concept for establishing ecotoxicologic priority lists for chemicals. *Chemosphere* No.1, pp 79-102.

Lande, S.S., Banerjee, S. (1981) Predicting aqueous solubility of organic nonelectrolytes from molar volume. *Chemosphere* 10, 751-759.

Landolt-Börnstein (1951) *Zahlenwerte und Funktionen aus Physik, Chemie, Astronomie, Geophysik, und Technik* (6th ed.) Vol. 1, Atom- und Molekularphysik, Part 3, Molekeln II. pp. 509-517, Springer-Verlag, Berlin.

Leahy, D.E. (1986) Intrinsic molecular volume as a measure of the cavity term in linear solvation energy relationships: octanol-water partition coefficients and aqueous solubilties. *J. Pharm. Sci.* 75, 629-636.

Lee, J-F., Crum, J.R., Boyd, S.A. (1989) Enhanced retention of organic contaminants by soils exchanged with organic cations. *Environ. Sci. Technol.* 23, 1365-1372.

Lee, R.F., Ryan, C. (1976) Biodegradation of petroleum hydrocarbons by marine microbes. In: *Proc. Int. Biodegradation Symp.* 3rd. 1975, pp.119-125.

Lee, R.F., Ryan, C. (1979) Microbial degradation of organochlorine compounds in estuarine waters and sediments. In: *Proceedings of the Workshop of Microbial Degradation of Pollutants in Marine Environments. EPA*-600/9-79-012. Washington D.C.

Leegwater, D.C. (1989) QSAR-analysis of acute toxicity of industrial pollutants to the guppy using molecular connectivity indices. *Aqua. Toxicol.* 15:157-168.

Leighton, D.T., Calo, J.M. (1981) Distribution coefficients of chlorinated hydrocarbons in dilute air-water systems for groundwater contamination applications. *J. Chem. Eng. Data* 26, 382-385.

Leo, A. (1985) *Medchem. Project. Issue* No.26, Pomona College, Claremont, CA.

Leo, A., Hansch, C., Elkins, D. (1971) Partition coefficients and their uses. *Chemical Reviews* 71, 525-616.

Lindenberg, M.A.B. (1956) Physicochimie des solutions-sur une relation simple entrele volume molieulaire et la solubilité dans l'eau des hydrocarbures et dérivés halogénés. *Seance Du* 17, 2057-2060.

Lo, J.M., Tseng, C.L., Yang, J.Y. (1986) Radiometric method for determining solubility of organic solvents in water. *Anal. Chem.* 58, 1596-1597.

Lu, P.Y., Metcalf, R. (1975) Environmental fate and biodegradability of benzenes derivatives as studied in a model aquatic ecosystem. *Environ. Health Perspec.* 10, 269-284.

Lyman, W.J. (1982) Adsorption coefficient for soils and sediments. In: *Handbook of Chemical Property Estimation Methods.* Lyman, W.J., Reehl, W.F., Rosenblatt, D.H. Editors, Chapter 4, Ann Arbor Sci., Michigan.

Lyman, W.J., Reehl, W.F., Rosenblatt, D.H. (1982) *Handbook on Chemical Property Estimation Methods.* Environmental Behavior of Organic Compounds. McGraw-Hill, New York.

Ma, K.C., Shiu, W.Y., Mackay, D. (1990) *A critically reviewed compilation of physical and chemical and persistence data for 110 selected EMPPL substances.* A Report Prepared for the Ontario Ministry of Environment, Water Resources Branch, Toronto, Ontario.

Mabey, W., Smith, J.H., Podoll, R.T., Johnson, H.L., Mill, T., Chou, T.W., Gate, J., Waight-Partridge, I., Jaber, H., Vandenberg, D. (1982) *Aquatic Fate Process for Organic Priority Pollutants. EPA Report,* No. 440/4-81-14.

Mackay, D. (1981) *Environmental and Laboratory Rates of Volatilization of Toxic Chemicals from Water.* pp.303-319.

Mackay, D. (1982) Correlation of bioconcentration factors. *Environ. Sci. Technol.* 16, 274-278.

Mackay, D., Bobra, A.M., Chan, D.W., Shiu, W.Y. (1982a) Vapor pressure correlation for low-volatility environmental chemicals. *Environ. Sci. Technol.* 16, 645-649.

Mackay, D., Bobra, A.M., Shiu, W.Y., Yalkowsky, S.H. (1980) Relationships between aqueous solubility and octanol-water partition coefficient. *Chemosphere* 9, 701-711.

Mackay, D., Paterson, S. (1991) Evaluating the multimedia fate of organic chemicals. A Level III fugacity model. *Environ. Sci. Technol.* 25, 427-436.

Mackay, D., Paterson, S., Chung, B., Neely, W.B. (1985) Evaluation of the environmental behavior of chemicals with a level III fugacity model. *Chemosphere* 14(3/4), 335-374.

Mackay, D., Shiu, W.Y. (1981) A critical review of Henry's law constants for chemicals of environmental interest. *J. Phys. Chem. Ref. Data* 10, 1175-1199.

Mackay, D., Shiu, W.Y. (1990) Physical-chemical properties and fate of volatile organic compounds: an application of the fugacity approach. In: *Significance and Treatment of Volatile Organic Compounds in Water Supplies.* Ram, N.M., Christman, R.F., Cantor, K.P. Editors, pp. 183-204., Lewis Publishers, Inc., Chelsea, Michigan.

Mackay, D., Shiu, W.Y., Bobra, A., Billinton, J., Chau, E., Yuen, A., Ng, C.,Szeto, F. (1982b) Volatilization of organic pollutants from water. *EPA* 600/3-82-019.

Mackay, D., Shiu, W.Y., Sutherland, R.P. (1979) Determination of air-water Henry's law constants for hydrophobic pollutants. *Environ. Sci. Technol.* 13, 333-337.

Mailhot, H. (1987) Prediction of algae bioaccumulation and uptake rate of nine organic compounds by ten physicochemical properties. *Environ. Sci. Technol.* 21, 1009-1013.

Marinucci, A.C., Bartha, R. (1979) Biodegradation of 1,2,3,-trichlorobenzene and 1,2,4-trichlorobenzene in soil and liquid enrichment culture. *Appl. Environ. Microbiol.* 38, 811-817.

Matter-Müller, C., Gujer, W., Giger, W. (1981) Transfer of volatile substances from water to the atmosphere. *Water Res.* 15, 1271-1279.

McDonald, R.A., Shrader, S.A., Stull, D.R. (1959) *J. Chem. Eng. Data* 4, 311-.

McDuffie, D. (1981) Estimation of octanol/water partition coefficients for organic pollutants using reversed-phase HPLC. *Chemosphere* 10, 73-83.

McGovern, E.W. (1943) Chlorohydrocarbon solvents. *Ind. Eng. Chem.* 35, 1230.

McKim, J., Schnieder, P., Veith, G. (1985) Absorption dynamics of organic chemical transport across trout gills as related to octanol-water partition coefficient. *Toxicol. Appl. Pharmacol.* 77, 1-10.

Mellan, I. (1970) *Industrial Solvents Handbook.* Noyes Data Corporation, Park Ridge, New Jersey, U.S.A.

Metcalf, R.L., Sanborn, J.R., Lu, P-Y, Nye, D. (1975) Laboratory model ecosystem studies of the degradation and fate of radiolabelled tri-, tetra-, and pentachlorobiphenyls compared with DDE. *Arch. Environ. Contam. Toxicol.* 3, 151-165.

Miller, M.M., Ghodbane, S., Wasik, S.P., Tewari, Y.B., Martire, D.E. (1984) Aqueous solubilities, octanol/water partition coefficients and entropies of melting of chlorinated benzenes and biphenyls. *J. Chem. Eng. Data* 29, 184-190.

Miller, M.M., Wasik, S.P., Huang, G.L., Shiu, W.Y., Mackay, D. (1985) Relationships between octanol-water partition coefficient and aqueous solubility. *Environ. Sci. Technol.* 19, 522-529.

Mills, W.B., Dean, J.D., Porcella, D.B., Gherini, S.A., Hudson, R.J.M., Frick, W.E., Rupp, G.L., Bowie, G.L. (1982). *Water Quality Assessment: A Screening Procedure for Toxic and Conventional Pollutants. Part 1, EPA*-600/6-82-004a.

Miyake, K., Tereda, H. (1982) Determination of partition coefficients of very hydrophobic compounds by high-performance liquid chromatography on glyceryl-coated controlled-pore glass. *J. Chromatogr.* 240, 9-20.

Neely, W.B. (1979) Estimating rate constants for the uptake and clearance by fish. *Environ. Sci. Technol.* 13, 1506-1510.

Neely, W.B. (1980) A method for selecting the most appropriate environmental experiments on a new chemical. In: *Dynamics, Exposure and Hazard Assessment of Toxic Chemicals.* R. Haque, Ed., pp.287-298., Ann Arbor Sci. Publ. Ann Arbor, Michigan.

Neely, W.B. (1982) Organizing data for environmental studies. *Environ. Toxicol. Chem.* 1, 259-266.

Neely, W.B. (1984) An analysis of aquatic toxicity data: Water solubility and acute LC50 fish data. *Chemosphere* 13, 813-819.

Neely, W.B., Branson, D.R. Blau, G.E. (1974) Partition coefficient to measure bioconcentration potential of organic chemicals in fish. *Environ. Sci. Technol.* 8, 1113-1115.

Neuhauser, E.F., Loehr, R.C., Malecki, M.R., Milligan, D.L., Durkin, P.R. (1985) The toxicity of selected organic chemicals to earthworm *eisenia fetida. J. Environ. Qual.* 14(3), 383-388.

Niimi, A.J., Cho, C.Y. (1980) Uptake of hexachlorobenzene (HCB) from feed by rainbow trout *(salmo gairdneri). Bull. Environ. Toxicol.* 24, 834-837.

Niimi, A.J., Palazzo, V. (1985) Temperature effect on the elimination of pentachlorophenol, hexachlorobenzene and mirex by rainbow trout *(salmo gairdneri). Water Res.* 19(2), 205-207.

Nirmalakhandan, N.N., Speece, R.E. (1988a) Prediction of aqueous solubility of organic chemicals based on molecular structure. *Environ. Sci. Technol.* 22, 328-338.

Nirmalakhandan, N.N., Speece, R.E. (1988b) QSAR model for predicting Henry's law constant. *Environ. Sci. Technol.* 22, 1349-1357.

OECD (1979) OECD Environmental Committee Chemicals Group, *OECD Chemical Testing Programme Expert Group, Physical Chemical Final Report Volume I, Part 1 and Part 2, Summary of the OECD Laboratory Intercomparison Testing Programme Part I-On the Physico-chemical Properties.* p.33. Dec., 1979, Berlin.

OECD (1981) *OECD Guidelines for Testing of Chemicals.* Organization for Economic Co-operation and Development. OECD, Paris.

Oliver, B.G. (1984) The relationship between bioconcentration factor in rainbow trout and physical-chemical properties for some halogenated compounds. In: *QSAR in Environmental Toxicology.* Kaiser, K.L.E. (Ed.), pp. 300-317. D. Reidel Publishing Company.

Oliver, B.G. (1985) Desorption of chlorinated hydrocarbons from spiked and anthropogenically contaminated sediments. *Chemosphere* 14, 1087-1106.

Oliver, B.G. (1987a) Biouptake of chlorinated hydrocarbons from laboratory-spiked and field sediments by oligochaete worms. *Environ. Sci. Technol.* 21, 785-790.

Oliver, B.G. (1987b) Fate of some chlorobenzenes from the Niagara River in Lake Ontario. In: *Sources and Fates of Aquatic Pollutants.* Hite, R.A., Eisenreich, S.J., Eds., pp. 471-489. Advances in Chemistry Series 216, Am. Chem. Soc., Washington, D.C.

Oliver, B.G. (1987c) Partitioning relationships for chlorinated organics between water and particulates in the St. Clair, Detroit and Niagara Rivers. In: *QSAR in Environmental Toxicology* - II, K.L.E. Kaiser, Ed., pp.251-260, D. Reidel Publishing Co.

Oliver, B.G., Charlton, M.N. (1984) Chlorinated organic contaminants on settling particulates in the Niagara River vicinity of Lake Ontario. *Environ. Sci. Technol.* 18, 903-908.

Oliver, B.G., Nicol, K.D. (1982) Chlorobenzenes in sediments, water and selected fish from Lakes Superior, Huron, Erie and Ontario. *Environ. Sci. Technol.* 16, 532-536.

Oliver, B.G., Nimii, A.J. (1983) Bioconcentration of chlorobenzenes from water by rainbow trout: correlations with partition coefficients and environmental residues. *Environ. Sci. Technol.* 17, 287-291.

Oliver, B.G., Niimi, A.J. (1985) Bioconcentration factors of some halogenated organics for rainbow trout: limitations in their use for prediction of environmental residues. *Environ. Sci. Technol.* 19, 842-849.

Olsen, R.L., Davis, A. (1990) Predicting the fate and transport of organic compounds in groundwater. *Hazard. Mat. Control* 3, 40-64.

Opperhuizen, A., Van Develde, E.W., Gobas, F.A.P.C., Liem, D.A.K., Van der Steen, J.M., Hutzinger, O. (1985) Relationship between bioconcentration in fish and steric factors of hydrophobic chemicals. *Chemosphere* 14, 1871-1896.

Opperhuizen, A. (1986) Bioconcentration of hydrophobic chemicals in fish. In: *Aquatic Toxicology and Environmental Fate*: Nineth Volume. ASTM STP 921. Poston, T.M., Purdy, R., Eds., pp. 304-315. American Society for Testing and Materials, Philadelphia.

Pankow, J.F. (1986) Magnitude of artifacts caused by bubbles and headspace in the determination of volatile compounds in water. *Anal. Chem.* 58, 1821-1826.

Pankow, J.F., Isabelle, L.M., Asher, W.E. (1984) Trace organic compounds in rain. 1. Sample design and analysis by adsorption/thermal desorption (ATD). *Environ. Sci. Technol.* 18, 310-318.

Pankow, J.F., Rosen, M.E. (1984) The analysis of volatile compounds by purge and trap with whole column cryotrapping (WCC) on a fused silica capillary column. *J. High Resolut. Chromatogr. Commu.* 7, 504-508.

Pankow, J.F., Rosen, M.E. (1988) Determination of volatile compounds in water by purging directly to a capillary column with whole column cryotrapping. *Environ. Sci. Technol.* 22, 398-405.

Pankow, J.F. (1990) Minimization of volatilization losses during sampling and analysis of volatile organic compounds in water. In: *Significance and Treatment of Volatile Organic Compounds in Water Supplies.* Ram, N.M., Christman,R.F., Cantor, K.P., Eds., pp. 73-99. Lewis Publishers, Inc., Chelsea, Michigan.

Paya-Perez,B., Riaz, M., Larsen, B.R. (1991) Soil sorption of 20 PCB congeners and six chlorobenzenes. *Ecotoxicol. Environ. Safety* 21, 1-17.

Pereira, W.E., Rostad, C.E., Chiou, C.T., Brinton, T.I., Barber, II, L.B., Demcheck, D.K., Demas, C.R. (1988) Contamination of estuarine water, biota and sediment by halogenated organic compounds: a field study. *Environ. Sci. Technol.* 22, 772-778.

Perry, R.A., Atkinson, R., Pitts, J.N. (1977) Kinetics and mechanisms of the gas phase reaction of the hydroxy radicals with aromatic hydrocarbons over temperature range 296-473°K. *J. Phys. Chem.* 81, 296-304.

Pierotti, C., Deal, C., Derr, E. (1959) Activity coefficient and molecular structure. *Ind. Eng. Chem. Fundam.* 51, 95-101.

Pirsch, J. (1956) Beitrag zur frage der gittercrafte organischer molekule. *Mikrochimica Acta.* 1-6, 992-1004.

Politzki, G.R., Bieniek, D., Lahaniatis, E.S., Scheunert, I., Klein, W., Korte, F. (1982) Determination of vapour pressures of nine organic chemicals adsorbed on silica gel. *Chemosphere* 11, 1217-1229.

Prausnitz, J.M. (1969). *Molecular Thermodynamic of Fluid-Phase Equilibria.* Prentice Hall, Englewood Cliffs, N.J.

Radding, S.B., Liu, D.H., Johnson, H.L., Mill, T. (1977) *Review of the Environmental Fate of Selected Chemicals.* U.S. Environmental Protection Agency Report No. *EPA*-560/5-77-003.

Reid, R.C., Prausnitz, J.M., Sherwood, T.K. (1977) *The Properties of Gases and Liquids.* 3rd ed. McGraw Hill, New York.

Reid, R.C., Prausnitz, J.M., Poling, B.E. (1987) *The Properties of Gases and Liquids.* 4th ed., McGraw Hill, New York.

Reischl, A., Reissinger, M., Thoma, H., Hutzinger, O. (1989) Uptake and accumulation of PCDD/F in terrestrial plants: basic considerations. *Chemosphere* 19, 467-474.

Rekker, R.F. (1977) *The Hydrophobic Fragmental Constants. Its Derivation and Application, a Means of Characterizing Membrane Systems.* Elsevier Sci. Publ. Co., Oxford, England.

Riddick, J.A. et al., (1986) *Organic Solvents: Physical Properties and Methods of Purification.* 4th Edn. J. Wiley & Sons, New York, N.Y.

Rippen, G., Ilgenstein, M., Klöpffer, W., Poreniski, H.J. (1982) Screening of the adsorption behavior of new chemicals: natural soils and model adsorbents. *Ecotox. Environ. Safety* 6, 236-245.

Rippen, G., Frank, R. (1986) In: *Hexachlorobenzene: Proceedings of an International Symposium.* Morris, C.R., Cabral, J.R.P. (Eds.), pp. 45-52. IARC, Lyon, France.

Ryan, J.A., Bell, R.M., Davidson, J.M., O'Connor, G.A. (1988) Plant uptake of non-ionic organic chemicals from soils. *Chemosphere* 17, 2299-2323.

Sabljic, A. (1984) Predictions of the nature and strength of soil sorption of organic pollutants by molecular topology. *J. Agric. Food Chem.* 32, 243-246.

Sabljic, A. (1987a) On the prediction of soil sorption coefficients of organic pollutants from molecular structure: application of molecular topology model. *Environ. Sci. Technol.* 21, 358-366.

Sabljic, A. (1987b) Nonempirical modeling of environmental distribution and toxicity of major organic pollutants. In: *QSAR in Environmental Toxicology* - II. Kaiser, K.L.E., Ed., pp 309-322, D. Reidel Publ. Co., Dordrecht, Netherlands.

Sanemasa, I., Miyazaki, Y., Arakawa, S.,Deguchi, T. (1987) The solubility of benzene-hydrocarbon binary mixtures in water. *Bull. Chem. Soc. Jpn.* 60, 517-523.

Sangster, J. (1989) Octanol-water partition coefficients of simple organic compounds. *J. Phys. Chem. Ref. Data* 18, 1111-1230.

Sarna, L.P., Hodge, P.E., Webster, G.R.B. (1984) Octanol-water partition coefficients of chlorinated dioxins and dibenzofurans by reversed-phase HPLC using several C_{18} columns. *Chemosphere* 13, 975-983.

Sato, A., Nakajima, T. (1979) A structure-activity reationship of some chlorinated hydrocarbons. *Arch. Environ. Health* 34, 69-75.

Schmidt-Bleek, F., Haberland, W., Klein, A.W., Caroli, S. (1982) Steps towards environment hazard assessment of new chemicals. *Chemosphere* 11, 383-415.

Schwarz, F.P. (1980) Measurement of the solubilities of slightly soluble organic liquids in water by elution chromatography. *Anal. Chem.* 52, 10-15.

Schwarz, F.P., Miller, J. (1980) Determination of the aqueous solubilities of organic liquids in water by elution chromatography. *Anal. Chem.* 52, 2161-2164.

Schwarzenbach, R.P., Molnar-Kubica, E., Giger, W., Wakeham, S.G. (1979) Distribution, residence time and fluxes of tetrachloroethylene and 1,4-dichlorobenzene in Lake Zurich, Switzerland. *Environ. Sci. Technol.* 13(11), 1367-1373.

Schwarzenbach, R.P., Westall, J. (1981) Transport of nonpolar compounds from surface water to groundwater. Laboratory sorption studies. *Environ. Sci. Technol.* 11, 1360-1367.

Sears, G.W., Hopke, E.R. (1949) Vapor pressures of naphthalene, anthracene, and hexachlorobenzene in a low pressure region. *J. Am. Chem. Soc.* 71, 1632-1634.

Seidell, A. (1941) *Solubilities of Organic Compounds.* Vol.2, Van Nostrand, New York.

Seip, H.M., Alstad, J., Carlberg, G.E., Martinsen, K., Skaane, P. (1986) Measurement of solubility of organic compounds in soils. *Sci. Total Environ.* 50, 87-101.

Shiu, W.Y., Mackay, D. (1986) A critical review of aqueous solubilities, vapor pressures, Henry's law constants and octanol-water partition coefficients of the polychlorinated biphenyls. *J. Phys. Chem. Data* 15, 911-929.

Shiu, W.Y., Gobas, F.P.A.C., Mackay, D. (1987) Physical-chemical properties of three congeneric series of chlorinated aromatic hydrocarbons. In: *QSAR in Envrionmental Toxicology.* II. Kaiser, K.L.E., Ed., pp. 347-362, D. Reidel Publishing Co., Dordrecht, Netherlands.

Shiu, W.Y., Ma, K.C., Mackay, D. (1990) Solubilities of pesticides in water. Part 1, Environmental physical chemistry and Part 2, Data compilation. *Reviews Environ. Contam. Toxicol.* 116, 1-187.

Simmons, P., Branson, D., Bailey, R.I. (1976) *1,2,4-Trichlorobenzene-biodegradable or not?* Canadian Assoc. of Textile colourists and Chemists. International Technical Conference. Quebec, Canada.

Smith, A.D., Bharath, A., Mullard, C., Orr, D., McCarthy, L.S., Ozburn, G.W. (1990) Bioconcentration kinetics of some chlorinated benzenes and chlorinated phenols in American flagfish, Jordanella floridae (Goode and Bean). *Chemosphere* 20, 379-386.

Smith, J.H., Mabey, W.R., Bahonos, N., Holt, B.R., Lee, S.S., Chou, T.W., Venberger, D.C., Mill, T. (1978) *Environmental pathways of selected chemicals in fresh water systems: Part II. Laboratory Studies.* Interagency Energy-Environment Research Program Report. EPA-600/7-78-074. Environmental Research Laboratory Office of Research and Development. U.S. Environment Protection Agency, Athens, Georgia 30605, p. 304.

Stauffer, T.B., MacIntyre, W.G. (1986) Sorption of low-polarity organic compounds on oxide minerals and aquifer material. *Environ. Toxicol. Chem.* 5, 949-955.

Stauffer, T.B., MacIntyre, W.G., Wickman, D.C. (1989) Sorption fo nonpolar organic chemicals on low-carbon-content aquifer materials. *Environ. Toxicol. Chem.* 8, 845-852.

Steen, W.C., Karickhoff, S.W. (1981) Biosorption of hydrophobic organic pollutants by mixed microbial populations. *Chemosphere* 10, 27-32.

Stephen, H., Stephen, Y. (1963). *Solubilities of Inorganic and Organic Compounds.* Vol. 1 & 2, Pergamon Press, Oxford.

Stephenson, R.M., Malanowski, A. (1987) *Handbook of the Thermodynamics of Organic Compounds.* Elsevier, New York.

Stull, D.R. (1947) *Vapor pressure of pure substances. Organic compounds. Ind. Eng. Chem.* 39, 517-540.

Suntio, L.R., Shiu, W.Y., Mackay, D., Seiber, J.N., Glofelty, D. (1988) Critical review of Henry's law constants. *Rev. Environ. Contam. Toxicol.* 103, 1-59.

Suntio, L.R., Shiu, W.Y., Mackay, D. (1988) A review of the nature and properties of chemicals present in pulp mill effluents. *Chemosphere* 17, 1249-1290.

Tabak, H.H., et al. (1964) Microbial metabolism of aromatic compounds. I. Decomposition of phenolic compounds and aromatic hydrocarbons by phenol-adapted bacteria. *J. Bacteriol.* 87, 910-919.

Tabak, H.H., Quave, S.A., Mashni, C.I., Barth, E.F. (1981) Biodegradability studies with organic priority pollutant compounds. *J. Water Pollut. Control. Fed.* 53, 1503-1518.

Tadokoro, H., Tomita, Y. (1987) The relationship between bioaccumulation and lipid content of fish. In: *QSAR in Environmental Toxicology.* II. Kaiser, K.L.E., Ed., pp. 363-373, D. Reidel Publ. Co., Dorhrect, Holland.

Taft, R.W., Abraham, M.H., Famini, G.R., Doherty, R.M., Abboud, J.M., Kamlet M.J. (1985) Solubility properties in polymers and biological media 5: An analysis of the physicochemical properties which influence octanol-water partition coefficients of aliphatic and aromatic solutes. *J. Pharm. Sci.* 74, 807-814.

Tewari, Y.B., Miller, M.M., Wasik, S.P., Martire, D.E. (1982) Aqueous solubility and octanol/water partition coefficient of organic compounds at 25.0 °C. *J. Chem. Eng. Data* 27, 451-454.

The Merck Index. *An Encyclopedia of Chemicals, Drugs and Biologicals.* (1989). Budavari, S., Ed., The Merck & Co., Inc., Rahway, N.J., 11th edition.

The Merck Index. *An Encyclopedia of Chemicals, Drugs and Biologicals* (1983). Windholz, M., Ed., The Merck and Co., Inc., Rahway, N.J., U.S.A., 10th edition.

Thibodeaux, L.J. (1979) *Chemodynamics.* John Wiley & Sons, New York.

Thomann, R.V. (1989) Bioaccumulation model of organic chemical distribution in aquatic food chains. *Environ. Sci. Technol.* 23, 699-707.

Travis, C.C., Arms, A.D. (1988) Bioconcentration of organics in beef, milk, and vegetation. *Environ. Sci. Technol.* 22, 271-174.

Tsonopoulos, C., Prausnitz, J.M. (1971) Activity coefficients of aromatic solutes in dilute aqueous solutions. *Ind. Eng. Chem. Fundam.* 10, 593-600.

Valsaraj, K.T. (1988) On the physico-chemical aspects of partitioning of nonpolar hydrophobic organics at the air-water interface. *Chemosphere* 17(5), 875-887.
Valsaraj, K.T., Thibodeaux, L.J. (1989) Relationships between micelle-water and octanol-water partition constants for hydrophobic organics of environmental interest. *Water Res.* 23(2), 183-189.
Valvani S.C., Yalkowsky, S.H. (1980) Solubility and partitioning in drug design. In: *Physical Chemical Properties of Drug Medical Research Series.* Vol. 10. Yalkowsky, S.H., Sinkinla, A.A., Valvani, S.C. Eds., pp. 201-229. Marcel Dekker Inc., New York, N.Y.
Valvani, S.C., Yalkowsky, S.H., Roseman, T.J. (1981) Solubility and Partitioning. IV. Aqueous solubility and octanol/water partition coefficient of liquid electrolytes. *J. Pharm. Sci.* 70, 502-507.
Van der Linden, A.C. (1978) Degradation of oil in the marine environment. *Dev. Biodegradation Hydrocarbons* 1, 165-200.
Van Hoogen, G., Opperhuizen, A. (1988) Toxicokinetics of chlorobenzenes in fish. *Environ. Toxicol. Chem.* 7, 213-219.
Veith, G.D., Austin, N.M., Morris, R.T. (1979a) A rapid method for estimating log P for organic chemicals. *Water Res.* 13, 43-47.
Veith, G.D., Defor, D.L. Bergstedt, B.V. (1979b) Measuring and estimating the bioconcentration factor of chemicals in fish. *J. Fish Res. Board Can.* 26, 1040-1048.
Veith, G.D., Macek, K.J., Petrocelli, S.R., Carroll, J. (1980) An evaluation of using partition coefficients and water solubility to estimate bioconcentration factors for organic chemicals in fish. *Aquatic Toxicology, ASTM STP* 707, Eaton, J.G., Parrish, P.R., Hendricks, A.C., Eds., pp 116-129, Amer. Soc. for Testing and Materials, Philadelphia.
Veith, G.D., Call, D.J., Brooke, L.T. (1983) Structure toxicity relationships for the fathead minnow, Pimephales promelas, narcotic industrial chemicals. *Can. J. Fish Aquat. Sci.* 40, 743-748.
Vesala, A. (1974) Thermodynamics of transfer of nonelectrolytes from light to heavy water. I. Linear free energy correlations of free energy of transfer with solubility and heat of melting of a nonelectrolyte. *Acta Chem. Scand.* 28A(8), 839-845.
Verschueren,K. (1977) *Handbook of Environmental Data on Organic Chemicals.* Van Nostrand Reinhold, New York.
Verschueren, K. (1983) *Handbook of Environmental Data on Organic Chemicals.* 2nd ed. Van Nostrand Reinhold Co., New York.
Voice, T.C., Weber, W.J., Jr. (1985) Sorbent concentration effects in liquid/solid partitioning. *Environ. Sci. Technol.* 19(9), 789-796.
Wakeham S.G., Davis, A.C., Karas, J. (1983) Mesocosm experiments to determine the fate and persistence of volatile organic compounds in coastal seawater. *Environ. Sci. Technol.* 17, 611-617.
Ware, W., West, W. (1977) *Investigation of Selected Potential Environmental Contaminants-Halogenated Benzenes.* EPA 560/2-77-004.

Wasik, S.P., Miller, M.M., Teware, Y.B., May, W.E., Sonnefeld, W.J., DeVoe, H., Zoller, W.H. (1983) Determination of the vapor pressure, aqueous solubility, and octanol/water partition coefficient of hydrophobic substances by coupled generator column/liquid chromatographic methods. *Residue Rev.* 85, 29-42.

Wateral, H., Tanaka, M., Suzuki, N. (1982) Determination of partition coefficients of halobenzenes in heptane/water and 1-octanol/water systems and comparison with the scaled particle calculation. *Anal. Chem.* 54, 702-705.

Wauchope, R.D., Getzen, F.W. (1972) Temperature dependence of solubilities in water and heats of fusion of solid aromatic hydrocarbons. *J. Chem. Eng. Data* 17, 38-41.

Weast, R.C., Ed. (1972-73) *Handbook of Chemistry and Physics,* 53th ed. CRC Press, Cleveland.

Weast, R. (1976-77) *Handbook of Chemistry and Physics.* 57th ed., CRC Press, Boca Raton, Florida.

Weast, R.C., Ed. (1983-84). *Handbook of Chemistry and Physics,* 64th ed., CRC Press, Boca Raton, Florida.

Weil, L., Dure, G., Quentin, K.L. (1974) Solubility in water of insecticide chlorinated hydrocarbons and polychlorinated biphenyls in view of water pollution. *Z. Wasser Abwasser Forsch.* 7, 169-175.

Windholz, M., Ed. (1983) *The Merck Index. An Encyclopedia of Chemicals, Drugs and Biologicals.* 10th Edition, The Merck & Co. Inc., Rahway, New Jersey.

Wing, J., Johnston, W.H. (1957) The solubility of water in aromatic halides. *J. Am. Chem. Soc.* 79, 864-865.

Wolfe, D.A. (1987) Interactions of spilled oil with suspended materials and sediments in aquatic systems. In: *Fate and Effects of Sediment-Bound Chemicals in Aquatic Systems.* Dickson, K.L., Maki, A.W., Brungs, W.A., Eds., pp. 299-316. Pergamon Press, New York.

Wong, P.T.S., Chau, Y.K., Rhamey, J.S., Docker, M. (1984) Relationship between water solubility of chlorobenzenes and their effects on a fresh water green algae. *Chemosphere* 13, 991-996.

Yalkowsky, S.H. (1979) Estimation of entropies of fusion of organic compounds. *Ind. Eng. Chem. Fundam.* 18, 108-111.

Yalkowsky, S.H., Orr, R.J., Valvani, S.C. (1979) Soliblity and partitioning. 3. The solubility of halobenzenes in water. *Ind. Eng. Chem. Fundam.* 18, 351-353.

Yalkowsky, S.H., Valvani, S.C. (1979) Solubilities and partitioning. 2. Relationships between aqueous solubilities, partition coefficients, and molecular surface areas of rigid aromatic hydrocarbons. *J. Chem. Eng. Data* 24, 127-129.

Yalkowsky, S.H., Valvani, S.C. (1980) Solubility and partitioning. I. Solubility of nonelectrolytes in water. *J. Pharm. Sci.* 69, 912-922.

Yalkowsky, S.H., Valvani, S.C., Mackay, D. (1983) Estimation of the aqueous solubility of some aromatic compounds. *Residue Rev.* 85, 43-55.

Yurteri, C., Ryan, D.F., Callow, J.J., Gurol, M.D. (1987) The effect of chemical composition of water on Henry's law constant. *J. Water Pollut. Control Fed.* 59, 950-056.

Yoshida, K., Shigeoka, T., Yamauchi, F. (1983a) Relationship between molar refraction and n-octanol/water partition coefficient. *Ecotox. Environ. Safety* 7, 558-565.

Yoshida, K., Shigeoka, T., Yamauchi, F. (1983b) Non-steady state equilibrium model for the preliminary prediction of the fate of chemicals in the environment. *Ecotox. Environ. Safety* 7, 179-190.

Zaroogian, G.E., Heltshe, J.F., Johnson, M. (1985) Estimation of bioconcentration in marine species using structure-activity models. *Envriron. Toxicol. Chem.* 4, 3-12.

Zimmerli, B., Marek, B. (1974) Moldelversuche Zur kontamination von lebensmitten mit pestiziden via gasphase. *Mitt. Gebiete Lebensm. Hyg.* 65, 55-64.

Zoeteman, B.C.J., Harmsen, K.M., Linders, J.B.H.J. (1980) Persistent organic pollutants in river water and groundwater of the Netherlands. *Chemosphere* 9, 231-249.

Zoeteman, B.C.J., De Greef, E., Brinkmann, F.J.J. (1981) Persistency of organic contaminants in groundwater. Lessons from soil pollution incidents in the Netherlands. *Sci. Total. Environ.* 21, 187-202.

4. Polychlorinated Biphenyls (PCBs)

4.1 List of Chemicals and Data Compilations:

4.1.1 PCB Congeners

Common Name: Biphenyl
Synonym: PCB-0, diphenyl
Chemical Name: biphenyl
CAS Registry No: 92-52-4
Molecular Formula: $C_{12}H_{10}$
Molecular Weight: 154.21
Melting Point (°C):

71 (Weast 1972,1973; Mackay et al. 1980, Bruggeman et al. 1982; Burkhard et al. 1985a; Shiu et al. 1987; Opperhuizen et al. 1988)

87 (Banerjee et al. 1990)

Boiling Point (°C):

255.9 (Weast 1972,1973)

246.0 (Shiu & Mackay 1986)

Density (g/cm^3 at 20°C): 0.866
Molar Volume (cm^3/mol):

184.6 (LeBas method, Miller et al. 1985; Shiu & Mackay 1986; Shiu et al. 1987)

0.920 (intrinsic volume, $V_I/100$, Kamlet et al. 1988; Hawker 1989b,1990; De Bruijn & Hermens 1990)

Molecular Volume (A^3):

207.0 (Opperhuizen et al. 1988)

181.87, 163.7, 207.0 (De Bruijn & Hermens 1990)

Total Surface Area, TSA (A^2):

182.0 (Yalkowsky & Valvani 1979)

192.2 (Mackay et al. 1980; Shiu et al. 1987)

189.53 (Burkhard 1984)

184.77 (planar, Doucette 1985)

192.3 (nonplanar, Doucette 1985; Doucette & Andren 1988)

195.2 (Shiu & Mackay 1986)

224.1 (planar, shorthand, Opperhuizen et al. 1988)

184.43 (planar, Hawker & Connell 1988a)

216.44, 189.44, 224.1 (De Bruijn & Hermens 1990)

Heat of Fusion, kcal/mol:

4.18 (Miller et al. 1984; Shiu & Mackay 1986)

Entropy of Fusion, cal/mol K (e.u.):

12.20 (Miller et al. 1984; Shiu & Mackay 1986)

Fugacity Ratio, F (calculated, assuming ΔS_{fusion} = 13.5 e.u.):

0.35 (Mackay et al. 1980,1983; Shiu & Mackay 1986)

0.352 (Shiu et al. 1987)

Water Solubility (g/m^3 or mg/L at 25°C):

5.94 (shake flask-UV, Andrew & Keefer 1949)

7.48 (shake flask-UV, Bohon & Claussen 1951; quoted, Mackay & Wolkoff 1973;

Mackay & Leinonen 1975; Kilzer et al. 1979; Shiu et al. 1990)
7.50 (calculated-molar volume, Lindenberg 1956; quoted, Horvath 1982)
3.87 (UV, Sahyun 1966)
7.08 (shake flask-UV, Wauchope & Getzen 1972)
7.0 (Hutzinger et al. 1974)
7.45 (shake flask-GC, Eganhouse & Calder 1976; quoted, Shiu et al. 1990)
7.0 (shake flask-fluorescence, Mackay & Shiu 1977; quoted, Chiou et al. 1982; Shiu et al. 1990)
7.50 (quoted, Verschueren 1977,1983; Neely 1980)
8.5 (shake flask-Nephelometric, Hollifield 1979)
7.48 (Kilzer et al. 1979; quoted, Sklarew & Girvin 1987)
7.50 (quoted, Kenaga & Goring 1980; Kenaga 1980)
7.45 (quoted, Mackay et al. 1980; 1983a)
7.05, 8.28 (observed, estimated, Yalkowsky & Valvani 1980)
8.09 (TLC-RT, Bruggeman et al. 1982)
20.3 (quoted, subcooled liquid, Chiou et al. 1982; Chiou 1985)
21.4 (quoted, subcooled liquid, Mackay et al. 1983a)
7.21 (quoted lit. average, Yalkowsky et al. 1983; quoted, Erickson 1986)
6.80 (Neely 1983)
2.82 (calculated-UNIFAC, Arbuckle 1983)
23.6 (calculated-HPLC-k', converted from reported γ_w, Hafkenscheid & Tomlinson 1983)
6.71 (gen. col.-GC/ECD, Miller et al. 1984,1985; quoted, Hawker 1989b)
7.09 (Pearlman et al. 1984)
22.82 (calculated-TSA, subcooled liquid, Burkhard et al. 1985b)
7.10, 8.08 (quoted, calculated-UNIFAC, Banerjee 1985)
7.12, 20.3 (quoted, subcooled liquid, Chiou & Block 1986)
7.0 (selected, Shiu & Mackay 1986; Shiu et al. 1987)
7.50 (quoted, Chou & Griffin 1987)
20.0 (selected, subcooled liquid, Shiu et al. 1987)
8.09 (calculated-UNIFAC, converted from log γ, Arbuckle 1986)
19.2, 8.0 (exptl., calculated-UNIFAC, converted from log γ, Burkhard & Kuehl 1986)
7.05 (vapor-saturation, Akiyoshi et al. 1987)
8.37 (calculated-UNIFAC, Banerjee & Howard 1988)
7.2 (gen. col.-HPLC, Billington et al. 1988)
8.09, 10.13 (quoted, calculated average of HPLC-RI, Brodsky & Ballschmiter 1988)
7.0 (quoted, Formica et al. 1988)
9.30 (quoted, Metcalfe et al. 1988)
16.9 (quoted, subcooled liq., Hawker 1989b)
7.05 (quoted, Nirmalakhandan & Speece 1989)
1.90 (calculated- χ , Nirmalakhandan & Speece 1989)

Vapor Pressure (Pa at 25°C):

1.30 (effusion method, Bright 1951)
0.031 (manometery, Augood et al. 1953; quoted, Bidleman 1984)
1.273 (effusion method, Bradley & Cleasby 1953; quoted, Bidleman 1984)
0.97 (Seki & Suzuki 1953)
1.03 (Aihara 1959)
7.60 (selected, Mackay & Wolkoff 1973; Mackay & Leinonen 1975; quoted, Kilzer et al. 1979; Bopp 1983)
1.41 (effusion method, Radchenko & Kitiagorodskii 1974; quoted, Bidleman 1984)
7.705 (subcooled liquid, Weast 1976-77)
1.866 (calculated-SxHLC, Mackay et al. 1979; quoted, Bidleman 1984)
1.293 (selected, Neely 1980)
1.33 (Neely 1981)
1.40 (Swann et al. 1983)
1.1-5.6 (selected, Mackay et al. 1983a)
3.8 (selected, subcooled liquid, Mackay et al. 1983a)
7.7 (selected, subcooled liquid, Bopp 1983)
1.27 (selected, Neely 1983; quoted, Erickson 1986)
5.608 (selected, subcooled liquid, Bidleman 1984)
7.04, 6.22 (subcooled liquid, GC-RT, Bildleman 1984)
5.61, 6.62 (subcooled liq., selected, GC-RT, Foreman & Bidleman 1985)
1.19 (gas saturation, Burkhard et al. 1984; quoted, Sklarew & Gievin 1987)
1.01, 1.80 (quoted exptl, calculated-P/C, Burkhard et al. 1985a)
0.423, 0.703, 0.594 (calculated-MW, GC-RI, χ, Burkhard et al. 1985a)
2.03 (subcooled liq., calculated, GC-RI, Burkhard et al. 1985b)
2.43 (selected, Shiu & Mackay 1986; Shiu et al. 1987; Sklarew & Girvin 1987)
6.9 (selected, subcooled liquid, Shiu et al. 1987)
4.90 (quoted, Metcalfe et al. 1988)
1.186, 0.639 (quoted, calculated-UNIFAC, Banerjee et al. 1990)
5.610, 5.0 (quoted, Hinckley et al. 1990)

Henry's Law Constant (Pa m^3/mol):

157 (calculated-P/C, Mackay & Leinonen 1975)
41.34 (batch stripping, Mackay et al. 1979)
30.4 (batch stripping, Mackay et al. 1980; Mackay & Shiu 1981)
28.0 (calculated-P/C, Mackay & Shiu 1981)
66.27 (Dow Chemicals, Neely 1982)
30-66 (calculated-P/C, Mackay et al. 1983a)
27.0 (selected, Mackay et al. 1983)
69.37 (calculated-P/C, Bopp 1983)
34.65 (calculated-UNIFAC, Arbuckle 1983)
13.68 (calculated-P/C, Burkhard et al. 1985b)
53.5 (calculated-P/C, Shiu & Mackay 1986; Shiu et al. 1987)

124 (calculated- χ , Nirmalakhandan & Speece 1988)
19.57 (wetted-wall column, Fendinger & Glotfelty 1990)

Octanol/Water Partition Coefficient, log K_{OW}:

3.16 (shake flask-UV, Rogers & Cammarata 1969; quoted, Sangster 1989)
3.19 (calculated-molecular orbital indices, Rogers & Cammarata 1969)
4.09 (shake flask, Leo et al. 1971; Hansch & Leo 1979; quoted, Chiou et al. 1982)
4.04 (shake flask, Hansch et al. 1973)
4.17, 4.09, 3.16, 4.04 (Neely et al. 1974; Hansch & Leo 1979)
3.75 (HPLC-RT, Veith et al. 1979)
4.09 (quoted, Veith et al. 1979; Veith & Kosian 1983)
4.03 (calculated-f const., Yalkowsky & Valvani 1979,1980; quoted, Arbuckle 1983)
4.04 (shake flask-HPLC, Banerjee et al. 1980; quoted, Sangster 1989)
3.88 (selected, Kenaga & Goring 1980; Chou & Griffin 1987)
4.14 (calculated, Mackay et al. 1980)
4.10 (TLC, Bruggeman et al. 1982; quoted, Erickson 1986)
4.06 (quoted of Hansch & Leo 1979, Hammers et al. 1982; Haky & Young 1984)
3.70 (HPLC-RT, Woodburn 1982)
3.16-4.09, 3.91 (shake flask, range, average, Eadsforth & Moser, 1983; quoted, Sangster 1989)
3.91-4.15, 4.05 (HPLC, range, average, Eadsforth & Moser 1983)
4.03 (calculated-HPLC-k', Hafkenscheid & Tomlinson 1983)
4.10 (quoted, Mackay et al. 1983; Kaiser 1983)
3.79 (calculated-f const., Yalkowsky et al. 1983)
4.09, 4.23 (quoted, calculated-molar fraction, Yoshida et al. 1983)
4.15 (calculated-TSA, Burkhard 1984)
3.76 (generator column-GC/ECD, Miller et al. 1984,1985; quoted, Sklarew & Girvin 1987; Hawker 1989b; Sangster 1989)
3.89 (generator column-HPLC, Woodburn et al. 1984; quoted, Burkhard & Kuehl 1986; Sklarew & Girvin 1987; Sangster 1989)
3.70 (HPLC-RT, Woodburn et al. 1984)
3.79 (HPLC-RT, Rapaport & Eisenreich 1984; quoted, Sklarew & Girvin 1987)
4.11 - 4.13 (HPLC-RV, Garst 1984)
4.09 (quoted exptl., Garst 1984)
4.10 (HPLC-RV, Garst & Wilson 1984)
3.88 (quoted, Freitag et al. 1985)
3.90 (selected, Shiu & Mackay 1986; Shiu et al. 1987; Sklarew & Girvin 1987)
4.06 (quoted, Tomlinson & Hafkenscheid 1986)
3.63, 4.00 (HPLC-k', calculated, De Kock & Lord 1987)
3.89 (generator column-GC, Doucette & Andren 1987,1988)
4.03, 4.42 (calculated-π, TSA, Doucette & Andren 1987)
4.04, 3.62 (quoted, calculated-UNIFAC, Banerjee & Howard 1988)
3.94, 4.04 (quoted, calculated average of HPLC-RI, Brodsky & Ballschmiter 1988)

4.03, 3.79, 3.69, 3.88, 4.35, 4.52 (calculated-π, f-const., HPLC-RT, MW, χ , TSA, Doucette & Andren 1988)
4.33 (quoted, Formica et al. 1988)
3.90, 3.99, 3.98 (calculated-f const., π, solvatochromic p., Kamlet et al. 1988)
4.09 (calculated-TSA, Hawker & Connell 1988a & b)
4.30 (Metcalfe et al. 1988)
3.90 (quoted, Hawker 1989b)
3.88 (correlated, Isnard & Lambert 1989)
4.008 (slow stirring, De Bruijn et al. 1989; De Bruijn & Hermens 1990)
4.10 (calculated-π, De Bruijn et al. 1989)
3.98 (recommended, Sangster 1989)
4.09 (quoted, Hawker 1990)

Bioconcentration Factor, log BCF:
2.64 (trout, calculated-k_1/k_2, Neely et al. 1974; quoted, Connell & Hawker 1988; Hawker 1990)
3.12 (rainbow trout, Veith et al. 1979; Veith & Kosian 1983)
2.53 (fish, flowing water, Kenaga & Goring 1980; Kenaga 1980)
2.30 (calculated-S, Kenaga 1980)
2.10 (calculated- χ , Koch 1983)
2.73, 2.45, 3.41 (selected, algae, fish, activated sludge, Halfon & Reggiani 1986)
3.0 (fish, selected, Metcalfe et al. 1988)

Sorption Partition Coefficient, log K_{OC}:
3.15 (calculated, Kenaga 1980)
3.95 (suspended particulate-matter, calculated-K_{OW}, Burkhard 1984)
3.57, 3.77 (Lake Erie at 9.6 mg/L DOC, Landrum et al. 1984)
5.58, 4.04 (Huron River at 7.8 mg/L DOC, Landrum et al. 1984)
3.40 (calculated, soil, Chou & Griffin 1987)
3.23 (soils, batch equilibration adsorption coefficient, Kishi et al. 1990)

Half-Lives in the Environment:
Air:
Surface water: half-life in river water estimated to be 1.5 days (Bailey et al. 1983).
Groundwater:
Sediment:
Soil:
Biota: estimated half-life from fish in simulated ecosystem to be 29 hours (Neely 1980).

Environmental Fate Rate Constants or Half-Lives:
Volatilization/Evaporation: half life of evaporation from water depth of 1 meter estimated to be 7.52 hours (Mackay & Leinonen 1975) and rate of volatilization, 0.92 g/m^2h (Mackay 1986; Metcalfe et al. 1988).

Photolysis:
Hydrolysis:
Oxidation: rate constant for reaction with OH radicals in the gas phase using relative rate technique compared with cyclohexane determined to be 8.5×10^{-12} cm^3 $molecule^{-1}$ sec^{-1} at 295°K for a 24-hours average OH radical concentration of 5×10^5 cm^{-3} (Atkinson & Aschmann 1985); room temperature rate constants for reaction with OH radicals in the gas phase calculated to be 7.9×10^{-12} cm^3 $molecule^{-1}$ sec^{-1} and 5.8 to 8.2×10^{-12} cm^3 $molecule^{-1}$ sec^{-1} with a calculated tropospheric lifetime of 3 days (Atkinson 1987).
Biodegradation: 100% degraded by activated sludge in 47-hour cycle (Monsanto Co. 1972); microbial degradation with pseudo first-order rate constant of 109 $year^{-1}$ in the water column and 1090 $year^{-1}$ in the sediment (Wong & Kaiser 1975; quoted, Neely 1981); rate of biodegradation in water from Port Valdez estimated to be 9.3-9.8 nmol/L-day with an initial biphenyl concentration of 4.4-4.7 μmol/L (data of Aug. 1977, Reichardt et al. 1981) and 3.2 nmol/L-day with initial concentration of 2.9 μmol/L (data of Aug. 1978, Reichardt et al. 1981); half-life of biodegradation estimated to be 1.5 days by using water dieaway test (Bailey et al. 1983).
Biotransformation:
Bioconcentration, Uptake (k_1) and Elimination (k_2) Rate Constants or Half-Lives:

k_1:	6.79 $hour^{-1}$	(trout muscle, Neely et al. 1974; Neely 1979)
k_2:	0.0155 $hour^{-1}$	(trout muscle, Neely et al. 1974)
k_1:	6.8 $hour^{-1}$	(trout, quoted, Hawker & Connell 1985)
$1/k_2$:	65 hour	(trout, quoted, Hawker & Connell 1985)
log k_1:	2.21 day^{-1}	(fish, Connell & Hawker 1988)
log $1/k_2$:	0.43 day	(fish, Connell & Hawker 1988)
$1/k_2$:	65 hour	(trout, Hawker & Connell 1988b)
log k_2:	-0.43 day^{-1}	(fish, quoted, Thomann 1989)

Common Name: 2-Chlorobiphenyl
Synonym: PCB-1, o-chlorobiphenyl
Chemical Name: 2-chlorobiphenyl
CAS Registry No: 2051-60-7
Molecular Formula: $C_{12}H_9Cl$
Molecular Weight: 188.65
Melting Point (°C):
34 (Beaven et al. 1961; Hutzinger et al. 1974; NAS 1979; Mackay et al. 1980; Bruggeman et al. 1982; Burkhard et al. 1985a; Opperhuizen et al. 1988)
32-32.5 (Wallnöfer et al. 1973)
34, 27 (selected, calculated, Abramowitz & Yalkowsky 1990)
Boiling Point (°C):
274 (Weast 1972,1973)
285 (calculated, Mackay et al. 1982; Shiu & Mackay 1986)
Density (g/cm^3 at 20°C): 0.9837
Molar Volume (cm^3/mol):
205.5 (LeBas method, Shiu & Mackay 1986)
1.010 (intrinsic, $V_I/100$, Kamlet et al. 1988; Hawker 1989b; De Bruijn & Hermens 1990)
Molecular Volume (A^3):
215.2 (planar, shorthand, Opperhuizen et al. 1988)
195.97, 176.71, 217.9 (De Bruijn & Hermens 1990)
Total Surface Area, TSA (A^2):
208.40 (Mackay et al. 1980; Shiu & Mackay 1986)
204.49 (Burkhard 1984)
195.32 (planar, Doucette 1985)
208.3 (nonplanar, Doucette 1985; Doucette & Andren 1987; 1988)
206.9 (Sabljic 1987)
232.2 (planar, shorthand, Opperhuizen et al. 1988)
195.45 (planar, Hawker & Connell 1988a)
233.74, 202.42, 232.3 (De Bruijn & Hermens 1990)
195 (Abramowitz & Yalkowsky 1990)
Heat of Fusion, kcal/mol:
3.66 (Miller et al. 1984; Shiu & Mackay 1986)
Entropy of Fusion, cal/mol K (e.u.):
12.0 (Miller et al. 1984; Shiu & Mackay 1986)
Fugacity Ratio, F (assuming ΔS_{fusion} = 13.5 e.u.):
0.817 (Mackay et al. 1980)
0.814 (Shiu & Mackay 1986)

Water Solubility (g/m^3 or mg/L at 25°C):
1.60 (Webb 1970)
0.90 (Hoover 1971)

5.90 (shake flask-GC/ECD, Wallnöfer et al. 1973; quoted, Hutzinger et al. 1974; NAS 1979; Erickson 1986)
4.13 (generator column-GC/ECD, Weil et al. 1974)
5.76 (calculated-TSA, Mackay et al. 1980)
4.12 (TLC-RT, Bruggeman et al. 1982)
5.08 (subcooled liq., shake flask-GC/ECD, Chiou et al. 1983)
2.73 (calculated-K_{OW}, Yalkowsky et al. 1983)
7.80 (Neely 1983)
5.06 (generator column-GC/ECD, Miller et al. 1984,1985; quoted, Hawker 1989b)
6.06 (calculated-TSA, subcooled liquid, Burkhard et al. 1985b)
5.08, 49.6 (quoted, calculated-K_{OW} & HPLC-RT, Chiou et al. 1986)
5.50 (selected, Shiu & Mackay 1986)
5.90 (quoted, Chou & Griffin 1987)
4.13, 5.79 (quoted, calculated average of HPLC-RI, Brodsky & Ballschmiter 1988)
5.97 (quoted, subcooled liq., Hawker 1989b)
4.42, 0.685 (quoted, calculated- χ , Nirmalakhanden & Speece 1989)
5.97, 4.73 (quoted, calculated-TSA, Abramowitz & Yalkowsky 1990)
0.77, 2.32 (quoted average of Brodsky & Ballschmiter 1988, correlated- χ , Patil 1991)

Vapor Pressure (Pa at 25°C):
1.53 (subcooled liquid, Stull 1947)
1.84 (subcooled liquid, Weast 1976-77)
1.12 (Neely 1981,1983; quoted, Erickson 1986)
1.84 (quoted, subcooled liquid, Bidleman 1984)
1.892, 2.56 (subcooled liquid, GC-RT, Bidleman 1984)
1.53 (quoted exptl., Burkhard et al. 1985a)
0.355, 0.755, 0.260 (calculated-MW, GC-RI, χ, Burkhard et al. 1985a)
0.926 (calculated, GC-RI, subcooled liquid, Burkhard et al. 1985b)
1.84, 2.20 (subcooled liq., quoted, GC-RT, Foreman & Bidleman 1985)
1.38 (P_L calculated from P_S & F, Shiu & Mackay 1986)
2.04 (selected, Shiu & Mackay 1986; quoted, Sklarew & Girvin 1987)
2.50 (selected, subcooled liq., Shiu & Mackay 1986)
1.84 (quoted, subcooled liq., Hinckley et al. 1990)

Henry's Law Constant (Pa m^3/mol):
28.9 (calculated-P/C, Burkhard et al. 1985b)
70.1 (calculated-P/C, Shiu & Mackay 1986)
74.6 (Dow Chemical, Neely 1982)

Octanol/Water Partition Coefficient, log K_{OW}:
4.54 (shake flask-GC, Tulp & Hutzinger 1978)
4.80 (Hansch & Leo 1979)
4.35 (HPLC-RT, Veith et al. 1979)

4.59 (shake flask-GC, Bruggeman et al. 1982; quoted, Sangster 1989)
4.56 (TLC-RT, Bruggeman et al. 1982; quoted, Erickson 1986)
3.75 (HPLC-RT, Woodburn 1982)
4.51 (shake flask-GC, Chiou et al. 1983)
4.50 (gen. col.-GC, Miller et al. 1984,1985; quoted, Sklarew & Girvin 1987; Hawker 1989b; Sangster 1989)
4.56 (calculated-f const., Yalkowsky et al. 1983)
4.56 (calculated-TSA, Burkhard 1984)
4.38 (gen. col.-HPLC, Woodburn et al. 1984; quoted, Sklarew & Girvin 1987; Sangster 1989)
3.90 (HPLC-RT, Rapaport & Eisenreich 1984)
4.60 (calculated-π, Rapaport & Eisenreich 1984; quoted, Sklarew & Girvin 1987)
4.80 (calculated-π, Woodburn et al. 1984)
5.7, 4.44 (quoted, HPLC-RP/MS, Burkhard & Kuehl 1986)
4.30 (selected, Shiu & Mackay 1986; quoted, Sklarew & Girvin 1987)
4.17 (calculated-S, Chou & Griffin 1987)
4.38 (generator column-GC/ECD, Doucette & Andren 1987,1988)
4.74, 4.80 (calculated-π, TSA, Doucette & Andren 1987)
4.50, 4.42 (quoted, calculated average of HPLC-RI, Brodsky & Ballschmiter 1988)
4.74, 4.56, 3.92, 4.44, 4.81, 4.90 (calculated-π, f-const., HPLC-RT, MW, χ , TSA, Doucette & Andren 1988)
4.46 (calculated-TSA, Hawker & Connell 1988a)
4.46 (calculated-solvatochromic parameters, Kamlet et al. 1988)
4.30, 4.70 (quoted, calculated-π or f const., Kamlet et al. 1988)
4.531 (slow stirring-GC, De Bruijn et al. 1989; De Bruijn & Hermens 1990)
4.56 (calculated-π, De Bruijn et al. 1989)
4.30 (quoted, Hawker 1989b)
4.52 (recommended, Sangster 1989)

Bioconcentration Factor, log BCF:

Sorption Partition Coefficient, log K_{OC}:
4.35 (suspended particulate matter, calculated-K_{OW}, Burkhard 1984)
3.47 (calculated, soil, Chou & Griffin 1987)

Sorption Partition Coefficient, log K_{OM}:
3.23 (soil org. matter, sorption isotherm, Chiou et al. 1983)
3.23, 3.96 (quoted, calculated- χ , Sabljic 1984)

Half-Lives in the Environment:
Air:
Surface water: half-life in Lake Michigan, 1.4 days (Neely 1983).
Groundwater:
Sediment:

Soil:
Biota:

Environmental Fate Rate Constants or Half-Lives:

Volatilization:

Photolysis: calculated sunlight photolysis rate constant in surface water at 40°L in summer is $1.1\text{-}3.7\times10^{-4}$ day^{-1} and with a half-life of 18 years (Dulin et al. 1986).

Hydrolysis:

Oxidation: rate constant for the reaction with OH radicals in the gas phase using relative rate technique compared with cyclohexane determined to be 2.9×10^{-12} cm^3 $molecule^{-1}$ sec^{-1} at 295°K for a 24-hours average OH radical concentration of 5×10^5 cm^{-3} (Atkinson & Aschmann 1985); room temperature rate constant $k\times10^{12}$ cm^3 $molecule^{-1}$ sec^{-1} for reaction with the OH radicals in the gas phase calculated to be 1.6 to the C_6H_4CL ring (ring B), 1.2 (ring A), and 2.8 (observed ring A + ring B) with a calculated tropospheric lifetime of 5-11 days (Atkinson 1987).

Biodegradation: half-life of biodegradation by bacteria estimated to be 100 hours (Wong & Kaiser 1975); within 7 hours by Alkaligenes sp. strain Y-42 from lake sediments (Furukawa & Matsumura 1976; quoted, Pal et al. 1980); microbial degradation with pseudo first-order rate constant of 63 $year^{-1}$ in the water column and 630 $year^{-1}$ in the sediment (Wong & Kaiser 1975; quoted, Neely 1981); biodegradation rate in water from Port Valdez estimated to be 4.1 nmol/L-day with an initial concentration of 1.5 μmol/L (data of Aug. 1977, Reichardt et al. 1981) and 1.2 nmol/L-day with initial concentration of 4.5 μmol/L (data of Aug. 1978, Reichardt et al. 1981); time for 50% degradation of an initial concentration of 1-100 μg/L by river dieaway test is about 2-3.5 days (Bailey et al. 1983); the degradation rate at 30 μg ml^{-1} was 1.1 μg ml^{-1} day^{-1} under culture conditions include river water as supportive medium and mixed bacterial cultures obtained from river sediments (Kong & Sayler 1983).

Biotransformation:

Bioconcentration, Uptake (k_1) and Elimination (k_2) Rate Constants:

Common Name: 3-Chlorobiphenyl
Synonym: PCB-2, m-chlorobiphenyl
Chemical Name: 3-chlorobiphenyl
CAS Registry No: 2051-61-8
Molecular Formula: $C_{12}H_9Cl$
Molecular Weight: 188.65
Melting Point (°C):
16.5 (Hutzinger et al. 1974)
25 (Mackay et al. 1980; Opperhuizen et al. 1988)
16 (Burkhard et al. 1985a)
25, 27 (selected, calculated, Abramowitz & Yalkowsky 1990)
Boiling Point (°C):
284 (Weast 1972,1973)
285 (calculated, Mackay et al. 1982; Shiu & Mackay 1986)
Density (g/cm^3 at 20°C): 0.9837
Molar Volume (cm^3/mol):
205.5 (LeBas method, Shiu & Mackay 1986)
1.010 (intrinsic, $V_I/100$, Kamlet et al. 1988)
Molecular Volume (A^3):
224.0 (planar, shorthand, Opperhuizen et al. 1988)
Total Surface Area, TSA (A^2):
210.02 (Mackay et al. 1980)
207.05 (Burkhard 1984)
202.3 (planar, Doucette 1985)
209.87 (nonplanar, Doucette 1985; Doucette & Andren 1987; 1988)
210.0 (Shiu & Mackay 1986)
237.0 (planar, shorthand, Opperhuizen et al. 1988)
201.95 (planar, Hawker & Connell 1988a)
202 (Abramowitz & Yalkowsky 1990)
Heat of Fusion, kcal/mol:
3.66 (Miller et al. 1984; Shiu & Mackay 1986)
Fugacity Ratio, F (assuming ΔS_{fusion} = 13.5 e.u.):
1.0 (Shiu & Mackay 1986)

Water Solubility (g/m^3 or mg/L at 25°C):
3.50 (shake flask-GC/ECD, Wallnöfer et al. 1973; Hutzinger et al. 1974; quoted, NAS 1979; Erickson 1986)
1.30 (generator column-GC/ECD, Weil et al. 1974; quoted, Chiou et al. 1982; Chiou & Block 1986)
6.22 (calculated-TSA, Mackay et al. 1980)
1.305 (TLC-RT, Bruggeman et al. 1982)
1.84 (quoted lit. average, Yalkowsky et al. 1983)
1.20 (quoted, Neely 1983)

4.72 (subcooled liq., calculated-TSA, Burkhard et al. 1985b)
2.50 (selected, Shiu & Mackay 1986)
3.56 (quoted, Chou & Griffin 1987)
1.31, 0.77 (quoted, calculated average of HPLC-RI, Brodsky & Ballschmiter 1988)
3.63 (gen. col.-HPLC, Billington et al. 1988)
1.31, 0.685 (quoted, calculated-χ, Nirmalakhanden & Speece 1989)
2.37, 2.37 (quoted, calculated-TSA, Abramowitz & Yalkowsky 1990)
0.77, 2.32 (quoted average of Brodsky & Ballschmiter 1988, calculated- χ , Patil 1991)

Vapor Pressure (Pa at 25°C):
0.723 (extrapolated, Antoine eqn., Weast 1972, 1973; quoted, Shiu & Mackay 1986)
1.01, 0.98 (subcooled liq., GC-RT, Bidleman 1984)
0.723 (quoted exptl., Burkhard et al. 1985a; quoted, Shiu & Mackay 1986)
0.435, 0.366, 0.260 (calculated-MW, GC-RI, χ, Burkhard et al. 1985a)
0.362 (subcooled liq., calculated, GC-RI, Burkhard et al. 1985b)
0.997 (subcooled liq., GC-RT, Foreman & Bidleman 1985)
0.20 (quoted from Bidleman 1984, Erickson 1986)
1.00 (selected, subcooled liq., Shiu & Mackay 1986)

Henry's Law Constant (Pa m^3/mol):
14.49 (calculated-P/C, Burkhard et al. 1985b)
75.55 (calculated-P/C, Shiu & Mackay 1986)
62.11 (Dow Chemical, Neely 1982)

Octanol/Water Partition Coefficient, log K_{OW}:
4.54 (shake flask-GC/ECD, Tulp & Hutzinger 1978)
4.35 (HPLC, Veith et al. 1979)
4.80 (Hansch & Leo 1979)
4.71 (shake flask-GC/ECD, Bruggeman et al. 1982; quoted, Kaiser 1983)
4.72 (TLC-RT, Bruggeman et al. 1982; quoted, Erickson 1986; Sangster 1989)
4.95 (quoted, Chiou et al. 1982 from Tulp & Hutzinger 1978)
4.35 (HPLC-RT, Woodburn 1982)
4.56 (calculated-f constant, Yalkowsky et al. 1983)
4.72 (calculated, Neely 1983)
4.62 (calculated-TSA, Burkhard 1984)
4.58 (gen. col.-HPLC, Woodburn et al. 1984; quoted, Sklarew & Girvin 1987; Sangster 1989)
3.75 (HPLC-RT, Woodburn et al. 1984)
4.42 (HPLC-RT, Rapaport & Eisenreich 1984; quoted, Sklarew & Girvin 1987)
4.60 (calculated-π, Rapaport & Eisenreich 1984)
4.80 (calculated-π, Woodburn et al. 1984)
4.60 (selected, Shiu & Mackay 1986; quoted, Sklarew & Girvin 1987)
4.34 (calculated-S, Chou & Griffin 1987)

4.58 (gen. col.-GC/ECD, Doucette & Andren 1987,1988)
4.74, 4.84 (calculated-π, TSA, Doucette & Andren 1987)
4.65, 4.66 (quoted, calculated average of HPLC-RI, Brodsky & Ballschmiter 1988)
4.74, 4.56, 4.44, 4.80, 4.94 (calculated-π, f-const., MW, χ , TSA, Doucette & Andren 1988)
4.69 (calculated-TSA, Hawker & Connell 1988a)
4.58 (recommended, Sangster 1989)
4.66 (quoted average of Brodsky & Ballschmiter 1988, calculated- χ , Patil 1991)

Bioconcentration Factor, log BCF:

Sorption Partition Coefficient, log K_{OC}:
4.42 (suspended particulate matter, calculated-K_{OW}, Burkhard 1984)
3.62 (calculated-S, soil, Chou & Griffin 1987)

Half-Lives in the Environment:
Air:
Surface water:
Groundwater:
Sediment:
Soil:
Biota:

Environmental Fate Rate Constants or Half-Lives:
Volatilization:
Photolysis:
Hydrolysis:
Oxidation: rate constant for reaction with OH radicals in the gas phase using relative rate technique compared with cyclohexane determined to be 5.4×10^{-12} cm^3 $molecule^{-1}$ sec^{-1} at 295°K for a 24-hours average OH radical concentration of 5×10^5 cm^{-3} (Atkinson & Aschmann 1985); room temperature rate constant $k\times10^{12}$ cm^3 $molecule^{-1}$ sec^{-1} for reaction with the OH radicals in the gas phase calculated to be 2.4 to the C_6H_4CL ring (ring B), 1.2 (ring A) and 5.2 (ring A + Ring B) with a calculated tropospheric lifetime of 5-11 days (Atkinson 1987).
Biodegradation: half-life of biodegradation by Alkaligenes sp. strain Y-42 from lake sediments estimated to be within 7 hours (Furukawa & Matsumara 1976; selected, Pal et al. 1980); biodegradation rate in water from Port Valdez estimated to be 2.6 nmol/L-day with an initial concentration of 3.6 μmol/L (data of Aug. 1977, Reichardt et al. 1981); time for biodegradation of 50% initial concentration of 1-100 μg/L is about 3-4 days by river dieaway test (Bailey et al. 1983); the degradation rate at 30 $\mu g\ ml^{-1}$ was 1.6 $\mu g\ ml^{-1}\ day^{-1}$ under culture conditions include river water as supportive medium and mixed bacterial cultures obtained from river sediments (Kong & Sayler 1983).

Biotransformation:
Bioconcentration, Uptake (k_1) and Elimination (k_2) Rate Constants:

k_1: 0.29 day^{-1} (golden orfe, Sugiura et al. 1979)
k_1: 0.13 day^{-1} (carp, Sugiura et al. 1979)
k_1: 0.11 day^{-1} (brown trout, Sugiura et al. 1979)
k_1: 0.11 day^{-1} (guppy, Sugiura et al. 1979)

Common Name: 4-Chlorobiphenyl
Synonym: PCB-3, p-chlorobiphenyl
Chemical Name: 4-chlorobiphenyl
CAS Registry No: 2051-62-9
Molecular Formula: $C_{12}H_9Cl$
Molecular Weight: 188.65
Melting Point (°C):

77.7 (Gomberg & Bachman 1924; NAS 1979; Mackay et al. 1980; Bruggeman et al. 1982; Burkhard et al. 1985a; Opperhuizen et al. 1988)
75.7-77 (Wallnöfer et al. 1973)
78, 77 (selected, calculated, Abramowitz & Yalkowsky 1990)

Boiling Point (°C):

291 (Weast 1972,1973)
285 (calculated, Mackay et al. 1982; Shiu & Mackay 1986)

Density (g/cm³ at 20°C): 0.9837
Molar Volume (cm³/mol):

205.5 (LeBas method, Shiu & Mackay 1986)
1.010 (intrinsic, $V_I/100$, Kamlet et al. 1988)

Molecular Volume (A³):

224.0 (planar, shorthand, Opperhuizen et al. 1988)

Total Surface Area, TSA (A²):

210.0 (Mackay et al. 1980; Shiu & Mackay 1986)
207.22 (Burkhard 1984)
202.3 (planar, Doucette 1985)
209.9 (nonplanar, Doucette 1985; Doucette & Andren 1987; 1988)
224.0 (calculated- χ , Sabljic 1987)
237.0 (planar, shorthand, Opperhuizen et al. 1988)
202.12 (planar, Hawker & Connell 1988a)
202.0 (Abramowitz & Yalkowsky 1990)

Heat of Fusion, kcal/mol:
Fugacity Ratio, F (assuming ΔS_{fusion} = 13.5 e.u.):

0.30 (Shiu & Mackay 1986)
0.282 (Gobas et al. 1987)

Water Solubility (g/m³ or mg/L at 25°C):

1.0 (Webb 1970)
0.40 (Hoover, 1971)
1.19 (shake flask-GC/ECD, Wallnöfer et al. 1973; Hutzinger et al. 1974; quoted, NAS 1979; Erickson 1986)
0.90 (generator column-GC/ECD, Weil et al. 1974)
1.65 (Branson 1977; quoted, Kenaga & Goring 1980; Kenaga 1980)
0.0151 (shake flask-GC/ECD, Lee et al. 1979)
1.87 (calculated-TSA, Mackay et al. 1980)

1.30 (generator column-HPLC/UV, Billington 1982)
0.902 (TLC-RT, Bruggeman et al. 1982)
1.47 (generator column-GC/ECD, Stolzenburg & Andren 1983)
0.824 (quoted lit. average, Yalkowsky et al. 1983)
4.62 (subcooled liq., calculated-TSA, Burkhard et al. 1985b)
1.20 (selected, Shiu & Mackay 1986)
1.19 (quoted, Chou & Griffin 1987)
1.30 (gen. col.-HPLC, Billington et al. 1988)
0.90, 0.88 (quoted, calculated average of HPLC-RI, Brodsky & Ballschmiter 1988)
1.207 (generator column-GC/ECD, Dunnivant & Elzerman 1988)
0.88, 0.67 (quoted, calculated-χ, Nirmalakhanden & Speece 1989)
0.81 (Eadie et al. 1990)
1.19, 0.75 (quoted, calculated-TSA, Abramowitz & Yalkowsky 1990)
0.88, 2.32 (quoted average of Brodsky & Ballschmiter 1988, calculated- χ , Patil 1991)

Vapor Pressure (Pa at 25°C):
1.41 (subcooled liq., Weast 1972-1973)
0.172 (selected, Mackay et al. 1979)
0.57 (P_L calculated from P_S using F, Mackay et al. 1979)
0.61 (Neely 1981,1983; quoted, Erickson 1986)
0.92, 0.942 (subcooled liquid, GC-RT, Bidleman 1984)
1.41 (quoted, subcooled liquid, Bidleman 1984)
0.175 (gas saturation, Burkhard et al. 1984; quoted, Sklarew & Girvin 1987)
0.131, 0.0979, 0.114 (calculated-MW, GC-RI, χ, Burkhard et al. 1985a)
0.320 (subcooled liq., calculated, GC-RI, Burkhard et al. 1985b)
0.907, 0.931 (subcooled liquid, quoted, GC-RT, Foreman & Bidleman 1985)
0.270 (selected, Shiu & Mackay 1986; quoted, Sklarew & Girvin 1987)
0.90 (selected, subcooled liquid, Shiu & Mackay 1986)
1.41 (quoted, subcooled liq., Hinckley et al. 1990)

Henry's Law Constant (Pa m^3/mol):
58.06 (batch stripping, Atlas et al. 1982; Dow Chemical, Neely 1982))
55.73 (calculated-P/C, Burkhard et al. 1985b)
42.56 (calculated-P/C, Shiu & Mackay 1986)
34.14 (batch stripping, Dunnivant & Elzerman 1988)
34.14 (Dunnivant et al. 1988)
24.39 (wetted-wall col.-GC, Fendinger & Glotfelty 1990)

Octanol/Water Partition Coefficient, log K_{OW}:
4.90 (Branson 1977; quoted, Kenaga & Goring 1980; Chou & Griffin 1987)
4.26 (Sugiura et al. 1978)
4.80 (Hansch & Leo 1979)
2.59 (NAS 1979)

4.61 (shake flask-GC, Bruggeman et al. 1982; quoted, Kaiser 1983; Sangster 1989)
4.69 (TLC-RT, Bruggeman et al. 1982; quoted, Erickson 1986)
4.34 (HPLC-RT, Woodburn 1982)
4.56 (calculated-f const., Yalkowsky et al. 1983)
4.34 (calculated, Neely 1983)
4.63 (calculated-TSA, Burkhard 1984)
4.49 (generator column-HPLC, Woodburn et al. 1984; quoted, Sklarew & Girvin 1987; Sangster 1989)
4.34 (HPLC-RT, Woodburn et al. 1984)
4.80 (calculated- π, Woodburn et al. 1984)
4.40 (HPLC-RT, Rapaport & Eisenreich 1984; quoted, Sklarew & Girvin 1987)
4.40 (calculated- π, Rapaport & Eisenreich 1984)
4.50 (selected, Shiu & Mackay 1986; quoted, Gobas et al. 1987; Sklarew & Girvin 1987)
4.49 (generator column-GC/ECD, Doucette & Andren 1987,1988)
4.74, 4.84 (calculated-π, TSA, Doucette & Andren 1987)
4.55, 4.63 (quoted, calculated average of HPLC-RI, Brodsky & Ballschmiter 1988)
4.74, 4.56, 4.44, 4.80, 4.94 (calculated-π, f const., MW, χ ,TSA, Doucette & Andren 1988)
4.69 (calculated-TSA, Hawker & Connell 1988a)
4.26 (Isnard & Lambert 1988)
4.61 (recommended, Sangster 1989)
4.63, 4.71 (quoted average of Brodsky & Ballschmiter 1988, calculated- χ , Patil 1991)

Bioconcentration Factor, log BCF:
2.08 (killifish, Goto et al. 1978)
2.68 (NAS 1979)
2.77 (estimated, fish, flowing water, Kenaga & Goring 1980; Kenaga 1980)
2.67 (calculated-S, Kenaga 1980)
2.69 (quoted, Bysshe 1982)
3.88 (fish, normalized, lipid basis, Tadokoro & Tomita 1987)
2.77 (quoted, fish, Isnard & Lambert 1988)

Sorption Partition Coefficient, log K_{OC}:
3.52 (calculated, Kenaga 1980)
4.43 (suspended particulate matter, calculated-K_{OW}, Burkhard 1984)
3.90 (soil, calculated-S, Chou & Griffin 1987)
4.71 (organic carbon, Eadie et al. 1990)
Sorption Partition Coefficient, log K_{OM}:
4.02 (natural sediment, Eadie et al. 1990)

Half-Lives in the Environment:
Air:

Surface water: half-life in Lake Michigan, 4.9 days (Neely 1983).
Groundwater:
Sediment:
Soil:
Biota:

Environmental Fate Rate Constants or Half-Lives:

Volatilization:

Photolysis: calculated sunlight photolysis rate constant in surface water at 40°L in summer is 0.115 to 2.3×10^{-4} day^{-1} with a half-life of 8.2 years (Dulin et al. 1986).

Hydrolysis:

Oxidation: rate constant for reaction with OH radicals in the gas phase using relative rate technique compared with cyclohexane determined to be 3.9×10^{-12} cm^3 $molecule^{-1}$ sec^{-1} at 295°K for a 24-hours average OH radical concentration of 5×10^5 cm^{-3} (Atkinson & Aschmann 1985); room temp. rate constant $k \times 10^{12}$ cm^3 $molecule^{-1}$ sec^{-1} for reaction with OH radicals in the gas phase calculated to be 2.6 to the C_6H_4CL ring (ring B), 1.2 (ring A) and 3.8 (observed for ring A + ring B) with a calculated tropospheric lifetime of 5-11 days (Atkinson 1987).

Biodegradation: half-life of biodegradation by bacteria estimated to be 175 hours (Wong & Kaiser 1975; quoted, Pal et al. 1980); within 7 hours by Alkaligenes sp. strain Y-42 from lake sediments (Furukawa & Matsumura 1976; quoted, Pal et al. 1980); microbial degradation with pseudo first-order rate constant of 38 $year^{-1}$ in the water column and 380 $year^{-1}$ in the sediment (Wong & Kaiser 1975; quoted, Neely 1981); degradation rate in water from Port Valdez estimated to be 3.1 nmol/L-day with an initial concentration of 2.9 μmol/L (data of Aug. 1977, Reichardt et al. 1981); time for 50% degradation of an initial concentration of 1-100 μg/L by river dieaway test is about 2-5 days (Bailey et al. 1983); the degradation rate at 30 $\mu g\ ml^{-1}$ was 2.0 $\mu g\ ml^{-1}\ day^{-1}$ under culture conditions include river water as supportive medium and mixed bacteria cultures obtained from river sediments (Kong & Sayler 1983).

Biotransformation:

Bioconcentration, Uptake (k_1) and Elimination (k_2) Rate Constants:

Common Name: 2,2'-Dichlorobiphenyl
Synonym: PCB-4
Chemical Name: 2,2'-dichlorobiphenyl
CAS Registry No: 13029-08-8
Molecular Formula: $C_{12}H_8Cl_2$
Molecular Weight: 223.1
Melting Point (°C):
- 60.5 (Hutzinger et al. 1974; NAS 1979)
- 61 (Mackay et al. 1980; Bruggeman et al. 1982; Burkhard et al. 1985a; Opperhuizen et al. 1988)
- 61, 46 (selected, calculated, Abramowitz & Yalkowsky 1990)

Boiling Point (°C):
- 312 (calculated, Mackay et al. 1982; Shiu & Mackay 1986)

Density (g/cm^3 at 20°C): 1.0536
Molar Volume (cm^3/mol):
- 226.4 (LeBas method, Shiu & Mackay 1986)
- 1.10 (intrinsic, $V_I/100$, Kamlet et al. 1988; De Bruijn & Hermens 1990)

Molecular Volume (A^3):
- 228.8 (planar, shorthand, Opperhuizen et al. 1988)
- 209.9, 189.65, 228.8 (De Bruijn & Hermens 1990)

Total Surface Area, TSA (A^2):
- 224.4 (Mackay et al. 1980)
- 224.4 (Yalkowsky et al. 1983)
- 219.41 (Burkhard 1984)
- 199.57 (planar, Doucette 1985)
- 224.32 (nonplanar, Doucette 1985; Doucette & Andren 1987; 1988)
- 224.2 (Shiu & Mackay 1986)
- 200.8 (planar, Hawker & Connell 1988a)
- 240.5 (nonplanar, shorthand, Opperhuizen et al. 1988)
- 250.06, 215.44, 240.5 (De Bruijn & Hermens 1990)
- 201.0 (Abramowitz & Yalkowsky 1990)

Heat of Fusion, kcal/mol:
Fugacity Ratio, F (assuming ΔS_{fusion} = 13.5 e.u.):
- 0.442 (Mackay et al. 1980; Shiu & Mackay 1986)

Water Solubility (g/m^3 or mg/L at 25°C):
- 1.0 (Webb 1970)
- 0.900 (Hoover 1971)
- 1.50 (shake flask-GC/ECD, Wallnöfer et al. 1973; Hutzinger et al. 1974; quoted, NAS 1979; Erickson 1986)
- 0.79 (generator column-GC/ECD, Weil et al. 1974; quoted, Geyer et al. 1980)
- 0.0212 (shake flask-GC/ECD, Lee et al. 1979)
- 1.00 (calculated-TSA, Mackay et al. 1980)

0.791 (TLC-RT, Bruggeman et al. 1982)
1.86 (20°C, subcooled liq., shake flask-GC/ECD, Chiou et al. 1982,1983; Chiou 1985; Chiou & Block 1986)
0.997 (quoted lit. average, Yalkowsky et al. 1983)
0.891 (quoted, Garten & Trabalka 1983)
1.70 (subcooled liq., calculated-TSA, Burkhard et al. 1985b)
1.00 (selected, Shiu & Mackay 1986)
1.50 (quoted, Chou & Griffin 1987)
1.124 (20°C, subcooled liq., calculated-mole fraction of Aroclor mixtures, Murphy et al. 1987)
0.792, 0.418 (quoted, calculated average of HPLC-RI, Brodsky & Ballschmiter 1988)
1.207 (gen. col.-GC/ECD, Dunnivant & Elzerman 1988)
0.792, 0.239 (quoted, calculated-χ, Nirmalakhanden & Speece 1989)
1.118, 2.23 (quoted, calculated-TSA, Abramowitz & Yalkowsky 1990)
0.425, 0.544 (quoted average of Brodsky & Ballschmiter 1988, calculated- χ, Patil 1991)

Vapor Pressure (Pa at 25°C):
0.133 (Knudsen-effusion technique, extrapolated from 37-54.92°C, Smith et al. 1964)
0.36 (quoted, Neely 1981)
0.82 (P_L calculated from P_S using F, Neely 1981)
0.133 (quoted, Neely 1983; Erickson 1986)
0.134 (quoted exptl. value, Burkhard et al. 1985a)
0.0691, 0.189, 0.114 (calculated-MW, GC-RI, χ, Burkhard et al. 1985a)
0.424 (subcooled liquid, calculated, GC-RI, Burkhard et al. 1985a)
0.326, 0.335 (subcooled liquid, GC-RT, Foreman & Bidleman 1985)
0.60 (selected, subcooled liquid, Shiu & Mackay 1986)
0.260 (selected, Shiu & Mackay 1986; quoted, Sklarew & Girvin 1987)
0.152 (20°C, subcooled liq., calculated-mole fraction of Aroclor mixtures, Murphy et al. 1987)
0.184 (calculated, S x HLC, Dunnivant & Elzerman 1988)
0.415 (subcooled liquid, calculated-M.P., Dunnivant & Elzerman 1988)

Henry's Law Constant (Pa m^3/mol):
22.3 (calculated-P/C, Murphy et al. 1983)
55.73 (calculated-P/C, Burkhard et al. 1985b)
59.17 (calculated-P/C, Shiu & Mackay 1986)
22.29 (selected, Eisenreich 1987)
30.2 (20°C, calculated-P/C, Murphy et al. 1987)
34.14 (batch stripping, Dunnivant & Elzerman 1988)
34.14 (batch stripping, Dunnivant et al. 1988)
23.3 (wetted wall col.-GC, Brunner et al. 1990)

Octanol/Water Partition Coefficient, log K_{OW}:

4.04 (HPLC-RT, Sugiura et al. 1978)
5.70 (HPLC-RT, Sugiura et al. 1979)
5.51 (Hansch & Leo 1979)
3.76 (selected, NAS 1979)
4.00 (HPLC-k', McDuffie 1981)
5.0 (shake flask-GC, Bruggeman et al. 1982)
5.02 (TLC-RT, Bruggeman et al. 1982; quoted, Erickson 1986)
3.55 (HPLC-RT, Woodburn 1982)
4.80 (shake flask-GC/ECD, Chiou et al. 1983; Chiou 1985; Chiou & Block 1986)
5.32 (calculated-f const., Yalkowsky et al. 1983)
4.96 (calculated, Neely 1983)
4.52 (quoted, Garten & Trabalka 1983)
4.96 (calculated-TSA, Burkhard 1984)
4.0 (quoted, Geyer et al. 1984)
4.90 (generator column-GC/ECD, Woodburn et al. 1984)
3.55 (HPLC-RT, Woodburn et al. 1984)
5.51 (calculated-π, Woodburn et al. 1984)
3.63 (HPLC-RT, Rapaport & Eisenreich 1984)
4.89 (calculated-π, Rpaport & Eisenreich 1984; quoted, Eisenreich 1987)
4.02 (quoted, Freitag et al. 1985)
4.90 (selected, Shiu & Mackay 1986; quoted, Gobas & Mackay 1987; Clark et al. 1990)
4.60 (calculated-S, Chou & Griffin 1987)
5.46, 5.18 (calculated-π, TSA, Doucette & Andren 1987)
4.90 (generator column-GC/ECD, Doucette & Andren 1987,1988)
4.90, 4.73 (quoted, calculated average of HPLC-RI, Brodsky & Ballschmiter 1988)
5.46, 5.32, 5.10, 5.24, 5.28 (calculated-π, f const., MW, χ, TSA, Doucette & Andren 1988)
4.65 (calculated-TSA, Hawker & Connell 1988a)
4.52 (Isnard & Lambert 1988)
5.02 (calculated-π, De Bruijn et al. 1989)
4.965 (slow stirring-GC, De Bruijn et al. 1989; De Bruijn & Hermens 1990)
4.72, 5.11 (quoted average of Brodsky & Ballschmiter 1988, calculated- χ, Patil 1991)

Bioconcentration Factor, log BCF:

2.04 (killifish, Goto et al. 1978)
3.81 (NAS 1979)
3.72, 2.95, 3.60, 3.26 (golden orfe, carp, brown trout, guppy, Sugiura et al. 1979)
-1.40, -1.30 (adipose tissue of male, female Albino rats, Geyer et al. 1980)
2.05 (fish, flowing water, Garten & Trabalka 1983)
2.45 (fish, microcosm condition, Garten & Trabalka 1983)
-1.35 (rodents, Garten & Trabalka 1983)

3.43, 2.90 (algae, calculated, Geyer et al. 1984)

3.43, 3.38, 3.80 (algae, fish, act. sludge, Freitag et al. 1984,1985; quoted, Halfon & Reggiani 1986)

3.85 (fish, normalized, lipid basis, Tadokoro & Tomita 1987)

2.05 (quoted, fish, Isnard & Lambert 1988)

Sorption Partition Coefficient, log K_{OC}:

4.76 (suspended particulate matter, calculated-K_{OW}, Burkhard 1984)

3.84 (soil, calculated-S, Chou & Griffin 1987)

Sorption Partition Coefficient, log K_{OM}:

3.68 (soil org. matter, sorption isotherm, Chiou et al. 1983)

3.68, 4.18 (quoted, calculated- χ , Sabljic 1984)

Half-Lives in the Environment:

Air:

Surface water: half life in Lake Michigan, 34.5 days (Neely 1983).

Groundwater:

Sediment:

Soil:

Biota: half-life in rainbow trout, 40 days and in its muscle, 20 days (Niimi & Oliver 1983).

Environmental Fate Rate Constants or Half-Lives:

Volatilization: estimated half-life of evaporation with an initial concentration of 0.1 ppm from 4.5 cm depth of water solution in a glass dish at 24 °C is 3.1 hours and 0.4 hours with stirring of the solution; the experimental half-life of evaporation under same condition is 3.9 hours and 0.9 hours with stirring of the solution (Chiou et al. 1979).

Photolysis:

Hydrolysis:

Oxidation: room temperature rate constant for reaction with OH radicals in the gas phase calculated to be 1.4×10^{-12} cm^3 $molecule^{-1}$ sec^{-1} with a calculated tropospheric lifetime of 8-17 days (Atkinson 1987).

Biodegradation: Biodegraded fairly quickly by Alkaligenes sp. strain Y-42 but small residue was detected after 7 hours (Furukawa & Matsumura 1976; quoted, Pal et al. 1980). Microbial degradation with pseudo first-order rate constant of 0.65 $year^{-1}$ in the water column and 6.5 $year^{-1}$ in the sediment (Furukawa et al. 1978; quoted, Neely 1981).

Biotransformation:

Bioconcentration, Uptake (k_1) and Elimination (k_2) Rate Constants:

k_2: 0.027 day^{-1} (10°C, sandworm, Goerke & Erst 1977; quoted, Waid 1986)

k_1: 0.29 day^{-1} (golden orfe, Sugiura et al. 1979)

k_1: 0.13 day^{-1} (carp, Sugiura et al. 1979)

k_1: 0.11 day^{-1} (brown trout, Sugiura et al. 1979)
k_1: 0.11 day^{-1} (guppy, Sugiura et al. 1979)
k_1: 122 day^{-1} (rainbow trout, calculated, Gobas & Mackay 1987)
k_2: 0.014 day^{-1} (rainbow trout, calculated, Gobas & Mackay 1987)
k_2: 0.017 day^{-1} (rainbow trout, Niimi & Oliver 1983; quoted, Clark et al. 1990)

Common Name: 2,3-Dichlorobiphenyl
Synonym: PCB-5
Chemical Name: 2,3-dichlorobiphenyl
CAS Registry No: 16605-91-7
Molecular Formula: $C_{12}H_8Cl_2$
Molecular Weight: 223.1
Melting Point (°C): 28.0
Boiling Point (°C): 172 (at 4000 Pa, Erickson 1986)
Density (g/cm^3):
Molar Volume (cm^3/mol):
226.4 (LeBas method, Shiu & Mackay 1986)
Molecular Volume (A^3):
Total Surface Area, TSA (A^2):
210.86 (planar, Doucette 1985)
223.87 (nonplanar, Doucette 1985)
210.34 (planar, Hawker & Connell 1988)
Heat of Fusion, kcal/mol:
Fugacity Ratio, F (assuming ΔS_{fusion} = 13.5 e.u.):
0.935

Water Solubility (g/m^3 or mg/L at 25°C):
1.70 (subcooled liq., calculated-TSA, Burkhard et al. 1985b)
1.00 (calculated average of HPLC-RI, Brodsky & Ballschmiter 1988)
1.00, 0.658 (quoted average of Brodsky & Ballschmiter 1988, calculated- χ , Patil 1991)

Vapor Pressure (Pa at 25°C):
0.147, 0.144, 0.114 (calculated-MW, GC-RI, χ , Burkhard et al. 1985a)
0.151 (subcooled liq., calculated-GC-RT, Burkhard et al. 1985b)

Henry's Law Constant (Pa m^3/mol):
19.56 (calculated-P/C, Burkhard et al. 1985b)
28.57 (calculated- χ , Sabljic & Güsten 1989)
23.30 (wetted-wall col.-GC, Brunner et al. 1990)

Octanol/Water Partition Coefficient, log K_{OW}:
5.20 (TLC, Bruggeman et al. 1982; quoted, Oliver & Niimi 1985)
5.36 (quoted, Kaiser 1983)
4.99 (calculated average of HPLC-RI, Brodsky & Ballschmiter 1988)
4.97 (calculated-TSA, Hawker & Connell 1988a; quoted, Clark et al. 1990)
5.0 (quoted, Hawker & Connell 1988b)
5.08 (quoted, Hawker 1990)
4.99, 5.11 (quoted average of Brodsky & Ballschmiter 1988, calculated- χ , Patil 1991)

Bioconcentration Factor, log BCF:

3.08 (oyster, quoted, Vreeland 1974)

4.11 (rainbow trout. Oliver & Niimi 1985)

4.41, 4.11 (rainbow trout, kinetic k_1/k_2, steady state, Oliver & Niimi 1985)

Sorption Partition Coefficient, log K_{OC}:

Half-Lives in the Environment:

Air:

Surface water:

Groundwater:

Sediment:

Soil:

Biota: half-life in rainbow trout, 61 days (Niimi & Oliver 1983; Oliver & Niimi 1985), and its muscle, 26 days (Niimi & Oliver 1983).

Environmental Fate Rate Constants or Half-Lives:

Volatilization:

Photolysis:

Hydrolysis:

Oxidation:

Biodegradation:

Biotransformation:

Bioconcentration, Uptake (k_1) and Elimination (k_2) Rate Constants:

k_2: 0.011 day^{-1} (rainbow trout, Niimi & Oliver 1983; quoted, Clark et al. 1990)

k_1: 300 day^{-1} (rainbow trout, Oliver & Niimi 1985)

k_2: 0.011 day^{-1} (rainbow trout, Oliver & Niimi 1985)

log $1/k_2$: 2.0, 2.2 hour (fish, quoted, calculated-K_{OW}, Hawker & Connell 1988b)

Common Name: 2,4-Dichlorobiphenyl
Synonym: PCB-7
Chemical Name: 2,4-dichlorobiphenyl
CAS Registry No: 33284-50-3
Molecular Formula: $C_{12}H_8Cl_2$
Molecular Weight: 223.1
Melting Point (°C):
24.1-24.4 (Dikerman & Weiss 1957; Weingarten 1961; Hutzinger et al. 1974)
24.4 (Mackay et al. 1980; Bruggeman et al. 1982; Opperhuizen et al. 1988)
24.0 (Burkhard et al. 1985a)
24, 51 (quoted, calculated, Abramowitz & Yalkowsky 1990)
Boiling Point (°C):
312 (calculated, Mackay et al. 1982; Shiu & Mackay 1986)
Density (g/cm³ at 20°C): 1.0536
Molar Volume (cm³/mol):
226.4 (LeBas method, Shiu & Mackay 1986)
1.10 (intrinsic, $V_I/100$, Kamlet et al. 1988)
Molecular Volume (A³):
234.9 (planar, shorthand, Opperhuizen et al. 1988)
Total Surface Area, TSA (A²):
226.0 (Mackay et al. 1980; Yalkowsky et al. 1983)
222.19 (Burkhard 1984)
212.86 (planar, Doucette 1985)
225.85 (nonplanar, Doucette 1985)
245.2 (planar, shorthand, Opperhuizen et al. 1988)
213.14 (planar, Hawker & Connell 1988a)
213.0 (Abramowitz & Yalkowsky 1990)
Heat of Fusion, kcal/mol:
Fugacity Ratio, F (assuming ΔS_{fusion} = 13.5 e.u.):
1.0 (Shiu & Mackay 1986)

Water Solubility (g/m³ or mg/L at 25°C):
1.40 (shake flask-GC/ECD, Wallnöfer et al. 1973; Hutzinger et al. 1974; quoted, NAS 1979; Erickson 1986; Sawhney 1987)
0.637 (shake flask-GC/ECD, Haque & Schmedding 1975; quoted, Haque et al. 1980)
2.025 (calculated-TSA, Mackay et al. 1980)
1.13 (generator column-HPLC/UV, Billington 1982)
0.629 (TLC-RT, Bruggeman et al. 1982)
1.408 (quoted, lit. average, Yalkowsky et al. 1983)
1.31 (subcooled liquid, calculated-TSA, Burkhard et al. 1985b)
1.25 (selected, Shiu & Mackay 1986)
1.40 (quoted, Chou & Griffin 1987)
0.613 (20°C, calculated-mole fraction of Aroclor mixtures, Murphy et al. 1987)

1.13 (gen. col.-HPLC, Billington et al. 1988)
0.629, 0.614 (quoted, calculated average of HPLC-RI, Brodsky & Ballschmiter 1988)
1.148 (generator column-GC/ECD, Dunnivant & Elzerman 1988)
1.41, 0.239 (quoted, calculated-χ, Nirmalakhanden & Speece 1989)
1.41 (quoted, Isnard & Lambert 1989)
0.163 (calculated-mole fraction of Aroclor mixtures, Murphy et al. 1987)
1.118, 0.560 (quoted, calculated-TSA, Abramowitz & Yalkowsky 1990)
0.614, 0.658 (quoted average of Brodsky & Ballschmiter 1988, calculated- χ , Patil 1991)

Vapor Pressure (Pa at 25°C):
0.184, 0.321 (subcooled liquid, GC-RT, Bidleman 1984)
0.175 (subcooled liquid, Burkhard et al. 1984)
0.157, 0.179, 0.0499 (calculated-MW, GC-RI, χ, Burkhard et al. 1985a)
0.175 (subcooled liquid, calculated, GC-RI, Burkhard et al. 1985b)
0.210, 0.216, 0.243 (subcooled liquid, GC-RT, Foreman & Bidleman 1985)
0.25 (selected, Shiu & Mackay 1986)
0.24 (quoted from Bidleman 1984, Erickson 1986)
0.100 (20°C, calculated-mole fraction of Aroclor mixtures, Murphy et al. 1987)
0.181 (subcooled liquid, calculated- S x HLC, Dunnivant & Elzerman 1988)

Henry's Law Constant (Pa m^3/mol):
96.66 (batch stripping, Atlas et al. 1982)
29.80 (calculated-P/C, Burkhard et al. 1985b)
45.39 (calculated-P/C, Shiu & Mackay 1986)
36.37 (20°C, calculated-P/C, Murphy et al. 1987)
35.26 (batch stripping, Dunnivant & Elzerman 1988; Dunnivant et al. 1988)
28.37 (wetted-wall col.-GC, Brunner et al. 1990)
24.78 (wetted-wall column-GC, Fendinger & Glotfelty 1990)

Octanol/Water Partition Coefficient, log K_{OW}:
5.15 (TLC-RT, Bruggeman et al. 1982; quoted, Erickson 1986)
5.00 (quoted, Kaiser 1983)
5.23 (calculated-f const., Yalkowsky et al. 1983)
5.04 (calculated-TSA, Burkhard 1984)
4.67 (HPLC-RT, Rapaport & Eisenreich 1984)
5.30 (calculated-π, Rapaport & Eisenreich 1984; quoted, Sklarew & Girvin 1987)
5.14 (quoted, Burkhard & Kuehl 1986)
5.00 (selected, Shiu & Mackay 1986)
4.62 (calculated-S, Chou & Griffin 1987)
5.30 (quoted, Eisenreich 1987)
5.16 (calculated average of HPLC-RI, Brodsky & Ballschmiter 1988)
5.07 (calculated-TSA, Hawker & Connell 1988)

5.23 (quoted, Isnard & Lambert 1989)
5.15, 5.10 (quoted average of Brodsky & Ballschmiter 1988, calculated- χ , Patil 1991)

Bioconcentration Factor, log BCF:

Sorption Partition Coefficient, log K_{OC}:
4.83 (suspended particulate matter, calculated-K_{OW}, Burkhard 1984)

Half-Lives in the Environment:
Air:
Surface water:
Groundwater:
Sediment:
Soil:
Biota:

Environmental Fate Rate Constants or Half Lives:
Volatilization:
Photolysis: calculated sunlight photolysis rate constant at 40° latitude in winter is < 2×10^{-8} sec^{-1} and 5.7×10^{-11} sec^{-1}, with a half-life > 400 days (Dulin et al. 1986).
Hydrolysis:
Oxidation: room temperature rate constant for reaction with OH radicals in the gas phase calculated to be 1.8×10^{-12} cm^3 $molecule^{-1}$ sec^{-1} with a calculated tropospheric lifetime of 8-17 days (Atkinson 1987).
Biodegradation: biodegraded fairly quickly by Alkaligenes sp. strain Y-42 from lake sediments but small amount residue was detected after 7 hours (Furukawa & Matsumura 1976; quoted, Pal et al. 1980).
Biotransformation:
Bioconcentration, Uptake (k_1) and Elimination (k_2) Rate Constants:

Common Name: 2,4'-Dichlorobiphenyl
Synonym: PCB-8
Chemical Name: 2,4'-dichlorobiphenyl
CAS Registry No: 34883-43-7
Molecular Formula: $C_{12}H_8Cl_2$
Molecular Weight: 223.1
Melting Point (°C):
42.5-43.5 (Wallnöfer et al. 1973)
46 (Hutzinger et al. 1974)
44.5-46.0 (NAS 1979)
42 (Mackay et al. 1980; Burkhard et al. 1985a; Opperhuizen et al. 1988)
43, 51 (quoted, calculated, Abramowitz & Yalkowsky 1990)
Boiling Point (°C):
312 (calculated, Mackay et al. 1982; Shiu & Mackay 1986)
Density (g/cm³ at 20°C): 1.0536
Molar Volume (cm³/mol):
226.4 (LeBas method, Shiu & Mackay 1986)
1.10 (intrinsic, $V_I/100$, Kamlet et al. 1988)
Molecular Volume (A³):
234.9 (planar, shorthand, Opperhuizen et al. 1988)
Total Surface Area, TSA (A²):
226.01 (Mackay et al. 1980; Yalkowsky et al. 1983; Shiu & Mackay 1986)
222..14 (Burkhard 1984)
212.85 (planar, Doucette 1985)
225.9 (nonplanar, Doucette 1985; Doucette & Andren 1988)
224.0 (Sabljic 1987)
213.14 (planar, Hawker & Connell 1988a)
245.2 (planar, shorthand, Opperhuizen et al. 1988)
213.0 (Abramowitz & Yalkowsky 1990)
Heat of Fusion, kcal/mol:
Fugacity Ratio, F (assuming ΔS_{fusion} = 13.5 e.u.):
0.66 (Mackay et al. 1980; Shiu & Mackay 1986)

Water Solubility (g/m³ or mg/L at 25°C):
0.50 (Webb 1970; Hoover 1971)
1.88 (shake flask-GC/ECD, Wallnöfer et al. 1973; Hutzinger et al. 1974; quoted, NAS 1979; Erickson 1986; Sawhney 1987)
0.637 (shake flask-GC, Haque & Schmedding 1975)
0.62 (generator column-GC/ECD, Weil et al. 1974)
0.139 (shake flask-GC/ECD, Lee et al. 1979)
1.325 (calculated-TSA, Mackay et al. 1980)
1.17 (20°C, subcooled liq., shake flask-GC/ECD, Chiou et al. 1983; Chiou 1985; Chiou & Block 1986)

0.706 (quoted lit. average, Yalkowsky et al. 1983)
0.706, 0.245 (quoted, calculated-UNIFAC, Banerjee 1985)
1.314 (subcooled liquid, calculated-TSA, Burkhard et al. 1985b)
1.00 (selected, Shiu & Mackay 1986)
1.26 (quoted, Chou & Griffin 1987)
0.538 (20°C, subcooled liq., calculated-mole fraction of Aroclor mixtures, Murphy et al. 1987)
0.769 (calculated average of HPLC-RI, Brodsky & Ballschmiter 1988)
0.220 (calculated-UNIFAC, Banerjee & Howard 1988)
0.614, 0.239 (quoted, calculated-χ, Nirmalakhanden & Speece 1989)
1.118, 0.560 (quoted, calculated-TSA, Abramowitz & Yalkowsky 1990)
0.774, 0.658 (quoted average of Brodsky & Ballschmiter 1988, calculated- χ , Patil 1991)

Vapor Pressure (Pa at 25°C):
0.0755 (quoted exptl. value, Burkhard et al. 1985a)
0.279 (calculated-SxHLC, Burkhard et al. 1985a)
0.104, 0.0998, 0.0499 (calculated-MW, GC-RI, χ, Burkhard et al. 1985a)
0.147 (subcooled liquid, calculated, GC-RI, Burkhard et al. 1985b)
0.157, 0.143 (subcooled liquid, GC-RT, Foreman & Bidleman 1985)
0.0685 (20°C, subcooled liq., calculated-mole fraction of Aroclor mixtures, Murphy et al. 1987)

Henry's Law Constant (Pa m^3/mol):
96.7 (batch stripping, Atlas et al. 1982; Murphy et al. 1983)
24.93 (calculated-P/C, Burkhard et al. 1985b)
45.39 (calculated-P/C, Shiu & Mackay 1986)
22.29 (quoted, Eisenreich 1987)
28.37 (20°C, calculated-P/C, Murphy et al. 1987)
31.31 (calculated-χ, Sabljic & Gusten 1989)

Octanol/Water Partition Coefficient, log K_{OW}:
5.51 (Hansch & Leo 1979)
4.48 (HPLC-RT, Woodburn 1982)
5.32 (calculated-f const., Yalkowsky et al. 1983; quoted, Erickson 1986)
5.20 (calculated, Neely 1983)
5.10 (shake flask-GC, Chiou et al. 1983; Chiou 1985; Chiou & Block 1986)
5.03 (calculated-TSA, Burkhard 1984)
5.14 (generator column-GC/ECD, Woodburn et al. 1984; quoted, Sklarew & Girvin 1987)
5.51 (calculated-π, Woodburn et al. 1984)
4.47 (HPLC-RT, Rapaport & Eisenreich 1984)
5.10 (calculated-π, Rapaport & Eisenreich 1984; quoted, Eisenreich 1987; Sklarew

& Girvin 1987)

4.66 (calculated-S, Chou & Griffin 1987)

5.46, 5.22 (calculated-π, TSA, Doucette & Andren 1987)

5.14 (gen. column-GC/ECD, Doucette & Andren 1987,1988)

5.14, 4.81 (quoted, calculated-UNIFAC, Banerjee & Howard 1988)

5.14, 5.09 (quoted, calculated average of HPLC-RI, Brodsky & Ballschmiter 1988)

5.07 (calculated-TSA, Hawker & Connell 1988a)

5.46, 5.32, 5.10, 5.24, 5.32 (calculated-π, f const., MW, χ ,TSA, Doucette & Andren 1988)

5.09, 5.10 (quoted average of Brodsky & Ballschmiter 1988, calculated- χ , Patil 1991)

Bioconcentration Factor, log BCF:

3.83, 3.55, 3.99 (algae, fish, act. sludge, Freitag et al. 1984,1985; quoted, Halfon & Reggiani 1986)

Sorption Partition Coefficient, log K_{OC}:

3.90 (Haque & Schmedding 1976; quoted, Chou & Griffin 1987)

4.83 (suspended particulate matter, calculated-K_{OW}, Burkhard 1984)

5.90 (Lake Michigan water column, Swackhamer & Armstrong 1987)

4.89 (calculated after Karickhoff et al. 1979, Capel & Eisenreich 1990)

4.86 (calculated after Schwarzenbach & Westall 1981, Capel & Eisenreich 1990)

Sorption Partition Coefficient, log K_{OM}:

3.89 (soil org. matter, sorption isotherm, Chiou et al. 1983)

3.90, 4.17 (quoted, calculated- χ , Sabljic 1984)

Half-Lives in the Environment:

Air:

Surface water:

Groundwater:

Sediment:

Soil:

Biota:

Environmental Fate Rate Constants or Half-Lives:

Volatilization:

Photolysis:

Hydrolysis:

Oxidation: room temperature rate constant for reaction with OH radicals in the gas phase calculated to be 1.4×10^{-12} cm^3 $molecule^{-1}$ sec^{-1} with a calculated tropospheric lifetime of 8-17 days (Atkinson 1987).

Biodegradation:

Biotransformation:

Bioconcentration, Uptake (k_1) and Elimination (k_2) Rate Constants:

Common Name: 2,5-Dichlorobiphenyl
Synonym: PCB-9
Chemical Name: 2,5-dichlorobiphenyl
CAS Registry No: 34883-39-1
Molecular Formula: $C_{12}H_8Cl_2$
Molecular Weight: 223.1
Melting Point (°C):

22-23 (Weingarten 1962)
23 (Mackay et al. 1980; Bruggeman et al. 1982; Burkhard et al. 1985a; Opperhuizen et al. 1988)
36 (calculated, Abramowitz & Yalkowsky 1990)

Boiling Point (°C):

171 (15mm Hg, Erickson 1986))
312 (calculated, Mackay et al. 1982; Shiu & Mackay 1986)

Density (g/cm^3 at 20°C): 1.0536
Molar Volume (cm^3/mol):

226.4 (LeBas method, Miller et al. 1985; Shiu & Mackay 1986)
1.100 (intrinsic volume: $V_I/100$, Kamlet et al. 1988; Hawker 1990)

Molecular Volume (A^3):

234.9 (planar, shorthand, Opperhuizen et al. 1988)

Total Surface Area, TSA (A^2):

227.6 (Mackay et al. 1980; Yalkowsky et al. 1983; Shiu & Mackay 1986)
222.01 (Burkhard 1984)
212.85 (planar, Doucette 1985)
225.85 (nonplanar, Doucette 1985; Doucette & Andren 1988)
212.97 (planar, Hawker & Connell 1988a)
245.2 (planar, shorthand, Opperhuizen et al. 1988)
213.0 (Abramowitz & Yalkowsky 1990)

Heat of Fusion, kcal/mol:
Fugacity Ratio, F (assuming ΔS_{fusion} = 13.5 e.u.):

1.0 (Shiu & Mackay 1986)

Water Solubility (g/m^3 or mg/L at 25°C):

0.58 (generator column-GC/ECD, Weil et al. 1974)
2.028 (shake flask-GC, Chiou et al. 1977)
0.58, 2.09 (quoted, calculated-TSA, Mackay et al. 1980)
0.19 (gen. col.-GC/ECD, Bruggeman et al. 1981)
0.587 (TLC-RT, Bruggeman et al. 1982)
0.587 (selected lit. average, Yalkowsky et al. 1983; quoted, Erickson 1986)
0.200 (Bruggeman Ph.D. Thesis 1983 in Opperhuizen et al. 1988)
1.940 (generator column-GC/ECD, Miller et al. 1984; 1985)
1.33 (subcooled liquid, calculated-TSA, Burkhard et al. 1985b)
2.0 (selected, Shiu & Mackay 1986)

0.58 (quoted, Opperhuizen 1986)
0.587, 0.726 (quoted, calculated average of HPLC-RI, Brodsky & Ballschmiter 1988)
1.115 (generator column-GC/ECD, Dunnivant & Elzerman 1988)
1.95 (quoted, Isnard & Lambert 1988, 1989)
1.943 (quoted, subcooled liquid, Hawker 1989b)
0.56, 0.239 (quoted, calculated-χ, Nirmalakhanden & Speece 1989)
2.23, 0.888 (quoted, calculated-TSA, Abramowitz & Yalkowsky 1990)

Vapor Pressure (Pa at 25°C):
0.0775 (Augood et al. 1953)
0.184 (subcooled liquid, Yalkowsky et al. 1983)
0.184 (subcooled liq., GC-RT, Bidleman 1984; quoted, Erickson 1986)
0.157, 0.202, 0.114 (calculated-MW, GC-RI, χ, Burkhard et al. 1985a)
0.198 (subcooled liq., calculated, GC-RI, Burkhard et al. 1985b)
0.231, 0.232 (subcooled liq., GC-RT, Foreman & Bidleman 1985)
0.180 (selected, Shiu & Mackay 1986)
0.197 (subcooled liquid, calculated-SxHLC, Dunnivant & Elzerman 1988)

Henry's Law Constant (Pa m^3/mol):
33.13 (calculated-P/C, Burkhard et al. 1985b)
20.1 (calculated-P/C, Shiu & Mackay 1986)
39.31 (batch stripping, Dunnivant & Elzerman 1988)
39.31 (batch stripping, Dunnivant et al. 1988)

Octanol/Water Partition Coefficient, log K_{OW}:
5.18 (TLC-RT, Bruggeman et al. 1982; quoted, Mackay & Hughes 1984; Oliver & Niimi 1985; Erickson 1986)
5.16 (generator column-GC/ECD, Miller et al. 1984,1985; quoted, Sklarew & Girvin 1987)
5.2 (quoted, Bruggeman et al. 1984; Opperhuizen 1986)
5.37 (quoted, Kaiser 1983)
5.03 (calculated-TSA, Burkhard 1984)
4.67 (HPLC-RT, Rapaport & Eisenreich 1984)
5.30 (calculated-π, Rapaport & Eisenreich 1984; quoted, Sklarew & Girvin 1987)
5.30 (quoted, Hawker & Connell 1985; Thomann 1989)
5.10 (selected, Shiu & Mackay 1986; quoted, Gobas & Mackay 1987; Gobas et al. 1987; Sklarew & Girvin 1987; Clark et al. 1990)
5.16, 5.10 (quoted, calculated average of HPLC-RI, Brodsky & Ballschmiter 1988)
5.06 (calculated-TSA, Hawker & Connell 1988a; Connell & Hawker 1988; Hawker 1990)
5.10 (quoted, Hawker & Connell 1988b; Hawker 1989b,1990; Clark et al. 1990)
5.10, 5.30, 5.08 (quoted, calculated-π or f const., solvatochromic parameters, Kamlet et al. 1988)

5.16 (quoted exptl., Doucette & Andren 1988)

5.46, 5.32, 5.10, 5.24, 5.32 (calculated-π, f const., MW, χ , TSA, Doucette & Andren 1988)

5.16 (quoted, Isnard & Lambert 1988, 1989)

Bioconcentration Factor, log BCF:

5.68 (goldfish, 3% lipid by wt., Bruggeman et al. 1981)

5.45 (goldfish, 10% lipid dry wt. in food, Bruggeman et al. 1981)

5.72 (guppy, 3.5% lipid by wt., Bruggeman et al. 1982,1984; quoted, Gobas et al. 1987)

4.14, 3.90 (goldfish, exptl., correlated, Mackay & Hughes 1984)

4.38 (rainbow trout, kinetic, Oliver & Niimi 1985)

4.00 (rainbow trout, steady state, Oliver & Niimi 1985)

4.26 (guppy, Gobas et al. 1987)

4.15, 4.92 (quoted, Connell & Hawker 1988; Hawker 1990)

4.00 (quoted, fish, Isnard & Lambert 1988)

Sorption Partition Coefficient, log K_{OC}:

4.83 (suspended particulate matter, calculated-K_{OW}, Burkhard 1984)

Half-Lives in the Environment:

Air:

Surface water:

Groundwater:

Sediment:

Soil:

Biota: half-life in guppy, 6.5 days (Bruggeman et al. 1984); half-life in rainbow trout, 85 days (Niimi & Oliver 1983; Oliver & Niimi 1985), and its muscle, 56 days (Niimi & Oliver 1983).

Environmental Fate Rate Constants or Half-Lives:

Volatilization:

Photolysis:

Hydrolysis:

Oxidation: room temperature rate constant for reaction with OH radicals in the gas phase calculated to be 1.8×10^{-12} cm^3 $molecule^{-1}$ sec^{-1} with a calculated tropospheric lifetime of 8-17 days (Atkinson 1987).

Biodegradation:

Biotransformation:

Bioconcentration, Uptake (k_1) and Elimination (k_2) Rate Constants:

k_1: 920 day^{-1} (23°C, goldfish, 3 % lipid, Bruggeman et al. 1981; quoted, Waid 1986)

k_2: 0.066 day^{-1} (23°C, goldfish, 3 % lipid, Bruggeman et al. 1981; quoted

Waid 1986; Hawker & Connell 1988; Clark et al. 1990)

k_2: 0.008 day^{-1} (rainbow trout, Niimi & Oliver 1983; quoted, Clark et al. 1990)

k_1: 190 day^{-1} (guppy, Bruggeman et al. 1984)

k_2: 0.11 day^{-1} (guppy, Bruggeman et al. 1984; quoted, Clark et al. 1990)

k_2: 0.066, 0.0523 day^{-1} (goldfish, exptl., calculated, Mackay & Hughes 1984)

k_1: 280 day^{-1} (rainbow trout, Oliver & Niimi 1985)

k_2: 0.0082 day^{-1} (rainbow trout, Oliver & Niimi 1985)

k_1: 38.3, 49.2 hour^{-1} (goldfish, guppy, quoted, Hawker & Connell 1985)

$1/k_2$: 360, 220 hour (goldfish, guppy, quoted, Hawker & Connell 1985)

k_1: 1200 day^{-1} (guppy, Opperhuizen 1986)

k_1: 122 day^{-1} (rainbow trout, calculated, Gobas & Mackay 1987)

k_2: 0.0089 day^{-1} (rainbow trout, calculated, Gobas & Mackay 1987)

$\log k_1$: 2.96, 3.07 day^{-1} (fish, quoted, Connell & Hawker 1988)

$\log 1/k_2$: 0.96, 1.18 day (fish, quoted, Connell & Hawker 1988)

$\log 1/k_2$: 2.1, 2.3 hour (fish, quoted, calculated-K_{OW}, Hawker & Connell 1988b)

$\log k_2$: -0.96, -1.17 day^{-1} (fish, quoted, Thomann 1989)

Common Name: 2,6-Dichlorobiphenyl
Synonym: PCB-10
Chemical Name: 2,6-dichlorobiphenyl
CAS Registry No: 33146-45-1
Molecular Formula: $C_{12}H_8Cl_2$
Molecular Weight: 223.1
Melting Point (°C):
35-36 (Hutzinger et al. 1971)
36 (Burkhard et al. 1985a; Opperhuizen et al. 1988)
35, 46 (quoted, calculated, Abramowitz & Yalkowsky 1990)
Boiling Point (°C):
312 (calculated, Mackay et al. 1982; Shiu & Mackay 1986)
Density (g/cm^3):
Molar Volume (cm^3/mol):
226.4 (LeBas method, Miller et al. 1985; Shiu & Mackay 1986)
1.10 (intrinsic volume: $V_I/100$, Kamlet et al. 1988; Hawker 1989b,1990; De Bruijn & Hermens 1990)
Molecular Volume (A^3):
228.8 (planar, shorthand, Opperhuizen et al. 1988)
209.69, 189.66, 228.8 (De Bruijn & Hermens 1990)
Total Surface Area, TSA (A^2):
219.36 (Burkhard 1984)
205.87 (planar, Doucette 1985)
224.32 (nonplanar, Doucette 1985; Doucette & Andren 1988)
206.46 (planar, Hawker & Connell 1988a)
240.5 (planar, shorthand, Opperhuizen et al. 1988)
249.51, 214.95, 240.5 (De Bruijn & Hermens 1990)
206.0 (Abramowitz & Yalkowsky 1990)
Heat of Fusion, kcal/mol:
3.01 (Miller et al. 1984)
Entropy of Fusion, cal/mol K (e.u.):
9.80 (Miller et al. 1984; Shiu & Mackay 1986)
Fugacity Ratio, F (assuming ΔS_{fusion} = 13.5 e.u.):
0.801 (Shiu & Mackay 1986)

Water Solubility (g/m^3 or mg/L at 25°C):
1.452 (shake flask-GC, Chiou et al. 1977)
1.390 (generator column-GC/ECD, Miller et al. 1984,1985; quoted, Hawker 1989b)
1.704 (subcooled liquid, calculated-TSA, Burkhard et al. 1985b)
1.40 (selected, Shiu & Mackay 1986)
0.336 (calculated average of HPLC-RI, Brodsky & Ballschmiter 1988)
2.41 (gen. col.-GC/ECD, Dunnivant & Elzerman 1988)
0.540 (generator column-GC/ECD, Opperhuizen et al. 1988)

1.65 (quoted, subcooled liquid, Hawker 1989b)
1.38 (quoted, Isnard & Lambert 1989)
0.523, 0.239 (quoted, calculated-χ, Nirmalakhanden & Speece 1989)
1.408, 1.408 (quoted, calculated-TSA, Abramowitz & Yalkowsky 1990)

Vapor Pressure (Pa at 25°C):
0.122, 0.288, 0.114 (calculated-MW, GC-RI, χ, Burkhard et al. 1985a)
0.365 (subcooled liq., calculated, GC-RI, Burkhard et al. 1985b)
0.336, 0.371 (subcooled liquid, GC-RT, Foreman & Bidleman 1985)

Henry's Law Constant (Pa m^3/mol):
47.83 (calculated-P/C, Burkhard et al. 1985b)
47.61 (calculated- χ , Sabljic & Güsten 1989)

Octanol/Water Partition Coefficient, log K_{OW}:
5.30 (quoted, Kaiser 1983)
4.96 (calculated-TSA, Burkhard 1984)
4.05, 5.31 (HPLC-RT, calculated-π, Rapaport & Eisenreich 1984; quoted, Sklarew & Girvin 1987)
4.93 (gen. col.-GC/ECD, Miller et al. 1984,1985; quoted, Sklarew & Girvin 1987)

5.00 (selected, Shiu & Mackay 1986; quoted, Sklarew & Girvin 1987)
4.93, 4.99 (quoted, calculated average of HPLC-RI, Brodsky & Ballschmiter 1988)
4.84 (calculated-TSA, Hawker & Connell 1988a)
4.95 (calculated-solvatochromic parameters, Kamlet et al. 1988)
4.93 (quoted exptl., Doucette & Andren 1988)
5.46, 5.23, 5.10, 5.24, 5.28 (calculated-π, f const., MW, χ , TSA, Doucette & Andren 1988)
5.0 (quoted, Hawker 1989b)
4.93 (quoted, Isnard & Lambert 1989)
4.982 (slow stripping-GC, De Bruijn et al. 1989; De Bruijn & Hermens 1990)
5.02 (calculated-π, De Bruijn et al. 1989)

Bioconcentration Factor, log BCF:

Sorption Partition Coefficient, log K_{OC}:
4.76 (suspended particulate matter, calculated-K_{OW}, Burkhard 1984)

Half-Lives in the Environment:
Air:
Surface water:
Groundwater:
Sediment:

Soil:
Biota:

Environmental Fate Rate Constants or Half-Lives:
Volatilization:
Photolysis:
Hydrolysis:
Oxidation: room temperature rate constant for reaction with OH radicals in the gas phase calculated to be 1.8×10^{-12} cm^3 $molecule^{-1}$ sec^{-1} with a tropospheric lifetime of 8-17 days (Atkinson 1987).
Biodegradation:
Biotransformation:
Bioconcentration, Uptake (k_1) and Elimination (k_2) Rate Constants:

Common Name: 3,3'-Dichlorobiphenyl
Synonym: PCB-11
Chemical Name: 3,3'-dichlorobiphenyl
CAS Registry No: 2050-67-1
Molecular Formula: $C_{12}H_8Cl_2$
Molecular Weight: 223.1
Melting Point (°C):
        29
        29, 72 (quoted, calculated, Abramowitz & Yalkowsky 1990)
Boiling Point (°C):
        322-324 (Weast 1972,1973)
        312 (calculated, Mackay et al. 1982; Shiu & Mackay 1986)
Density (g/cm^3):
Molar Volume (cm^3/mol):
        226.4 (LeBas method, Shiu & Mackay 1986; Shiu et al. 1987)
Molecular Volume (A^3):
Total Surface Area, TSA (A^2):
        224.49 (Burkhard 1984)
        219.82 (planar, Doucette 1985)
        227.43 (nonplanar, Doucette 1985)
        219.47 (planar, Hawker & Connell 1988a)
        219.0 (Abramowitz & Yalkowsky 1990)
Heat of Fusion, kcal/mol:
Fugacity Ratio, F (assuming ΔS_{fusion} = 13.5 e.u.):
        0.912 (Shiu & Mackay 1986)

Water Solubility (g/m^3 or mg/L at 25°C):
        1.057 (subcooled liquid, calculated-TSA, Burkhard et al. 1985b)
        0.078 (calculated average of HPLC-RI, Brodsky & Ballschmiter 1988)
        0.354 (generator column-GC/ECD, Dunnivant & Elzerman 1988)
        0.223 (calculated-TSA, Abramowitz & Yalkowsky 1990)
        0.079, 0.658 (quoted average of Brodsky & Ballschmiter 1988, calculated- χ , Patil 1991)

Vapor Pressure (Pa at 25°C):
        0.0267 (gas saturation-GC/ECD, Westcott et al. 1981)
        0.0267 (GC-RT, Westcott & Bidleman 1981; Westcott et al. 1981)
        0.0290 (P_L calculated from P_S using F, Westcott et al. 1981)
        0.0865, 0.0952 (subcooled liquid, GC-RT, Bidleman 1984)
        0.0645 (subcooled liquid, Burkhard 1984)
        0.143, 0.0612, 0.114 (calculated-MW, GC-RI, χ, Burkhard et al. 1985a)
        0.0258 (quoted, exptl., Burkhard et al. 1985a)
        0.0646 (subcooled liq., calculated, GC-RI, Burkhard et al. 1985b)

0.091, 0.076, 0.0907 (subcooled liquid, GC-RT, Foreman & Bidleman 1985)
0.091 (quoted from Bidleman 1984, Erickson 1986)
0.09 (selected, subcooled liquid, Shiu & Mackay 1986)
0.0367 (calculated-SxHLC, Dunnivant & Elzerman 1988)
0.0413 (subcooled liquid, calculated-MP, Dunnivant & Elzerman 1988)

Henry's Law Constant (Pa m^3/mol):
13.58 (calculated-P/C, Burkhard et al. 1985b)
23.61 (batch stripping, Dunnivant & Elzerman 1988)
23.61 (batch stripping, Dinnivant et al. 1988)

Octanol/Water Partition Coefficient, log K_{OW}:
5.30 (shake flask-GC/ECD, Bruggeman et al. 1982)
5.34 (TLC-RT, Bruggeman et al. 1982; quoted, Erickson 1986)
5.10 (calculated-TSA, Burkhard 1984)
5.30 (selected, Shiu & Mackay 1986; quoted, Sklarew & Girvin 1987; Clark et al. 1990)
5.30, 5.27 (quoted, calculated average of HPLC-RI, Brodsky & Ballschmiter 1988)
5.28 (calculated-TSA, Hawker & Connell 1988a)
5.27, 5.10 (quoted average of Brodsky & Ballschmiter 1988, calculated- χ , Patil 1991)

Bioconcentration Factor, log BCF:

Sorption Partition Coefficient, log K_{OC}:
4.90 (suspended particulate matter, calculated-K_{OW}, Burkhard 1984)

Half-Lives in the Environment:
Air:
Surface water:
Groundwater:
Sediment:
Soil:
Biota: half-life in rainbow trout, 5 days and its muscle < 5 days (Niimi & Oliver 1983).

Environmental Fate Rate Constants or Half-Lives:
Volatilization:
Photolysis:
Hydrolysis:
Oxidation: room temperature rate constant for reaction with OH radicals in the gas phase calculated to be 2.7×10^{-12} cm^3 $molecule^{-1}$ sec^{-1} with a calculated tropospheric lifetime of 8-17 days (Atkinson 1987).
Biodegradation:
Biotransformation:

Bioconcentration, Uptake (k_1) and Elimination (k_2) Rate Constants:

k_2: 0.1385 day^{-1} (rainbow trout, Niimi & Oliver 1983; quoted, Clark et al. 1990)

Common Name: 3,4-Dichlorobiphenyl
Synonym: PCB-12
Chemical Name: 3,4-dichlorobiphenyl
CAS Registry No: 2974-92-7
Molecular Formula: $C_{12}H_8Cl_2$
Molecular Weight: 223.1
Melting Point (°C):
49-50 (Weingarten 1961)
50, 66 (quoted, calculated, Abramowitz & Yalkowsky 1990)
Boiling Point (°C):
195-200 (Weast 1972,1973)
312 (calculated, Mackay et al. 1982; Shiu & Mackay 1986)
Density (g/cm^3):
Molar Volume (cm^3/mol):
226.4 (LeBas method, Miller et al. 1985; Shiu & Mackay 1986)
Molecular Volume (A^3):
Total Surface Area, TSA (A^2):
222.82 (Burkhard 1984)
217.84 (planar, Doucette 1985)
225.41 (nonplanar, Doucette 1985; Doucette & Andren 1987; 1988)
217.73 (planar, Hawker & Connell 1988a)
218.0 (Abramowitz & Yalkowsky 1990)
Heat of Fusion, kcal/mol:
Fugacity Ratio, F (assuming ΔS_{fusion} = 13.5 e.u.):
0.566 (Shiu & Mackay 1986)

Water Solubility (g/m^3 or mg/L at 25°C):
1.233 (calculated-TSA, subcooled liquid, Burkhard et al. 1985b)
0.090 (calculated average of HPLC-RI, Brodsky & Ballschmiter 1988)
0.00791 (generator column-GC/ECD, Dunnivant & Elzerman 1988)
0.0138 (subcooled liquid, calculated-MP, Dunnivant & Elzerman 1988)
0.280 (calculated-TSA, Abramowitz & Yalkowsky 1990)
0.091, 0.658 (quoted average of Brodsky & Ballschmiter 1988, calculated- χ , Patil 1991)

Vapor Pressure (Pa at 25°C):
0.0532 (subcooled liquid, Burkhard 1984)
0.0888, 0.0313, 0.0499 (calculated-MW, GC-RI, χ, Burkhard et al. 1985a)
0.0532 (subcooled liq., calculated, GC-RT, Burkhard et al. 1985b)
0.078, 0.062 (subcooled liq., GC-RT, Foreman & Bidleman 1985)
0.00074 (calculated-SxHLC, Dunnivant & Elzerman 1988)
0.00129 (calculated-MP, Dunnivant & Elzerman 1988)

Henry's Law Constant (Pa m^3/mol):

9.60 (calculated-P/C, Burkhard et al. 1985b)
20.77 (batch stripping, Dunnivant & Elzerman 1988)
20.77 (batch stripping, Dunnivant et al. 1988)
14.18 (wetted-wall col.-GC, Brunner et al. 1990)

Octanol/Water Partition Coefficient, log K_{OW}:

5.51 (Hansch & Leo 1979)
5.10 (HPLC-RT, Woodburn 1982,1984)
5.51 (calculated-π, Woodburn et al. 1984)
5.05 (calculated-TSA, Burkhard 1984)
5.29 (gen. col.-GC, Woodburn et al. 1984; quoted, Sklarew & Girvin 1987)
5.30 (selected, Shiu & Mackay 1986; quoted, Sklarew & Girvin 1987)
5.46, 5.22 (calculated-π, TSA, Doucette & Andren 1987)
5.29 (generator column-GC/ECD, Doucette & Andren 1987)
5.29, 5.23 (quoted, calculated average of HPLC-RI, Brodsky & Ballschmiter 1988)
5.46, 5.32, 5.10, 5.24, 5.31 (calculated-π, f const., MW, χ, TSA, Doucette & Andren 1988)
5.22 (calculated-TSA, Hawker & Connell 1988a)
5.23, 5.10 (quoted average of Brodsky & Ballschmiter 1988, calculated-χ, Patil 1991)

Bioconcentration Factor, log BCF:

Sorption Partition Coefficient, log K_{OC}:

4.85 (suspended particulate matter, calculated-K_{OW}, Burkhard 1984)

Half-Lives in the Environment:

Air:
Surface water:
Groundwater:
Sediment:
Soil:
Biota:

Environmental Fate Rate Constants or Half-Lives:

Volatilization:
Photolysis:
Hydrolysis:
Oxidation: room temperature rate constant for reaction with OH radicals in the gas phase calculated to be 1.8×10^{-12} cm^3 $molecule^{-1}$ sec^{-1} with a calculated tropospheric lifetime of 8-17 days (Atkinson 1987).
Biodegradation: 100% degraded by Nocardia strain NCIB 10603 within one week (Baxter et al. 1975; quoted, Pal et al. 1980).

Biotransformation:
Bioconcentration, Uptake (k_1) and Elimination (k_2) Rate Constants:

Common Name: 3,5-Dichlorobiphenyl
Synonym: PCB-14
Chemical Name: 3,5-dichlorobiphenyl
CAS Registry No: 31883-41-5
Molecular Formula: $C_{12}H_8Cl_2$
Molecular Weight: 223.1
Melting Point (°C):
 36 (Hinkel & Hay 1928)
 31-32 (Weingarten 1961; Hutzinger et al. 1974)
 34 (Burkhard et al. 1985a)
 31, 72 (quoted, calculated, Abramowitz & Yalkowsky 1990)
Boiling Point (°C):
 166 (10mm Hg)
 312 (calculated, Mackay et al. 1982; Shiu & Mackay 1986)
Density (g/cm^3):
Molar Volume (cm^3/mol):
 226.4 (LeBas method, Shiu & Mackay 1986)
Molecular Volume (A^3):
Total Surface Area, TSA (A^2):
 224.57 (Burkhard 1984)
 219.82 (planar, Doucette 1985)
 227.39 (nonplanar, Doucette 1985)
 219.47 (planar, Hawker & Connell 1988a)
 219.0 (Abramowitz & Yalkowsky 1990)
Heat of Fusion, kcal/mol:
Fugacity Ratio, F (assuming ΔS_{fusion} = 13.5 e.u.):
 0.872 (Shiu & Mackay 1986)

Water Solubility (g/m^3 or mg/L at 25°C):
 1.05 (subcooled liq., calculated-TSA, Burkhard et al. 1985b)
 0.04 (calculated average of HPLC-RI, Brodsky & Ballschmiter 1988)

Vapor Pressure (Pa at 25°C):
 0.128, 0.0662, 0.114 (calculated-MW, GC-RI, χ , Burkhard et al. 1985a)
 0.0785 (subcooled liq., calculated-GC-RI, Burkhard et al. 1985b)
 0.126, 0.117 (subcooled liquid, GC-RT, Foreman & Bidleman 1985)

Henry's Law Constant (Pa m^3/mol):
 16.72 (calculated-P/C, Burkahrd et al. 1985b)
 49.55 (calculated- χ , Sabljic & Güsten 1989)

Octanol/Water Partition Coefficient, log K_{OW}:

5.37 (shake flask-GC/ECD, Bruggeman et al. 1982)

5.10 (calculated-TSA, Burkhard 1984)

5.20 (quoted, Oliver & Niimi 1985)

5.40 (selected, Shiu & Mackay 1986; quoted, Sklarew & Girvin 1987; Clark et al. 1990)

5.37, 5.41 (quoted, calculated average of HPLC-RI, Brodsky & Ballschmiter 1988)

5.28 (calculated-TSA, Hawker & Connell 1988a)

5.30 (quoted, Hawker & Connell 1988b)

Bioconcentration Factor, log BCF:

3.79 (rainbow trout, steady state, Oliver & Niimi 1985)

3.82 (rainbow trout, kinetic, Oliver & Niimi 1985)

Sorption Partition Coefficient, log K_{OC}:

4.90 (suspended particulate matter, calculated-K_{OW}, Burkhard 1984)

Half-Lives in the Environment:

Air:

Surface water:

Groundwater:

Sediment:

Soil:

Biota: half-life in rainbow trout, 15 days (Niimi & Oliver 1983; Oliver & Niimi 1985), in its muscle 14 days (Niimi & Oliver 1983).

Environmental Fate Rate Constants or Half-Lives:

Volatilization:

Photolysis:

Hydrolysis:

Oxidation: room temperature rate constant for reaction with OH radicals in the gas phase calculated to be 2.9×10^{-12} cm^3 $molecule^{-1}$ sec^{-1} with a calculated tropospheric lifetime of 8-17 days (Atkinson 1987).

Biodegradation:

Biotransformation:

Bioconcentration, Uptake (k_1) and Elimination (k_2) Rate Constants:

k_1: 310 day^{-1} (rainbow trout, Oliver & Niimi 1985)

k_2: 0.046 day^{-1} (rainbow trout, Oliver & Niimi 1985)

log $1/k_2$: 1.3, 2.5 hour (fish, selected, calculated-K_{OW}, Hawker & Connell 1988)

$1/k_2$: 21.7 day (rainbow trout, Clark et al. 1990)

Common Name: 4,4'-Dichlorobiphenyl
Synonym: PCB-15
Chemical Name: 4,4'-dichlorobiphenyl
CAS Registry No: 2050-68-2
Molecular Formula: $C_{12}H_8Cl_2$
Molecular Weight: 223.1
Melting Point (°C):

148-149 (McKillop et al. 1968; Hutzinger et al. 1974)
148-149 (Wallnöfer et al. 1973)
149-150 (NAS 1979)
149 (Mackay et al. 1980; Bruggeman et al. 1982; Burkhard et al. 1985a; Opperhuizen et al. 1988)
149, 135 (quoted, calculated, Abramowitz & Yalkowsky 1990)

Boiling Point (°C):

289-315 (Weast 1972,1973)
312 (calculated, Mackay et al. 1982; Shiu & Mackay 1986)

Density (g/cm^3 at 20°C): 1.0536
Molar Volume (cm^3/mol):

226.4 (LeBas method, Miller et al. 1985; Shiu & Mackay 1986; Gobas & Mackay 1987)
1.10 (intrinsic, $V_I/100$, Kamlet et al. 1988)

Molecular Volume (A^3):

241.0 (planar, shorthand, Opperhuizen et al. 1988)

Total Surface Area, TSA (A^2):

227.6 (Mackay et al. 1980; Shiu & Mackay 1986)
224.88 (Burkhard 1984)
219.82 (planar, Doucette 1985)
227.4 (nonplanar, Doucette 1985; Doucette & Andren 1988)
219.81 (planar, Hawker & Connell 1988a)
249.9 (planar, shorthand, Opperhuizen et al. 1988)
220.0 (Abramowitz & Yalkowsky 1990)

Heat of Fusion, kcal/mol:
Fugacity Ratio, F (assuming ΔS_{fusion} = 13.5 e.u.):

0.059 (Shiu & Mackay 1986)
0.0546 (Gobas et al. 1987)

Water Solubility (g/m^3 or mg/L at 25°C):

0.050 (Webb 1970)
0.060 (Hoover 1971)
0.080 (shake flask-GC/ECD, Wallnöfer et al. 1973; Hutzinger et al. 1974; quoted, NAS 1979; Erickson 1986; Sawhney 1987)
0.056 (generator column-GC/ECD, Weil et al. 1974)
0.062 (20°C, shake flask-GC, Chiou et al. 1977; Freed et al. 1977; quoted, Kenaga &

Goring 1980; Kenaga 1980; Mackay et al. 1980; Sklarew & Girvin 1987)

0.104 (calculated-TSA, Mackay et al. 1980)
0.046 (generator column-HPLC/UV, Billington 1982)
0.056 (TLC-RT, Bruggeman et al. 1982)
0.065 (gen. col.-HPLC, Huang 1983)
0.060 (quoted lit. average, Yalkowsky et al. 1983)
1.02 (subcooled liquid, calculated-TSA, Burkhard et al. 1985b)
0.060 (selected, Shiu & Mackay 1986)
0.080 (quoted, Chou & Griffin 1987)
0.058 (gen. col.-HPLC, Billington et al. 1988)
0.056, 0.094 (quoted, calculated average of HPLC-RI, Brodsky & Ballschmiter 1988)
0.0363 (generator column-GC/ECD, Dunnivant & Elzerman 1988)
0.0603 (selected, Isnard & Lambert 1989)
0.054, 0.239 (quoted, calculated-χ, Nirmalakhanden & Speece 1989)
0.056, 0.045 (quoted, calculated-TSA, Abramowitz & Yalkowsky 1990)
0.095, 0.674 (quoted average of Brodsky & Ballschmiter 1988, calculated- χ , Patil 1991)

Vapor Pressure (Pa at 25°C):

0.00253 (Knudsen-effusion technique, extrapolated, Smith et al. 1964)
0.043 (P_L calculated from P_S using F, Smith et al. 1964)
0.00253 (quoted, Neely 1983; Erickson 1986)
0.071, 0.084 (subcooled liquid, GC-RT, Bidleman 1984)
0.0508 (subcooled liquid, Burkhard 1984)
0.075, 0.059 (subcooled liquid, GC-RT, Foreman & Bidleman 1985)
0.00263 (quoted exptl., Burkhard et al. 1985a)
0.0093, 0.00313, 0.0219 (calculated-MW, GC-RI, χ, Burkhard et al. 1985a)
0.0508 (subcooled liq., calculated, GC-RI, Burkhard et al. 1985b)
0.080 (selected, subcooled liquid, Shiu & Mackay 1986)
0.035 (subcooled liq., calculated, Murphy et al. 1987)
0.00328 (calculated-SxHLC, Dunnivant & Elzerman 1988)
0.0548 (subcooled liquid, calculated-MP, Dunnivant & Elzerman 1988)

Henry's Law Constant (Pa m^3/mol):

30.4 (calculated-P/C, Murphy et al. 1983)
11.04 (calculated-P/C, Burkhard et al. 1985b)
14.69 (calculated, Coates & Elzerman 1986)
17.0 (calculated-P/C, Shiu & Mackay 1986)
20.16 (batch stripping, Dunnivant & Elzerman 1988)
20.16 (batch stripping, Dunnivant et al. 1988)
9.66 (wetted-wall column-GC/ECD, Fendinger & Glotfelty 1990)

Octanol/Water Partition Coefficient, log K_{OW}:

5.58 (shake flask-GC, Chiou et al. 1977; quoted, Kenaga & Goring 1980; Mackay et al. 1980; Oliver & Niimi 1984; Sklarew & Girvin 1987)
5.51 (calculated-π, Tulp & Hutzinger 1978)
5.17 (HPLC-RT, Sugiura et al. 1978)
5.51 (Hansch & Leo 1979)
5.57 (calculated, Mackay et al. 1980a)
5.36 (shake flask-GC, Bruggeman et al. 1982)
5.28 (TLC-RT, Bruggeman et al. 1982; quoted, Erickson 1986)
4.92 (HPLC-RT, Woodburn 1982,1984)
4.77 (calculated, Neely 1983)
5.32 (calculated-f const., Yalkowsky et al. 1983)
5.11 (calculated-TSA, Burkhard 1984)
5.33 (generator column-GC/ECD, Woodburn et al. 1984; quoted, Sklarew & Girvin 1987)
5.51 (calculated-π, Woodburn et al. 1984)
4.82 (HPLC-RT, Rapaport & Eisenreich 1984; quoted, Sklarew & Girvin 1987)
5.57 (calculated-π, Rapaport & Eisenreich 1984)
5.30 (selected, Shiu & Mackay 1986; quoted, Gobas et al. 1987; Sklarew & Girvin 1987)
5.54 (calculated-S, Chou & Griffin 1987)
5.46, 5.25 (calculated-π, TSA, Doucette & Andren 1987)
5.33 (generator column-GC/ECD, Doucette & Andren 1987)
5.35, 5.23 (quoted, calculated average of HPLC-RI, Brodsky & Ballschmiter 1988)
5.30 (calculated-TSA, Hawker & Connell 1988a)
5.53 (quoted exptl. value, Doucette & Andren 1988)
5.46, 5.32, 5.10, 5.23, 5.36 (calculated-π, f const., MW, χ , TSA, Doucette & Andren 1988)
5.58 (quoted, Isnard & Lambert 1989)
5.23, 5.10 (quoted average of Brodsky & Ballschmiter 1988, calculated- χ , Patil 1991)

Bioconcentration Factor, log BCF:

2.97 (killifish, Goto et al. 1978)
2.33 (fish, flowing water, Kenaga & Goring 1980; Kenaga 1980; quoted, Bysshe 1982)
3.47 (calculated-S, Kenaga 1980)
3.58 (rainbow trout, highest value-nonequilibrated, Oliver & Niimi 1984)
4.10 (Sabljic 1987)
4.27 (picea omorika, Reischl et al. 1989 from Reischl 1988)

Sorption Partition Coefficient, log K_{OC}:

4.30 (calculated, Kenaga 1980)
4.91 (suspended particulate matter, calculated-K_{OW}, Burkhard 1984)
5.65 (EPA-B2 river sediment, Coates & Elzerman 1986)

Half-Lives in the Environment:

Air:

Surface water: half life in Lake Michigan, 57.5 days (Neely 1983).

Groundwater:

Sediment:

Soil:

Biota: half-life in picea omorika, 29 days (Reischl et al. 1989 from Reischl 1988).

Environmental Fate Rate Constants or Half-Lives:

Volatilization: estimated evaporation half-life for an initial concentration of 0.03 ppm in a 4.5 cm depth of water solution in a glass dish is 4.5 hours and 1.7 hours with stirring of the solution; while experimental observed half-life under the same condition is 4.0 hours and 1.5 hours with stirring of the solution (Chiou et al. 1979).

Photolysis:

Hydrolysis:

Oxidation: room temperature rate constant for reaction with OH radicals in the gas phase calculated to be 1.4×10^{-12} cm^3 $molecule^{-1}$ sec^{-1} with a calculated tropospheric lifetime of 8-17 days (Atkinson 1987).

Biodegradation: 50-80% biodegraded by Alkaligenes sp. strain Y-42 from lake sediments within 7-hour period (Furukawa & Matsumura 1976; quoted, Pal et al. 1980).

Biotransformation:

Bioconcentration, Uptake (k_1) and Elimination (k_2) Rate Constants:

Common Name: 2,2',3-Trichlorobiphenyl
Synonym: PCB-16
Chemical Name: 2,2',3-trichlorobiphenyl
CAS Registry No: 38444-78-9
Molecular Formula: $C_{12}H_7Cl_3$
Molecular Weight: 257.54
Melting Point (°C):
28.1-28.8 (Weingarten 1961; Hutzinger et al. 1974; NAS 1979)
28.5 (Burkhard et al. 1985a)
28, 32 (quoted, calculated, Abramowitz & Yalkowsky 1990)
Boiling Point (°C):
337 (calculated, Mackay et al. 1982; Shiu & Mackay 1986)
Density (g/cm^3):
Molar Volume (cm^3/mol):
247.3 (LeBas method, Miller et al. 1985)
247.3 (LeBas method, Shiu & Mackay 1986)
1.190 (intrinsic volume: $V_I/100$, Kamlet et al. 1988)
Molecular Volume (A^3):
Total Surface Area, TSA (A^2):
234.30 (Burkhard 1984)
251.11 (planar, Doucette 1985)
239.86 (nonplanar, Doucette 1985)
215.69 (planar, Hawker & Connell 1988a)
216.0 (Abramowitz & Yalkowsky 1990)
Heat of Fusion, kcal/mol:
Fugacity Ratio, F (assuming ΔS_{fusion} = 13.5 e.u.):
0.934 (Shiu & Mackay 1986)

Water Solubility (g/m^3 or mg/L at 25°C):
0.505 (subcooled liq., calculated-TSA, Burkhard et al. 1985b)
0.293 (20°C, subcooled liq., calculated-mole fraction of Aroclor mixtures, Murphy et al. 1987)
0.205 (calculated of HPLC-RI, Brodsky & Ballschmiter 1988)
0.814 (calculated-TSA, Abramowitz & Yalkowsky 1990)
0.205, 0.174 (quoted of Brodsky & Ballschmiter 1988, calculated- χ , Patil 1991)

Vapor Pressure (Pa at 25°C):
0.0522, 0.066, 0.033 (calculated-MW,GC-RI,χ, Burkhard et al. 1985a)
0.069 (subcooled liq., calculated-GC-RI, Burkhard et al. 1985b)
0.0538, 0.060 (subcooled liq., GC-RT, Foreman & Bidleman 1985)
0.0275 (20°C, subcooled liq., calculated-mole fraction of Aroclor mixtures, Murphy et al. 1987)

Henry's Law Constant (Pa m^3/mol):

35.16 (calculated-P/C, Burkhard et al. 1985b)
80.0 (calculated-P/C, Shiu & Mackay 1986)
24.11 (20°C, calculated-P/C, Murphy et al. 1987)
81.77 (batch stripping, Atlas et al. 1982)
28.07 (calculated-χ, Sabljic & Güsten 1989)
20.27 (wetted-wall col.-GC, Brunner et al. 1990)

Octanol/Water Partition Coefficient, log K_{OW}:

5.36 (calculated-TSA, Burkhard 1984)
4.15 (HPLC-RT, Rapaport & Eisenreich 1984)
5.31 (calculated-π, Rapaport & Eisenreich 1984; quoted, Sklarew & Girvin 1987)
5.60 (selected, Shiu & Mackay 1986; quoted, Sklarew & Girvin 1987)
5.12 (calculated-HPLC-RI, Brodsky & Ballschmiter 1988)
5.16 (calculated-TSA, Hawker & Connell 1988a)
5.12, 5.49 (quoted of Brodsky & Ballschmiter 1988, calculated- χ , Patil 1991)

Bioconcentration Factor, log BCF:

Sorption Partition Coefficient, log K_{OC}:

5.16 (suspended particulate matter, Burkhard 1984)

Half-Lives in the Environment:

Air:
Surface water:
Groundwater:
Sediment:
Soil:
Biota:

Environmental Fate Rate Constants or Half-Lives:

Volatilization:
Photolysis:
Hydrolysis:
Oxidation:
Biodegradation: 50% degraded by Nocardia strain NCIB 10603 within 7 days (Baxter et al. 1975; quoted, Pal et al. 1980).
Biotransformation:
Bioconcentration, Uptake (k_1) and Elimination (k_2) Rate Constants:

Common Name: 2,2',5-Trichlorobiphenyl
Synonym: PCB-18
Chemical Name: 2,2',5-trichlorobiphenyl
CAS Registry No: 37680-65-2
Molecular Formula: $C_{12}H_7Cl_3$
Molecular Weight: 257.54
Melting Point (°C):
43-44 (Hutzinger et al. 1974; NAS 1979; Erickson 1986)
44 (Mackay et al. 1980; Bruggeman et al. 1982; Burkhard et al. 1985a; Opperhuizen et al. 1988)
44, 20 (quoted, calculated, Abramowitz & Yalkowsky 1990)
Boiling Point (°C):
337 (calculated, Mackay et al. 1982; Shiu & Mackay 1986)
Density (g/cm^3 at 20°C): 1.1485
Molar Volume (cm^3/mol):
247.3 (LeBas method, Miller et al. 1985)
247.3 (LeBas methjod, Shiu & Mackay 1986)
1.190 (intrinsic volume: $V_I/100$, Kamlet et al. 1988; Hawker 1990)
Molecular Volume (A^3):
245.8 (planer, shorthand, Opperhuizen et al. 1988)
Total Surface Area, TSA (A^2):
241.98 (Mackay et al. 1980)
236.93 (Burkhard 1984)
217.10 (planar, Doucette 1985)
241.84 (nonplanar, Doucette 1985; Doucette & Andren 1987; 1988))
218.32 (planar, Hawker & Connell 1988a)
253.4 (planar, shorthand, Opperhuizen et al. 1988)
218.0 (Abramowitz & Yalkowsky 1990)
Heat of Fusion, kcal/mol:
Fugacity Ratio, F (assuming ΔS_{fusion} = 13.5 e.u.):
0.648 (Mackay et al. 1980)
0.651 (Shiu & Mackay 1986)

Water Solubility (g/m^3 or mg/L at 25°C):
0.640 (shake flask-GC/ECD, Weil et al. 1974; quoted, Könemann 1981)
0.248 (shake flask-GC/ECD, Haque & Schmedding 1975; quoted, Haque et al. 1980)
0.016 (radioactive isotope-^{14}C labelled, Metcalf et al. 1975)
0.0614 (shake flask-GC/ECD, Lee et al. 1979)
0.016 (NAS 1979)
0.085 (Kenaga & Goring 1980; Kenaga 1980)
0.407 (calculated-TSA, Mackay et al. 1980)
0.110 (GC/ECD, Bruggeman et al. 1981)
0.103 (TLC-RT, Bruggeman et al. 1982)

0.135 (quoted lit. average, Yalkowsky et al. 1983; selected, Erickson 1986)
0.402 (subcooled liq., calculated-TSA, Burkhard et al. 1985b)
0.40 (selected, Shiu & Mackay 1986)
0.299 (20°C, subcooled liq., calculated-mole fraction of Aroclor mixtures, Murphy et al. 1987)
0.647, 0.115 (quoted, calculated average of HPLC-RI, Brodsky & Ballschmiter 1988)
0.510 (generator column-GC/ECD, Dunnivant & Elzerman 1988)
0.780 (subcooled liquid, calculated-MP, Dunnivant & Elzerman 1988)
0.45 (quoted, Isnard & Lambert 1989)
0.0618, 0.0814 (quoted, calculated- χ , Nirmalakhanden & Speece 1989)
0.408, 0.814 (quoted, calculated-TSA, Abramowitz & Yalkowsky 1990)
0.115, 0.174 (quoted average of Brodsky & Ballschmiter 1988, calculated- χ , Patil 1991)

Vapor Pressure (Pa at 25°C):
0.267 (Neely 1981)
0.412 (P_S calculated from P_L using F, Neely 1981)
0.012 (Neely 1983; quoted, Erickson 1986)
0.0904 (subcooled liquid, Burkhard 1984)
0.0367, 0.0605, 0.033 (calculated-MW,GC-RI,χ, Burkhard et al. 1985a)
0.0897 (calculated-SxHLC, Burkhard et al. 1985a)
0.0904 (subcooled liquid, GC-RT, Burkhard et al. 1985b)
0.0776, 0.0833 (subcooled liquid, GC-RT, Foreman & Bidleman 1985)
0.143 (selected, Shiu & Mackay 1986; quoted, Sklarew & Girvin 1987)
0.22 (selected, subcooled liquid, Shiu & Mackay 1986)
0.0352 (20°C, subcooled liq., calculated-mole fraction of Aroclor mixtures, Murphy et al. 1987)
0.0762 (calculated-SxHLC, Dunnivant & Elzerman 1988)
0.117 (subcooled liquid, calculated-MP, Dunnivant & Elzerman 1988)

Henry's Law Constant (Pa m^3/mol):
101.53 (23°C, batch stripping-GC, Atlas et al. 1982)
20.26 (calculated-P/C, Murphy et al. 1983)
58.06 (calculated-P/C, Burkhard et al. 1985b)
20.26 (20°C, batch stripping, Oliver 1985)
92.21 (calculated-P/C, Shiu & Mackay 1986)
30.30 (20°C, calculated-P/C, Murphy et al. 1987)
38.5 (batch stripping, Dunnivant & Elzerman 1988)
25.33 (wetted-wall column-GC, Brunner et al. 1990)

Octanol/Water Partition Coefficient, log K_{OW}:
3.89 (radiolabelled-^{14}C, Metcalf et al. 1975)
6.20 (calculated-π, Tulp & Hutzinger 1978)

6.22 (shake flask, Hansch & Leo 1979)
3.89 (NAS 1979)
5.88 (calculated after Rekker 1977, Könemann 1981)
5.63 (calculated, Neely 1983)
5.64 (TLC-RT, Bruggeman et al. 1982; quoted, Erickson 1986)
4.34 (HPLC-RT, Woodburn 1982)
5.59 (gen. col.-GC/ECD, Woodburn 1982; quoted, Mackay & Hughes 1984)
5.40 (quoted, Voice et al. 1983; Voice & Weber 1985)
6.00 (calculated-f const., Yalkowsky et al. 1983)
5.44 (calculated-TSA, Burkhard 1984)
5.60 (quoted, Oliver & Charlton 1984)
4.39 (HPLC-RT, Rapaport & Eisenreich 1984)
5.55 (calculated-π, Rapaport & Eisenreich; quoted, Sklarew & Girvin 1987)
5.60 (generator column-GC/ECD, Woodburn et al. 1984; quoted, Sklarew & Girvin 1987)
4.34 (HPLC-RT, Woodburn et al. 1984)
5.60 (shake flask-GC, Chiou et al. 1985; quoted, Oliver & Niimi 1985)
5.60 (selected, Shiu & Mackay 1986; quoted, Sklarew & Girvin 1987; Clark et al. 1990)
5.59 (quoted, Hawker & Connell 1985; Thomann 1989)
5.55 (quoted, Eisenreich 1987)
4.97, 5.68 (HPLC-k', calculated, De Kock & Lord, 1987)
5.60 (generator column-GC/ECD, Doucette & Andren 1987,1988)
6.16, 5.60 (calculated-π, TSA, Doucette & Andren 1987)
5.60 (calculated-chlorine substituent, Oliver 1987a & c)
6.16, 6.0, 5.44, 5.65, 5.71 (calculated-π, f const. MW, χ , TSA, Doucette & Andren 1988)
5.24 (calculated-TSA, Hawker & Connell 1988a)
5.20 (quoted, Hawker & Connell 1988b)
5.60 (quoted, Isnard & Lambert 1989)
5.24 (quoted, Hawker 1990)

Bioconcentration Factor, log BCF:
3.86, 3.76, 2.91, 3.81 (algae, snail, mosquito, fish, ^{14}C-labelled, Metcalf et al. 1975)
1.72 (green sunfish, 15 days in static water, Sanborn et al. 1975)
4.08 (NAS 1979)
3.39 (calculated-S, Kenaga 1980)
5.52 (goldfish, 10% lipid by wt. in food, Bruggeman et al. 1981)
5.83 (goldfish, 3% lipid, Bruggeman et al. 1981)
4.30, 4.31 (goldfish, exptl., correlated, Mackay & Hughes 1984)
4.91 (rainbow trout, kinetic, Oliver & Niimi 1985)
4.23 (rainbow trout, steady state, Oliver & Niimi 1985)
5.77 (rainbow trout, field data, Oliver & Niimi 1985)

4.30 (fish, calculated-C_B/C_W or k_1/k_2, Connell & Hawker 1988; Hawker 1990)
3.75 (picea omorika, Reischl et al. 1989 from Reischl 1988)
4.23 (fish, Isnard & Lambert 1989)
6.87 (rainbow trout, quoted, Thomann 1989)

Sorption Partition Coefficient, log K_{OC}:
4.23 (calculated, Kenaga 1980)
5.24 (suspended particulate matter, Burkhard 1984)
5.4 (field data, Oliver & Charlton 1984)
5.5 (Niagara River-org. matter, Oliver & Charlton 1984)
5..2 (calculated-K_{OW}, Oliver & Charlton 1984)
4.5 (sediment, Voice & Weber 1985)
5.1-6.3, 5.5 (suspended sediment, average, Oliver 1987a)
7.0 (algae > 50 μm, Oliver 1987a)
4.57 (natural solids, Chin & Weber 1989)
5.34 (calculated after Karickhoff et al. 1979, Capel & Eisenreich 1990)
4.49 (calculated after Schwarzenbach & Westall 1981, Capel & Eisenreich 1990)
4.85, 4.15 (Adrich humic acid substrate with methyl salicylate, organic polymers present in Huron River water, Chin et al. 1990)

Half-Lives in the Environment:
Air:
Surface water: half-life in Lake Michigan, 43.1 days (Neely 1983).
Groundwater:
Sediment:
Soil:
Biota: half-life in rainbow trout, 190 days and its muscle 86 days (Niimi & Oliver 1983); in fish, 190 days (Niimi & Oliver 1983; Oliver & Niimi 1985); in worms at 8°C, 26 days (Oliver 1987c); in omorika, 25 days (Reischl et al. 1989).

Environmental Fate Rate Constants or Half-Lives:
Volatilization:
Photolysis:
Hydrolysis:
Oxidation:
Biodegradation: 50-80% being degraded by alkaligenes sp. strain Y-42 from lake sediments within 7-hour period (Furukawa & Matsumura 1976; quoted, Pal et al. 1980).
Biotransformation:
Bioconcentration, Uptake (k_1) and Elimination (k_2) Rate Constants:
k_1: 950 day^{-1} (23°C, goldfish, 3% lipid content, Bruggeman et al. 1981; quoted, Waid 1986)
k_2: 0.048 day^{-1} (23°C, goldfish, 3% lipid content, Bruggeman et al.

1981; quoted, Waid 1986; Clark et al. 1990)

k_2: 0.0037 day^{-1} (rainbow trout, Niimi & Oliver 1983; quoted, Clark 1990)

k_2: 0.048, 0.0372 day^{-1} (goldfish, exptl., correlated, Mackay & Hughes 1984)

k_1: 300 day^{-1} (rainbow trout, Oliver & Niimi 1985)

k_2: 0.0037 day^{-1} (rainbow trout, Oliver & Niimi 1985)

k_1: 39.6 hour^{-1} (goldfish, quoted, Hawker & Connell 1985)

$1/k_2$: 500 hour (goldfish, quoted, Hawker & Connell 1985)

log k_1: 2.98 day^{-1} (goldfish, quoted, Connell & Hawker 1988b)

log $1/k_2$: 1.32 day (goldfish, quoted, Connell & Hawker 1988b; Thomann 1989)

log $1/k_2$: 2.4, 2.4 day (fish, quoted, calculated-K_{OW}, Hawker & Connell 1988b).

Common Name: 2,3,3'-Trichlorobiphenyl
Synonym: PCB-20
Chemical Name: 2,3,3'-trichlorobiphenyl
CAS Registry No: 38444-84-7
Molecular Formula: $C_{12}H_7Cl_3$
Molecular Weight: 257.54
Melting Point (°C):
    58 (calculated, Abramowitz & Yalkowsky 1990)
Boiling Point (°C):
    337 (calculated, Mackay et al. 1982; Shiu & Mackay 1986)
Density (g/cm^3):
Molar Volume (cm^3/mol):
    247.3 (LeBas method, Miller et al. 1985)
    247.3 (LeBas method, Shiu & Mackay 1986)
Molecular Volume (A^3):
Total Surface Area, TSA (A^2):
    236.89 (Burkhard 1984)
    228.39 (planar, Doucette 1985)
    241.44 (nonplanar, Doucette 1985)
    227.86 (planar, Hawker & Connell 1988a)
    228.0 (Abramowitz & Yalkowsky 1990)
Heat of Fusion, kcal/mol:
Fugacity Ratio, F (assuming ΔS_{fusion} = 13.5 e.u.):

Water Solubility (g/m^3 or mg/L at 25°C):
    0.402 (subcooled liq., calculated-TSA, Burkhard et al. 1985b)
    0.162 (calculated-TSA, Abramowitz & Yalkowsky 1990)
    0.174 (calculated- χ , Patil 1991)

Vapor Pressure (Pa at 25°C):
    0.033 (calculated- χ , Burkhard et al. 1985a)
    0.0566, 0.0283, 0.0330 (subcooled liq., calculated-MW,GC-RI, χ , Burkhard et al. 1985a)
    0.0269 (subcooled liq., calculated-GC-RI, Burkhard et al. 1985b)
    0.0285, 0.249 (subcooled liq., GC-RT, Foreman & Bidleman 1985)

Henry's Law Constant (Pa m^3/mol):
    81.77 (batch stripping, Atlas et al. 1982)
    17.23 (calculated-P/C, Burkhard et al. 1985b)
    82.0 (selected, Shiu & Mackay 1986)
    30.7 (calculated- χ , Sabljic & Güsten 1989)
    16.21 (wetted-wall col.-GC, Brunner et al. 1990)

Octanol/Water Partition Coefficient, log K_{OW}:

5.43 (calculated-TSA, Burkhard 1984)

4.99, 5.57 (HPLC-RT, calculated-π, Rapaport & Eisenreich 1984)

5.60 (selected, Shiu & Mackay 1986)

5.57 (calculated-TSA, Hawker & Connell 1988a)

5.60, 5.49 (quoted of Shiu & Mackay 1986, calculated- χ , Patil 1991)

Bioconcentration Factor, log BCF:

Sorption Partition Coefficient, log K_{OC}:

5.23 (suspended particulate matter, Burkhard 1984)

Half-Lives in the Environment:

Air:

Surface water:

Groundwater:

Sediment:

Soil:

Biota:

Environmental Fate Rate Constants or Half-Lives:

Volatilization:

Photolysis:

Hydrolysis:

Oxidation:

Biodegradation:

Biotransformation:

Bioconcentration, Uptake (k_1) and Elimination (k_2) Rate Constants:

Common Name: 2,3,4-Trichlorobiphenyl
Synonym: PCB-21
Chemical Name: 2,3,4-trichlorobiphenyl
CAS Registry No: 55702-46-0
Molecular Formula: $C_{12}H_7Cl_3$
Molecular Weight: 257.54
Melting Point (°C):
101-102 (Hutzinger et al. 1974)
102 (Burkhard et al. 1985a)
79 (calculated, Abramowitz & Yalkowsky 1990)
Boiling Point (°C):
Density (g/cm^3):
Molar Volume (cm^3/mol):
247.3 (LeBas method, Shiu & Mackay 1986)
1.19 (intrinsic volume: V_I/100, De Bruijn & Hermens 1990)
Molecular Volume (A^3):
223.67, 203.54, 244.5 (De Bruijn & Hermens 1990)
Total Surface Area, TSA (A^2):
235.16 (Burkhard 1984)
226.40 (planar, Doucette 1985)
239.41 (nonplanar, Doucette 1985)
226.11 (planar, Hawker & Connell 1988a)
226.0 (Abramowitz & Yalkowsky 1990)
266.22, 231.81, 253.1 (De Bruijn & Hermens 1990)
Heat of Fusion, kcal/mol:
Fugacity Ratio, F (assuming ΔS_{fusion} = 13.5 e.u.): 0.173

Water Solubility (g/m^3 or mg/L at 25°C):
0.469 (calculated-TSA, Burkhard et al. 1985b)
0.169 (calculated average of HPLC-RI, Brodsky & Ballschmiter 1988)
0.103 (calculated-TSA, Abramowitz & Yalkowsky 1990)
0.170, 0.174 (quoted average of Brodsky & Ballschmiter 1988, calculated- χ , Patil 1991)

Vapor Pressure (Pa at 25°C):
0.0098, 0.00489, 0.0219 (calculated-MW, GC-RI, χ , Burkhard et al. 1985a)
0.0269 (subcooled liq., calculated, GC-RI, Burkhard et al. 1985b)

Henry's Law Constant (Pa m^3/mol):
14.79 (calculated-P/C, Burkhard et al. 1985b)
21.38 (calculated- χ , Sabljic & Güsten 1989)

Octanol/Water Partition Coefficient, log K_{OW}:

5.88 (calculated after Rekker 1977, Könemann 1981)
5.77 (calculated-π, Bruggeman et al. 1982)
5.39 (calculated-TSA, Burkhard 1984)
5.68 (calculated average of HPLC-RI, Brodsky & Ballschmiter 1988)
5.51 (calculated-TSA, Hawker & Connell 1988a)
5.77 (calculated-π, De Bruijn et al. 1989)
5.86 (slow stirring-GC, De Bruijn et al. 1989)
5.86 (slow stirring-GC, De Bruijn & Hermens 1990)
5.68, 5.49 (quoted average of Brodsky & Ballschmiter 1988, calculated- χ , Patil 1991)

Bioconcentration Factor, log BCF:

Sorption Partition Coefficient, log K_{OC}

5.19 (Suspended particulate matter, Burkhard 1984)

Half-Lives in the Environment:

Air:
Surface water:
Groundwater:
Sediment:
Soil:
Biota:

Environmental Fate Rate Constants or Half-Lives:

Volatilization:
Photolysis:
Hydrolysis:
Oxidation:
Biodegradation:
Biotransformation:
Bioconcentration, Uptake (k_1) and Elimination (k_2) Rate Constants:

Common Name: 2,3',5-Trichlorobiphenyl
Synonym: PCB-26
Chemical Name: 2,3',5-trichlorobiphenyl
CAS Registry No: 38444-81-4
Molecular Formula: $C_{12}H_7Cl_3$
Molecular Weight: 257.54
Melting Point (°C):
    40.0-40.5
    46 (calculated, Abramowitz & Yalkowsky 1990)
Boiling Point (°C):
Density (g/cm³):
Molar Volume (cm³/mol):
    247.3 (LeBas method, Shiu & Mackay 1986)
Molecular Volume (A³):
Total Surface Area, TSA (A²):
    239.52 (Burkhard 1984)
    230.37 (planar, Doucette 1985)
    243.42 (nonplanar, Doucette 1985)
    230.49 (planar, Hawker & Connell 1988a)
    230.0 (Abramowitz & Yalkowsky 1990)
Heat of Fusion, kcal/mol:
Fugacity Ratio, F (assuming ΔS_{fusion} = 13.5 e.u.): 0.703

Water Solubility (g/m³ or mg/L at 25°C):
    0.319 (subcooled liquid, calculated-TSA, Burkhard et al. 1985b)
    0.184 (calculated average of HPLC-RI, Brodsky & Ballschmiter 1988)
    0.253 (gen. col.-GC/ECD, Dunnivant & Elzerman 1988)
    0.138 (20°C, calculated-mole fraction of Aroclor mixtures, Murphy et al. 1987)
    0.162 (calculated-TSA, Abramowitz & Yalkowsky 1990)
    0.187, 0.178 (quoted average of Brodsky & Ballschmiter 1988, calculated- χ , Patil 1991)

Vapor Pressure (Pa at 25°C):
    0.0353 (subcooled liq., Burkhard et al. 1984)
    0.0402, 0.0262, 0.033 (calculated-MW,GC-RI,χ, Burkhard et al. 1985a)
    0.0363 (subcooled liq., calculated-GC-RI, Burkhard et al. 1985b)
    0.0411, 0.0412 (subcooled liq., GC-RT, Foreman & Bidleman 1985)
    0.0182 (20°C, subcooled liq., calculated-mole fraction of Aroclor mixtures, Murphy et al. 1987)
    0.0323 (calculated-SxHLC, Dunnivant & Elzerman 1988)
    0.0459 (subcooled liquid, calculated-M.P., Dunnivant & Elzerman 1988)

Henry's Law Constant (Pa m^3/mol):

28.37 (calculated, Burkhard et al. 1985b)
34.34 (20°C, calculated-P/C, Murphy et al. 1987)
32.93 (batch stripping, Dunnivant et al. 1988)
32.93 (batch stripping, Dunnivant & Elzerman 1988)
20.27 (wetted-wall col.-GC, Brunner et al. 1990)

Octanol/Water Partition Coefficient, log K_{OW}:

5.88 (calculated after Rekker 1977, Könemann 1981)
5.51 (calculated-TSA, Burkhard 1984)
5.18, 5.76 (HPLC-RT, calculated-π, Rapaport & Eisenreich 1984)
5.50 (selected, Shiu & Mackay 1986)
5.76 (quoted, Eisenreich 1987)
5.65 (calculated average of HPLC-RI, Brodsky & Ballschmiter 1988)
5.66 (calculated-TSA, Hawker & Connell 1988a)
5.65, 5.48 (quoted average of Brodsky & Ballschmiter 1988, calculated- χ , Patil 1991)

Bioconcentration Factor, log BCF:

Sorption Partition Coefficient, log K_{OC}:

5.31 (suspended particulate matter, Burkhard 1984)

Half-Lives in the Environment:

Air:
Surface water:
Groundwater:
Sediment:
Soil:
Biota:

Environmental Fate Rate Constants or Half-Lives:

Volatilization:
Photolysis:
Hydrolysis:
Oxidation:
Biodegradation: biodegraded fairly quickly by Alkaligenes sp. strain Y-42 from lake sediments but small residue was detected after 7 hours (Furukawa & Matsumura 1976; quoted, Pal et al. 1980).
Biotransformation:
Bioconcentration, Uptake (k_1) and Elimination (k_2) Rate Constants:

Common Name: 2,4,4'-Trichlorobiphenyl
Synonym: PCB-28
Chemical Name: 2,4,4'-trichlorobiphenyl
CAS Registry No:7012-37-5
Molecular Formula: $C_{12}H_7Cl_3$
Molecular Weight: 257.54
Melting Point (°C):

57-58 (Hutzinger et al. 1971,1974; NAS 1979)
57-57.5 (Wallnöfer et al. 1973)
57.0 (Mackay et al. 1980; Bruggeman et al. 1982; Burkhard et al. 1985a; Opperhuizen et al. 1988; Ballschmiter & Wittlinger 1991)
57, 74 (quoted, calculated, Abramowitz & Yalkowsky 1990)

Boiling Point (°C):

206-207 (Sengupta 1966)
337.0 (calculated, Mackay et al. 1982; Shiu & Mackay 1986)

Density (g/cm³ at 20°C): 1.1485
Molar Volume (cm³/mol):

247.3 (LeBas method, Miller et al. 1985)
247.3 (LeBas method, Shiu & Mackay 1986)

Molecular Volume (A³):

248.2 (planar, shorthand, Opperhuizen et al. 1988)

Total Surface Area, TSA (A²):

243.58 (Mackay et al. 1980; Shiu & Mackay 1986)
239.83 (Burkhard 1984)
230.83 (planar, Hawker & Connell 1988a)
258.1 (planar, shorthand, Opperhuizen et al. 1988)
231.0 (Abramowitz & Yalkowsky 1990)

Heat of Fusion, kcal/mol:
Fugacity Ratio, F (assuming ΔS_{fusion} = 13.5 e.u.):

0.482 (Mackay et al. 1980)
0.484 (Shiu & Mackay 1986)

Water Solubility (g/m³ or mg/L at 25°C):

0.085 (shake flask-GC/ECD, Wallnöfer et al. 1973; Hutzinger et al. 1974; quoted, NAS 1979; Kenaga & Goring 1980; Kenaga 1980; Erickson 1986)
0.260 (gen. col.-GC/ECD, Weil et al. 1974)
0.119 (Dexter & Pavlou 1978)
0.085 (quoted, NAS 1979; Chou & Griffin 1987)
0.266 (calculated-TSA, Mackay et al. 1980; quoted, Ballschmiter & Wittlinger 1991)
0.263 (TLC-RT, Bruggeman et al. 1982)
0.27 (20°C, subcooled liq., shake flask-GC/ECD, Chiou et al. 1983; Chiou 1985; Chiou & Block 1986)
0.148 (quoted lit. average, Yalkowsky et al. 1983)

0.142 (gen. col.-HPLC/UV, Huang 1983)
0.163 (gen. col.-GC/ECD, Miller et al. 1984)
0.312 (subcooled liq., calculated-TSA, Burkhard et al. 1985b)
0.116 (shake flask-GC/ECD, Chiou et al. 1986)
0.16 (selected, Shiu & Mackay 1986)
0.143 (20°C, subcooled liq., calculated-mole fraction of Aroclor mixture, Murphy et al. 1987)
0.264, 0.153 (quoted, calculated average of HPLC-RI, Brodsky & Ballschmiter 1988)
0.067 (22°C, gen. col.-GC, Opperhuizen et al. 1988)
0.117 (gen. col.-GC, Dunnivant & Elzerman 1988)
0.0662, 0.0814 (quoted; calculated- χ , Nirmalakhanden & Speece 1989)
0.162, 0.0814 (quoted, calculated-TSA, Abramowitz & Yalkowsky 1990)
0.204 (quoted from Brodsky et al. 1988, Ballschmiter & Wittlinger 1991)
0.155, 0.178 (quoted average of Brodsky & Ballschmiter 1988, calculated- χ , Patil 1991)

Vapor Pressure (Pa at 25°C):
0.0277 (subcooled liquid, Burkhard 1984)
0.0273, 0.014, 0.00957 (calculated-MW,GC-RI,χ, Burkhard et al. 1985a)
0.0277 (subcooled liq., GC-RI, Burkhard et al. 1985b)
0.0339, 0.0340 (subcooled liq., GC-RT, Foreman & Bidleman 1985)
0.0149 (20°C, subcooled liq.,calculated-mole fraction of Aroclor mixtures, Murphy et al. 1987)
0.0145 (calculated-SxHLC, Dunnivant & Elzerman 1988)
0.0304 (subcooled liquid, calculated-M.P., Dunnivant & Elzerman 1988)
0.026 (quoted from Wittlinger et al. 1990, Ballschmiter & Wittlinger 1991)
0.034 (quoted from Bidleman 1984, Ballschmiter & Wittlinger 1991)

Henry's Law Constant (Pa m^3/mol):
22.80 (calculated-P/C, Burkhard et al. 1985b)
26.75 (20°C, calculated-P/C, Murphy et al. 1987)
32.02 (batch stripping, Dunnivant & Elzerman 1988; quoted, Ballschmiter & Wittlinger 1991)
20.27 (wetted-wall col.-GC, Brunner et al. 1990)
33.0 (quoted from Wittlinger et al. 1990, Ballschmiter & Wittlinger 1991)

Octanol/Water Partition Coefficient, log K_{OW}:
5.62 (shake flask, McDuffie 1981)
5.74 (TLC-RT, Bruggeman et al. 1982)
5.62 (shake flask-GC/ECD, Chiou et al. 1983; Chiou 1985; Chiou & Block 1986)
6.00 (calculated-f const., Yalkowsky et al. 1983)
5.51 (calculated-TSA, Burkhard 1984)
5.11 (HPLC-RT, Rapaport & Eisenreich 1984)

5.69 (calculated-π, Rapaport & Eisenreich 1984; quoted Sklarew & Girvin 1987)
5.60 (selected, Shiu & Mackay 1986; quoted, Sklarew & Girvin 1987)
5.51 (calculated-S, Chou & Griffin 1987)
5.69 (quoted, Eisenreich 1987)
5.71 (calculated average of HPLC-RI, Brodsky & Ballschmiter 1988)
5.67 (calculated-TSA, Hawker & Connell 1988a)
5.50 (quoted, Brodsky et al. 1988; Ballschmiter & Wittlinger 1991)
5.71, 5.48 (quoted average of Brodsky & Ballschmiter 1988, calculated- χ , Patil 1991)

Bioconcentration Factor, log BCF:
3.39 (calculated-S, Kenaga 1980)

Sorption Partition Coefficient, log K_{OC}:
4.23 (calculated, Kenaga 1980)
4.38 (soil org. matter, sorption isotherm, Chiou et al. 1983)
5.31 (suspended particulate matter, Burkhard 1984)
5.28 (suspended solids-Lake Superior, Baker et al. 1986)
5.30, 4.59 (suspended solids-Lake superior, calculated, Baker et al. 1986)
4.40, 3.54 (Sanhedron soil, Suwannee River, humic acid, Chiou et al. 1986,1987)
3.89, 3.57 (Sanhedron soil, Suwannee River, fulvic acid, Chiou et al. 1986,1987)
4.84 (Fluka-Tridom humic acid, Chiou et al. 1987)
4.24 (Calcasieu River humic extract, Chiou et al. 1987)
3.53 (Suwannee River water sample, Chiou et al. 1987)
3.57 (Sopchoppy River water sample, Chiou et al. 1987)
5.48 (calculated after Karickhoff et al. 1979, Capel & Eisenreich 1990)
4.63 (calculated after Schwarzenbach & Westall 1981, Capel & Eisenreich 1990)
4.21 (calculated-polymaleic acid, Chin & Weber 1990)

Half-Lives in the Environment:
Air:
Surface water:
Groundwater:
Sediment:
Soil:
Biota:

Environmental Fate Rate Constants or Half-Lives:
Volatilization:
Photolysis: measured sunlight photolysis rate constant in surface water at 40°L in winter is 2.2×10^{-8} sec^{-1} with a half-life of 133 days and a calculated rate constant of 6×10^{-8} sec^{-1} (Dulin et al. 1986).
Hydrolysis:
Oxidation:

Biodegradation: half-life of biodegradation by Alkaligenes sp. strain Y-42 from lake sediments to be within 7 hours (Furukawa & Matsumura 1976; quoted, Pal et al. 1980).

Biotransformation:

Bioconcentration, Uptake (k_1) and Elimination (k_2) Rate Constants:

Common Name: 2,4,5-Trichlorobiphenyl
Synonym: PCB-29
Chemical Name: 2,4,5-trichlorobiphenyl
CAS Registry No: 15862-07-4
Molecular Formula: $C_{12}H_7Cl_3$
Molecular Weight: 257.54
Melting Point (°C):
    78-79 (General Aniline & Film corp. 1942; Hutzinger et al. 1974)
    78.0 (Mackay et al. 1980; Bruggeman et al. 1982; Opperhuizen et al. 1988)
    79.0 (Burkhard et al. 1985a)
    78, 73 (quoted, calculated, Abramowitz & Yalkowsky 1990)
Boiling Point (°C):
Density (g/cm^3 at 20°): 1.1485
Molar Volume (cm^3/mol):
    247.3 (LeBas method, Miller et al. 1985)
    247.3 (LeBas method, Shiu & Mackay 1986)
    1.190 (intrinsic volume: $V_I/100$, Kamlet et al. 1988; Hawker 1989b; De Bruijn & Hermens 1990)
Molecular Volume (A^3):
    248.2 (planar, shorthand, Opperhuizen et al. 1988)
    223.71, 203.65, 248.2 (De Bruijn & Hermens 1990)
Total Surface Area, TSA (A^2):
    241.59 (Mackay et al. 1980; Shiu & Mackay 1986)
    237.79 (Burkhard 1984)
    228.39 (planar, Doucette 1985)
    241.39 (nonplanar, Doucette 1985)
    241.39 (Doucette & Andren 1988)
    228.74 (planar, Hawker & Connell 1988a)
    255.6 (planar, shorthand, Opperhuizen et al. 1988)
    267.02, 233.71, 255.6 (De Bruijn & Hermens 1990)
    229.0 (Abramowitz & Yalkowsky 1990)
Heat of Fusion, kcal/mol:
    5.45 (Miller et al. 1984)
Entropy of Fusion, cal/mol K (e.u.):
    15.60 (Miller et al. 1984; Shiu & Mackay 1986)
Fugacity Ratio, F (assuming ΔS_{fusion} = 13.5 e.u.):
    0.30 (Mackay et al. 1980; Shiu & Mackay 1986)

Water Solubility (g/m^3 or mg/L at 25°C):
    0.092 (gen. col.-GC/ECD, Weil et al. 1974; quoted, Könemann 1981)
    0.119 (Dexter & Pavlou 1978)
    0.193 (calculated-TSA, Mackay et al. 1980)
    0.094 (TLC-RT, Bruggeman et al. 1982)

0.140 (gen. col.-HPLC, Billington 1982)
0.142 (gen. col.-HPLC, Huang 1983)
0.0914 (quoted lit. average, Yalkowsky et al. 1983; quoted, Erickson 1986)
0.373 (subcooled liquid, calculated-TSA, Burkhard et al. 1985b)
0.162 (gen. col.-GC/ECD, Miller et al. 1984,1985)
0.140 (selected, Shiu & Mackay 1986; quoted, Hawker 1989b)
0.094, 0.115 (quoted, calculated average of HPLC-RI, Brodsky & Ballschmiter 1988)
0.632 (quoted, subcooled liquid, Hawker 1989b)
0.162 (quoted, Isnard & Lambert 1989)
0.0914, 0.0814 (quoted, calculated-χ, Nirmalakhanden & Speece 1989)
0.129, 0.103 (quoted, calculated-TSA, Abramowitz & Yalkowsky 1990)

Vapor Pressure (Pa at 25°C):
0.0443 (subcooled liq.,GC-RT, Bidleman 1984)
0.0165, 0.0112, 0.0219 (calculated-MW,GC-RI,χ, Burkhard et al. 1985a)
0.0320 (subcooled liq., calculated-GC-RI, Burkhard et al. 1985b)
0.0453, 0.0464, 0.0443 (subcooled liq., GC-RT, Foreman & Bidleman 1985)
0.132 (selected, Shiu & Mackay 1986; quoted, Sklarew & Girvin 1987)
0.044 (subcooled liq., selected, Shiu & Mackay 1986)

Henry's Law Constant (Pa m^3/mol):
25.33 (calculated-P/C, Burkhard et al. 1985b)
24.29 (calculated-P/C, Shiu & Mackay 1986)
27.05 (calculated-χ, Sabljic & Güsten 1989)
20.27 (wetted-wall col.-GC, Brunner et al. 1990)

Octanol/Water Partition Coefficient, log K_{OW}:
6.22 (Hansch & Leo 1979)
5.88 (calculated after Rekker 1977, Könemann 1981)
5.77 (TLC-RT, Bruggeman et al. 1982; quoted, Erickson 1986)
5.77 (calculated-π, Bruggeman et al. 1982)
5.86 (HPLC-RT, Woodburn 1982)
5.46 (calculated-TSA, Burkhard 1984)
5.51 (gen. col.-GC/ECD, Miller et al. 1984,1985; quoted, Sklarew & Girvin 1987)
5.99 (calculated-f const., Yalkowsky et al. 1983)
5.81 (gen. col.-GC, Woodburn et al. 1984; quoted, Sklarew & Girvin 1987)
5.86 (HPLC-RT, Woodburn et al. 1984)
6.22 (calculated-π, Woodburn et al. 1984)
5.67 (HPLC-RT, Rapaport & Eisenreich 1984)
6.25 (calculated-π, Rapaport & Eisenreich 1984; quoted, Sklarew & Girvin 1987)
5.70, 5.66 (quoted, HPLC-RP/MS, Burkhard & Kuehl 1986)
5.60 (selected, Shiu & Mackay 1986; quoted, Gobas et al. 1987,1989; Sklarew & Girvin 1987; Hawker 1989b)

5.81 (gen. col.-GC/ECD, Doucette & Andren 1987)
6.17, 5.59 (calculated-π, TSA, Doucette & Andren 1987)
5.68, 5.81 (quoted, calculated average of HPLC-RI, Brodsky & Ballschmiter 1988)
6.17, 5.99, 5.47, 5.44, 5.65, 5.69 (calculated-π, f const., HPLC-RT, MW, χ , TSA, Doucette & Andren 1988)
5.60 (calculated-TSA, Hawker & Connell 1988a)
5.52 (calculated-solvatochromic p., Kamlet et al. 1988)
5.60, 6.15 (quoted, calculated-π or f const., Kamlet et al. 1988)
5.86 (slow stirring-GC, De Bruijn et al. 1988)
5.77 (calculated-π, De Bruijn et al. 1989)
5.60 (quoted, Hawker 1989b)
5.77 (quoted, reported as PCB-30, Thomann 1989)
5.51 (quoted, Isnard & Lambert 1989)
5.901 (slow stirring-GC, De Bruijn et al. 1989; De Bruijn & Hermens 1990)

Bioconcentration Factor, log BCF:
4.97 (guppy, lipid wt. based, Gobas et al. 1989)
5.41 (guppy, corr. lipid wt. based, Gobas et al. 1989)

Sorption Partition Coefficient, log K_{OC}:
5.26 (suspended particulate matter, Burkhard 1984)

Half-Lives in the Environment:
Air:
Surface water:
Groundwater:
Sediment:
Soil:
Biota:

Environmental Fate Rate Constants or Half-Lives:
Volatilization:
Photolysis:
Hydrolysis:
Oxidation:
Biodegradation: half-life of biodegradation by Alkaligenes sp. strain Y-42 from lake sediments estimated to be less than 7 hours (Furukawa & Matsumura 1976; quoted, Pal et al. 1980).
Biotransformation:
Bioconcentration, Uptake (k_1) and Elimination (k_2) Rate Constants:
log k_2: -1.68 day^{-1} (fish, quoted, Hawker & Connell 1985; Thomann 1989)

Common Name: 2,4,6-Trichlorobiphenyl
Synonym: PCB-30
Chemical Name: 2,4,6-trichlorobiphenyl
CAS Registry No: 35693-92-6
Molecular Formula: $C_{12}H_7Cl_3$
Molecular Weight: 257.54
Melting Point (°C):

62.5 (Augwood et al. 1953; Hutzinger et al. 1974)
46.0 (Saeki et al. 1971)
61.9 (Burkhard et al. 1985a)
61.3 (Opperhuizen et al. 1988)
63, 70 (quoted, calculated, Abramowitz & Yalkowsky 1990)

Boiling Point (°C):

337 (calculated, Mackay et al. 1982; Shiu & Mackay 1986)

Density (g/cm³):
Molar Volume (cm³/mol):

247.3 (LeBas method, Miller et al. 1985)
247.3 (LeBas method, Shiu & Mackay 1986)
1.19 (intrinsic volume: $V_I/100$, Kamlet et al. 1988; Hawker 1989b; De Bruijn & Hermens 1990)

Molecular Volume (A³):

245.8 (planar, shorthand, Opperhuizen et al. 1988)
223.61, 203.12, 245.8 (De Bruijn & Hermens 1990)

Total Surface Area, TSA (A²):

237.05 (Burkhard 1984)
223.39 (planar, Doucette 1985)
241.84 (nonplanar, Doucette 1985)
241.84 (Doucette & Andren 1987, 1988)
224.16 (planar, Hawker & Connell 1988a)
252.7 (planar, shorthand, Opperhuizen et al. 1988)
266.37, 230.96, 252.7 (De Bruijn & Hermens 1990)
224.0 (Abramowitz & Yalkowsky 1990)

Heat of Fusion, kcal/mol:

3.94 (Miller et al. 1984)

Entropy of Fusion, cal/mol K (e.u.):

11.8 (Miller et al. 1984; Shiu & Mackay 1986)

Fugacity Ratio, F (assuming ΔS_{fusion} = 13.5 e.u.):

0.427 (Shiu & Mackay 1986)

Water Solubility (g/m³ or mg/L at 25°C):

0.226 (gen. col.-GC/ECD, Miller et al. 1984,1985)
0.40 (subcooled liq., calculated-TSA, Burkhard et al. 1985b)
0.20 (selected, Shiu & Mackay 1986)

0.047 (calculated average of HPLC-RI, Brodsky & Ballschmiter 1988)
0.187 (gen. col.-GC/ECD, Doucette & Andren 1988)
0.252, 0.243 (gen. col.-GC, Dunnivant & Elzerman 1988)
0.592 (subcooled liquid, calculated-M.P., Dunnivant & Elzerman 1988)
0.469 (quoted, Hawker 1989b)
0.224 (quoted, Isnard & Lambert 1989)
0.224, 0.0814 (quoted, calculated-χ, Nirmalakhanden & Speece 1989)
0.205, 0.162 (quoted, calculated-TSA, Abramowitz & Yalkowsky 1990)

Vapor Pressure (Pa at 25°C):
0.031 (Augood 1953; quoted, Bidleman 1984))
0.0306 (Weast 1972,1973)
0.0955, 0.144 (subcooled liq., GC-RT, Bidleman 1984)
0.030 (quoted, subcooled liquid, Bidleman 1984)
0.0946 (subcooled liquid, Burkhard 1984)
0.0244, 0.0421, 0.0219 (calculated-MW,GC-RI,χ, Burkhard et al. 1985a)
0.111, 0.135, 0.117 (subcooled liq., GC-RT, Foreman & Bidleman 1985)
0.0946 (subcooled liq.,calculated-GC-RI, Burkhard et al. 1985b)
0.117 (quoted from Bidleman 1984, Erickson 1986)
0.0384 (selected, Shiu & Mackay 1986; quoted, Sklarew & Girvin 1987; Hawker 1989b)
0.090 (selected, subcooled liq., Shiu & Mackay 1986)
0.064 (calculated-SxHLC, Dunnivant & Elzerman 1988)
0.152 (subcooled liquid, calculated-M.P., Dunnivant & Elzerman 1988)
0.030 (quoted subcooled liq., Hinckley et al. 1990)

Henry's Law Constant (Pa m^3/mol):
61.40 (calculated-P/C, Burkhard et al. 1985b)
49.51 (calculated-P/C, Shiu & Mackay 1986)
65.76 (batch stripping, Dunnivant & Elzerman 1988)
65.76 (batch stripping, Dunnivant et al. 1988)

Octanol/Water Partition Coefficient, log K_{OW}:
5.88 (calculated after Rekker 1977, Könemann1981)
5.44 (calculated-TSA, Burkhard 1984)
5.47 (gen. col.-GC/ECD, Miller et al. 1984,1985; quoted, Sklarew & Girvin 1987)
5.50 (selected, Shiu & Mackay 1986; quoted, Sklarew & Girvin 1987; Hawker 1989b)
5.57 (gen. col.-GC/ECD, Doucette & Andren 1987,1988)
6.17, 5.60 (calculated-π, TSA, Doucette & Andren 1987)
5.47, 5.62 (quoted, calculated average of HPLC-RI, Brodsky & Ballschmiter 1988)
6.17, 5.99, 5.44, 5.65, 5.71 (calculated-π, f const., MW, χ , TSA, Doucette & Andren 1988)

5.44 (calculated-TSA, Hawker & Connell 1988a)
5.59 (calculated-solvatochromic p., Kamlet et al. 1988)
5.50 (quoted, Kamlet et al. 1988; Hawker 1989b)
5.47 (quoted, Isnard & Lambert 1989)
5.61 (calculated-π, De Bruijn et al. 1989)
5.711 (slow stirring, De Bruijn et al. 1989; De Bruijn & Hermens 1990)

Bioconcentration Factor, log BCF:

Sorption Partition Coefficient, log K_{OC}:
5.24 (suspended particulates, Burkhard 1984)

Half-Lives in the Environment:
Air:
Surface water:
Groundwater:
Sediment:
Soil:
Biota:

Environmental Fate Rate Constants or Half-Lives:
Volatilization:
Photolysis:
Hydrolysis:
Oxidation:
Biodegradation:
Biotransformation:
Bioconcentration, Uptake (k_1) and Elimination (k_2) Rate Constants:

Common Name: 2,4',5-Trichlorobiphenyl
Synonym: PCB-31
Chemical Name: 2,4',5-trichlorobiphenyl
CAS Registry No: 16606-02-3
Molecular Formula: $C_{12}H_7Cl_3$
Molecular Weight: 257.54
Melting Point (°C):
67 (Bellavita 1935; Hutzinger et al. 1974; NAS 1979)
65 (Burkhard et al. 1985a)
63.5-64.5 (selected, Erickson 1986)
64 (Opperhuizen et al. 1988)
67, 61 (quoted, calculated, Abramowitz & Yalkowsky 1990)
Boiling Point (°C):
337 (calculated, Mackay et al. 1982; Shiu & Mackay 1986)
Density (g/cm^3):
Molar Volume (cm^3/mol):
247.3 (LeBas method, Miller et al. 1985)
247.3 (LeBas method, Shiu & Mackay 1986)
1.190 (intrinsic volume: $V_I/100$, Kamlet et al. 1988; Hawker 1990)
Molecular Volume (A^3):
251.9 (planar, shorthand, Opperhuizen et al. 1988)
Total Surface Area, TSA (A^2):
239.66 (Burkhard 1984)
230.37 (planar, Doucette 1985)
243.39 (nonplanar, Doucette 1985)
243.39 (Doucette & Andren 1988)
230.66 (planar, Hawker & Connell 1988a)
258.1 (planar, shorthand, Opperhuizen et al. 1988)
231.0 (Abramowitz & Yalkowsky 1990)
Heat of Fusion, kcal/mol:
Fugacity Ratio, F (assuming ΔS_{fusion} = 13.5 e.u.):
0.384 (Shiu & Mackay 1986)

Water Solubility (g/m^3 or mg/L at 25°C):
0.11 (Kilzer et al. 1979; quoted, Sklarew & Girvin 1987)

0.11 (quoted, Geyer et al. 1980)
0.075 (GC/ECD, Bruggeman et al. 1981)
0.317 (subcooled liq., calculated-TSA, Burkhard et al. 1985b)
0.170 (calculated average of HPLC-RI, Brodsky & Ballschmiter 1988)
0.090 (gen. col., Opperhuizen et al. 1988)
0.0935, 0.0814 (quoted, calculated- χ , Nirmalakhanden & Speece 1989)
0.143 (20°C, subcooled liq., calculated-mole fraction of Aroclor mixtures, Murphy et

al. 1987)

0.163 (calculated-TSA, Abramowitz & Yalkowsky 1990)

0.170, 0.178 (quoted average of Brodsky & Ballschmiter 1988, calculated- χ , Patil 1991)

Vapor Pressure (Pa at 25°C):

0.0341, 0.0474 (subcooled liq., GC-RT, Bidleman 1984)

0.0227, 0.0132, 0.00957 (calculated-MW,GC-RI,χ, Burkhard et al. 1985a)

0.0313 (subcooled liq., calculated-GC-RI, Burkhard et al. 1985b)

0.0373, 0.0346, 0.0403 (subcooled liq., GC-RT, Foreman & Bidleman 1985)

0.040 (quoted from Bidleman 1984, Erickson 1986)

0.015 (selected, Shiu & Mackay 1986; quoted, Sklarew & Girvin 1987)

0.0149 (20°C, subcooled liq., calculated-mole fraction of Aroclor mixtures, Murphy et al. 1987)

Henry's Law Constant (Pa m^3/mol):

94.13 (batch stripping, Atlas et al. 1982)

20.26 (calculated, Murphy et al. 1983)

25.43 (calculated-P/C, Burkhard et al. 1985b)

26.75 (20°C, calculated-P/C, Murphy et al. 1987)

20.27 (quoted, Eisenreich 1987)

28.47 (calculated- χ , Sabljic & Güsten 1989)

19.25 (wetted-wall col.-GC, Brunner et al. 1990)

Octanol/Water Partition Coefficient, log K_{OW}:

6.22 (Hansch & Leo 1979)

5.88 (calculated after Rekker 1977, Könemann 1981)

5.30 (HPLC-RT, Woodburn 1982; Woodburn et al.1984)

5.77 (TLC-RT, Bruggeman et al. 1982; quoted, Erickson 1986)

5.56 (quoted, Garten & Trabalka 1983)

5.51 (calculated-TSA, Burkhard 1984)

5.11 (HPLC-RT, Rapaport & Eisenreich 1984)

5.69 (calculated-π, Rapaport & Eisenreich 1984; quoted, Sklarew & Girvin 1987)

5.79 (gen. col.-GC/ECD, Woodburn et al. 1984; quoted, Sklarew & Girvin 1987)

6.22 (calculated-π, Woodburn et al. 1984)

5.70 (selected, Shiu & Mackay 1986; quoted, Sklarew & Girvin 1987; Clark et al. 1990)

5.79, 5.68 (quoted, calculated average of HPLC-RI, Brodsky & Ballschmiter 1988)

5.77 (quoted, Hawker & Connell 1985; Thomann 1989)

5.69 (quoted, Eisenreich 1987)

5.6 (calculated-chlorine substituent, Oliver 1987a & c)

5.79 (gen. col.-GC/ECD, Doucette & Andren 1987,1988)

6.17, 5.63 (calculated-π, TSA, Doucette & Andren 1987)

6.17, 6.00, 5.44, 5.64, 5.74 (calculated-π, f const., MW, χ, TSA, Doucette & Andren 1988)
5.67 (calculated-TSA, Hawker & Connell 1988a; quoted, Connell & Hawker 1988; Hawker 1990)
5.68, 5.48 (quoted average of Brodsky & Ballschmiter 1988, calculated- χ, Patil 1991)

Bioconcentration Factor, log BCF:
3.45 (fish, Korte et al. 1978)
-0.30, -0.22 (adipose tissue of male, female Albino rats, Geyer et al. 1980)
6.15 (goldfish, 3% lipid, Bruggeman et al. 1981)
5.98 (goldfish, 10% lipid content in food, Bruggeman et al. 1981)
3.95, 3.95, 4.51 (algae, fish, activated sludge, Freitag et al. 1984,1985; selected, Halfon & Reggiani 1986)
3.66 (salmon fry in humic water-steady state, Carlberg et al. 1986)
3.83 (salmon fry in lake water-steady state, Carlberg et al. 1986)
4.62 (fish, calculated-C_B/C_W or k_1/k_2 Connell & Hawker 1988,; Hawker 1990)

Sorption Partition Coefficient, log K_{OC}:
5.31 (suspended particulates, Burkhard 1984)
5.5-6.3, 5.9 (suspended sediment, average, Oliver 1987a)
6.8 (algae > 50 μm, Oliver 1987a)
5.48 (calculated after Karickhoff et al. 1979, Capel & Eisenreich 1990)
4.63 (calculated after Schwarzenbach & Westall 1981, Capel & Eisenreich 1990)

Half-Lives in the Environment:
Air:
Surface water:
Groundwater:
Sediment:
Soil:
Biota: half-life in rainbow trout, 196 days and its muscle, 81 days (Niimi & Oliver 1983); in worms at 8°C, 30 days (Oliver 1987c).

Environmental Fate Rate Constants or Half-Lives:
Volatilization:
Photolysis:
Hydrolysis:
Oxidation:
Biodegradation: 50-80% being degraded by Alkaligenes sp. strain Y-42 from lake sediments within 7-hour period (Furukawa & Matsumura 1976; quoted, Pal et al. 1980).
Biotransformation:
Bioconcentration, Uptake (k_1) and Elimination (k_2) Rate Constants:

k_1: 890 day^{-1} (23°C, goldfish, 3% lipid content, Bruggeman et al. 1981; quoted, Waid 1986)

k_2: 0.021 day^{-1} (23°C, goldfish, 3% lipid content, Bruggeman et al. 1981; quoted, Waid 1986; Clark et al. 1990)

k_2: 0.0035 day^{-1} (rainbow trout, Niimi & Oliver 1983; quoted, Clark et al. 1990)

k_1: 37.1 $hour^{-1}$ (goldfish, quoted, Hawker & Connell 1985)

$1/k_2$: 1142 hour (goldfish, quoted, Hawker & Connell 1985)

log k_1: 2.95 day^{-1} (fish, quoted, Connell & Hawker 1988)

log $1/k_2$: 1.68 day (fish, quoted, Connell & Hawker 1988; Thomann 1989)

Common Name: 2',3,4-Trichlorobiphenyl
Synonym: PCB-33
Chemical Name: 2',3,4-trichlorobiphenyl
CAS Registry No: 38444-86-9
Molecular Formula: $C_{12}H_7Cl_3$
Molecular Weight: 257.54
Melting Point (°C):
65-66 (Mascarelli & Gatti 1933; NAS 1979)
54 (Bellavita 1935)
60.1-60.4 (Weingarten 1961)
60-60.5 (Wallnöfer et al. 1973)
60.0 (Hutzinger et al. 1974; Mackay et al. 1980; Burkhard et al. 1985a)
60, 73 (quoted, calculated, Abramowitz & Yalkowsky 1990)
Boiling Point (°C):
337 (calculated, Mackay et al. 1982; Shiu & Mackay 1986)
Density (g/cm^3 at 20°C): 1.1485
Molar Volume (cm^3/mol):
247.3 (LeBas method, Miller et al. 1985)
247.3 (LeBas method, Shiu & Mackay 1986)
1.19 (intrinsic volume: V_I/100, De Bruijn & Hermens 1990)
Molecular Volume (A^3):
248.8 (planar, shorthand, Opperhuizen et al. 1988)
223.88, 203.62, 248.2 (De Bruijn & Hermens 1990)
Total Surface Area, TSA (A^2):
241.59 (Mackay et al. 1980; Shiu & Mackay 1986)
237.84 (Burkhard 1984)
228,39 (planar, Doucette 1985)
241.40 (nonplanar, Doucette 1985)
228.75 (planar, Hawker & Connell 1988a)
255.6 (planar, shorthand, Opperhuizen et al. 1988)
267.36, 233.68, 255.6 (De Bruijn & Hermens 1990)
229.0 (Abramowitz & Yalkowsky 1990)
Heat of Fusion, kcal/mol:
Fugacity Ratio, F (assuming ΔS_{fusion} = 13.5 e.u.):
0.452 (Shiu & Mackay 1986)

Water Solubility (g/m^3 or mg/L at 25°C):
0.078 (shake flask-GC/ECD, Wallnöfer et al. 1973; Hutzinger et al. 1974; quoted, NAS 1979; Erickson 1986)
0.078 (quoted, NAS 1979; Chou & Griffin 1987)
0.291 (calculated-TSA, Mackay et al. 1980)
0.0796 (quoted lit. average, Yalkowsky et al. 1983)
0.371 (subcooled liq., calculated-TSA, Burkhard et al. 1985b; quoted, Capel et al.

1991)

0.080 (selected, Shiu & Mackay 1986)

0.157 (calculated average of HPLC-RI, Brodsky & Ballschmiter 1988)

0.0778, 0.0814 (quoted, calculated-χ, Nirmalakhanden & Speece 1989)

0.133 (20°C, subcooled liq., calculated-mole fraction of Aroclor mixtures, Murphy et al. 1987)

0.0814, 0.103 (quoted, calculated-TSA, Abramowitz & Yalkowsky 1990)

0.159, 0.174 (quoted average of Brodsky & Ballschmiter 1988, calculated- χ , Patil 1991)

Vapor Pressure (Pa at 25°C):

0.0046 (Augood et al. 1953)

0.0107 (GC-RT, Westcott & Bidleman 1981; quoted, Erickson 1986)

0.0133 (gas saturation, Westcott & Bidleman 1981; Westcott et al. 1981)

0.030 (P_L calculated from P_S using F, Westcott et al. 1981)

0.0255, 0.0115, 0.0145 (calculated-MW,GC-RI, χ , Burkhard et al. 1985a)

0.0243 (subcooled liq., calculated-GC-RI, Burkhard et al. 1985b; quoted, Capel et al. 1991)

0.0264, 0.0219 (subcooled liq., GC-RT, Foreman & Bidleman 1985)

0.0136 (selected, Shiu & Mackay 1986; quoted, Sklarew & Girvin 1987)

0.030 (subcooled liq., Shiu & Mackay 1986)

0.0119 (20°C, subcooled liq., calculated-mole fraction of Aroclor mixtures, Murphy et al. 1987)

0.0137, 0.0272 (quoted, calculated-UNIFAC, Banerjee et al. 1990)

Henry's Law Constant (Pa m^3/mol):

15.20 (calculated, Murphy et al. 1983)

16.92 (calculated-P/C, Burkhard et al. 1985b; quoted, Capel et al. 1991)

43.67 (calculated-P/C, Shiu & Mackay 1986)

15.20 (quoted, Eisenreich 1987)

22.70 (20°C, calculated-P/C, Murphy et al. 1987)

21.99 (calculated, Sabljic & Güsten 1989)

Octanol/Water Partition Coefficient, log K_{OW}:

5.77 (calculated-π, Bruggeman et al. 1982)

6.10 (calculated, Neely 1983; quoted, Erickson 1986)

6.00 (calculated-f const., Yalkowsky et al. 1983)

5.46 (calculated-TSA, Burkhard 1984)

6.03 (quoted, Hawker & Connell 1986)

5.80 (selected, Shiu & Mackay 1986; quoted, Sklarew & Girvin 1987)

5.54 (calculated-S, Chou & Griffin 1987)

5.57 (quoted, Eisenreich 1987; Sklarew & Girvin 1987)

5.71 (calculated average of HPLC-RI, Brodsky & Ballschmiter 1988)

5.60 (calculated-TSA, Hawker & Connell 1988; quoted, Capel et al. 1991)
5.872 (slow stirring-GC, De Bruijn et al. 1989)
5.77 (calculated-π, De Bruijn et al. 1989)
5.872 (slow stirring-GC, De Bruijn & Hermens 1990)
5.71, 5.48 (quoted average of Brodsky & Ballschmiter 1988, calculated- χ , Patil 1991)

Bioconcentration Factor, log BCF:
3.79 (oyster,Vreeland 1974; quoted, Hawker & Connell 1986)

Sorption Partition Coefficient, log K_{OC}:
5.26 (suspended particulate matter, Burkhard 1984)
4.64 (soil, calculated-S, Chou & Griffin 1987)

Half-Lives in the Environment:
Air:
Surface water:
Groundwater:
Sediment:
Soil:
Biota:

Environmental Fate Rate Constants or Half-Lives:
Volatilization:
Photolysis:
Hydrolysis:
Oxidation:
Biodegradation:
Biotransformation:
Bioconcentration, Uptake (k_1) and Elimination (k_2) Rate Constants:

Common Name: 3,3',4-Trichlorobiphenyl
Synonym: PCB-35
Chemical Name: 3,3',4-trichlorobiphenyl
CAS Registry No: 37680-69-6
Molecular Formula: $C_{12}H_7Cl_3$
Molecular Weight: 257.54
Melting Point (°C):
    87.0 (Burkhard et al. 1985a)
    88.0 (Opperhuizen et al. 1988)
    73.0 (calculated, Abramowitz & Yalkowsky 1990)
Boiling Point (°C):
Density (g/cm^3):
Molar Volume (cm^3/mol):
    247.3 (LeBas method, Miller et al. 1985)
    247.3 (LeBas method, Shiu & Mackay 1986)
Molecular Volume (A^3):
    254.3 (planar, shorthand, Opperhuizen et al. 1988)
Total Surface Area, TSA (A^2):
    243.3 (Mackay et al. 1980)
    240.26 (Burkhard 1984)
    228.39 (planar, Doucette 1985)
    241.40 (nonplanar, Doucette 1985)
    235.25 (planar, Hawker & Connell 1988a)
    260.3 (planar, shorthand, Opperhuizen et al. 1988)
    235.0 (Abramowitz & Yalkowsky 1990)
Heat of Fusion, kcal/mol:
Fugacity Ratio, F (assuming ΔS_{fusion} = 13.5 e.u.): 0.244

Water Solubility (g/m^3 or mg/L at 25°C):
    0.301 (subcooled liq., calculated-TSA, Burkhard et al. 1985b)
    0.0155, 0.0814 (quoted, calculated- χ , Nirmalakhanden & Speece 1989)
    0.0514 (calculated-TSA, Abramowitz & Yalkowsky 1990)

Vapor Pressure (Pa at 25°C):
    0.0138, 0.00246, 0.00957 (calculated-MW, GC-RI, χ , Burkhard et al. 1985a)
    0.00949 (subcooled liq., calculated-GC-RI, Burkhard et al. 1985b)
    0.014, 0.0105 (subcooled liq., GC-RT, Foreman & Bidleman 1985)

Henry's Law Constant (Pa m^3/mol):
    8.13 (calculated-P/C, Burkhard et al. 1985b)
    22.49 (calculated- χ , Sabljic & Güsten 1989)

Octanol/Water Partition Coefficient, log K_{OW}:

5.53 (calculated-TSA, Burkhard 1984)

5.82 (calculated-TSA, Hawker & Connell 1988a)

Bioconcentration Factor, log BCF:

Sorption Partition Coefficient, log K_{OC}:

5.33 (suspended particulate matter, Burkhard 1984)

Half-Lives in the Environment:

Air:

Surface water:

Groundwater:

Sediment:

Soil:

Biota:

Environmental Fate Rate Constants or Half-Lives:

Volatilization:

Photolysis:

Hydrolysis:

Oxidation:

Biodegradation:

Biotransformation:

Bioconcentration, Uptake (k_1) and Elimination (k_2) Rate Constants or Half-Lives:

Common Name: 3,4,4'-Trichlorobiphenyl
Synonym: PCB-37
Chemical Name: 3,4,4'-trichlorobiphenyl
CAS Registry No: 38444-90-5
Molecular Formula: $C_{12}H_7Cl_3$
Molecular Weight: 257.54
Melting Point (°C):
86.8-87.8 (Weingarten 1961)
88.0 (Mackay et al. 1980; Bruggeman et al. 1982)
88, 83 (quoted, calculated, Abramowitz & Yalkowsky 1990)
Boiling Point (°C):
337 (calculated, Mackay et al. 1982; Shiu & Mackay 1986)
Density (g/cm³ at 20°C): 1.2024
Molar Volume (cm³/mol):
247.3 (LeBas method, Miller et al. 1985)
247.3 (LeBas method, Shiu & Mackay 1986)
Molecular Volume (A³):
Total Surface Area, TSA (A²):
243.18 (Mackay et al. 1980; Shiu & Mackay 1986)
240.49 (Burkhard 1984)
235.36 (planar, Doucette 1985)
242.94 (nonplanar, Doucette 1985)
235.42 (planar, Hawker & Connell 1988a)
235.0 (Abramowitz & Yalkowsky 1990)
Heat of Fusion, kcal/mol:
Fugacity Ratio, F (assuming ΔS_{fusion} = 13.5 e.u.):
0.244 (Shiu & Mackay 1986)

Water Solubility (g/m³ or mg/L at 25°C):
0.0152 (gen. col.-GC/ECD, Weil et al. 1974)
0.135 (calculated-TSA, Mackay et al. 1980)
0.0152 (TLC-RT, Bruggeman et al. 1982)
0.0152 (quoted lit. average, Yalkowsky et al. 1983; quoted, Erickson 1986)
0.296 (subcooled liq., calculated-TSA, Burkhard et al. 1985b)
0.015 (selected, Shiu & Mackay 1986)
0.072 (20°C, subcooled liq., calculated-mole fraction of Aroclor mixtures, Murphy et al. 1987)
0.0152, 0.0103 (quoted, calculated average of HPLC-RI, Brodsky & Ballschmiter 1988)
0.0162, 0.0408 (quoted, calculated-TSA, Abramowitz & Yalkowsky 1990)

Vapor Pressure (Pa at 25°C):
0.0566, 0.00897, 0.00419 (subcooled liq., calculated-MW,GC-RI, χ , Burkhard et al. 1985a)

0.0084 (subcooled liq., calculated-GC-RI, Burkhard et al. 1985b)
0.0127, 0.0094 (subcooled liq., GC-RT, Foreman & Bidleman 1985)
0.00454 (20°C, subcooled liq., calculated-mole fraction of Aroclor mixtures, Murphy et al. 1987)

Henry's Law Constant (Pa m^3/mol):
84.21 (batch stripping, Atlas et al. 1982; Shiu & Mackay 1986)
7.34 (calculated-P/C, Burkhard et al. 1985b)
15.40 (20°C, calculated-P/C, Murphy et al. 1987)
14.59 (calculated- χ , Sabljic & Güsten 1989)
10.13 (wetted-wall col.-GC, Brunner et al. 1990)

Octanol/Water Partition Coefficient, log K_{OW}:
5.90 (TLC-RT, Bruggeman et al. 1982; quoted, Erickson 1986)
6.00 (calculated-f const., Yalkowsky et al. 1983)
5.53 (calculated-TSA, Burkhard 1984)
4.94 (HPLC-RT, Rapaport & Eisenreich 1984)
5.56 (calculated-π, Rapaport & Eisenreich 1984; quoted, Sklarew & Girvin 1987)
5.90 (selected, Shiu & Mackay 1986; quoted, Sklarew & Girvin 1987)
5.78 (calculated average of HPLC-RI, Brodsky & Ballschmiter 1988)
5.83 (calculated-TSA, Hawker & Connell 1988a)

Bioconcentration Factor, log BCF:

Sorption Partition Coefficient, log K_{OC}:
5.33 (suspended particulate matter, Burkhard 1984)
4.81 (calculated after Karickhoff et al. 1979, Capel & Eisenreich 1990)
4.81 (calculated after Schwarzenbach & Westall 1981, Capel & Eisenreich 1990)

Half-Lives in the Environment:
Air:
Surface water:
Groundwater:
Sediment:
Soil:
Biota:

Environmental Fate Rate Constants or Half-Lives:
Volatilization:
Photolysis:
Hydrolysis:
Oxidation:
Biodegradation:

Biotransformation:
Bioconcentration, Uptake (k_1) and Elimination (k_2) Rate Constants:

Common Name: 2,2',3,3'-Tetrachlorobiphenyl
Synonym: PCB-40
Chemical Name: 2,2',3,3'-tetrachlorobiphenyl
CAS Registry No: 38444-93-8
Molecular Formula: $C_{12}H_6Cl_4$
Molecular Weight: 291.99
Melting Point (°C):
119.5-121.5 (Wallnöfer et al. 1973; Hutzinger et al. 1974; Erickson 1986)
121.0 (Mackay et al. 1980; Burkhard et al. 1985a; Opperhuizen et al. 1988)
121, 90 (quoted, calculated, Abramowitz & Yalkowsky 1990)
Boiling Point (°C):
360 (calculated, Mackay et al. 1982; Shiu & Mackay 1986)
Density (g/cm^3 at 20°C): 1.2024
Molar Volume (cm^3/mol):
268.2 (LeBas method, Shiu & Mackay 1986)
1.280 (intrinsic volume: V_I/100, Kamlet et al. 1988; De Bruijn & Hermens 1990)
Molecular Volume (A^3):
255.4 (planar, shorthand, Opperhuizen et al. 1988)
237.40, 216.37, 255.4 (De Bruijn & Hermens 1990)
Total Surface Area, TSA (A^2):
255.61 (Mackay et al. 1980; Shiu & Mackay 1986)
249.19 (Burkhard 1984)
230.65 (planar, Doucette 1985)
255.40 (nonplanar, Doucette 1985)
230.58 (planar, Hawker & Connell 1988a)
261.3 (planar, shorthand, Opperhuizen et al. 1988)
282.28, 244.05, 261.3 (De Bruijn & Hermens 1990)
231.0 (Abramowitz & Yalkowsky 1990)
Heat of Fusion, kcal/mol:
Fugacity Ratio, F (assuming ΔS_{fusion} = 13.5 e.u.):
0.113 (Shiu & Mackay 1986)

Water Solubility (g/m^3 or mg/L at 25°C):
0.034 (shake flask-GC/ECD, Wallnöfer et al. 1973; Hutzinger et al. 1974; quoted, NAS 1979; Erickson 1986)
0.034 (quoted, NAS 1979; Chou & Griffin 1987)
0.026 (calculated-TSA, Mackay et al. 1980)
0.030 (quoted lit. average, Yalkowsky et al. 1983)
0.161 (subcooled liq., calculated-TSA, Burkhard et al. 1985b)
0.030 (selected, Shiu & Mackay 1986)
0.161 (quoted, Eisenreich 1987)
0.0807 (20°C, subcooled liq., calculated-mole fraction of Aroclor mixtures, Murphy et al. 1987)

0.044 (calculated average of HPLC-RI, Brodsky & Ballschmiter 1988)
0.0156 (gen. col.-GC/ECD, Dunnivant & Elzerman 1988)
0.0136 (subcooled liquid, calculated-MP, Dunnivant & Elzerman 1988)
0.0343, 0.0273 (quoted, calculated-χ, Nirmalakhanden & Speece 1989)
0.034 (quoted, Isnard & Lambert 1989)
0.0292, 0.0583 (quoted, calculated-TSA, Abramowitz & Yalkowsky 1990)
0.044, 0.044 (quoted average of Brodsky & Ballschmiter 1988, calculated- χ , Patil 1991)

Vapor Pressure (Pa at 25°C):
0.0098 (subcooled liq., GC-RT, Bidleman 1984; quoted, Erickson 1986)
0.00229, 0.00134, 0.00957 (calculated-MW,GC-RI,χ, Burkhard et al. 1985a)
0.0112 (subcooled liq., calculated-GC-RI, Burkhard et al. 1985b; quoted, Eisenreich 1987)
0.00887, 0.00861, 0.0098 (subcooled liq., GC-RT, Foreman & Bidleman 1985)
0.00255 (selected, Shiu & Mackay 1986)
0.002 (subcooled liq., selected, Shiu & Mackay 1986)
0.00452 (20°C, subcooled liq., calculated-mole fraction of Aroclor mixtures, Murphy et al. 1987)
0.00109 (calculated-SxHLC, Dunnivant & Elzerman 1988)
0.00956 (subcooled liquid, Dunnivant & Elzerman 1988)

Henry's Law Constant (Pa m^3/mol):
20.27 (calculated-P/C, Burkhard et al. 1985b)
12.16 (batch stripping, Oliver 1985)
21.94 (calculated-P/C, Shiu & Mackay 1986)
20.27 (quoted, Eisenreich 1987)
16.31 (20°C, calculated-P/C, Murphy et al. 1987)
20.47 (batch stripping, Dunnivant & Elzerman 1988)
20.47 (batch stripping, Dunnivant et al. 1988)
10.13 (wetted-wall col.-GC, Brunner et al. 1990)

Octanol/Water Partition Coefficient, log K_{OW}:
4.63 (HPLC-RT, Sugiura et al. 1978)
6.67 (calculated-f const., Yalkowsky et al. 1983; quoted, Erickson 1986)
5.77 (calculated-TSA, Burkhard 1984)
4.54 (HPLC-RT, Rapaport & Eisenreich 1984)
5.56 (calculated-π, Rapaport & Eisenreich 1984; quoted, Sklarew & Girvin 1987)
5.80 (shake flask-GC, Chiou 1985; Oliver & Niimi 1985)
5.60 (selected, Shiu & Mackay 1986; quoted, Sklarew & Girvin 1987; Clark et al. 1990)
5.81 (calculated-S, Chou & Griffin 1987)
4.56 (quoted, Eisenreich 1987)

5.90 (calculated-chlorine substituents, Oliver 1987c)
5.67 (calculated average of HPLC-RI, Brodsky & Ballschmiter 1988)
5.55 (gen. col.-GC/ECD, Hawker & Connell 1988)
5.66 (calculated-TSA, Hawker & Connell 1988a)
5.70 (quoted, Hawker & Connell 1988b)
5.80 (quoted, Isnard & Lambert 1989)
5.80 (quoted, Thomann 1989)
6.26 (calculated-π, De Bruijn et al. 1989)
6.178 (slow stirring-GC, De Bruijn et al. 1989)
6.178 (slow stirring-GC, De Bruijn & Hermens 1990)
5.67, 5.88 (quoted average of Brodsky & Ballschmiter 1988, calculated- χ , Patil 1991)

Bioconcentration Factor, log BCF:
3.08 (killifish, Goto et al. 1978)
4.69 (rainbow trout, kinetic, Oliver & Niimi 1985)
4.23 (rainbow trout, steady state, Oliver & Niimi 1985)
5.38 (rainbow trout, field data, Oliver & Niimi 1985)
4.38, 4.23 (worms, fish, Oliver 1987c)
6.48 (rainbow trout, quoted, Thomann 1989)
4.23 (fish, quoted, Isnard & Lambert 1988,1989)

Sorption Partition Coefficient, log K_{OC}:
5.57 (suspended particulate matter, Burkhard 1984)

Half-Lives in the Environment:
Air:
Surface water:
Groundwater:
Sediment:
Soil:
Biota: half-life in rainbow trout, 107 days and its muscle, 61 days (Niimi & Oliver 1983); in worms at 8°C, 29 days (Oliver 1987c).

Environmental Fate Rate Constants or Half-Lives:
Volatilization:
Photolysis:
Hydrolysis:
Oxidation:
Biodegradation:
Biotransformation:
Bioconcentration, Uptake (k_1) and Elimination (k_2) Rate Constants:
k_2: 0.0065 day^{-1} (rainbow trout, Niimi & Oliver 1983; quoted, Clark et al. 1990)

k_1: 320 day^{-1} (rainbow trout, Oliver & Niimi 1985)
k_2: 0.0065 day^{-1} (rainbow trout, Oliver & Niimi 1985)
log $1/k_2$: 2.2, 2.9 hour (fish, quoted, calculated-K_{OW}, Hawker & Connell 1988b).

Common Name: 2,2',3,5'-Tetrachlorobiphenyl
Synonym: PCB-44
Chemical Name: 2,2',3,5'-tetrachlorobiphenyl
CAS Registry No: 41464-39-5
Molecular Formula: $C_{12}H_6Cl_4$
Molecular Weight: 291.99
Melting Point (°C):
46.5-47 (Hutzinger et al. 1974; Erickson 1986)
49-50 (NAS 1979)
47.0 (Mackay et al. 1980; Burkhard et al. 1985a; Opperhuizen et al. 1988)
47, 42 (quoted, calculated, Abramowitz & Yalkowsky 1990)
Boiling Point (°C):
360 (calculated, Mackay et al. 1982; Shiu & Mackay 1986)
Density (g/cm^3 at 20°C): 1.2024
Molar Volume (cm^3/mol):
268.2 (LeBas method, Miller et al. 1985)
268.2 (LeBas method, Shiu & Mackay 1986)
Molecular Volume (A^3):
259.1 (planar, shorthand, Opperhuizen et al. 1988)
Total Surface Area, TSA (A^2):
257.58 (Mackay et al. 1980; Shiu & Mackay 1986)
254.75 (Burkhard 1984)
232.63 (planar, Doucette, 1985)
257.43 (nonplanar, Doucette 1985)
233.21 (planar, Hawker & Connell 1988a)
263.8 (planar, shorthand, Opperhuizen et al. 1988)
233.0 (Abramowitz & Yalkowsky 1990)
Heat of Fusion, kcal/mol:
Fugacity Ratio, F (assuming ΔS_{fusion} = 13.5 e.u.):
0.608 (Mackay et al. 1980; Shiu & Mackay 1986)

Water Solubility (g/m^3 or mg/L at 25°C):
0.170 (gen. col.-GC/ECD, Wallnöfer et al. 1973; Weil et al. 1974; Hutzinger et al. 1974; quoted, NAS 1979; Erickson 1986)
0.170 (quoted, NAS 1979; Chou & Griffin 1987)
0.121 (calculated-TSA, Mackay et al. 1980)
0.080 (gen. col.-HPLC/UV, Bellington 1982)
0.172 (quoted lit. average, Yalkowsky et al. 1983)
0.131 (subcooled liq., calculated-TSA, Burkhard et al. 1985b; quoted, Capel et al. 1991)
0.10 (selected, Shiu & Mackay 1986)
0.131 (quoted, Eisenreich 1987)
0.10 (20°C, subcooled liq., calculated-mole fraction of Aroclor mixtures, Murphy et

al. 1987)

0.08 (gen. col.-HPLC, Billington et al. 1988)

0.036 (calculated average of HPLC-RI, Brodsky & Ballschmiter 1988)

0.0172, 0.0273 (quoted, calculated-χ, Nirmalakhanden & Speece 1989)

0.0923, 0.146 (quoted, calculated-TSA, Abramowitz & Yalkowsky 1990)

0.036, 0.044 (quoted average of Brodsky & Ballschmiter 1988, calculated- χ , Patil 1991)

Vapor Pressure (Pa at 25°C):

0.0124, 0.00943, 0.00957 (calculated-MW,GC-RI,χ, Burkhard et al. 1985a)

0.0147 (subcooled liq.,calculated-GC-RI, Burkhard et al. 1985b; quoted, Capel et al. 1991)

0.0128, 0.013 (subcooled liq., GC-RT, Foreman & Bidleman 1985)

0.0152 (quoted, Eisenreich 1987)

0.0064 (20°C, subcooled liq., calculated-mole fraction of Aroclor mixtures, Murphy et al. 1987)

Henry's Law Constant (Pa m^3/mol):

79.28 (batch stripping, Atlas et al. 1982)

24.32 (calculated, Murphy et al. 1983)

32.83 (calculated-P/C, Burkhard et al. 1985b; quoted, Capel et al. 1991)

19.15 (20°C, calculated-P/C, Murphy et al. 1987)

25.43 (calculated- χ , Sabljic & Gusten 1989)

Octanol/Water Partition Coefficient, log K_{OW}:

6.67 (calculated-f const., Yalkowsky et al. 1983; quoted, Erickson 1986)

5.84 (calculated-TSA, Burkard 1984)

4.79 (HPLC-RT, Rapaport & Eisenreich 1984)

5.81 (calculated-π, Rapaport & Eisenreich; quoted, Sklarew & Girvin 1987)

5.81 (quoted, Hawker & Connell 1986)

6.00 (selected, Shiu & Mackay 1986; quoted, Sklarew & Girvin 1987)

5.29 (calculated-S, Chou & Griffin 1987)

5.81 (quoted, Eisenreich 1987)

5.73 (calculated average of HPLC-RI, Brodsky & Ballschmiter 1988)

5.75 (calculated-TSA, Hawker & Connell 1988b; quoted, Capel et al. 1991)

5.73, 5.88 (quoted average of Brodsky & Ballschmiter 1988, calculated- χ , Patil 1991)

Bioconcentration Factor, log BCF:

4.04 (oyster, Vreeland 1974; quoted, Hawker & Connell 1986)

Sorption Partition Coefficient, log K_{OC}:

5.64 (suspended particulate matter, Burkhard 1984)

4.43 (soil, calculated-S, Chou & Griffin 1987)

5.60 (calculated after Karickhoff et al. 1979, Capel & Eisenreich 1990)
4.67 (calculated after Schwarzenbach & Westall 1981, Capel & Eisenreich 1990)

Half-Lives in the Environment:
Air:
Surface water:
Groundwater:
Sediment:
Soil:
Biota:

Environmental Fate Rate Constants or Half-Lives:
Volatilization:
Photolysis:
Hydrolysis:
Oxidation:
Biodegradation:
Biotransformation:
Bioconcentration, Uptake (k_1) and Elimination (k_2) Rate Constants:

Common Name: 2,2',4,4'-Tetrachlorobiphenyl
Synonym: PCB-47
Chemical Name: 2,2',4,4'-tetrachlorobiphenyl
CAS Registry No: 2437-79-8
Molecular Formula: $C_{12}H_6Cl_4$
Molecular Weight: 291.99
Melting Point (°C):
83 (Ullman 1904; Fichter & Adler 1926)
41-42 (Hall & Minhaj 1957; Wallnöfer et al. 1973)
41.0 (Hutzinger et al. 1971)
41, 83 (Hutzinger et al. 1974; Mackay et al. 1980)
42 (Opperhuizen et al. 1988)
83, 93 (quoted, calculated, Abramowitz & Yalkowsky 1990)
Boiling Point (°C):
360 (calculated, Mackay et al. 1982; Shiu & Mackay 1986)
Density (g/cm^3 at 20°C): 1.2024
Molar Volume (cm^3/mol):
268.2 (LeBas method, Miller et al. 1985)
268.2 (LeBas method, Shiu & Mackay 1986)
Molecular Volume (A^3):
262.8 (planar, shorthand, Opperhuizen et al. 1988)
Total Surface Area, TSA (A^2):
259.56 (Mackay et al. 1980; Shiu & Mackay 1986)
254.75 (Burkhard 1984)
234.62 (planar, Doucette 1985)
259.38 (nonplanar, Doucette 1985)
236.19 (planar, Hawker & Connell 1988a)
266.3 (planar, shorthand, Opperhuizen et al. 1988)
236.0 (Abramowitz & Yalkowsky 1990)
Heat of Fusion, kcal/mol:
Fugacity Ratio, F (assuming ΔS_{fusion} = 13.5 e.u.):
0.268 (Shiu & Mackay 1986)

Water Solubility (g/m^3 or mg/L at 25°C):
0.068 (Wallnöfer et al. 1973; Hutzinger et al. 1974; quoted, NAS 1979; Erickson 1986)
0.99 (subcooled liquid, Johnstone et al. 1974)
0.115 (calculated-TSA, Mackay et al. 1980)
0.067 (quoted lit. average, Yalkowsky et al. 1983)
0.103 (subcooled liq., calculated-TSA, Burkhard et al. 1985b)
0.090 (selected, Shiu & Mackay 1986)
0.066 (quoted, Chou & Griffin 1987)
0.102 (quoted, Eisenreich 1987)

0.017 (calculated average of HPLC-RI, Brodsky & Ballschmiter 1988)
0.0541 (22°C, gen. col.-GC, Opperhuizen et al. 1988)
0.0544, 0.0273 (quoted, calculated- χ , Nirmalakhanden & Speece 1989)
0.017 (quoted, Isnard & Lambert 1988, 1989)
0.0923, 0.0367 (quoted, calculated-TSA, Abramowitz & Yalkowsky 1990)
0.017, 0.045 (quoted average of Brodsky & Ballschmiter 1988, calculated- χ , Patil 1991)

Vapor Pressure (Pa at 25°C):
0.0115 (Neely 1983)
0.0101 (calculated-SxHLC, Burkhard et al. 1985a)
0.0142, 0.0111, 0.00419 (calculated-MW,GC-RI, χ , Burkhard et al. 1985a)
0.0151 (subcooled liq., calculated-GC-RI, Burkhard et al. 1985b)
0.0152, 0.0156 (subcooled liq., GC-RT, Foreman & Bidleman 1985)
0.020 (selected, subcooled liq., Shiu & Mackay 1986)
0.0152 (quoted, Eisenreich 1987)

Henry's Law Constant (Pa m^3/mol):
42.86 (calculated-P/C, Burkhard et al. 1985b)
17.38 (calculated-P/C, Shiu & Mackay 1986)
42.56 (quoted, Eisenreich 1987)
44.48 (calculated- χ , Sabljic & Güsten 1989)
19.25 (wetted-wall col.-GC/ECD, Brunner et al. 1990)

Octanol/Water Partition Coefficient, log K_{OW}:
5.68 (Tulp & Hutzinger 1978)
5.2 (HPLC-k', McDuffie 1981; quoted, Chou & Griffin 1987)
7.13, 7.01 (calculated-π, f-const., McDuffie 1981)
6.44 (calculated, Neely 1983)
6.67 (calculated-f const., Yalkowsky et al. 1983; quoted, Erickson 1986)
5.92 (calculated-TSA, Burkhard 1984)
5.27 (HPLC-RT, Rapaport & Eisenreich 1984)
6.29 (calculated-π, Rapaport & Eisenreich 1984; quoted, Sklarew & Girvin 1987)
5.90 (selected, Shiu & Mackay 1986; quoted, Sklarew & Girvin 1987)
6.29 (quoted, Eisenreich 1987)
6.17, 6.12 (HPLC-k', calculated, De Kock & Lord 1987)
5.94 (calculated average of HPLC-RI, Brodsky & Ballschmiter 1988)
5.85 (calculated-TSA, Hawker & Connell 1988a)
6.11 (quoted, Isnard & Lambert 1988, 1989)
5.94, 5.87 (quoted average of Brodsky & Ballschmiter 1988, calculated- χ , Patil 1991)

Bioconcentration Factor, log BCF:
3.98 (rainbow trout muscle, steady state, Branson et al. 1975; quoted, Waid 1986)

4.09 (rainbow trout, Neely et al. 1974)
3.95 (rainbow trout, Branson et al. 1975; quoted, NAS 1979)
4.86 (fish, quoted, Isnard & Lambert 1988,1989)

Sorption Partition Coefficient, log K_{OC}:
5.72 (suspended particulate matter, calculated-K_{OW}, Burkhard 1984)
4.68 (soil, calculated-S, Chou & Griffin 1987)

Half-Lives in the Environment:
Air:
Surface water: half life in Lake Michigan, 49.2 days (Neely 1983).
Groundwater:
Sediment:
Soil:
Biota: half-life in rainbow trout muscle, 28 days (Branson et al. 1975; selected, Waid 1986).

Environmental Fate Rate Constants or Half-Lives:
Volatilization:
Photolysis: calculated sunlight photolysis in surface water at 40°L in summer is 0.055-0.553 day^{-1} with a half-life of 13 days and 2.1×10^{-7} sec^{-1} in winter with a half-life of 170 days, while the observed rate constant in winter is 5×10^{-8} sec^{-1} (Dulin et al. 1986).
Hydrolysis:
Oxidation:
Biodegradation: no degradation observed after 98 days incubation by river dieaway test (Bailey et al. 1983).

Bioconcentration, Uptake (k_1) and Elimination (k_2) Rate Constants:
k_1: 286 day^{-1} (10-12°C, rainbow trout muscle, Branson et al. 1975; selected, Waid 1986)
k_2: 0.030 day^{-1} (10-12°C, rainbow trout muscle, Branson et al. 1975; selected, Waid 1986)

Common Name: 2,2',4,5'-Tetrachlorobiphenyl
Synonym: PCB-49
Chemical Name: 2,2',4,5'-tetrachlorobiphenyl
CAS Registry No: 41464-40-8
Molecular Formula: $C_{12}H_6Cl_4$
Molecular Weight: 291.99
Melting Point (°C):
66-68.5 (Webb & McCall 1972; NAS 1979)
64-66 (Hutzinger et al. 1974; Erickson 1986)
66.0 (Burkhard et al. 1985a)
87.0 (Opperhuizen et al. 1988)
64, 45 (quoted, calculated, Abramowitz & Yalkowsky 1990)
Boiling Point (°C):
360 (calculated, Mackay et al. 1982; Shiu & Mackay 1986)
Density (g/cm^3):
Molar Volume (cm^3/mol):
268.2 (LeBas method, Shiu & Mackay 1986)
1.280 (intrinsic volume: V_I/100, Kamlet et al. 1988; De Bruijn & Hermens 1990)
Molecular Volume (A^3):
262.8 (planar, shorthand, Opperhuizen et al. 1988)
237.75, 216.53, 262.8 (De Bruijn & Hermens 1990)
Total Surface Area, TSA (A^2):
259.6 (Mackay et al. 1980)
254.59 (Burkhard 1984)
234.62 (planar, Doucette 1985)
259.41 (nonplanar, Doucette 1985)
259.41 (Doucette & Andren 1988)
236.01 (planar, Hawker & Connell 1988a)
266.3 (planar, shorthand, Opperhuizen et al. 1988)
283.80, 247.28, 266.3 (De Bruijn & Hermens 1990)
236.0 (Abramowitz & Yalkowsky 1990)
Heat of Fusion, kcal/mol:
Fugacity Ratio, F (assuming ΔS_{fusion} = 13.5 e.u.):
0.411 (Shiu & Mackay 1986)

Water Solubility (g/m^3 or mg/L at 25°C):
0.0164 (gen. col.-GC/ECD, Miller et al. 1984,1985)
0.105 (subcooled liq., calculated-TSA, Burkhard et al. 1985b; quoted, Eisenreich 1987)
0.0781 (20°C, subcooled liq., calculated-mole fraction of Aroclor mixtures, Murphy et al. 1987)
0.0160 (selected, Shiu & Mackay 1986; quoted, Isnard & Lambert 1989; Hawker 1989b)
0.022 (calculated average of HPLC-RI, Brodsky & Ballschmiter 1988)
0.0164, 0.0273 (quoted, calculated- χ , Nirmalakhanden & Speece 1989)

0.0146, 0.923 (quoted, calculated-TSA, Abramowitz & Yalkowsky 1990)
0.022, 0.045 (quoted average of Brodsky & Ballschmiter 1988, calculated- χ , Patil 1991)

Vapor Pressure (Pa at 25°C):
0.00113 (calculated-SxHLC, Burkhard et al. 1985a)
0.00801, 0.00708, 0.00278 (calculated-MW,GC-RI, χ , Burkhard et al. 1985a)
0.0170 (subcooled liq., calculated-GC-RI, Burkhard et al. 1985b; quoted, Eisenreich 1987)
0.0167, 0.0162 (subcooled liq., GC-RT, Foreman & Bidleman 1985)
0.00742 (20°C, subcooled liq., calculated-mole fraction of Aroclor mixtures, Murphy et al. 1987)

Henry's Law Constant (Pa m³/mol):
20.27 (calculated, Murphy et al. 1983)
47.72 (calculated-P/C, Burkhard et al. 1985b)
27.96 (20°C, calculated-P/C, Murphy et al. 1987)
37.90 (calculated- χ , Sabljic & Güsten 1989)
21.28 (wetted-wall col.-GC, Brunner et al. 1990)

Octanol/Water Partition Coefficient, log K_{OW}:
6.23 (calculated-π, Bruggeman et al. 1982)
5.73 (gen. col.-GC/ECD, Miller et al. 1984,1985; quoted, Kamlet et al. 1988; Hawker 1989b)
5.92 (calculated-TSA, Burkhard 1984)
5.20 (HPLC-RT, Rapaport & Eisenreich 1984)
6.22 (calculated-π, Rapaport & Eisenreich 1984; quoted, Sklarew & Girvin 1987)
6.10 (selected, Shiu & Mackay 1986; quoted, Sklarew & Girvin 1987)
6.22 (quoted, Eisenreich 1987)
5.73, 5.87 (quoted, calculated average of HPLC-RI, Brodsky & Ballschmiter 1988)
5.85 (calculated-TSA, Hawker & Connell 1988)
5.95 (calculated-solvatochromic p., Kamlet et al. 1988)
5.73 (quoted exptl., Doucette & Andren 1988)
6.88, 6.67, 5.89, 6.03, 6.13 (calculated-π, f const., MW, χ , TSA, Doucette & Andren 1988)
5.73 (quoted, Isnard & Lambert 1989)
6.23 (calculated-π, De Bruijn et al. 1989)
6.361 (slow stirring-GC, De Bruijn et al. 1989; De Bruijn & Hermens 1990)
5.87, 5.87 (quoted average of Brodsky & Ballschmiter 1988, calculated- χ , Patil 1991)

Bioconcentration Factor, log BCF:

Sorption Partition Coefficient, log K_{OC}:

5.71 (suspended particulate matter, calculated-K_{OW}, Burkhard 1984)

Half-Lives in the Environment:

Air:
Surface water:
Groundwater:
Sediment:
Soil:
Biota:

Environmental Fate Rate Constants or Half-Lives:

Volatilization:
Photolysis:
Hydrolysis:
Oxidation:
Biodegradation:
Biotransformation:
Bioconcentration, Uptake (k_1) and Elimination (k_2) Rate Constants:

Common Name: 2,2',4,6-Tetrachlorobiphenyl
Synonym: PCB-50
Chemical Name: 2,2'4,6-tetrachlorobiphenyl
CAS Registry No: 62796-65-8
Molecular Formula: $C_{12}H_6Cl_4$
Molecular Weight: 291.99
Melting Point (°C):
45.0 (calculated, Abramowitz & Yalkowsky 1990)
Boiling Point (°C):
Density (g/cm^3):
Molar Volume (cm^3/mol):
Molecular Volume (A^3):
Total Surface Area, TSA (A^2):
256.18 (Burkhard 1984)
227.64 (planar, Doucette 1985)
257.83 (nonplanar, Doucette 1985)
229.51 (planar, Hawker & Connell 1988a)
230 (Abramowitz & Yalkowsky 1990)
Heat of Fusion, kcal/mol:
Fugacity Ratio, F (assuming ΔS_{fusion} = 13.5 e.u.):

Water Solubility (g/m^3 or mg/L at 25°C):
0.092 (subcooled liq., calculated-TSA, Burkhard et al. 1985b)
0.034 (calculated average of HPLC-RI, Brodsky & Ballschmiter 1988)
0.1842 (calculated-TSA, Abramowitz & Yalkowsky 1990)
0.034, 0.044 (quoted average of Brodsky & Ballschmiter 1988, calculated- χ , Patil 1991)

Vapor Pressure (Pa at 25°C):
0.0204, 0.0451, 0.00957 (calculated-MW,GC-RI, χ , Burkhard et al. 1985a)
0.0433 (subcooled liq., calculated-GC-RI, Burkhard et al. 1985b)

Henry's Law Constant (Pa m^3/mol):
76.80 (batch stripping, Atlas et al. 1982)
137.8 (calculated-P/C, Burkhard et al. 1985b)
58.57 (calculated- χ , Sabljic & Güsten 1989)

Octanol/Water Partition Coefficient, log K_{OW}:
5.96 (calculated-TSA, Burkhard 1984)
5.75 (calculated average of HPLC-RI, Brodsky & Ballschmiter 1988)
5.63 (calculated-TSA, Hawker & Connell 1988a)
5.75, 5.87 (quoted average of Brodsky & Ballschmiter 1988, calculated- χ , Patil 1991)

Bioconcentration Factor, log BCF:

4.26, 3.50, 3.81 (algae, fish, activated sludge, Freitag et al. 1984,1985; quoted, Halfon & Reggiani 1986)

Sorption Partition Coefficient, log K_{OC}:

5.76 (suspended particulate matter, calculated-K_{OW}, Burkhard 1984)

Half-Lives in the Environment:

Air:
Surface water:
Groundwater:
Sediment:
Soil:
Biota:

Environmental Fate Rate Constants or Half-Lives:

Volatilization:
Photolysis:
Hydrolysis:
Oxidation:
Biodegradation:
Biotransformation:
Bioconcentration, Uptake (k_1) and Elimination (k_2) Rate Constants:

Common Name: 2,2',4,6'-Tetrachlorobiphenyl
Synonym: PCB-51
Chemical Name: 2,2',4,6'-tetrachlorobiphenyl
CAS Registry No: 58194-04-7
Molecular Formula: $C_{12}H_6Cl_4$
Molecular Weight: 291.99
Melting Point (°C):
 45.0 (calculated, Abramowitz & Yalkowsky 1990)
Boiling Point (°C):
Density (g/cm^3):
Molar Volume (cm^3/mol):
 268.2 (LeBas method, Miller et al. 1985)
Molecular Volume (A^3):
 262.8 (planar, shorthand, Opperhuizen et al. 1988)
Total Surface Area, TSA (A^2):
 256.32 (Burkhard 1984)
 227.64 (planar, Doucette 1985)
 257.83 (nonplanar, Doucette 1985)
 229.51 (planar, Hawker & Connell 1988a)
 230.0 (Abramowitz & Yalkowsky 1990)
Heat of Fusion, kcal/mol:
Fugacity Ratio, F (assuming ΔS_{fusion} = 13.5 e.u.):

Water Solubility (g/m^3 or mg/L at 25°C):
 0.0911 (subcooled liq., calculated-TSA, Burkhard et al. 1985b)
 0.065 (calculated-HPLC-RI, Brodsky & Ballschmiter 1988)
 0.1842 (calculated-TSA, Abramowitz & Yalkowsky 1990)
 0.065, 0.044 (quoted of Brodsky & Ballschmiter 1988, calculated- χ , Patil 1991)

Vapor Pressure (Pa at 25°C):
 0.0204, 0.033, 0.00957 (calculated-MW,GC-RI,χ, Burkhard et al. 1985a)
 0.0315 (subcooled liq., calculated-GC-RI, Burkhard et al. 1985b)

Henry's Law Constant (Pa m^3/mol):
 76.80 (batch stripping, Atlas et al. 1982)
 101.2 (calculated-P/C, Burkhard et al. 1985b)
 49.04 (calculated- χ , Sabljic & Güsten 1989)

Octanol/Water Partition Coefficient, log K_{OW}:
 5.96 (calculated-TSA, Burkhard 1984)
 5.51 (calculated-HPLC-RI, Brodsky & Ballschmiter 1988)
 5.63 (calculated-TSA, Hawker & Connell 1988a)
 5.51, 5.88 (quoted of Brodsky & Ballschmiter 1988, calculated- χ , Patil 1991)

Bioconcentration Factor, log BCF:

Sorption Partition Coefficient, log K_{OC}:

5.76 (suspended particulate matter, calculated-K_{OW}, Burkhard 1984)

Half-Lives in the Environment:

Air:

Surface water:

Groundwater:

Sediment:

Soil:

Biota:

Environmental Fate Rate Constants or Half-Lives:

Volatilization:

Photolysis:

Hydrolysis:

Oxidation:

Biodegradation:

Biotransformation:

Bioconcentration, Uptake (k_1) and Elimination (k_2) Rate Constants:

Common Name: 2,2',5,5'-Tetrachlorobiphenyl
Synonym: PCB-52
Chemical Name: 2,2',5,5'-tetrachlorobiphenyl
CAS Registry No: 35693-99-3
Molecular Formula: $C_{12}H_6Cl_4$
Molecular Weight: 291.99
Melting Point (°C):
87-89 (Webb & McCall 1972; Hutzinger et al. 1974; NAS 1979)
86.5-87 (Safe & Hutzinger 1972)
86-87 (Wallnöfer et al. 1973)
87.0 (Mackay et al. 1980; Burkhard et al. 1985a; Opperhuizen et al. 1988)
87, 66 (quoted, calculated, Abramowitz & Yalkowsky 1990)
Boiling Point (°C):
360 (calculated, Mackay et al. 1982; Shiu & Mackay 1986)
Density (g/cm^3 at 20°C): 1.2024
Molar Volume (cm^3/mol):
268.2 (LeBas method, Shiu & Mackay 1986; Shiu et al. 1987)
1.28 (intrinsic volume: $V_I/100$, Hawker 1990)
Molecular Volume (A^3):
262.8 (planar, shorthand, Opperhuizen et al. 1988)
Total Surface Area, TSA (A^2):
259.57 (Mackay et al. 1980)
254.42 (Burkhard 1984)
234.62 (planar, Doucette 1985)
259.41 (nonplanar, Doucette 1985)
258.2 (Sabljic 1987)
235.84 (planar, Hawker & Connell 1988a)
266.3 (planar, shorthand, Opperhuizen et al. 1988)
236.0 (Abramowitz & Yalkowsky 1990)
Heat of Fusion, kcal/mol:
Entropy of Fusion, cal/mol K (e.u.):
11.0 (Hinkley et al. 1990)
Fugacity Ratio, F (assuming ΔS_{fusion} = 13.5 e.u.):
0.243 (Mackay et al. 1980; Shiu & Mackay 1986)

Water Solubility (g/m^3 or mg/L at 25°C):
0.046 (shake flask-GC/ECD, Wallnöfer et al. 1973; Hutzinger et al. 1974; Tulp & Hutzinger 1978; quoted, Nas 1979; Landrum et al. 1984; Erickson 1986)
0.0265 (shake flask-GC/ECD, Haque & Schmedding 1975; selected, Chiou et al. 1979; Haque et al. 1980; Bruggeman et al. 1984; Miller et al. 1984; McKim et al. 1985)
0.016 (LSC, Metcalf et al. 1975)
0.0060 (16.5°C, shake flask-GC/ECD, Wiese & Griffin 1978; quoted, McKim et al.

1985)

0.0223 (shake flask-GC/ECD, Lee et al. 1979)

0.041 (calculated-TSA, Mackay et al. 1980; quoted, Ballschmiter & Wittlinger 1991)

0.055 (gen. col.-GC/ECD, Bruggeman et al. 1981)

0.074 (Bruggeman Ph.D. Thesis 1983 in Opperhuizen et al. 1988)

0.0365 (gen. col.-HPLC/UV, Huang 1983)

0.0184 (calculated-K_{OW}, Yalkowsky et al. 1983)

0.184, 0.039 (quoted, calculated-UNIFAC, Banerjee 1985)

0.027 (gen. col.-GC/ECD, Miller et al. 1984; 1985)

0.0153 (quoted, Opperhuizen 1986)

0.106 (subcooled liquid, calculated-TSA, Burkhard et al. 1985b; quoted, Eisenreich 1987; Capel et al. 1991)

0.030 (selected, Shiu & Mackay 1986)

0.36 (quoted, Chou & Griffin 1987)

0.113 (20°C, subcooled liq., calculated-mole fraction of Aroclor mixtures, Murphy et al. 1987)

0.029 (calculated average of HPLC-RI, Brodsky & Ballschmiter 1988)

0.0041 (calculated-UNIFAC, Banerjee & Howard 1988)

0.170 (gen. col.-HPLC, Billington et al. 1988)

0.0237 (Brodsky et al. 1988; quoted, Ballschmiter & Wittlinger 1991)

0.110 (gel. col., Dunnivant & Elzerman 1988)

0.461 (subcooled liquid, calculated-M.P., Dunnivant & Elzerman 1988)

0.0153 (22°C, gen. col.-GC/ECD, Opperhuizen et al. 1988)

0.027 (quoted, Isnard & Lambert 1988,1989)

0.0153, 0.0273 (quoted, calculated- χ , Nirmalakhanden & Speece 1989)

0.0292, 0.0582 (quoted, calculated-TSA, Abramowitz & Yalkowsky 1990)

0.0161 (Eadie et al. 1990)

0.029, 0.045 (quoted average of Brodsky & Ballschmiter 1988, calculated- χ , Patil 1991)

0.031 (quoted, Paya-Perez et al. 1991)

Vapor Pressure (Pa at 25°C):

0.00734 (GC-RT, Westcott & Bidleman 1981)

0.0071 (GC-RT, Westcott et al. 1981)

0.0203 (P_L calculated from P_S using F, Westcott et al. 1981)

0.00253 (gas saturation, Westcott et al. 1981)

0.104 (quoted, Neely 1981)

0.428 (P_L calculated from P_S using F, Neely 1981)

0.00493 (Neely 1983; quoted, Erickson 1986)

0.0203 (P_L calculated from P_S using F, Neely 1983)

0.0159, 0.0229 (subcooled liquid, GC-RT, Bidleman 1984)

0.0104 (subcooled liquid, quoted, Bidleman 1984)

0.0192 (subcooled liquid, Burkhard 1984)

0.00427 (calculated-SxHLC, Burkhard et al. 1985a)
0.00497 (selected, exptl., Burkhard et al. 1985b)
0.00497, 0.00492, 0.00957 (calculated-MW, GC-RI, χ, Burkhard et al. 1985a)
0.0193 (subcooled liquid, calculated-GC-RI, Burkhard et al. 1985b; quoted, Eisenreich 1987; Capel et al. 1991)
0.0184, 0.0173, 0.0191 (subcooled liq., GC-RT, Foreman & Bidleman 1985)
0.0049 (selected, Shiu & Mackay 1986; quoted, Sklarew & Girvin 1987)
0.0020 (selected subcooled liq., Shiu & Mackay 1986)
0.00904 (20°C, subcooled liq., calculated-mole fraction of Aroclor mixtures, Murphy et al. 1987)
0.0127 (calculated-SxHLC, Dunnivant & Elzerman 1988)
0.0546 (subcooled liquid, calculated-M.P., Dunnivant & Elzerman 1988)
0.0104, 0.008 (quoted, subcooled liq., Hinckley et al. 1990)
0.013 (Wittlinger et al. 1990; quoted, Ballschmiter & Wittlinger 1991)
0.016 (quoted from Bidleman 1984, Ballschmiter & Wittlinger 1991)

Henry's Law Constant (Pa m^3/mol):
31.41-53.7 (calculated-P/C, Westcott et al. 1981)
14.1-53.7 (calculated-P/C, Westcott & Bidleman 1981)
94.15 (batch stripping, Atlas et al. 1982)
22.29 (calculated-P/C, Murphy et al. 1983)
26.34 (calculated-P/C, Murphy et al. 1984)
53.20 (calculated-P/C, Burkhard et al. 1985b; quoted, Eisenreich 1987; Capel et al. 1991)
2.53 (batch stripping, Hassett & Milicic 1985)
12.16 (batch stripping, Oliver 1985)
47.59 (calculated, Shiu & Mackay 1986)
24.11 (20°C, subcooled liq., calculated-mole fraction of Aroclor mixtures, Murphy et al. 1987)
34.65 (batch stripping, Dunnivant & Elzerman 1988; quoted, Ballschmiter & Wittlinger 1991)
34.65 (batch stripping, Dunnivant et al. 1988)
20.27 (wetted-wall col.-GC, Brunner et al. 1990)
160 (Wittlinger et al. 1990; quoted, Ballschmiter & Wittlinger 1991)

Octanol/Water Partition Coefficient, log K_{OW}:
3.91 (radiolabelled-^{14}C, Metcalf et al. 1975)
5.81 (Hansch & Leo 1979; Chiou et al. 1982; Chiou & Block 1986)
6.90 (calculated-π, Tulp & Hutzinger 1978)
3.91 (NAS 1979)
6.26 (TLC-RT, Bruggeman et al. 1982; quoted, Bruggeman et al. 1984; Landrum et al. 1984; Mackay & Hughes 1984; Opperhuizen 1986; Erickson 1986)
5.81 (quoted, Chiou et al. 1982; Chiou & Block 1986)

6.12 (calculated, Neely 1983)
6.67 (calculated-f const., Yalkowsky et al. 1983)
5.91 (calculated-TSA, Burkhard 1984)
5.07 (HPLC-RT, Rapaport & Eisenreich 1984)
6.09 (calculated-π, Rapaport & Eisenreich 1984; quoted, Sklarew & Girvin 1987)
6.10 (quoted, Oliver & Chalton 1984)
6.0 (quoted, McKim et al. 1985)
6.09 (quoted, Hawker & Connell 1985,1986)
6.09 (quoted, Eisenreich 1987)
6.10 (selected, Shiu & Mackay 1986; quoted, Gobas et al. 1987; Sklarew & Girvin 1987; Banerjee & Baughman 1991; Clark et al. 1990)
5.78 (calculated-S, Chou & Griffin 1987)
5.9 (calculated-chlorine substituent, Oliver 1987a & c)
5.79 (calculated average of HPLC-RI, Brodsky & Ballschmiter 1988)
5.80 (Brodsky et al. 1988; quoted, Ballschmiter & Wittlinger 1991)
6.09, 6.02 (quoted, calculated-UNIFAC, Banerjee & Howard 1988)
5.84 (calculated-TSA, Hawker & Connell 1988a; Connell & Hawker 1988; Hawker 1990; quoted, Capel et al. 1991)
5.80 (quoted, Isnard & Lambert 1988,1989)
6.09 (quoted, Thomann 1989)
5.79, 5.87 (quoted average of Brodsky & Ballschmiter 1988, calculated-χ, Patil 1991)

Bioconcentration Factor, log BCF:
3.87 (oyster, Vreeland 1974; quoted, Hawker & Connell 1986)
4.26, 4.60, 4.02, 4.07 (algae, snail, mosquito, fish, Metcalf et al. 1975)
2.66 (green sunfish, 15 days in static water, Sanborn et al. 1975)
6.21 (goldfish, 3% lipid, Bruggeman et al. 1981)
6.07 (goldfish, 10% lipid dry wet in food, Bruggeman et al. 1981)
6.38 (guppy, 3.5% lipid, Bruggeman et al. 1982,1984; quoted, Gobas et al. 1987)
4.69, 4.98 (goldfish, exptl., correlated, Mackay & Hughes 1984)
5.30 (rainbow trout, kinetic, Oliver & Niimi 1985)
4.26 (rainbow trout, steady state, Oliver & Niimi 1985)
6.28 (rainbow trout, field data, Oliver & Niimi 1985)
4.92 (guppy, Gobas et al. 1987; quoted, Banerjee & Baughman 1991)
4.26, 4.69 (guppy, goldfish, calculated-C_B/C_W, or k_1/k_2, Connell & Hawker 1988; Hawker 1990)
4.26 (fish, quoted, Isnard & Lambert 1988,1989)
6.38 (guppy, lipid wt. based, Gobas et al. 1989)
5.76 (guppy, corr. lipid wt. based, Gobas et al. 1989)
7.38 (rainbow trout, quoted, Thomann 1989)
4.25 (guppy, estimated, Banerjee & Baughman 1991)

Sorption Partition Coefficient, log K_{OC}:

4.67 (soil, Chiou et al. 1979)
5.91 (suspended particulate matter, calculated-K_{OW}, Burkhard 1984)
5.6 (field data, Oliver & Charlton 1984)
5.5 (Niagara River-org. matter, Oliver & Charlton 1984)
5.7 (calculated-K_{OW}, Oliver & Charlton 1984)
3.87, 4.36 (Huron River at 7.8 mg C/L DOC, Landrum et al. 1984)
4.87 (dissolved humic acid, gas purging-LSC, Hassett & Milicic 1985)
5.35 (suspended solids-Lake Superior, Baker et al. 1986)
5.70, 4.87 (suspended solids-Lake Superior, calculated, Baker et al. 1986)
5.0-6.4, 5.9 (suspended sediment, range, average, Oliver 1987a)
7.0 (algae > 50 μm, Oliver 1987a)
6.12 (Lake Michigan water column, Swackhamer & Armstrong 1987)
4.65 (calculated, Bahnick & Doucette 1988)
3.48 (12 lakes/streams in S. Ontario at 1.6-26.5 mg C/L, Evans 1988)
4.35 (calculated-polymaleic acid, Chin & Weber 1989)
5.88 (calculated after Karickhoff et al. 1979, Capel & Eisenreich 1990)
4.88 (calculated after Schwarzenbach & Westall 1981, Capel & Eisenreich 1990)
3.83 (organic carbon, Eadie et al. 1990)
5.00 (Aldrich humic acid from soil & water samples, Jota & Hassett 1991)
4.38-4.81 (humic acid from soil & water samples, Jota & Hassett 1991)
2.89-3.93 (fulvic acid & dissolved organic matter samples, Jota & Hassett 1991)
5.41 (soil, Paya-Perez et al. 1991)

Sorption Partition Coefficient, log K_{OM}:

4.67 (Haque & Schmedding 1976; Chou & Griffin 1987)
4.67, 4.62 (quoted, calculated- χ , Sabljic 1984)
3.88 (natural sediment, Eadie et al. 1990)

Half-Lives in the Environment:

Air:
Surface water: half-life in Lake Michigan, 19.7 days (Neely 1983).
Groundwater:
Sediment:
Soil:
Biota: half-life in female rainbow trouts , 1.76 years and in males, 1.43 years (Guiney et al. 1980); in rainbow trout, 500 days (Niimi & Oliver 1983; Oliver & Niimi 1985); and its muscle, 99 days (Niimi & Oliver 1983); in guppy, 46 days (Bruggeman et al. 1984); in worms at 8°C, 43 days (Oliver 1987c).

Environmental Fate Rate Constants or Half-Lives:

Volatilization: half-life of evaporation from an initial concentration of 0.005 ppm in a glass dish of 4.5 cm depth of water solution at 24°C is measured to be 2.8 hours and 0.68 hours with stirring of the solution (Chiou et al. 1979).

Photolysis:
Hydrolysis:
Oxidation:
Biodegradation: microbial degradation with pseudo first-order rate constant of 0.1 $year^{-1}$ in the water column and 1.0 $year^{-1}$ in the sediment (Furukawa et al. 1978; quoted, Neely 1981).
Biotransformation:
Bioconcentration, Uptake (k_1) and Elimination (k_2) Rate Constants:

k_2: 0.008 day^{-1} (11°C, rainbow trout, Guiney et al. 1977; quoted, Waid 1986)
k_2: 0.003 day^{-1} (10-11°C, rainbow trout eggs & sac fry, Guiney et al. 1980; quoted, Waid 1986)
k_1: 740 day^{-1} (23°C, goldfish, 3% lipid content, Bruggeman et al. 1981; quoted, Waid 1986)
k_2: 0.015 day^{-1} (23°C, goldfish, 3% lipid content, Bruggeman et al. 1981; quoted, Waid 1986; selected, Clark et al. 1990)
k_2: 0.0014 day^{-1} (rainbow trout, Niimi & Oliver 1983; quoted, Clark et al. 1990)
k_1: 1200 day^{-1} (guppy, Bruggeman et al. 1984)
k_2: 0.015 day^{-1} (guppy, Bruggeman et al. 1984; quoted, Clark et al. 1990)
k_2: 0.015, 0.0134 day^{-1} (goldfish, exptl., correlated, Mackay & Hughes 1984)
k_1: 2800 day^{-1} (rainbow trout, Oliver & Niimi 1985)
k_2: 0.0014 day^{-1} (rainbow trout, Oliver & Niimi 1985)
k_1: 30.8 $hour^{-1}$ (goldfish, quoted, Hawker & Connell 1985)
$1/k_2$: 1600 hour (goldfish, quoted, Hawker & Connell 1985)
k_1: 50.0 $hour^{-1}$ (guppy, quoted, Hawker & Connell 1985)
$1/k_2$: 1600 hour (guppy, quoted, Hawker & Connell 1985)
k_1: 1100 day^{-1} (guppy, Opperhuizen 1986)
log k_1: 2.87 day^{-1} (fish, quoted, Connell & Hawker 1988)
log $1/k_2$: 1.82 day (fish, quoted, Connell & Hawker 1988; Thomann 1989)
log $1/k_2$: 2.9, 3.0 hour (fish, quoted, calculated-K_{OW}, Hawker & Connell 1988b).
$1/k_2$: 30.3 day (guppy, quoted, Clark et al. 1990)
$1/k_2$: 61.7, 102 day (guppy, Gobas et al. 1989; quoted, Clark et al. 1990)

Common Name: 2,2',5,6'-Tetrachlorobiphenyl
Synonym: PCB-53
Chemical Name: 2,2',5,6'-tetrachlorobiphenyl
CAS Registry No: 41464-41-9
Molecular Formula: $C_{12}H_6Cl_4$
Molecular Weight: 291.99
Melting Point (°C):
   103-104.5 (Hutzinger et al. 1974)
   30.0 (calculated, Abramowitz & Yalkowsky 1990)
Boiling Point (°C):
Density (g/cm^3):
Molar Volume (cm^3/mol):
   268.2 (LeBas method, Shiu & Mackay 1986)
Molecular Volume (A^3):
Total Surface Area, TSA (A^2):
   256.15 (Burkhard 1984)
   227.64 (planar, Doucette 1985)
   257.83 (nonplanar, Doucette 1985)
   229.34 (planar, Hawker & Connell 1988a)
   229.0 (Abramowitz & Yalkowsky 1990)
Heat of Fusion, kcal/mol:
Fugacity Ratio, F (assuming ΔS_{fusion} = 13.5 e.u.): 0.165

Water Solubility (g/m^3 or mg/L at 25°C):
   0.0923 (subcooled liq., calculated-TSA, Burkhard et al. 1985b)
   0.109 (20°C, subcooled liq., calculated-mole fraction of Aroclor mixtures, Murphy et al. 1987)
   0.065 (calculated average of HPLC-RI, Brodsky & Ballschmiter 1988)
   0.0476 (gen. col.-GC/ECD, Dunnivant & Elzerman 1988)
   0.288 (subcooled liquid, calculated-MP, Dunnivant & Elzerman 1988)
   0.292 (calculated-TSA, Abramowitz & Yalkowsky 1990)
   0.065, 0.044 (quoted average of Brodsky & Ballschmiter 1988, calculated- χ , Patil 1991)

Vapor Pressure (Pa at 25°C):
   0.00493 (Neely 1983; quoted, Erickson 1986)
   0.0273 (subcooled liquid, GC-RT, Bidleman 1984)
   0.0356 (subcooled liquid Burkhard 1984)
   0.0204, 0.0372, 0.00957 (calculated-MW, GC-RI, χ, Burkhard et al. 1985a)
   0.0356 (subcooled liquid, calculated-GC-RI, Burkhard et al. 1985b)
   0.0268, 0.0331, 0.0273 (subcooled liq., GC-RT, Foreman & Bidleman 1985)
   0.011 (20°C, subcooled liq., calculated-mole fraction of Aroclor mixtures, Murphy et al. 1987)

0.00671 (calculated-SxHLC, Dunnivant & Elzerman 1988)
0.0405 (subcooled liquid, calculated-M.P., Dunnivant & Elzerman 1988)

Henry's Law Constant (Pa m^3/mol):
30.40 (calculated-P/C, Murphy et al. 1983)
112.5 (calculated-P/C, Burkhard et al. 1985b)
28.67 (20°C, calculated-P/C, Murphy et al. 1987)
41.14 (batch stripping, Dunnivant & Elzerman 1988)
41.14 (batch stripping, Dunnivant et al. 1988)

Octanol/Water Partition Coefficient, log K_{OW}:
5.96 (calculated-TSA, Burkhard 1984)
5.90 (calculated-chlorine substituents, Oliver 1987c)
5.55 (calculated average of HPLC-RI, Brodsky & Ballschmiter 1988)
5.46 (gen. col.-GC, Hawker & Connell 1988a)
5.62 (calculated-TSA, Hawker & Connell 1988a; quoted, Clark et al. 1990)
5.55, 5.87 (quoted average of Brodsky & Ballschmiter 1988, calculated- χ , Patil 1991)

Bioconcentration Factor, log BCF:

Sorption Partition Coefficient, log K_{OC}:
5.76 (suspended particulate matter, calculated-K_{OW}, Burkhard 1984)

Half-Lives in the Environment:
Air:
Surface water:
Groundwater:
Sediment:
Soil:
Biota: half-life in rainbow trout, 365 days and its muscle 107 days (Niimi & Oliver 1983); in worms at 8°C, 30 days (Oliver 1987c).

Environmental Fate Rate Constants or Half-Lives:
Volatilization:
Photolysis:
Hydrolysis:
Oxidation:
Biodegradation:
Biotransformation:
Bioconcentration, Uptake (k_1) and Elimination (k_2) Rate Constants:
k_2: 0.0019 day^{-1} (rainbow trout, Niimi & Oliver 1983; selected, Clark et al. 1990)

Common Name: 2,2',6,6'-Tetrachlorobiphenyl
Synonym: PCB-54
Chemical Name: 2,2',6,6'-tetrachlorobiphenyl
CAS Registry No: 15968-05-5
Molecular Formula: $C_{12}H_6Cl_4$
Molecular Weight: 291.99
Melting Point (°C):
198 (Van Roosmalen 1934; Hutzinger et al. 1974; Burkhard et al. 1985a; Opperhuizen et al. 1988)
101.0 (calculated, Abramowitz & Yalkowsky 1990)
Boiling Point (°C):
360 (calculated, Mackay et al. 1982; Shiu & Mackay 1986)
Density (g/cm³):
Molar Volume (cm³/mol):
268.2 (LeBas method, Shiu & Mackay 1986)
1.28 (intrinsic volume: $V_I/100$, De Bruijn & Hermens 1990)
Molecular Volume (A³):
237.0 (planar, shorthand, Opperhuizen et al. 1988)
237.46, 215.32, 237.0, 250.6 (De Bruijn & Hermens 1990)
Total Surface Area, TSA (A²):
254.88 (Burkhard 1984)
214.36 (planar, Doucette 1985)
256.30 (nonplanar, Doucette 1985)
217.18 (planar, Hawker & Connell 1988a)
246.7 (planar, shorthand, Opperhuizen et al. 1988)
281.79, 238.98, 248.3, 256.9 (De Bruijn & Hermens 1990)
217.0 (Abramowitz & Yalkowsky 1990)
Heat of Fusion, kcal/mol:
Fugacity Ratio, F (assuming ΔS_{fusion} = 13.5 e.u.): 0.0195

Water Solubility (g/m³ or mg/L at 25°C):
0.102 (subcooled liquid, calculated-TSA, Burkhard et al. 1985b)
0.184 (calculated average of HPLC-RI, Brodsky & Ballschmiter 1988)
0.0027 (22°C, gen. col.-GC/ECD, Opperhuizen et al. 1988)
0.616 (subcooled liquid, calculated-M.P., Dunnivant & Elzerman 1988)
0.00273, 0.0273 (quoted, calculated-χ, Nirmalakhanden & Speece 1989)
0.0119 (gen. col.-GC/ECD, Dunnivant & Elzerman 1988)
0.1842 (calculated-TSA, Abramowitz & Yalkowsky 1990)
0.184, 0.044 (quoted average of Brodsky & Ballschmiter 1988, calculated- χ , Patil 1991)

Vapor Pressure (Pa at 25°C):

0.0659 (subcooled liquid, Burkhard 1984)
0.000396, 0.00132, 0.0219 (calculated-MW,GC-RI,χ, Burkhard et al. 1985a)
0.0659 (subcooled liquid, calculated-GC-RI, Burkhard et al. 1985b)
0.0392, 0.0517 (GC-RT, subcooled liq., Foreman & Bidleman 1985)
0.00227 (calculated-SxHLC, Dunnivant & Elzerman 1988)
0.0405 (subcooled liquid, calculated-M.P., Dunnivant & Elzerman 1988)

Henry's Law Constant (Pa m^3/mol):

188.5 (calculated-P/C, Burkhard et al. 1985b)
15.0 (calculated, Coates & Elzerman 1986)
55.73 (batch stripping, Dunnivant & Elzerman 1988; Dunnivant et al. 1988)
20.27 (wetted-wall col.-GC/ECD, Brunner et al. 1990)

Octanol/Water Partition Coefficient, log K_{OW}:

6.63 (calculated after Rekker 1977, Könemann 1981)
4.16 (HPLC-k', McDuffie 1981)
7.13 (calculated-π, McDuffie 1981)
7.01 (calculated-f const., McDuffie 1981)
5.94 (TLC-RT, Bruggeman et al. 1982; quoted, Erickson 1986)
5.94 (calculated-π, Bruggeman et al. 1982; De Bruijn et al. 1989)
5.92 (calculated-TSA, Burkhard 1984)
5.90 (selected, Shiu & Mackay 1986; quoted, Sklarew & Girvin 1987)
5.24 (calculated average of HPLC-RI, Brodsky & Ballschmiter 1988)
5.48 (gen. col.-GC, Hawker & Connell 1988a)
5.21 (calculated-TSA, Hawker & Connell 1988a)
5.94 (calculated-π, De Bruijn et al. 1989)
5.936 (slow stirring-GC, De Bruijn et al. 1989)
5.936 (slow stirring-GC, De Bruijn & Hermens 1990)
5.24, 5.88 (quoted average of Brodsky & Ballschmiter 1988, calculated- χ , Patil 1991)

Bioconcentration Factor, log BCF:

Sorption Partition Coefficient, log K_{OC}:

5.72 (suspended particulate matter, calculated-K_{OW}, Burkhard 1984)
5.63 (river sediment, Coates & Elzerman 1986)
4.79 (correlated literature values in soils, Sklarew & Girvin 1987)

Half-Lives in the Environment:

Air:
Surface water:
Groundwater:
Sediment:

Soil:
Biota:

Environmental Fate Rate Constants or Half-Lives:
Volatilization:
Photolysis:
Hydrolysis:
Oxidation:
Biodegradation:
Biotransformation:
Bioconcentration, Uptake (k_1) and Elimination (k_2) Rate Constants or Half-Lives:

Common Name: 2,3,4,4'-Tetrachlorobiphenyl
Synonym: PCB-60
Chemical Name: 2,3,4,4'-tetrachlorobiphenyl
CAS Registry No: 33025-41-1
Molecular Formula: $C_{12}H_6Cl_4$
Molecular Weight: 291.99
Melting Point (°C):
128 (Weast 1972,1973)
142 (Saeki et al. 1971; Hutzinger et al. 1974; NAS 1979; Burkhard et al. 1985a)
142, 102 (quoted, calculated, Abramowitz & Yalkowsky 1990)
Boiling Point (°C):
360 (calculated, Mackay et al. 1982; Shiu & Mackay 1986)
Density (g/cm^3 at 20°C): 1.2024
Molar Volume (cm^3/mol):
268.2 (LeBas method, Shiu & Mackay 1986)
Molecular Volume (A^3):
265.2 (planar, shorthand, Opperhuizen et al. 1988)
Total Surface Area, TSA (A^2):
259.15 (Mackay et al. 1980)
252.80 (Burkhard 1984)
243.93 (planar, Doucette 1985)
256.94 (nonplanar, Doucette 1985)
243.80 (planar, Hawker & Connell 1988a)
268.5 (planar, shorthand, Opperhuizen et al. 1988)
244.0 (Abramowitz & Yalkowsky 1990)
Heat of Fusion, kcal/mol:
Fugacity Ratio, F (assuming ΔS_{fusion} = 13.5 e.u.):
0.0695 (Shiu & Mackay 1986)

Water Solubility (g/m^3 or mg/L at 25°C):
0.058 (shake flask-GC/ECD, Wallnöfer et al. 1973; Hutzinger et al. 1974; quoted, Erickson 1986)
0.0168 (calculated-TSA, Mackay et al. 1980)
0.121 (calculated-TSA, subcooled liq., Burkhard et al. 1985b)
0.0389 (20°C, subcooled liq., calculated-mole fraction of Aroclor mixtures, Murphy et al. 1987)
0.041 (calculated average of HPLC-RI, Brodsky & Ballschmiter 1988)
0.0359, 0.0273 (Nirmalakhanden & Speece 1989)
0.0146 (calculated-TSA, Abramowitz & Yalkowsky 1990)
0.040, 0.045 (quoted average of Brodsky & Ballschmiter 1988, calculated- χ , Patil 1991)

Vapor Pressure (Pa at 25°C):

0.00142, 0.000319, 0.00184 (calculated-MW,GC-RI,χ, Burkhard et al. 1985a)
0.00427 (subcooled liq., calculated-GC-RI, Burkhard et al. 1985b)
0.00527, 0.00414, (subcooled liq., GC-RT, Foreman & Bidleman 1985)
0.00217 (20°C, subcooled liq., calculated-mole fraction of Aroclor mixtures, Murphy et al. 1987)

Henry's Law Constant (Pa m^3/mol):

84.20 (batch stripping, Atlas et al. 1982)
10.34 (calculated-P/C, Burkhard et al. 1985b)
16.41 (20°C, calculated-P/C, Murphy et al. 1987)
15.40 (calculated- χ , Sabljic & Güsten 1989)

Octanol/Water Partition Coefficient, log K_{OW}:

5.87 (calculated-TSA, Burkhard 1984)
5.33 (HPLC-RT, Rapaport & Eisenreich 1984)
5.84 (calculated-π, Rapaport & Eisenreich 1984; quoted, Sklarew & Girvin 1987)
5.90 (selected, Shiu & Mackay 1986; quoted, Sklarew & Girvin 1987)
6.25 (calculated average of HPLC-RI, Brodsky & Ballschmiter 1988)
6.11 (calculated-TSA, Hawker & Connell 1988a)
6.24, 5.88 (quoted average of Brodsky & Ballschmiter 1988, calculated- χ , Patil 1991)

Bioconcentration Factor, log BCF:

Sorption Partition Coefficient, log K_{OC}:

5.67 (suspended particulate matter, calculated-K_{OW}, Burkhard 1984)

Half-Lives in the Environment:

Air:
Surface water:
Groundwater:
Sediment:
Soil:
Biota:

Environmental Fate Rate Constants or Half-Lives:

Volatilization:
Photolysis:
Hydrolysis:
Oxidation:
Biodegradation:
Biotransformation:
Bioconcentration, Uptake (k_1) and Elimination (k_2) Rate Constants:

Common Name: 2,3,4,5-Tetrachlorobiphenyl
Synonym: PCB-61
Chemical Name: 2,3,4,5-tetrachlorobiphenyl
CAS Registry No: 33284-53-6
Molecular Formula: $C_{12}H_6Cl_4$
Molecular Weight: 291.99
Melting Point (°C):
92-92.5 (McBee et al. 1955; Hutzinger et al. 1974)
92.0 (Mackay et al. 1980; Bruggeman et al. 1982; Burkhard et al. 1985a; Opperhuizen et al. 1988)
92, 95 (quoted, calculated, Abramowitz & Yalkowsky 1990)
Boiling Point (°C):
360 (calculated, Mackay et al. 1982; Shiu & Mackay 1986)
Density (g/cm^3 at 20°C): 1.2024
Molar Volume (cm^3/mol):
268.2 (LeBas method, Shiu & Mackay 1986)
1.28 (intrinsic volume: $V_I/100$, Kamlet et al. 1988; Hawker 1989b; De Bruijn & Hermens 1990)
Molecular Volume (A^3):
257.8 (planar, shorthand, Opperhuizen et al. 1988)
237.44, 217.02, 257.8 (De Bruijn & Hermens 1990)
Total Surface Area, TSA (A^2):
255.21 (Mackay et al. 1980; Shiu & Mackay 1986)
250.76 (Burkhard 1984)
241.94 (planar, Doucette 1985)
254.95 (nonplanar, Doucette 1985; Doucette & Andren 1988)
241.72 (planar, Hawker & Connell 1988a)
263.5 (planar, shorthand, Opperhuizen et al. 1988)
282.21, 246.66, 263.5 (De Bruijn & Hermens 1990)
242.0 (Abramowitz & Yalkowsky 1990)
Heat of Fusion, kcal/mol:
6.02 (Miller et al. 1984)
Entropy of Fusion, cal/mol K (e.u.):
16.60 (Miller et al. 1984)
Fugacity Ratio, F (assuming ΔS_{fusion} = 13.5 e.u.):
0.217 (Mackay et al. 1980)
0.218 (Shiu & Mackay 1986)

Water Solubility (g/m^3 or mg/L at 25°C):
0.0192 (shake flask-GC/ECD, Wallnöfer et al. 1973; Weil et al. 1974)
0.0209 (shake flask-GC/ECD, Haque & Schmedding 1975)
0.0525 (calculated-TSA, Mackay et al. 1980)
0.0193 (TLC-RT, Bruggeman et al. 1982)

0.0099 (gen. col.-HPLC/UV, Bellington 1982)
0.0193 (quoted lit. average, Yalkowsky et al. 1983; quoted, Erickson 1986)
0.0209 (gen. col.-GC/ECD, Miller et al. 1984,1985)
0.142 (subcooled liquid, calculated-TSA, Burkhard et al. 1985b)
0.0099 (gen. col.-HPLC, Billington et al. 1988)
0.020 (selected, Shiu & Mackay 1986; quoted, Hawker 1989b)
0.0193, 0.023 (quoted, calculated average of HPLC-RI, Brodsky & Ballschmiter 1988)
0.014 (gen. col.-GC, Dunnivant & Elzerman 1988)
0.0648 (subcooled liquid, calculated-M.P., Dunnivant & Elzerman 1988)
0.130 (quoted, subcooled liquid, Hawker 1989b)
0.0193, 0.0273 (quoted, calculated- χ , Nirmalakhanden & Speece 1989)
0.0209 (quoted, Isnard & Lambert 1989)
0.01842, 0.01842 (quoted, calculated-TSA, Abramowitz & Yalkowsky 1990)

Vapor Pressure (Pa at 25°C):
0.00443, 0.0013, 0.00957 (calculated-MW,GC-RT, χ , Burkhard et al. 1985a)
0.00558 (subcooled liquid, calculated-GC-RI, Burkhard et al. 1985b)

Henry's Law Constant (Pa m^3/mol):
11.45 (calculated-P/C, Burkhard et al. 1985b)
17.53 (calculated- χ , Sabljic & Güsten 1989)

Octanol/Water Partition Coefficient, log K_{OW}:
5.78 (HPLC-RT, Sugiura et al. 1979)
6.39 (TLC-RT, Bruggeman et al. 1982; quoted, Erickson 1986)
6.74 (calculated-f const., Yalkowsky et al. 1983)
5.81 (calculated-TSA, Burkhard 1984)
5.72 (gen. col.-GC/ECD, Miller et al. 1984,1985; quoted, Burkhard & Kuehl 1986; Sklarew & Girvin 1987)
5.90 (selected, Shiu & Mackay 1986; Gobas et al. 1987; quoted, Sklarew & Girvin 1987; Hawker 1989b; Clark et al. 1990)
5.72, 6.44 (quoted, calculated average of HPLC-RI, Brodsky & Ballschmiter 1988)
6.18 (gen. col.-GC/ECD, Hawker & Connell 1988a)
6.04 (calculated-TSA, Hawker & Connell 1988a)
5.92 (calculated-solvatochromic p., Kamlet et al. 1988)
5.91, 6.74 (quoted, calculated-π or f const., Kamlet et al. 1988)
5.72 (quoted exptl., Doucette & Andren 1988)
6.88, 6.74, 5.89, 6.04, 6.02 (calculated-π, f const., MW, χ , TSA, Doucette & Andren 1988)
5.91 (quoted, Hawker 1989b)
5.72 (quoted, Isnard & Lambert 1989)
6.39 (calculated-π, De Bruijn etal. 1989)
6.406 (slow stirring-GC, De Bruijn et al. 1989; De Bruijn & Hermens 1990)

Bioconcentration Factor, log BCF:

4.29, 3.57, 3.94, 3.90 (golden orfe, carp, brown trout, guppy, Sugiura et al. 1979)

Sorption Partition Coefficient, log K_{OC}:

5.61 (suspended particulate matter, calculated-K_{OW}, Burkhard 1984)

Half-Lives in the Environment:

Air:

Surface water:

Groundwater:

Sediment:

Soil:

Biota: half-life in rainbow trout, 312 days and its muscle 93 days (Niimi & Oliver 1983).

Environmental Fate Rate Constants or Half-Lives:

Volatilization:

Photolysis:

Hydrolysis:

Oxidation:

Biodegradation:

Biotransformation:

Bioconcentration, Uptake (k_1) and Elimination (k_2) Rate Constants:

k_1: 0.038 day^{-1} (golden orfe, Sugiura et al. 1979)

k_2: 0.0022 day^{-1} (rainbow trout, Niimi & Oliver 1983; selected, Clark et al. 1990)

Common Name: 2,3',4,4'-Tetrachlorobiphenyl
Synonym: PCB-66
Chemical Name: 2,3',4,4'-tetrachlorobiphenyl
CAS Registry No: 32598-10-0
Molecular Formula: $C_{12}H_6Cl_4$
Molecular Weight: 291.99
Melting Point (°C):
124 (Saeki et al. 1971; Hutzinger et al. 1974)
127.-127.5 (Webb & McCall 1972; NAS 1979)
127-128 (Wallnöfer et al. 1973)
128 (Mackay et al. 1980; Opperhuizen et al. 1988)
126 (Burkhard et al. 1985a)
124, 96 (quoted, calculated, Abramowitz & Yalkowsky 1990)
Boiling Point (°C):
360 (calculated, Mackay 1982; Shiu & Mackay 1986)
Density (g/cm³):
Molar Volume (cm³/mol):
268.2 (LeBas method, Shiu & Mackay 1986)
Molecular Volume (A³):
265.2 (planar, shorthand, Opperhuizen et al. 1988)
Total Surface Area, TSA (A²):
259.15 (Mackay et al. 1980; Shiu & Mackay 1986)
255.42 (Burkhard 1984)
245.91 (planar, Doucette 1985)
258.99 (nonplanar, Doucette 1985)
246.44 (planar, Hawker & Connell 1988a)
268.5 (planar, shorthand, Opperhuizen et al. 1988)
246.0 (Abramowitz & Yalkowsky 1990)
Heat of Fusion, kcal/mol:
Fugacity Ratio, F (assuming ΔS_{fusion} = 13.5 e.u.):
0.105 (Shiu & Mackay 1986)

Water Solubility (g/m³ or mg/L at 25°C):
0.058 (shake flask-GC/ECD, Wallnöfer et al.; Hutzinger et al. 1974; quoted, Erickson 1986)
0.058 (quoted, NAS 1979; Chou & Griffin 1987)
0.0168 (calculated-TSA, Mackay et al. 1980)
0.098 (subcooled liq., calculated-TSA, Burkhard et al. 1985b; quoted, Eisenreich 1987; Capel et al. 1991)
0.040 (selected, Shiu & Mackay 1986)
0.0368 (20°C, subcooled liq., calculated-mole fraction of Aroclor mixtures, Murphy et al. 1987)
0.068 (calculated-HPLC-RI, Brodsky & Ballschmiter 1988)

0.0359, 0.0273 (quoted, calculated- χ , Nirmalakhanden & Speece 1989)
0.0046, 0.0116 (quoted, calculated-TSA, Abramowitz & Yalkowsky 1990)
0.068, 0.045 (quoted average of Brodsky & Ballschmiter 1988, calculated- χ , Patil 1991)

Vapor Pressure (Pa at 25°C):
0.00616 (subcooled liquid, GC-RT, Bidleman 1984; quoted, Erickson 1986)
0.00204, 0.000494, 0.00122 (calculated-MW, GC-RI, χ , Burkhard et al. 1985a)
0.00459 (subcooled liq., calculated-GC-RI, Burkhard et al. 1985b; quoted, Eisenreich 1987; Capel et al. 1991)
0.00569, 0.00507, 0.00616 (subcooled liq., GC-RT, Foreman & Bidleman 1985)
0.00252 (20°C, subcooled liq., calculated-mole fraction of Aroclor mixtures, Murphy et al. 1987)

Henry's Law Constant (Pa m^3/mol):
84.20 (calculated-P/C, Murphy et al. 1983)
13.68 (calculated-P/C, Burkhard et al. 1985b; quoted, Capel et al. 1991)
14.18 (quoted, Eisenreich 1987)
20.37 (20°C, calculated-P/C, Murphy et al. 1987)
25.84 (calculated- χ , Sabljic & Güsten 1989)

Octanol/Water Partition Coefficient, log K_{OW}:
6.67 (calculated-f const., Yalkowsky et al. 1983; quoted, Erickson 1986)
5.94 (calculated-TSA, Burkhard 1984)
5.80, 6.31 (HPLC-RT, calculated-π, Rapaport & Eisenreich 1984)
5.45 (quoted, Rapaport & Eisenreich 1985; Sklarew & Girvin 1987)
5.80 (selected, Shiu & Mackay 1986; quoted, Sklarew & Girvin 1987; Clark et al. 1990)
5.64 (calculated-S, Chou & Griffin 1987)
6.31 (quoted, Eisenreich 1987)
5.90 (calculated-chlorine substituents, Oliver 1987c)
5.98 (calculated-HPLC-RI, Brodsky & Ballschmiter 1988)
6.31 (gen. col.-GC-RT, Hawker & Connell 1988a)
6.20 (calculated-TSA, Hawker & Connell 1988a; quoted, Capel et al. 1991)
5.98, 5.87 (quoted of Brodsky & Ballschmiter 1988, calculated- χ , Patil 1991)

Bioconcentration Factor, log BCF:
4.45 (worms, Oliver 1987c)

Sorption Partition Coefficient, log K_{OC}:
5.74 (suspended particulate matter, calculated-K_{OW}, Burkhard 1984)
4.72 (soil, calculated-S, Chou & Griffin 1987)
5.23 (calculated after Karickhoff et al. 1979, Capel & Eisenreich 1990)

4.90 (calculated after Schwarzenbach & Westall 1981, Capel & Eisenreich 1990)

Half-Lives in the Environment:

Air:

Surface water:

Groundwater:

Sediment:

Soil:

Biota: half-life in rainbow trout, 670 days and its muscle 108 days (Niimi & Oliver 1983); in worms at 8°C, 33 days (Oliver 1987c).

Environmental Fate Rate Constants or Half-Lives:

Volatilization:

Photolysis:

Hydrolysis:

Oxidation:

Biodegradation:

Biotransformation:

Bioconcentration, Uptake (k_1) and Elimination (k_2) Rate Constants:

k_2: 0.001 day^{-1} (rainbow trout, Niimi & Oliver 1983; selected, Clark et al. 1990)

Common Name: 2,3',4',5-Tetrachlorobiphenyl
Synonym: PCB-70
Chemical Name: 2,3',4',5-tetrachlorobiphenyl
CAS Registry No: 32598-11-1
Molecular Formula: $C_{12}H_6Cl_4$
Molecular Weight: 291.99
Melting Point (°C):
104 (Bellavita 1934; Hutzinger et al. 1974; Mackay et al. 1980; Burkhard et al. 1985a; Opperhuizen et al. 1988)
104-107 (Wallnöfer et al. 1973)
104-105 (NAS 1979)
104, 83 (quoted, calculated, Abramowitz & Yalkowsky 1990)
Boiling Point (°C):
360 (calculated, Mackay et al. 1982; Shiu & Mackay 1986)
Density (g/cm^3 at 20°C): 1.2024
Molar Volume (cm^3/mol):
268.2 (LeBas method, Shiu & Mackay 1986)
1.28 (intrinsic volume: $V_I/100$, Hawker 1990)
Molecular Volume (A^3):
265.2 (planar, shorthand, Opperhuizen et al. 1988)
Total Surface Area, TSA (A^2):
259.15 (Mackay et al. 1980; Shiu & Mackay 1986)
255.25 (Burkhard 1984)
245.91 (planar, Doucette 1985)
258.97 (nonplanar, Doucette 1985)
246.26 (planar, Hawker & Connell 1988a)
268.5 (planar, shorthand, Opperhuizen et al. 1988)
246.0 (Abramowitz & Yalkowsky 1990)
Heat of Fusion, kcal/mol:
Fugacity Ratio, F (assuming ΔS_{fusion} = 13.5 e.u.): 0.165

Water Solubility (g/m^3 or mg/L at 25°C):
0.041 (shake flask-GC/ECD, Wallnöfer et al. 1973; Hutzinger et al. 1974; quoted, Erickson 1986)
0.041 (quoted, NAS 1979; Chou & Griffin 1987)
0.029 (calculated-TSA, Mackay et al. 1980)
0.022 (gen. col.-GC, Bruggeman et al. 1981)
0.0412 (selected lit. average, Yalkowsky et al. 1983)

0.016 (Bruggeman Ph.D. Thesis 1983 in Opperhuizen et al. 1988)
0.099 (subcooled liq., calculated-TSA, Burkhard et al. 1985b; quoted, Eisenreich 1987; Capel et al. 1991)
0.0091 (gen. col.-GC, Opperhuizen et al. 1985)

0.041 (selected, Shiu & Mackay 1986)
0.0362 (20°C, subcooled liq., calculated-mole fraction of Aroclor mixtures, Murphy et al. 1987)
0.059 (calculated average of HPLC-RI, Brodsky & Ballschmiter 1988)
0.016 (quoted, Bruggeman Ph.D. Thesis in Opperhuizen et al. 1988)
0.0164, 0.0273 (quoted, calculated-χ, Nirmalakhanden & Speece 1989)
0.01842 (calculated-TSA, Abramowitz & Yalkowsky 1990)
0.060, 0.045 (quoted average of Brodsky & Ballschmiter 1988, calculated- χ , Patil 1991)

Vapor Pressure (Pa at 25°C):
0.00544, 0.00642 (subcooled liq.,GC-RT, Bidleman 1984)
0.00769 (calculated-SxHLC, Burkhard et al. 1985a)
0.00337, 0.000919, 0.00278 (calculated-MW,GC-RI,χ, Burkhard et al. 1985a)
0.00519 (subcooled liquid, calculated-GC-RI, Burkhard et al. 1985b; quoted, Eisenreich 1987; Capel et al. 1991)
0.00526, 0.00551, 0.00591 (subcooled liq., GC-RT, Foreman & Bidleman 1985)
0.006 (selected, subcooled liquid, Shiu & Mackay 1986)
0.000587 (quoted from Bidleman 1984; Erickson 1986)
0.00236 (20°C, subcooled liq., calculated-mole fraction of Aroclor mixtures, Murphy et al. 1987)

Henry's Law Constant (Pa m^3/mol):
20.26 (calculated-P/C, Murphy et al. 1983)
15.30 (calculated-P/C, Burkhard et al. 1985b; quoted, Eisenreich 1987; Capel et al. 1991)
19.05 (20°C, calculated-P/C, Murphy et al. 1987)
19.15 (calculated- χ , Sabljic & Güsten 1989)
10.13 (wetted-wall col.-GC, Brunner et al. 1990)

Octanol/Water Partition Coefficient, log K_{OW}:
5.95 (shake flask-GC, Tulp & Hutzinger 1978)
5.20 (HPLC-k', McDuffie 1981)
7.13 (calculated-π, McDuffie 1981)
7.01 (calculated-f const., McDuffie 1981)
6.39 (TLC-RT, Bruggeman et al. 1982; quoted, Mackay & Hughes 1984; Erickson 1986)
6.67 (calculated-f const., Yalkowsky et al. 1983)
5.93 (calculated-TSA, Burkhard 1984)
5.72 (HPLC-RT, Rapaport & Eisenreich 1984)
6.23 (calculated-π, Rapaport & Eisenreich 1984; quoted, Sklarew & Girvin 1987)
6.23 (quoted, Hawker & Connell 1985; Thomann 1989)
6.18 (HPLC-RT, Opperhuizen et al. 1985)

5.90 (selected, Shiu & Mackay 1986; quoted, Gobas et al. 1987; Sklarew & Girvin 1987; Clark et al. 1990)
5.75 (calculated-S, Chou & Griffin 1987)
6.23 (quoted, Eisenreich 1987; Thomann 1989)
5.90 (calculated-chlorine substituents, Oliver 1987a)
6.23 (calculated average of HPLC-RI, Brodsky & Ballschmiter 1988)
6.20 (calculated-TSA, Hawker & Connell 1988a; Connell & Hawker 1988, Hawker 1990; selected, Capel et al. 1991)
6.22, 5.87 (quoted average of Brodsky & Ballschmiter 1988, calculated- χ , Patil 1991)

Bioconcentration Factor, log BCF:
6.15 (goldfish, 3% lipid, Bruggeman et al. 1981)
6.20 (goldfish, 3% lipid, calculated-K_{OW}, Bruggeman et al. 1981)
4.62, 5.11 (goldfish, exptl., correlated, Mackay & Hughes 1984)
4.32, 5.50 (guppy, lipid phase, Opperhuizen et al. 1985)
4.92 (guppy, Gobas et al. 1987)
4.62, 4.32 (fish, calculated-C_B/C_W or k_1/k_2, Connell & Hawker 1988; Hawker 1990)

Sorption Partition Coefficient, log K_{OC}:
5.73 (suspended particulate matter, calculated$_{OW}$, Burkhard 1984)
5.6-6.8, 6.3 (suspended sediment, average, Oliver 1987a)
7.20 (algae > 50 μm, Oliver 1987a)
4.81 (soil, calculated-S, Chou & Griffin 1987)
4.76 (correlated literature values in soils, Sklarew & Girvin 1987)
6.04 (calculated after Karickhoff et al. 1979, Capel & Eisenreich 1990)
5.52 (calculated after Schwarzenbach & Westall 1981, Capel & Eisenreich 1990)

Half-Lives in the Environment:
Air:
Surface water:
Groundwater:
Sediment:
Soil:
Biota: half-life in guppy, 38.6 days (Opperhuizen et al. 1985).

Environmental Fate Rate Constants or Half-Lives:
Volatilization:
Photolysis:
Hydrolysis:
Oxidation:
Biodegradation:
Biotransformation:
Bioconcentration, Uptake (k_1) and Elimination (k_2) Rate Constants:

k_1: 420 day^{-1} (23°C, goldfish, 3% lipid content, Bruggeman et al. 1981; quoted, Waid 1986)
k_2: 0.01 day^{-1} (23°C, goldfish, 3% lipid content, Bruggeman et al. 1981; quoted, Waid 1986; Clark et al. 1990)
k_2: 0.01, 0.0104 day^{-1} (goldfish, exptl., correlated, Mackay & Hughes 1984)
k_1: 380 day^{-1} (guppy, Opperhuizen et al. 1985)
k_2: 0.018 day^{-1} (guppy, Opperhuizen et al. 1985; quoted, Clark et al. 1990)
k_1: 17.5 hour^{-1} (goldfish, quoted, Hawker & Connell 1985)
$1/k_2$: 2400 hour (goldfish, quoted, Hawker & Connell 1985,1988b)
log k_1: 2.58, 2.62 day^{-1} (fish, quoted, Connell & Hawker 1988)
log 1/k2: 1.74, 2.0 day (fish, quoted, Connell & Hawker 1988)
log k_2: -2.0 day^{-1} (fish, quoted, Thomann 1989)

Common Name: 2,4,4',6-Tetrachlorobiphenyl
Synonym: PCB-75
Chemical Name: 2,4,4',6-tetrachlorobiphenyl
CAS Registry No: 32598-12-2
Molecular Formula: $C_{12}H_6Cl_4$
Molecular Weight: 291.99
Melting Point (°C):
93.0 (calculated, Abramowitz & Yalkowsky 1990)
Boiling Point (°C):
Density (g/cm^3):
Molar Volume (cm^3/mol):
268.2 (LeBas method, Shiu & Mackay 1986)
Molecular Volume (A^3):
Total Surface Area, TSA (A^2):
254.69 (Burkhard 1984)
240.92 (planar, Doucette 1985)
259.38 (nonplanar, Doucette 1985)
241.85 (planar, Hawker & Connell 1988a)
242.0 (Abramowitz & Yalkowsky 1990)
Heat of Fusion, kcal/mol:
Fugacity Ratio, F (assuming ΔS_{fusion} = 13.5 e.u.):

Water Solubility (g/m^3 or mg/L at 25°C):
0.104 (subcooled liq., calculated-TSA, Burkhard et al. 1985b)
0.013 (calculated average of HPLC-RI, Brodsky & Ballschmiter 1988)
0.091 (gen. col.-GC/ECD, Dunnivant & Elzerman 1988)
0.0184 (calculated-TSA, Abramowitz & Yalkowsky 1990)
0.016, 0.045 (quoted average of Brodsky & Ballschmiter 1988, calculated- χ , Patil 1991)

Vapor Pressure (Pa at 25°C):
0.0204, 0.0158, 0.00419 (subcooled liq., calculated-MW,GC-RI, χ , Burkhard et al. 1985a)
0.0150 (subcooled liq., calculated-GC-RI, Burkhard et al. 1985b)
0.0179, 0.0202 (subcooled liq., GC-RT, Foreman & Bidleman 1985)

Henry's Law Constant (Pa m^3/mol):
42.25 (calculated-P/C, Burkhard et al. 1985b)
42.25 (batch stripping, Dunnivant & Elzerman 1988)
55.32 (calculated- χ , Sabljic & Güsten 1989)

Octanol/Water Partition Coefficient, log K_{OW}:

5.92 (calculated-TSA, Burkhard 1984)
6.03 (calculated average of HPLC-RI, Brodsky & Ballschmiter 1988)
6.05 (calculated, Hawker & Connell 1988a)
6.03, 5.87 (quoted average of Brodsky & Ballschmiter 1988, calculated- χ , Patil 1991)

Bioconcentration Factor, log BCF:

Sorption Partition Coefficient, log K_{OC}:

5.72 (suspended particulate matter, calculated-K_{OW}, Burkhard 1984)

Half-Lives in the Environment:

Air:
Surface water:
Groundwater:
Sediment:
Soil:
Biota:

Environmental Fate Rate Constants or Half-Lives:

Volatilization:
Photolysis:
Hydrolysis:
Oxidation:
Biodegradation:
Biotransformation:
Bioconcentration, Uptake (k_1) and Elimination (k_2) Rate Constants:

Common Name: 3,3',4,4'-Tetrachlorobiphenyl
Synonym: PCB-77
Chemical Name: 3,3',4,4'-tetrachlorobiphenyl
CAS Registry No: 32598-13-3
Molecular Formula: $C_{12}H_6Cl_4$
Molecular Weight: 291.99
Melting Point (°C):
    173 (Van Roosmalen 1934; Hutzinger et al. 1974; Erickson 1986)
    182-184 (Wallöfer et al. 1973; Hutzinger et al. 1974)
    180.0 (Mackay et al. 1980; Bruggeman et al. 1982; Burkhard et al. 1985a; Opperhuizen et al. 1988)
    180, 173 (quoted, calculated, Abramowitz & Yalkowsky 1990)
Boiling Point (°C):
    360 (calculated, Mackay et al. 1982; Shiu & Mackay 1986)
Density (g/cm^3 at 20°C): 1.2024
Molar Volume (cm^3/mol):
    268.2 (LeBas method, Shiu & Mackay 1986)
    1.28 (intrinsic volume: $V_I/100$, De Bruijn & Hermens 1990)
Molecular Volume (A^3):
    267.6 (planar, shorthand, Opperhuizen et al. 1988)
    237.70, 217.52, 267.6 (De Bruijn & Hermens 1990)
Total Surface Area, TSA (A^2):
    258.76 (Mackay et al. 1980)
    256.02 (Burkhard 1984)
    250.91 (planar, Doucette 1985)
    258.52 (nonplanar, Doucette 1985)
    251.02 (planar, Hawker & Connell 1988a)
    270.7 (planar, shorthand, Opperhuizen et al. 1988)
    283.72, 251.94, 270.7 (De Bruijn & Hermens 1990)
    251.0 (Abramowitz & Yalkowsky 1990)
Heat of Fusion, kcal/mol:
Entropy of Fusion, cal/mol K (e.u.):
    20.40 (Shiu & Mackay 1986)
Fugacity Ratio, F (assuming ΔS_{fusion} = 13.5 e.u.):
    0.029 (Shiu & Mackay 1986)

Water Solubility (g/m^3 or mg/L at 25°C):
    0.175 (shake flask-GC/ECD, Wallnöfer et al. 1973; Hutzinger et al. 1974; quoted, Erickson 1986)
    0.00075 (gen. col.-GC/ECD, Weil et al. 1974)
    0.180 (quoted, NAS 1979; Chou & Griffin 1987)
    0.0174 (calculated-TSA, Mackay et al. 1980)
    0.00075 (TLC-RT, Bruggeman et al. 1982)

0.0114 (quoted lit. average, Yalkowsky et al. 1983)
0.0931 (subcooled liquid, calculated-TSA, Burkhard et al. 1985b)
0.001 (selected, Shiu & Mackay 1986)
0.000569 (gen. col.-GC/ECD, Dickhut et al. 1986)
0.00301 (quoted, Dickhut et al. 1986 from Burkhard et al. 1985b)
0.00075, 0.00099 (quoted, calculated average of HPLC-RI, Brodsky & Ballschmiter 1988)
0.0018 (22°C, gen. col.-GC/ECD, Opperhuizen et al. 1988)
0.00055 (gen. col.-GC, Dunnivant & Elzerman 1988)
0.0161 (subcooled liquid, calculated-M.P., Dunnivant & Elzerman 1988)
0.0114, 0.0273 (quoted, calculated-χ, Nirmalakhanden & Speece 1989)
0.00923, 0.00292 (quoted, calculated-TSA, Abramowitz & Yalkowsky 1990)

Vapor Pressure (Pa at 25°C):
0.000307 (calculated, Neely 1983; quoted, Erickson 1986)
0.00219, 0.00196 (subcooled liq., GC-RT, Bidleman 1984)
0.0014 (subcooled liquid, Burkhard 1984)
5.97×10^{-4}, 4.46×10^{-5}, 8.04×10^{-4} (calculated-MW,GC-RT,χ, Burkhard et al. 1985a)
0.0014 (subcooled liquid, calculated-GC-RI, Burkhard et al. 1985b)
0.00213, 0.00144, 0.00207 (subcooled liq., GC-RT, Foreman & Bidleman 1985)
0.002 (selected, subcooled liq., Shiu & Mackay 1986)
5.88×10^{-5} (selected, Shiu & Mackay 1986)
1.82×10^{-5} (calculated-SxHLC, Dunnivant & Elzerman 1988)
5.26×10^{-4} (subcooled liquid, calculated-M.P., Dunnivant & Elzerman 1988)

Henry's Law Constant (Pa m^3/mol):
4.37 (calculated-P/C, Burkhard et al. 1985b)
1.72 (calculated-P/C, Shiu & Mackay 1986)
9.52 (batch stripping, Dunnivant & Elzerman 1988)
9.52 (batch stripping, Dunnivant et al. 1988)

Octanol/Water Partition Coefficient, log K_{OW}:
6.04 (HPLC-RT, Sugiura et al. 1979)
6.52 (TLC-RT, Bruggeman et al. 1982; quoted, Erickson 1986)
6.67 (calculated-f const., Yalkowsky et al. 1983)
5.95 (calculated-TSA, Burkhard 1984)
5.62, 5.62 (HPLC-RT, calculated-π, Rapaport & Eisenreich 1984)
6.10 (selected, Shiu & Mackay 1986; quoted, Sklarew & Girvin 1987; Clark et al. 1990)
5.27 (calculated-S, Chou & Griffin 1987)
6.77, 6.12 (HPLC-k', calculated, De Kock & Lord 1987)
5.81 (quoted, Geyer et al. 1987)
6.29 (calculated average of HPLC-RI, Brodsky & Ballschmiter 1988)

6.21 (gen. col.-GC, Hawker & Connell 1988a)
6.36 (calculated-TSA, Hawker & Connell 1988a)
6.52 (calculated-π, De Bruijn et al. 1989)
6.63 (slow stirring-GC, De Bruijn et al. 1989; De Bruijn & Hermens 1990)

Bioconcentration Factor, log BCF:
3.46 (killifish, Goto et al. 1978)
3.90, 3.24, 3.63, 4.15 (golden orfe, carp, brown trout, guppy, Sugiura et al. 1979)
2.77, 2.63 (human fat in lipid, wet wt. basis, calculated-K_{OW}, Geyer et al. 1987)

Sorption Partition Coefficient, log K_{OC}:
5.75 (suspended particulate matter, calculated-K_{OW}, Burkhard 1984)
4.41 (soil, calculated, Chou & Griffin 1987)

Half-Lives in the Environment:
Air:
Surface water: half-life in Lake Michigan, 805 days (Neely 1983).
Groundwater:
Sediment:
Soil:
Biota: half-life in rainbow trout, 44 days and its muscle, 29 days (Niimi & Oliver 1983).

Environmental Fate Rate Constants or Half-Lives:
Volatilization:
Photolysis:
Hydrolysis:
Oxidation:
Biodegradation:
Biotransformation:
Bioconcentration, Uptake (k_1) and Elimination (k_2) Rate Constants:
k_1: 0.029, 0.0047 day^{-1} (golden orfe, guppy, Sugiura et al. 1979
k_1: 0.029 day^{-1} (golden orfe, Sugiura et al. 1979)
k_1: 0.0047 day^{-1} (guppy, Sugiura et al. 1979)
k_2: 0.0157 day^{-1} (rainbow trout, Niimi & Oliver 1983; quoted, Clark et al. 1990)

Common Name: 3,3',5,5'-Tetrachlorobiphenyl
Synonym: PCB-80
Chemical Name: 3,3',5,5'-tetrachlorobiphenyl
CAS Registry No: 33284-52-5
Molecular Formula: $C_{12}H_6Cl_4$
Molecular Weight: 291.99
Melting Point (°C):
164 (Van Roosmalen 1934; Hutzinger et al. 1974; Burkhard et al. 1985a; Erickson 1986)
164, 127 (quoted, calculated, Abramowitz & Yalkowsky 1990)
Boiling Point (°C):
360 (calculated, Mackay et al. 1982; Shiu & Mackay 1986)
Density (g/cm^3):
Molar Volume (cm^3/mol):
268.2 (LeBas method, Shiu & Mackay 1986)
Molecular Volume (A^3):
Total Surface Area, TSA (A^2):
259.47 (Burkhard 1984)
254.88 (planar, Doucette 1985)
262.52 (nonplanar, Doucette 1985)
254.51 (planar, Hawker & Connell 1988a)
255.0 (Abramowitz & Yalkowsky 1990)
Heat of Fusion, kcal/mol:
Fugacity Ratio, F (assuming ΔS_{fusion} = 13.5 e.u.):
0.0421 (Shiu & Mackay 1986)

Water Solubility (g/m^3 or mg/L at 25°C):
0.0712 (subcooled liq., calculated-TSA, Burkhard et al. 1985b)
0.00057 (calculated average of HPLC-RI, Brodsky & Ballschmiter 1988)
0.00124 (gen. col.-GC/ECD, Dunnivant & Elzerman 1988)
0.0295 (subcooled liquid, calculated-M.P., Dunnivant & Elzerman 1988)
0.00292 (calculated-TSA, Abramowitz & Yalkowsky 1990)

Vapor Pressure (Pa at 25°C):
0.000859, 0.000139, 0.0219 (calculated-MW,GC-RI,χ, Burkhard et al. 1985a)
0.00305 (subcooled liq., calculated-GC-RI, Burkhard et al. 1985b)
0.00547, 0.00511 (subcooled liq., GC-RT, Foreman & Bidleman 1985)

Henry's Law Constant (Pa m^3/mol):
12.46 (calculated-P/C, Burkhard et al. 1985b)
63.02 (calculated- χ , Sabljic & Güsten 1989)

Octanol/Water Partition Coefficient, log K_{OW}:

6.85 (HPLC-RT, Sugiura et al. 1978)
6.58 (TLC-RT, Bruggeman et al. 1982; quoted, Erickson 1986)
6.05 (calculated-TSA, Burkhard 1984)
6.10 (selected, Shiu & Mackay 1986; quoted, Sklarew & Girvin 1987)
6.60 (calculated average of HPLC-RI, Brodsky & Ballschmiter 1988)
6.48 (calculated-TSA, Hawker & Connell 1988a)

Bioconcentration Factor, log BCF:

Sorption Partition Coefficient, log K_{OC}:

5.85 (suspended particulate matter, calculated-K_{OW}, Burkhard 1984)

Half-Lives in the Environment:

Air:
Surface water:
Groundwater:
Sediment:
Soil:
Biota:

Environmental Fate Rate Constants or Half-Lives:

Volatilization:
Photolysis:
Hydrolysis:
Oxidation:
Biodegradation:
Biotransformation:
Bioconcentration, Uptake (k_1) and Elimination (k_2) Rate Constants:

Common Name: 2,2',3,3',5-Pentachlorobiphenyl
Synonym: PCB-83
Chemical Name: 2,2',3,3',5-pentachlorobiphenyl
CAS Registry No: 60145-20-2
Molecular Formula: $C_{12}H_5Cl_5$
Molecular Weight: 326.43
Melting Point (°C):
65 (calculated, Abramowitz & Yalkowsky 1990)
Boiling Point (°C):
381 (calculated, Mackay et al. 1982; Shiu & Mackay 1986)
Density (g/cm^3):
Molar Volume (cm^3/mol):
289.1 (LeBas method, Shiu & Mackay 1986)
Molecular Volume (A^3):
Total Surface Area, TSA (A^2):
266.71 (Burkhard 1984)
248.17 (planar, Doucette 1985)
272.93 (nonplanar, Doucette 1985)
248.10 (planar, Hawker & Connell 1988a)
248.0 (Abramowitz & Yalkowsky 1990)
Heat of Fusion, kcal/mol:
Fugacity Ratio, F (assuming ΔS_{fusion} = 13.5 e.u.):

Water Solubility (g/m^3 or mg/L at 25°C):
0.0045 (gen. col.-GC/ECD, Weil et al. 1974)
0.023 (shake flask-GC/ECD, Wallnöfer et al. 1973; Tulp & Hutzinger 1978)
0.046 (subcooled liq., calculated-TSA, Burkhard et al. 1985b)
0.0282 (20°C, subcooled liq., calculated-mole fraction of Aroclor mixtures, Murphy et al. 1987)
0.0260 (calculated-TSA, Abramowitz & Yalkowsky 1990)

Vapor Pressure (Pa at 25°C):
0.00735, 0.00324, 0.00278 (calculated-MW,GC-RI,χ, subcooled liq., Burkhard et al. 1985a)
0.00299 (subcooled liq., calculated-GC-RI, Burkhard et al. 1985b; quoted, Eisenreich 1987)
0.00274, 0.00303 (subcooled liq., GC-RT, Foreman & Bidleman 1985)
0.00154 (20°C, subcooled liq., calculated-mole fraction of Aroclor mixtures, Murphy et al. 1987)

Henry's Law Constant (Pa m^3/mol):
21.28 (calculated-P/C, Burkhard et al. 1985b; quoted, Eisenreich 1987)
16.62 (20°C, calculated-P/C, Murphy et al. 1987)

26.65 (calculated- χ , Sabljic & Güsten 1989)

Octanol/Water Partition Coefficient, log K_{OW}:

6.24 (calculated-TSA, Burkhard 1984)

6.26 (calculated-TSA, Hawker & Connell 1988a)

Bioconcentration Factor, log BCF:

Sorption Partition Coefficient, log K_{OC}:

6.04 (suspended particulate matter, calculated-K_{OW}, Burkhard 1984)

4.748 (calculated- χ , reported as log K_h at 5 mg C/L, Sabljic et al. 1989)

Half-Lives in the Environment:

Air:

Surface water:

Groundwater:

Sediment:

Soil:

Biota:

Environmental Fate Rate Constants or Half-Lives:

Volatilization:

Photolysis:

Hydrolysis:

Oxidation:

Biodegradation:

Biotransformation:

Bioconcentration, Uptake (k_1) and Elimination (k_2) Rate Constants:

Common Name: 2,2',3,4,5-Pentachlorobiphenyl
Synonym: PCB-86
Chemical Name: 2,2'3,4,5-pentachlorobiphenyl
CAS Registry No: 55312-69-1
Molecular Formula: $C_{12}H_5Cl_5$
Molecular Weight: 326.43
Melting Point (°C):
100 (Mackay et al. 1980; Burkhard et al. 1985a; Opperhuizen et al. 1988)
100, 79 (quoted, calculated, Abramowitz & Yalkowsky 1990)
Boiling Point (°C):
Density (g/cm³):
Molar Volume (cm³/mol):
289.1 (LeBas method, Shiu & Mackay 1986)
Molecular Volume (A³):
268.7 (planar, shorthand, Opperhuizen et al. 1988)
Total Surface Area, TSA (A²):
271.18 (Mackay et al. 1980; Shiu & Mackay 1986)
265.68 (Burkhard 1984)
246.19 (planar, Doucette 1985)
270.94 (nonplanar, Doucette 1985)
247.07 (planar, Hawker & Connell 1988a)
274.2 (planar, shorthand, Opperhuizen et al. 1988)
247.0 (Abramowitz & Yalkowsky 1990)
Heat of Fusion, kcal/mol:
Fugacity Ratio, F (assuming ΔS_{fusion} = 13.5 e.u.):
0.181 (Mackay et al. 1980)
0.182 (Shiu & Mackay 1986)

Water Solubility (g/m³ or mg/L at 25°C):
0.0098 (gen. col.-GC/ECD, Weil et al. 1974; quoted, Erickson 1986)
0.0098, 0.0133 (quoted, calculated-TSA, Mackay et al. 1980)
0.0349 (gen. col.-HPLC/UV, Huang 1983)
0.0099 (quoted lit. average, Yalkowsky et al. 1983)
0.0496 (subcooled liq., calculated-TSA, Burkhard et al. 1985b)
0.0098 (quoted, Erickson 1986 from Mackay et al. 1980)
0.020 (selected, Shiu & Mackay 1986)
0.034 (gen. col.-HPLC, Billington et al. 1988)
0.0044 (calculated average of HPLC-RI, Brodsky & Ballschmiter 1988)
0.00986, 0.0090 (quoted, calculated- χ , Nirmalakhanden & Speece 1989)
0.0206, 0.0206 (quoted, calculated-TSA, Abramowitz & Yalkowsky 1990)
0.0044, 0.011 (quoted average of Brodsky & Ballschmiter 1988, calculated- χ , Patil 1991)

Vapor Pressure (Pa at 25°C):

0.00933 (quoted, Neely 1981)
0.051 (P_L calculated from P_S using F, Neely 1981)
0.00643 (subcooled liq., Neely 1981)
0.000077 (calculated, Neely 1983; quoted, Erickson 1986)
0.00133, 0.00245, 0.00184 (calculated-MW,GC-RI,χ, Burkhard et al. 1985a)
0.0128 (subcooled liq., calculated-GC-RI, Burkhard et al. 1985b)
0.051 (selected, subcooled liquid, Shiu & Mackay 1986)

Henry's Law Constant (Pa m^3/mol):

84.1 (calculated-P/C, Burkhard et al. 1985b)
17.23 (calculated- χ , Sabljic & Güsten 1989)

Octanol/Water Partition Coefficient, log K_{OW}:

6.38 (Neely 1983; quoted, Erickson 1986)
7.49 (calculated-f const., Yalkowsky et al. 1983)
6.22 (calculated-TSA, Burkhard 1984)
6.38 (calculated-π, Woodburn et al. 1984)
6.37 (quoted, Hawker & Connell 1986)
6.20 (selected, Shiu & Mackay 1986; quoted, Sklarew & Girvin 1987)
6.39 (calculated average of HPLC-RI, Brodsky & Ballschmiter 1988)
6.23 (calculated-TSA, Hawker & Connell 1988a)
6.38, 6.26 (quoted average of Brodsky & Ballschmiter 1988, calculated- χ , Patil 1991)

Bioconcentration Factor, log BCF:

4.43 (oyster, Vreeland 1974; quoted, Hawker & Connell 1986)

Sorption Partition Coefficient, log K_{OC}:

6.02 (suspended particulate matter, calculated-K_{OW}, Burkhard 1984)
4.770 (humic substances, calculated- χ , reported as log K_h, Sabljic et al. 1989)

Half-Lives in the Environment:

Air:
Surface water: half-life in Lake Michigan, 108 days (Neely 1983).
Groundwater:
Sediment:
Soil:
Biota:

Environmental Fate Rate Constants or Half-Lives:

Volatilization:
Photolysis:

Hydrolysis:

Oxidation:

Biodegradation: microbial degradation with pseudo first-order rate constant Of 0.005 $year^{-1}$ in the water column and 0.05 $year^{-1}$ in the sediment (Furukawa et al. 1978; quoted, Neely 1981).

Biotransformation:

Bioconcentration, Uptake (k_1) and Elimination (k_2) Rate Constants:

Common Name: 2,2',3,4,5'-Pentachlorobiphenyl
Synonym: PCB-87
Chemical Name: 2,2',3,4,5'-pentachlorobiphenyl
CAS Registry No: 38380-02-8
Molecular Formula: $C_{12}H_5Cl_5$
Molecular Weight: 326.43
Melting Point (°C):
111.5-113 (Hutzinger et al. 1974; NAS 1979; Erickson 1986)
112 (Mackay et al. 1980; Bruggeman et al. 1982; Burkhard et al. 1985a; Opperhuizen et al. 1988)
114, 73 (quoted, calculated, Abramowitz & Yalkowsky 1990)
Boiling Point (°C):
381 (calculated, Mackay et al. 1982; Shiu & Mackay 1986)
Density (g/cm^3 at 20°C): 1.2803
Molar Volume (cm^3/mol):
289.1 (LeBas method, Shiu & Mackay 1986)
Molecular Volume (A^3):
272.4 (planar, shorthand, Opperhuizen et al. 1988)
Total Surface Area, TSA (A^2):
273.19 (Mackay et al. 1980)
267.57 (Burkhard 1984)
248.17 (planar, Doucette 1985)
272.97 (nonplanar, Doucette 1985)
273.6 (Shiu & Mackay 1986)
275.3 (calculated- χ , Sabljic 1987)
248.99 (planar, Hawker & Connell 1988a)
276.7 (planar, shorthand, Opperhuizen et al. 1988)
249.0 (Abramowitz & Yalkowsky 1990)
Heat of Fusion, kcal/mol:
Fugacity Ratio, F (assuming ΔS_{fusion} = 13.5 e.u.):
0.132 (Mackay et al. 1980)
0.138 (Shiu & Mackay 1986)

Water Solubility (g/m^3 or mg/L at 25°C):
0.022 (shake flask-GC/ECD, Wallnöfer et al. 1973)
0.022 (Hutzinger et al. 1974; quoted, Erickson 1986)
0.0045 (gen. col.-GC/ECD, Weil et al. 1974)
0.022 (quoted, NAS 1979; Chou & Griffin 1987)
0.0086 (calculated-TSA, Mackay et al. 1980)
0.0045 (TLC-RT, Bruggeman et al. 1982)
0.0101 (quoted lit. average, Yalkowsky et al. 1983)
0.0431 (subcooled liq., calculated-TSA, Burkhard et al. 1985b; quoted, Eisenreich 1987; Capel et al. 1991)

0.0040 (selected, Shiu & Mackay 1986)
0.0294 (20°C, subcooled liq., calculated-mole fraction of Aroclor mixtures, Murphy et al. 1987)
0.0045, 0.0071 (quoted, calculated average of HPLC-RI, Brodsky & Ballschmiter 1988)
0.0045, 0.009 (quoted, calculated- χ , Nirmalakhanden & Speece 1989)
0.0041, 0.0164 (quoted, calculated-TSA, Abramowitz & Yalkowsky 1990)
0.0045 (quoted, Paya-Perez et al. 1991)
0.0071, 0.011 (quoted average of Brodsky & Ballschmiter 1988, calculated- χ , Patil 1991)

Vapor Pressure (Pa at 25°C):
0.00226 (subcooled liq., GC-RT, Bidleman 1984)
0.00141 (calculated-SxHLC, Burkhard et al. 1985a)
0.00101, 0.000392, 0.000804 (calculated-MW,GC-RI, χ , Burkhard et al. 1985a)
0.00141, 0.00262 (subcooled liq., calculated-GC-RI, Burkhard et al. 1985b; selected, Eisenreich 1987; Capel et al. 1991)
0.00261, 0.00248, 0.00226 (subcooled liq., GC-RT, Foreman & Bidleman 1985)
0.000304 (selected, Shiu & Mackay 1986)
0.0023 (selected, subcooled liq., Shiu & Mackay 1986)
0.00116 (20°C, subcooled liq., calculated-mole fraction of Aroclor mixtures, Murphy et al. 1987)

Henry's Law Constant (Pa m^3/mol):
33.44 (calculated, Murphy et al. 1983)
19.86 (calculated-P/C, Burkhard et al. 1985b; quoted, Capel et al. 1991)
24.81 (calculated-P/C, Shiu & Mackay 1986)
12.87 (20°C, calculated-P/C, Murphy et al. 1987)
20.27 (quoted, Eisenreich 1987)
18.24 (calculated- χ , Sabljic & Güsten 1989)
7.50 (wetted-wall col.-GC, Brunner et al. 1990)

Octanol/Water Partition Coefficient, log K_{OW}:
6.85 (TLC-RT, Bruggeman et al. 1982; quoted, Erickson 1986)
7.43 (calculated-f const., Yalkowsky et al. 1983)
6.27 (calculated-TSA, Burkhard 1984)
5.45 (HPLC-RT, Rapaport & Eisenreich 1984)
6.37 (calculated-π, Rapaport & Eisenreich 1984; quoted, Sklarew & Girvin 1987)
6.50 (selected, Shiu & Mackay 1986; quoted, Sklarew & Girvin 1987)
5.94 (calculated-S, Chou & Griffin 1987)
6.37 (quoted, Eisenreich 1987)
6.23 (calculated average of HPLC-RI, Brodsky & Ballschmiter 1988)
6.29 (calculated-TSA, Hawker & Connell 1988; quoted, Capel et al. 1991)
6.23, 6.26 (quoted average of Brodsky & Ballschmiter 1988, calculated- χ , Patil 1991)

Bioconcentration Factor, log BCF:

4.43 (oyster, Vreeland 1974)

Sorption Partition Coefficient, log K_{OC}:

4.54 (Koch 1983)
6.07 (suspended particulate matter, calculated-K_{OW}, Burkhard 1984)
4.98 (soil, calculated-S, Chou & Griffin 1987)
4.88 (calculated, Bahnick & Doucette 1988)
4.761, 4.748 (humic substances of 5 mg C/L, selected, calculated- χ , reported as log K_h, Sabljic et al. 1989)
4.76, 4.87, 4.85, 3.75 (humic substance in concn. of 5,10, 20, 40 mg C/L, reported as log K_h, Lara & Ernst 1989)
6.18 (calculated after Karickhoff et al. 1979, Capel & Eisenreich 1990)
5.08 (calculated after Schwarzenbach & Westall 1981, Capel & Eisenreich 1990)
5.73 (soil, Paya-Perez et al. 1991)

Sorption Partition Coefficient, log K_{OM}:

4.50, 4.85 (selected, calculated- χ , Sabljic 1984)

Half-Lives in the Environment:

Air:
Surface water:
Groundwater:
Sediment:
Soil:
Biota: half-life in rainbow trout, 155 days and its muscle, 62 days (Niimi & Oliver 1983).

Environmental Fate Rate Constants or Half-Lives:

Volatilization:
Photolysis:
Hydrolysis:
Oxidation:
Biodegradation:
Biotransformation:
Bioconcentration, Uptake (k_1) and Elimination (k_2) Rate Constants:

k_2: 0.0045 day^{-1} (rainbow trout, Niimi & Oliver 1983; quoted, Clark et al. 1990)
k_s: 0.049 hour^{-1} (uptake of mayfly-sediment model II, Gobas et al. 1989)
k_t: 0.013 hour^{-1} (depuration of mayfly-sediment model II, Gobas et al. 1989)

Common Name: 2,2',3,4,6-Pentachlorobiphenyl
Synonym: PCB-88
Chemical Name: 2,2',3,4,6-pentachlorobiphenyl
CAS Registry No: 55215-17-3
Molecular Formula: $C_{12}H_5Cl_5$
Molecular Weight: 326.43
Melting Point (°C):

100 (Mackay et al. 1980; Burkhard et al. 1985a; Opperhuizen et al. 1988)
100, 99 (quoted, calculated, Abramowitz & Yalkowsky 1990)

Boiling Point (°C):

381 (calculated, Mackay et al. 1982; Shiu & Mackay 1986)

Density (g/cm^3 at 20°C): 1.2803
Molar Volume (cm^3/mol):

289.1 (LeBas method, Shiu & Mackay 1986)

Molecular Volume (A^3):

259.5 (planar, shorthand, Opperhuizen et al. 1988)

Total Surface Area, TSA (A^2):

271.6 (Mackay et al. 1980; Shiu & Mackay 1986)
269.15 (Burkhard 1984)
241.19 (planar, Doucette 1985)
271.39 (nonplanar, Doucette 1985)
242.48 (planar, Hawker & Connell 1988a)
264.4 (planar, shorthand, Opperhuizen et al. 1988)
242.0 (Abramowitz & Yalkowsky 1990)

Heat of Fusion, kcal/mol:
Fugacity Ratio, F (assuming ΔS_{fusion} = 13.5 e.u.):

0.182 (Mackay et al. 1980)

Water Solubility (g/m^3 or mg/L at 25°C):

0.012 (gen. col.-GC/ECD, Weil et al. 1974)
0.0129 (calculated-TSA, Mackay et al. 1980)
0.0121 (quoted lit. average, Yalkowsky et al. 1983; quoted, Erickson 1986)
0.0385 (subcooled liq., calculated-TSA, Burkhard et al. 1985b)
0.012 (selected, Shiu & Mackay 1986)
0.0119, 0.009 (quoted, calculated- χ , Nirmalakhanden & Speece 1989)
0.0130, 0.0206 (quoted, calculated-TSA, Abramowitz & Yalkowsky 1990)
0.012, 0.011 (quoted of Mackay & Shiu 1986, calculated- χ , Patil 1991)

Vapor Pressure (Pa at 25°C):

0.00133, 0.00309, 0.00278 (calculated-MW,GC-RI,χ, Burkhard et al. 1985a)
0.0161 (subcooled liq., calculated-GC-RI, Burkhard et al. 1985b)

Henry's Law Constant (Pa m^3/mol):

136.8 (calculated-P/C, Burkhard et al. 1985b)
34.65 (calculated- χ , Sabljic & Güsten 1989)

Octanol/Water Partition Coefficient, log K_{OW}:

7.51 (calculated-f const., Yalkowsky et al. 1983; quoted, Erickson 1986)
6.31 (calculated-TSA, Burkhard 1984)
6.50 (selected, Shiu & Mackay 1986; quoted, Sklarew & Girvin 1987)
6.07 (calculated-TSA, Hawker & Connell 1988a)
6.50, 6.26 (quoted of Mackay & Shiu 1986, calculated- χ , Patil 1991)

Bioconcentration Factor, log BCF:

Sorption Partition Coefficient, log K_{OC}:

6.11 (suspended particulate matter, calculated-K_{OW}, Burkhard 1984)
4.611 (calculated- χ , reported as log K_h at 5 mg C/L, Sabljic et al. 1989)

Half-Lives in the Environment:

Air:
Surface water:
Groundwater:
Sediment:
Soil:
Biota:

Environmental Fate Rate Constants or Half-Lives:

Volatilization:
Photolysis:
Hydrolysis:
Oxidation:
Biodegradation:
Biotransformation:
Bioconcentration, Uptake (k_1) and Elimination (k_2) Rate Constants:

Common Name: 2,2',3,5',6-Pentachlorobiphenyl
Synonym: PCB-95
Chemical Name: 2,2',3,5',6-pentachlorobiphenyl
CAS Registry No: 38379-99-6
Molecular Formula: $C_{12}H_5Cl_5$
Molecular Weight: 326.4
Melting Point (°C):
98.5-100 (Hutzinger et al. 1974; NAS 1979; Erickson 1986)
79.0 (calculated, Abramowitz & Yalkowsky 1990)
Boiling Point (°C):
381 (calculated, Mackay et al. 1982; Shiu & Mackay 1986)
Density (g/cm^3):
Molar Volume (cm^3/mol):
289.1 (LeBas method, Shiu & Mackay 1986)
Molecular Volume (A^3):
Total Surface Area, TSA (A^2):
271.0 (Burkhard 1984)
243.18 (planar, Doucette 1985)
273.41 (nonplanar, Doucette 1985)
244.23 (planar, Hawker & Connell 1988a)
244.0 (Abramowitz & Yalkowsky 1990)
Heat of Fusion, kcal/mol:
Fugacity Ratio, F (assuming ΔS_{fusion} = 13.5 e.u.): 0.182

Water Solubility (g/m^3 or mg/L at 25°C):
0.0336 (subcooled liq., calculated-TSA, Burkhard 1985b)
0.0541 (20°C, subcooled liq., calculated-mole fraction of Aroclor mixtures, Murphy et al. 1987)
0.021 (calculated average of HPLC-RI, Brodsky & Ballschmiter 1988)
0.0259 (calculated-TSA, Abramowitz & Yalkowsky 1990)
0.021, 0.011 (quoted average of Brodsky & Ballschmiter 1988, calculated- χ , Patil 1991)

Vapor Pressure (Pa at 25°C):
0.00735, 0.00905, 0.00278 (subcooled liq., calculated-MW, GC-RI, χ , Burkhard et al. 1985a)
0.00849 (subcooled liq., GC-RI, Burkhard et al. 1985b)
0.00537, 0.00744 (subcooled liquid, GC-RT, Foreman & Bidleman 1985)
0.00335 (20°C, subcooled liq., calculated-mole fraction of Aroclor mixtures, Murphy et al. 1987)

Henry's Law Constant (Pa m^3/mol):
82.78 (calculated-P/C, Burkhard 1985b)

20.06 (20°C, calculated-P/C, Murphy et al. 1987)
29.38 (calculated- χ , Sabljic & Güsten 1989)

Octanol/Water Partition Coefficient, log K_{OW}:
6.36 (calculated-TSA, Burkhard 1984)
5.18 (HPLC-RT, Rapaport & Eisenreich 1984)
6.55 (calculated-π, Rapaport & Eisenreich 1984; quoted, Sklarew & Girvin 1987)
6.40 (selected, Shiu & Mackay 1986; quoted, Sklarew & Girvin 1987)
6.20 (calculated-chlorine substituents, Oliver 1987a)
5.92 (calculated average of HPLC-RI, Brodsky & Ballschmiter 1988)
6.13 (calculated-TSA, Hawker & Connell 1988a)
5.92, 6.26 (quoted average of Brodsky & Ballschmiter 1988, calculated- χ , Patil 1991)

Bioconcentration Factor, log BCF:

Sorption Partition Coefficient, log K_{OC}:
6.16 (suspended particulate matter, calculated-K_{OW}, Burkhard 1984)
5.3-6.8, 6.1 (suspended sediment, average, Oliver 1987a)
7.40 (algae > 50 μm, Oliver 1987a)
6.30 (Lake Michigan water column, Swackhamer & Armstrong 1987)
4.60, 4.66, 4.61, 3.70 (humic substances, in concn. 5, 10, 20, 40 mg C/L, reported as log K_h, Lara & Ernst 1989)
4.603, 4.589 (humic substances of 5 mg C/L, quoted, calculated- χ , reported as log K_h, Sabljic et al. 1989)
5.55 (soil, Paya-Perez et al. 1991)

Half-Lives in the Environment:
Air:
Surface water:
Groundwater:
Sediment:
Soil:
Biota:

Environmental Fate Rate Constants or Half-Lives:
Volatilization:
Photolysis:
Hydrolysis:
Oxidation:
Biodegradation:
Biotransformation:
Bioconcentration, Uptake (k_1) and Elimination (k_2) Rate Constants:

Common Name: 2,2',4,4',5-Pentachlorobiphenyl
Synonym: PCB-99
Chemical Name: 2,2',4,4',5-pentachlorobiphenyl
CAS Registry No: 38380-01-7
Molecular Formula: $C_{12}H_5Cl_5$
Molecular Weight: 326.43
Melting Point (°C):
81.0 (calculated, Abramowitz & Yalkowsky 1990)
Boiling Point (°C):
Density (g/cm^3):
Molar Volume (cm^3/mol):
Molecular Volume (A^3):
Total Surface Area, TSA (A^2):
270.35 (Burkhard 1984)
250.16 (planar, Doucette 1985)
274.92 (nonplanar, Doucette 1985)
251.79 (planar, Hawker & Connell 1988a)
252.0 (Abramowitz & Yalkowsky 1990)
Heat of Fusion, kcal/mol:
Fugacity Ratio, F (assuming ΔS_{fusion} = 13.5 e.u.):

Water Solubility (g/m^3 or mg/L at 25°C):
0.0353 (subcooled liq., calculated-TSA, Burkhard et al. 1985b; quoted, Capel et al. 1991)
0.0222 (20°C, subcooled liq., calculated-mole fraction of Aroclor mixtures, Murphy et al. 1987)
0.0037 (calculated-HPLC-RI, Brodsky & Ballschmiter 1988)
0.0103 (calculated-TSA, Abramowitz & Yalkowsky 1990)
0.0037, 0.011 (quoted of Brodsky & Ballschmiter 1988, calculated- χ , Patil 1991)

Vapor Pressure (Pa at 25°C):
0.0029 (subcooled liquid, GC-RT, Bidleman 1984; quoted, Erickson 1986)
0.00735, 0.00342, 0.000532 (subcooled liq., calculated-MW, GC-RI, χ , Burkhard et al. 1985a)
0.00316 (subcooled liq., calculated-GC-RI, Burkhard et al. 1985b; quoted, Eisenreich 1987; Capel et al. 1991)
0.00328, 0.00375, 0.00293 (subcooled liquid, GC-RT, Foreman & Bidleman 1985)
0.00147 (20°C, subcooled liq., calculated-mole fraction of Aroclor mixtures, Murphy et al. 1987)

Henry's Law Constant (Pa m^3/mol):
29.28 (calculated-P/C, Burkhard et al. 1985b; quoted, Eisenreich 1987; Capel et al. 1991)

21.68 (20°C, calculated-P/C, Murphy et al. 1987)
30.50 (calculated- χ , Sabljic & Güsten 1989)
7.90 (wetted-wall col.-GC/ECD, Brunner et al. 1990)

Octanol/Water Partition Coefficient, log K_{OW}:
6.11 (selected, Garten & Trabalka 1983)
6.34 (calculated-TSA, Burkhard 1984)
6.29 (HPLC-RT, Rapaport & Eisenreich 1984)
7.21 (calculated-π, Rapaport & Eisenreich 1984; quoted, Sklarew & Girvin 1987)
6.60 (selected, Shiu & Mackay 1986; quoted, Sklarew & Girvin 1987)
7.21 (quoted, Eisenreich 1987)
6.41 (calculated-HPLC-RI, Brodsky & Ballschmiter 1988)
6.39 (calculated-TSA, Hawker & Connell 1988a; quoted, Capel et al. 1991)
6.41, 6.26 (quoted of Brodsky & Ballschmiter 1988, calculated- χ , Patil 1991)

Bioconcentration Factor, log BCF:
4.09 (fish, microcosm, Garten & Trabalka 1983)

Sorption Partition Coefficient, log K_{OC}:
6.14 (suspended particulate matter, calculated-K_{OW}, Burkhard 1984)
4.726 (calculated- χ , reported as log K_h at 5 mg C/L, Sabljic et al. 1989)
7.00 (calculated after Karickhoff et al. 1979, Capel & Eisenreich 1990)
5.68 (calculated after Schwarzenbach & Westall 1981, Capel & Eisenreich 1990)

Half-Lives in the Environment:
Air:
Surface water:
Groundwater:
Sediment:
Soil:
Biota:

Environmental Fate Rate Constants or Half-Lives:
Volatilization:
Photolysis:
Hydrolysis:
Oxidation:
Biodegradation:
Biotransformation:
Bioconcentration, Uptake (k_1) and Elimination (k_2) Rate Constants:

Common Name: 2,2',4,4',6-Pentachlorobiphenyl
Synonym: PCB-100
Chemical Name: 2,2'4,4'6-pentachlorobiphenyl
CAS Registry No: 39485-83-1
Molecular Formula: $C_{12}H_5Cl_5$
Molecular Weight: 326.4
Melting Point (°C):
Boiling Point (°C):
Density (g/cm³):
Molar Volume (cm³/mol):
289.1 (LeBas method, Shiu & Mackay 1986)
Molecular Volume (A³):
Total Surface Area, TSA (A²):
273.83 (Burkhard 1984)
245.16 (planar, Doucette 1985)
275.37 (nonplanar, Doucette 1985)
247.20 (planar, Hawker & Connell 1988a)
Heat of Fusion, kcal/mol:
Fugacity Ratio, F (assuming ΔS_{fusion} = 13.5 e.u.):

Water Solubility (g/m³ or mg/L at 25°C):
0.0275 (subcooled liq., calculated-TSA, Burkhard et al. 1985b; quoted, Eisenreich 1987)
0.031 (unpublished data of Weil 1978; quoted, Kilzer et al. 1979; Geyer et al. 1980)
0.0071 (calculated average of HPLC-RI, Brodsky & Ballschmiter 1988)
0.0071, 0.011 (quoted average of Brodsky & Ballschmiter 1988, calculated- χ , Patil 1991)

Vapor Pressure (Pa at 25°C):
0.00735, 0.00872, 0.00184 (subcooled liq., calculated-MW, GC-RI, χ, Burkhard et al 1985a)
0.00818 (subcooled liq., calculated-GC-RT, Burkhard et al. 1985b; quoted, Eisenreich 1987)

Henry's Law Constant (Pa m³/mol):
97.27 (calculated-P/C, Burkhard et al. 1985b; quoted, Eisenreich 1987)
62.62 (calculated- χ , Sabljic & Güsten 1989)

Octanol/Water Partition Coefficient, log K_{OW}:
5.89 (quoted, Garten & Trabalka 1983)
6.44 (calculated-TSA, Burkhard 1984)
5.5 (quoted, Geyer et al. 1984; Freitag et al. 1985)
6.23 (calculated average of HPLC-RI, Brodsky & Ballschmiter 1988)

6.23 (calculated-TSA, Hawker & Connell 1988a)

6.23, 6.26 (quoted average of Brodsky & Ballschmiter 1988, calculated- χ , Patil 1991)

Bioconcentration Factor, log BCF:

1.11, 1.06 (adipose tissue of male, female Albino rats, Geyer et al. 1980)

0.32 (rodent, Garten & Trabalka 1983)

4.06, 3.91 (algae, calculated, Geyer et al. 1984)

4.06, 3.37, 4.44 (algae, fish, activated sludge, Freitag et al. 1984,1985; quoted, Halfon & Reggiani 1986)

Sorption Partition Coefficient, log K_{OC}:

6.24 (suspended particulate matter, calculated-K_{OW}, Burkhard 1984)

4.567 (calculated- χ , reported as log K_h at 5 mg C/L, Sabljic et al. 1989)

Half-Lives in the Environment:

Air:

Surface water:

Groundwater:

Sediment:

Soil:

Biota:

Environmental Fate Rate Constants or Half-Lives:

Volatilization:

Photolysis:

Hydrolysis:

Oxidation:

Biodegradation:

Biotransformation:

Bioconcentration, Uptake (k_1) and Elimination (k_2) Rate Constants:

k_2: 0.010 day^{-1} (10°C, sandworm, Goerke & Ernst 1977; quoted, Waid 1986)

Common Name: 2,2'4,5,5'-Pentachlorobiphenyl
Synonym: PCB-101
Chemical Name: 2,2',4,5,5'-pentachlorobiphenyl
CAS Registry No: 37680-72-3
Molecular Formula: $C_{12}H_5Cl_5$
Molecular Weight: 326.43
Melting Point (°C):
76.5-77.5 (Hutzinger et al. 1974; NAS 1979; Erickson 1986)
77.0 (Mackay et al. 1980; Bruggeman et al. 1982; Burkhard et al. 1985a; Opperhuizen et al. 1988; Ballschmiter & Wittlinger 1991)
77, 67 (quoted, calculated, Abramowitz & Yalkowsky 1990)
Boiling Point (°C):
381 (calculated, Mackay et al. 1982; Shiu & Mackay 1986)
Density (g/cm³ at 20°C): 1.2803
Molar Volume (cm³/mol):
289.1 (LeBas method, Shiu & Mackay 1986)
1.370 (intrinsic volume: V_I/100, Kamlet et al. 1988; Hawker 1989b)
Molecular Volume (A³):
276.1 (planar, shorthand, Opperhuizen et al. 1988)
Total Surface Area, TSA (A²):
275.15 (Mackay et al. 1980)
270.20 (Burkhard 1984)
250.16 (planar, Doucette 1985)
269.2 (Shiu & Mackay 1986)
274.95 (nonplanar, Doucette 1985; Doucette & Andren 1988)
275.3 (calculated- χ , Sabljic 1987)
251.62 (planar, Hawker & Connell 1988a)
276.7 (planar, shorthand, Opperhuizen et al. 1988)
252.0 (Abramowitz & Yalkowsky 1990)
Heat of Fusion, kcal/mol:
4.49 (Miller et al. 1984)
Entropy of Fusion, cal/mol K (e.u.):
12.80 (Miller et al. 1984; selected, Hinckley et al. 1990)
Fugacity Ratio, (assuming ΔS_{fusion} = 13.5 e.u.):
0.306 (Mackay et al. 1980)
0.318 (Shiu & Mackay 1986)

Water Solubility (g/m³ or mg/L at 25°C):
0.031 (shake flask-GC/ECD, Wallnöfer et al. 1973; Hutzinger et al. 1974; quoted, Erickson 1986)
0.0103 (shake flask-GC/ECD, Haque & Schmedding 1975; Chiou et al. 1977; quoted, Haque et al. 1980; Sklarew & Girvin 1987)
0.0042 (gen. col.-GC/ECD, Weil et al. 1974; quoted, Lara & Ernst 1989)

0.019 (LSC, Metcalf et al. 1975)
0.010 (24°C, shake flask-GC/ECD, Chiou et al. 1977; Freed et al. 1977; selected, Mackay et al. 1980a)
0.00424 (16.5°C, shake flask-GC/ECD, Wiese & Griffin 1978)
0.031 (quoted, NAS 1979; Chou & Griffin 1987)
0.01 (quoted, Kenaga & Goring 1980; Kenaga 1980)
0.0163 (calculated-TSA, Mackay et al. 1980; quoted, Ballschmiter & Wittlinger 1991)
0.0042 (TLC-RT, Bruggeman et al. 1982)
0.0130 (quoted lit. average, Yalkowsky et al. 1983)
0.0005 (gen. col.-HPLC, Swann et al. 1983)
0.004 (RP-HPLC-RT, Swann et al. 1983)
0.0194 (gen. col.-GC/ECD, Miller et al. 1984,1985; quoted, Hawker 1989b)
0.0356 (subcooled liq., calculated-TSA, Burkhard et al. 1985b; quoted, Eisenreich 1987; Capel et al. 1991)
0.0154 (gen. col.-GC/ECD, Dickhut et al. 1986)
0.0115 (quoted, Dickhut et al. 1986 from Burkhard et al. 1985b)
0.011 (shake flask-GC/ECD, Chiou et al. 1986,1991)
0.010 (selected, Shiu & Mackay 1986)
0.0263 (20°C, calculated-mole fraction of Aroclor mixtures, Murphy et al. 1987)
0.010, 0.020, 0.021 (reported lit., Sawhney 1987)
0.0042, 0.0056 (quoted, calculated average of HPLC-RI, Brodsky & Ballschmiter 1988)
0.00392 (Brodsky et al. 1988; quoted, Ballschmiter & Wittlinger 1991)
0.00674 (gen. col.-GC/ECD, Dunnivant & Elzerman 1988)
0.0222 (subcooled liquid, calculated-M.P., Dunnivant & Elzerman 1988)
0.0594 (quoted, subcooled liquid, Hawker 1989b)
0.0042, 0.009 (quoted, calculated- χ , Nirmalakhanden & Speece 1989)
0.010 (quoted, Isnard & Lambert 1988,1989)
0.0103, 0.0164 (quoted, calculated-TSA, Abramowitz & Yalkowsky 1990)

Vapor Pressure (Pa at 25°C):
0.00123 (GC-RT, Westcott et al. 1981; quoted, Erickson 1986)
0.0039 (P_L calculated from P_S using F, Westcott et al. 1981)
0.00096 (gas saturation-GC/ECD, Westcott et al. 1981)
0.0016 (GC-RT, Westcott & Bidleman 1981)
0.0031 (P_L calculated from P_S using F, Westcott & Bidleman 1981)
0.00336, 0.00402 (subcooled liq.,GC-RT, Bidleman 1984)
0.00315 (subcooled liquid, quoted, Bidleman 1984; Ballschmiter& Wittlinger 1991)
0.00358 (subcooled liquid, Burkhard 1984)
0.00358 (subcooled liq.,calculated-GC-RI, Burkhard et al. 1985b; quoted, Eisenreich 1987; Capel et al. 1991)
0.00225, 0.00118, 0.000804 (calculated-MW,GC-RI, χ , Burkhard et al. 1985a)
0.00361, 0.00403, 0.00367 (subcooled liq., GC-RT, Foreman & Bidleman 1985)
0.00109 (selected, Shiu & Mackay 1986; quoted, Sklarew & Girvin 1987)

0.0035 (selected, subcooled liquid, Shiu & Mackay 1986)
0.00142 (20°C, subcooled liq., calculated-mole fraction of Aroclor mixtures, Murphy et al. 1987)
0.000527 (calculated-SxHLC, Dunnivant & Elzerman 1988)
0.00173 (subcooled liquid, calculated-M.P., Dunnivant & Elzerman 1988)
0.00111, 0.00152 (quoted, calculated-UNIFAC, Banerjee et al. 1990)
0.00315, 0.00296 (quoted, subcooled liq., Hinckley et al. 1990)
0.0026 (Wittlinger et al. 1990; quoted, Ballschmiter & Wittlinger 1991)

Henry's Law Constant (Pa m^3/mol):
11.46-35.46 (calculated-P/C, Westcott et al. 1981)
32.73 (calculated-P/C, Burkhard et al. 1985b; quoted, Eisenreich 1987; Capel et al. 1991)
7.09 (20°C, batch stripping, Oliver 1985)
35.48 (calculated-P/C, Shiu & Mackay 1986)
18.14 (20°C, calculated-P/C, Murphy et al. 1987)
25.43 (batch stripping, Dunnivant & Elzerman 1988; Dunnivant et al. 1988; quoted, Ballschmiter & Wittlinger 1991)
9.12 (wetted-wall col.-GC/ECD, Brunner et al. 1990)
217 (Wittlinger et al. 1990; quoted, Ballschmiter & Wittlinger 1991)

Octanol/Water Partition Coefficient, log K_{OW}:
4.12 (radiolabelled-^{14}C, Metcalf et al. 1975)
6.11 (shake flask-GC/ECD, Chiou et al. 1977; Freed et al. 1977; Chiou et al. 1982; Chiou 1985; Chiou & Block 1986; quoted, Mackay et al. 1980a; Sklarew & Girvin 1987)
7.64 (calculated-π, Tulp & Hutzinger 1978)
7.64 (Hansch & Leo 1979)
4.20 (NAS 1979)
6.44 (HPLC-RT, Veith et al. 1979)
7.24 (calculated, Mackay et al. 1980a)
6.85 (TLC-RT, Bruggeman et al. 1982; quoted, Erickson 1986)
6.11 (selected, Garten & Trabalka 1983)
6.42 (HPLC-RT, Swann et al. 1983)
7.43 (calculated-f const., Yalkowsky et al. 1983)
6.34 (calculated-TSA, Burkhard 1984)
5.92 (gen. col.-GC/ECD, Miller et al. 1984; quoted, Sklarew & Girvin 1987)
6.15 (HPLC-k', Rapaport & Eisenreich 1984)
7.07 (calculated-π, Rapaport & Eisenreich 1984; quoted, Sklarew & Girvin 1987)
7.64 (HPLC-RT, Woodburn et al. 1984)
7.64 (calculated-π, Woodburn et al. 1984)
6.50 (gen. col.-HPLC, Woodburn et al. 1984; quoted, Sklarew & Girvin 1987)
7.1 (quoted, Oliver & Charlton 1984)

6.18 (quoted, Burkhard & Kuehl 1986; Geyer et al. 1987)
6.40 (selected, Shiu & Mackay 1986; quoted, Sklarew & Girvin 1987; Clark et al. 1990)
5.84 (calculated-S, Chou & Griffin 1987)
7.07 (quoted, Eisenreich 1987)
6.88, 7.0 (HPLC-k', calculated, De Kock & Lord 1987)
6.20 (calculated-chlorine substituents, Oliver 1987a)
6.50 (gen. col.-GC/ECD, Doucette & Andren 1987,1988)
7.60, 6.39 (calculated-group contribution, TSA, Doucette & Andren 1987)
6.30, 6.30 (quoted, calculated average of HPLC-RI, Brodsky & Ballschmiter 1986)
6.40 (Brodsky et al. 1988; quoted, Ballschmiter & Wittlinger 1991)
6.38 (calculated-TSA, Hawker & Connell 1988a; quoted, Capel et al. 1991)
6.62 (calculated-solvatochromic p., Kamlet et al. 1988)
6.40 (quoted, Hawker & Connell 1989b)
7.60, 7.43, 5.88, 6.32, 6.39, 6.50 (calculated-π, f const., HPLC-RT, MW, χ, TSA, Doucette & Andren 1988)
5.92 (quoted, Isnard & Lambert 1988,1989)
6.10 (quoted, Thomann 1989)

Bioconcentration Factor, log BCF:
3.18 (green sunfish, 15 days in static water, Sanborn et al. 1975)
3.74, 4.78, 4.08, 4.24 (algae, snail, fish, mosquito, Metcalf et al. 1975)
4.66 (fish, flowing water, Kenaga & Goring; Kenaga 1980)
3.92, 3.60 (calculated-S, K_{OC}, Kenaga 1980)
4.09 (fish, microcosm, Garten & Trabalka 1983)
>5.40 (rainbow trout, kinetic, Oliver & Niimi 1985)
4.15 (rainbow trout, steady state, Oliver & Niimi 1985)
6.92 (rainbow trout, field data, Oliver & Niimi 1985)
2.73, 2.60 (human fat of lipid, wet wt. basis, calculated-K_{OW}, Geyer et al. 1987)
8.02 (rainbow trout, quoted, Thomann 1989)
4.15 (fish, quoted, Isnard & Lambert 1988)

Sorption Partition Coefficient, log K_{OC}:
4.63 (Kenaga & Goring 1980; selected, Lyman 1982; Hodson & Williams 1988)
4.74, 4.80 (estimated-S, K_{OW}, Lyman 1982)
5.13, 4.67 (estimated-BCF, Lyman 1982)
4.70 (soil, slurry method, Swann et al. 1983)
5.45 (calculated-HPLC-RT, Swann et al. 1983)
6.14 (suspended particulate matter, calculated-K_{OW}, Burkhard 1984)
5.6 (field data, Oliver & Charlton 1984)
5.5 (Niagara River-org. matter, Oliver & Charlton 1984)
6.6 (calculated-K_{OW}, Oliver & Charlton 1984)
5.65 (suspended solids-Lake Superior, Baker et al. 1986)

6.68, 5.58 (suspended solids-Lake Superior, calculated, Baker et al. 1986)
4.87, 4.07 (Sanhedron soil, Suwannee River, humic acid, GC/ECD, Chiou et al. 1986,1987)
4.12, 4.10 (Sanhedron soil, Suwannee River, fulvic acid, Chiou et al. 1986,1987)
4.89 (soil, calculated-S, Chou & Griffin 1987)
5.1-6.7, 6.2 (suspended sediment, average, Oliver 1987a)
7.20 (algae > 50 μm, Oliver 1987a)
5.41 (Aldrich humic acid Na salt, Chiou et al. 1987)
5.41 (Fluka-Tridon humic acid, Chiou et al. 1987)
4.81 (Calcasieu River humic extract, Chiou et al. 1987)
4.09 (Suwannee River water sample, Chiou et al. 1987)
4.01 (Sopchoppy River water sample, Chiou et al. 1987)
6.25 (Lake Michigan water column, Swackhamer & Armstrong 1987)
4.77, 4.86, 4.80, 3.86 (humic substances, in concn. of 5, 10, 20, 40 mg C/L, reported as log K_h, Lara & Ernst 1989)
4.772, 4.726 (humic substances of 5 mg C/L, quoted, calculated- χ , reported as log K_h, Sabljic et al. 1989)
5.86 (calculated after Karickhoff et al. 1979, Capel & Eisenreich 1990)
5.58 (calculated after Schwarzenbach & Westall 1981, Capel & Eisenreich 1990)
5.67 (soil, Paya-Perez et al. 1991)

Sorption Partition Coefficient, log K_{OM}:
4.63, 4.84 (quoted, calculated- χ , Sabljic 1984)

Half-Lives in the Environment:
Air:
Surface water:
Groundwater:
Sediment:
Soil:
Biota: half-life in rainbow trout, >1000 days (Niimi & Oliver 1983; Oliver & Niimi 1985), and its muscle, 85 days (Niimi & Oliver 1983).

Environmental Fate Rate Constants or Half-Lives:
Volatilization:
Photolysis:
Hydrolysis:
Oxidation:
Biodegradation: degradation rate, 1.5×10^{-8} nmol cell^{-1} hour by species of Alcaligenes and Acinetobacter (Furukawa et al. 1978, selected, NAS 1979).
Biotransformation:
Bioconcentration, Uptake (k_1) and Elimination (k_2) Rate Constants:
k_2: >0.0007 day^{-1} (rainbow trout, Niimi & Oliver 1983; quoted, Clark et al. 1990)

k_1: 180 day^{-1} (rainbow trout, Oliver & Niimi 1985)
k_2: 0.0007 day^{-1} (rainbow trout, Oliver & Niimi 1985)
k_s: 0.049 $hour^{-1}$ (uptake of mayfly-sediment model II, Gobas et al. 1989)
k_t: 0.014 $hour^{-1}$ (depuration of mayfly-sediment model II, Gobas et al. 1989)
log $1/k_2$: >3.1, 3.6 hour (fish, quoted, calculated-K_{OW}, Hawker & Connell 1988b).

Common Name: 2,2',4,6,6'-Pentachlorobiphenyl
Synonym: PCB-104
Chemical Name: 2,2',4,6,6'-pentachlorobiphenyl
CAS Registry No: 56558-16-8
Molecular Formula: $C_{12}H_5Cl_5$
Molecular Weight: 326.43
Melting Point (°C):
 91.0 (calculated, Abramowitz & Yalkowsky 1990)
Boiling Point (°C):
Density (g/cm^3):
Molar Volume (cm^3/mol):
 289.1 (LeBas method, Shiu & Mackay 1986)
Molecular Volume (A^3):
Total Surface Area, TSA (A^2):
 272.57 (Burkhard 1984)
 231.88 (planar, Doucette 1985)
 273.82 (nonplanar, Doucette 1985)
 234.87 (planar, Hawker & Connell 1988a)
 235.0 (Abramowitz & Yalkowsky 1990)
Heat of Fusion, kcal/mol:
Entropy of Fusion, cal/mol K (e.u.):
Fugacity Ratio, F (assuming ΔS_{fusion} = 13.5 e.u.):

Water Solubility (g/m^3 or mg/L at 25°C):
 0.030 (subcooled liq., calculated-TSA, Burkhard et al. 1985b)
 0.0156 (generator column-GC, Dunnivant & Elzerman 1988)
 0.0411 (calculated-TSA, Abramowitz & Yalkowsky 1990)

Vapor Pressure (Pa at 25°C):
 0.017 (subcooled liquid, Burkhard 1984)
 0.00735, 0.018, 0.00419 (subcooled liq., calculated-MW,GC-RI,χ, Burkhard et al. 1985a)
 0.0170 (subcooled liq., calculated-GC-RI, Burkhard et al. 1985b)
 0.00434 (subcooled liq., GC-RT, Foreman & Bidleman 1985)
 0.00434 (calculated-SxHLC, Dunnivant & Elzerman 1988)

Henry's Law Constant (Pa m^3/mol):
 185.4 (calculated-P/C, Burkhard et al. 1985b)
 90.9 (batch stripping, Dunnivant & Elzerman 1988;Dunnivant et al. 1988)

Octanol/Water Partition Coefficient, log K_{OW}:
 6.40 (calculated-TSA, Burkhard 1984)
 5.37 (gen. col.-GC-RT, Hawker & Connell 1988a)

5.81 (calculated-TSA, Hawker & Connell 1988a; quoted, Clark et al. 1990)

Bioconcentration Factor, log BCF:

Sorption Partition Coefficient, log K_{OC}:

6.20 (suspended particulate matter, calculated-K_{OW}, Burkhard 1984)

4.431 (humic substances, calculated- χ , Sabljic et al. 1989)

Half-Lives in the Environment:

Air:

Surface water:

Groundwater:

Sediment:

Soil:

Biota: half-life in rainbow trout, >1000 days and its muscle, 101 days (Niimi & Oliver 1983).

Environmental Fate Rate Constants or Half-Lives:

Volatilization:

Photolysis:

Hydrolysis:

Oxidation:

Biodegradation:

Biotransformation:

Bioconcentration, Uptake (k_1) and Elimination (k_2) Rate Constants:

k_2: >0.0007 day^{-1} (rainbow trout, Niimi & Oliver 1983; quoted, Clark et al. 1990)

Common Name: 2,3,3',4',6-Pentachlorobiphenyl
Synonym: PCB-110
Chemical Name: 2,3,3',4',6-pentachlorobiphenyl
CAS Registry No: 38380-03-9
Molecular Formula: $C_{12}H_5Cl_5$
Molecular Weight: 326.4
Melting Point (°C):
    79 (calculated, Abramowitz & Yalkowsky 1990)
Boiling Point (°C):
Density (g/cm^3):
Molar Volume (cm^3/mol):
    289.1 (LeBas method, Shiu & Mackay 1986)
Molecular Volume (A^3):
Total Surface Area, TSA (A^2):
    267.48 (Burkhard 1984)
    254.78 (planar, Doucette 1985)
    272.97 (nonplanar, Doucette 1985)
    254.65 (planar, Hawker & Connell 1988a)
    255.0 (Abramowitz & Yalkowsky 1990)
Heat of Fusion, kcal/mol:
Entropy of Fusion, cal/mol K (e.u.):
Fugacity Ratio, F (assuming ΔS_{fusion} = 13.5 e.u.):

Water Solubility (g/m^3 or mg/L at 25°C):
    0.0434 (subcooled liq., calculated-TSA, Burkhard et al. 1985b)
    0.0288 (20°C, subcooled liq., calculated-mole fraction of Aroclor mixtures, Murphy et al. 1987)
    0.0073 (calculated-HPLC-RI, Brodsky & Ballschmiter 1988)
    0.0082 (calculated-TSA, Abramowitz & Yalkowsky 1990)
    0.0073, 0.011 (quoted of Brodsky & Ballschmiter 1988, calculated- χ , Patil 1991)

Vapor Pressure (Pa at 25°C):
    0.00735, 0.00248, 0.000804 (subcooled liq., calculated-MW, GC-RI, χ , Burkhard et al. 1985a)
    0.00228 (subcooled liq., calculated-GC-RI, Burkhard et al. 1985b)
    0.00182, 0.00199 (subcooled liq., GC-RT, Foreman & Bidleman 1985)
    9.48×10^{-4} (20°C, subcooled liq., calculated-mole fraction of Aroclor mixtures, Murphy et al. 1987)

Henry's Law Constant (Pa m^3/mol):
    37.48 (calculated-P/C, Murphy et al. 1983)
    17.12 (calculated-P/C, Burkhard et al. 1985b)
    10.74 (20°C, calculated-P/C, Murphy et al. 1987)

19.15 (calculated- χ , Sabljic & Güsten 1989)

Octanol/Water Partition Coefficient, log K_{OW}:

6.27 (calculated-TSA, Burkhard 1984)
6.53 (quoted, Rapaport & Eisenreich 1985; Sklarew & Girvin 1987)
6.20 (calculated-chlorine substituent, Oliver 1987a)
6.20 (calculated-HPLC-RI, Brodsky & Ballschmiter 1988)
6.48 (calculated-TSA, Hawker & Connell 1988a)
6.23, 6.26 (quoted of Brodsky & Ballschmiter 1988, calculated- χ , Patil 1991)

Bioconcentration Factor, log BCF:

Sorption Partition Coefficient, log K_{OC}:

6.06 (suspended particulate matter, calculated-K_{OW}, Burkhard 1984)
5.6-6.8, 6.4 (suspended sediment, average, Oliver 1987a)
7.70 (algae > 50μm, Oliver 1987a)
4.72, 4.80, 4.77, 3.79 (humic substances, in concn. of 5, 10, 20, 40 mg C/L, reported as log K_h, Lara & Ernst 1989)
4.717, 4.748 (humic substances of 5 mg C/L, quoted, calculated- χ , Sabljic et al. 1989)
6.32 (calculated after Karickhoff et al. 1979, Capel & Eisenreich 1990)
5.20 (calculated after Schwarzenbach & Westall 1981, Capel & Eisenreich 1990)
5.71 (soil, Paya-Perez et al. 1991)

Half-Lives in the Environment:

Air:
Surface water:
Groundwater:
Sediment:
Soil:
Biota:

Environmental Fate Rate Constants or Half-Lives:

Volatilization:
Photolysis:
Hydrolysis:
Oxidation:
Biodegradation:
Biotransformation:
Bioconcentration, Uptake (k_1) and Elimination (k_2) Rate Constants:

Common Name: 2,3,4,5,6-Pentachlorobiphenyl
Synonym: PCB-116
Chemical Name: 2,3,4,5,6-pentachlorobiphenyl
CAS Registry No: 18259-05-7
Molecular Formula: $C_{12}H_5Cl_5$
Molecular Weight: 326.43
Melting Point (°C):
124 (Mackay et al. 1980; Bruggeman et al. 1982; Opperhuizen et al. 1988)
123 (Burkhard et al. 1985a; Erickson 1986)
124, 121 (quoted, calculated, Abramowitz & Yalkowsky 1990)
Boiling Point (°C):
381 (calculated, Mackay et al. 1982; Shiu & Mackay 1986)
Density (g/cm^3 at 20°C): 1.2803
Molar Volume (cm^3/mol):
289.1 (LeBas method, Shiu & Mackay 1986)
1.370 (intrinsic volume: $V_I/100$, Kamlet et al. 1988; Hawker 1989b; De Bruijn & Hermens 1990)
Molecular Volume (A^3):
265.4 (planar, shorthand, Opperhuizen et al. 1988)
251.05, 229.79, 265.0 (De Bruijn & Hermens 1990)
Total Surface Area, TSA (A^2):
269.22 (Mackay et al. 1980; Shiu & Mackay 1986)
262.99 (Burkhard 1984)
250.51 (planar, Doucette 1985)
268.95 (nonplanar, Doucette 1985; Doucette & Andren 1988)
250.10 (planar, Hawker & Connell 1988a)
271.7 (planar, shorthand, Opperhuizen et al. 1988)
250.0 (Abramowitz & Yalkowsky 1990)
Heat of Fusion, kcal/mol:
5.21 (Miller et al. 1984)
Entropy of Fusion, cal/mol K (e.u.):
13.10 (Miller et al. 1984)
12.86 (Shiu & Mackay 1986)
Fugacity Ratio, F (assuming ΔS_{fusion} = 13.5 e.u.):
0.105 (Mackay et al. 1980; Shiu & Mackay 1986)

Water Solubility (g/m^3 or mg/L at 25°C):
0.0068 (gen. col.-GC/ECD, Weil et al. 1974)
0.0207 (shake flask-GC/ECD, Dexter & Pavlou 1978)
0.00904 (calculated-TSA, Mackay et al. 1980)
0.0068 (TLC-RT, Bruggeman et al. 1982)
0.00682 (quoted lit. average, Yalkowsky et al. 1983; quoted, Erickson 1986)
0.00548 (gen. col.-GC/ECD, Miller et al. 1984,1985; quoted, Hawker 1989b)

0.0607 (subcooled liq., calculated-TSA, Burkhard et al. 1985b)
0.00554, 0.44 (quoted, calculated-K_{OW} & HPLC-RT, Chiou et al. 1986)
0.008 (selected, Shiu & Mackay 1986)
0.0036 (calculated average of HPLC-RI, Brodsky & Ballschmiter 1988)
0.00401 (gen. col.-GC/ECD, Dunnivant & Elzerman 1988)
0.0375 (subcooled liquid, calculated-M.P., Dunnivant & Elzerman 1988)
0.0136 (22°C, gen. col.-GC/ECD, Opperhuizen et al. 1988)
0.0494 (quoted, subcooled liquid, Hawker 1989b)
0.0055 (quoted, Isnard & Lambert 1989)
0.00136, 0.009 (quoted, calculated-χ, Nirmalakhanden & Speece 1989)
0.00517, 0.00517 (quoted, calculated-TSA, Abramowitz & Yalkowsky 1990)

Vapor Pressure (Pa at 25°C):
0.000788, 0.000394, 0.00419 (calculated-MW,GC-RI,χ, Burkhard et al. 1985a)
0.00341 (subcooled liq., calculated-GC-RI, Burkhard et al. 1985b)

Henry's Law Constant (Pa m^3/mol):
18.34 (calculated-P/C, Burkhard et al. 1985b)
23.41 (calculated-χ, Sabljic et al. 1989)

Octanol/Water Partition Coefficient, log K_{OW}:
6.85 (TLC-RT, Bruggeman et al. 1982; quoted, Erickson 1986)
7.49 (calculated-f const., Yalkowsky et al. 1983)
6.14 (calculated-TSA, Burkhard 1984)
6.30 (gen. col.-GC/ECD, Miller et al. 1984,1985; quoted, Sklarew & Girvin 1987; Hawker 1989b)
6.30 (selected, Shiu & Mackay 1986; quoted, Sklarew & Girvin 1987; Clark et al. 1990)
6.44 (calculated average of HPLC-RI, Brodsky & Ballschmiter 1988)
6.33 (calculated-TSA, Hawker & Connell 1988a)
6.45 (calculated-solvatochromic p., Kamlet et al. 1988)
6.30, 7.49 (quoted, calculated-π or f const., Kamlet et al. 1988)
6.30 (quoted exptl., Doucette & Andren 1988)
7.60, 7.43, 6.32, 6.40, 6.36 (calculated-π, f const., MW, χ, TSA, Doucette & Andren 1988)
6.30 (quoted, Isnard & Lambert 1989)
6.754 (slow stirring-GC/ECD, De Bruijn et al. 1989)
6.85 (calculated-π, De Bruijn et al. 1989)
6.754 (slow stirring-GC/ECD, De Bruijn & Hermens 1990)

Bioconcentration Factor, log BCF:

Sorption Partition Coefficient, log K_{OC}:

5.94 (suspended particulate matter, calculated-K_{OW}, Burkhard 1984)

4.791 (humic substances, calculated- χ , Sabljic et al. 1989)

Half-Lives in the Environment:

Air:

Surface water:

Groundwater:

Sediment:

Soil:

Biota: half-life in rainbow trout, >1000 days and its muscle, 100 days (Niimi & Oliver 1983).

Environmental Fate Rate Constants or Half-Lives:

Volatilization:

Photolysis:

Hydrolysis:

Oxidation:

Biodegradation:

Biotransformation:

Bioconcentration, Uptake (k_1) and Elimination (k_2) Rate Constants:

k_2: >0.0007 day^{-1} (rainbow trout, Niimi & Oliver 1983; quoted, Clark et al. 1990)

Common Name: 2,2',3,3',4,4'-Hexachlorobiphenyl
Synonym: PCB-128
Chemical Name: 2,2',3,3',4,4'-hexachlorobiphenyl
CAS Registry No: 38380-07-3
Molecular Formula: $C_{12}H_4Cl_6$
Molecular Weight: 360.88
Melting Point (°C):
150 (Mackay et al. 1980; Bruggeman et al. 1982; Burkhard et al. 1985a; Opperhuizen et al. 1988)
145.5-146.6 (Erickson 1986)
150, 151 (quoted, calculated, Abramowitz & Yalkowsky 1990)
Boiling Point (°C):
400 (calculated, Mackay et al. 1982; Shiu & Mackay 1986)
Density (g/cm^3 at 20°C): 1.3482
Molar Volume (cm^3/mol):
310 (LeBas method, Shiu & Mackay 1986)
1.460 (intrinsic volume: V_I/100, Kamlet et al. 1988; Hawker 1989b; De Bruijn & Hermens 1990)
Molecular Volume (A^3):
282.4 (planar, shorthand, Opperhuizen et al. 1988)
265.25, 243.26, 282.0 (De Bruijn & Hermens 1990)
Total Surface Area, TSA (A^2):
286.77 (Mackay et al. 1980; Shiu & Mackay 1986)
280.70 (Burkhard 1984)
261.73 (planar, Doucette 1985)
286.49 (nonplanar, Doucette 1985; Doucette & Andren 1988)
292.4 (Sabljic 1987)
262.13 (planar, Hawker & Connell 1988a)
282.1 (planar, shorthand, Opperhuizen et al. 1988)
314.91, 273.83, 282.1 (De Bruijn & Hermens 1990)
262.0 (Abramowitz & Yalkowsky 1990)
Heat of Fusion, kcal/mol:
6.89 (Miller et al. 1984)
Entropy of Fusion, cal/mol K (e.u.):
16.40 (Miller et al. 1984; Shiu & Mackay 1986)
Fugacity Ratio, F (assuming ΔS_{fusion} = 13.5 e.u.):
0.0582 (Mackay et al. 1980; Shiu & Mackay 1986)

Water Solubility (g/m^3 or mg/L at 25°C):
0.00044 (gen. col.-GC/ECD, Weil et al. 1974)
0.00099 (shake flask-GC/ECD, Dexter & Pavlou 1978)
0.00132 (calculated-TSA, Mackay et al. 1980)
0.00056 (TLC-RT, Bruggeman et al. 1982)

0.00044 (quoted lit. average, Yalkowsky et al. 1983; quoted, Erickson 1986)
0.000285(gen. col.-GC/ECD, Miller et al. 1984,1985)
0.0189 (subcooled liq., calculated-TSA, Burkhard et al. 1985b; quoted, Capel et al. 1991)
0.0006 (selected, Shiu & Mackay 1986; quoted, Hawker 1989b)
0.0067 (20°C, subcooled liq., calculated-mole fraction of Aroclor mixtures, Murphy et al. 1987)
0.00044, 0.0017 (quoted, calculated average of HPLC-RI, Brodsky & Ballschmiter 1988)
7.28×10^{-5} (calculated-UNIFAC, Banerjee & Howard 1988)
0.0023 (gen. col.-GC, Dunnivant & Elzerman 1988)
0.0055 (subcooled liquid, calculated-M.P., Dunnivant & Elzerman 1988)
0.00949 (quoted, subcooled liquid, Hawker 1989b)
0.000282(quoted, Isnard & Lambert 1989)
0.00444, 0.00293 (quoted, calculated- χ, Nirmalakhanden & Speece 1989)
0.000572, 0.000906 (quoted, calculated-TSA, Abramowitz & Yalkowsky 1990)
0.0017, 0.0024 (quoted average of Brodsky & Ballschmiter 1988, calculated- χ, Patil 1991)

Vapor Pressure (Pa at 25°C):
0.000341 (subcooled liq., GC-RT, Bidleman 1984; quoted, Erickson 1986)
0.000359 (subcooled liquid, Burkhard 1984)
0.000154, 0.0000231, 0.0000676 (calculated-MW,GC-RI,χ, Burkhard et al. 1985a)
0.0000508 (calculated-SxHLC, Burkhard et al. 1985a)
0.000359 (subcooled liq., calculated-GC-RI. Burkhard et al. 1985b; quoted, Capel et al. 1991)
0.000359 (subcooled liq., Burkhard et al. 1985)
0.000367, 0.000294, 0.000341 (subcooled liq.,GC-RT, Foreman & Bidleman 1985)
0.0000198 (selected, Shiu & Mackay 1986)
0.00034 (selected, subcooled liq., Shiu & Mackay 1986)
0.0000983 (20°C, subcooled liquid, calculated-mole fraction of Aroclor mixtures, Murphy et al. 1987)
2.94×10^{-6} (calculated-SxHLC, Dunnivant & Elzerman 1988)
4.66×10^{-5} (subcooled liquid, calculated-M.P., Dunnivant & Elzerman 1988)
0.000367, 0.000294, 0.000341 (GC-RT, Foreman & Bidleman 1985)

Henry's Law Constant (Pa m^3/mol):
50.66 (calculated, Murphy et al. 1983)
6.85 (calculated-P/C, Burkhard et al. 1985b; quoted, Capel et al. 1991)
11.91 (calculated-P/C, Shiu Mackay 1986)
5.78 (20°C, calculated-P/C, Murphy et al. 1987)
3.04 (batch stripping, Dunnivant & Elzerman 1988)
3.04 (batch stripping, Dunnivant et al. 1988)
1.32 (wetted-wall col.-GC, Brunner et al. 1990)

Octanol/Water Partition Coefficient, log K_{OW}:

7.44 (TLC-RT, Bruggeman et al. 1982; quoted, Erickson 1986)
8.18 (calculated-f const., Yalkowsky et al. 1983)
6.66 (quoted, Garten & Trabalka 1983)
6.62 (calculated-TSA, Burkhard 1984)
6.98 (gen. col.-GC/ECD, Miller et al. 1984,1985; quoted, Sklarew & Girvin 1987)
6.14 (HPLC-RT, Rapaport & Eisenreich 1984)
6.96 (calculated-π, Rapaport & Eisenreich 1984; quoted, Sklarew & Girvin 1987)
6.98, 6.28 (quoted, HPLC-RP/MS, Burkhard et al. 1985c)
8.31 (calculated-π, Burkhard et al. 1985c)
7.00 (selected, Shiu & Mackay 1986; quoted, Sklarew & Girvin 1987; Hawker 1989b)
6.50 (calculated-chlorine substituents, Oliver 1987c)
6.98, 6.68 (quoted, calculated average of HPLC-RI, Brodsky & Ballschmiter 1988)
6.74 (calculated-TSA, Hawker & Connell 1988a; quoted, Clark et al. 1990; Capel et al. 1991)
6.98 (gen. col.-GC/ECD, Doucette & Andren 1988)
8.31, 8.18, 6.71, 6.73, 6.78 (calculated-π, f-const., MW, χ, TSA, Doucette & Andren 1988)
6.81 (calculated-solvatochromic p., Kamlet et al. 1988)
7.00, 7.97 (quoted, calculated-π or f const., Kamlet et al. 1988)
6.96, 7.24 (quoted, calculated-UNIFAC, Banerjee & Howard 1988)
7.44 (calculated-π, DeBruijn et al. 1989)
7.0 (selected, Hawker 1989b)
6.93 (selected, Isnard & Lambert 1989)
7.321 (slow stirring-GC/ECD, De Bruijn et al. 1989; De Bruijn & Hermens 1990)

Bioconcentration Factor, log BCF:

1.0 (poultry, Garten & Trabalka 1983)

Sorption Partition Coefficient, log K_{OC}:

6.42 (suspended particulate matter, calculated-K_{OW}, Burkhard 1984)
4.28 (worms, Oliver 1987c)
5.259 (calculated- χ, reported as log K_h at 5 mg C/L, Sabljic et al. 1989)

Sorption Partition Coefficient, log K_{OM}:

5.05, 5.09 (quoted, calculated- χ, Sabljic 1984)

Half-Lives in the Environment:

Air:
Surface water:
Groundwater:
Sediment:
Soil:
Biota: half-life in rainbow trout, >1000 days and its muscle, 89 days (Niimi & Oliver

1983).

Environmental Fate Rate Constants or Half-Lives:

Volatilization:

Photolysis:

Hydrolysis:

Oxidation:

Biodegradation:

Biotransformation:

Bioconcentration, Uptake (k_1) and Elimination (k_2) Rate Constants:

k_2: >0.0007 day^{-1} (rainbow trout, Niimi & Oliver 1983; quoted, Clark et al. 1990)

Common Name: 2,2',3,3',4,5-Hexachlorobiphenyl
Synonym: PCB-129
Chemical Name: 2,2',3,3',4,5-hexachlorobiphenyl
CAS Registry No: 55215-18-4
Molecular Formula: $C_{12}H_4Cl_6$
Molecular Weight: 360.88
Melting Point (°C):
    85 (Mackay et al. 1980; Bruggeman et al. 1982; Burkhard et al. 1985a; Opperhuizen et al. 1988)
    85, 102 (quoted, calculated, Abramowitz & Yalkowsky 1990)
Boiling Point (°C):
    400 (calculated, Mackay et al. 1982; Shiu & Mackay 1986)
Density (g/cm^3 at 20°C): 1.3482
Molar Volume (cm^3/mol):
    310.0 (LeBas method, Shiu & Mackay 1986)
Molecular Volume (A^3):
    286.1 (planar, shorthand, Opperhuizen et al. 1988)
Total Surface Area, TSA (A^2):
    286.78 (Mackay et al. 1980; Shiu & Mackay 1986)
    280.57 (Burkhard 1984)
    261.73 (planar, Doucette 1985)
    286.49 (nonplanar, Doucette 1985)
    261.96 (planar, Hawker & Connell 1988a)
    282.1 (planar, shorthand, Opperhuizen et al. 1988)
    262.0 (Abramowitz & Yalkowsky 1990)
Heat of Fusion, kcal/mol:
Entropy of Fusion, cal/mol K (e.u.):
Fugacity Ratio, F (assuming ΔS_{fusion} = 13.5 e.u.):
    0.256 (Shiu & Mackay 1986)

Water Solubility (g/m^3 or mg/L at 25°C):
    0.00085 (gen. col.-GC/ECD, Weil et al. 1974)
    0.00581 (calculated-TSA, Mackay et al. 1980)
    0.00085 (quoted lit. average, Yalkowsky et al. 1983; quoted, Erickson 1986)
    0.0190 (subcooled liq., calculated-TSA, Burkhard et al. 1985b; quoted, Eisenreich 1987)
    0.0006 (selected, Shiu & Mackay 1986)
    0.0117 (20°C, subcooled liq., calculated-mole fraction of Aroclor mixtures, Murphy et al. 1987)
    0.0014 (calculated average of HPLC-RI, Brodsky & Ballschmiter 1988)
    0.00582 (gen. col.-GC/ECD, Dunnivant & Elzerman 1988)
    0.00846, 0.00293 (quoted, calculated- χ , Nirmalakhanden & Speece 1989)
    0.000572, 0.00287 (quoted, calculated-TSA, Abramowitz & Yalkowsky 1990)

0.0014, 0.0024 (quoted average of Brodsky & Ballschmiter 1988, calculated- χ , Patil 1991)

Vapor Pressure (Pa at 25°C):

0.000675, 0.000576, 0.000233 (calculated-MW,GC-RI, χ , Burkhard et al. 1985a)
0.00208 (subcooled liq., calculated-GC-RI, Burkhard et al. 1985b)
0.00213 (quoted, Eisenreich 1987)

Henry's Law Constant (Pa m^3/mol):

39.52 (calculated-P/C, Burkhard et al. 1985b; quoted, Eisenreich 1987)
8.61 (calculated- χ , Sabljic & Güsten 1989)
2.94 (wetted-wall col.-GC, Brunner et al. 1990)

Octanol/Water Partition Coefficient, log K_{OW}:

8.26 (calculated-f const., Yalkowsky et al. 1983; quoted, Erickson 1986)
6.62 (calculated-TSA, Burkhard 1984)
6.50 (HPLC-RT, Rapaport & Eisenreich 1984)
7.32 (calculated-π, Rapaport & Eisenreich 1984; quoted, Sklarew & Girvin 1987)
7.30 (selected, Shiu & Mackay 1986; quoted, Sklarew & Girvin 1987)
7.32 (quoted, Eisenreich 1987)
6.76 (calculated average of HPLC-RI, Brodsky & Ballschmiter 1988)
6.73 (calculated-TSA, Hawker & Connell 1988a)
6.76, 6.64 (quoted average of Brodsky & Ballschmiter 1988, calculated- χ ,Patil 1991)

Bioconcentration Factor, log BCF:

Sorption Partition Coefficient, log K_{OC}:

6.42 (suspended particulate matter, calculated-K_{OW}, Burkhard 1984)
5.259 (calculated- χ , reported as log K_h at 5 mg C/L, Sabljic 1989)

Half-Lives in the Environment:

Air:
Surface water:
Groundwater:
Sediment:
Soil:
Biota:

Environmental Fate Rate Constants or Half-Lives:

Volatilization:
Photolysis:
Hydrolysis:
Oxidation:

Biodegradation:
Biotransformation:
Bioconcentration, Uptake (k_1) and Elimination (k_2) Rate Constants:

Common Name: 2,2',3,3',5,6-Hexachlorobiphenyl
Synonym: PCB-134
Chemical Name: 2,2',3,3',5,6-hexachlorobiphenyl
CAS Registry No: 52704-70-8
Molecular Formula: $C_{12}H_4Cl_6$
Molecular Weight: 360.88
Melting Point (°C):
100 (Mackay et al. 1980; Burkhard et al. 1985a)
100, 114 (quoted, calculated, Abramowitz & Yalkowsky 1990)
Boiling Point (°C):
400 (calculated, Mackay et al. 1982; Shiu & Mackay 1986)
Density (g/cm^3 at 20°C): 1.3482
Molar Volume (cm^3/mol):
310.0 (LeBas method, Shiu & Mackay 1986)
Molecular Volume (A^3):
276.9 (planar, shorthand, Opperhuizen et al. 1988)
Total Surface Area, TSA (A^2):
287.19 (Mackay et al. 1980; Shiu & Mackay 1986)
283.17 (Burkhard 1984)
256.73 (planar, Doucette 1985)
286.93 (nonplanar, Doucette 1985; Doucette & Andren 1988)
256.49 (planar, Hawker & Connell 1988a)
287.3 (planar, shorthand, Opperhuizen et al. 1988)
256.0 (Abramowitz & Yalkowsky 1990)
Heat of Fusion, kcal/mol:
Entropy of Fusion, cal/mol K (e.u.):
Fugacity Ratio, F:
0.181 (Mackay et al. 1980; Shiu & Mackay 1986)

Water Solubility (g/m^3 or mg/L at 25°C):
0.00091 (gen. col.-GC/ECD, Weil et al. 1974; quoted, Lara & Ernst 1989)
0.00399 (calculated-TSA, Mackay et al. 1980)
0.00091 (quoted lit. average, Yalkowsky et al. 1983; quoted, Erickson 1986)
0.0160 (subcooled liq., calculated-TSA, Burkhard et al. 1985b)
0.0004 (selected, Shiu & Mackay 1986)
0.0130 (20°C, subcooled liq., calculated-mole fraction of Aroclor mixtures, Murphy et al. 1987)
0.0081 (calculated-HPLC-RI, Brodsky & Ballschmiter 1988)
0.000907, 0.00293 (calculated- χ , Nirmalakhanden & Speece 1989)
0.00036, 0.0036 (quoted, calculated-TSA, Abramowitz & Yalkowsky 1990)
0.0081, 0.0024 (quoted average of Brodsky & Ballschmiter 1988, calculated- χ , Patil 1991)

Vapor Pressure (Pa at 25°C):

0.000146 (calculated-SxHLC, Burkhard et al. 1985a)

0.00246 (subcooled liq., calculated-GC-RI, Burkhard et al. 1985b)

0.00048, 0.000483, 0.000804 (subcooled liq., calculated-MW,GC-RI,χ, Burkhard et al. 1985a)

0.00127, 0.00185 (subcooled liq., GC-RT, Foreman & Bidleman 1985)

0.00036 (20°C, subcooled liq., calculated-mole fraction of Aroclor mixtures, Murphy et al. 1987)

Henry's Law Constant (Pa m^3/mol):

57.76 (Murphy et al. 1983)

55.53 (calculated-P/C, Burkhard et al. 1985b)

9.83 (20°C, calculated-P/C, Murphy et al. 1987)

20.67 (calculated-χ, Sabljic & Güsten 1989)

4.96 (wetted-wall col.-GC, Brunner et al. 1990)

Octanol/Water Partition Coefficient, log K_{OW}:

8.18 (calculated, Yalkowsky et al. 1983; quoted, Erickson 1986)

6.69 (calculated-TSA, Burkhard 1984)

6.70 (selected, Shiu & Mackay 1986)

6.81 (gen. col.-GC/ECD,. Doucette & Andren 1987)

8.31 (calculated-π, Doucette & Andren 1987)

6.75 (calculated-TSA, Doucette & Andren 1987)

6.20 (calculated-HPLC-RI, Brodsky & Ballschmiter 1988)

6.55 (calculated-TSA, Hawker & Connell 1988a)

6.20, 6.64 (quoted of Brodsky & Ballschmiter 1988, calculated- χ , Patil 1991)

Bioconcentration Factor, log BCF:

Sorption Partition Coefficient, log K_{OC}:

6.49 (suspended particulate matter, calculated-K_{OW}, Burkhard 1984)

5.18, 5.16, 5.15, 4.41 (humic substances, in concn. of 5, 10, 20, 40 mg C/L, reported as log K_h, Lara & Ernst 1989)

5.178, 5.100 (humic substances, quoted, calculated- χ , reported as log K_h at 5 mg C/L, Sabljic et al. 1989)

Half-Lives in the Environment:

Air:

Surface water:

Groundwater:

Sediment:

Soil:

Biota:

Environmental Fate Rate Constants or Half-Lives:

Volatilization:

Photolysis:

Hydrolysis:

Oxidation:

Biodegradation:

Biotransformation:

Bioconcentration, and Uptake (k_1) and Elimination (k_2) Rate Constants:

Common Name: 2,2',3,3',6,6'-Hexachlorobiphenyl
Synonym: PCB-136
Chemical Name: 2,2',3,3',6,6'-hexachlorobiphenyl
CAS Registry No: 38411-22-2
Molecular Formula: $C_{12}H_4Cl_6$
Molecular Weight: 360.88
Melting Point (°C):
114-114.5 (NAS 1979)
112.2 (Opperhuizen et al. 1988)
114, 111 (quoted, calculated, Abramowitz & Yalkowsky 1990)
Boiling Point (°C):
400 (calculated, Mackay et al. 1982; Shiu & Mackay 1986)
Density (g/cm³):
Molar Volume (cm³/mol):
310.0 (LeBas method, Shiu & Mackay 1986)
1.460 (intrinsic volume: $V_I/100$, Kamlet et al. 1988; Hawker 1989b; De Bruijn & Hermens 1990)
Molecular Volume (A³):
263.6 (planar, shorthand, Opperhuizen et al. 1988)
264.89, 242.58, 263.6, 277.2 (De Bruijn & Hermens 1990)
Total Surface Area, TSA (A²):
284.68 (Burkhard 1984)
245.44 (planar, Doucette 1985)
287.38 (nonplanar, Doucette 1985; Doucette & Andren 1988)
246.95 (planar, Hawker & Connell 1988a)
269.5 (planar, shorthand, Opperhuizen et al. 1988)
313.80, 268.56, 269.1, 277.7 (De Bruijn & Hermens 1990)
247.0 (Abramowitz & Yalkowsky 1990)
Heat of Fusion, kcal/mol:
5.07 (Miller et al. 1984)
Entropy of Fusion, cal/mol K (e.u.):
13.10 (Miller et al. 1984)
Fugacity Ratio, F (assuming ΔS_{fusion} = 13.5 e.u.):
0.138 (Shiu & Mackay 1986)

Water Solubility (g/m³ or mg/L at 25°C):
0.00099 (shake flask-GC/ECD, Dexter & Pavlou 1978)
0.00603 (gen. col.-GC/ECD, Miller et al. 1984)
0.0145 (subcooled liq., calculated-TSA, Burkhard et al. 1985b; quoted, Eisenreich 1987)
0.00451 (gen. col.-GC/ECD, Dickhut et al. 1986)
0.00210 (quoted, Dickhut et al. 1986 from Burkhard et al. 1985b)
0.00080 (selected, Shiu & Mackay 1986)

0.0202 (20°C, subcooled liq., calculated-mole fraction of Aroclor mixtures, Murphy et al. 1987)
0.0414 (quoted, subcooled liquid, Hawker 1989b)
0.006, 0.00293 (quoted, calculated- χ , Nirmalakhanden & Speece 1989)
0.00603 (quoted, Isnard & Lambert 1989)
0.00072, 0.0114 (quoted, calculated-TSA, Abramowitz & Yalkowsky 1990)

Vapor Pressure (Pa at 25°C):
0.000349, 0.000531, 0.000804 (calculated-MW,GC-RI, χ , Burkhard et al. 1985a)
0.00374 (subcooled liq., calculated-GC-RI, Burkhard et al. 1985b; quoted, Eisenreich 1987)
0.00156, 0.00292 (subcooled liq.-GC-RT, Foreman & Bidleman 1985)
0.00127 (20°C, subcooled liq., calculated-mole fraction of Aroclor mixtures, Murphy et al. 1987)

Henry's Law Constant (Pa m^3/mol):
93.22 (calculated-P/C, Burkhard et al. 1985b; quoted, Eisenreich 1987)
22.79 (20°C, calculated-P/C, Murphy et al. 1987)
25.54 (calculated- χ , Sabljic & Güsten 1989)
8.92 (wetted-wall col.-GC, Brunner et al. 1990)

Octanol/Water Partition Coefficient, log K_{OW}:
8.35 (Hansch & Leo 1979)
6.0 (quoted, Garten & Trabalka 1983)
6.73 (calculated-TSA, Burkhard 1984)
6.81 (gen. col.-HPLC, Woodburn et al. 1984; quoted, Sklarew & Girvin 1987)
6.63 (gen. col.-GC/ECD, Miller et al. 1984; quoted, Sklarew & Girvin 1987)
4.91 (HPLC-RT, Rapaport & Eisenreich 1984)
6.51 (calculated-π, Rapaport & Eisenreich 1984; quoted, Sklarew & Girvin 1987)
6.81, 5.48 (quoted, HPLC-RP/MS, Burkhard et al. 1985c)
8.31 (calculated-π, Burkhard et al. 1985c)
6.70 (selected, Shiu & Mackay 1986; quoted, Sklarew & Girvin 1987)
6.51 (quoted, Eisenreich 1987)
6.81 (gen. col.-GC/ECD, Doucette & Andren 1987,1988)
8.31, 6.68 (calculated-π, TSA, Doucette & Andren 1987)
8.31, 8.18, 6.71, 6.73, 6.80 (calculated-π, f const., MW, χ , TSA, Doucette & Andren 1988)
5.76 (gen. col.-GC/ECD, Hawker & Connell 1988a)
6.22 (calculated-TSA, Hawker & Connell 1988a)
7.118 (slow stirring-GC, De Bruijn et al. 1989)
7.18 (calculated-π, De Bruijn et al. 1989)
6.70 (quoted, Hawker 1989b)
6.63 (quoted, Isnard & Lambert 1989)

7.118 (slow stirring-GC, De Bruijn et al. 1989; De Bruijn & Hermens 1990)

Bioconcentration Factor, log BCF:

0.86 (poultry, Garten & Trabalka 1983)

Sorption Partition Coefficient, log K_{OC}:

6.53 (suspended particulate matter, calculated-K_{OW}, Burkhard 1984)

6.53 (suspended solids, 0.7 mg/L, 43.2% OC-Lake Michigan, Voice & Weber 1985)

5.68 (suspended solids, 6.5 mg/L, 14.8% OC-Lake Michigan, Voice & Weber 1985)

4.95, 5.05, 4.95, 4.27 (humic substances in concn. of 5, 10, 20, 40 mg C/L, reported as log K_h, Lara & Ernst 1989)

4.952, 4.942 (humic substances, quoted, calculated- χ , reported as log K_h at 5 mg C/L, Sabljic et al. 1989)

Half-Lives in the Environment:

Air:

Surface water:

Groundwater:

Sediment:

Soil:

Biota:

Environmental Fate Rate Constants or Half-Lives:

Volatilization:

Photolysis:

Hydrolysis:

Oxidation:

Biodegradation:

Biotransformation:

Bioconcentration, Uptake (k_1) and Elimination (k_2) Rate Constants:

Common Name: 2,2',3,4,4',5'-Hexachlorobiphenyl
Synonym: PCB-138
Chemical Name: 2,2',3,4,4',5'-hexachlorobiphenyl
CAS Registry No: 35065-28-2
Molecular Formula: $C_{12}H_4Cl_6$
Molecular Weight: 360.9
Melting Point (°C):
79 (Hutzinger et al. 1974)
78.5-80 (NAS 1979)
79, 109 (quoted, calculated, Abramowitz & Yalkowsky 1990)
Boiling Point (°C):
400 (calculated, Mackay et al. 1982; Shiu & Mackay 1986)
Density (g/cm^3):
Molar Volume (cm^3/mol):
310 (LeBas method, Shiu & Mackay 1986)
Molecular Volume (A^3):
Total Surface Area, TSA (A^2):
283.29 (Burkhard 1984)
263.71 (planar, Doucette 1985)
288.52 (nonplanar, Doucette 1985)
264.76 (planar, Hawker & Connell 1988a)
265.0 (Abramowitz & Yalkowsky 1990)
Heat of Fusion, kcal/mol:
Entropy of Fusion, cal/mol K (e.u.):
Fugacity Ratio, F (assuming ΔS_{fusion} = 13.5 e.u.): 0.286

Water Solubility (g/m^3 or mg/L at 25°C):
0.0159 (subcooled liq., calculated-TSA, Burkhard et al. 1985b; quoted, Eisenreich 1987; Capel et al. 1991)
0.00729 (20°C, subcooled liq., calculated-mole fraction of Aroclor mixtures, Murphy et al. 1987)
0.0015 (calculated average of HPLC-RI, Brodsky & Ballschmiter 1988)
0.0017 (Brodsky et al. 1988; quoted, Ballschmiter & Wittlinger 1991)
0.00181 (calculated-TSA, Abramowitz & Yalkowsky 1990)
0.0015, 0.0024 (quoted average of Brodsky & Ballschmiter 1988, calculated- χ , Patil 1991)

Vapor Pressure (Pa at 25°C):
0.000506, 0.000563 (subcooled liquid, GC-RT, Bidleman 1984; quoted, Ballschmiter & Wittlinger 1991)
0.000774, 0.000158, 0.0000674 (calculated-MW, GC-RI, χ , Burkhard et al. 1985a)
0.000487 (subcooled liq., calc.-GC-RI, Burkhard et al. 1985b; quoted, Eisenreich 1987; Capel et al. 1991)

0.00051, 0.000496, 0.000534 (subcooled liq., GC-RT, Foreman & Bidleman 1985)
0.000533 (quoted from Bidleman 1984, Erickson 1986)
0.0005 (selected, subcooled liquid, Shiu & Mackay 1986)
0.000147 (20°C, subcooled liq., calculated-mole fraction of Aroclor mixtures, Murphy et al. 1987)
0.00033 (Wittlinger et al. 1990; quoted, Ballschmiter & Wittlinger 1991)

Henry's Law Constant (Pa m^3/mol):
48.64 (calculated-P/C, Murphy et al. 1983)
11.04 (calculated-P/C, Burkhard et al. 1985b; quoted, Eisenreich 1987; Capel et al. 1991)
7.60 (20°C, calculated-P/C, Murphy et al. 1987)
10.84 (calculated- χ , Sabljic & Güsten 1989)
2.13 (wetted-wall col.-GC, Brunner et al. 1990)
69.0 (Wittlinger et al. 1990; quoted, Ballschmiter & Wittlinger 1991)

Octanol/Water Partition Coefficient, log K_{OW}:
6.69 (calculated-TSA, Burkhard 1984)
6.62 (HPLC-RT, Rapaport & Eisenreich 1984)
7.44 (calculated-π, Rapaport & Eisenreich 1984; quoted, Sklarew & Girvin 1987)
7.44 (quoted, Eisenreich 1987)
6.50 (calculated-chlorine substituents, Oliver 1987a)
6.73 (calculated average of HPLC-RI, Brodsky & Ballschmiter 1988)
6.70 (Brodsky et al. 1988; quoted, Ballschmiter & Wittlinger 1991)
6.83 (calculated-TSA, Hawker & Connell 1988a; quoted, Capel et al. 1991)
6.73, 6.64 (quoted average of Brodsky & Ballschmiter 1988, calculated- χ , Patil 1991)

Bioconcentration Factor, log BCF:

Sorption Partition Coefficient, log K_{OC}:
6.49 (suspended particulate matter, calculated-K_{OW}, Burkhard 1984)
5.8-7.3, 6.6 (suspended sediment , average, Oliver 1987a)
7.60 (algae > 50 μm, Oliver 1987a)
6.65 (Lake Michigan water column, Swackhamer & Armstrong 1987)
5.21, 5.22, 5.17, 4.60 (humic substances, in concn. of 5, 10, 20, 40 mg C/L, reported as log K_h, Lara & Ernst 1989)
5.207, 5.241 (humic substances, quoted, calculated- χ , reported as log K_h at 5 mg C/L, Sabljic et al. 1989)
5.93 (soil, Paya-Perez et al. 1991)

Half-Lives in the Environment:
Air:
Surface water:

Groundwater:
Sediment:
Soil:
Biota:

Environmental Fate Rate Constants or Half-Lives:
Volatilization:
Photolysis:
Hydrolysis:
Oxidation:
Biodegradation:
Biotransformation:
Bioconcentration, Uptake (k_1) and Elimination (k_2) Rate Constants:
k_s: 0.049 $hour^{-1}$ (uptake of mayfly-sediment model II, Gobas et al. 1989)
k_t: 0.008 $hour^{-1}$ (depuration of mayfly-sediment model II, Gobas et al. 1989)

Common Name: 2,2',4,4',5,5'-Hexachlorobiphenyl
Synonym: PCB-153
Chemical Name: 2,2',4,4',5,5'-hexachlorobiphenyl
CAS Registry No: 35065-27-1
Molecular Formula: $C_{12}H_4Cl_6$
Molecular Weight: 360.88
Melting Point (°C):
102-104 (Wallöfer et al. 1973)
103-104 (NAS 1979)
103.0 (Mackay et al. 1980;Bruggeman et al. 1982; Burkhard et al. 1985a; Opperhuizen et al. 1988)
103, 139 (quoted, calculated, Abramowitz & Yalkowsky 1990)
Boiling Point (°C):
400 (calculated, Mackay et al. 1982; Shiu & Mackay 1986)
Density (g/cm³):
Molar Volume (cm³/mol):
310.0 (LeBas method, Shiu & Mackay 1986)
1.460 (intrinsic volume: V_I/100, Hawker 1990)
Molecular Volume (A³):
289.4 (planar, shorthand, Opperhuizen et al. 1988)
Total Surface Area, TSA (A²):
290.78 (Mackay et al. 1980; Shiu & Mackay 1986)
285.92 (Burkhard 1984)
265.70 (planar, Doucette 1985)
290.50 (nonplanar, Doucette 1985; Doucette & Andren 1987; 1988)
292.4 (calculated- χ, Sabljic 1987)
267.39 (planar, Hawker & Connell 1988a)
287.1 (planar, shorthand, Opperhuizen et al. 1988)
267.0 (Abramowitz & Yalkowsky 1990)
Heat of Fusion, kcal/mol:
Entropy of Fusion, cal/mol K (e.u.):
Fugacity Ratio, F (assuming ΔS_{fusion} = 13.5 e.u.):
0.169 (Mackay et al. 1980; Shiu & Mackay 1986)

Water Solubility (g/m³ or mg/L at 25°C):
0.0088 (shake flask-GC/ECD, Wallnöfer et al. 1973)
0.0012 (gen. col.-GC/ECD, Weil et al. 1974; quoted, Lara & Ernst 1989)
0.000953 (shake flask-GC/ECD, Haque & Schmedding 1975; quoted, Chiou et al. 1979; Haque et al. 1980; Bruggeman et al. 1984; Sklarew & Girvin 1987; Chou & Griffin 1987)
0.00095 (24°C, shake flask-GC/ECD, Chiou et al. 1977; Freed et al. 1977; quoted, Kenaga & Goring 1980; Kenaga 1980; Mackay et al. 1980a; Sklarew & Girvin 1987)

0.00105 (16.5°C, shake flask-GC/ECD, Wiese & Griffin 1978)
0.088 (NAS 1979)
0.00278 (calculated-TSA, Mackay et al. 1980; quoted, Ballschmiter & Wittlinger 1991)
0.0012 (Bruggeman et al. 1982; Opperhuizen 1986)
0.00131 (quoted lit. average, Yalkowsky et al. 1983; quoted, Erickson 1986)
0.00133, 0.000069 (quoted, calculated-UNIFAC, Banerjee 1985)
0.0134 (subcooled liq., calculated-TSA, Burkhard et al. 1985b; quoted, Eisenreich 1987; Capel et al. 1991)
0.001 (selected, Shiu & Mackay 1986)
0.00914 (20°C, subcooled liq., calculated-mole fraction of Aroclor mixtures, Murphy et al. 1987)
0.0012, 0.0012 (quoted, calculated average of HPLC-RI, Brodsky & Ballschmiter 1988)
0.0009 (Brodsky et al. 1988; quoted, Ballschmiter & Wittlinger 1991)
0.0000528 (calculated-UNIFAC, Banerjee & Howard 1988)
0.00086 (gen. col.-GC/ECD, Dunnivant & Elzerman 1988)
0.00516 (subcooled liquid, calculated-MP, Dunnivant & Elzerman 1988)
0.00845 (gen. col.-GC/ECD, Doucette & Andren 1988)
0.00115 (22°C, generator column-GC/ECD, Opperhuizen et al. 1988)
0.0012 (quoted, Isnard & Lambert 1989)
0.00117, 0.00293 (selected, calculated- χ , Nirmalakhanden & Speece 1989)
0.00091, 0.00072 (selected, calculated-TSA, Abramowitz & Yalkowsky 1990)
0.00094 (quoted, Eadie et al. 1990)
0.0012, 0.0024 (quoted average of Brodsky & Ballschmiter 1988, calculated- χ , Patil 1991)
0.0011 (quoted, Paya-Perez et al. 1991)

Vapor Pressure (Pa at 25°C):
0.00068, 0.000719 (subcooled liq., GC-RT, Bidleman 1984; quoted, Ballschmiter & Wittlinger 1991)
0.00663 (subcooled liquid, Burkhard 1984)
0.000457 (calculated-SxHLC, Burkhard et al. 1985a)
0.000448, 0.000124, 0.0000676 (calculated-MW,GC-RI, χ , Burkhard et al. 1985a)
0.000663 (subcooled liq., calculated-GC-RI, Burkhard et al. 1985b; quoted, Eisenreich 1987; Capel et al. 1991)
0.000708, 0.000813, 0.0007 (subcooled liq., GC-RT, Foreman & Bidleman 1985)
0.000457 (subcooled liquid, GC-RT, Burkhard et al. 1985b)
0.000693 (quoted from Bidleman 1984, Erickson 1986)
0.000119 (selected, Shiu & Mackay 1986)
0.00070 (selected, subcooled liq., Shiu & Mackay 1986)
0.000253 (20°C, subcooled liquid, calculated-mole fraction of Aroclor mixtures, Murphy et al. 1987)
0.0000324 (calculated-SxHLC, Dunnivant & Elzerman 1988)
0.000193 (subcooled liquid, calculated-M.P., Dunnivant & Elzerman 1988)

0.00056 (Wittlinger et al. 1990; quoted, Ballschmiter & Wittlinger 1991)

Henry's Law Constant (Pa m^3/mol):

35.46 (calculated, Murphy et al. 1983)
12.46 (batch stripping, Coates 1984)
17.93 (calculated-P/C, Burkhard et al. 1985b; quoted, Capel et al. 1991)
6.08 (20°C, batch stripping, Oliver 1985)
42.9 (calculated-P/C, Shiu & Mackay 1986)
18.24 (quoted, Eisenreich 1987)
10.03 (20°C, calculated-P/C, Murphy et al. 1987)
13.37 (batch stripping, Dunnivant & Elzerman 1988; Dunnivant et al. 1988; quoted, Ballschmiter & Wittlinger 1991)
2.33 (wetted-wall column-GC, Brunner et al. 1990)
224 (Wittlinger et al. 1990; quoted, Ballschmiter & Wittlinger 1991)

Octanol/Water Partition Coefficient, log K_{OW}:

6.72 (shake flask-GC/ECD, Chiou et al. 1977; Freed et al. 1977; Chiou et al. 1982; Chiou & Block 1986; quoted, Mackay et al. 1980a; Schwarzenbach & Westall 1981; Chou & Griffin 1987; Sklarew & Girvin 1987)
8.35 (calculated-π, Tulp & Hutzinger 1978)
8.35 (Hansch & Leo 1979)
4.48 (NAS 1979)
6.34 (shake flask-GC/ECD, Karickhoff et al. 1979)
6.57 (quoted, Kenaga & Goring 1980)
8.06 (calculated, Mackay et al. 1980a)
7.44 (TLC-RT, Bruggeman et al. 1982; 1984; quoted, Opperhuizen 1986; Erickson 1986)
6.72 (quoted, Garten & Trabalka 1983)
6.70 (quoted, Voice et al. 1983; Voice & Weber 1985)
8.18 (calculated-f const., Yalkowsky et al. 1983)
6.77 (calculated-TSA, Burkhard 1984)
7.40 (extrapolated from lit. values, Coates 1984)
7.8 (quoted, Oliver & Charlton 1984)
6.93 (HPLC-RT, Rapaport & Eisenreich 1984)
7.75 (calculated-π, Rapaport & Eisenreich 1984; quoted, Sklarew & Girvin 1987)
6.90 (gen. col.-HPLC, Woodburn et al. 1984; quoted, Sklarew & Girvin 1987; Banerjee & Baughman 1991)
8.35 (calculated-π, Woodburn et al. 1984)
6.68 (HPLC/MS, Burkhard et al. 1985c)
8.31 (calculated-π, Burkhard et al. 1985)
7.20 (quoted, Di Toro et al. 1985)
6.65 (quoted, Burkhard et al. 1985c; Burkhard & Kuehl 1986; Geyer et al. 1987)
6.90 (selected, Shiu & Mackay 1986; quoted, Gobas et al. 1987; Sklarew & Girvin

1987; Clark et al. 1990; Banerjee & Baughman 1991)

7.75 (quoted, Hawker & Connell 1986; Eisenreich 1987)

7.69, 7.71 (HPLC-k', calculated, De Kock & Lord 1987)

6.50 (calculated-chlorine substituent, Oliver 1987)

6.90 (gen. col.-GC, Doucette & Andren 1987,1988)

8.31, 6.75 (calculated-π, TSA, Doucette & Andren 1987)

6.90, 6.62 (quoted, calculated average of HPLC-RI, Brodsky & Ballschmiter 1988)

6.80 (Brodsky et al. 1988; quoted from Ballschmiter & Wittlinger 1991)

6.92 (calculated-TSA, Hawker & Connell 1988a; quoted, Connell & Hawker 1988; Hawker 1990; Capel et al. 1991)

8.31, 8.18, 6.46, 6.71, 6.73, 6.87 (calculated-π, f const., HPLC-RT, MW, χ, TSA, Doucette & Andren 1988)

6.70 (quoted, Evans & Landrum 1989)

6.68 (quoted, Isnard & Lambert 1989)

6.80, 6.64 (quoted average of Brodsky & Ballschmiter 1988, calculated- χ, Patil 1991)

6.81 (quoted, Paya-Perez et al. 1991)

Bioconcentration Factor, log BCF:

4.68 (oyster, Vreeland 1974; quoted, Hawker & Connell 1986)

4.66 (fish, flowing water, Kenaga & Goring 1980; Kenaga 1980)

4.48, 5.23 (calculated-S, K_{OC}, Kenaga 1980)

4.66 (quoted, Bysshe 1982)

5.23, 3.76 (amphipods, clams, Lynch et al. 1982)

5.03, 4.88, 4.65 (algae, snail, aquatic earthworm, Lynch et al. 1982)

4.82, 4.63 (crayfish, fish, Lynch et at. 1982)

4.00, 4.72, 3.77 (calculated-S, calculated-C, calculated-K_{OW}, Lynch et al. 1982)

0.99 (poultry, Garten & Trabalka 1983)

6.78 (guppy, 3.5% lipid, Bruggeman et al. 1982,1984; quoted, Gobas et al. 1987)

4.84 (rainbow trout, calculated-centrifugal water concentrations, Muir et al. 1985)

5.87 (rainbow trout, calculated-K_{OW}, Oliver & Niimi 1985)

7.0 (rainbow trout, field data, Oliver & Niimi 1985)

2.60, 2.48 (human fat of lipid, wet wt. basis, calculated-K_{OW}, Geyer et al. 1987)

5.32 (guppy, Gobas et al. 1987; selected, Banerjee & Baughman 1991)

4.57 (selenastrum capricornutum, Mailhot 1987)

4.48 (worms, Oliver 1987c)

4.84, 5.32 (fish, calculated-C_B/C_W or k_1/k_2, Connell & Hawker 1988; Hawker 1990)

3.85 (hexagenia limbata, Landrum & Poore 1988)

5.01 (pontoporela hoyi, calculated, Evans & Landrum 1989)

5.06 (guppy, estimated, Banerjee & Baughman 1991)

Sorption Partition Coefficient, log K_{OC}:

5.62 (Haque & Schmedding 1976)

5.34 (soil, exptl., Chiou et al. 1979; quoted, Schwarzenbach & Westall 1981)

6.08 (natural sediment, sorption isotherm, Karickhoff et al. 1979; quoted, Kenaga & Goring 1980; Chou & Griffin 1987)
5.33 (calculated-K_{OW}, Schwarzenbach & Westall 1981; quoted, Voice & Weber 1985)
6.51 (calculated-K_{OW}, Schwarzenbach & Westall 1981 from Karickhoff 1981)
5.62 (soil/sediment, quoted, Karickhoff 1981)
6.43, 6.42, 5.33 (calculated-K_{OW}, C_L, C_S, Karickhoff 1981)
6.51 (sediment, calculated-K_{OW}, Lynch et al. 1982)
6.57 (suspended particulate matter, calculated-K_{OW}, Burkhard 1984)
5.6 (field data, Oliver & Charlton 1984)
5.5 (Niagara River-org. matter, Oliver & Charlton 1984)
7.3 (calculated-K_{OW}, Oliver & Charlton 1984)
6.61 (sediment/pore water-Saginaw Bay, Di Toro et al. 1985)
5.40 (Offshore Grand Haven sediment, Voice & Weber 1985)
7.56, 7.68 (river sediment, Coates & Elzerman 1986)
5.575 (correlated literature values in soils, Sklarew & Girvin 1987)
5.8-7.3, 6.6 (suspended sediment, average, Oliver 1987a)
7.20 (algae > 50 μm, Oliver 1987a)
6.48 (Lake Michigan water column, Swackhamer & Armstrong 1987)
4.75 (12 lakes/streams in S. Ontario at 1.6-26.5 mg C/L, Evans 1988)
5.26, 5.25, 5.19, 4.62 (humic substances, in concn. of 5, 10, 20, 40 mg C/L, reported as log K_h, Lara & Ernst 1989)
5.258, 5.222 (humic substances, reported as log K_h at 5 mg C/L, Sabljic et al. 1989)
7.54 (calculated after Karickhoff et al. 1979, Capel & Eisenreich 1990)
6.08 (calculated after Schwarzenbach & Westall 1981, Capel & Eisenreich 1990)
6.76 (organic carbon, Eadie et al. 1990)
5.86 (soil, Paya-Perez et al. 1991)

Sorption Partition Coefficient, log K_{OM}:
4.42 (natural sediment, Eadie et al. 1990)

Half-Lives in the Environment:
Air:
Surface water: 25-53 min in aqueous solution purged at a flow rate of 1 L/min (Coates 1984).
Groundwater:
Sediment:
Soil:
Biota: half-life in rainbow trout, >1000 days and its muscle, 77 days (Niimi & Oliver 1983); in worms at 8°C, 170 days (Oliver 1987c); in guppy, 175 days (Bruggeman et al. 1984); in pontoporela hoyi, 45.6 days (Evans & Landrum 1989).

Environmental Fate Rate Constants or Half-Lives:
Volatilization:

Photolysis:
Hydrolysis:
Oxidation:
Biodegradation:
Biotransformation:
Bioconcentration, Uptake (k_1) and Elimination (k_2) Rate Constants:

k_2: >0.0007 day^{-1} (rainbow trout, Niimi & Oliver 1983; quoted, Clark et al. 1990)
k_1: 800 day^{-1} (guppy, Bruggeman et al. 1984)
k_2: 0.004 day^{-1} (guppy, Bruggeman et al. 1984; quoted, Clark et al. 1990)
k_1: 461 day^{-1} (rainbow trout, total ^{14}C in whole fish-wet weight, Muir et al. 1985)
k_2: 0.008 day^{-1} (rainbow trout, total ^{14}C in whole fish-wet weght, Muir et al. 1985)
k_1: 1100 day^{-1} (guppy, Opperhuizen 1986)
k_1: 63.2 $hour^{-1}$ (10-20°C, hexagenia limbata, Landrum & Poore 1988)
k_2:0.009 $hour^{-1}$ (10-20°C, hexagenia limbata, Landrum & Poore 1988)
k_s:0.049 $hour^{-1}$ (uptake of mayfly-sediment model II, Gobas et al. 1989)
k_t:0.009 $hour^{-1}$ (depuration of mayfly-sediment model II, Gobas et al. 1989)
log k_1: 2.90, 2.66 day^{-1} (fish, quoted, Connell & Hawker 1990)
log $1/k_2$: 2.39, 2.10 day (fish, quoted, Connell & Hawker 1990)

Common Name: 2,2',4,4',6,6'-Hexachlorobiphenyl
Synonym: PCB-155
Chemical Name: 2,2',4,4',6,6'-hexachlorobiphenyl
CAS Registry No: 33979-03-2
Molecular Formula: $C_{12}H_4Cl_6$
Molecular Weight: 360.88
Melting Point (°C):
112.5-114
114 (Mackay et al. 1980; Opperhuizen et al. 1988)
113 (Burkhard et al. 1985a)
114, 150 (quoted, calculated, Abramowitz & Yalkowsky 1990)
Boiling Point (°C):
400 (calculated, Mackay et al. 1982; Shiu & Mackay 1986)
Density (g/cm^3 at 20°C): 1.3482
Molar Volume (cm^3/mol):
310.0 (LeBas method, Shiu & Mackay 1986)
1.460 (intrinsic volume: V_I/100, Kamlet et al. 1988; Hawker 1989b; De Bruijn & Hermens 1990)
Molecular Volume (A^3):
271.0 (planar, shorthand, Opperhuizen et al. 1988)
265.22, 242.26, 271.0, 284.6 (De Bruijn & Hermens 1990)
Total Surface Area, TSA (A^2):
291.50 (Mackay et al. 1980; Shiu & Mackay 1986)
290.22 (Burkhard 1984)
249.41 (planar, Doucette 1985)
291.36 (nonplanar, Doucette 1985; Doucette & Andren 1988)
252.56 (planar, Hawker & Connell 1988a)
272.5 (planar, shorthand, Opperhuizen et al. 1988)
315.30, 271.16, 274.1, 282.7 (De Bruijn & Hermens 1990)
253.0 (Abramowitz & Yalkowsky 1990)
Heat of Fusion, kcal/mol:
4.18 (Miller et al. 1984)
Entropy of Fusion, cal/mol K (e.u.):
10.80 (Miller et al. 1984; Shiu & Mackay 1986)
Fugacity Ratio, F (assuming ΔS_{fusion} = 13.5 e.u.):
0.131 (Mackay et al. 1980; Shiu & Mackay 1986)

Water Solubility (g/m^3 or mg/L at 25°C):
0.00090 (shake flask-GC/ECD, Wallnöfer et al. 1973; Weil et al. 1974)
0.00204 (calculated-TSA, Mackay et al. 1980)
0.00091 (TLC-RT, Bruggeman et al. 1982)
0.00091 (quoted lit. average, Yalkowsky et al. 1983; quoted, Erickson 1986)
0.00041 (gen. col.-GC/ECD, Miller et al. 1984,1985; quoted, Hawker 1989b)

0.0101 (subcooled liq., calculated-TSA, Burkhard et al. 1985b)
0.0007 (selected, Shiu & Mackay 1986)
0.0000528 (calculated-UNIFAC, Banerjee & Howard 1988)
0.00091, 0.0027 (quoted, calculated average of HPLC-RI, Brodsky & Ballschmiter 1988)
0.0023 (gen. col.-GC/ECD, Dunnivant & Elzerman 1988)
0.0167 (subcooled liquid, calculated-M.P., Dunnivant & Elzerman 1988)
0.00208 (quoted, subcooled liquid, Hawker 1989b)
0.00109 (22°C, gen. col.-GC/ECD, Opperhuizen et al. 1988)
0.000407 (quoted, Isnard & Lambert 1989)
0.00109, 0.00293 (quoted, calculated- χ , Nirmalakhanden & Speece 1989)
0.00072, 0.00228 (quoted, calculated-TSA, Abramowitz & Yalkowsky 1990)
0.0027, 0.0025 (quoted average of Brodsky & Ballschmiter 1988, calculated- χ , Patil 1991)

Vapor Pressure (Pa at 25°C):
0.0016 (GC-RT, Westcott et al. 1981)
0.0122 (P_L calculated from P_S using F, Westcott et al. 1981)
0.00173 (Westcott & Bidleman 1981; quoted, Erickson 1986)
0.00443 (subcooled liquid, Burkhard 1984)
0.000357, 0.000641, 0.000804 (calculated-MW,GC-RI, χ , Burkhard et al. 1985a)
0.00443 (subcooled liq., calculated-GC-RI, Burkhard et al. 1985b)
0.00159 (selected, Shiu & Mackay 1986; quoted, Sklarew & Girvin 1987)
0.012 (selected, subcooled liq., Shiu & Mackay 1986)
0.000476 (calculated-SxHLC, Dunnivant & Elzerman 1988)
0.000354 (subcooled liquid, calculated-M.P., Dunnivant & Elzerman 1988)

Henry's Law Constant (Pa m^3/mol):
11.65 (batch stripping, Coates 1984)
157.0 (calculated-P/C, Burkhard et al. 1985b)
12.46 (calculated, Coates & Elzerman 1986)
817.9 (calculated-P/C, Shiu & Mackay 1986)
76.5 (batch stripping, Dunnivant & Elzerman 1988)
76.5 (batch stripping, Dunnivant et al. 1988)

Octanol/Water Partition Coefficient, log K_{OW}:
6.70 (shake flask-GC, Chiou et al. 1977; selected, Oliver & Niimi 1985)
6.34 (shake flask-GC, Karickhoff et al. 1979)
6.37 (HPLC-k', McDuffie 1981)
7.12 (TLC, Bruggeman et al. 1982; quoted, Erickson 1986)
6.10 (quoted, Garten & Trabalka 1983)
8.18 (calculated-f const., Yalkowsky et al. 1983)
6.88 (calculated-TSA, Burkhard 1984)
7.55 (gen. col.-GC/ECD, Miller et al. 1984,1985; quoted, Sklarew & Girvin 1987)

7.55, 6.39 (selected, HPLC-RP/MS, Burkhard et al. 1985c)
8.31 (calculated-π, Burkhard et al. 1985c)
6.01 (HPLC-k', Tomlinson & Hafkanscheid 1986)
7.00 (selected, Shiu & Mackay 1986; quoted, Sklarew & Girvin 1987; Hawker 1989b; Clark et al. 1990)
6.50 (calculated-chlorine substituents, Oliver 1987c)
7.75, 7.24, 8.31 (quoted, calculated-UNIFAC, π or f const., Banerjee & Howard 1988)
7.55, 6.54 (quoted, calculated average of HPLC-RI, Brodsky & Ballschmiter 1988)
6.41 (calculated-TSA, Hawker & Connell 1988a)
6.40 (quoted, Hawker & Connell 1988b)
7.20 (calculated-solvatochromic p., Kamlet et al. 1988)
7.20, 8.18 (quoted, calculated-π or f const., Kamlet et al. 1988)
7.55 (quoted exptl., Doucette & Andren 1988)
8.31, 8.18, 6.71, 6.73, 6.89, (calculated-π, f const., MW, χ, TSA, Doucette & Andren 1988)
6.70 (quoted, Isnard & Lambert 1988,1989)
7.12 (calculated-π, De Bruijn et al. 1989)
7.287 (slow stirring-GC, De Bruijn et al. 1989; De Bruijn & Hermens 1990)
6.70 (quoted, reported as 2,3,6,2',4',6'-hexachlorobiphenyl, Thomann 1989)
6.54, 6.64 (quoted average of Brodsky & Ballschmiter 1988, calculated- χ, Patil 1991)

Bioconcentration Factor, log BCF:
1.02 (poultry, Garten & Trabalka 1983)
>4.93 (rainbow trout, kinetic, Oliver & Niimi 1985)
3.68 (rainbow trout, steady state, Oliver & Niimi 1985)
4.53 (worms, Oliver 1987c)
3.68 (fish, selected, Isnard & Lambert 1988)
8.09 (quoted, Thomann 1989)

Sorption Partition Coefficient, log K_{OC}:
6.08 (exptl., Karickhoff et al. 1979; Karickhoff 1981)
5.95, 7.28, 5.95 (calculated-K_{OW}, C_L, C_S, Karickhoff 1981)
6.68 (suspended particulate matter, calculated-K_{OW}, Burkhard 1984)
4.905 (calculated- χ, reported as log K_h at 5 mg C/L, Sabljic et al. 1989)

Half-Lives in the Environment:
Air:
Surface water:
Groundwater:
Sediment:
Soil:
Biota: half-life in rainbow trout, >1000 days (Niimi & Oliver 1983; Oliver & Niimi 1985), and its muscle, 77 days (Niimi & Oliver 1983).

Environmental Fate Rate Constants or Half-Lives:

Volatilization:

Photolysis:

Hydrolysis:

Oxidation:

Biodegradation:

Biotransformation:

Bioconcentration, Uptake (k_1) and Elimination (k_2) Rate Constants:

k_2: >0.0007 day^{-1} (rainbow trout, Niimi & Oliver 1983; quoted, Clark et al. 1990)

k_1: 60 day^{-1} (rainbow trout, Oliver & Niimi 1985; Thomann 1989)

k_2: >0.0007 day^{-1} (rainbow trout, Oliver & Niimi 1985)

log $1/k_2$: 3.1, 3.6 hour (fish, quoted, calculated-K_{OW}, Hawker & Connell 1988b)

$1/k_2$: 141 day (guppy, Gobas et al. 1989; quoted, Clark et al. 1990)

Common Name: 3,3',4,4',5,5'-Hexachlorobiphenyl
Synonym: PCB-169
Chemical Name: 3,3',4,4',5,5'-hexachlorobiphenyl
CAS Registry No: 32774-16-6
Molecular Formula: $C_{12}H_4Cl_6$
Molecular Weight: 360.88
Melting Point (°C):

201-202.0
202 (calculated, Abramowitz & Yalkowsky 1990)

Boiling Point (°C):
Density (g/cm^3):
Molar Volume (cm^3/mol):

310.0 (LeBas method, Shiu & Mackay 1986)
1.46 (intrinsic volume: $V_I/100$, De Bruijn & Hermens 1990)

Molecular Volume (A^3):

265.20, 244.49, 294.2 (De Bruijn & Hermens 1990)

Total Surface Area, TSA (A^2):

287.18 (Burkhard 1984)
281.99 (planar, Doucette 1985)
289.64 (nonplanar, Doucette 1985)
282.23 (planar, Hawker & Connell 1988a)
281.99 (planar, Doucette 1985)
289.64 (nonplanar, Doucette 1985)
315.84, 281.93, 291.5 (De Bruijn & Hermens 1990)
282.0 (Abramowitz & Yalkowsky 1990)

Heat of Fusion, kcal/mol:
Entropy of Fusion, cal/mol K (e.u.):
Fugacity Ratio, F (assuming ΔS_{fusion} = 13.5 e.u.): 0.0178

Water Solubility (g/m^3 or mg/L at 25°C):

0.01230 (subcooled liq., calculated-TSA, Burkhard et al. 1985b)
0.000504 (calculated average of HPLC-RI, Brodsky & Ballschmiter 1988)
0.000036 (calculated-TSA, Abramowitz & Yalkowsky 1990)
0.00051, 0.0025 (quoted average of Brodsky & Ballschmiter 1988, calculated- χ , Patil 1991)

Vapor Pressure (Pa at 25°C):

4.7×10^{-5}, 1.08×10^{-6}, 2.96×10^{-5} (calculated-MW,GC-RI, χ , Burkhard et al. 1985a)
5.36×10^{-5} (subcooled liq., calculated-GC-RI, Burkhard et al. 1985b)

Henry's Law Constant (Pa m^3/mol):

1.57 (calculated-P/C, Burkhard et al. 1985b)
5.98 (calculated- χ , Sabljic & Güsten 1989)

Octanol/Water Partition Coefficient, log K_{OW}:

6.80 (calculated-TSA, Burkhard 1984)
7.55 (calculated average of HPLC-RI, Brodsky & Ballschmiter 1988)
7.42 (calculated-TSA, Hawker & Connell 1988a)
7.76 (calculated-π, De Bruijn et al. 1989)
7.408 (slow stirring-GC/ECD, De Bruijn et al. 1989; De Bruijn & Hermens 1990)
7.55, 6.64 (quoted average of Brodsky & Ballschmiter 1988, calculated- χ , Patil 1991)

Bioconcentration Factor, log BCF:

Sorption Partition Coefficient, log K_{OC}:

6.60 (suspended particulate matter, calculated-K_{OW}, Burkhard 1984)

Half-Lives in the Environment:

Air:
Surface water:
Groundwater:
Sediment:
Soil:
Biota:

Environmental Fate Rate Constants or Half-Lives:

Volatilization:
Photolysis:
Hydrolysis:
Oxidation:
Biodegradation:
Biotransformation:
Bioconcentration, Uptake (k_1) and Elimination (k_2) Rate Constants:

Common Name: 2,2',3,3',4,4',5-Heptachlorobiphenyl
Synonym: PCB-170
Chemical Name: 2,2',3,3',4,4',5-heptachlorobiphenyl
CAS Registry No: 35065-30-6
Molecular Formula: $C_{12}H_3Cl_7$
Molecular Weight: 395.3
Melting Point (°C):
  135 (Hutzinger et al. 1974)
  134.5-135.5 (NAS 1979)
  132 (calculated, Abramowitz & Yalkowsky 1990)
Boiling Point (°C):
Density (g/cm^3):
Molar Volume (cm^3/mol):
Molecular Volume (A^3):
Total Surface Area, TSA (A^2):
  296.30 (Burkhard 1984)
  277.27 (planar, Doucette 1985)
  302.03 (nonplanar, Doucette 1985)
  277.74 (planar, Hawker & Connell 1988a)
  278.0 (Abramowitz & Yalkowsky 1990)
Heat of Fusion, kcal/mol:
Entropy of Fusion, cal/mol K (e.u.):
Fugacity Ratio, F (assuming ΔS_{fusion} = 13.5 e.u.): 0.0807

Water Solubility (g/m^3 or mg/L at 25°C):
  0.00767 (subcooled liq., Burkhard et al. 1985b; quoted, Eisenreich 1987; Capel et al. 1991)
  0.00347 (20°C, subcooled liq., calculated-mole fraction of Aroclor mixtures, Murphy et al. 1987)
  0.000504 (calculated-HPLC-RI, Brodsky & Ballschmiter 1988)
  0.000395 (calculated-TSA, Abramowitz & Yalkowsky 1990)
  0.000498, 0.00052 (quoted average of Brodsky & Ballschmiter 1988, calculated- χ , Patil 1991)

Vapor Pressure (Pa at 25°C):
  0.0000837 (subcooled liquid, GC-RT, Bidleman 1984; quoted, Erickson 1986)
  0.000078, 0.0000337, 0.00000858 (calculated-MW, GC-RI, χ , Burkhard et al. 1985a)
  0.000372 (subcooled liq., calculated-GC-RI, Burkhard et al. 1985b; quoted, Eisenreich 1987; Capel et al. 1991)
  0.0000873, 0.0000811, 0.0000837 (subcooled liq., GC-RT, Foreman & Bidleman 1985)
  0.0000132 (20°C, subcooled liq., calculated-mole fraction of Aroclor mixtures, Murphy et al. 1987)

Henry's Law Constant (Pa m^3/mol):

19.25 (calculated-P/C, Burkhard 1985b; quoted, Eisenreich 1987; Capel et al. 1991)
1.52 (20°C, calculated-P/C, Murphy et al. 1987)
0.91 (wetted-wall col.-GC, Brunner et al. 1990)

Octanol/Water Partition Coefficient, log K_{OW}:

7.05 (calculated-TSA, Burkhard 1984)
7.08 (calculated-HPLC-RI, Brodsky & Ballschmiter 1988)
7.27 (calculated-TSA, Hawker & Connell 1988a; quoted, Capel et al. 1991)
7.08, 7.03 (quoted of Brodsky & Ballschmiter 1988, calculated- χ , Patil 1991)

Bioconcentration Factor, log BCF:

Sorption Partition Coefficient, log K_{OC}:

6.85 (suspended particulate matter, calculated-K_{OW}, Burkhard 1984)
5.63, 5.48, 5.42, 4.99 (humic substances, in concn. of 5, 10, 20, 40 mg C/L, reported as log K_h, Lara & Ernst 1989)
5.632, 5.675 (humic substances, quoted, calculated- χ , reported as log K_h at 5 mg C/L, Sabljic et al. 1989)

Half-Lives in the Environment:

Air:
Surface water:
Groundwater:
Sediment:
Soil:
Biota:

Environmental Fate Rate Constants or Half-Lives:

Volatilization:
Photolysis:
Hydrolysis:
Oxidation:
Biodegradation:
Biotransformation:
Bioconcentration, Uptake (k_1) and Elimination (k_2) Rate Constants:

Common Name: 2,2',3,3',4,4',6-Heptachlorobiphenyl
Synonym: PCB-171
Chemical Name: 2,2',3,3',4,4',6-heptachlorobiphenyl
CAS Registry No: 52663-71-5
Molecular Formula: $C_{12}H_3Cl_7$
Molecular Weight: 395.32
Melting Point (°C):
 122.4 (Weast 1972,1973; Opperhuizen et al. 1988)
 122, 152 (quoted, calculated, Abramowitz & Yalkowsky 1990)
Boiling Point (°C):
 417 (calculated, Mackay et al. 1982; Shiu & Mackay 1986)
Density (g/cm^3):
Molar Volume (cm^3/mol):
 330.9 (LeBas method, Shiu & Mackay 1986)
 1.55 (intrinsic volume: V_I/100, Kamlet et al. 1988; Hawker 1989b)
Molecular Volume (A^3):
 286.5 (planar, shorthand, Opperhuizen et al. 1988)
Total Surface Area, TSA (A^2):
 299.79 (Burkhard 1984)
 272.28 (planar, Doucette 1985)
 302.48 (nonplanar, Doucette 1985; Doucette & Andren 1988)
 273.15 (planar, Hawker & Connell 1988a)
 289.9 (planar, shorthand, Opperhuizen et al. 1988)
 273.0 (Abramowitz & Yalkowsky 1990)
Heat of Fusion, kcal/mol:
 4.85 (Miller et al. 1984)
Entropy of Fusion, cal/mol K (e.u.):
 12.20 (Miller et al. 1984; Shiu & Mackay 1986)
Fugacity Ratio, F (assuming ΔS_{fusion} = 13.5 e.u.):
 0.109 (Mackay et al. 1980; Shiu & Mackay 1986)

Water Solubility (g/m^3 or mg/L at 25°C):
 0.00624 (shake flask-GC/ECD, Dexter & Pavlou 1978)
 0.00217 (gen. col.-GC/ECD, Miller et al. 1984,1985)
 0.00625 (subcooled liq., calculated-TSA, Burkhard et al. 1985b)
 0.0020 (selected, Shiu & Mackay 1986; selected, Hawker 1989b)
 0.00412 (20°C, subcooled liq., calculated-mole fraction of Aroclor mixtures, Murphy et al. 1987)
 9.81×10^{-6} (calculated-UNIFAC, Banerjee & Howard 1988)
 0.000714 (calculated average of HPLC-RI, Brodsky & Ballschmiter 1988)
 0.0165 (quoted, subcooled liquid, Hawker 1989b)
 0.00219 (quoted, Isnard & Lambert 1989)
 0.000217, 0.00095 (quoted, calculated- χ , Nirmalakhanden & Speece 1989)

0.00198, 0.000395 (quoted, calculated-TSA, Abramowitz & Yalkowsky 1990)

Vapor Pressure (Pa at 25°C):

0.000187, 0.000298 (subcooled liq., GC-RT, Bidleman 1984)
0.000105, 0.000571, 0.0000196 (calculated-MW,GC-RI, χ , Burkhard et al. 1985a)
0.000469 (subcooled liq., calculated-GC-RI, Burkhard et al. 1985b)
0.000257, 0.000317, 0.000236 (subcooled liq., GC-RT, Foreman & Bidleman 1985)
0.00024 (quoted from Bidleman 1984, Erickson 1986)
0.0000273 (selected, Shiu & Mackay 1986)
0.00025 (selected, subcooled liq., Shiu & Mackay 1986)

Henry's Law Constant (Pa m^3/mol):

29.79 (calculated-P/C, Burkhard et al. 1985b)
5.40 (calculated-P/C, Shiu & Mackay 1986)

Octanol/Water Partition Coefficient, log K_{OW}:

7.14 (calculated-TSA, Burkhard 1984)
6.68 (gen. col.-GC/ECD, Miller et al. 1984,1985; quoted, Sklarew & Girvin 1987)
6.70 (selected, Shiu & Mackay 1986; quoted, Sklarew & Girvin 1987; Kamlet et al. 1988; Hawker 1989b)
6.70 (calculated-chlorine substituents, Oliver 1987c)
6.68, 7.85 (quoted, calculated-UNIFAC, Banerjee & Howard 1988)
6.68, 6.99 (quoted, calculated average of HPLC-RI, Brodsky & Ballschmiter 1988)
6.68 (quoted, Doucette & Andren 1988)
9.03, 8.94, 7.06, 7.04, 7.16 (calculated-π, f const., MW, χ , TSA, Doucette & Andren 1988)
7.11 (calculated-TSA, Hawker & Connell 1988a)
7.34 (calculated-solvatochromic p., Kamlet et al. 1988)
6.68 (quoted, Isnard & Lambert 1989)

Bioconcentration Factor, log BCF:

Sorption Partition Coefficient, log K_{OC}:

6.94 (suspended particulate matter, calculated-K_{OW}, Burkhard 1984)
5.516 (calculated- χ , reported as log K_h at 5 mg C/L, Sabljic et al. 1989)

Half-Lives in the Environment:

Air:
Surface water:
Groundwater:
Sediment:
Soil:
Biota: half-life in worms at 8°C, 260 days (Oliver 1987c).

Environmental Fate Rate Constants or Half-Lives:

- Volatilization:
- Photolysis:
- Hydrolysis:
- Oxidation:
- Biodegradation:
- Biotransformation:
- Bioconcentration, Uptake (k_1) and Elimination (k_2) Rate Constants:

Common Name: 2,2'3,4,4',5,5'-Heptachlorobiphenyl
Synonym: PCB-180
Chemical Name: 2,2',3,4,4',5,5'-heptachlorobiphenyl
CAS Registry No: 35065-29-3
Molecular Formula: $C_{12}H_3Cl_7$
Molecular Weight: 395.32
Melting Point (°C):

109-110.0
99 (calculated, Abramowitz & Yalkowsky 1990)

Boiling Point (°C): 240-280 (20mm Hg)
Density (g/cm^3):
Molar Volume (cm^3/mol):

330.9 (LeBas method, Shiu & Mackay 1986)

Molecular Volume (A^3):
Total Surface Area, TSA (A^2):

298.90 (Burkhard 1984)
279.25 (planar, Doucette 1985)
304.06 (nonplanar, Doucette 1985)
280.37 (planar, Hawker & Connell 1988a)
280.0 (Abramowitz & Yalkowsky 1990)

Heat of Fusion, kcal/mol:
Entropy of Fusion, cal/mol K (e.u.):
Fugacity Ratio, F (assuming ΔS_{fusion} = 13.5 e.u.):

0.144

Water Solubility (g/m^3 or mg/L at 25°C):

0.00656 (subcooled liq., calculated-TSA, Burkhard et al. 1985b; quoted, Eisenreich 1987; Capel et al. 1991)
0.00385 (20°C, subcooled liq., calculated-mole fraction of Aroclor mixtures, Murphy et al. 1987)
0.00031 (calculated-HPLC-RI, Brodsky & Ballschmiter 1988)
0.00031 (Brodsky et al. 1988; quoted, Ballschmiter & Wittlinger 1991)
0.00063 (calculated-TSA, Abramowitz & Yalkowsky 1990)
0.00031, 0.00053 (quoted of Brodsky & Ballschmiter 1988, calculated- χ , Patil 1991)

Vapor Pressure (Pa at 25°C):

0.000081 (Verlag Chemie 1983; quoted, Ballschmiter & Wittlinger 1991)
0.00013, 0.000129 (subcooled liquid, GC-RT, Bidleman 1984; quoted, Erickson 1986)
0.000138, 0.0000807, 0.00000858 (calculated-MW, GC-RI, χ , Burkhard et al. 1985a)
0.000506 (subcooled liq., calculated-GC-RI, Burkhard 1985b; quoted, Eisenreich 1987)
0.000121, 0.000143, 0.000129 (subcooled liq., GC-RT, Foreman & Bidleman 1985)
0.0000314 (20°C, subcooled liq., calculated-mole fraction of Aroclor mixtures, Murphy et al. 1987)

Henry's Law Constant (Pa m^3/mol):

30.40 (calculated-P/C, Burkhard et al. 1985b; quoted, Eisenreich 1987; Capel et al. 1991)
3.24 (20°C, calculated-P/C, Murphy et al. 1987)
1.013 (wetted-wall col.-GC, Brunner et al. 1990)
102 (Wittlinger et al.; quoted, Ballschmiter & Wittlinger 1991)

Octanol/Water Partition Coefficient, log K_{OW}:

7.12 (calculated-TSA, Burkhard 1984)
6.70 (calculated-chlorine substituents, Oliver 1987a)
7.21 (calculated-HPLC-RI, Brodsky & Ballschmiter 1988)
7.20 (Brodsky et al. 1988; quoted, Ballschmiter & Wittlinger 1991)
7.36 (calculated-TSA, Hawker & Connell 1988a; selected, Capel et al. 1991)
7.21, 7.02 (quoted of Brodsky & Ballschmiter 1988, calculated- χ, Patil 1991)

Bioconcentration Factor, log BCF:

Sorption Partition Coefficient, log K_{OC}:

6.92 (suspended particulate matter, calculated-K_{OW}, Burhard 1984)
6.2-7.4, 6.9 (suspended sediments, range, average, Oliver 1987a)
7.30 (algae > 50 μm, Oliver 1987a)
6.51 (Lake Michigan water column, Swackhamer & Armstrong 1987)
5.73, 5.54, 5.50, 5.09 (humic substances, in concn. of 5, 10, 20, 40 mg C/L, reported as log K_h, Lara & Ernst 1989)
5.732, 5.659 (humic substances, selected, calculated- χ , reported as log K_h at 5 mg C/L, Sabljic et al. 1989)
6.97 (calculated after Karickhoff et al. 1979, Capel & Eisenreich 1990)
5.66 (calculated after Schwarzenbach & Westall 1981, Capel & Eisenreich 1990)
5.78 (soil, Paya-Perez et al. 1991)

Half-Lives in the Environment:

Air:
Surface water:
Groundwater:
Sediment:
Soil:
Biota:

Environmental Fate Rate Constants or Half-Lives:

Volatilization:
Photolysis:
Hydrolysis:
Oxidation:

Biodegradation:
Biotransformation:
Bioconcentration, Uptake (k_1) and Elimination (k_2) Rate Constants:

k_s: 0.049 $hour^{-1}$ (uptake of mayfly-sediment model II, Gobas et al. 1989)

k_t: 0.008 $hour^{-1}$ (depuration of mayfly-sediment model II, Gobas et al. 1989)

Common Name: 2,2',3,4,5,5',6-Heptachlorobiphenyl
Synonym: PCB-185
Chemical Name: 2,2',3,4,5,5',6-heptachlorobiphenyl
CAS Registry No: 52712-05-7
Molecular Formula: $C_{12}H_3Cl_7$
Molecular Weight: 395.32
Melting Point (°C):
149 (Bruggeman et al. 1982; Opperhuizen et al. 1988)
149, 133 (quoted, calculated, Abramowitz & Yalkowsky 1990)
Boiling Point (°C):
417 (calculated, Mackay et al. 1982; Shiu & Mackay 1986)
Density (g/cm^3 at 20°C): 1.3702
Molar Volume (cm^3/mol):
330.9 (LeBas method, Shiu & Mackay 1986)
1.55 (intrinsic volume: V_I/100, Kamlet et al. 1988)
Molecular Volume (A^3):
298.8 (planar, shorthand, Opperhuizen et al. 1988)
Total Surface Area, TSA (A^2):
302.78 (Mackay et al. 1980; Shiu & Mackay 1986)
299.74 (Burkhard 1984)
272.25 (planar, Doucette 1985)
302.51 (nonplanar, Doucette 1985)
309.5 (calculated- χ , Sabljic 1987)
272.98 (planar, Hawker & Connell 1988a)
298.8 (planar, shorthand, Opperhuizen et al. 1988)
273.0 (Abramowitz & Yalkowsky 1990)
Heat of Fusion, kcal/mol:
Entropy of Fusion, cal/mol K (e.u.):
Fugacity Ratio, F (assuming ΔS_{fusion} = 13.5 e.u.):
0.0595 (Mackay et al. 1980; Shiu & Mackay 1986)

Water Solubility (g/m^3 or mg/L at 25°C):
0.00047 (gen. col.-GC/ECD, Weil et al. 1974)
0.000402 (calculated-TSA, Mackay et al. 1980)
0.000475 (TLC-RT, Bruggeman et al. 1982)
0.000475 (quoted lit. average, Yalkowsky et al. 1983; quoted Erickson 1986)
0.000625 (subcooled liq., calculated-TSA, Burkhard et al. 1985b; quoted, Eisenreich 1987)
0.00045 (selected, Shiu & Mackay 1986)
0.00546 (20°C, subcooled liq., calculated-mole fraction of Aroclor mixtures, Murphy et al. 1987)
0.000475, 0.000664 (quoted, calculated average of HPLC-RI, Brodsky & Ballschmiter 1988)

0.000464, 0.00095 (quoted, calculated- χ , Nirmalakhanden & Speece 1989)
0.000498, 0.000627 (quoted, calculated-TSA, Abramowitz & Yalkowsky 1990)
0.00070, 0.00052 (quoted average of Brodsky & Ballschmiter 1988, calculated- χ , Patil 1991)

Vapor Pressure (Pa at 25°C):
5.66×10^{-5}, 4.75×10^{-5}, 6.76×10^{-5} (calculated-MW,GC-RI, χ , Burkhard et al. 1985a)
7.28×10^{-4} (subcooled liq., calculated-GC-RI, Burkhard et al. 1985b; quoted, Eisenreich 1987)
3.21×10^{-4}, 4.78×10^{-4} (subcooled liq., GC-RT, Foreman & Bidleman 1985)

Henry's Law Constant (Pa m^3/mol):
46.0 (calculated-P/C, Burkhard et al. 1985b; quoted, Eisenreich 1987)
1.62 (wetted-wall col.-GC, Brunner et al. 1990)

Octanol/Water Partition Coefficient, log K_{OW}:
7.93 (TLC-RT, Bruggeman et al. 1982)
8.94 (calculated-f const., Yalkowsky et al. 1983)
7.14 (calculated-TSA, Burkhard 1984)
7.00 (selected, Shiu & Mackay 1986; quoted, Sklarew & Girvin 1987)
6.99 (calculated average of HPLC-RI, Brodsky & Ballschmiter 1988)
7.11 (calculated-TSA, Hawker & Connell 1988a)
6.99, 7.03 (quoted average of Brodsky & Ballschmiter 1988, calculated- χ , Patil 1991)

Bioconcentration Factor, log BCF:
4.36 (picea omorika, Reischl et al. 1989 from Reischl 1988)

Sorption Partition Coefficient, log K_{OC}:
5.95 (Koch 1983)
6.94 (suspended particulate matter, calculated-K_{OW}, Burkhard 1984)
5.516 (calculated- χ , reported as log K_h at 5 mg C/L, Sabljic et al. 1989)
5.33 (calculated, Bahnick & Doucette 1988)
6.43 (calculated after Karickhoff et al. 1979, Capel & Eisenreich 1990)
5.28 (calculated after Schwarzenbach & Westall 1981, Capel & Eisenreich 1990)

Sorption Partition Coefficient, log K_{OM}:
5.95, 5.31 (selected, calculated- χ , Sabljic 1984)

Half-Lives in the Environment:
Air:
Surface water:
Groundwater:
Sediment:
Soil:

Biota: 48 days in picea omorika (Reischl et al. 1989 from Reischl 1988).

Environmental Fate Rate Constants or Half-Lives:

Volatilization:

Photolysis:

Hydrolysis:

Oxidation:

Biodegradation:

Biotransformation:

Bioconcentration, Uptake (k_1) and Elimination (k_2) Rate Constants:

$t_{1/2}$: 48 days (picea omorika, Reischl et al. 1989 from Reischl 1988).

Common Name: 2,2',3,4',5,5'6-Heptachlorobiphenyl
Synonym: PCB-187
Chemical Name: 2,2',3,4',5,5',6-heptachlorobiphenyl
CAS Registry No: 52663-68-0
Molecular Formula: $C_{12}H_3Cl_7$
Molecular Weight: 395.32
Melting Point (°C):
149.0 (Mackay et al. 1980; Burkhard et al. 1985a)
139.0 (calculated, Abramowitz & Yalkowsky 1990)
Boiling Point (°C):
417 (calculated, Mackay et al. 1982; Shiu & Mackay 1986)
Density (g/cm^3):
Molar Volume (cm^3/mol):
330.9 (LeBas method, Shiu & Mackay 1986)
Molecular Volume (A^3):
Total Surface Area, TSA (A^2):
302.78 (calculated-TSA, Mackay et al. 1980)
301.61 (Burkhard 1984)
274.26 (planar, Doucette 1985)
304.50 (nonplanar, Doucette 1985)
274.89 (planar, Hawker & Connell 1988a)
275.0 (Abramowitz & Yalkowsky 1990)
Heat of Fusion, kcal/mol:
Entropy of Fusion, cal/mol K (e.u.):
Fugacity Ratio, F (assuming ΔS_{fusion} = 13.5 e.u.):
0.0593

Water Solubility (g/m^3 or mg/L at 25°C):
0.00047 (gen. col.-GC/ECD, Weil et al. 1974)
0.000402 (calculated-TSA, Mackay et al. 1980)
0.00561 (subcooled liq., calculated-TSA, Burkhard et al. 1985b)
0.00451 (20°C, subcooled liq., calculated-mole fraction of Aroclor mixtures, Murphy et al. 1987)
0.083 (calculated-HPLC-RI, Brodsky & Ballschmiter 1988)
0.000395 (calculated-TSA, Abramowitz & Yalkowsky 1990)

Vapor Pressure (Pa at 25°C):
3.05×10^{-4} (subcooled liquid, GC-RT, Bidleman 1984; quoted, Erickson 1986)
5.66×10^{-5}, 3.92×10^{-5}, 2.96×10^{-5} (calculated-MW, GC-RI, χ , Burkhard et al. 1985a)
5.98×10^{-4} (subcooled liq., calculated-GC-RI, Burkhard et al. 1985b)
3.47×10^{-4}, 5.74×10^{-4}, 3.05×10^{-4} (subcooled liq., GC-RT, Foreman & Bidleman 1985)
9.42×10^{-5} (20°C, subcooled liq., calculated-mole fraction of Aroclor mixtures, Murphy et al. 1987)

Henry's Law Constant (Pa m^3/mol):

42.15 (calculated-P/C, Burkhard et al. 1985b)

8.41 (20°C, calculated-P/C, Murphy et al. 1987)

Octanol/Water Partition Coefficient, log K_{OW}:

7.19 (calculated-TSA, Burkhard 1984)

6.92 (calculated-HPLC-RI, Brodsky & Ballschmiter 1988)

7.17 (calculated-TSA, Hawker & Connell 1988a)

Bioconcentration Factor, log BCF:

Sorption Partition Coefficient, log K_{OC}:

6.99 (suspended particulate matter, calculated-K_{OW}, Burkhard 1984)

5.51, 5.40, 5.33, 4.90 (humic substances, in concn. of 5, 10, 20, 40 mg C/L, reported as log K_h, Lara & Ernst 1989)

5.510, 5.501 (humic substances, selected, calculated- χ , reported as log K_h at 5 mg C/L, Sabljic et al. 1989)

Half-Lives in the Environment:

Air:

Surface water:

Groundwater:

Sediment:

Soil:

Biota:

Environmental Fate Rate Constants or Half-Lives:

Volatilization:

Photolysis:

Hydrolysis:

Oxidation:

Biodegradation:

Biotransformation:

Bioconcentration, Uptake (k_1) and Elimination (k_2) Rate Constants:

Common Name: 2,2',3,3',4,4',5,5'-Octachlorobiphenyl
Synonym: PCB-194
Chemical Name: 2,2',3,3',4,4',5,5'-octachlorobiphenyl
CAS Registry No: 35694-08-7
Molecular Formula: $C_{12}H_2Cl_8$
Molecular Weight: 429.77
Melting Point (°C):

152-123 (Tas & DeVos 1971)
159-160 (Binns & Suschitzby 1971)
156-157 (Safe & Hutzinger 1972)
158-159 (Wallnöfer et al. 1973)
159 (Mackay et al. 1980; Bruggeman et al. 1982; Burkhard 1985a; Opperhuizen et al. 1988)
159, 185 (quoted, calculated, Abraomowitz & Yalkowsky 1990)

Boiling Point (°C):

432 (calculated, Mackay et al. 1982; Shiu & Mackay 1986)

Density (g/cm^3): 1.507 (at 22°C)
Molar Volume (cm^3/mol):

351.8 (LeBas method, Shiu & Mackay 1986)
1.64 (intrinsic volume: V_I/100, Kamlet et al. 1988; Hawker 1990)

Molecular Volume (A^3):

308.6 (planar, shorthand, Opperhuizen et al. 1988)

Total Surface Area, TSA (A^2):

317.92 (Mackay et al. 1980)
311.87 (Burkhard 1984)
292.81 (planar, Doucette 1985)
317.62 (nonplanar, Doucette 1985)
293.34 (planar, Hawker & Connell 1988a)
302.9 (planar, shorthand, Opperhuizen et al. 1988)
293 (Abramowitz & Yalkowsky 1990)

Heat of Fusion, kcal/mol:
Entropy of Fusion, cal/mol K (e.u.):
Fugacity Ratio, F (assuming ΔS_{fusion} = 13.5 e.u.):

0.0474 (Mackay et al. 1980; Shiu & Mackay 1986)

Water Solubility (g/m^3 or mg/L at 25°C):

0.0072 (shake flask-GC/ECD, Wallnöfer et al. 1973; quoted, NAS 1979)
0.000272 (gen. col.-GC/ECD, Weil et al. 1974; quoted, Bruggeman et al. 1984; Lara & Ernst 1989)
0.000101 (calculated-TSA, Mackay et al. 1980)
0.000277 (TLC-RT, Bruggeman et al. 1982)
0.00140 (quoted lit. average, Yalkowsky et al. 1983; Erickson 1986)
0.00349 (subcooled liq., calculated-TSA, Burkhard et al. 1985b; quoted, Eisenreich

1987)

0.0002 (selected, Shiu & Mackay 1986)

0.00027 (quoted, Opperhuizen 1986)

0.00 (20°C, subcooled liquid, calculated-mole fraction of Aroclor mixtures, Murphy et al. 1987)

0.000277, 0.000086 (quoted, calculated-HPLC-RI, Brodsky & Ballschmiter 1988)

0.00124 (22°C, gen. col.-GC/ECD, Opperhuizen et al. 1988)

0.000124, 0.00029 (quoted, calculated- χ , Nirmalakhanden & Speece 1989)

0.000215, 0.000027 (quoted, calculated-TSA, Abromowitz & Yalkowsky 1990)

0.000086, 0.00011 (quoted of Brodsky & Ballschmiter 1988, calculated- χ , Patil 1991)

Vapor Pressure (Pa at 25°C):

$1.63x10^{-5}$, $2.03x10^{-5}$, $1.09x10^{-6}$ (calculated-MW, GC-RI, χ , Burkhard et al. 1985a)

$3.86x10^{-4}$ (subcooled liq., calculated-GC-RI, Burkhard et al. 1985b; selected, Eisenreich 1987)

$2.07x10^{-5}$, $1.79x10^{-5}$ (subcooled liq., GC-RT, Foreman & Bidleman 1985)

Henry's Law Constant (Pa m^3/mol):

47.52 (calculated-P/C, Burkhard et al. 1985b; quoted, Eisenreich 1987)

10.13 (wetted-wall col.-GC, Brunner et al. 1990)

Octanol/Water Partition Coefficient, log K_{OW}:

8.68 (TLC-RT, Bruggeman et al. 1982; quoted, Bruggeman et al. 1984; selected, Opperhuizen 1986; Erickson 1986)

9.69 (calculated-f const., Yalkowsky et al. 1983)

7.47 (calculated-TSA, Burkhard 1984)

7.10 (selected, Shiu & Mackay 1986; quoted, Gobas et al. 1987; Sklarew & Girvin 1987; Clark et al. 1990; Banerjee & Baughman 1991)

6.90 (calculated-chlorine substituents, Oliver 1987a &c)

7.62 (calculated-HPLC-RI, Brodsky & Ballschmiter 1988)

7.67 (gen. col.-GC-RT, Hawker & Connell 1988a)

7.80 (calculated-TSA, Hawker & Connell 1988a; quoted, Connell & Hawker 1988; Hawker 1990)

6.90 (calculated-chlorine substituent, Oliver 1987)

7.62, 7.41 (quoted of Brodsky & Ballschmiter 1988, calculated- χ , Patil 1991)

Bioconcentration Factor, log BCF:

5.81 (guppy, 3.5% lipid, Bruggeman et al. 1984; quoted, Gobas et al. 1987)

4.35 (guppy, Gobas et al. 1987; quoted, Banerjee & Baughman 1991)

4.18 (worms, Oliver 1987c)

4.35 (guppy, calculated-C_B/C_W, or k_1/k_2, Connell & Hawker 1988; Hawker 1990)

4.81 (guppy, estimated, Banerjee & Baughman 1991)

Sorption Partition Coefficient, log K_{OC}:

7.27 (suspended particulate matter, calculated-K_{OW}, Burkhard 1984)

6.5-7.1, 6.8 (suspended sediment, average, Oliver 1987a)

7.80 (algae > 50 μm, Oliver 1987a)

5.94, 5.72, 5.68, 5.36 (humic substances, in concn. of 5, 10, 20, 40 mg C/L, reported as log K_h, Lara & Ernst 1989)

5.943, 6.016 (humic substances, quoted, calculated- χ , reported as log K_h at 5 mg C/L, Sabljic et al. 1989)

Half-Lives in the Environment:

Air:

Surface water:

Groundwater:

Sediment:

Soil:

Biota: half-life in rainbow trout, >1000 days and its muscle, 78 days (Niimi & Oliver 1983); in guppy, 100 days (Bruggeman et al. 1984); in worms at 8°C, 220 days (Oliver 1987c).

Environmental Fate Rate Constants or Half-Lives:

Volatilization:

Photolysis:

Hydrolysis:

Oxidation:

Biodegradation:

Biotransformation:

Bioconcentration, Uptake (k_1) and Elimination (k_2) Rate Constants:

k_2: >0.0007 day^{-1} (rainbow trout, Niimi & Oliver 1983; quoted, Clark et al. 1990)

k_1: 150 day^{-1} (guppy, Bruggeman et al. 1984)

k_2: 0.007 day^{-1} (guppy, Bruggeman et al. 1984; quoted, Clark et al. 1990)

k_1: 1000 day^{-1} (guppy, Opperhuizen 1986)

log k_1: 2.18 day^{-1} (fish, quoted, Connell & Hawker 1988)

log $1/k_2$: 2.15 day (fish, quoted, Connell & Hawker 1988)

Common Name: 2,2',3,3',5,5',6,6'-Octachlorobiphenyl
Synonym: PCB-202
Chemical Name: 2,2',3,3',5,5',6,6'-octachlorobiphenyl
CAS Registry No:2136-99-4
Molecular Formula: $C_{12}H_2Cl_8$
Molecular Weight: 429.77
Melting Point (°C):
161 (Van Roosmalen 1934; Burkhard et al. 1985a; Erickson 1986)
162 (Mackay et al. 1980; Bruggeman et al. 1982; Opperhuizen et al. 1988)
162, 193 (quoted, calculated, Abramowitz & Yalkowsky 1990)
Boiling Point (°C):
432 (calculated, Mackay et al. 1982; Shiu & Mackay 1986)
Density (g/cm^3 at 20°C): 1.507
Molar Volume (cm^3/mol):
351.8 (LeBas method, Shiu & Mackay 1986)
1.640 (intrinsic volume: $V_I/100$, Kamlet et al. 1988; Hawker 1989b; De Bruijn & Hermens 1990)
Molecular Volume (A^3):
292.0 (planar, shorthand, Opperhuizen et al. 1988)
292.55, 268.70, 290.2, 303.8 (De Bruijn & Hermens 1990)
Total Surface Area, TSA (A^2):
318.73 (Mackay et al.1980)
314.47 (Burkhard 1984)
276.52 (planar, Doucette 1985)
318.47 (nonplanar, Doucette 1985; Doucette & Andren 1987,1988)
276.73 (planar, Hawker & Connell 1988a)
293.3 (planar, shorthand, Opperhuizen et al. 1988)
346.09, 295.68, 289.9, 298.5 (De Bruijn & Hermens 1990)
277.0 (Abramowitz & Yalkowsky 1990)
Heat of Fusion, kcal/mol:
5.45 (Miller et al. 1984)
Entropy of Fusion, cal/mol K (e.u.):
12.60 (Miller et al. 1984; Shiu & Mackay 1986; selected, Hinckley et al. 1990)
Fugacity Ratio, F (assuming ΔS_{fusion} = 13.5 e.u.):
0.0443 (Mackay et al. 1980; Shiu & Mackay 1986)

Water Solubility (g/m^3 or mg/L at 25°C):
0.00018 (gen. col.-GC/ECD, Weil et al. 1974; quoted, Lara & Ernst 1989)
0.000088 (calculated-TSA, Mackay et al. 1980)
0.00018 (TLC-RT, Bruggeman et al. 1982)
0.000179 (quoted lit. average, Yalkowsky et al. 1983; quoted, Erickson 1986)
0.000393 (gen. col.-GC/ECD, Miller et al. 1984,1985)
0.00306 (subcooled liq., calculated-TSA, Burkhard et al. 1985b)

0.000147 (gen. col.-GC/ECD, Dickhut et al. 1986)
0.00017 (quoted, Dickhut et al. 1986 from Burkhard et al. 1985b)
0.0003 (quoted, Shiu & Mackay 1986; Hawker 1989b)
0.00018, 0.00039 (quoted, calculated average of HPLC-RI, Brodsky & Ballschmiter 1988)
0.00697 (quoted, Hawker 1989b)
0.000389 (quoted, Isnard & Lambert 1989)
0.000179, 0.00029 (quoted, calculated- χ , Nirmalakhaden & Speece 1989)
0.00027, 0.000108 (quoted, calculated-TSA, Abramowitz & Yalkowsky 1990)

Vapor Pressure (Pa at 25°C):
$2.83x10^{-5}$, $6.43x10^{-4}$ (P_S, P_L calculated from P_S using F, Burkhard et al. 1984)
$2.89x10^{-5}$ (gas saturation, Burkhard et al. 1985a)
$5.4x10^{-4}$ (subcooled liq., calculated, GC-RI, Burkhard et al. 1985b)
$1.55x10^{-5}$, $2.69x10^{-5}$, $2.96x10^{-5}$ (calculated-MW,GC-RI, χ , Burkhard et al. 1985a)
$1.70x10^{-4}$, $3.91x10^{-4}$ (subcooled liq., GC-RT, Foreman & Bidleman 1985)
$2.66x10^{-5}$ (Shiu & Mackay 1986)
$6.0x10^{-4}$ (subcooled liquid, Shiu & Mackay 1986)
$2.89x10^{-5}$, $5.96x10^{-6}$ (quoted, calculated-UNIFAC, Banerjee et al. 1990)
$6.59x10^{-4}$, $5.26x10^{-4}$ (subcooled liq., quoted, GC-RT, Hinckley et al. 1990)

Henry's Law Constant (Pa m^3/mol):
75.79 (calculated-P/C, Burkhard et al. 1985b)
38.08 (calculated-P/C, Shiu & Mackay 1986)
1.82 (wetted-wall col.-GC, Brunner et al. 1990)

Octanol/Water Partition Coefficient, log K_{OW}:
9.77 (Hansch & Leo 1979)
8.42 (TLC-RT, Bruggeman et al. 1982; quoted, Erickson 1986)
9.69 (calculated-f const., Yalkowsky et al. 1983)
7.54 (calculated-TSA, Burkhard 1984)
7.11 (gen. col.-GC/ECD, Miller et al. 1984,1985; quoted, Sklarew & Girvin 1987)
7.14 (gen. col.-HPLC, Woodburn et al. 1984; quoted, Sklarew & Girvin 1987)
9.77 (calculated-π, Woodburn et al. 1984)
7.10 (selected, Shiu & Mackay 1986; quoted, Sklarew & Girvin 1987; Hawker 1989b)
7.12 (gen. col.-GC/ECD, Doucette & Andren 1987,1988)
9.73, 7.42 (calculated-π, TSA, Doucette & Andren 1987)
7.12, 7.16 (quoted, calculated average of HPLC-RI, Brodsky & Ballschmiter 1988)
7.21 (gen. col.-GC-RT, Hawker & Connell 1988a)
7.24 (calculated-TSA, Hawker & Connell 1988a)
7.67 (HPLC-RT, Hawker & Connell 1988)
8.12 (calculated-solvatochromic p., Kamlet et al. 1988)
7.10, 9.73 (quoted, calculated-π or f const., Kamlet et al. 1988)

9.73, 9.69, 7.39, 7.34, 7.54 (calculated-π, f const., MW, χ ,TSA, Doucette & Andren 1988)
7.729 (slow stirring-GC, DeBruijn et al. 1989)
8.42 (calculated-π, De Bruijn et al. 1989)
7.11 (quoted, Isnard & Lambert 1989)
7.729 (slow stirring-GC, De Bruijn & Hermens 1990)

Bioconcentration Factor, log BCF:

Sorption Partition Coefficient, log K_{OC}:
7.34 (suspended particulate matter, calculated-K_{OW}, Burkhard 1984)
5.61, 5.46, 5.41, 4.99 (humic substances, in concn. of 5, 10, 20, 40 mg C/L, reported as log K_h, Lara & Ernst 1989)
5.610, 5.699 (humic substances, selected, calculated- χ , reported as log K_h at 5 mg C/L, Sabljic et al. 1989)

Half-Lives in the Environment:
Air:
Surface water:
Groundwater:
Sediment:
Soil:
Biota:

Environmental Fate Rate Constants or Half-Lives:
Volatilization:
Photolysis:
Hydrolysis:
Oxidation:
Biodegradation:
Biotransformation:
Bioconcentration, Uptake (k_1) and Elimination (k_2) Rate Constants:

Common Name: 2,2',3,3',4,4',5,5',6-Nonachlorobiphenyl
Synonym: PCB-206
Chemical Name: 2,2',3,3',4,4',5,5',6-nonachlorobiphenyl
CAS Registry No: 40186-72-9
Molecular Formula: $C_{12}HCl_9$
Molecular Weight: 462.21
Melting Point (°C):
 204.5-206.5 (Hutzinger et al. 1974)
 206.0 (Bruggeman et al. 1982; Opperhuizen et al. 1988)
 206, 193 (quoted, calculated, Abramowitz & Yalkowsky 1990)
Boiling Point (°C):
 445 (calculated, Mackay et al. 1982; Shiu & Mackay 1986)
Density (g/cm^3 at 20°C): 1.507
Molar Volume (cm^3/mol):
 372.7 (LeBas method, Shiu & Mackay 1986)
 1.73 (intrinsic volume: $V_I/100$, Kamlet et al. 1988)
Molecular Volume (A^3):
 309.1 (planar, shorthand, Opperhuizen et al. 1988)
Total Surface Area, TSA (A^2):
 331.9 (Mackay et al. 1980; Shiu & Mackay 1986)
 328.46 (Burkhard 1984)
 301.37 (planar, Doucette 1985)
 331.62 (nonplanar, Doucette 1985)
 301.73 (planar, Hawker & Connell 1988a)
 306.9 (planar, shorthand, Opperhuizen et al. 1988)
 302.0 (Abramowitz & Yalkowsky 1990)
Heat of Fusion, kcal/mol:
 5.40 (Miller et al. 1984)
Entropy of Fusion, cal/mol K (e.u.):
 11.9 (Miller et al. 1984; Shiu & Mackay 1986)
Fugacity Ratio, F:
 0.016 (Mackay et al. 1980; Shiu & Mackay 1986)

Water Solubility (g/m^3 or mg/L at 25°C):
 0.000112 (gen. col.-GC/ECD, Weil et al. 1974; quoted, Lara & Ernst 1989)
 0.000012 (calculated-TSA, Mackay et al. 1980)
 0.00011 (TLC-RT, Bruggeman et al. 1982)
 0.00011 (quoted lit. average, Yalkowsky et al. 1983; quoted, Erickson 1986)
 0.000112 (Miller et al. 1984)
 0.0000315 (quoted, Dickhut 1986 from Burkhard et al. 1985b)
 0.000025 (gen. col.-GC/ECD, Dickhut et al. 1986)
 0.00011 (selected, Shiu & Mackay 1986)
 0.00 (20°C, subcooled liquid, calculated-mole fraction of Aroclor mixtures, Murphy

et al. 1987)

0.00011, 0.000031 (quoted, calculated average of HPLC-RI, Brodsky & Ballschmiter 1988)

0.000078 (22°C, gen. col.-GC/ECD, Opperhuizen et al. 1988)

0.0000788, 0.000095 (Nirmalakhanden & Speece 1989)

0.000116, 0.0000116 (quoted, calculated-TSA, Abramowitz & Yalkowsky 1990)

0.000031, 0.000021 (quoted average of Brodsky & Ballschmiter 1988, calculated- χ , Patil 1991)

Vapor Pressure (Pa at 25°C):

$2.01X10^{-6}$, $1.88x10^{-6}$, $2.09x10^{-7}$ (calculated-MW,GC-RI,χ, Burkhard et al. 1985a)

$1.034x10^{-4}$ (subcooled liq., calculated-GC-RI, Burkhard et al. 1985b)

$1.08x10^{-5}$, $1.53x10^{-5}$ (subcooled liq., GC-RT, Foreman & Bidleman 1985)

0.00 (20°C, subcooled liquid, calculated-mole fraction of Aroclor mixtures, Murphy et al. 1987)

Henry's Law Constant (Pa m^3/mol):

27.66 (calculated-P/C, Burkhard et al. 1985b)

0.00 (20°C, calculated-P/C, Murphy et al. 1987)

Octanol/Water Partition Coefficient, log K_{OW}:

9.14 (TLC-RT, Bruggeman et al. 1982; quoted, Erickson 1986)

10.44 (calculated-f const., Yalkowsky et al. 1983)

7.92 (calculated-TSA, Burkhard 1984)

7.20 (selected, Shiu & Mackay 1986; quoted, Sklarew & Girvin 1987; Clark et al. 1990)

7.94 (calculated average of HPLC-RI, Brodsky & Ballschmiter 1988)

8.09 (calculated-TSA, Hawker & Connell 1988a)

7.94, 7.80 (quoted average of Brodsky & Ballschmiter 1988, calculated- χ , Patil 1991)

Bioconcentration Factor, log BCF:

Sorption Partition Coefficient, log K_{OC}:

7.72 (suspended particulate matter, calculated-K_{OW}, Burkhard 1984)

6.15, 5.92, 5.83, 5.69 (humic substances, in concn. of 5, 10, 20, 40 mg C/L, reported as log K_h, Lara & Ernst 1989)

6.152, 6.133 (humic substances, selected, calculated- χ , reported as log K_h at 5 mg C/L, Sabljic et al. 1989)

Half-Lives in the Environment:

Air:

Surface water:

Groundwater:

Sediment:

Soil:

Biota: half-life in rainbow trout, >1000 days and its muscle, 84 days (Niimi & Oliver 1983).

Environmental Fate Rate Constants or Half-Lives:

Volatilization:

Photolysis:

Hydrolysis:

Oxidation:

Biodegradation:

Biotransformation:

Bioconcentration, Uptake (k_1) and Elimination (k_2) Rate Constants:

k_2: >0.0007 day^{-1} (rainbow trout, Niimi & Oliver 1983; quoted, Clark et al. 1990)

Common Name: 2,2',3,3',4,4',5,6,6'-Nonachlorobiphenyl
Synonym: PCB-207
Chemical Name: 2,2',3,3',4,4',5,6,6'-nonachlorobiphenyl
CAS Registry No: 52663-79-3
Molecular Formula: $C_{12}HCl_9$
Molecular Weight: 462.21
Melting Point (°C):
161 (calculated, Abramowitz & Yalkowsky 1990)
Boiling Point (°C):
Density (g/cm³):
Molar Volume (cm³/mol):
372.7 (LeBas method, Shiu & Mackay 1986)
Molecular Volume (A³):
Total Surface Area, TSA (A²):
329.15 (Burkhard 1984)
290.07 (planar, Doucette 1985)
332.03 (nonplanar, Doucette 1985)
291.48 (planar, Hawker & Connell 1988a)
291.0 (Abramowitz & Yalkowsky 1990)
Heat of Fusion, kcal/mol:
Fugacity Ratio, F (assuming ΔS_{fusion} = 13.5 e.u.):

Water Solubility (g/m³ or mg/L at 25°C):
0.00167 (subcooled liquid, calculated-TSA, Burkhard et al. 1985b)
0.000039 (calculated average of HPLC-RI, Brodsky & Ballschmiter 1988)
0.000058 (calculated-TSA, Abramowitz & Yalkowsky 1990)
0.000036, 0.000021 (quoted average of Brodsky & Ballschmiter 1988, calculated- χ, Patil 1991)

Vapor Pressure (Pa at 25°C):
1.24×10^{-4}, 1.47×10^{-4}, 4.77×10^{-7} (subcooled liq., calculated-MW,GC-RI, χ, Burkhard et al. 1985a)
1.30×10^{-4} (subcooled liq., calculated-GC-RI, Burkhard et al. 1985b)
3.17×10^{-5}, 4.99×10^{-5} (subcooled liq., GC-RT, Foreman & Bidleman 1985)

Henry's Law Constant (Pa m³/mol):
35.97 (calculated-P/C, Burkhard et al. 1985b)

Octanol/Water Partition Coefficient, log K_{OW}:
7.94 (calculated-TSA, Burkhard 1984)
7.88 (calculated average of HPLC-RI, Brodsky & Ballschmiter 1988)
7.52 (gen. col.-GC-RT, Hawker & Connell 1988a)
7.74 (calculated-TSA, Hawker & Connell 1988a)

7.88, 7.80 (quoted average of Brodsky & Ballschmiter 1988, calculated- χ , Patil 1991)

Bioconcentration Factor, log BCF:

Sorption Partition Coefficient, log K_{OC}:

7.74 (suspended particulate matter, calculated-K_{OW}, Burkhard 1984)

5.98, 5.77, 5.67, 5.44 (humic substances, in concn. of 5, 10, 20, 40 mg C/L, reported as log K_h, Lara & Ernst 1989)

5.984, 5.974 (humic substances, selected, calculated- χ , reported as log K_h at 5 mg C/L, Sabljic et al. 1989)

Half-Lives in the Environment:

Air:

Surface water:

Groundwater:

Sediment:

Soil:

Biota:

Environmental Fate Rate Constants or Half-Lives:

Volatilization:

Photolysis:

Hydrolysis:

Oxidation:

Biodegradation:

Biotransformation:

Bioconcentration, Uptake (k_1) and Elimination (k_2) Rate Constants:

Common Name: 2,2',3,3',4,5,5',6,6'-Nonachlorobiphenyl
Synonym: PCB-208
Chemical Name: 2,2',3,3',4,5,5',6,6'-nonachlorobiphenyl
CAS Registry No: 52663-77-1
Molecular Formula: $C_{12}HCl_9$
Molecular Weight: 462.21
Melting Point (°C):
 182.0 (Opperhuizen et al. 1988)
Boiling Point (°C):
 445 (calculated, Mackay et al. 1982; Shiu & Mackay 1986)
Density (g/cm³):
Molar Volume (cm³/mol):
 372.7 (LeBas method, Shiu & Mackay 1986)
 1.73 (intrinsic volume: $V_I/100$, Kamlet et al. 1988; Hawker 1989b)
Molecular Volume (A³):
 299.8 (planar, shorthand, Opperhuizen et al. 1988)
Total Surface Area, TSA (A²):
 328.33 (Burkhard 1984)
 290.07 (planar, Doucette 1985)
 332.02 (nonplanar, Doucette 1985)
 290.59 (planar, Hawker & Connell 1988a)
 301.2 (planar, shorthand, Opperhuizen et al. 1988)
 291.0 (Abramowitz & Yalkowsky 1990)
Heat of Fusion, kcal/mol:
Entropy of Fusion, cal/mol K (e.u.):
 11.82 (Shiu & Mackay 1986)
Fugacity Ratio, F (calculated, assuming ΔS_{fusion} = 13.5 e.u.):
 0.0276 (Shiu & Mackay 1986)

Water Solubility (g/m³ or mg/L at 25°C):
 0.000018 (gen. col.-GC/ECD, Miller et al. 1984,1985; quoted, Hawker 1989b)
 0.00174 (subcooled liq., calculated-TSA, Burkhard et al. 1985b)
 0.000018 (selected, Shiu & Mackay 1986)
 0.000057 (calculated average of HPLC-RI, Brodsky & Ballschmiter 1988)
 0.000422 (quoted, subcooled liquid, Hawker 1989b)
 1.74×10^{-7} (calculated-UNIFAC, Banerjee & Howard 1988)
 0.000018 (quoted, Isnard & Lambert 1989)
 0.00018, 0.000095 (quoted, calculated- χ , Nirmalakhanden & Speece 1989)
 0.000018, 0.000037 (quoted, calculated-TSA, Abramowitz & Yalkowsky 1990)

Vapor Pressure (Pa at 25°C):
 3.41×10^{-6}, 3.78×10^{-6}, 1.09×10^{-6} (calculated-MW,GC-RI, χ , Burkhard et al. 1985a)
 1.22×10^{-4} (subcooled liq., calculated-GC-RI, Burkhard et al. 1985b)

$3.08x10^{-5}$, $6.62x10^{-5}$ (subcooled liq., GC-RT, Foreman & Bidleman 1985)

Henry's Law Constant (Pa m^3/mol):

32.53 (calculated-P/C, Burkhard et al. 1985b)

Octanol/Water Partition Coefficient, log K_{OW}:

7.92 (calculated-TSA, Burkhard 1984)

8.16 (gen. col.-GC/ECD, Miller et al. 1984,1985; quoted, Sklarew & Girvin 1987; Hawker 1989b)

8.16, 9.05, 10.4 (quoted, calculated-UNIFAC, π or f const., Banerjee & Howard 1988)

8.16 (selected, Shiu & Mackay 1986; quoted, Sklarew & Girvin 1987; Kamlet et al. 1988; Hawker 1989b)

8.16, 7.77 (quoted, calculated average of HPLC-RI, Brodsky & Ballschmiter 1988)

7.71 (calculated-TSA, Hawker & Connell 1988a)

8.52 (calculated-solvatochromic p. , Kamlet et al. 1988)

10.44 (calculated-π or f const., Kamlet et al. 1988)

8.16 (quoted, Doucette & Andren 1988)

10.45, 10.44, 7.68, 7.60, 7.87 (calculated-π, f const., MW, χ , TSA, Doucette & Andren 1988)

8.16 (quoted, Isnard & Lambert 1989)

Bioconcentration Factor, log BCF:

Sorption Partition Coefficient, log K_{OC}:

7.72 (suspended particulate matter, calculated-K_{OW}, Burkhard 1984)

5.974 (calculated- χ , reported as log K_h at 5 mg C/L, Sabljic et al. 1989)

Half-Lives in the Environment:

Air:

Surface water:

Groundwater:

Sediment:

Soil:

Biota:

Environmental Fate Rate Constants or Half-Lives:

Volatilization:

Photolysis:

Hydrolysis:

Oxidation:

Biodegradation:

Biotransformation:

Bioconcentration, Uptake (k_1) and Elimination (k_2) Rate Constants:

Common Name: 2,2',3,3',4,4',5,5',6,6'-Decachlorobiphenyl
Synonym: PCB-209
Chemical Name: 2,2',3,3',4,4',5,5',6,6'-decachlorobiphenyl
CAS Registry No: 2051-24-3
Molecular Formula: $C_{12}Cl_{10}$
Molecular Weight: 498.66
Melting Point (°C):

310 (Van Roosmalen 1934)
305-306 (Hutzinger et al. 1971)
300-310 (sublimes, Wallnöfer et al. 1973)
305.0 (Mackay et al. 1980; Bruggeman et al. 1982; Shiu et al. 1987; Opperhuizen et al. 1988)
305, 256 (quoted, calculated, Abramowitz & Yalkowsky 1990)

Boiling Point (°C):

456 (calculated, Mackay et al. 1982; Shiu & Mackay 1986)

Chlorine Content: 71.18% (Hutzinger et al. 1974)
Density (g/cm^3 at 20°C): 1.507
Molar Volume (cm^3/mol):

393.6 (LeBas method, Shiu & Mackay 1986; Shiu et al. 1987)
1.820 (intrinsic volume: V_I/100, Kamlet et al. 1988; Hawker 1989b,1990; De Bruijn & Hermens 1990)

Molecular Volume (A^3):

309.4 (planar, shorthand, Opperhuizen et al. 1988)
319.91, 295.17, 309.4, 323.0 (De Bruijn & Hermens 1990)

Total Surface Area, TSA (A^2):

345.92 (Mackay et al. 1980; Shiu et al. 1987)
342.14 (Burkhard 1984)
303.63 (planar, Doucette 1985)
345.59 (nonplanar, Doucette 1985; Doucette & Andren 1987,1988)
304.45 (planar, Hawker & Connell 1988a)
309.1 (planar, shorthand, Opperhuizen et al. 1988)
375.95, 320.94, 305.7, 314.3 (De Bruijn & Hermens 1990)
304.0 (Abramowitz & Yalkowsky 1990)

Heat of Fusion, kcal/mol:

6.86 (Miller et al. 1984)

Entropy of Fusion, cal/mol K (e.u.):

11.80 (Miller et al. 1984; Shiu & Mackay 1986)
11.80 (Hinckley et al. 1990)

Fugacity Ratio, F (calculated, assuming ΔS_{fusion} = 13.5 e.u.):

0.0017 (Mackay et al. 1980; Shiu & Mackay 1986)
0.002 (Mackay et al. 1983)
0.00167 (Shiu et al. 1987)

Water Solubility (g/m^3 or mg/L at 25°C):

0.015 (shake flask-GC/ECD, Wallnöfer et al. 1973; Tulp & Hutzinger 1978; selected, NAS 1979)

1.6×10^{-5} (generator column-GC/ECD, Weil et al. 1974; quoted, Bruggeman et al. 1984; Opperhuizen 1986; Lara & Ernst 1989)

4.0×10^{-7} (calculated-TSA, Mackay et al. 1980)

1.6×10^{-5} (quoted, Bruggeman et al. 1982)

0.00002 (quoted, Mackay et al. 1983)

0.0002 (quoted, subcooled liquid, Mackay et al. 1983)

4.9×10^{-4} (quoted lit. average, Yalkowsky et al. 1983; Erickson 1986)

7.43×10^{-6} (gen. col.-GC/ECD, Miller et al. 1984,1985)

1.09×10^{-3} (subcooled liq., calculated-TSA, Burkhard et al. 1985b)

2.12×10^{-8} (calculated-UNIFAC, converted from log γ, Arbuckle 1986)

4.1×10^{-4}, 4.6×10^{-4} (exptl., calculated-UNIFAC, converted from log γ, Burkhard et al. 1986)

7.6×10^{-4} (Mackay 1986; Metcalfe et al. 1988)

6.48×10^{-6} (gen. co.-GC/ECD, Dickhut et al. 1986)

4.05×10^{-6} (Dickhut et al. 1986 from Burkhard et al. 1985b)

1.2×10^{-6} (selected, Shiu & Mackay 1986; Shiu et al. 1987)

0.0072 (selected, subcooled liquid, Shiu et al. 1987)

1.6×10^{-5}, 1.04×10^{-5} (quoted, calculated average of HPLC-RI, Brodsky & Ballschmiter 1988)

2.1×10^{-5} (22°C, gen. col.-GC/ECD, Opperhuizen et al. 1988)

0.00203 (quoted, Hawker 1989b)

7.4×10^{-6} (quoted, Isnard & Lambert 1989)

2.08×10^{-5}, 3×10^{-5} (quoted, calculated- χ , Nirmalakhanden & Speece 1989)

1.25×10^{-6}, 2.5×10^{-6} (quoted, calculated-TSA, Abramowitz & Yalkowsky 1990)

Vapor Pressure (Pa at 25°C):

1.4×10^{-5} (calculated-volatilization rate, Dobbs & Cull 1982)

1.39×10^{-6} (subcooled liquid, GC-RT, Bidleman 1984)

0.00004 (quoted, subcooled liquid, Mackay et al. 1983)

5.17×10^{-8}, 3.04×10^{-5} (P_S, P_L calculated from P_S using F, Burkhard et al. 1984)

5.3×10^{-8} (extrapolated, gas saturation, Burkhard et al. 1985a)

7.59×10^{-8},5.36×10^{-8},4.01×10^{-8} (calculated-MW,GC-RI, χ , Burkhard et al. 1985a)

2.75×10^{-5} (subcooled liq., calculated-GC-RI, Burkhard et al. 1985b)

5.58×10^{-6},1.32×10^{-5} (subcooled liq., GC-RT, Foreman & Bidleman 1985)

5.0×10^{-8} (selected, Shiu & Mackay 1986; Shiu et al. 1987)

3.0×10^{-5} (selected, subcooled liquid, Shiu & Mackay 1986; Shiu et al. 1987)

1.4×10^{-6} (selected, Mackay 1986; Metcalfe et al. 1988)

5.29×10^{-8}, 2.2×10^{-8} (quoted, calculated-UNIFAC, Banerjee et al. 1990)

5.14×10^{-6}, 1.44×10^{-6} (subcooled liq., quoted, GC-RT, Hinckley et al. 1990)

Henry's Law Constant (Pa m^3/mol):

- 100 (estimated, Mackay et al. 1983)
- 12.46 (calculated-P/C, Burkhard et al. 1985b)
- 20.84 (calculated-P/C, Shiu & Mackay 1986; Shiu et al. 1987)

Octanol/Water Partition Coefficient, log K_{OW}:

- 11.19 (Hansch & Leo 1979)
- 5.28 (NAS 1979)
- 9.60 (TLC-RT, Bruggeman et al. 1982; 1984; quoted, Opperhuizen 1986; Erickson 1986)
- 9.60 (quoted, Mackay et al. 1983)
- 11.20 (calculated-f const., Yalkowsky et al. 1983)
- 8.29 (calculated-TSA, Burkhard 1984)
- 8.26 (gen. col.-GC/ECD, Miller et al. 1984,1985; quoted, Sklarew & Girvin 1987; Hawker 1989b)
- 8.20 (gen. col.-HPLC, Woodburn et al. 1984; quoted, Sklarew & Girvin 1987)
- 11.19 (calculated-π, Woodburn et al. 1984)
- 8.30 (Mackay 1986; Metcalfe et al. 1988)
- 8.26 (selected, Shiu & Mackay 1986; Shiu et al. 1987; quoted, Geyer et al. 1987; Sklarew & Girvin 1987; Kamlet et al. 1988; Clark et al. 1990; Banerjee & Baughman 1991)
- 8.20 (slow stirring, Brooke et al. 1986)
- 8.60 (estimated-solubility, Brooke et al. 1986)
- 8.23 (quoted, Burkhard & Kuehl 1986)
- 8.20 (gen. col.-GC/ECD, Doucette & Andren 1987,1988; quoted, Geyer et al. 1987)
- 8.23, 8.36 (quoted, calculated average of HPLC-RI, Brodsky & Ballschmiter 1988)
- 11.16, 8.06 (calculated-π, TSA, Doucette & Andren 1988)
- 8.18 (calculated-TSA, Hawker & Connell 1988a; Connell & Hawker 1988)
- 8.92 (calculated-solvatochromic p., Kamlet et al. 1988)
- 11.20 (calculated-π or f const., Kamlet et al. 1988)
- 8.26, 9.45, 11.2 (selected, calculated-UNIFAC, π or f const., Banerjee & Howard 1988)
- 11.16, 11.2, 7.73, 7.95, 7.85, 8.20 (calculated-π, f const., HPLC-RT, MW, χ , TSA, Doucette & Andren 1988)
- 8.26 (quoted, Isnard & Lambert 1989; Hawker 1989b)
- 9.60 (calculated-π, De Bruijn et al. 1989)
- 8.274 (slow stirring-GC/ECD, De Bruijn et al. 1989; De Bruijn & Hermens 1990)
- 8.18 (quoted, Hawker 1990)

Bioconcentration Factor, log BCF:

- 5.48 (guppy, 3.5% extractable lipid, Bruggeman et al. 1984; quoted, Gobas et al. 1987)
- 7.0 (fish, quoted, Mackay 1986; Metcalfe et al. 1988)

1.48, 1.41 (human fat of lipid basis, calculated-K_{OW}, Geyer et al. 1987)
1.38, 1.32 (human fat of wet wt. basis, calculated-K_{OW}, Geyer et al. 1987)
4.02 (guppy, Gobas et al. 1987; quoted, Banerjee & Baughman 1991)
4.02 (guppy, calculated-C_B/C_W or k_1/k_2, Connell & Hawker 1988; Hawker 1990)
5.07 (guppy, estimated, Banerjee & Baughman 1991)

Sorption Partition Coefficient, log K_{OC}:
8.09 (suspended particulate matter, calculated-K_{OW}, Burkhard 1984)
6.19, 5.99, 5.83, 5.61 (humic substances, in concn. of 5, 10, 20, 40 mg C/L, reported as log K_h, Lara & Ernst 1989)
6.186, 6.169 (humic substances, selected, calculated- χ , reported as log K_h at 5 mg C/L, Sabljic et al. 1989)

Half-Lives in the Environment:
Air:
Surface water:
Groundwater:
Sediment:
Soil:
Biota: half-life in rainbow trout, >1000 days and its muscle, 122 days (Niimi & Oliver 1983); in guppy, 175 days (Bruggeman et al. 1984).

Environmental Fate Rate Constants or Half-Lives:
Volatilization/Evaporation: 8.5×10^{-7} g/m^2h (Mackay 1986; Metcalfe et al. 1988).
Photolysis:
Hydrolysis:
Oxidation:
Biodegradation:
Biotransformation:
Bioconcentration, Uptake (k_1) and Elimination (k_2) Rate Constants:
k_2: >0.0007 day^{-1} (rainbow trout, Niimi & Oliver 1983; quoted, Clark et al. 1990)
k_1: 40 day^{-1} (guppy, Bruggeman et al. 1984)
k_2: 0.004 day^{-1} (guppy, Bruggeman et al. 1984; quoted, Clark et al. 1990)
k_1: 600 day^{-1} (guppy, Opperhuizen 1986)
log k_1: 1.60 day^{-1} (fish, quoted, Connell & Hawker 1988)
log $1/k_2$: 2.39 day (fish, quoted, Connell & Hawker 1988)

4.1.2 Isomer Groups

Common Name: Monochlorobiphenyl
Synonym: Dowtherm G
Chemical Name: monochlorobiphenyl
CAS Registry No: 27323-18-8
Molecular Formula: $C_{12}H_9CL$
No. of Isomers: 3
Molecular Weight: 189
Melting Point (°C):
25-77.9 (Shiu et al. 1987)
Boiling Point (°C): 285
Chlorine Content: 18.79% (Hutzinger et al. 1974)
Density (g/cm^3): 1.1
Molar Volume (cm^3/mol):
205.5 (LeBas method, Shiu & Mackay 1986; Shiu et al. 1987)
Molecular Volume (A^3):
Total Surface Area, TSA (A^2):
208.4-210 (Shiu et al. 1987)
Heat of Fusion, kcal/mol:
Fugacity Ratio, F (calculated, assuming ΔS_{fusion} = 13.5 e.u.):
0.3-1.0 (25°C, Mackay et al. 1983)
0.301-1.0 (Shiu et al. 1987)

Water Solubility (g/m^3 or mg/L at 25°C):
0.06-1.5 (selected, Mackay et al. 1983)
7.20 (selected, subcooled liq., Mackay et al. 1983a)
0.795-6.17, 4.08 (exptl. range, calculated-UNIFAC, converted from log γ, Burkhard et al. 1986)
1.2-5.5 (selected, Shiu et al. 1987)
2.5-6.73 (selected, subcooled liquid, Shiu et al. 1987)
1.2-9.5 (selected, Formica et al. 1988)
4.0 (selected, Metcalfe et al. 1988)

Vapour Pressure (Pa at 25°C):
1.1-5.6 (selected, Mackay et al. 1983a)
2.30 (subcooled liquid, Mackay et al. 1983a)
1.32 (selected, subcooled liq., Bopp 1983)
1.10 (average, liquid, Mackay 1986; Metcalfe et al. 1988)
0.271-2.04 (selected, solid, Shiu et al. 1987)
0.9-2.5 (selected, subcooled liquid, Shiu et al. 1987)

Henry's Law Constant (Pa m^3/mol):
79.3 (calculated-P/C, Bopp 1983)
58-74 (calculated, Mackay et al. 1983a)

60.0 (selected, Mackay et al. 1983a & b)
42.56-75.55 (calculated, Shiu et al. 1987)

Octanol/Water Partition Coefficient, log K_{OW}:
4.66 (selected, Mackay et al. 1983b)
4.70 (selected, Mackay 1986; Metcalfe et al. 1988)
4.3-4.6 (selected, Shiu et al. 1987)
4.73 (calculated-no. Cl atoms, Formica et al. 1988)

Bioconcentration Factor, log BCF:
3.40 (fish, selected, Mackay 1986; Metcalfe et al. 1988)

Sorption Partition Coefficient, log K_{OC}:

Half-Lives in the Environment:
Air: atmospheric photodegradation, 0.62-1.4 days (Dilling et al. 1983).
Surface water: 1.4-4.9 days in Lake Michigan (Neely 1983); 2-3 days for river water (Bailey et al. 1983).
Groundwater:
Sediment:
Soil:
Biota:

Environmental Fate Rate Constants or Half-Lives:
Volatilization/Evaporation: 0.25 g/m^2h (Mackay 1986; Metcalfe et al. 1988).
Photolysis: half-life of photodegradation in the atmosphere, 0.62-1.4 days (Dilling et al. 1983).
Hydrolysis:
Oxidation: room temp. rate constant for reaction with hydroxy radicals in the gas phase calculated to be 3.1 to 4.7×10^{-12} cm^3 $molecule^{-1}$ sec^{-1} with a calculated tropospheric lifetime of 5-11 days (Atkinson 1987).
Biodegradation: rate of degradation using species of Alcaligenes and Acinetobacter, 7×10^{-8} nmol $cell^{-1}$ $hour^{-1}$ (Furukawa et al. 1978; selected, NAS 1979); time for 50% biodegradation of an initial concentration of 1-100 $\mu g/L$ by river dieaway test is about 2-5 days (Bailey et al. 1983).
Biotransformation:
Bioconcentration, Uptake (k_1) and Elimination (k_2) Rate Constants:

Common Name: Dichlorobiphenyl
Synonym:
Chemical Name: dichlorobiphenyl
CAS Registry No: 25512-42-9
Molecular Formula: $C_{12}H_8Cl_2$
No. of Isomers: 12
Molecular Weight: 223.1
Melting Point (°C):

25-77.8
24.4-149 (Shiu & Mackay 1986; Shiu et al. 1987; quoted, Metcalfe et al. 1988)

Boiling Point (°C):

312 (Shiu & Mackay 1986)

Chlorine Content: 31.77%
Density (g/cm^3 at 20°C): 1.30
Molar Volume (cm^3/mol):

226.4 (LeBas method, Shiu & Mackay 1986; Shiu et al. 1987)

Molecular Volume (A^3):
Total Surface Area, TSA (A^2):

224.2-227.6 (Mackay et al. 1980; Yalkowsky et al. 1983; Shiu et al. 1987)

Heat of Fusion, kcal/mol:
Fugacity Ratio, F (calculated, assuming ΔS_{fusion} = 13.5 e.u.):

0.06-1.5 (Mackay et al. 1983)
0.059-1.0 (Shiu et al. 1987)

Water Solubility (g/m^3 or mg/L at 25°C):

0.22 (quoted, Bopp 1983)
0.06-1.5 (quoted, Mackay et al. 1983a)
2.20 (quoted, subcooled liq., Mackay et al. 1983a)
0.898-1.96, 1.60 (exptl. range, calculated-UNIFGAC, converted from log γ, Burkhard & Kuehl 1986)
1.6 (Mackay 1986; quoted, Metcalfe et al. 1988)
0.06-2.0 (selected, Shiu et al. 1987)
1.02-2.26 (selected, subcooled liquid, Shiu et al. 1987)

Vapor Pressure (Pa at 25°C):

0.223 (quoted, subcooled liquid, Bopp 1983)
0.60 (quoted, subcooled liq., Mackay et al. 1983a)
0.03-0.36 (Mackay et al. 1983)
0.24 (average, liquid, Mackay 1986; Metcalfe et al. 1988)
0.0018-0.279 (selected, Shiu et al. 1987)
0.008-0.60 (selected, subcooled liquid, Shiu et al. 1987)

Henry's Law Constant (Pa m^3/mol):

153.6 (calculated-P/C, Bopp 1983)
60.0 (Mackay et al. 1983a & b)
97.0 (calculated-P/C, Mackay et al. 1983a)
17-92.2 (calculated, Shiu et al. 1987)

Octanol/Water Partition Coefficient, log K_{OW}:

5.19 (Mackay et al. 1983a & b)
5.10 (Mackay 1986; Metcalfe et al. 1988)
4.9-5.3 (selected, Shiu et al. 1987)
5.13 (calculated-chlorine atoms, Formica et al. 1988)

Bioconcentration Factor, log BCF:

3.89 (biota, Mackay et al. 1983b)
3.80 (fish, selected, Mackay 1986; Metcalfe et al. 1988)
4.10 (calculated- χ , Koch 1983)

Sorption Partition Coefficient, log K_{OC}:

Half-Lives in the Environment:

Air:
Surface water: 2-3 days (Bailey et al. 1983).
Groundwater:
Sediment:
Soil:
Biota:

Environmental Fate Rate Constants or Half Lives:

Volatilization/Evaporation: 0.065 g/m^2h (selected, Mackay 1986; Metcalfe et al. 1988).
Photolysis:
Hydrolysis:
Oxidation: room temp. rate constant for reaction with hydroxy radicals in the gas phase calculated to be 1.4 to 2.9×10^{-12} cm^3 $molecule^{-1}$ sec^{-1} with a calculated tropospheric lifetime of 8-17 days (Atkinson 1987).
Biodegradation: rate of degradation using species of Alcaligenes and Acinetobacter, 6×10^{-8} nmol $cell^{-1}$ $hour^{-1}$ (Furukawa et al. 1978; quoted, NAS 1979); half-life of degradation estimated to be 2-3 days using river water dieaway test (Bailey et al. 1983).
Biotransformation:
Bioconcentration, Uptake (k_1) and Elimination (k_2) Rate Constants:

Common Name: Trichlorobiphenyl
Synonym:
Chemical Name:trichlorobiphenyl
CAS Registry No: 25323-68-6
Molecular Formula: $C_{12}H_7Cl_3$
No. of Isomers: 24
Molecular Weight: 257.5
Melting Point (°C):
44-87 (Shiu et al. 1987)
28-87 (Shiu & Mackay 1986; Metcalfe et al. 1988)
Boiling Point (°C):
337 (average, Shiu & Mackay 1986; Metcalfe et al. 1988)
Chlorine Content: 41.4%
Density (g/cm^3):
Molar Volume (cm^3/mol):
247.3 (LeBas method, Shiu & Mackay 1986; Shiu et al. 1987)
Molecular Volume (A^3):
Total Surface Area, TSA (A^2):
241.6-243.6 (Shiu et al. 1987)
Heat of Fusion, kcal/mol:
Fugacity Ratio, F (calculated, assuming ΔS_{fusion} = 13.5 e.u.):
0.24-0.65 (Mackay et al. 1983)
0.244-0.651 (Shiu et al. 1987)

Water Solubility (g/m^3 or mg/L at 25°C):
0.05 (Neely 1980)
0.15-0.64 (Mackay et al. 1983)
0.67 (subcooled liq., Mackay et al. 1983)
0.0654-1.09 (exptl. range, calculated-UNIFAC, converted from log γ, Burkhard & Kuehl 1986)
0.65 (Mackay 1986; Metcalfe et al. 1988)
0.015-0.40 (selected, Shiu et al. 1987)
0.015-0.4 (Formica et al. 1988)

Vapor Pressure (Pa at 25°C):
0.200 (Neely 1980)
0.0375 (subcooled liq., Bopp 1983)
0.01-0.27 (Mackay et al. 1983)
0.20 (subcooled liq., Mackay et al. 1983)
0.054 (average, liquid, Mackay 1986; Metcalfe et al. 1988)
0.0136-0.143 (selected, Shiu et al. 1987)
0.003-0.022 (selected, subcooled liquid, Shiu et al. 1987)

Henry's Law Constant (Pa m^3/mol):

81.76 (calculated-P/C, Bopp 1983)
82-102 (calculated-P/C, Mackay et al. 1983)
77 (selected, Mackay et al. 1983)
24.3-92.2 (calculated-P/C, Shiu et al. 1987)

Octanol/Water Partition Coefficient, log K_{OW}:

5.76 (Mackay et al. 1983)
5.50 (Mackay 1986; Metcalfe et al. 1988)
5.5-5.9 (selected, Shiu et al. 1987)
5.53 (calculated-chlorine atoms, Formica et al. 1988)

Bioconcentration Factor, log BCF:

4.20 (fish, Mackay 1986; Metcalfe et al. 1988)
4.70 (calculated- χ , Koch 1983)

Sorption Partition Coefficient, log K_{OC}:

Sorption Partition Coefficient, log K_P:

3.34 (lake sediment, calculated, Formica et al. 1988)
3.50 (calculated- χ , Koch 1983)

Half-Lives in the Environment:

Air:
Surface water:
Groundwater:
Sediment:
Soil:
Biota: estimated half-life from fish in simulated ecosystem to be 134 hours (Neely 1980).

Environmental Fate Rate Constants or Half Lives:

Volatilization/Evaporation: 0.017 g/m^{2}h (Mackay 1986; Metcalfe et al. 1988).
Photolysis:
Hydrolysis:
Oxidation: room temp. rate constant for reaction with hydroxy radicals in the gas phase calculated to be 0.7 t0 1.6×10^{-12} cm^3 molecule^{-1} sec^{-1} with a calculated tropospheric lifetime of 14-30 days (Atkinson 1987).
Biodegradation: rate of degradation using species of Alcaligenes and Acinetobacter, 5×10^{-8} nmol cell^{-1} hour^{-1} (Furukawa et al. 1978; quoted, NAS 1979).
Biotransformation:
Bioconcentration, Uptake (k_1) and Elimination (k_2) Rate Constants:

Common Name: Tetrachlorobiphenyl
Synonym:
Chemical Name: tetrachlorobiphenyl
CAS Registry No: 26914-33-0
Molecular Formula: $C_{12}H_6Cl_4$
No. of Isomers: 42
Molecular Weight: 292
Melting Point (°C):
 47-180 (Shiu & Mackay 1986; Shiu et al. 1987; Metcalfe et al. 1988)
Boiling Point (°C):
 360 (average, Shiu & Mackay 1986; Metcalfe et al. 1988)
Chlorine Content: 48.6%
Density (g/cm^3 at 20°C): 1.5
Molar Volume (cm^3/mol):
 268.2 (LeBas method, Shiu & Mackay 1986; Shiu et al. 1987)
Molecular Volume (A^3):
Total Surface Area, TSA (A^2):
 255.6-259.6
Heat of Fusion, kcal/mol:
Fugacity Ratio, F (calculated, assuming ΔS_{fusion} = 13.5 e.u.):
 0.1-0.68 (Mackay et al. 1983)
 0.029-0.606 (Shiu et al. 1987)

Water Solubility (g/m^3 or mg/L at 25°C):
 0.05 (Neely 1980)
 0.0008-0.17 (Mackay et al. 1983)
 0.017 (McCall et al. 1983)
 0.02-0.0955, 0.224 (exptl. range, calculated-UNIFAC, converted from log γ, Burkhard & Kuehl 1986)
 0.26 (Mackay 1986; Metcalfe et al. 1988)
 0.0043-0.10 (selected, Shiu et al. 1987)
 0.039-0.38 (selected, subcooled liquid, Shiu et al. 1987)
 0.001-0.1 (Formica et al. 1988)

Vapor Pressure (Pa at 25°C):
 0.0653 (Neely 1980)
 0.0064 (subcooled liq., Bopp 1983)
 0.003-0.104 (Mackay et al. 1983)
 0.06 (subcooled liq., Mackay et al. 1983)
 0.0653 (McCall et al. 1983)
 0.012 (Mackay 1986; Metcalfe et al. 1988)
 5.9×10^{-5}-0.0054 (selected, Shiu et al. 1987)
 0.002 (selected, subcooled liquid, Shiu et al. 1987)

Henry's Law Constant (Pa m^3/mol):

34.69 (calculated-P/C, Bopp 1983)
75-94 (calculated-P/C, Mackay et al. 1983)
76.0 (Mackay et al. 1983)
1.72-47.59 (calculated-P/C, Shiu et al. 1987)

Octanol/Water Partition Coefficient, log K_{OW}:

6.35 (suggested, Mackay et al. 1983)
5.90 (Mackay 1986, Metcalfe et al. 1988)
5.6-6.5 (selected, Shiu et al. 1987)
5.93 (calculated-chlorine atoms, Formica et al. 1988)

Bioconcentration Factor, log BCF:

3.98 (pinfish, Branson et al. 1975; quoted, Waid 1986)
3.95 (trout, Branson et al. 1975; quoted, NAS 1979)
4.79 (McCall et al. 1983)
4.60 (fish, Mackay 1986; Metcalfe et al. 1988)
5.30 (calculated- χ , Koch 1983)

Sorption Partition Coefficient, log K_{OC}:

4.51 (correlated, McCall et al. 1983)
3.43-5.11 (correlated of literature values in high clay soils, Sklarew & Girvin 1987)

Half-Lives in the Environment:

Air:
Surface water:
Groundwater:
Sediment:
Soil: volatilization half-life from an Ottawa sand estimated to be 10 days (Haque et al. 1974; selected, Pal et al. 1980).
Biota: estimated half-life from fish in simulated ecosystem to be 139 hours (Neely 1980).

Environmental Fate Rate Constants or Half-Lives:

Volatilization/Evaporation: 4.2×10^{-3} g/m^2h (Mackay 1986; Metcalfe et al. 1988).
Photolysis:
Hydrolysis:
Oxidation: room temp. rate constant for reaction with hydroxy radicals in the gas phase calculated to be 0.4 to 0.9×10^{-12} cm^3 $molecule^{-1}$ sec^{-1} with a calculated tropospheric lifetime of 25-60 days (Atkinson 1987).
Biodegradation: rate of degradation for both rings substituted with chlorine using species of Alcaligenes and Acinetobacter, 2.5×10^{-8} nmol $cell^{-1}$ $hour^{-1}$ (Furukawa et al. 1978; selected, NAS 1979); no degradation by river dieaway test after 98 days of incubation (Bailey et al. 1983).

Biotransformation:

Bioconcentration, Uptake (k_1) and Elimination (k_2) Rate Constants:

Common Name: Pentachlorobiphenyl
Synonym:
Chemical Name: pentachlorobiphenyl
CAS Registry No: 25429-29-2
Molecular Formula: $C_{12}H_5Cl_5$
No. of Isomers: 46
Molecular Weight: 326
Melting Point (°C):
-23.5-124 (Shiu et al. 1987)
76.5-124 (Shiu & Mackay 1986; Metcalfe et al. 1988)
Boiling Point (°C):
381 (average, Shiu & Mackay 1986; Metcalfe et al. 1988)
Chlorine Content: 54.4%
Density (g/cm^3 at 20°C): 1.50
Molar Volume (cm^3/mol):
289.1 (LeBas method, Shiu & Mackay 1986)
Molecular Volume (A^3):
Total Surface Area, TSA (A^2):
269.2-275.2 (Shiu et al. 1987)
Heat of Fusion, kcal/mol:
Fugacity Ratio, F (calculated, assuming ΔS_{fusion} = 13.5 e.u.):
0.1-0.31 (Mackay et al. 1983)
0.105-0.311 (Shiu et al. 1987)

Water Solubility (g/m^3 or mg/L at 25°C):
0.01 (Neely 1980)
0.004-0.03 (Mackay et al. 1983)
0.072 (subcooled liq., Mackay et al. 1983)
0.0338-0.0525, 0.0829 (experimental range, calculated-UNIFAC, converted from log γ, Burkhard & Kuehl 1986)
0.099 (Mackay 1986; Metcalfe et al. 1988)
0.004-0.02 (selected, Shiu et al. 1987)
0.03-0.11 (selected, subcooled liquid, Shiu et al. 1987)
0.004-0.02 (Formica et al. 1988)
0.024 (Mackay 1989)

Vapor Pressure (Pa at 25°C):
0.01026 (Neely 1980)
0.00111 (subcooled liq., Bopp 1983)
0.015 (subcooled liq., Mackay et al. 1983)
0.004-0.03 (Mackay et al. 1983)
0.0026 (Mackay 1986; Metcalfe et al. 1988)
0.000304-0.0093 (selected, Shiu et al. 1987)

0.0023-0.051 (selected, subcooled liquid, Shiu et al. 1987)

Henry's Law Constant (Pa m^3/mol):
68.0 (suggested, Mackay et al. 1983)
17.34 (calculated-P/C, Bopp 1983)
24.8-151.4 (selected, Shiu et al. 1987)
12.2 (calculated, Mackay 1989)

Octanol/Water Partition Coefficient, log K_{OW}:
6.85 (suggested, Mackay et al. 1983)
6.33 (calculated-chlorine atoms, Formica et al. 1988)
6.30 (Mackay 1986; Metcalfe et al. 1988)
6.2-6.5 (selected, Shiu et al. 1987)
6.60 (Mackay 1989)

Bioconcentration Factor, log BCF:
5.0 (fish, Mackay 1986; Metcalfe et al. 1988)
5.30 (Mackay 1989)
5.90 (calculated- χ , Koch 1983)

Sorption Partition Coefficient, log K_{OC}:
6.21 (calculated, $0.41K_{OW}$, Mackay 1989)

Sorption Partition Coefficient, log K_P:
4.15 (lake sediment, calculated-K_{OW}, f_{OC}, Formica et al. 1988)
4.51 (calculated- χ , Koch 1983)

Half-Lives in the Environment:
Air: half-life for atmospheric photodegradation, 0.62-1.4 days (Dilling et al. 1983).
Surface water:
Groundwater:
Sediment:
Soil: volatilization half-life from an Ottawa sand estimated to be 25 days (Haque et al. 1974; selected, Pal et al. 1980).
Biota: estimated half-life from fish in simulated ecosystem to be 226 hours (Neely 1980).

Environmental Fate Rate Constants and Half Lives:
Volatilization /Evaporation: 1.0×10^{-3} g/m^2h (Mackay 1986; Metcalfe et al. 1988).
Photolysis:
Hydrolysis:
Oxidation: room temp. rate constant for reaction with hydroxy radicals in the gas phase calculated to be 0.2 to 0.4×10^{-12} cm^3 $molecule^{-1}$ sec^{-1} with a calculated tropospheric lifetime of 60-120 days (Atkinson 1987).

Biodegradation:
Biotransformation:
Bioconcentration, Uptake (k_1) and Elimination (k_2) Rate Constants:

Common Name: Hexachlorobiphenyl
Synonym:
Chemical Name: hexachlorobiphenyl
CAS Registry No: 26601-64-9
Molecular Formula: $C_{12}H_4Cl_6$
No. of Isomers: 42
Molecular Weight: 361
Melting Point (°C):
85-160 (Shiu et al. 1987)
77-150 (Shiu & Mackay 1986; Metcalfe et al. 1988)
Boiling Point (°C):
400 (average, Shiu & Mackay 1986; Metcalfe et al. 1988)
Chlorine Content: 62.77% (Hutzinger et al. 1974)
Density (g/cm^3 at 20°C): 1.60
Molar Volume (cm^3/mol):
310.0 (LeBas method, Shiu & Mackay 1986; Shiu et al. 1987)
Molecular Volume (A^3):
Total Surface Area, TSA (A^2):
287.2-291.5 (Shiu & Mackay 1986)
Heat of Fusion, kcal/mol:
Fugacity Ratio, F (calculated, assuming ΔS_{fusion} = 13.5 e.u.):
0.06-0.25 (Mackay et al. 1983)
0.0582-0.256 (Shiu & Mackay 1986)

Water Solubility (g/m^3 or mg/L at 25°C):
0.0004-0.01 (Mackay et al. 1983)
0.021 (subcooled liq., Mackay et al. 1983)
0.00303-0.0504, 0.00297 (experimental range, calculated-UNIFAC, converted from log γ, Burkhard & Kuehl 1986)
0.038 (Mackay 1986; Metcalfe et al. 1988)
0.0004-0.001 (selected, Shiu et al. 1987)
0.0022-0.01 (selected, subcooled liquid, Shiu et al. 1987)
0.0004-0.005 (Formica et al. 1988)
0.0035 (Mackay & Paterson 1991)

Vapor Pressure (Pa at 25°C):
0.000182 (subcooled liq., Bopp 1983)
0.0016 (Mackay et al. 1983)
0.005 (subcooled liq., Mackay et al. 1983)
5.8×10^{-4} (Mackay 1986; Metcalfe et al. 1988)
2.0×10^{-5}-1.59×10^{-3} (selected, Shiu et al. 1987)
7.0×10^{-4}-0.012 (selected, subcooled liquid, Shiu et al. 1987)
0.0005 (Mackay & Paterson 1991)

Henry's Law Constant (Pa m^3/mol):

86.0 (Mackay et al. 1983)

6.70 (calculated-P/C, Bopp 1983)

11.9-81.8 (calculated-P/C, Shiu et al. 1987)

Octanol/Water Partition Coefficient, log K_{OW}:

6.7-7.3 (selected, Shiu et al. 1987)

6.70 (Mackay 1986; Metcalfe et al. 1988)

6.80 (Mackay & Paterson 1991)

Bioconcentration Factor, log BCF:

6.50 (calculated- χ , Koch 1983)

5.39 (fish, selected, Mackay 1986; Metcalfe et al. 1988)

4.57 (green alga, Mailhot 1987)

Sorption Partition Coefficient, log K_{OC}:

4.785-6.869 (correlated literature values in high clay soils, Sklarew & Girvin 1987)

Half-Lives in the Environment:

Air:

Surface water:

Groundwater:

Sediment:

Soil: volatilization half-life from an Ottawa sand estimated to be 40 days (Haque et al. 1974; selected, Pal et al. 1980).

Biota:

Environmental Fate Rate Constants or Half-Lives:

Volatilization/Evaporation: 2.5×10^{-4} g/m^2h (Mackay 1986; Metcalf et al. 1988).

Photolysis:

Hydrolysis:

Oxidation:

Biodegradation: degradation rate constants estimated to be 1.5×10^{-5} $hour^{-1}$ in water, soil and sediment (Mackay & Patterson 1991).

Biotransformation:

Bioconcentration, Uptake (k_1) and Elimination (k_2) Rate Constants:

Common Name: Heptachlorobiphenyl
Synonym:
Chemical Name: heptachlorobiphenyl
CAS Registry No: 28655-71-2
Molecular Formula: $C_{12}H_3Cl_7$
No. of Isomers: 24
Molecular Weight: 395.3
Melting Point (°C):
122.4-149 (Shiu & Mackay 1986; Shiu et al. 1987; Metcalfe et al. 1988)
Boiling Point (°C):
417 (average, Shiu & Mackay 1986; Metcalfe et al. 1988)
Chlorine Content: 62.77% (Hutzinger et al. 1974)
Density (g/cm^3 at 20°C): 1.70
Molar Volume (cm^3/mol):
330.9 (LeBas method, Shiu & Mackay 1986)
Molecular Volume (A^3):
Total Surface Area, TSA (A^2):
302.9 (Shiu et al. 1987)
Heat of Fusion, kcal/mol:
Fugacity Ratio, F (calculated, assuming ΔS_{fusion} = 13.5 e.u.):
0.06 (Mackay et al. 1983)
0.0596-0.109 (Shiu et al. 1987)

Water Solubility (g/m^3 or mg/L at 25°C):
0.0005 (Mackay et al. 1983)
0.006 (subcooled liq., Mackay et al. 1983)
0.00816-0.0205, 0.011 (experimental range, calculated-UNIFAC, converted from log γ, Burkhard & Kuehl 1986)
0.014 (Mackay 1986; Metcalfe et al. 1988)
0.00045-0.002 (selected, Shiu et al. 1987)
0.0076-0.018 (selected, subcooled liquid, Shiu et al. 1987)
0.00046-0.002(Formica et al. 1988)

Vapor Pressure (Pa at 25°C):
0.0015 (subcooled liq., Mackay et al. 1983)
1.3×10^{-4} (Mackay 1986; Metcalfe et al. 1988)
2.73×10^{-5} (selected, Shiu et al. 1987)
2.5×10^{-4} (selected, subcooled liquid, Shiu et al. 1987)

Henry's Law Constant (Pa m^3/mol):
100.0 (Mackay et al. 1983)
5.4 (calculated-P/C, Shiu et al. 1987)

Octanol/Water Partition Coefficient, log K_{OW}:

7.1 (Mackay 1986; Metcalfe et al. 1988)

6.7-7.0 (selected, Shiu et al. 1987)

Bioconcentration Factor, log BCF:

7.10 (calculated- χ , Koch 1983)

5.80 (fish, Mackay 1986; Metcalfe et al. 1988)

Sorption Partition Coefficient, log K_{OC}:

Half-Lives in the Environment:

Air:

Surface water:

Groundwater:

Sediment:

Soil:

Biota:

Environmental Fate Rate Constants or Half-Lives:

Volatilization/Evaporation: 6.2×10^{-5} g/m^2h (Mackay 1986; Metcalfe et al. 1988).

Photolysis:

Hydrolysis:

Oxidation:

Biodegradation:

Biotransformation:

Bioconcentration, Uptake (k_1) and Elimination (k_2) Rate Constants:

Common Name: Octachlorobiphenyl
Synonym:
Chemical Name: octachlorobiphenyl
CAS Registry No: 31472-83-0
Molecular Formula: $C_{12}H_2Cl_8$
No. of Isomers: 12
Molecular Weight: 430
Melting Point (°C):
159-162 (Shiu & Mackay 1986; Shiu et al. 1987; Metcalfe et al. 1988)
Boiling Point (°C):
432 (average, Shiu & Mackay 1986; Metcalfe et al. 1988)
Chlorine Content: 65.98% (Hutzinger et al. 1974)
Density (g/cm^3 at 20°C): 1.7
Molar Volume (cm^3/mol):
351.8 (LeBas method, Shiu & Mackay 1986; Shiu et al. 1987)
Molecular Volume (A^3):
Total Surface Area, TSA (A^2):
317.9-318.7
Heat of Fusion, kcal/mol:
Fugacity Ratio, F (calculated, assuming ΔS_{fusion} = 13.5 e.u.):
0.04 (Mackay et al. 1983)
0.0443-0.0474 (Shiu et al. 1987)

Water Solubility (g/m^3 or mg/L at 25°C):
0.0002-0.007 (Mackay et al. 1983)
0.020 (subcooled liq., Mackay et al. 1983)
0.00345-0.006, 0.00378 (experimental range, calculated-UNIFAC, converted from log γ, Burkhard & Kuehl 1986)
0.0055 (Mackay 1986; Metcalfe et al. 1988)
0.0002-0.0003 (selected, Shiu et al. 1987)
0.004-0.0068 (selected, subcooled liquid, Shiu et al. 1987)

Vapor Pressure (Pa at 25°C):
2.8×10^{-5} (Mackay 1986; Metcalfe et al. 1988)
2.66×10^{-5} (selected, Shiu et al. 1987)
6.0×10^{-3} (selected, subcooled liquid, Shiu et al. 1987)

Henry's Law Constant (Pa m^3/mol):
100.0 (suggested, Mackay et al. 1983)
38.08 (calculated-P/C, Shiu et al. 1987)

Octanol/Water Partition Coefficient, log K_{OW}:
8.55 (Mackay et al. 1983)

7.50 (Mackay 1986; Metcalfe et al. 1988)
7.1 (selected, Shiu et al. 1987)

Bioconcentration Factor, log BCF:
6.20 (fish, Mackay 1986; Metcalfe et al. 1988)

Sorption Partition Coefficient, log K_{OC}:

Half-Lives in the Environment:
Air:
Surface water:
Groundwater:
Sediment:
Soil:
Biota:

Environmental Fate Rate Constants or Half-Lives:
Volatilization/Evaporation: 1.5×10^{-5} g/m^2h (Mackay 1986; Metcalfe et al. 1988).
Photolysis:
Hydrolysis:
Oxidation:
Biodegradation:
Biotransformation:
Bioconcentration, Uptake (k_1) and Elimination (k_2) Rate Constants:

Common Name: Nonachlorobiphenyl
Synonym:
Chemical Name: nonachlorobiphenyl
CAS Registry No: 53742-07-7
Molecular Formula: $C_{12}HCl_9$
No. of Isomers: 3
Molecular Weight: 464.2
Melting Point (°C):
182.8-206 (Shiu & Mackay 1986; Shiu et al. 1987; Metcalfe et al. 1988)
Boiling Point (°C):
445 (average, Shiu & Mackay 1986; Metcalfe et al. 1988)
Chlorine Content: 68.73% (Hutzinger et al. 1974)
Density (g/cm^3 at 20°C): 1.80
Molar Volume (cm^3/mol):
372.7 (LeBas method, Shiu & Mackay 1986; Shiu et al. 1987)
Molecular Volume (A^3):
Total Surface Area, TSA (A^2):
331.9 (Shiu et al. 1987)
Heat of Fusion, kcal/mol:
Fugacity Ratio, F (calculated, assuming ΔS_{fusion} = 13.5 e.u.):
0.016 (Mackay et al. 1983)
0.0163-0.0276 (Shiu et al. 1987)

Water Solubility (g/m^3 or mg/L at 25°C):
0.0001 (Mackay et al. 1983)
0.0007 (subcooled liq., Mackay et al. 1983)
0.000678-0.00148, 0.00145 (exptl. range, calculated-UNIFAC, converted from log γ, Burkhard & Kuehl 1986)
0.002 (Mackay 1986; Metcalfe et al. 1988)
0.000018-0.00011 (selected, Shiu et al. 1987)
0.00065-0.0068 (selected, subcooled liquid, Shiu et al. 1987)

Vapor Pressure (Pa at 25°C):
0.00015 (subcooled liq., Mackay et al. 1983)
6.3×10^{-6} (Mackay 1986; Metcalfe et al. 1988)

Henry's Law Constant (Pa m^3/mol):
100.0 (estimated, Mackay et al. 1983)

Octanol/Water Partition Coefficient, log K_{OW}:
9.14 (Mackay et al. 1983)
7.9 (Mackay 1986; Metcalfe et al. 1988)
7.2-8.16 (selected, Shiu et al. 1987)

Bioconcentration Factor, log BCF:

6.60 (fish, Mackay 1986; Metcalfe et al. 1988)

Sorption Partition Coefficient, log K_{OC}:

Half-Lives in the Environment:

Air:

Surface water:

Groundwater:

Sediment:

Soil:

Biota:

Environmental Fate Rate Constants or Half-Lives:

Volatilization/Evaporation: 3.5×10^{-6} g/m^{2}h (Mackay 1986; Metcalfe et al. 1988).

Photolysis:

Hydrolysis:

Oxidation:

Biodegradation:

Biotransformation:

Bioconcentration, Uptake (k_1) and Elimination (k_2) Rate Constants:

4.1.3 Aroclor Mixtures

Common Name: Aroclor 1016
Synonym:
Chemical Name:
CAS Registry No: 12674-11-2
Molecular Formula:
Average Molecular Weight: 257
Physical state: mobile oil
Distillation Range (°C):
323-356 (NAS 1979; Brinkman & De Kock 1980)
Chlorine Content: 41%
Density (g/cm^3):
1.37 (20°C, Brinkman & De Kock 1980)
1.36-1.37 (25°C, NAS 1979)
1.33 (Mills et al. 1982)
1.40 (25°C, Mackay 1986)
Molar Volume (cm^3/mol):
Molecular Volume (A^3):
Total Surface Area, TSA (A^2):
Heat of Fusion, kcal/mol:
Fugacity Ratio, F:

Water Solubility (g/m^3 or mg/L at 25°C):
0.22-0.25 (estimated, Tucker et al. 1975)
0.906 (23°C, shake flask-GC/ECD, Griffin et al. 1978; quoted, Lee et al. 1979)
0.42 (shake flask-GC/ECD, Paris et al. 1978; Callahan et al. 1979; Mackay et al. 1980; Mills et al. 1982)
0.906 (shake flask-GC/ECD, Lee et al. 1979)
0.049, 0.490 (shake flask, nephelometric, Hollifield 1979)
0.085 (quoted, Kenaga & Goring 1980)
0.34 (quoted, Pal et al. 1980)
0.906 (23°C, shake flask-GC/ECD, Griffin & Chian 1981; quoted, Sklarew & Girvin 1987)
0.40-0.91 (quoted, Mackay et al. 1983)
0.84 (selected, Mackay 1986; Metcalfe et al. 1988)
0.332 (quoted, Chou & Griffin 1987)

Vapor Pressure (Pa at 25°C):
0.0533 (Monsanto Co. 1972; quoted, Callahan et al. 1979; Mabey et al. 1982; Mills et al. 1982)
0.060 (quoted, Mackay et al. 1983)
0.200 (quoted, subcooled liquid, Mackay et al. 1983)
0.10 (selected, Mackay 1986; Metcalfe et al. 1988)
0.12, 0.121 (GC-RT, Foreman & Bidleman 1985)

Henry's Law Constant (Pa m^3/mol):

1368 (calculated, Paris et al. 1978)
33.4 (calculated-P/C, Mabey et al. 1982)
77.0 (calculated, Mackay et al. 1983)

Octanol/Water Partition Coefficient, log K_{OW}:

>5.58 (Chiou et al. 1977; Callahan et al. 1979)
4.38 (shake flask-GC, Paris et al. 1978; quoted, Callahan et al. 1979)
5.88 (GC-RT, Veith et al. 1979b)
5.58 (HPLC-RT, Veith et al. 1979a; quoted, Kenaga & Goring 1980)
3.48 (Pal et al. 1980; quoted, Sklarew & Girvin 1987)
5.58 (quoted, Callahan et al. 1979; Mabey et al. 1982)
5.88 (quoted, Mackay 1982)
4.3-5.48 (quoted, Mills et al. 1982)
4.40-5.80 (quoted, Mackay et al. 1983; Mackay 1986; Metcalfe et al. 1988)
5.88 (quoted, Garten & Trabalka 1983)
5.31 (quoted, Chou & Griffin 1987)
4.38 (quoted, Ryan et al. 1988)

Bioconcentration Factor, log BCF:

3.80, 3.81, 3.74 (bacteria: Doe Run pond, Hickory Hills pond, USDA pond, Paris et al. 1978)
4.18 (fish in Hudson River, Skea et al. 1979; quoted, Waid 1986)
4.63 (fathead minnow, 32 day exposure, Veith et al. 1979b; Veith & Kosian 1983)
4.69, 3.81 (fish, flowing water, static water, Kenaga & Goring 1980)
4.69 (quoted, Bysshe 1982)
4.70 (microorganism, calculated-K_{OW}, Mabey et al. 1982)
4.63 (fish, quoted, Mackay 1982)
4.56 (fish, calculated-K_{OW}, Mackay 1982)
0.15 (rodents, Garten & Trabalka 1983)
4.63 (fish, Garten & Trabalka 1983)
4.63 (fathead minnow, quoted, Zaroogian et al. 1985)
4.24, 4.30 (oyster, quoted, Zaroogian et al. 1985)
3.11-4.5 (fish, selected, Mackay 1986; Metcalfe et al. 1988)

Sorption Partition Coefficient, log K_{OC}:

5.26 (calculated-K_{OW}, Mabey et al. 1982)
4.87 (calculated, Sklarew & Girvin 1987)
4.25 (soil, calculated-S, Chou & Griffin 1987)

Half-Lives in the Environment:

Air:

Surface water: 9.9 hours in 1 m^3 water of 1 meter deep (Paris et al. 1978; selected,

Callahan et al. 1979; Mills et al. 1982)

Groundwater:

Sediment:

Soil: >50 days (Ryan et al. 1988).

Sludge: estimated half-life to be 15 days for the volatilization from activated sludge under aerobic conditions (Tucker et al. 1975; quoted, Pal et al. 1980).

Biota: half-life in fish, about 1.2 years in Hudson River (Armstrong & Sloan 1985).

Environmental Fate Rate Constants or Half-Lives:

Volatilization/Evaporation: 0.031 g/m^2h (Mackay 1986; Metcalfe et al. 1988).

Photolysis:

Hydrolysis: not environmentally significant (Mabey et al. 1982).

Oxidation: calculated rate constant for singlet oxygen, $<<360\ M^{-1}\ h^{-1}$ and RO_2 (peroxy radical) $<<1$ (Mabey et al. 1982).

Biodegradation: 32.9% degraded by activated sludge in 47-hour cycle (Monsanto Co. 1972); 33% degraded by activated sludge for 48-hour exposure (Tucker et al. 1975; Versar Inc. 1979; quoted, Pal et al. 1980); rate constant by acclimated activated sludge is 0.2 day^{-1} with a half-life of 3.5 days (Callahan et al. 1979). 96% loss by degradation with Nocardia strain NCIB 10603 and 91% loss with NCIB 10643, both within 52 days; >98% loss with NCIB 10603 and >96% loss with NCIB 10643, both within 100 days (Baxter et al. 1975; quoted, Pal et al. 1980).

Biotransformation: rate constant estimated to be $3x10^{-9}$ to $3x10^{-12}$ ml $cell^{-1}\ h^{-1}$ for bacteria transformation in water (Mabey et al. 1982).

Bioconcentration, Uptake (k_1) and Elimination (k_2) Rate Constants:

Common Name: Aroclor 1221
Synonym:
Chemical Name:
CAS Registry No: 11104-28-2
Molecular Formula:
Average Molecular Weight: 192-200.7
Physical State: mobile oil
Distillation Range (°C):
  275-320 (NAS 1979; Brinkman & De Kock 1980)
Chlorine Content: 20.5-21.5%
Density (g/cm^3):
  1.182-1.19 (25°C, NAS 1979)
  1.18 (20°C, Brinkman & De Kock 1980)
  1.15 (Callahan et al. 1979; Mills et al. 1982)
Molar Volume (cm^3/mol):
Molecular Volume (A^3):
Total Surface Area, TSA (A^2):
Heat of Fusion, kcal/mol:
Fugacity Ratio, F:

Water Solubility (g/m^3 or mg/L at 25°C):
  5.00 (Zitko 1970, 1971)
  3.5 (23°C, shake flask-GC/ECD, Griffin et al. 1978; quoted, Chou & Griffin 1987)
  15.0 (Monsanto Co. 1972; quoted, Callahan et al. 1979; Pal et al. 1980; Mills et al. 1982)
  3.52 (shake flask-GC/ECD, Lee et al. 1979)
  0.590 (shake flask-nephelometric, Hollifield 1979)
  40.0 (calculated-K_{OW}, Maybey et al. 1982)
  3.50-15.0 (quoted, Mackay et al. 1983)

Vapor Pressure (Pa at 25°C):
  0.893 (Monsanto Co. 1972; quoted, Callahan et al. 1979; Mabey et al. 1982; Mills et al. 1982)
  0.93 (Pal et al. 1980)
  0.89 (quoted, Mackay et al. 1983)
  2.00 (quoted, subcooled liquid, Mackay et al. 1983)

Henry's Law Constant (Pa m^3/mol):
  0.750 (Hetling et al. 1978)
  17.23 (calculated-P/C, Mabey et al. 1982)
  60.0 (suggested value, Mackay et al. 1983)
  23.10 (calculated, Burkhard et al. 1985b)

Octanol/Water Partition Coefficient, log K_{OW}:

2.8 (Monsanto Co. 1972; quoted, Callahan et al. 1979)
2.81 (Kalmaz & Kalmaz 1979; quoted, Pal et al. 1980)
4.08 (Callahan et al. 1979; Mabey et al. 1982)
4.10-4.70 (quoted, Mackay et al. 1980)
2.78-4.0 (quoted, Mills et al. 1982)
4.09 (quoted, Chou & Griffin 1987)
4.09 (quoted, Ryan et al. 1988)

Bioconcentration Factor, log BCF:

3.34 (microorganism, calculated-K_{OW}, Mabey et al. 1982)

Sorption Partition Coefficient, log K_{OC}:

3.76 (sediment, calculated-K_{OW}, Mabey et al. 1982)
3.62 (soil, calculated-S, Chou & Griffin 1987)

Half-Lives in the Environment:

Air:
Surface water:
Groundwater:
Sediment:
Soil: >50 days (Ryan et al. 1988).
Sludge: estimated half-life of volatilization from activated sludge to be 12 days under aerobic conditions (Tucker et al. 1975; quoted, Pal et al. 1980).
Biota:

Environmental Fate Rate Constants or Half-Lives:

Volatilization: estimated rate of volatilization from liquid substrate to be 1.74 $mg/cm^2/hour$ at 100°C (Hutzinger et al. 1974; quoted, Pal et al. 1980).
Photolysis:
Hydrolysis: not environmentally significant (Mabey et al. 1982).
Oxidation: calculated rate constant for singlet oxygen, $<< 360\ M^{-1}\ h^{-1}$ and RO_2 (peroxy radical) $<< 1\ M^{-1}\ h^{-1}$ (Mabey et al. 1982).
Biodegradation: 80.6% degraded by activated sludge in 47-hour cycle (Monsanto Co. 1972; quoted, Pal et al. 1980); 81% degraded by activated sludge for 48-hour exposure (Versar Inc. 1979; quoted, Pal et al. 1980); biodegraded by acclimated activated sludge with a first-order rate constant of 0.8 day^{-1} and a half-life of 0.9 days (Callahan et al. 1979).
Biotransformation: rate of transformation for bacteria in water estimated to be 3×10^{-9} to 3×10^{-12} ml $cell^{-1}$ $hour^{-1}$ (Mabey et al. 1982).
Bioconcentration, Uptake (k_1) and Elimination (k_2) Rate Constants:

Common Name: Aroclor 1232
Synonym:
Chemical Name:
CAS Registry No: 11141-16-5
Molecular Formula:
Average Molecular Weight: 221-232.2
Physical State: mobile oil
Distillation Range (°C):
270-325 (NAS 1979; Brinkman & De Kock 1980)
Chlorine Content: 32%
Density (g/cm^3):
1.260 (20°C, Brinkman & De Kock 1980)
1.27-1.28 (25°C, NAS 1979)
1.24 (Callahan et al. 1979; Mills et al. 1982)
Molar Volume (cm^3/mol):
Molecular Volume (A^3):
Total Surface Area, TSA (A^2):
Heat of Fusion, kcal/mol:
Fugacity Ratio, F:

Water Solubility (g/m^3 or mg/L at 25°C):
1.45 (Monsanto Co. 1972; quoted, Callahan et al. 1979; Chou & Griffin 1987)
1.45 (quoted, Pal et al. 1980; Mackay et al. 1983)

Vapor Pressure (Pa at 25°C):
0.54 (Monsanto Co. 1972; Callahan et al. 1979; Mabey et al. 1982)
0.533 (quoted, Mills et al. 1982)
0.54 (quoted, Mackay et al. 1983)

Henry's Law Constant (Pa m^3/mol):
60.0 (suggested, Mackay et al. 1980)
1.14 (calculated-P/C, Mabey et al. 1982)

Octanol/Water Partition Coefficient, log K_{OW}:
3.2 (Monsanto Co. 1972; Callahan et al. 1979; Mabey et al. 1982)
4.54 (Tulp & Hutzinger 1978; Callahan et al. 1979)
3.23 (quoted, Pal et al. 1980)
3.18-4.48 (quoted, Mills et al. 1982)
4.10-5.20 (quoted, Mackay et al. 1983)
4.62 (calculated-S, Chou & Griffin 1987)
4.54 (quoted, Ryan et al. 1988)

Bioconcentration Factor, log BCF:

2.54 (microorganism, calculated-K_{OW}, Mabey et al. 1982)

Sorption Partition Coefficient, log K_{OC}:

2.89 (sediment, calculated-K_{OW}, Mabey et al. 1982)
3.85 (soil, calculated-S, Chou & Griffin 1987)

Half-Lives in the Environment:

Air:
Surface water:
Groundwater:
Sediment:
Soil: >50 days (Ryan et al. 1988).
Biota:

Environmental Fate Rate Constants or Half-Lives:

Volatilization:
Photolysis:
Hydrolysis: not environmentally significant (Mabey et al. 1982).
Oxidation: calculated rate constant for singlet oxygen << 360 M^{-1} and RO_2 (peroxy radical) << 1 M^{-1} h^{-1} (Mabey et al. 1982).
Biodegradation:
Biotransformation: estimated rate of transformation for bacteria in water to be $3x10^{-9}$ to $3x10^{-12}$ ml $cell^{-1}$ $hour^{-1}$ (Mabey et al. 1982).
Bioconcentration, Uptake (k_1) and Elimination (k_2) Rate Constants:

Common Name: Aroclor 1242
Synonym:
Chemical Name:
CAS Registry No: 53469-21-9
Molecular Formula:
Average Molecular Weight: 261-266.5
Physical State: mobile oil
Distillation Range (°C):
325-366 (NAS 1979; Brinkman & De Kock 1980; Mackay et al. 1986)
Chlorine Content: 42%
Density (g/cm^3):
1.38 (Brinkman & De Kock 1980)
1.30-1.39 (NAS 1979)
1.35 (Callahan et al. 1979; Mills et al. 1982)
1.40 (Mackay 1986; Metcalfe et al. 1988)
Molar Volume (cm^3/mol):
Molecular Volume (A^3):
Total Surface Area, TSA (A^2):
Heat of Fusion, kcal/mol:
Fugacity Ratio, F: 1.0

Water Solubility (g/m^3 or mg/L at 25°C):
0.20 (Monsanto Co. 1972; quoted, Hutzinger et al. 1974; Sawhney 1987)
0.20 (20°C, Nisbet & Sarofim 1972)
0.24 (quoted, Mackay & Wolkoff 1973; Mackay & Leinonen 1975; Brinkman & De Kock 1980; Geyer et al. 1980)
0.20 (Tucker et al. 1975)
0.045 (shake flask-GC, Lawrence & Tosine 1976)
0.085 (Branson 1977; Kenaga & Goring 1980)
0.703 (23°C, shake flask-GC/ECD, Griffin et al. 1978)
0.34 (shake flask-GC/ECD, Paris et al. 1978)
0.1329 (11.5°C, shake flask-GC/ECD, Dexter & Pavlou 1978)
0.23 (quoted, Callahan et al. 1979; Mabey et al. 1982)
0.23-0.703 (shake flask-GC, Lee et al. 1979)
0.100 (shake flask-nephelometric, Hollifield 1979)
0.703 (20°C, Griffin & Chian 1980; quoted, Sklarew & Girvin 1987)
0.24 (quoted, Pal et al. 1980)
0.25 (quoted, Eisenreich et al. 1981)
0.34-0.703 (quoted, Westcott et al. 1981)
0.1-0.3 (quoted, Mills et al. 1982)
0.35 (quoted, Richardson et al. 1983)
0.24 (quoted, Erickson 1986)
0.50 (selected, Mackay et al. 1986)

0.288 (quoted, Chou & Griffin 1987)
0.40 (quoted, Eisenreich 1987)
0.75 (selected, Mackay 1986; Metcalfe et al. 1988)
0.277 (20°C, calculated from mole fraction, Murphy et al. 1987)

Vapor Pressure (Pa at 25°C):
0.0133 (20°C, Nisbet & Sarofim 1972)
0.055 (quoted, Mackay & Wolkoff 1973; Mackay & Leinonen 1975; Bidleman & Christinsen 1979)
0.12 (20°C, extrapolated, Monsanto 1972; NAS 1979)
0.054 (quoted, Callahan et al. 1979)
0.054 (Pal et al. 1980; quoted, Sklarew & Girvin 1987)
0.0537 (quoted, Eisenreich et al. 1981)
0.054 (quoted, Westcott et al. 1981)
0.055 (quoted, Mackay & Leinonen 1975; Mackay et al. 1983)
0.0533 (quoted, Mills et al. 1982)
0.054 (quoted, Richardson et al. 1983)
0.040 (38°C, average, Wingender & Williams 1984)
0.076, 0.077 (GC-RT, Foreman & Bidleman 1985)
0.077 (quoted, Mackay et al. 1986)
0.0517 (quoted, Eisenreich 1987)
0.091 (selected, Mackay 1986; Metcalfe et al. 1988)
0.033 (20°C, calculated-mole fraction, Murphy et al. 1987)

Henry's Law Constant (Pa m^3/mol):
58.06 (calculated, Mackay & Leinonen 1975)
768 (calculated, Paris et al. 1978)
59.5 (Slinn et al. 1978)
56.74 (calculated-P/C, Eisenreich et al. 1981)
20.32-41.62 (Westcott et al. 1981)
20.27-41.54 (calculated, Westcott et al. 1981)
79.02 (batch stripping, Atlas et al. 1982; quoted, Eisenreich & Looney 1983; Atlas & Giam 1986)
34.69 (radiotracer-equilibration, Atlas et al. 1982; Atlas & Giam 1986)
200.6 (calculated-P/C, Mabey et al. 1982)
57.75 (quoted, Mills et al. 1982)
22.29 (direct concn. ratio-GC/ECD, Murphy et al. 1983)
40.3 (16°C, calculated-P/C, Richardson et al. 1983)
34.75 (calculated, Burkhard et al. 1985b)
50.0 (calculated, Mackay et al. 1986)
34.45 (calculated-P/C, Eisenreich 1987)
23.0 (20°C, quoted, Murphy et al. 1987 from Burkhard et al. 1985b)
28.31 (20°C, equilibrium concn. ratio, Murphy et al. 1987)

Octanol/Water Partition Coefficient, log K_{OW}:

4.11 (Callahan et al. 1979; Mabey et al. 1982)
5.58 (HPLC-RT, Veith et al. 1979a; quoted, Kenaga & Goring 1980)
0.703 (shake flask-GC/ECD, Lee et al. 1979)
3.54 (Pal et al. 1980; quoted, Sklarew & Girvin 1987)
4.0-5.6 (quoted, Mills et al. 1982)
4.50-5.80 (quoted, Mackay et al. 1983; Mackay 1986; Metcalfe et al. 1988; Eisenreich 1987)
5.29 (quoted, Chou & Griffin 1987)
4.11 (quoted, Ryan et al. 1988)

Bioconcentration Factor, log BCF:

3.92, 3.65, 3.46 (bacteria: Doe Run pond, Hickory Hills pond, USDA pond, Paris et al. 1978)
0.08, -0.22 (adipose tissue of male, female Albino rats, Geyer et al. 1980)
4.69, 3.81 (fish, flowing water, static water, Kenaga & Goring 1980)
4.69 (quoted, Bysshe 1982)
3.36 (microorganism, calculated-K_{OW}, Mabey et al. 1982)
0.30 (rodent, Garten & Trabalka 1983)
0.13 (poultry, Garten & Trabalka 1983)
-0.11 (sheep, Garten & Trabalka 1983)
-0.50 (small birds, Garten & Trabalka 1983)
-0.27 (swine, Garten & Trabalka 1983)
3.20-4.51 (fish, selected, Mackay 1986; Metcalfe et al. 1988)

Sorption Partition Coefficient, log K_{OC}:

3.80 (sediment, calculated-K_{OW}, Mabey et al. 1982)
3.36 (calculated, Sklarew & Girvin 1987)
4.09 (soil, calculated-S, Chou & Griffin 1987)

Half-Lives in the Environment:

Air:
Surface water: 12 hours (Paris et al. 1978); volatilization half-life estimated to be 12 hours at 1 meter depth in 1 m^3 of water (Mackay & Leinonen 1975; quoted, Pal et al. 1980; Mills et al. 1982).
Groundwater:
Sediment:
Soil: >50 days (Ryan et al. 1988).
Biota:

Environmental Fate Rate Constants or Half-Lives:

Volatilization/Evaporation: 0.23 $\mu g/m^2 day$ and with a half-life of 17 days (Baker et al. 1985); 0.029 $g/m^2 h$ (Mackay 1986; Metcalfe et al. 1988).

Photolysis:

Hydrolysis: not environmentally significant (Mabey et al. 1982).

Oxidation: calculated rate constant for singlet oxygen, $<< 360\ M^{-1}\ h^{-1}$ and RO_2 (peroxy radical), $<< 1\ M^{-1}\ h^{-1}$ (Mabey et al. 1982).

Biodegradation: 26.3% degraded by activated sludge in 47-hour cycle (Monsanto Co. 1972; quoted, Pal et al. 1980); 26% degraded by activated sludge for 48-hour exposure (Versar Inc. 1979; quoted, Pal et al. 1980); degraded by acclimated activated sludge with a first-order rate constant of 0.15 day^{-1} and a half-life of 4.5 days (Callahan et al. 1979). 88% loss by degradation with Nocardia strain NCIB 10603, 76% loss with NCIB 10643 both within 52 days; 95% loss with NCIB 10603 and 85% loss with NCIB 10643 both within 100 days (Baxter et al. 1975; quoted, Pal et al. 1980).

Biotransformation: estimated rate constant for bacteria transformation in water to be 3×10^{-9} to 3×10^{-12} ml $cell^{-1}$ $hour^{-1}$ (Mabey et al. 1982).

Bioconcentration, Uptake (k_1) and Elimination (k_2) Rate Constants:

Common Name: Aroclor 1248
Synonym:
Chemical Name:
CAS Registry No: 12672-29-6
Molecular Formula:
Average Molecular Weight: 288-299.5
Physical State: mobile oil
Distillation Range (°C):

340-375 (NAS 1979; Brinkman & De Kock 1980; Mackay et al. 1986)

Chlorine Content: 48%
Density (g/cm^3):

1.44 (20°C, Brinkman & De Kock 1980)
1.40-1.41 (NAS 1979)
1.41 (Callahan et al. 1979)
1.40 (Mackay 1986; Mills et al. 1982; Metcalfe et al. 1988)

Molar Volume (cm^3/mol):
Molecular Volume (A^3):
Total Surface Area, TSA (A^2):
Heat of Fusion, kcal/mol:
Fugacity Ratio, F:

Water Solubility (g/m^3 or mg/L at 25°C):

0.10 (20°C, Nisbet & Sarofim 1972)
0.10 (Monsanto Co. 1972; selected, Hutzinger et al. 1974; Sawhney 1987)
0.043 (26°C, Hutzinger et al. 1974)
0.054 (quoted, Mackay & Wolkoff 1973; Mackay & Leinonen 1975)
0.054 (quoted, Callahan et al. 1979; Pal et al. 1980; Mabey et al. 1982; Mills et al. 1982)
0.060 (shake flask-nephelometry, Hollifield 1979)
0.054 (NAS 1979; Chou & Griffin 1987)
0.052 (quoted, Brinkman & De Kock 1980)
0.054 (quoted, Mackay et al. 1983, 1986)
0.052 (quoted, Erickson 1986)
0.32 (Mackay 1986; Metcalfe et al. 1988)
0.056 (quoted, Eisenreich 1987)

Vapor Pressure (Pa at 25°C):

0.004 (20°C, Nisbet & Sarofim 1972)
0.066 (selected, Mackay & Wolkoff 1973; Mackay & Leinonen 1975; quoted, Mills et al. 1982)
0.11 (20°C, extrapolated, Monsanto 1972; quoted, NAS 1979)
0.017 (Branson 1977; Kenaga & Goring 1980)
0.066 (quoted, Callahan et al. 1979; Mabey et al. 1982; Mackay et al. 1983)

0.025, 0.024 (GC-RT, Foreman & Bidleman 1985)
0.024 (quoted, Mackay et al. 1986)
0.0085 (quoted, Eisenreich 1987)
0.023 (quoted, Mackay 1986; Metcalfe et al. 1988)

Henry's Law Constant (Pa m^3/mol):
355.7 (calculated, Mackay & Leinonen 1975)
372 (Slinn et al. 1978)
364 (calculated-P/C, Mabey et al. 1982)
86.0 (suggested, Mackay et al. 1983)
44.58 (calculated, Burkhard et al. 1985b; quoted, Eisenreich 1987)
50.0 (calculated, Mackay et al. 1986)

Octanol/Water Partition Coefficient, log K_{OW}:
5.75 (Hansch et al. 1973; Callahan et al. 1979; Kenaga & Goring 1980; Mabey et al. 1982; Chou & Griffin 1987)
6.11 (shake flask-GC, Chiou et al. 1977; quoted, Callahan et al. 1979)
6.11 (HPLC-RT, Veith et al. 1979a & b)
6.11 (quoted, Mackay 1982)
6.0 (quoted, Mills et al. 1982)
6.11 (quoted, Garten & Trabalka 1983)
5.8-6.3 (quoted, Mackay et al. 1983, 1986; Mackay 1986; Metcalfe et al. 1988; Eisenreich 1987)
5.6 (quoted, Ryan et al. 1988)
6.10 (quoted, Thomann 1989)

Bioconcentration Factor, log BCF:
4.42 (bluegill sunfish, Stalling & Meyer 1972)
4.75-4.79 (channel catfish, Mayer et al. 1977; quoted, Waid 1986)
5.08 (fathead minnow, DeFoe et al. 1978; quoted, Waid 1986)
4.85 (fathead minnow, 32-day exposure, Veith et al. 1979b)
4.86, 4.07 (fish, flowing water, static water, Kenaga & Goring 1980)
4.86 (quoted, Bysshe 1982)
3.86-4.42, 4.19 (mussel, range, average, Geyer et al. 1982)
4.86 (microorganism, calculated-K_{OW}, Mabey et al. 1982)
4.85 (quoted, Mackay 1982)
4.79 (fish, calculated-K_{OW}, Mackay 1982)
4.85 (fish, Garten & Trabalka 1983)
0.82 (poultry, Garten & Trabalka 1983)
0.72 (rodents, Garten & Trabalka 1983)
4.5-5.0 (fish, selected, Mackay 1986; Metcalfe et al. 1988)

Sorption Partition Coefficient, log K_{OC}:

5.44 (sediment, calculated-K_{OW}, Mabey et al. 1982)

4.74 (soil, calculated-S, Chou & Griffin 1987)

Half-Lives in the Environment:

Air:

Surface water: volatilization half-life estimated to be 10 hours at 1 meter depth in 1 m^3 water (Mackay & Leinonen 1975; quoted, Pal et al. 1980; Mills et al. 1982).

Groundwater:

Sediment:

Soil: >50 days (Ryan et al. 1988).

Biota:

Environmental Fate Rate Constants or Half-Lives:

Volatilization/Evaporation: 8.3×10^{-3} g/m^2h (Mackay 1986; Metcalfe et al. 1988).

Photolysis:

Hydrolysis: not environmentally significant (Mabey et al. 1982).

Oxidation: calculated rate constant for singlet oxygen, $<< 360$ M^{-1} h^{-1} and RO_2 (peroxy radical), $<< 1$ M^{-1} h^{-1} (Mabey et al. 1982).

Biodegradation:

Biotransformation: estimated bacteria transformation in water to be 3×10^{-9} to 3×10^{-12} ml $cell^{-1}$ $hour^{-1}$ (Mabey et al. 1982).

Bioconcentration, Uptake (k_1) and Elimination (k_2) Rate Constants:

log k_2: -1.92 day^{-1} (fish, quoted, Thomann 1989)

Common Name: Aroclor 1254
Synonym:
Chemical Name:
CAS Registry No: 11097-69-1
Molecular Formula:
Average Molecular Weight: 327-328.4
Physical State: viscous liquid
Distillation Range (°C):
365-390 (NAS 1979; Brinkman & De Kock 1980; Mackay et al. 1986)
Chlorine Content: 54%
Density (g/cm^3):
1.505 (Monsanto 1972)
1.49-1.50 (65°C, NAS 1979)
1.54 (20°C, Brinkman & De Kock 1980)
1.50 (Mills et al. 1982; Mackay 1986; Metcalfe et al. 1988)
Molar Volume (cm^3/mol):
Molecular Volume (A^3):
Total Surface Area, TSA (A^2):
Heat of Fusion, kcal/mol:
Fugacity Ratio, F:

Water Solubility (g/m^3 or mg/L at 25°C):
0.30 (Zitko 1971)
0.043 (26°C, Nelson et al. 1972)
0.050 (20°C, Nisbet & Sarofim 1972)
0.040 (Monsanto Co. 1972; selected, Hutzinger et al. 1974; Sawhney 1987)
0.012-0.07 (Mackay & Wolkoff 1973; Mackay & Leinonen 1975; Geyer et al. 1980)
0.056 (shake flask-GC, Haque et al. 1974)
0.0001 (shake flask-GC/ECD, Schoor 1975)
0.045 (shake flask-GC, Lawrence & Tosine 1976)
0.070 (23°C, shake flask-GC/ECD, Griffin et al. 1978)
0.0242 (11.5°C, shake flask-GC/ECD, Dexter & Pavlou 1978)
0.012 (quoted, Brinkman & De Kock 1980)
0.056 (quoted, Haque et al. 1980)
0.010 (quoted, Kenaga & Goring 1980)
0.031 (Callahan et al. 1979; Mabey et al. 1982)
0.070 (shake flask-GC/ECD, Lee et al. 1979)
0.057 (shake flask-nephelometry, Hollifield 1979)
0.07 (23°C, Griffin & Chian 1980; quoted, Sklarew & Girvin 1987)
0.012 (quoted, Giam et al. 1980; Pal et al. 1980)
0.0115 (quoted, Eisenreich et al. 1981)
0.045-0.07 (quoted, Westcott et al. 1981)
0.01-0.06 (quoted, Mills et al. 1982)

0.012-0.07 (quoted, Mackay et al. 1983, 1986)
0.012 (quoted, Erickson 1986)
0.042 (quoted, Chou & Griffin 1987)
0.035 (quoted, Eisenreich 1987)
0.14 (selected, Mackay 1986; Metcalfe et al. 1988)
0.043 (20°C, calculated-mole fraction, Murphy et al. 1987)

Vapor Pressure (Pa at 25°C):
0.00048 (20°C, Nisbet & Sarofim 1972)
0.0103 (Monsanto Co. 1972; Callahan et al. 1979; Mabey et al. 1982)
0.0103 (quoted, Mackay & Wolkoff 1973; Mackay & Leinonen 1975)
0.0103 (quoted, Bidleman & Christinsen 1979)
0.024 (20°C, extrapolated, Monsanto 1974; quoted, NAS 1979)
0.0103 (Callahan et al. 1979; Mabey et al. 1982; Mills et al. 1982)
0.0103 (quoted, Giam et al. 1980; Westcott et al. 1980)
0.0101 (quoted, Eisenreich et al. 1981)
0.004 (38°C, Average, Wingender & Williams 1984)
0.00435, 0.00424 (GC-RT, Foreman & Bidleman 1985)
0.043 (selected, Mackay et al. 1986)
0.00263 (quoted, Eisenreich 1987)
$6.7x10^{-3}$ (selected, Mackay 1986; Metcalfe et al. 1988)
0.00294 (20°C, calculated-mole fraction, Murphy et al. 1987)

Henry's Law Constant (Pa m^3/mol):
279.7 (calculated, Mackay & Leinonen 1975)
0.0993 (Murphy & Rzeszutko 1977; quoted, Eisenreich & Looney 1983)
273 (Slinn et al. 1978)
274 (calculated-P/C, Eisenreich et al. 1981)
0.0070 (Eisenreich et al. 1981a)
0.0142 (Doskey & Andren 1981; quoted, Eisenreich & Looney 1983)
47.57-74.08 (calculated-P/C, Westcott et al. 1981)
16.60 (radiotracer-equilibration, Atlas et al. 1982; Atlas & Giam 1986)
284 (quoted, Mills et al. 1982)
82.0 (suggested, Mackay et al. 1983)
21.0 (direct concn. ratio-GC/ECD, Murphy et al. 1983)
28.67 (calculated, Burkhard et al. 1985b; quoted, Eisenreich 1987)
50.0 (calculated, Mackay et al. 1986)
18.24 (20°C, selected, Murphy et al. 1987 from Burkhard et al. 1985b)
19.25 (20°C, equilibrium concn. ratio, Murphy et al. 1987)

Octanol/Water Partition Coefficient, log K_{OW}:
6.03 (Hansch et al. 1973; Callahan et al. 1979; Mabey et al. 1982)
6.47 (GC-RT, Veith et al. 1979b; Veith & Kosian 1983)

6.72 (HPLC-RT, Veith et al. 1979a)
6.04 (Callahan et al. 1979; Mabey et al. 1982)
4.08 (Pal et al. 1980; quoted, Sklarew & Girvin 1987)
6.47 (quoted, Mackay 1982)
6.0 (quoted, Mills et al. 1982)
6.47 (quoted, Garten & Trabalka 1983; Travis & Arms 1988)
6.1-6.8 (selected, Mackay et al. 1983, 1986)
6.47 (Zaroogian et al. 1985; quoted, Södergren 1987)
6.1-6.8 (selected, Mackay 1986; Metcalfe et al. 1988)
6.11 (quoted, Chou & Griffin 1987)
6.04 (quoted, Ryan et al. 1988)
6.50 (quoted, Thomann 1989)

Bioconcentration Factor, log BCF:
4.57 (spot fish, Hansen et al. 1971; quoted, Waid 1986)
4.85 (bluegill sunfish, Stalling & Mayer 1972)
4.75-4.79 (channel catfish, Mayer et al. 1977; quoted, Waid 1986)
5.08, 5.57, 6.08 (mysis, sculpins, pelagic fish, Veith et al. 1977)
5.00 (fathead minnow, 32-day exposure, Veith et al. 1979b; Veith & Kosian 1983)
5.0-5.22 (oyster, Hansen 1976; NAS 1979)
4.41 (shrimp, Hansen 1976; NAS 1979)
4.57 (estuarine fish, Hansen 1976; NAS 1979)
0.79, 0.78 (adipose tissue of male, female Albino rats, Geyer et al. 1980)
4.66, 4.08 (fish, flowing water, static water, Kenaga & Goring 1980)
4.66 (quoted, Bysshe 1982)
5.12 (microorganism, calculated-K_{OW}, Mabey et al. 1982)
5.00 (fish, quoted, Mackay 1982)
5.15 (fish, calculated-K_{OW}, Mackay 1982)
4.57 (fish, estuarine, Hansen 1976; NAS 1979)
0.53 (cow, Garten & Trabalka 1983)
4.70 (fish, Garten & Trabalka 1983)
0.77 (poultry, Garten & Trabalka 1983)
0.79 (rodents, Garten & Trabalka 1983)
0.18 (sheep, Garten & Trabalka 1983)
0.98 (small birds, Garten & Trabalka 1983)
0.03 (swine, Garten & Trabalka 1983)
5.0 (fathead minnow, quoted, Zaroogian et al. 1985)
4.80, 4.68 (oyster, quoted, Zaroogian et al. 1985)
4.8-5.51 (fish, quoted, Mackay 1986; Metcalfe et al. 1988)
5.52 (oyster, Södergren 1987)
-1.28 (beef, reported as biotransfer factor log B_b, Travis & Arms 1988)
-1.95 (milk, reported as biotransfer factor log B_m, Travis & Arms 1988)
-1.77 (vegetable, reported as biotransfer factor log B_v, Travis & Arms 1988)

7.21 (field data, laketrout, Thomann 1989)
6.9, 6.51, 6.67, 6.8 (field data, large-mouth bass, Thomann 1989)

Sorption Partition Coefficient, log K_{OC}:
- 6.0 (sediment/pore water samples of pond, Halter & Johnson 1977; selected, Di Toro et al. 1985)
- 5.72 (sediment, calculated-K_{OW}, Mabey et al. 1982)
- 5.44 (sediment/pore water samples-Lake Michigan, Eadie et al. 1983; selected, Di Toro et al. 1985)
- 6.65 (subsurface water/suspended solids, 56% OC-Lake Michigan, Voice & Weber 1985)
- 5.88 (pore water/sediment, 0.7% OC-Lake Michigan, Voice & Weber 1985)
- 5.61 (pore water/sediment, 1.7% OC-Lake Michigan, Voice & Weber 1985)
- 4.82 (pore water/sediment, 3.8% OC-Lake Michigan, Voice & Weber 1985)
- 6.62 (calculated, Sklarew & Girvin 1987)
- 4.81 (soil, calculated-S, Chou & Griffin 1987)

Half-Lives in the Environment:

Air:

Surface water: volatilization half-life from 1 meter depth in 1 m^3 water estimated to be 10 hours (Mackay & Leinonen 1975; quoted, Pal et al. 1980; Mills et al. 1982).

Groundwater:

Sediment:

Soil: volatilization half-life from an Ottawa sand estimated to be 15 days (Haque et al. 1974; quoted, Pal et al. 1980); half-life in soil, >50 days (Ryan et al. 1988).

Biota: half-life in plant surface, < 12 days (Pal et al. 1980); in guppies, 3.3 days and in cichlids, 5.1 days (Gooch & Hamdy 1982; quoted, Waid 1986).

Environmental Fate Rate Constants or Half-Lives:

Volatilization/Evaporation: volatilization rate estimated to be $2x10^{-6}$ g/cm_2d at 26°C and $8.6x10^{-5}$ g/cm^2d at 60°C (Haque et al. 1974); 0.10 $\mu g/m^2d$ with a half-life of 28 days (Baker et al. 1985); $2.7x10^{-3}$ g/m^2h (Mackay 1986; Metcalfe et al. 1988).

Photolysis:

Hydrolysis: not environmentally significant.

Oxidation: calculated rate constant for singlet oxygen, << 360 M^{-1} h^{-1} and RO_2 (peroxy radical), << 1 M^{-1} h^{-1} (Mabey et al. 1982).

Biodegradation: no reduction of concentration in the spilled transformer fluid contaminant of Aroclor was detected over a two-year period (Moein et al. 1976; quoted, Pal et al. 1980). 15.2% degraded by activated sludge in 47-hour cycle (Monsanto Co. 1972); 19% degraded by activated sludge for 48-hour exposure (Versar Inc. 1975); biodegradation with a first-order rate constant of 0.1 day^{-1} by acclimated activated sludge and a half-life of 7.0 days (Callahan et al. 1979).

Biotransformation: rate constant for bacteria transformation in water estimated to be $3x10^{-9}$ to $3x10^{-12}$ ml cell^{-1} hour^{-1} (Mabey et al. 1982).

Bioconcentration, Uptake (k_1) and Elimination (k_2) Rate Constants:

k_2: 0.023 day^{-1} (0 to 1 day), 0.086 day^{-1} (1 to 2 days), & 0.0899 day^{-1} (2 to 6 days) with a biological half-life of 5.5 days (mosquito larvae, Gooch & Hamdy 1982; selected, Waid 1986)

k_2: 0.131 day^{-1}, 0.137 day^{-1} with biological half-life of 4.7 days (guppies, Gooch & Hamdy 1982; quoted, Waid 1986)

k_2: 0.102 day^{-1} (first day), 0.057 day^{-1} (thereafter) with a biological half-life of 6.1 days (cichlids, Gooch & Hamdy 1982; quoted, Waid 1986)

Common Name: Aroclor 1260
Synonym:
Chemical Name:
CAS Registry No: 11096-82-5
Molecular Formula:
Average Molecular Weight: 372-375.7
Physical State: sticky resin
Distillation Range (°C):
385-420 (NAS 1979; Brinkman & De Kock 1980)
Chlorine Content: 60%
Density (g/cm^3):
1.62 (20°C, Brinkman & De Kock 1980)
1.55-1.56 (90°C, NAS 1979)
1.58 (Callahan et al. 1979; Mills et al. 1982)
1.60 (Mackay 1986; Metcalfe et al. 1988)
Molar Volume (cm^3/mol):
Molecular Volume (A^3):
Total Surface Area, TSA (A^2):
Heat of Fusion, kcal/mol:
Fugacity Ratio, F:

Water Solubility (g/m^3 or mg/L at 25°C):
0.025 (Monsanto CO. 1972)
0.025 (20°C, Nisbet & Sarofim 1972)
0.0027 (Mackay & Wolkoff 1973; quoted, Callahan et al. 1979; Geyer et al. 1980; Pal et al. 1980; Mabey et al. 1982; Mills et al. 1982; Chou & Griffin 1987)
0.080 (shake flask-nephelometry, Hollifield 1979)
0.003 (quoted, Brinkman & De Kock 1980; Mackay et al. 1983)
0.0027 (quoted, Richardson et al. 1983)
0.00304 (quoted, Eisenreich 1987)
0.0144 (20°C, calculated-mole fraction, Murphy et al. 1987)

Vapor Pressure (Pa at 25°C):
2.67×10^{-5} (38°C, Nisbet & Sarofim 1972)
0.0054 (quoted, Mackay & Wolkoff 1973; Mackay & Leinonen 1975)
0.012 (20 °C, extrapolated, Monsanto 1972; NAS 1979)
0.0054 (Callahan et al. 1979; Mabey et al. 1982)
0.0053 (Pal et al. 1980; Mills et al. 1982)
0.0054 (quoted, Mackay et al. 1983)
0.0054 (quoted, Richardson et al. 1983)
0.0004 (38°C, average, Wingender & Williams 1984)
0.00183, 0.00162 (GC-RT, Foreman & Bidleman 1985)
0.003 (quoted, Erickson 1986)

0.00064 (Mackay 1986; Metcalfe et al. 1988)
0.000284 (quoted, Eisenreich 1987)
0.000841 (20°C, calculated-mole fraction, Murphy et al. 1987)

Henry's Law Constant (Pa m^3/mol):
722.4 (calculated, Mackay & Leinonen 1975)
718 (Slinn et al. 1978)
719 (quoted, Mills et al. 1982)
88.0 (suggested, Mackay et al. 1983)
72.24 (16°C, calculated-P/C, Richardson et al. 1983)
34.04 (calculated, Burkhard et al. 1985b; quoted, Eisenreich 1987)
21.27 (20°C, quoted, Murphy et al. 1987 from Burkhard et al. 1985b)
17.23 (20°C, equilibrium concn. ratio, Murphy et al. 1987)

Octanol/Water Partition Coefficient, log K_{OW}:
7.14 (Hansch et al. 1973; Chiou et al. 1977)
6.11 (Chiou et al. 1977; Callahan et al. 1979)
7.15 (Callahan et al. 1979; Mabey et al. 1982)
6.91 (GC-RT, Veith et al. 1979a; Veith & Kosian 1983)
4.34 (Pal et al. 1980)
6.91 (quoted, Mackay 1982)
>6.0 (quoted, Mills et al. 1982)
6.30-7.50 (quoted, Mackay et al. 1983; Mackay 1986; Metcalfe et al. 1988; Eisenreich 1987)
6.61 (calculated-S, Chou & Griffin 1987)
6.91 (quoted, Geyer et al. 1987)
6.11 (quoted, Ryan et al. 1988)
6.90 (quoted, Thomann 1989)

Bioconcentration Factor, log BCF:
5.43 (fathead minnows, DeFoe et al. 1978; quoted, Waid 1986)
5.29 (fathead minnow, 32-day exposure, Veith et al. 1979a; Veith & Kosian 1983)
0.672 (adipose tissue of male albino rats, Geyer et al. 1980)
5.29 (fish, quoted, Mackay 1982)
6.11 (microorganism, calculated-K_{OW}, Mabey et al. 1982)
5.59 (fish, calculated-K_{OW}, Mackay 1982)
5.0-6.20 (fish, quoted, Mackay 1986; Metcalfe et al. 1988)
2.28-2.50 (human fat of lipid basis, calculated-K_{OW}, Geyer et al. 1987)
2.11-2.36 (human fat of wet wt. basis, calculated-K_{OW}, Geyer et al. 1987)
4.38 (rhabdosargus holubi, De Kock & Lord 1988)

Sorption Partition Coefficient, log K_{OC}:

6.83 (sediment, calculated-K_{OW}, Mabey et al. 1982)

5.54 (soil, calculated-S, Chou & Griffin 1987)

Half-Lives in the Environment:

Air:

Surface water: volatilization half-life estimated to be 10 hours at 1 meter depth of 1 m^3 of water (Mackay & Leinonen 1975; quoted, Pal et al. 1980; Mills et al. 1982).

River water: volatilization half-life estimated to be 52 days (Oloffs et al. 1972; selected, Pal et al. 1980).

Sediment:

Soil:

Biota: half-life in Rhabdosargus holubi, 50 days (De Kock & Lord 1988).

Environmental Fate Rate Constants or Half-Lives:

Volatilization/Evaporation: estimated evaporation rate from liquid surfaces at 100°C to be 0.009 mg/cm^2h (Hutzinger et al. 1974); half-life of evaporation from water depth of 1 meter estimated to be 7.53 hours (Mackay & Leinonen 1975); and rate of evaporation to be 2.9×10^{-4} g/m^2h (Mackay 1986; Metcalfe et al. 1988).

Photolysis:

Hydrolysis: not environmentally significant (Mabey et al. 1982).

Oxidation: calculated rate constant for singlet oxygen, $<< 360$ M^{-1} h^{-1} and RO_2 (peroxy radical), $<< 1$ M^{-1} h^{-1} (Mabey et al. 1982).

Biodegradation: no degradation over a 12-week period in natural water samples (Oloffs et al. 1972; quoted, Pal et al. 1980).

Biotransformation: rate of transformation for bacteria in water estimated to be 3×10^{-9} to 3×10^{-12} ml $cell^{-1}$ $hour^{-1}$ (Mabey et al. 1982).

Bioconcentration, Uptake (k_1) and Elimination (k_2) Rate Constants:

k_1: 332 day^{-1} (rhabdosargus holubi, De Kock & Lord 1988)

k_2: 0.014 day^{-1} (rhabdosargus holubi, De Kock & Lord 1988)

log k_2: -2.40 day^{-1} (fish, quoted, Thomann 1989)

4.2 Summary Tables and QSPR Plots

Table 4.1 Summary of selected physical-chemical properties of some PCB congeners

IUPAC no.	Congener	CAS no.	MW g/mol	mp, °C	bp, °C	Fugacity ratio, F at 25 °C	Molar vol., cm^3/mol V_M LeBas	$V_I/100$ intrinsic (a)	Total surface area, $Å^2$ TSA (b)	TSA nonplanar (c)	TSA planar (c)	TSA planar (d)	TSA planar (e)	TMV, $Å^3$ planar (e)
0	Biphenyl	92-54-2	154.2	71	255	0.352	184.6	0.92	192.45	192.34	184.77	184.43	224.1	207
1	2-	2051-60-7	188.7	34	274	0.817	205.4	1.01	208.44	208.33	195.32	195.45	232.3	215.2
2	3-	2051-61-8	188.7	25.1	284	1	205.4	1.01	210.02	209.87	202.3	201.95	237	224
3	4-	2051-62-9	188.7	77.9	291	0.301	205.4	1.01	210.02	209.87	202.3	202.12	237	224
4	2,2'-	13029-08-8	223.1	61		0.442	226.4	1.1	224.42	224.32	199.57	200.8	240.5	228.8
5	2,3-	16605-91-7	223.1							223.87	210.86	210.34		
7	2,4-	33284-50-3	223.1	24.4		1	226.4	1.1	226.01	225.85	212.85	213.14	245.2	234.9
8	2,4'-	34883-43-7	223.1	43		0.666	226.4	1.1	226.01	225.86	212.85	213.14	245.2	234.9
9	2,5-	34883-39-1	223.1	25.1		1	226.4			225.85	212.85	212.97	245.2	234.9
10	2,6-	33146-45-1	223.1	34.9		0.801	226.4			224.32	205.87	206.46	240.5	
11	3,3'-	2050-67-1	223.1	29	322	0.913	226.4			227.43	219.82	219.47		
12	3,4-	2974-92-7	223.1	49		0.579	226.4			225.41	217.84	217.73		
14	3,5-	34883-41-5	223.1	31		0.872				227.39	219.82	219.47		
15	4,4'-	2050-68-2	223.1	149	315	0.059	226.4	1.1	227.59	277.4	219.82	219.81	249.9	241
16	2,2'3-	38444-78-9	257.5	28		0.934	247.3	1.19		215.11	239.86	215.69		
18	2,2',5-	37680-65-2	257.5	44		0.651	247.3		241.98	217.7	241.84	218.32	253.4	245.8
20	2,3,3'-	38444-84-7	257.5				247.3			241.44	228.39	227.86		
21	2,3,4-	55702-46-0	257.5	102		0.173	247.3			241.41	226.4	226.11		

Table 4.1 (cont'd)

IUPAC no.	congener	CAS no.	MW g/mol	mp, °C	bp, °C	fugacity ratio, F	V_M LeBas	V_I/100 intrinsic	TSA, $Å^2$	TSA nonplanar	TSA planar	TSA planar	TSA planar	TMV, $Å^3$ planar
26	2,3',5-	38444-85-8	257.5	40.5		0.703	247.3			239.41	230.37	230.49		
28	2,4,4'-	7012-37-5	257.5	57		0.484	247.3		243.58	241.39	230.37	230.83	255.6	
29	2,4,5-	15862-07-4	257.5	78		0.3	247.3	1.19	241.59	241.39	228.39	228.74	255.6	248.2
30	2,4,6-	35693-92-6	257.5	62.5		0.427	247.3			241.84	223.39	224.16	252.7	245.8
31	2,4',5-	16862-07-4	257.5	67		0.384	247.3	1.19		234.39	230.37	230.66	258.1	251.9
33	2',3,4-	38444-86-9	257.5	60		0.452	247.3		214.59	241.4	228.39	228.75	255.6	248.8
35	3,3',4-	37680-69-6	257.5	87		0.244	247.3			241.4	228.39	235.25	260.3	254.3
37	3,4,4'-	28444-90-5	257.5	87		0.244	247.3	1.19	247.18	243.18	235.36	235.42		
40	2,2',3,3'-	38444-93-8	292	121		0.113	268.2	1.28	255.61	255.4	230.65	230.58	261.3	255.4
44	2,2',3,5'-	41464-39-5	292	47		0.606	268.2		257.8	257.43	232.63	233.21	263.8	259.1
47	2,2',4,4'-	2437-79-8	292	83		0.268	268.2		259.56	259.38	234.62	236.19	266.3	262.8
49	2,2',4, 5'-	41464-40-8	292	64		0.413	268.2	1.28		259.41	234.62	236.01	266.3	262.8
50	2,2',4,6-	62796-65-0	292				268.2			247.83	227.64	229.5		
51	2,2',4,6'-	68194-04-7	292				268.2			257.83	227.64	229.51		262.8
52	2,2',5,5'-	35693-99-3	292	87		0.244	268.2		259.57	259.41	234.62	235.84	266.3	
53	2,2',5,6'-	41464-41-9	292	104		0.165	268.2				227.64	257.83	229.34	
54	2,2',6,6'-	15968-05-5	292	198		0.01945	268.2			256.3	214.36	217.18	246.7	237
60	2,3,4,4'-	33025-41-1	292	142		0.0696	268.2			259.15	256.94	243.8	268.5	
61	2,3,4,5-	33284-53-6	292	92		0.218	268.2	1.28	255.21	254.95	241.94	241.72	263.5	
65	2,3,5,6-	33284-54-7	292	79		0.292	268.2			255.4	236.95	236.24		
66	2,3',4,4'-	32598-10-0	292	124		0.105	268.2			258.97	245.91	246.44		
70	2,3'4',5-	32598-11-1	292	104		0.165	268.2		259.15	258.97	245.91	246.26	268.5	
75	2,4,4',6-	32598-12-2	292				268.2			258.52	250.91	241.85		
77	3,3',4,4'-	32598-13-3	292	180		0.0294	268.2		258.76	258.52	250.91	251.01	270.7	265.2
80	3,3',5,5'-	33284-52-5	292	164		0.0422	268.2			262.52	254.88	254.51		

Table 4.1 (cont'd)

IUPAC no.	congener	CAS no.	MW g/mol	mp, °C	bp, °C	fugacity ratio, F	V_M LeBas	$V_I/100$ intrinsic	TSA	TSA nonplanar	TSA planar	TSA planar	TSA planar	TMV, Å^3 planar
83	2,2'3,3',5-	60145-20-2	326.4				289.1			272.93	248.17	248.1		
86	2,2',3,4,5-	65510-45-4	326.4	100		0.182	289.1		271.18	270.94	246.19	247.07	274.2	268.7
87	2,2',3,4,5'-	38380-02-8	326.4	114		0.132	289.1		273.19	272.97	248.17	248.99	276.7	272.4
88	2,2',3,4,6-	55215-17-3	326.4	100		0.182	289.1			271.39	241.19	242.48	264.4	259.5
95	2,2'3,5,6-	38379-99-6	326.4	100		0.182	289.1			273.41	243.18	244.23		
99	2,2',4,4',5-	38380-01-7	326.4				289.1			274.92	250.16	251.79		
100	2,2'4,4',6-	39485-83-1	326.4							275.37	245.16	247.2		
101	2,2',4,5,5'-	37680-73-2	326.4	76.5		0.311	289.1	1.37		274.95	250.16	251.62	276.7	276.1
104	2,2',4,6,6'-	56558-16-8	326.4				289.1			273.82	231.88	234.87		
105	2,3,3',4,4'-	32598-14-4	326.4	105		0.175	289.1			272.52	259.47	259.41		
110	2,3,3',4',6-	38380-03-9	326.4				289.1			272.98	254.48	254.65		
116	2,3,4,5,6-	18259-05-7	326.4	124		0.105	289.1	1.37	269.22	268.95	250.51	250.1	271.7	265.4
118	2,3',4,4',5-	31508-00-6	326.4	107		0.155	289.1			274.51	261.45	262.04		
124	2',3,4,5,5'-	70424-70-3	326.4				289.1			274.5	261.45	261.87		
128	2,2',3,3',4,4'-	38380-07-3	360.9	150		0.0582	310		286.77	286.49	261.73	261.96	282.1	282.4
129	2,2',3,3'4,5-	55215-18-4	360.9	85		0.256	310	1.46	286.78	286.49	261.73	261.96	282.1	286.1
134	2,2',3,3',5,6-	52704-70-8	360.9	100		0.182	310		287.19	286.93	256.73	256.49	287.3	276.9
136	2,2',3,3',6,6'-	38411-22-2	360.9	112.2		0.138	310			287.38	245.44	246.95	269.5	263.6
138	2,2',3,4,4',5'-	35065-28-2	360.9	80		0.286	310	1.46		288.52	263.71	264.76		
149	2,2',3,4',5',6-	38380-04-0	360.9	oil		1	310			288.96	258.72	260		
153	2,2',4,4',5,5'-	35065-27-1		103		0.17	310		290.78	290.5	265.7	267.39	287.1	289.4
155	2,2',4,4',6,6'-	33979-03-2	360.9	114		0.132	310	1.46	291.5	291.36	249.41	252.56	272.5	271
156	2,3,3',4,4',5-	38380-08-4	360.9				310			288.07	275.01	275.01		
169	3,3'4,4',5,5'-	32774-16-6	360.9	202		0.0178	310			289.64	281.99	282.23		

Table 4.1 (cont'd)

IUPAC no.	congener	CAS no.	MW g/mol	mp, °C	bp, °C	fugacity ratio, F	V_M LeBas	$V_I/100$ intrinsic (a)	TSA (b)	TSA, Å^2 nonplanar (c)	TSA planar (c)	TSA planar (d)	TSA planar (e)	TMV, Å^3 planar (e)
171	2,2',3,3',4,4',6-	52663-71-5	395.3	122.4		0.0109	330.9	1.55		302.48	272.28	273.15	289.9	286.5
180	2,2',3,4,4',5,5'-	35065-29-3	395.3	110		0.144	330.9			304.51	279.25	280.37		
185	2,2',3,4,5,5',6-	52712-05-7	395.3	149		0.0596	330.9	1.55	302.78	302.51	272.28	272.98	298.8	298.8
187	2,2',3,4',5,5',6-	52663-68-0	395.3				330.9	1.55		304.5	274.26	274.89		
194	2,2',3,3',4,4',5,5'-	35694-08-7	429.8	159		0.0474	351.8	1.64	317.92	317.62	292.81	293.34	302.9	308.6
202	2,2',3,3',5,5',6,6'-	2136-99-4	429.7	162		0.0443	351.8	1.64	318.72	318.47	276.52	276.73	293.3	292
206	2,2',3,3',4,4',5,5',6-	40186-72-9	464.2	206		0.0163	372.7	1.73	331.93	331.62	301.37	301.73	306.9	309.1
207	2,2',3,3',4,4',5,6,6'-	52663-79-3	464.2							332.03	290.07	291.48		
208	2,2',3,3',4,5,5',6,6'-	52663-77-1	464.2	182.8		0.0276	372.7	1.73		332.02	290.07	290.59	301.2	299.8
209	2,2',3,3',4,4',5,5',6,6'-	2051-24-3	498.7	305.9		0.00167	393.6		345.92	345.59	303.63	304.45	309.1	309.4

(a) Kamlet et al. (1988)
(b) Mackay et al. (1980)
(c) Doucette (1985)
(d) Hawker (1988)
(e) Opperhuizen et al. (1988)

Table 4.2 Summary of selected physical-chemical properties of some PCB congeners at 25 °C

IUPAC no.	Congener	Selected properties: Vapor pressure P^S, Pa	P_L, Pa	Solubility S g/m^3	C^S mmol/m^3	C_L mmol/m^3	log K_{OW}	Henry's law const., H Pa m^3/mol calcd., P/C
0	Biphenyl	1.3	3.69	7	45.39	129.7	3.9	53.5
1	2-	2.04	2.5	5.5	29.15	35.66	4.3	70.1
2	3-	1	1	2.5	13.25	13.24	4.6	75.55
3	4-	0.271	0.9	1.2	6.36	21.15	4.5	42.56
4	2,2'-	0.265	0.6	1	4.48	10.14	4.9	59.17
5	2,3-							
7	2,4-	0.254	0.25	1.25	5.6	5.51	5	45.39
8	2,4'-			1	4.48	6.73	5.1	
9	2,5-	0.18	0.18	2	8.960	8.950	5.1	20.1
10	2,6-			1.4	6.280	7.840	5	
11	3,3'-	0.027	0.03	0.354	1.587	1.738	5.3	17.26
12	3,4-			0.008				
14	3,5-	0.105	0.12					
15	4,4'-	0.0048	0.08	0.06	0.269	4.560	5.3	17
16	2,2'3-							
18	2,2',5-	0.143	0.22	0.4	1.550	2.390	5.6	92.21
20	2,3,3'-							
21	2,3,4-							
26	2,3',5-			0.251	0.975	1.387		
28	2,4,4'-			0.16	0.621	1.28	5.8	
29	2,4,5-	0.132	0.044	0.14	0.544	1.81	5.6	24.29
30	2,4,6-	0.0384	0.09	0.2	0.777	1.82	5.5	49.51
31	2,4',5-							
33	2',3,4-	0.0136	0.003	0.08	0.311	0.69	5.8	43.67
35	3,3',4-							
37	3,4,4'-			0.015	0.0582	0.24	5.9	

Table 4.2 (cont'd)

IUPAC no.	Congener	Vapor pressure P^S, Pa solid	P_L,Pa liquid	Solubility S g/m³	C_S mmol/m³	C_L mmol/m³	log K_{OW}	Henry's law const., H Pa m³/mol calcd., P/C
40	2,2',3,3'-	0.00225	0.002	0.03	0.103	0.91	5.6	21.94
44	2,2',3,5'-			0.1	0.342	0.565	6	
47	2,2',4,4'-	0.0054	0.002	0.09	0.308	1.15	5.9	17.38
49	2,2',4, 5'-			0.016	0.0548	0.133	6.1	
50	2,2',4,6-							
51	2,2',4,6'-							
52	2,2',5,5'-	0.0049	0.002	0.03	0.103	0.42	6.1	47.59
53	2,2',5,6'-						5.5	
54	2,2',6,6'-						5.48	
60	2,3,4,4'-						6.31	
61	2,3,4,5-			0.02	0.0685	0.314	5.9	
65	2,3,5,6-						5.94	
66	2,3',4,4'-			0.04	0.0147	1.3	5.8	
70	2,3'4',5-							
75	2,4,4',6-			0.091			6.21	
77	3,3',4,4'-	0.0000588	0.002	0.001	0.0342	1.165	6.5	1.72
80	3,3',5,5'-			0.0012	0.0041	0.0974		
83	2,2'3,3',5-							
86	2,2',3,4,5-	0.00927	0.051	0.02	0.0613	0.337	6.2	151.4
87	2,2',3,4,5'-	0.000304	0.0023	0.004	0.0123	0.0927	6.5	24.81
88	2,2',3,4,6-			0.012	0.0368	0.202	6.5	
95	2,2'3,5,6-							
99	2,2',4,4',5-							
100	2,2'4,4',6-							
101	2,2',4,5,5'-	0.00109	0.0035	0.01	0.0306	0.0986	6.4	35.48
104	2,2',4,6,6'-		0.00434	0.0156	0.0306	0.3103		13.98
105	2,3,3',4,4'-						6	
110	2,3,3',4',6-			0.004			6.3	
116	2,3,4,5,6-			0.008	0.0145	0.233	6.3	
118	2,3',4,4',5-							
124	2',3,4,5,5'-							

Table 4.2 (cont'd)

IUPAC no.	Congener	Vapor pressure P^S, Pa solid	P_L,Pa liquid	Solubility S g/m³	C^S mmol/m³	C_L mmol/m³	log K_{OW}	Henry's law const., H Pa m³/mol calcd., P/C
128	2,2',3,3',4,4'-	0.0000198	0.00034	0.0006	0.00166	0.0286	7	11.91
129	2,2',3,3'4,5-			0.0006	0.00166	0.0065	7.3	
134	2,2',3,3',5,6-			0.0004	0.00111	0.0061	7.3	
136	2,2',3,3',6,6'-			0.0008	0.00222	0.0161	6.7	
138	2,2',3,4,4',5'-							
149	2,2',3,4',5',6-							
153	2,2',4,4',5,5'-	0.000119	0.0007	0.001	0.00277	0.0163	6.9	42.9
155	2,2',4,4',6,6'-	0.00048	0.00363	0.002	0.0055	0.0420	7	86.616
156	2,3,3',4,4',5-							
169	3,3'4,4',5,5'-							
170	2,2',3,3',4,4',5-							
171	2,2',3,3',4,4',6-	0.0000273	0.00025	0.002	0.00506	0.046	6.7	5.4
180	2,2',3,4,4',5,5'-							
185	2,2',3,4,5,5',6-			0.00045	0.00114	0.0191	7	
187	2,2',3,4',5,5',6-							
194	2,2',3,3',4,4',5,5'-			0.0002	0.00047	0.0098	7.4	
202	2,2',3,3',5,5',6,6'-	0.0000266	0.0006	0.0003	0.0007	0.0158	7.1	38.08
206	2,2',3,3',4,4',5,5',6-	$1.96x10^{-7}$	$1.2x10^{-5}$	0.00011	0.000237	0.0146	7.2	82.20
207	2,2',3,3',4,4',5,6,6'-						7.52	
208	2,2',3,3',4,5,5',6,6'-			0.000018	0.000038	0.00141	8.16	
209	2,2',3,3',4,4',5,5',6,6'-	$5.02x10^{-8}$	0.00003	0.000001	0.000002	0.0144	8.26	20.84

Table 4.3 Summary of physical-chemical properties PCB isomer groups and Aroclor mixtures

PCB isomer groups	CAS no.	MW g/mol	Cl no. n_{Cl}	mp, °C	Fugacity Ratio, F range	V_M cm³/mol LeBas	TSA, Å² range
Biphenyl	92-52-4	154.2	0	71	0.352	184.6	192.2
Mono-	27323-18-8	188.7	1	25.1 - 78	0.299-1.0	205.5	208.4 - 210
Di-	25512-42-9	223.1	2	24.4 - 149	0.0594-1.0	226.4	224.2 - 227.6
Tri-	25323-68-6	257.5	3	28.1 - 102	0.173-0.932	247.3	241.6 - 243.6
Tetra-	26914-33-0	292.0	4	47 - 164	0.042-0.606	268.2	255.6 - 259.6
Penta-	25429-29-2	326.4	5	76.5 - 123	0.107-0.310	289.1	269.2 - 275.2
Hexa-	26601-64-9	360.9	6	70 - 201	0.0182-0.359	310	287.2 - 291.5
Hepta-	28655-71-2	395.3	7	109 - 162	0.0596-0.148	330.9	302.8
Octa-	31472-83-0	429.8	8	132 - 161	0.0452-0.087	351.8	317.9 - 318.7
Nona-	53742-07-7	464.2	9	205 - 206	0.0163-0.027	372.7	331.9
Deca-	2051-24-3	498.7	10	305	0.00167	393.6	345.9

Aroclor mixtures	CAS no.	MW g/mol	% Cl	no. of Cl/molecule	Fugacity ratio, F at 25 °C	Density g/cm³ at 25°C	Distillation range, °C
Aroclor 1016	12674-11-2	257	41	3	1.0	1.33	323 - 356
Aroclor 1221	111-042-82	192	20.5-21.5	1.15	1.0	1.15	275 - 320
Aroclor 1232	111-411-65	221	31.4-32.5	2.04	1.0	1.24	290 - 325
Aroclor 1242	534-692-19	261	42	3.1	1.0	1.35	325 - 366
Aroclor 1248	126-722-96	288	48	3.9	1.0	1.41	340 - 375
Aroclor 1254	110-976-91	327	54	4.96	1.0	1.5	365 - 390
Aroclor 1260	110-968-25	372	60	6.3	1.0	1.58	385 - 420

Table 4.4 Summary of physical-chemical properties of PCB isomer groups and Aroclor mixtures at 20 -25 °C

PCB isomer groups	Aqueous Solubility, range			Vapor presssure, range		Henry's law const., H	log K_{OW} range
	S, g/m^3	C_S, $mmol/m^3$	C_L, $mmol/m^3$	P_S, Pa	P_L, Pa	Pa m^3/mol	
Biphenyl	7.0	45.39	129.7	1.30	3.69	28.64	3.90
Mono-	1.21-5.50	6.36-29.15	1 13.24-35.66	0.271-2.04	0.9-2.5	42.56-75.55	4.3-4.60
Di-	0.060-2.0	0.269-8.96	4.56-10.14	0.0048-0.279	0.008-0.60	17.0-92.21	4.9-5.30
Tri-	0.015-0.40	0.0582-1.55	0.24-2.39	0.0136-0.143	0.003-0.22	24.29-92.21	5.5-5.90
Tetra-	0.0043-0.010	0.0147-0.342	0.133-1.30	0.000059-.0054	0.002	1.72-47.59	5.6-6.50
Penta-	0.004-0.020	0.0123-0.0613	0.093-0.337	0.000304-0.0093	0.0023-0.051	24.8-151.4	6.2-6.50
Hexa-	0.0004-0.0007	0.0011-0.002	0.0061-0.0286	0.000020-0.0015	0.0007- 0.012	11.9-818	6.7-7.30
Hepta-	0.000045-0.0002	0.00114-0.0051	0.0191-0.046	0.0000273	0.00025	5.40	6.7-7.0
Octa-	0.0002-0.0003	0.00047-0.0007	0.0098-0.0158	0.0000266	0.0006	38.08	7.10
Nona-	0.00018-0.0012	0.000038-0.00024	0.00141-0.0146	-	-	-	7.2-8.16
Deca-	0.000761	0.0000024	0.0144	0.00000005	0.00003	20.84	8.26

Aroclor mixtures	Solubility, range		Vapor pressure	Henry's law const., H	log K_{OW}
	S, g/m^3	C_L, $mmol/m^3$	P_L, Pa	Pa m^3/mol	
Aroclor 1016	0.22-0.84	0.856-0.216	0.06-0.2	70-900	4.4 - 5.8
Aroclor 1221	0.59-5.0	0.307-26.0	0.89-2.0	34-450	4.1 - 4.7
Aroclor 1232	1.45	6.56-2.0	0.54	82-270	4.5 - 5.2
Aroclor 1242	0.1-0.75	0.383-2.87	0.05-0.13	45-130	4.5 - 5.8
Aroclor 1248	0.1-0.5	0.347-1.74	0.0085-0.11	5-300	5.8 - 6.3
Aroclor 1254	0.01-0.30	0.306-0.92	0.008-0.02	20-260	6.1 - 6.8
Aroclor 1260	0.003-0.08	0.00806-0.215	0.0002-0.012	20-60	6.3 - 7.5

Table 4.5 Suggested half-life classes of polychlorinated biphenyls in various environmental compartments

Compounds	Air class	Water class	Soil class	Sediment class
Biphenyl	3	4	5	6
Monochloro-	4	7	8	8
Dichloro-	4	7	8	8
Trichloro-	5	8	9	9
Tetrachloro-	6	9	9	9
Pentachloro-	6	9	9	9
Hexachloro-	7	9	9	9
Heptachloro-	7	9	9	9
Octachloro-	8	9	9	9
Nonachloro-	8	9	9	9
Decachloro-	9	9	9	9

where,

Class	Mean half-life (hours)	Range (hours)
1	5	< 10
2	17 (~ 1 day)	10-30
3	55 (~ 2 days)	30-100
4	170 (~ 1 week)	100-300
5	550 (~ 3 weeks)	300-1,000
6	1700 (~ 2 months)	1,000-3,000
7	5500 (~ 8 months)	3,000-10,000
8	17000 (~ 2 years)	10,000-30,000
9	55000 (~ 6 years)	> 30,000

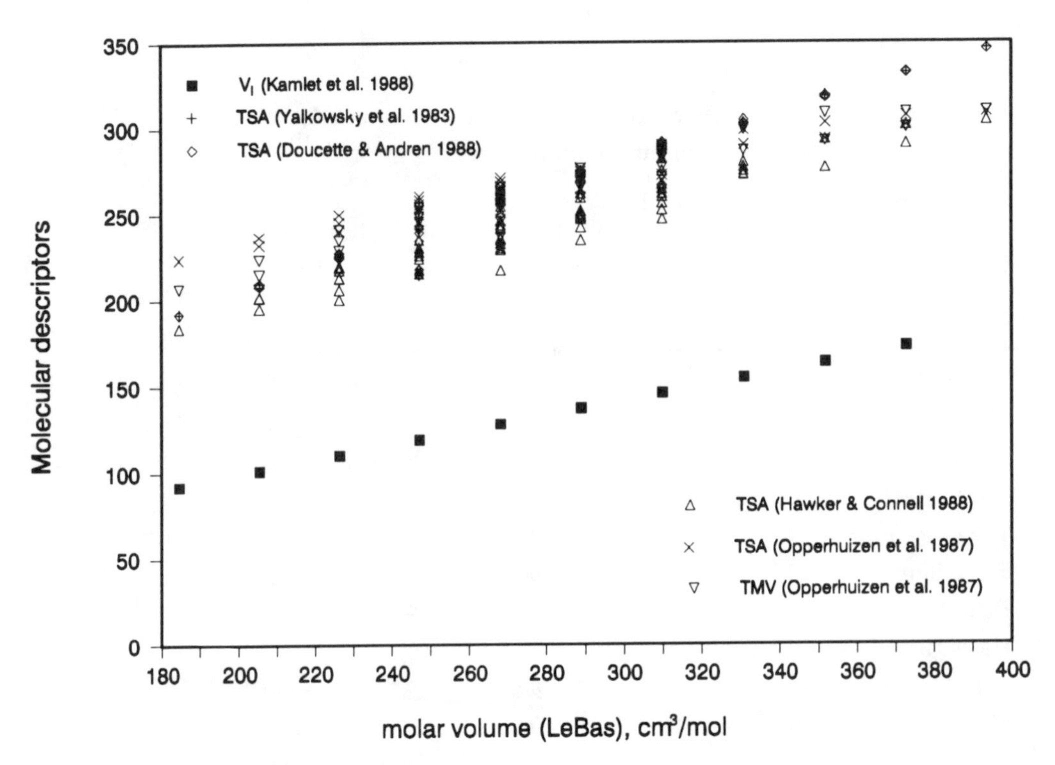

Figure 4.1 Plot of molecular descriptors versus LeBas molar volume

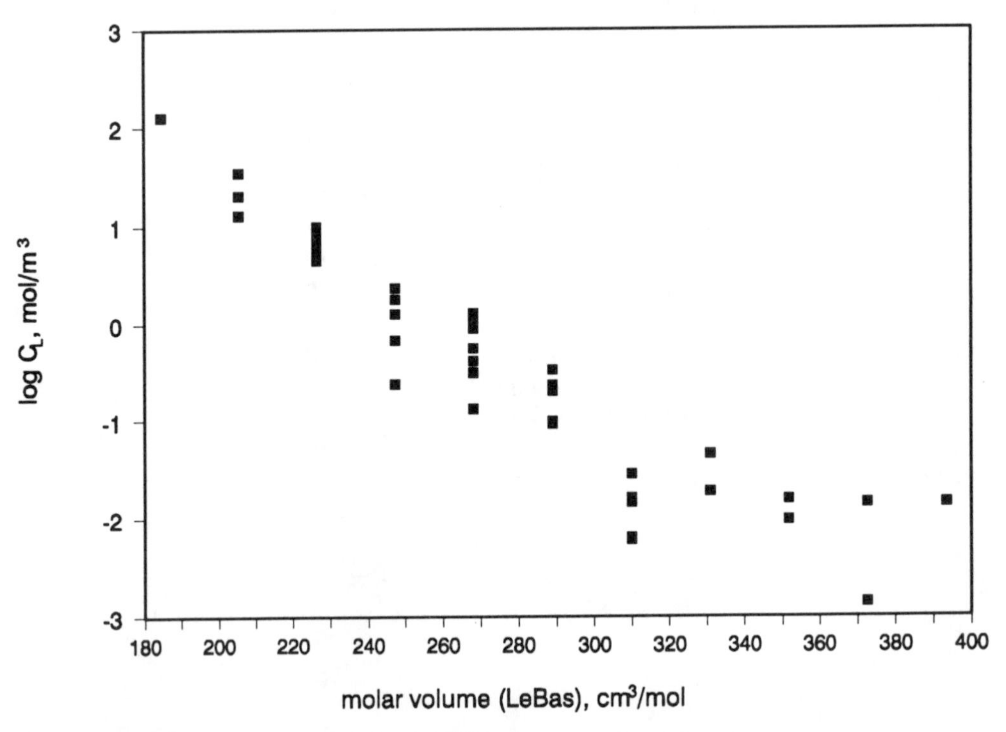

Figure 4.2 Plot of log C_L (liquid solubility) versus molar volume

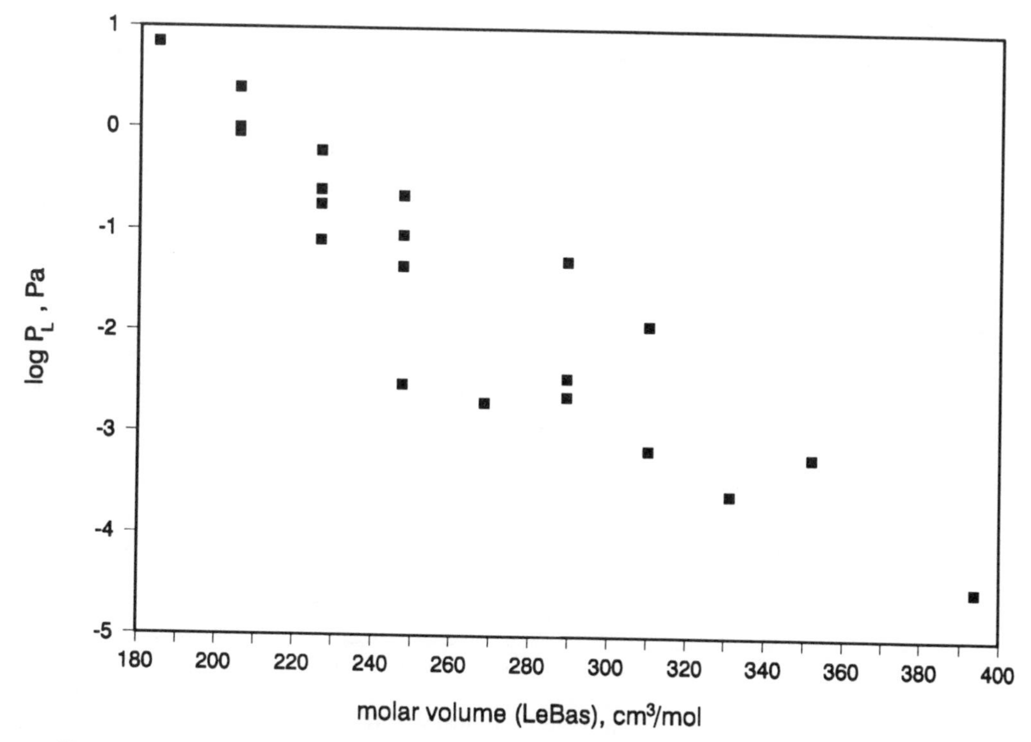

Figure 4.3 Plot of P_L (liquid vapor pressure) versus molar volume

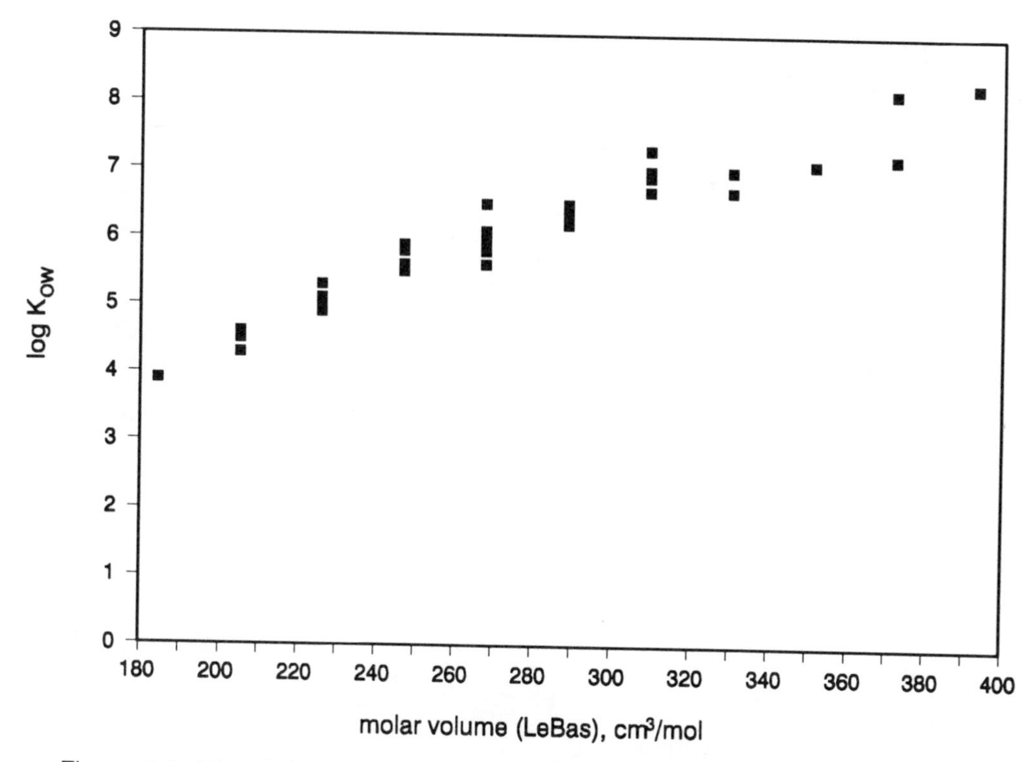

Figure 4.4 Plot of log K_{OW} versus molar volume

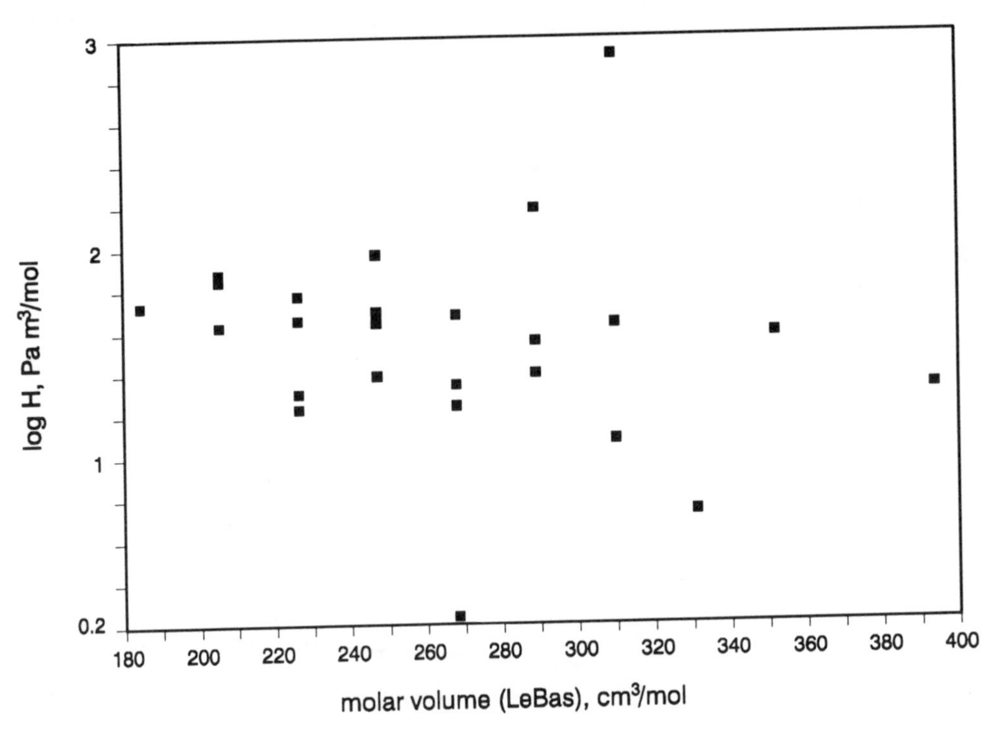

Figure 4.5 Plot of log H (Henry's law constant) versus molar volume

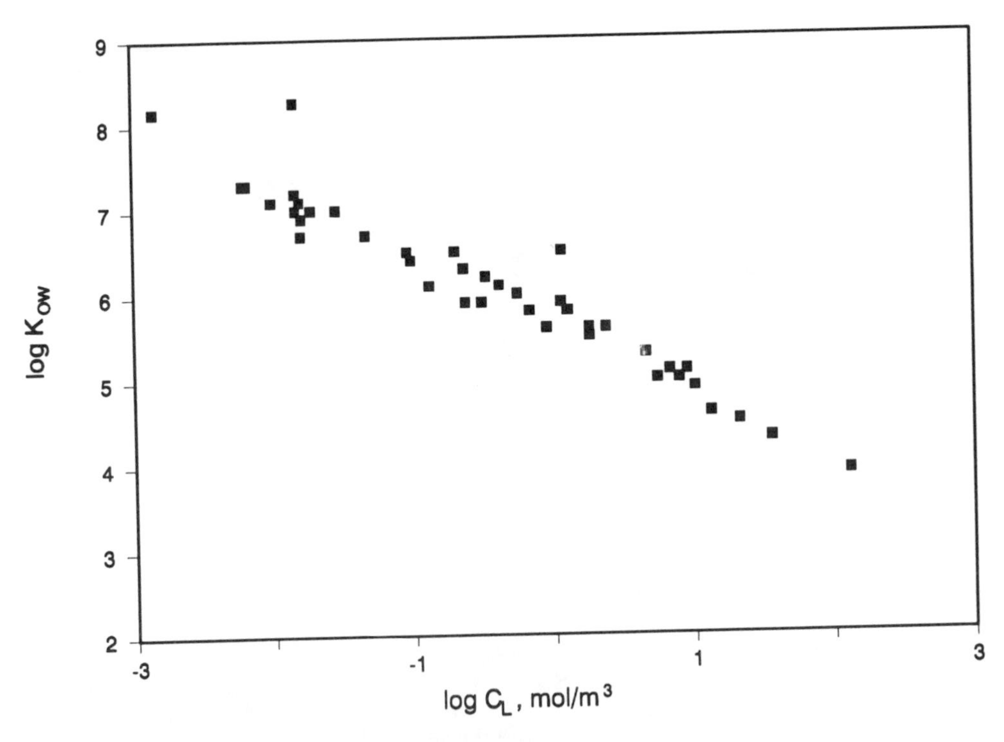

Figure 4.6 Plot log K_{OW} versus log C_L

4.3 Illustrative Fugacity Calculations: Level I, II and III

Chemical name: Biphenyl

Level I

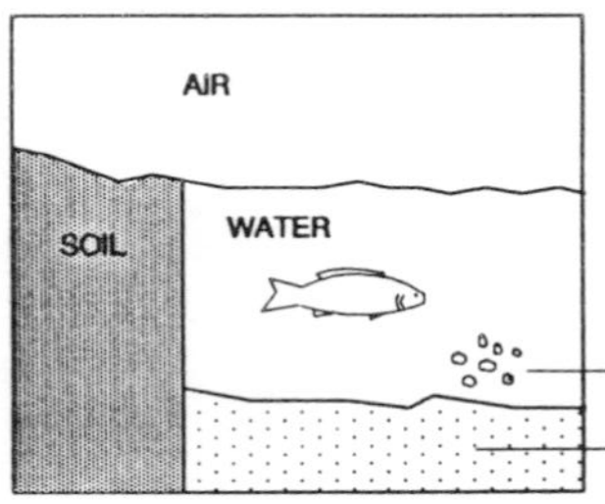

	Z	Concentration			Amount	Amount
	mol/m3 Pa	mol/m3	mg/L (or g/m3)	ug/g	kg	%
AIR	4.034E-04	2.681E-09	4.134E-07	3.487E-04	41341	41.341
WATER	3.492E-02	2.321E-07	3.578E-05	3.578E-05	7157	7.157
SOIL	5.459E+00	3.628E-05	5.594E-03	2.331E-03	50346	50.346
FISH	1.387E+01	9.217E-05	1.421E-02	1.421E-02	2.8425	2.84E-03
SUSPENDED SEDIMENT	3.412E+01	2.267E-04	3.496E-02	2.331E-02	34.962	3.50E-02
BOTTOM SEDIMENT	1.092E+01	7.255E-05	1.119E-02	4.662E-03	1118.79	1.12E+00
Total					100000	100

f = 6.646E-06 Pa

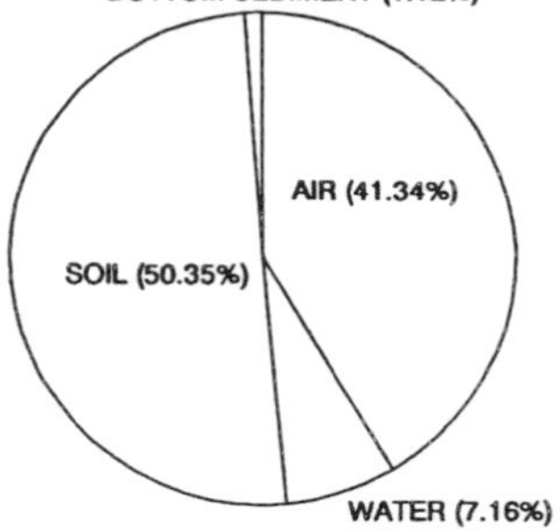

Distribution of mass

Level II

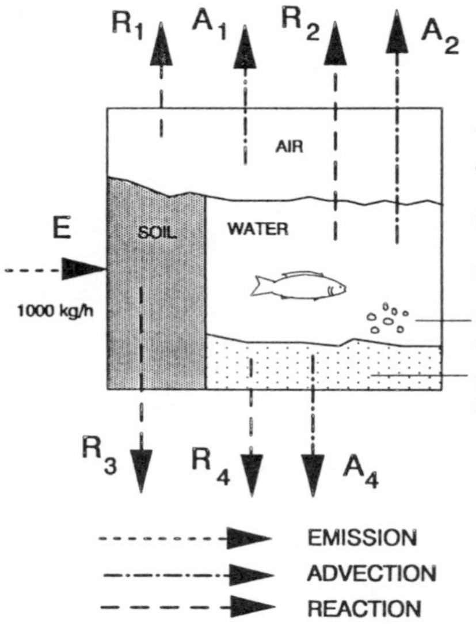

	Half-life h	D Values D(reaction)	D(advec'n)	Conc'n mol/m3	Loss Reaction kg/h	Loss Advection kg/h	Removal %
AIR	55	5.08E+08	4.03E+08	2.59E-09	5.03E+02	399.603	90.310
WATER	170	2.85E+07	6.98E+06	2.24E-07	2.82E+01	6.918	3.512
SOIL	550	6.19E+07		3.51E-05	6.13E+01		6.13E+00
FISH				8.91E-05			
SUSPENDED SEDIMENT				2.19E-04			
BOTTOM SEDIMENT	1700	4.45E+05	2.18E+04	7.01E-05	4.41E-01	2.16E-02	4.62E-02
Total		5.99E+08	4.10E+08		593.46	406.54	100
R + A			1.01E+09			1000	

f = 6.424E-06 Pa
Total amount = 96660 kg

Overall residence time = 96.66 h
Reaction time = 162.88 h
Advection time = 237.76 h

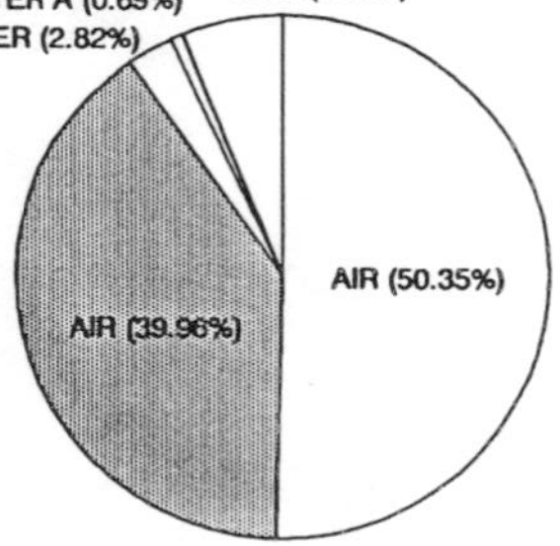

Distribution of removal rates

Level III

Chemical name: Biphenyl

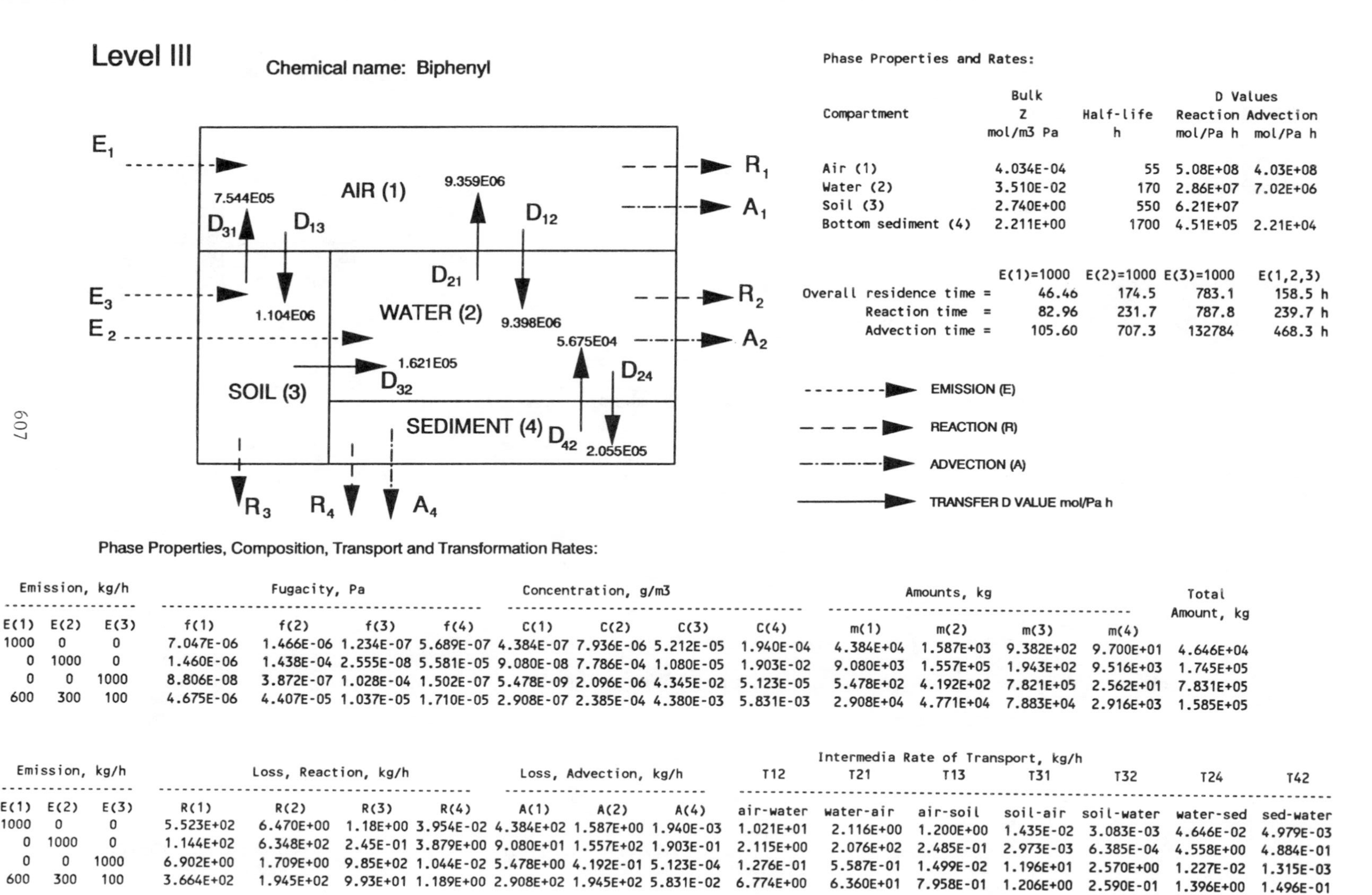

Phase Properties and Rates:

Compartment	Bulk Z mol/m3 Pa	Half-life h	D Values Reaction mol/Pa h	D Values Advection mol/Pa h
Air (1)	4.034E-04	55	5.08E+08	4.03E+08
Water (2)	3.510E-02	170	2.86E+07	7.02E+06
Soil (3)	2.740E+00	550	6.21E+07	
Bottom sediment (4)	2.211E+00	1700	4.51E+05	2.21E+04

	E(1)=1000	E(2)=1000	E(3)=1000	E(1,2,3)
Overall residence time =	46.46	174.5	783.1	158.5 h
Reaction time =	82.96	231.7	787.8	239.7 h
Advection time =	105.60	707.3	132784	468.3 h

Phase Properties, Composition, Transport and Transformation Rates:

Emission, kg/h			Fugacity, Pa				Concentration, g/m3				Amounts, kg				Total
E(1)	E(2)	E(3)	f(1)	f(2)	f(3)	f(4)	C(1)	C(2)	C(3)	C(4)	m(1)	m(2)	m(3)	m(4)	Amount, kg
1000	0	0	7.047E-06	1.466E-06	1.234E-07	5.689E-07	4.384E-07	7.936E-06	5.212E-05	1.940E-04	4.384E+04	1.587E+03	9.382E+02	9.700E+01	4.646E+04
0	1000	0	1.460E-06	1.438E-04	2.555E-08	5.581E-05	9.080E-08	7.786E-04	1.080E-05	1.903E-02	9.080E+03	1.557E+05	1.943E+02	9.516E+03	1.745E+05
0	0	1000	8.806E-08	3.872E-07	1.028E-04	1.502E-07	5.478E-09	2.096E-06	4.345E-02	5.123E-05	5.478E+02	4.192E+02	7.821E+05	2.562E+01	7.831E+05
600	300	100	4.675E-06	4.407E-05	1.037E-05	1.710E-05	2.908E-07	2.385E-04	4.380E-03	5.831E-03	2.908E+04	4.771E+04	7.883E+04	2.916E+03	1.585E+05

Emission, kg/h			Loss, Reaction, kg/h				Loss, Advection, kg/h			Intermedia Rate of Transport, kg/h						
										T12	T21	T13	T31	T32	T24	T42
E(1)	E(2)	E(3)	R(1)	R(2)	R(3)	R(4)	A(1)	A(2)	A(4)	air-water	water-air	air-soil	soil-air	soil-water	water-sed	sed-water
1000	0	0	5.523E+02	6.470E+00	1.18E+00	3.954E-02	4.384E+02	1.587E+00	1.940E-03	1.021E+01	2.116E+00	1.200E+00	1.435E-02	3.083E-03	4.646E-02	4.979E-03
0	1000	0	1.144E+02	6.348E+02	2.45E-01	3.879E+00	9.080E+01	1.557E+02	1.903E-01	2.115E+00	2.076E+02	2.485E-01	2.973E-03	6.385E-04	4.558E+00	4.884E-01
0	0	1000	6.902E+00	1.709E+00	9.85E+02	1.044E-02	5.478E+00	4.192E-01	5.123E-04	1.276E-01	5.587E-01	1.499E-02	1.196E+01	2.570E+00	1.227E-02	1.315E-03
600	300	100	3.664E+02	1.945E+02	9.93E+01	1.189E+00	2.908E+02	1.945E+02	5.831E-02	6.774E+00	6.360E+01	7.958E-01	1.206E+00	2.590E-01	1.396E+00	1.496E-01

Chemical name: 2-Chlorobiphenyl

Level I

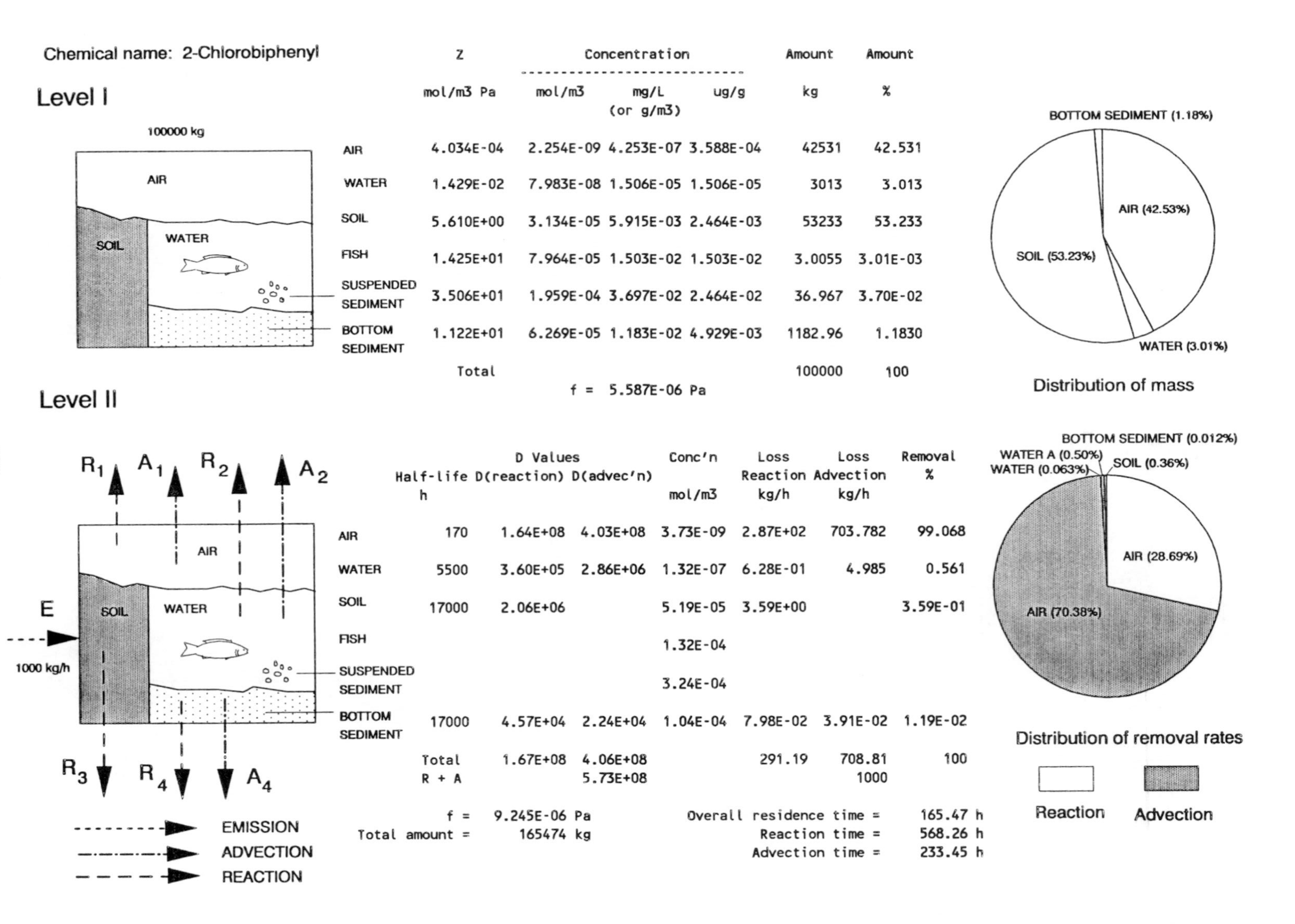

	Z	Concentration			Amount	Amount
	mol/m3 Pa	mol/m3	mg/L (or g/m3)	ug/g	kg	%
AIR	4.034E-04	2.254E-09	4.253E-07	3.588E-04	42531	42.531
WATER	1.429E-02	7.983E-08	1.506E-05	1.506E-05	3013	3.013
SOIL	5.610E+00	3.134E-05	5.915E-03	2.464E-03	53233	53.233
FISH	1.425E+01	7.964E-05	1.503E-02	1.503E-02	3.0055	3.01E-03
SUSPENDED SEDIMENT	3.506E+01	1.959E-04	3.697E-02	2.464E-02	36.967	3.70E-02
BOTTOM SEDIMENT	1.122E+01	6.269E-05	1.183E-02	4.929E-03	1182.96	1.1830
Total					100000	100

f = 5.587E-06 Pa

Distribution of mass

Level II

	Half-life h	D Values D(reaction)	D(advec'n)	Conc'n mol/m3	Loss Reaction kg/h	Loss Advection kg/h	Removal %
AIR	170	1.64E+08	4.03E+08	3.73E-09	2.87E+02	703.782	99.068
WATER	5500	3.60E+05	2.86E+06	1.32E-07	6.28E-01	4.985	0.561
SOIL	17000	2.06E+06		5.19E-05	3.59E+00		3.59E-01
FISH				1.32E-04			
SUSPENDED SEDIMENT				3.24E-04			
BOTTOM SEDIMENT	17000	4.57E+04	2.24E+04	1.04E-04	7.98E-02	3.91E-02	1.19E-02
Total		1.67E+08	4.06E+08		291.19	708.81	100
R + A			5.73E+08			1000	

f = 9.245E-06 Pa
Total amount = 165474 kg

Overall residence time = 165.47 h
Reaction time = 568.26 h
Advection time = 233.45 h

Distribution of removal rates

Level III

Chemical name: 2-Chlorobiphenyl

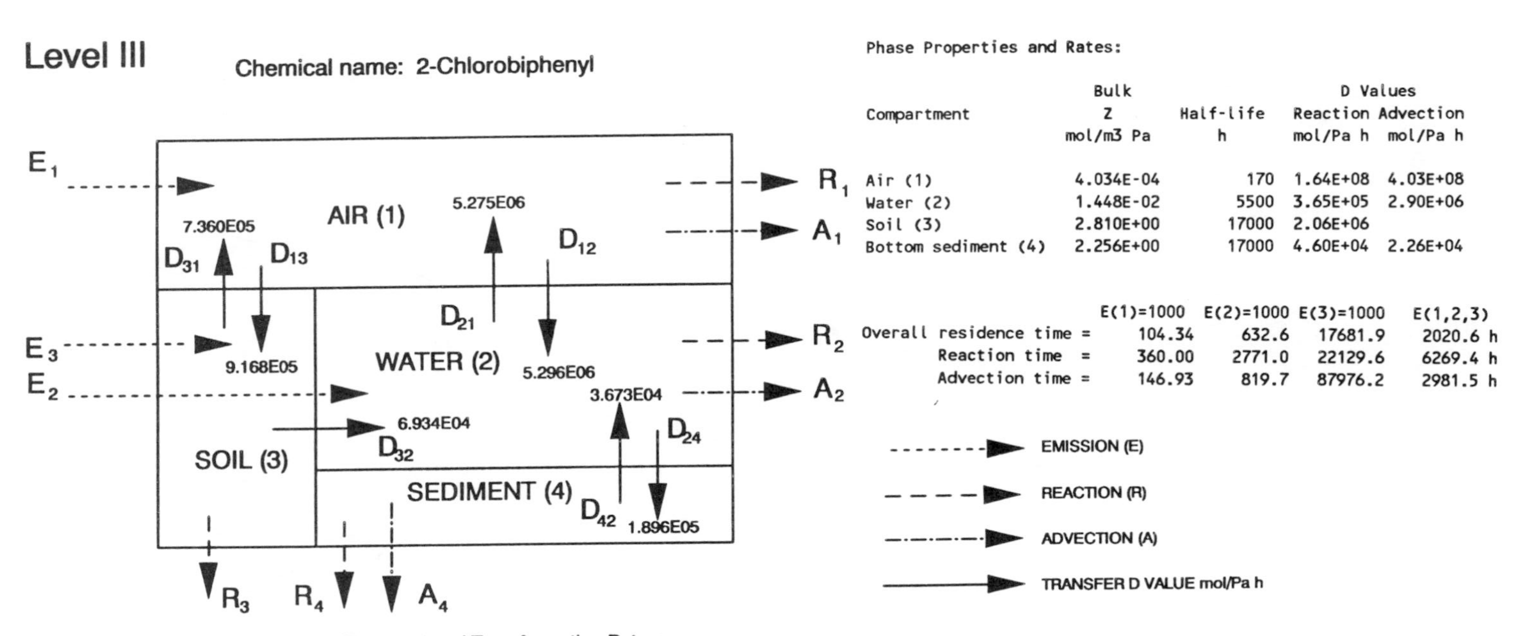

Phase Properties and Rates:

Compartment	Bulk Z mol/m3 Pa	Half-life h	D Values Reaction mol/Pa h	D Values Advection mol/Pa h
Air (1)	4.034E-04	170	1.64E+08	4.03E+08
Water (2)	1.448E-02	5500	3.65E+05	2.90E+06
Soil (3)	2.810E+00	17000	2.06E+06	
Bottom sediment (4)	2.256E+00	17000	4.60E+04	2.26E+04

	E(1)=1000	E(2)=1000	E(3)=1000	E(1,2,3)
Overall residence time =	104.34	632.6	17681.9	2020.6 h
Reaction time =	360.00	2771.0	22129.6	6269.4 h
Advection time =	146.93	819.7	87976.2	2981.5 h

EMISSION (E)

REACTION (R)

ADVECTION (A)

TRANSFER D VALUE mol/Pa h

Phase Properties, Compositions, Transport and Transformation Rates:

Emission, kg/h			Fugacity, Pa				Concentration, g/m3				Amounts, kg				Total
E(1)	E(2)	E(3)	f(1)	f(2)	f(3)	f(4)	C(1)	C(2)	C(3)	C(4)	m(1)	m(2)	m(3)	m(4)	Amount, kg
1000	0	0	9.287E-06	5.703E-06	2.970E-06	1.027E-05	7.070E-07	1.558E-05	1.574E-03	4.373E-03	7.070E+04	3.116E+03	2.834E+04	2.186E+03	1.043E+05
0	1000	0	5.658E-06	6.155E-04	1.809E-06	1.109E-03	4.307E-07	1.681E-03	9.592E-04	4.719E-01	4.307E+04	3.363E+05	1.727E+04	2.359E+05	6.326E+05
0	0	1000	2.521E-06	1.635E-05	1.849E-03	2.945E-05	1.919E-07	4.467E-05	9.804E-01	1.254E-02	1.919E+04	8.934E+03	1.765E+07	6.268E+03	1.768E+07
600	300	100	7.522E-06	1.897E-04	1.873E-04	3.417E-04	5.726E-07	5.182E-04	9.927E-02	1.454E-01	5.726E+04	1.036E+05	1.787E+06	7.272E+04	2.021E+06

Emission, kg/h			Loss, Reaction, kg/h				Loss, Advection, kg/h			Intermedia Rate of Transport, kg/h T12	T21	T13	T31	T32	T24	T42
E(1)	E(2)	E(3)	R(1)	R(2)	R(3)	R(4)	A(1)	A(2)	A(4)	air-water	water-air	air-soil	soil-air	soil-water	water-sed	sed-water
1000	0	0	2.882E+02	3.926E-01	1.16E+00	8.913E-02	7.070E+02	3.116E+00	4.373E-02	9.280E+00	5.677E+00	1.607E+00	4.125E-01	3.886E-02	2.041E-01	7.120E-02
0	1000	0	1.756E+02	4.237E+01	7.04E-01	9.618E+00	4.307E+02	3.363E+02	4.719E+00	5.654E+00	6.127E+02	9.788E-01	2.513E-01	2.368E-02	2.202E+01	7.684E+00
0	0	1000	7.824E+01	1.126E+00	7.19E+02	2.555E-01	1.919E+02	8.934E+00	1.254E-01	2.519E+00	1.628E+01	4.361E-01	2.568E+02	2.420E+01	5.850E-01	2.041E-01
600	300	100	2.334E+02	1.306E+01	7.28E+01	2.964E+00	5.726E+02	1.306E+01	1.454E+00	7.516E+00	1.888E+02	1.301E+00	2.601E+01	2.450E+00	6.787E+00	2.368E+00

Chemical name: 3-Chlorobiphenyl

Level I

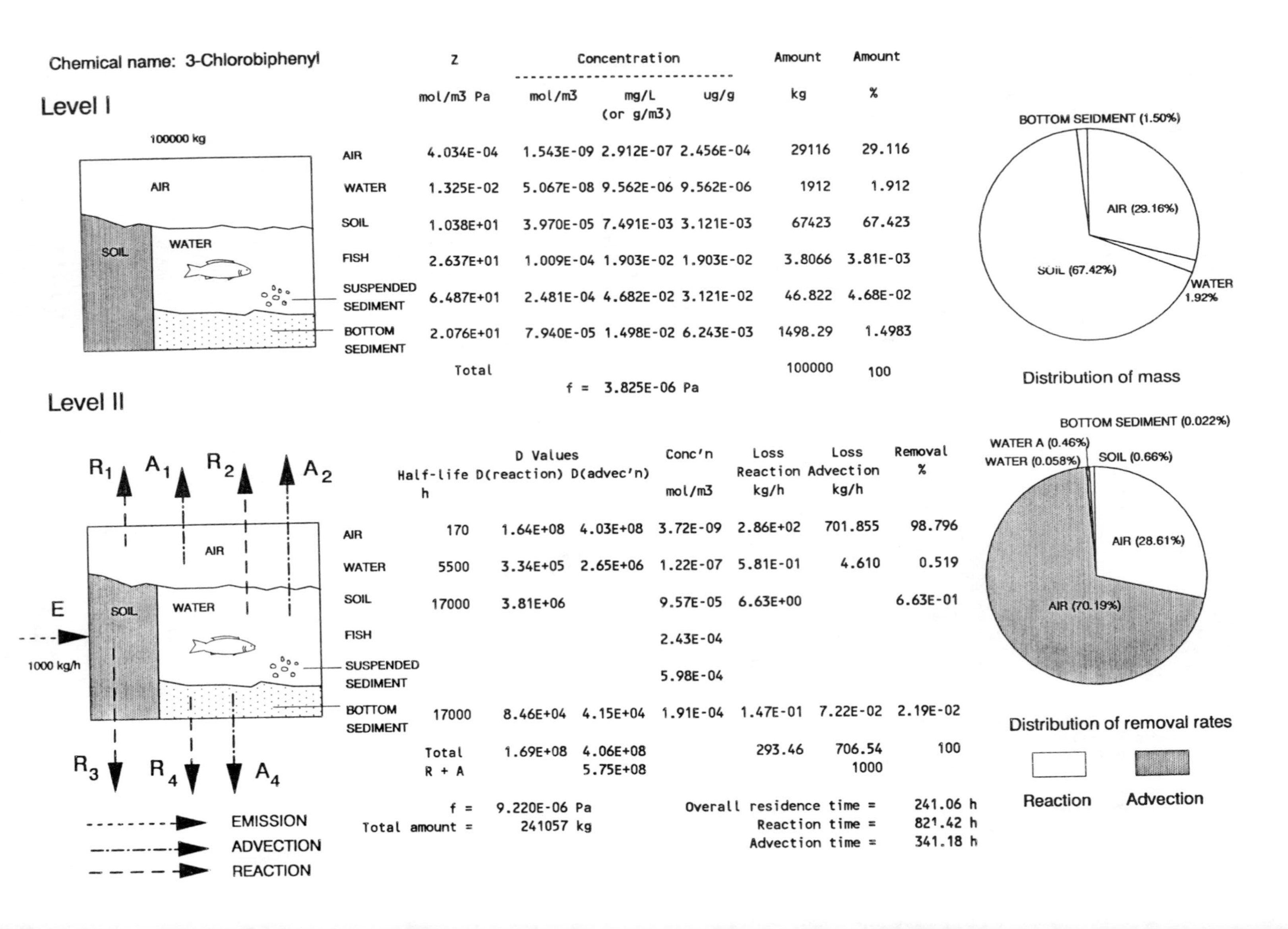

	Z	Concentration			Amount	Amount
	mol/m3 Pa	mol/m3	mg/L (or g/m3)	ug/g	kg	%
AIR	4.034E-04	1.543E-09	2.912E-07	2.456E-04	29116	29.116
WATER	1.325E-02	5.067E-08	9.562E-06	9.562E-06	1912	1.912
SOIL	1.038E+01	3.970E-05	7.491E-03	3.121E-03	67423	67.423
FISH	2.637E+01	1.009E-04	1.903E-02	1.903E-02	3.8066	3.81E-03
SUSPENDED SEDIMENT	6.487E+01	2.481E-04	4.682E-02	3.121E-02	46.822	4.68E-02
BOTTOM SEDIMENT	2.076E+01	7.940E-05	1.498E-02	6.243E-03	1498.29	1.4983
	Total				100000	100

f = 3.825E-06 Pa

Level II

	Half-life h	D Values D(reaction)	D(advec'n)	Conc'n mol/m3	Loss Reaction kg/h	Loss Advection kg/h	Removal %
AIR	170	1.64E+08	4.03E+08	3.72E-09	2.86E+02	701.855	98.796
WATER	5500	3.34E+05	2.65E+06	1.22E-07	5.81E-01	4.610	0.519
SOIL	17000	3.81E+06		9.57E-05	6.63E+00		6.63E-01
FISH				2.43E-04			
SUSPENDED SEDIMENT				5.98E-04			
BOTTOM SEDIMENT	17000	8.46E+04	4.15E+04	1.91E-04	1.47E-01	7.22E-02	2.19E-02
	Total	1.69E+08	4.06E+08		293.46	706.54	100
	R + A		5.75E+08			1000	

f = 9.220E-06 Pa
Total amount = 241057 kg

Overall residence time = 241.06 h
Reaction time = 821.42 h
Advection time = 341.18 h

Level III

Chemical name: 3-Chlorobiphenyl

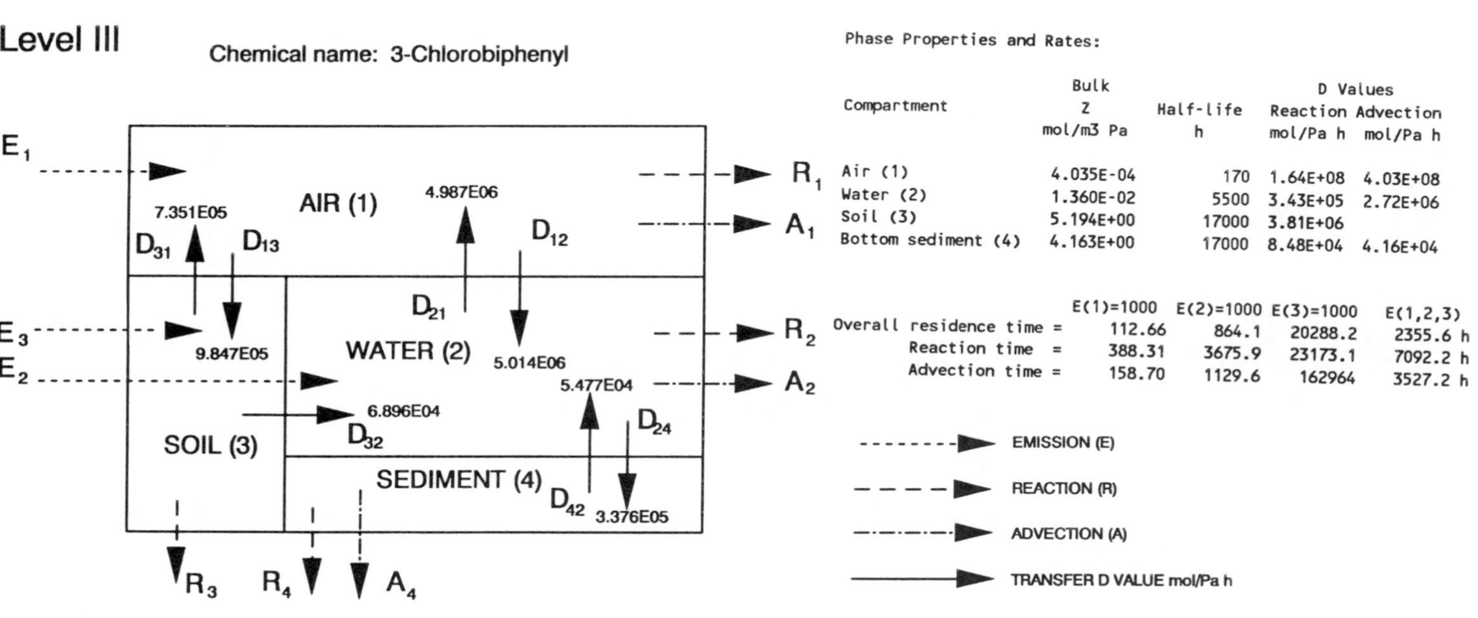

Phase Properties and Rates:

Compartment	Bulk Z mol/m3 Pa	Half-life h	D Values Reaction mol/Pa h	D Values Advection mol/Pa h
Air (1)	4.035E-04	170	1.64E+08	4.03E+08
Water (2)	1.360E-02	5500	3.43E+05	2.72E+06
Soil (3)	5.194E+00	17000	3.81E+06	
Bottom sediment (4)	4.163E+00	17000	8.48E+04	4.16E+04

	E(1)=1000	E(2)=1000	E(3)=1000	E(1,2,3)
Overall residence time =	112.66	864.1	20288.2	2355.6 h
Reaction time =	388.31	3675.9	23173.1	7092.2 h
Advection time =	158.70	1129.6	162964	3527.2 h

Phase Properties, Compositions, Transport and Transformation Rates:

Emission, kg/h			Fugacity, Pa				Concentration, g/m3				Amounts, kg				Total
E(1)	E(2)	E(3)	f(1)	f(2)	f(3)	f(4)	C(1)	C(2)	C(3)	C(4)	m(1)	m(2)	m(3)	m(4)	Amount, kg
1000	0	0	9.285E-06	5.636E-06	1.981E-06	1.050E-05	7.069E-07	1.446E-05	1.942E-03	8.247E-03	7.069E+04	2.893E+03	3.495E+04	4.124E+03	1.127E+05
0	1000	0	5.589E-06	6.431E-04	1.192E-06	1.198E-03	4.255E-07	1.650E-03	1.169E-03	9.409E-01	4.255E+04	3.300E+05	2.104E+04	4.705E+05	8.641E+05
0	0	1000	1.562E-06	1.051E-05	1.149E-03	1.957E-05	1.189E-07	2.696E-05	1.126E+00	1.537E-02	1.189E+04	5.392E+03	2.026E+07	7.686E+03	2.029E+07
600	300	100	7.404E-06	1.973E-04	1.164E-04	3.676E-04	5.637E-07	5.064E-04	1.141E-01	2.888E-01	5.637E+04	1.013E+05	2.054E+06	1.444E+05	2.356E+06

Emission, kg/h			Loss, Reaction, kg/h				Loss, Advection, kg/h			Intermedia Rate of Transport, kg/h T12	T21	T13	T31	T32	T24	T42
E(1)	E(2)	E(3)	R(1)	R(2)	R(3)	R(4)	A(1)	A(2)	A(4)	air-water	water-air	air-soil	soil-air	soil-water	water-sed	sed-water
1000	0	0	2.882E+02	3.645E-01	1.42E+00	1.681E-01	7.069E+02	2.893E+00	8.247E-02	8.786E+00	5.304E+00	1.725E+00	2.748E-01	2.578E-02	3.591E-01	1.085E-01
0	1000	0	1.734E+02	4.158E+01	8.58E-01	1.918E+01	4.255E+02	3.300E+02	9.409E+00	5.288E+00	6.051E+02	1.038E+00	1.654E-01	1.552E-02	4.097E+01	1.238E+01
0	0	1000	4.849E+01	6.794E-01	8.26E+02	3.133E-01	1.189E+02	5.392E+00	1.537E-01	1.478E+00	9.886E+00	2.903E-01	1.593E+02	1.495E+01	6.693E-01	2.023E-01
600	300	100	2.298E+02	1.276E+01	8.37E+01	5.886E+00	5.637E+02	1.276E+01	2.888E+00	7.006E+00	1.857E+02	1.376E+00	1.615E+01	1.515E+00	1.257E+01	3.799E+00

Chemical name: 4-Chlorobiphenyl

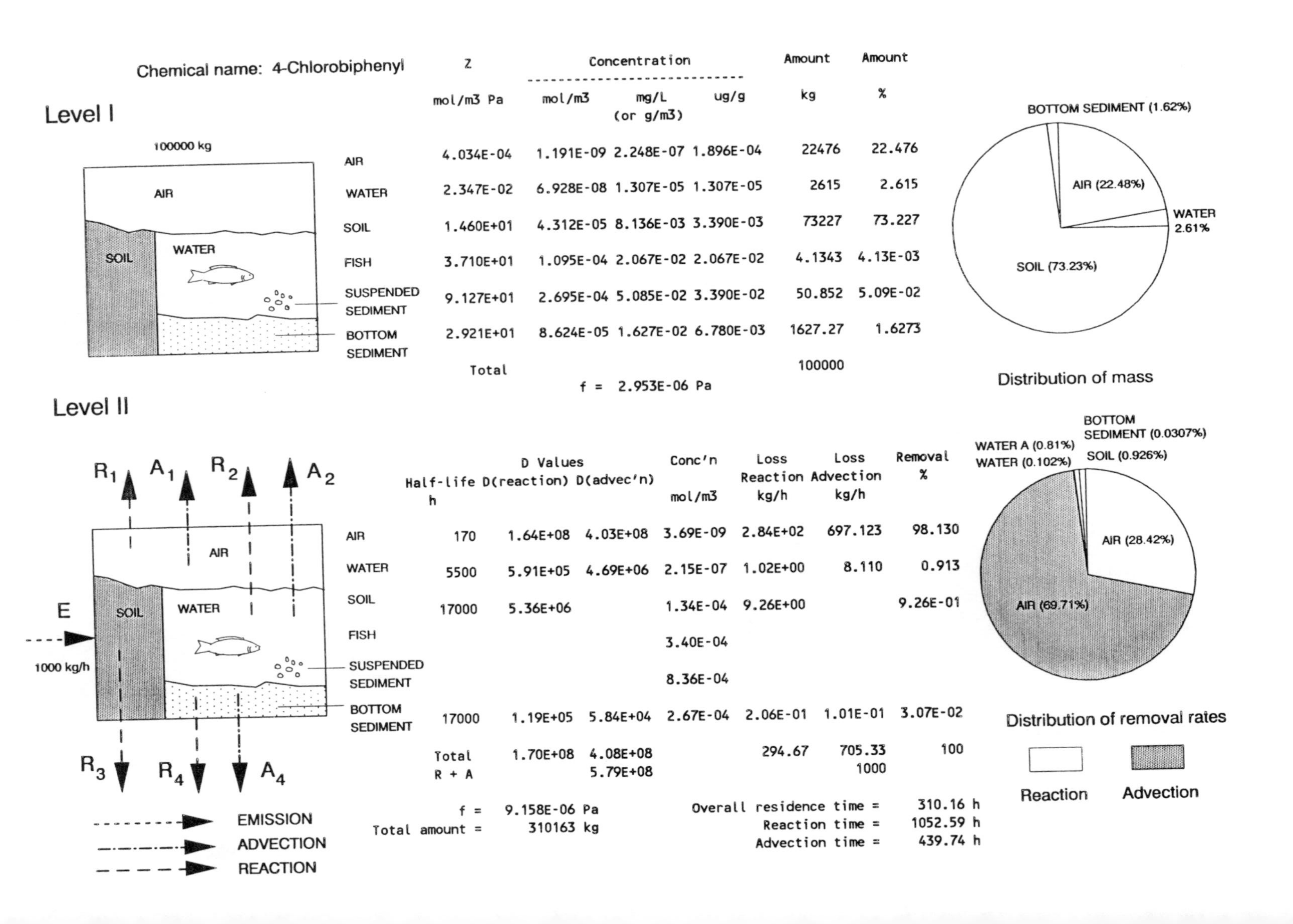

Level I

	Z mol/m3 Pa	Concentration mol/m3	mg/L (or g/m3)	ug/g	Amount kg	Amount %
AIR	4.034E-04	1.191E-09	2.248E-07	1.896E-04	22476	22.476
WATER	2.347E-02	6.928E-08	1.307E-05	1.307E-05	2615	2.615
SOIL	1.460E+01	4.312E-05	8.136E-03	3.390E-03	73227	73.227
FISH	3.710E+01	1.095E-04	2.067E-02	2.067E-02	4.1343	4.13E-03
SUSPENDED SEDIMENT	9.127E+01	2.695E-04	5.085E-02	3.390E-02	50.852	5.09E-02
BOTTOM SEDIMENT	2.921E+01	8.624E-05	1.627E-02	6.780E-03	1627.27	1.6273
	Total				100000	

f = 2.953E-06 Pa

Level II

	Half-life h	D Values D(reaction)	D(advec'n)	Conc'n mol/m3	Loss Reaction kg/h	Loss Advection kg/h	Removal %
AIR	170	1.64E+08	4.03E+08	3.69E-09	2.84E+02	697.123	98.130
WATER	5500	5.91E+05	4.69E+06	2.15E-07	1.02E+00	8.110	0.913
SOIL	17000	5.36E+06		1.34E-04	9.26E+00		9.26E-01
FISH				3.40E-04			
SUSPENDED SEDIMENT				8.36E-04			
BOTTOM SEDIMENT	17000	1.19E+05	5.84E+04	2.67E-04	2.06E-01	1.01E-01	3.07E-02
	Total	1.70E+08	4.08E+08		294.67	705.33	100
	R + A		5.79E+08			1000	

f = 9.158E-06 Pa
Total amount = 310163 kg

Overall residence time = 310.16 h
Reaction time = 1052.59 h
Advection time = 439.74 h

Level III

Chemical name: 4-Chlorobiphenyl

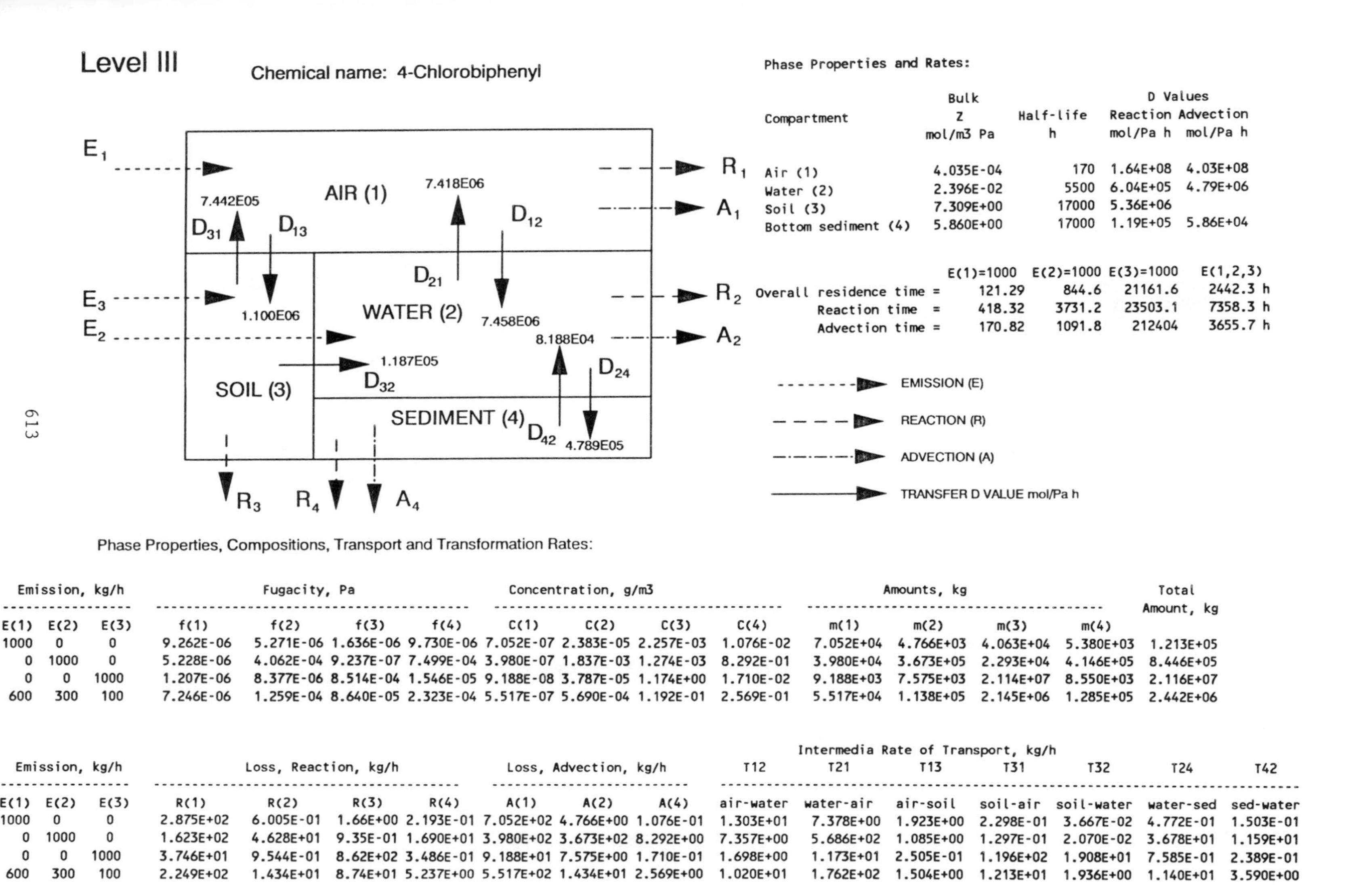

Phase Properties and Rates:

Compartment	Bulk Z mol/m3 Pa	Half-life h	D Values Reaction mol/Pa h	D Values Advection mol/Pa h
Air (1)	4.035E-04	170	1.64E+08	4.03E+08
Water (2)	2.396E-02	5500	6.04E+05	4.79E+06
Soil (3)	7.309E+00	17000	5.36E+06	
Bottom sediment (4)	5.860E+00	17000	1.19E+05	5.86E+04

	E(1)=1000	E(2)=1000	E(3)=1000	E(1,2,3)
Overall residence time =	121.29	844.6	21161.6	2442.3 h
Reaction time =	418.32	3731.2	23503.1	7358.3 h
Advection time =	170.82	1091.8	212404	3655.7 h

Phase Properties, Compositions, Transport and Transformation Rates:

Emission, kg/h			Fugacity, Pa				Concentration, g/m3				Amounts, kg				Total
E(1)	E(2)	E(3)	f(1)	f(2)	f(3)	f(4)	C(1)	C(2)	C(3)	C(4)	m(1)	m(2)	m(3)	m(4)	Amount, kg
1000	0	0	9.262E-06	5.271E-06	1.636E-06	9.730E-06	7.052E-07	2.383E-05	2.257E-03	1.076E-02	7.052E+04	4.766E+03	4.063E+04	5.380E+03	1.213E+05
0	1000	0	5.228E-06	4.062E-04	9.237E-07	7.499E-04	3.980E-07	1.837E-03	1.274E-03	8.292E-01	3.980E+04	3.673E+05	2.293E+04	4.146E+05	8.446E+05
0	0	1000	1.207E-06	8.377E-06	8.514E-04	1.546E-05	9.188E-08	3.787E-05	1.174E+00	1.710E-02	9.188E+03	7.575E+03	2.114E+07	8.550E+03	2.116E+07
600	300	100	7.246E-06	1.259E-04	8.640E-05	2.323E-04	5.517E-07	5.690E-04	1.192E-01	2.569E-01	5.517E+04	1.138E+05	2.145E+06	1.285E+05	2.442E+06

Emission, kg/h			Loss, Reaction, kg/h				Loss, Advection, kg/h			Intermedia Rate of Transport, kg/h T12	T21	T13	T31	T32	T24	T42
E(1)	E(2)	E(3)	R(1)	R(2)	R(3)	R(4)	A(1)	A(2)	A(4)	air-water	water-air	air-soil	soil-air	soil-water	water-sed	sed-water
1000	0	0	2.875E+02	6.005E-01	1.66E+00	2.193E-01	7.052E+02	4.766E+00	1.076E-01	1.303E+01	7.378E+00	1.923E+00	2.298E-01	3.667E-02	4.772E-01	1.503E-01
0	1000	0	1.623E+02	4.628E+01	9.35E-01	1.690E+01	3.980E+02	3.673E+02	8.292E+00	7.357E+00	5.686E+02	1.085E+00	1.297E-01	2.070E-02	3.678E+01	1.159E+01
0	0	1000	3.746E+01	9.544E-01	8.62E+02	3.486E-01	9.188E+01	7.575E+00	1.710E-01	1.698E+00	1.173E+01	2.505E-01	1.196E+02	1.908E+01	7.585E-01	2.389E-01
600	300	100	2.249E+02	1.434E+01	8.74E+01	5.237E+00	5.517E+02	1.434E+01	2.569E+00	1.020E+01	1.762E+02	1.504E+00	1.213E+01	1.936E+00	1.140E+01	3.590E+00

Chemical name: 2,2'-Dichlorobiphenyl

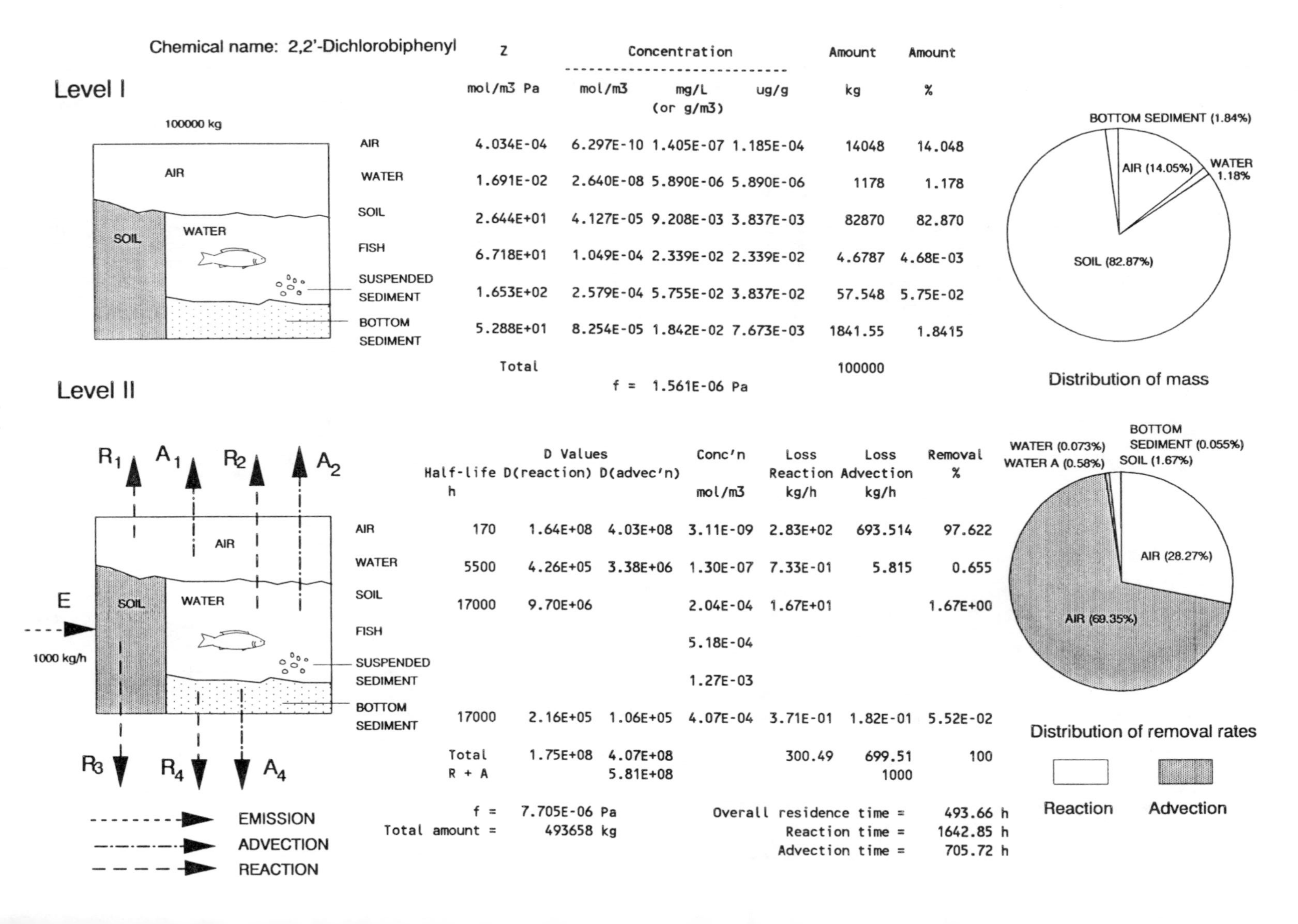

Level I

	Z	Concentration			Amount	Amount
	mol/m3 Pa	mol/m3	mg/L (or g/m3)	ug/g	kg	%
AIR	4.034E-04	6.297E-10	1.405E-07	1.185E-04	14048	14.048
WATER	1.691E-02	2.640E-08	5.890E-06	5.890E-06	1178	1.178
SOIL	2.644E+01	4.127E-05	9.208E-03	3.837E-03	82870	82.870
FISH	6.718E+01	1.049E-04	2.339E-02	2.339E-02	4.6787	4.68E-03
SUSPENDED SEDIMENT	1.653E+02	2.579E-04	5.755E-02	3.837E-02	57.548	5.75E-02
BOTTOM SEDIMENT	5.288E+01	8.254E-05	1.842E-02	7.673E-03	1841.55	1.8415
	Total				100000	

f = 1.561E-06 Pa

Level II

	Half-life h	D Values D(reaction)	D(advec'n)	Conc'n mol/m3	Loss Reaction kg/h	Loss Advection kg/h	Removal %
AIR	170	1.64E+08	4.03E+08	3.11E-09	2.83E+02	693.514	97.622
WATER	5500	4.26E+05	3.38E+06	1.30E-07	7.33E-01	5.815	0.655
SOIL	17000	9.70E+06		2.04E-04	1.67E+01		1.67E+00
FISH				5.18E-04			
SUSPENDED SEDIMENT				1.27E-03			
BOTTOM SEDIMENT	17000	2.16E+05	1.06E+05	4.07E-04	3.71E-01	1.82E-01	5.52E-02
	Total	1.75E+08	4.07E+08		300.49	699.51	100
	R + A		5.81E+08			1000	

f = 7.705E-06 Pa
Total amount = 493658 kg

Overall residence time = 493.66 h
Reaction time = 1642.85 h
Advection time = 705.72 h

Level III

Chemical name: 2,2'-Dichlorobiphenyl

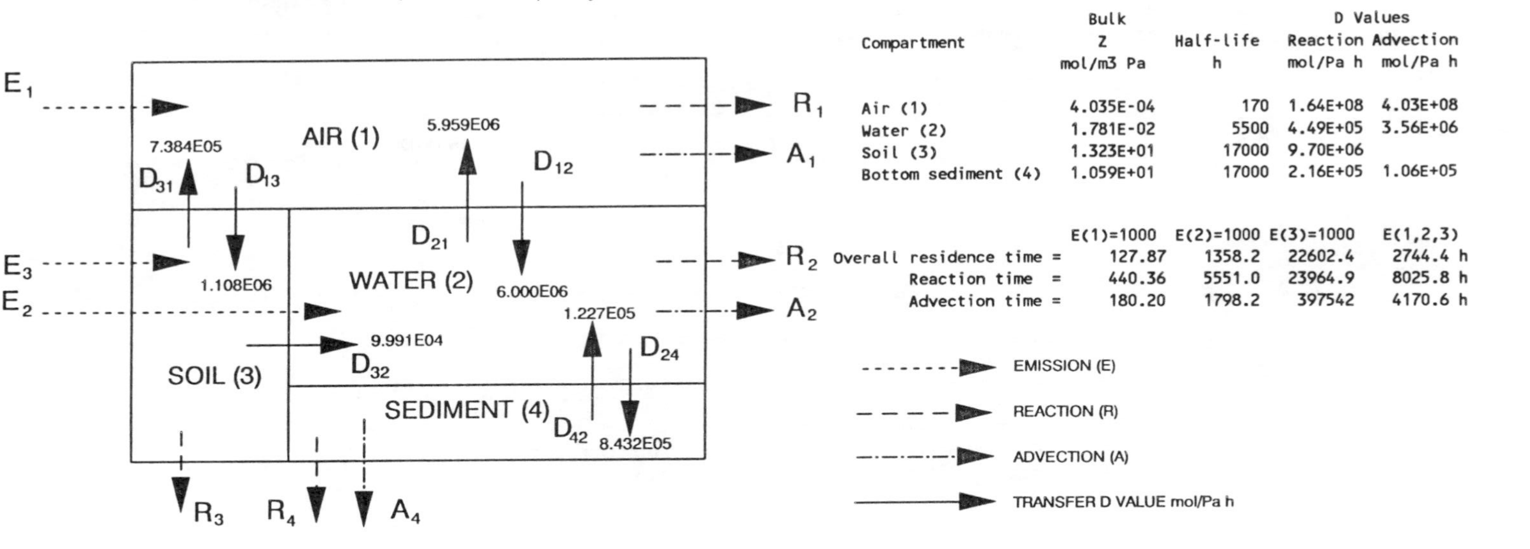

Phase Properties and Rates:

Compartment	Bulk Z mol/m3 Pa	Half-life h	D Values Reaction mol/Pa h	D Values Advection mol/Pa h
Air (1)	4.035E-04	170	1.64E+08	4.03E+08
Water (2)	1.781E-02	5500	4.49E+05	3.56E+06
Soil (3)	1.323E+01	17000	9.70E+06	
Bottom sediment (4)	1.059E+01	17000	2.16E+05	1.06E+05

	E(1)=1000	E(2)=1000	E(3)=1000	E(1,2,3)
Overall residence time =	127.87	1358.2	22602.4	2744.4 h
Reaction time =	440.36	5551.0	23964.9	8025.8 h
Advection time =	180.20	1798.2	397542	4170.6 h

Phase Properties, Compositions, Transport and Transformation Rates:

Emission, kg/h			Fugacity, Pa				Concentration, g/m3				Amounts, kg				Total Amount, kg
E(1)	E(2)	E(3)	f(1)	f(2)	f(3)	f(4)	C(1)	C(2)	C(3)	C(4)	m(1)	m(2)	m(3)	m(4)	
1000	0	0	7.841E-06	4.455E-06	8.240E-07	8.452E-06	7.059E-07	1.770E-05	2.431E-03	1.997E-02	7.059E+04	3.540E+03	4.376E+04	9.984E+03	1.279E+05
0	1000	0	4.417E-06	4.262E-04	4.641E-07	8.086E-04	3.976E-07	1.693E-03	1.369E-03	1.910E+00	3.976E+04	3.386E+05	2.465E+04	9.552E+05	1.358E+06
0	0	1000	5.910E-07	4.351E-06	4.252E-04	8.255E-06	5.320E-08	1.729E-05	1.255E+00	1.950E-02	5.320E+03	3.457E+03	2.258E+07	9.751E+03	2.260E+07
600	300	100	6.089E-06	1.310E-04	4.315E-05	2.485E-04	5.481E-07	5.203E-04	1.273E-01	5.870E-01	5.481E+04	1.041E+05	2.292E+06	2.935E+05	2.744E+06

Emission, kg/h			Loss, Reaction, kg/h				Loss, Advection, kg/h			Intermedia Rate of Transport, kg/h T12	T21	T13	T31	T32	T24	T42
E(1)	E(2)	E(3)	R(1)	R(2)	R(3)	R(4)	A(1)	A(2)	A(4)	air-water	water-air	air-soil	soil-air	soil-water	water-sed	sed-water
1000	0	0	2.877E+02	4.460E-01	1.78E+00	4.070E-01	7.059E+02	3.540E+00	1.997E-01	1.050E+01	5.922E+00	1.938E+00	1.357E-01	1.837E-02	8.380E-01	2.313E-01
0	1000	0	1.621E+02	4.267E+01	1.00E+00	3.894E+01	3.976E+02	3.386E+02	1.910E+01	5.912E+00	5.666E+02	1.092E+00	7.645E-02	1.034E-02	8.017E+01	2.213E+01
0	0	1000	2.169E+01	4.356E-01	9.21E+02	3.975E-01	5.320E+01	3.457E+00	1.950E-01	7.911E-01	5.784E+00	1.461E-01	7.004E+01	9.478E+00	8.185E-01	2.259E-01
600	300	100	2.234E+02	1.311E+01	9.34E+01	1.197E+01	5.481E+02	1.311E+01	5.870E+00	8.150E+00	1.741E+02	1.505E+00	7.109E+00	9.619E-01	2.464E+01	6.801E+00

Chemical name: 2,5-Dichlorobiphenyl

Level I

	Z	Concentration			Amount	Amount
	mol/m3 Pa	mol/m3	mg/L (or g/m3)	ug/g	kg	%
AIR	4.034E-04	1.524E-10	3.401E-08	2.869E-05	3401	3.401
WATER	4.980E-02	1.882E-08	4.198E-06	4.198E-06	839.62	0.8396
SOIL	1.234E+02	4.662E-05	1.040E-02	4.334E-03	93609	93.609
FISH	3.135E+02	1.184E-04	2.643E-02	2.643E-02	5.2851	5.29E-03
SUSPENDED SEDIMENT	7.712E+02	2.914E-04	6.501E-02	4.334E-02	65.006	6.50E-02
BOTTOM SEDIMENT	2.468E+02	9.324E-05	2.080E-02	8.668E-03	2080.21	2.0802
Total					100000	100

f = 3.778E-07 Pa

Level II

	Half-life h	D Values D(reaction)	D(advec'n)	Conc'n mol/m3	Loss Reaction kg/h	Loss Advection kg/h	Removal %
AIR	170	1.64E+08	4.03E+08	2.89E-09	262.76	644.59	90.735
WATER	5500	1.26E+06	9.96E+06	3.57E-07	2.0053	15.915	1.792
SOIL	17000	4.53E+07		8.84E-04	72.333		7.233
FISH				2.25E-03			
SUSPENDED SEDIMENT				5.52E-03			
BOTTOM SEDIMENT	17000	1.01E+06	4.94E+05	1.77E-03	1.607	0.789	0.240
Total		2.11E+08	4.14E+08		338.71	661.29	100
R + A			6.25E+08			1000	

f = 7.162E-06 Pa
Total amount = 1895538 kg

Overall residence time = 1895.54 h
Reaction time = 5596.35 h
Advection time = 2866.42 h

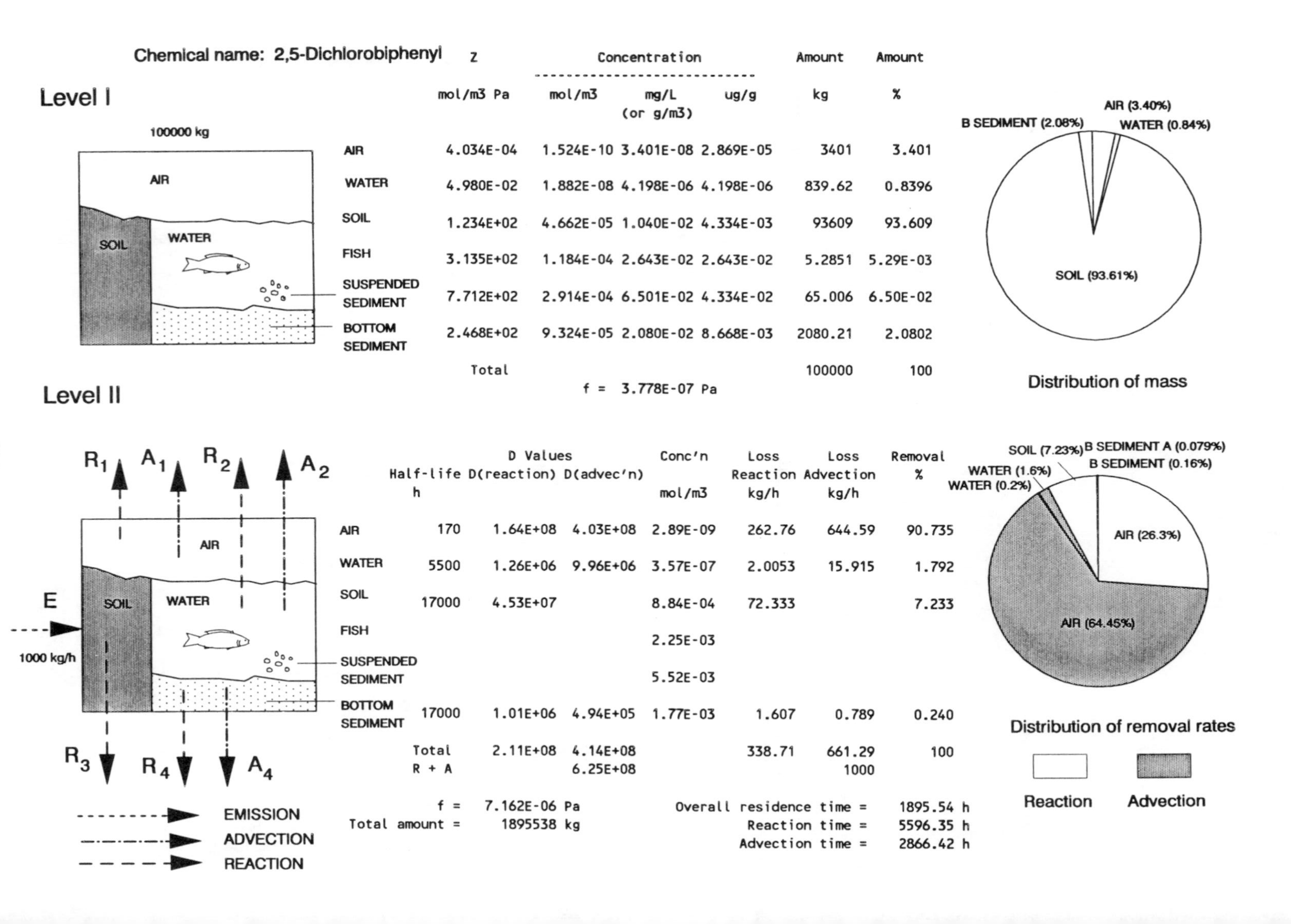

Level III

Chemical name: 2,5-Dichlorobiphenyl

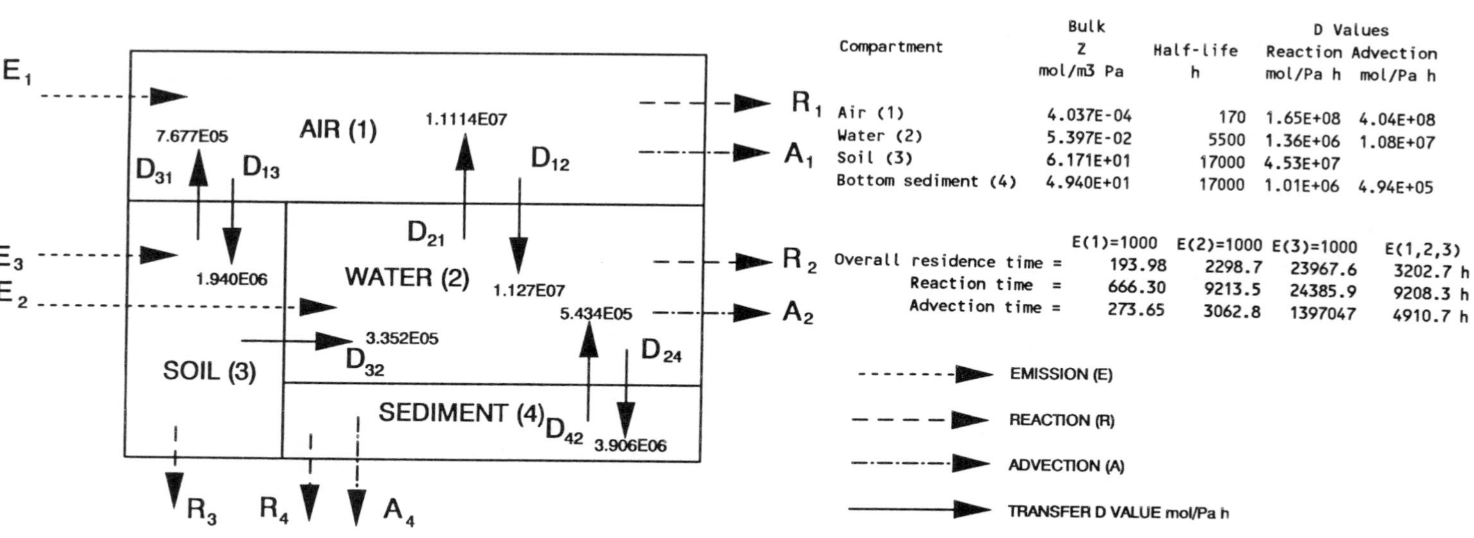

Phase Properties and Rates:

Compartment	Bulk Z mol/m3 Pa	Half-life h	D Values Reaction mol/Pa h	D Values Advection mol/Pa h
Air (1)	4.037E-04	170	1.65E+08	4.04E+08
Water (2)	5.397E-02	5500	1.36E+06	1.08E+07
Soil (3)	6.171E+01	17000	4.53E+07	
Bottom sediment (4)	4.940E+01	17000	1.01E+06	4.94E+05

	E(1)=1000	E(2)=1000	E(3)=1000	E(1,2,3)
Overall residence time =	193.98	2298.7	23967.6	3202.7 h
Reaction time =	666.30	9213.5	24385.9	9208.3 h
Advection time =	273.65	3062.8	1397047	4910.7 h

EMISSION (E)

REACTION (R)

ADVECTION (A)

TRANSFER D VALUE mol/Pa h

Phase Properties, Composition, Transport and Transformation Rates:

Emission, kg/h			Fugacity, Pa				Concentration, g/m3				Amounts, kg				Total Amount, kg
E(1)	E(2)	E(3)	f(1)	f(2)	f(3)	f(4)	C(1)	C(2)	C(3)	C(4)	m(1)	m(2)	m(3)	m(4)	
1000	0	0	7.773E-06	3.353E-06	3.252E-07	6.408E-06	7.001E-07	4.038E-05	4.477E-03	7.061E-02	7.001E+04	8.076E+03	8.059E+04	3.531E+04	1.940E+05
0	1000	0	3.311E-06	1.727E-04	1.385E-07	3.300E-04	2.982E-07	2.080E-03	1.907E-03	3.637E+00	2.982E+04	4.160E+05	3.432E+04	1.819E+06	2.299E+06
0	0	1000	1.526E-07	1.304E-06	9.664E-05	2.491E-06	1.374E-08	1.570E-05	1.331E+00	2.745E-02	1.374E+03	3.139E+03	2.395E+07	1.373E+04	2.397E+07
600	300	100	5.672E-06	5.396E-05	9.901E-06	1.031E-04	5.109E-07	6.498E-04	1.363E-01	1.136E+00	5.109E+04	1.300E+05	2.454E+06	5.681E+05	3.203E+06

Emission, kg/h			Loss, Reaction, kg/h				Loss, Advection, kg/h			Intermedia Rate of Transport, kg/h						
										T12	T21	T13	T31	T32	T24	T42
E(1)	E(2)	E(3)	R(1)	R(2)	R(3)	R(4)	A(1)	A(2)	A(4)	air-water	water-air	air-soil	soil-air	soil-water	water-sed	sed-water
1000	0	0	2.854E+02	1.018E+00	3.29E+00	1.439E+00	7.001E+02	8.076E+00	7.061E-01	1.955E+01	8.338E+00	3.365E+00	5.570E-02	2.432E-02	2.922E+00	7.768E-01
0	1000	0	1.215E+02	5.241E+01	1.40E+00	7.413E+01	2.982E+02	4.160E+02	3.637E+01	8.327E+00	4.294E+02	1.433E+00	2.372E-02	1.036E-02	1.505E+02	4.001E+01
0	0	1000	5.602E+00	3.956E-01	9.76E+02	5.595E-01	1.374E+01	3.139E+00	2.745E-01	3.838E-01	3.241E+00	6.605E-02	1.655E+01	7.226E+00	1.136E+00	3.020E-01
600	300	100	2.083E+02	1.637E+01	1.00E+02	2.316E+01	5.109E+02	1.637E+01	1.136E+01	1.427E+01	1.342E+02	2.456E+00	1.696E+00	7.403E-01	4.702E+01	1.250E+01

Chemical name: 4,4'-Dichlorobiphenyl

Level I

	Z mol/m3 Pa	Concentration mol/m3	Concentration mg/L (or g/m3)	Concentration ug/g	Amount kg	Amount %
AIR	4.034E-04	8.706E-11	1.942E-08	1.638E-05	1942	1.942
WATER	5.603E-02	1.209E-08	2.697E-06	2.697E-06	539	0.539
SOIL	2.200E+02	4.748E-05	1.059E-02	4.413E-03	95328	95.328
FISH	5.590E+02	1.206E-04	2.691E-02	2.691E-02	5.3821	5.38E-03
SUSPENDED SEDIMENT	1.375E+03	2.967E-04	6.620E-02	4.413E-02	66.200	6.62E-02
BOTTOM SEDIMENT	4.400E+02	9.495E-05	2.118E-02	8.827E-03	2118.41	2.1184
Total					100000	

f = 2.158E-07 Pa

Level II

	Half-life h	D Values D(reaction)	D Values D(advec'n)	Conc'n mol/m3	Loss Reaction kg/h	Loss Advection kg/h	Removal %
AIR	170	1.64E+08	4.03E+08	2.72E-09	2.48E+02	607.669	85.538
WATER	5500	1.41E+06	1.12E+07	3.78E-07	2.13E+00	16.879	1.901
SOIL	17000	8.07E+07		1.49E-03	1.22E+02		1.22E+01
FISH				3.77E-03			
SUSPENDED SEDIMENT				9.28E-03			
BOTTOM SEDIMENT	17000	1.79E+06	8.80E+05	2.97E-03	2.70E+00	1.33E+00	4.03E-01
Total		2.47E+08	4.16E+08		374.13	625.87	100
R + A			6.62E+08			1000	

f = 6.752E-06 Pa
Total amount = 3128729 kg

Overall residence time = 3128.73 h
Reaction time = 8362.76 h
Advection time = 4998.98 h

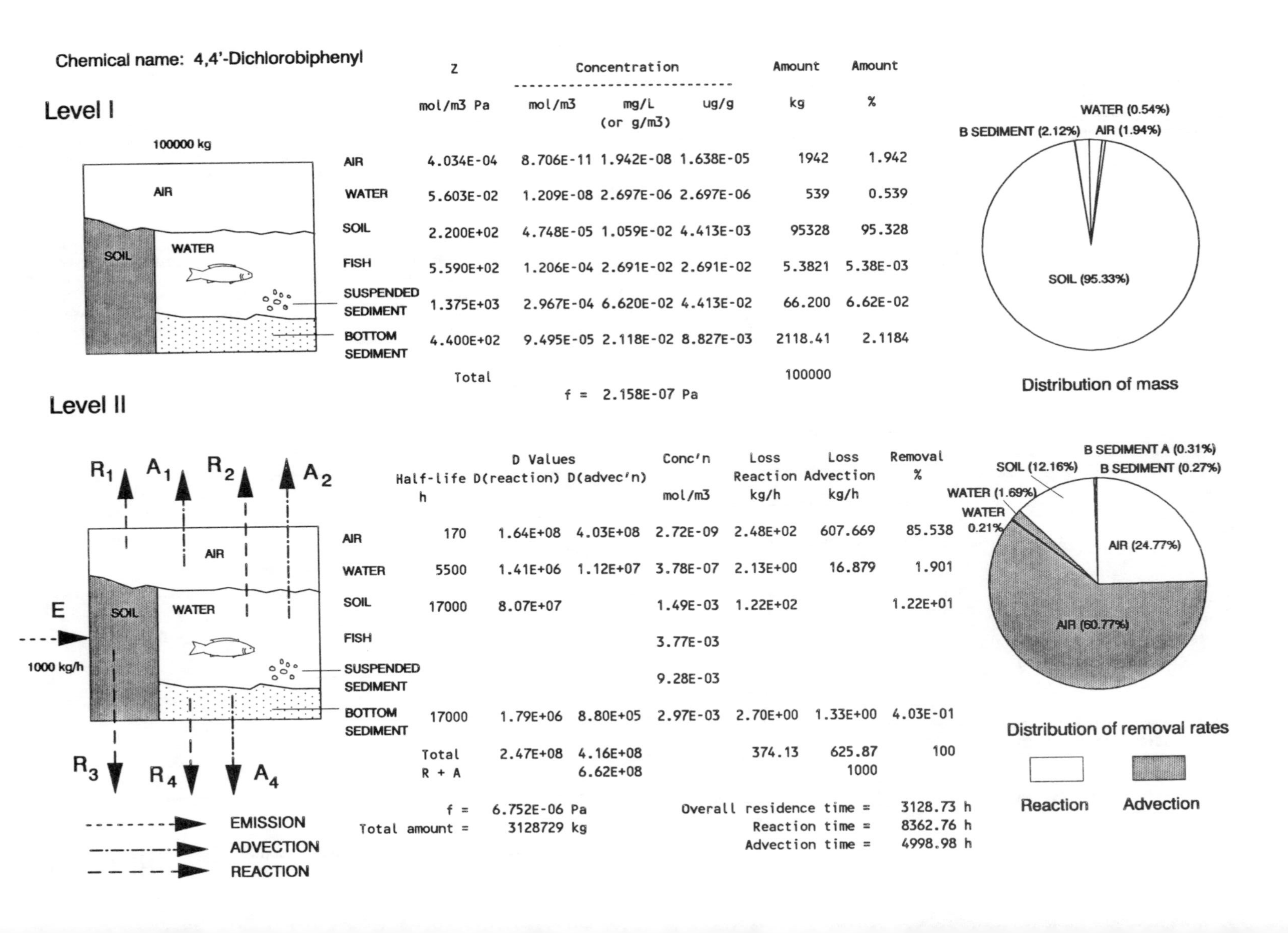

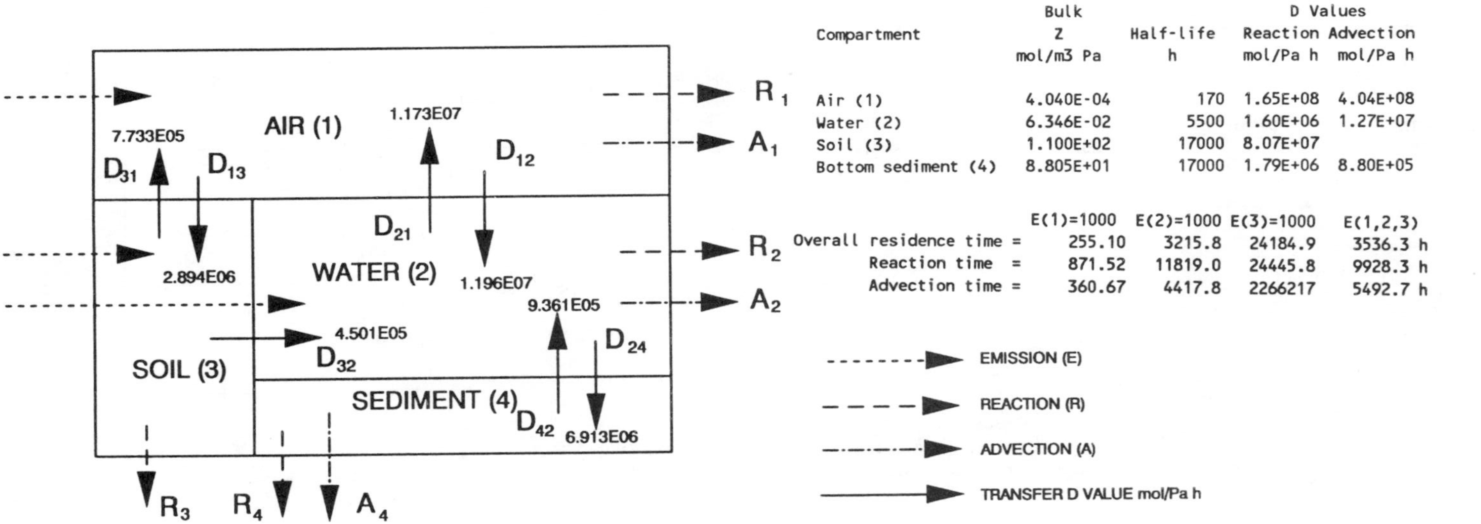

Phase Properties and Rates:

Compartment	Bulk Z mol/m3 Pa	Half-life h	D Values Reaction mol/Pa h	D Values Advection mol/Pa h
Air (1)	4.040E-04	170	1.65E+08	4.04E+08
Water (2)	6.346E-02	5500	1.60E+06	1.27E+07
Soil (3)	1.100E+02	17000	8.07E+07	
Bottom sediment (4)	8.805E+01	17000	1.79E+06	8.80E+05

	E(1)=1000	E(2)=1000	E(3)=1000	E(1,2,3)
Overall residence time =	255.10	3215.8	24184.9	3536.3 h
Reaction time =	871.52	11819.0	24445.8	9928.3 h
Advection time =	360.67	4417.8	2266217	5492.7 h

EMISSION (E)

REACTION (R)

ADVECTION (A)

TRANSFER D VALUE mol/Pa h

Phase Properties, Compositions, Transport and Transformation Rates:

Emission, kg/h			Fugacity, Pa				Concentration, g/m3				Amounts, kg				Total Amount, kg
E(1)	E(2)	E(3)	f(1)	f(2)	f(3)	f(4)	C(1)	C(2)	C(3)	C(4)	m(1)	m(2)	m(3)	m(4)	
1000	0	0	7.741E-06	2.976E-06	2.734E-07	5.713E-06	6.977E-07	4.214E-05	6.710E-03	1.122E-01	6.977E+04	8.428E+03	1.208E+05	5.611E+04	2.551E+05
0	1000	0	2.914E-06	1.450E-04	1.029E-07	2.783E-04	2.627E-07	2.053E-03	2.526E-03	5.467E+00	2.627E+04	4.106E+05	4.547E+04	2.733E+06	3.216E+06
0	0	1000	8.905E-08	8.245E-07	5.470E-05	1.583E-06	8.026E-09	1.167E-05	1.343E+00	3.109E-02	8.026E+02	2.335E+03	2.417E+07	1.554E+04	2.418E+07
600	300	100	5.528E-06	4.537E-05	5.665E-06	8.708E-05	4.982E-07	6.423E-04	1.390E-01	1.711E+00	4.982E+04	1.285E+05	2.503E+06	8.553E+05	3.536E+06

Emission, kg/h			Loss, Reaction, kg/h				Loss, Advection, kg/h			Intermedia Rate of Transport, kg/h T12	T21	T13	T31	T32	T24	T42
E(1)	E(2)	E(3)	R(1)	R(2)	R(3)	R(4)	A(1)	A(2)	A(4)	air-water	water-air	air-soil	soil-air	soil-water	water-sed	sed-water
1000	0	0	2.844E+02	1.062E+00	4.92E+00	2.287E+00	6.977E+02	8.428E+00	1.122E+00	2.066E+01	7.787E+00	4.998E+00	4.716E-02	2.745E-02	4.603E+00	1.193E+00
0	1000	0	1.071E+02	5.173E+01	1.85E+00	1.114E+02	2.627E+02	4.106E+02	5.467E+01	7.777E+00	3.794E+02	1.882E+00	1.775E-02	1.033E-02	2.242E+02	5.812E+01
0	0	1000	3.272E+00	2.942E-01	9.85E+02	6.336E-01	8.026E+00	2.335E+00	3.109E-01	2.377E-01	2.157E+00	5.750E-02	9.436E+00	5.493E+00	1.275E+00	3.305E-01
600	300	100	2.031E+02	1.619E+01	1.02E+02	3.486E+01	4.982E+02	1.619E+01	1.711E+01	1.475E+01	1.187E+02	3.569E+00	9.772E-01	5.689E-01	7.015E+01	1.819E+01

Chemical name: 2,2',5-Trichlorobiphenyl

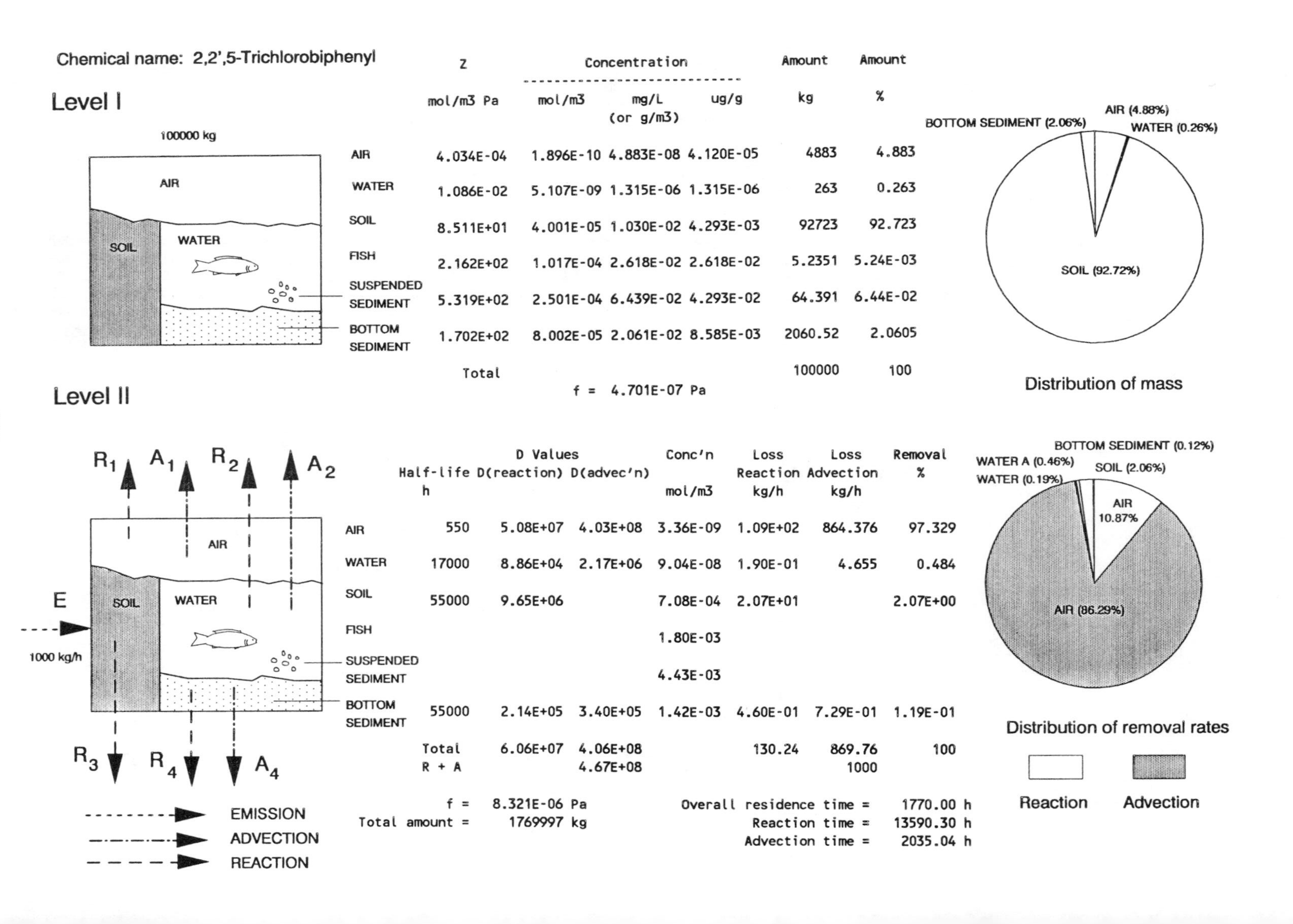

Level I

	Z	Concentration			Amount	Amount
	mol/m3 Pa	mol/m3	mg/L (or g/m3)	ug/g	kg	%
AIR	4.034E-04	1.896E-10	4.883E-08	4.120E-05	4883	4.883
WATER	1.086E-02	5.107E-09	1.315E-06	1.315E-06	263	0.263
SOIL	8.511E+01	4.001E-05	1.030E-02	4.293E-03	92723	92.723
FISH	2.162E+02	1.017E-04	2.618E-02	2.618E-02	5.2351	5.24E-03
SUSPENDED SEDIMENT	5.319E+02	2.501E-04	6.439E-02	4.293E-02	64.391	6.44E-02
BOTTOM SEDIMENT	1.702E+02	8.002E-05	2.061E-02	8.585E-03	2060.52	2.0605
Total					100000	100

f = 4.701E-07 Pa

Level II

	Half-life h	D Values D(reaction)	D(advec'n)	Conc'n mol/m3	Loss Reaction kg/h	Loss Advection kg/h	Removal %
AIR	550	5.08E+07	4.03E+08	3.36E-09	1.09E+02	864.376	97.329
WATER	17000	8.86E+04	2.17E+06	9.04E-08	1.90E-01	4.655	0.484
SOIL	55000	9.65E+06		7.08E-04	2.07E+01		2.07E+00
FISH				1.80E-03			
SUSPENDED SEDIMENT				4.43E-03			
BOTTOM SEDIMENT	55000	2.14E+05	3.40E+05	1.42E-03	4.60E-01	7.29E-01	1.19E-01
	Total	6.06E+07	4.06E+08		130.24	869.76	100
	R + A		4.67E+08			1000	

f = 8.321E-06 Pa
Total amount = 1769997 kg

Overall residence time = 1770.00 h
Reaction time = 13590.30 h
Advection time = 2035.04 h

Level III

Chemical name: 2,2'5-Trichlorobiphenyl

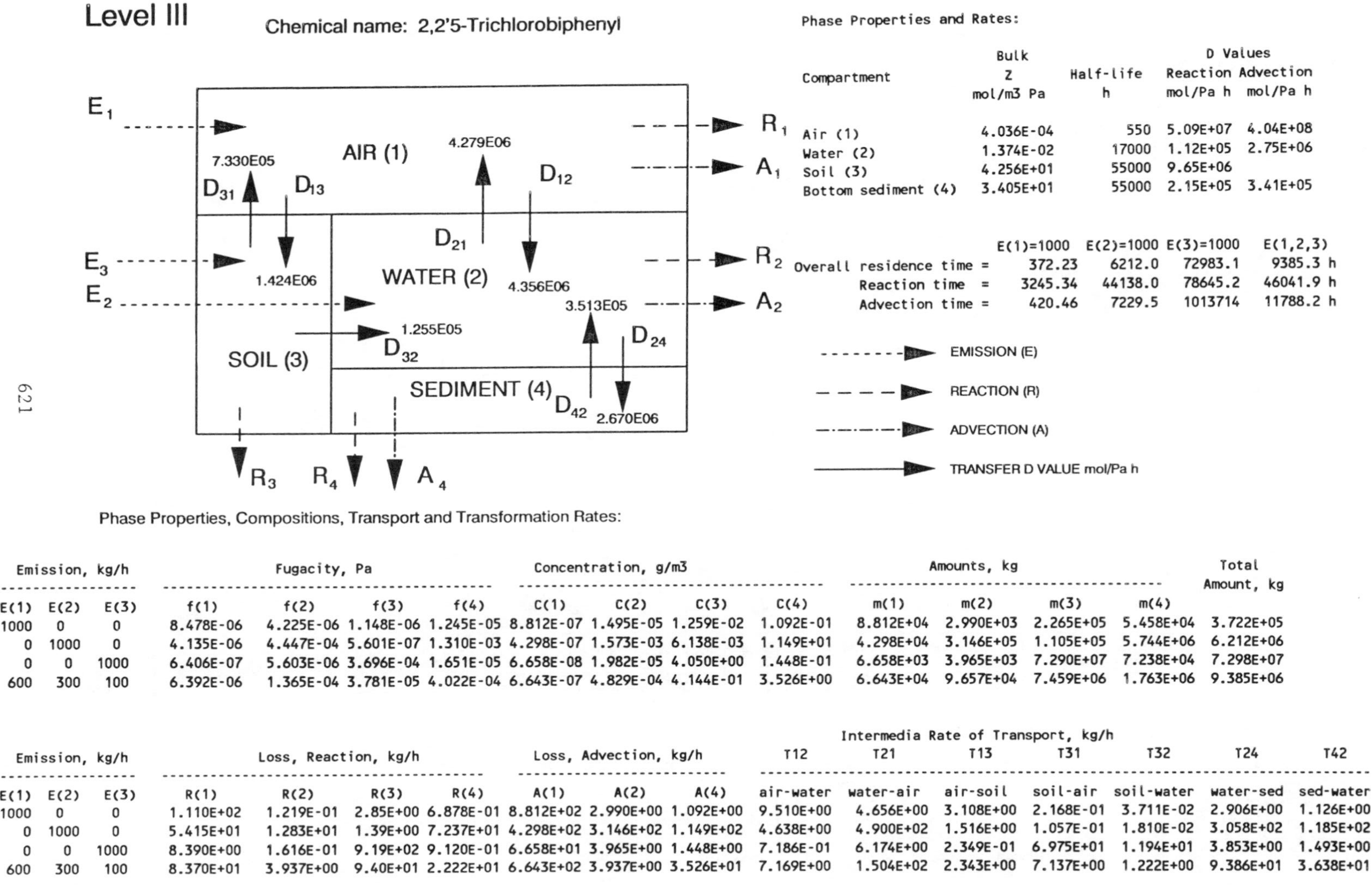

Phase Properties and Rates:

Compartment	Bulk Z mol/m3 Pa	Half-life h	D Values Reaction mol/Pa h	D Values Advection mol/Pa h
Air (1)	4.036E-04	550	5.09E+07	4.04E+08
Water (2)	1.374E-02	17000	1.12E+05	2.75E+06
Soil (3)	4.256E+01	55000	9.65E+06	
Bottom sediment (4)	3.405E+01	55000	2.15E+05	3.41E+05

	E(1)=1000	E(2)=1000	E(3)=1000	E(1,2,3)	
Overall residence time =	372.23	6212.0	72983.1	9385.3	h
Reaction time =	3245.34	44138.0	78645.2	46041.9	h
Advection time =	420.46	7229.5	1013714	11788.2	h

Phase Properties, Compositions, Transport and Transformation Rates:

Emission, kg/h			Fugacity, Pa				Concentration, g/m3				Amounts, kg				Total Amount, kg
E(1)	E(2)	E(3)	f(1)	f(2)	f(3)	f(4)	C(1)	C(2)	C(3)	C(4)	m(1)	m(2)	m(3)	m(4)	
1000	0	0	8.478E-06	4.225E-06	1.148E-06	1.245E-05	8.812E-07	1.495E-05	1.259E-02	1.092E-01	8.812E+04	2.990E+03	2.265E+05	5.458E+04	3.722E+05
0	1000	0	4.135E-06	4.447E-04	5.601E-07	1.310E-03	4.298E-07	1.573E-03	6.138E-03	1.149E+01	4.298E+04	3.146E+05	1.105E+05	5.744E+06	6.212E+06
0	0	1000	6.406E-07	5.603E-06	3.696E-04	1.651E-05	6.658E-08	1.982E-05	4.050E+00	1.448E-01	6.658E+03	3.965E+03	7.290E+07	7.238E+04	7.298E+07
600	300	100	6.392E-06	1.365E-04	3.781E-05	4.022E-04	6.643E-07	4.829E-04	4.144E-01	3.526E+00	6.643E+04	9.657E+04	7.459E+06	1.763E+06	9.385E+06

Emission, kg/h			Loss, Reaction, kg/h				Loss, Advection, kg/h			Intermedia Rate of Transport, kg/h T12	T21	T13	T31	T32	T24	T42
E(1)	E(2)	E(3)	R(1)	R(2)	R(3)	R(4)	A(1)	A(2)	A(4)	air-water	water-air	air-soil	soil-air	soil-water	water-sed	sed-water
1000	0	0	1.110E+02	1.219E-01	2.85E+00	6.878E-01	8.812E+02	2.990E+00	1.092E+00	9.510E+00	4.656E+00	3.108E+00	2.168E-01	3.711E-02	2.906E+00	1.126E+00
0	1000	0	5.415E+01	1.283E+01	1.39E+00	7.237E+01	4.298E+02	3.146E+02	1.149E+02	4.638E+00	4.900E+02	1.516E+00	1.057E-01	1.810E-02	3.058E+02	1.185E+02
0	0	1000	8.390E+00	1.616E-01	9.19E+02	9.120E-01	6.658E+01	3.965E+00	1.448E+00	7.186E-01	6.174E+00	2.349E-01	6.975E+01	1.194E+01	3.853E+00	1.493E+00
600	300	100	8.370E+01	3.937E+00	9.40E+01	2.222E+01	6.643E+02	3.937E+00	3.526E+01	7.169E+00	1.504E+02	2.343E+00	7.137E+00	1.222E+00	9.386E+01	3.638E+01

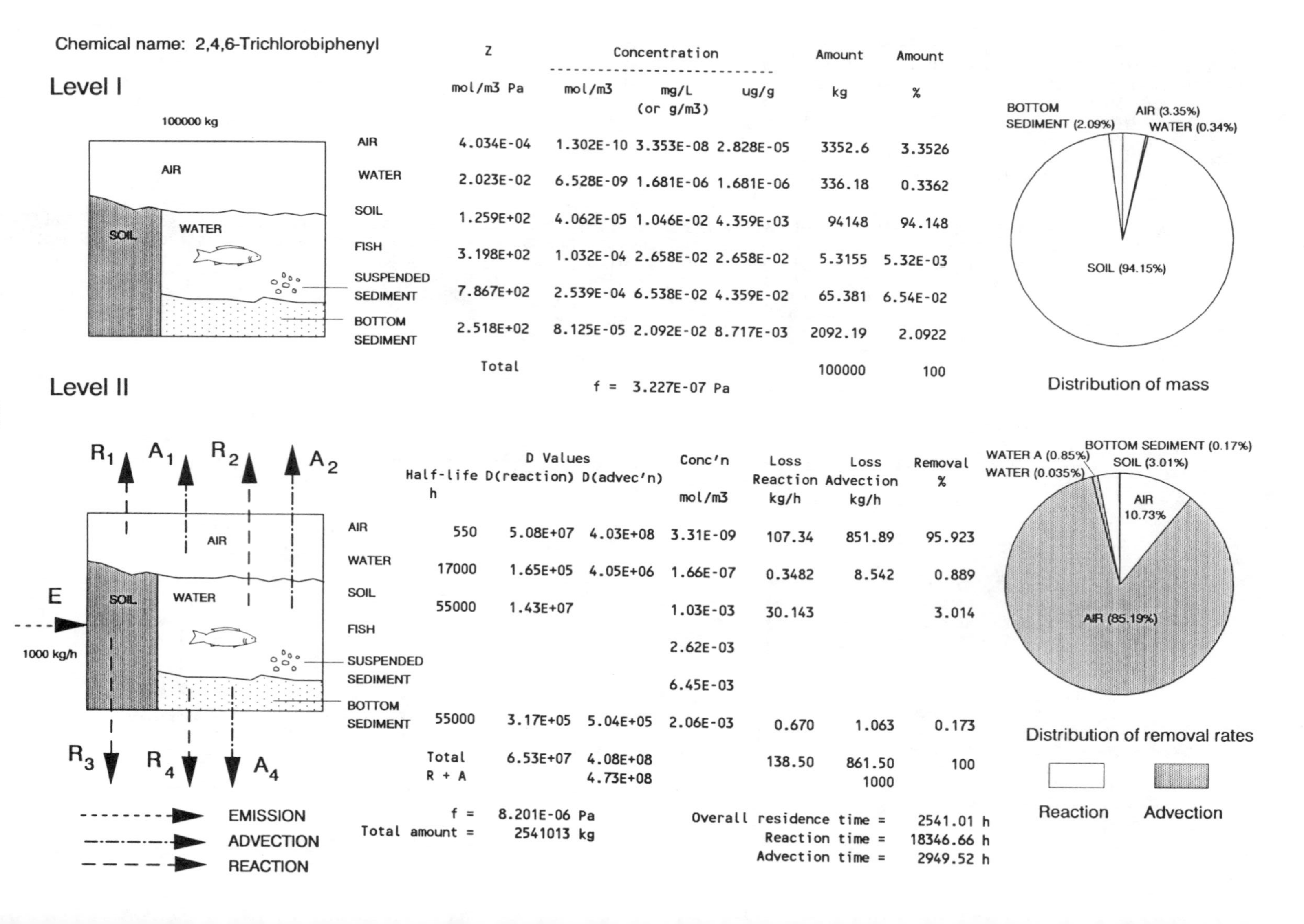

Chemical name: 2,4,6-Trichlorobiphenyl

Level I

100000 kg

	Z	Concentration			Amount	Amount
	mol/m3 Pa	mol/m3	mg/L (or g/m3)	ug/g	kg	%
AIR	4.034E-04	1.302E-10	3.353E-08	2.828E-05	3352.6	3.3526
WATER	2.023E-02	6.528E-09	1.681E-06	1.681E-06	336.18	0.3362
SOIL	1.259E+02	4.062E-05	1.046E-02	4.359E-03	94148	94.148
FISH	3.198E+02	1.032E-04	2.658E-02	2.658E-02	5.3155	5.32E-03
SUSPENDED SEDIMENT	7.867E+02	2.539E-04	6.538E-02	4.359E-02	65.381	6.54E-02
BOTTOM SEDIMENT	2.518E+02	8.125E-05	2.092E-02	8.717E-03	2092.19	2.0922
	Total				100000	100

f = 3.227E-07 Pa

Level II

	Half-life h	D Values D(reaction)	D Values D(advec'n)	Conc'n mol/m3	Loss Reaction kg/h	Loss Advection kg/h	Removal %
AIR	550	5.08E+07	4.03E+08	3.31E-09	107.34	851.89	95.923
WATER	17000	1.65E+05	4.05E+06	1.66E-07	0.3482	8.542	0.889
SOIL	55000	1.43E+07		1.03E-03	30.143		3.014
FISH				2.62E-03			
SUSPENDED SEDIMENT				6.45E-03			
BOTTOM SEDIMENT	55000	3.17E+05	5.04E+05	2.06E-03	0.670	1.063	0.173
	Total R + A	6.53E+07	4.08E+08 4.73E+08		138.50	861.50 1000	100

f = 8.201E-06 Pa
Total amount = 2541013 kg

Overall residence time = 2541.01 h
Reaction time = 18346.66 h
Advection time = 2949.52 h

Level III

Chemical name: 2,4,6-Trichlorobiphenyl

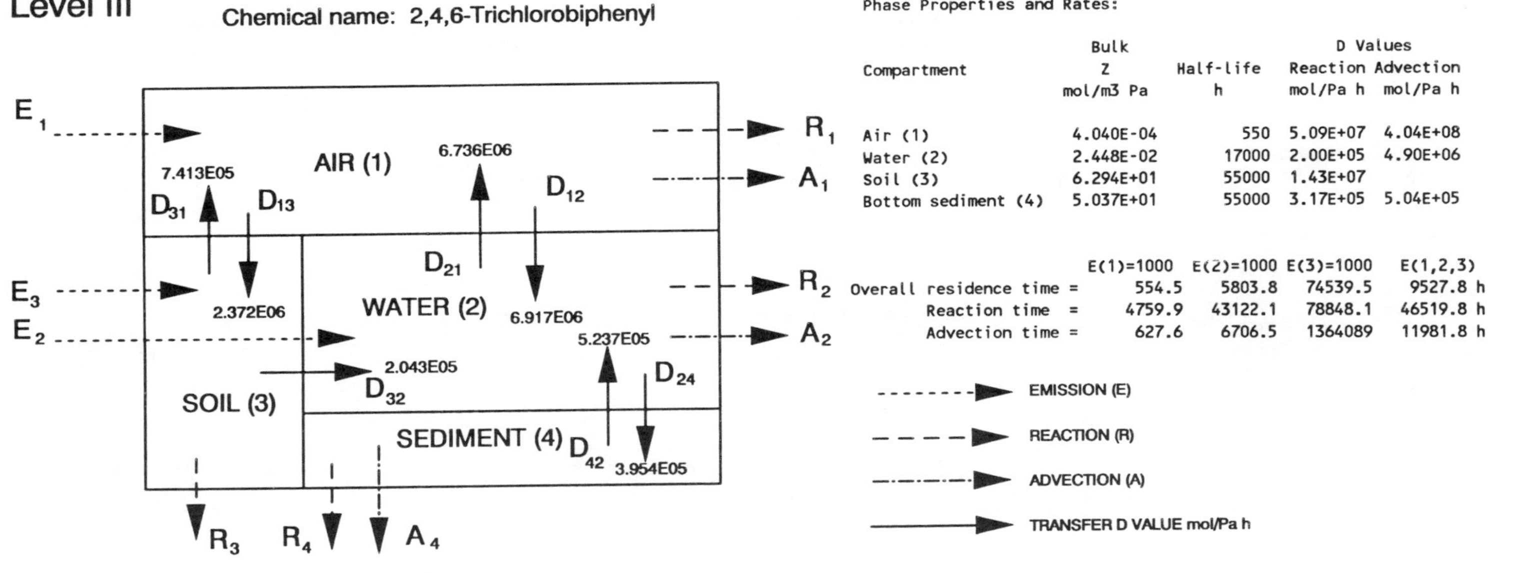

Phase Properties and Rates:

Compartment	Bulk Z mol/m3 Pa	Half-life h	D Values Reaction mol/Pa h	D Values Advection mol/Pa h
Air (1)	4.040E-04	550	5.09E+07	4.04E+08
Water (2)	2.448E-02	17000	2.00E+05	4.90E+06
Soil (3)	6.294E+01	55000	1.43E+07	
Bottom sediment (4)	5.037E+01	55000	3.17E+05	5.04E+05

	E(1)=1000	E(2)=1000	E(3)=1000	E(1,2,3)
Overall residence time =	554.5	5803.8	74539.5	9527.8 h
Reaction time =	4759.9	43122.1	78848.1	46519.8 h
Advection time =	627.6	6706.5	1364089	11981.8 h

Phase Properties, Compositions, Transport and Transformation Rates:

Emission, kg/h			Fugacity, Pa				Concentration, g/m3				Amounts, kg				Total Amount, kg
E(1)	E(2)	E(3)	f(1)	f(2)	f(3)	f(4)	C(1)	C(2)	C(3)	C(4)	m(1)	m(2)	m(3)	m(4)	
1000	0	0	8.429E-06	4.112E-06	1.314E-06	1.209E-05	8.767E-07	2.592E-05	2.129E-02	1.568E-01	8.767E+04	5.184E+03	3.833E+05	7.840E+04	5.545E+05
0	1000	0	3.986E-06	2.746E-04	6.212E-07	8.073E-04	4.146E-07	1.731E-03	1.007E-02	1.047E+01	4.146E+04	3.461E+05	1.812E+05	5.235E+06	5.804E+06
0	0	1000	4.640E-07	3.885E-06	2.552E-04	1.142E-05	4.826E-08	2.449E-05	4.136E+00	1.482E-01	4.826E+03	4.898E+03	7.446E+07	7.408E+04	7.454E+07
600	300	100	6.299E-06	8.522E-05	2.650E-05	2.506E-04	6.552E-07	5.372E-04	4.294E-01	3.250E+00	6.552E+04	1.074E+05	7.730E+06	1.625E+06	9.528E+06

Emission, kg/h			Loss, Reaction, kg/h				Loss, Advection, kg/h			Intermedia Rate of Transport, kg/h T12	T21	T13	T31	T32	T24	T42
E(1)	E(2)	E(3)	R(1)	R(2)	R(3)	R(4)	A(1)	A(2)	A(4)	air-water	water-air	air-soil	soil-air	soil-water	water-sed	sed-water
1000	0	0	1.105E+02	2.113E-01	4.83E+00	9.878E-01	8.767E+02	5.184E+00	1.568E+00	1.501E+01	7.132E+00	5.149E+00	2.508E-01	6.911E-02	4.186E+00	1.630E+00
0	1000	0	5.224E+01	1.411E+01	2.28E+00	6.596E+01	4.146E+02	3.461E+02	1.047E+02	7.099E+00	4.762E+02	2.435E+00	1.186E-01	3.268E-02	2.795E+02	1.089E+02
0	0	1000	6.081E+00	1.997E-01	9.38E+02	9.335E-01	4.826E+01	4.898E+00	1.482E+00	8.265E-01	6.739E+00	2.834E-01	4.872E+01	1.343E+01	3.956E+00	1.541E+00
600	300	100	8.256E+01	4.380E+00	9.74E+01	2.047E+01	6.552E+02	4.380E+00	3.250E+01	1.122E+01	1.478E+02	3.848E+00	5.058E+00	1.394E+00	8.677E+01	3.379E+01

Chemical name: 2',3,4-Trichlorobiphenyl

Level I

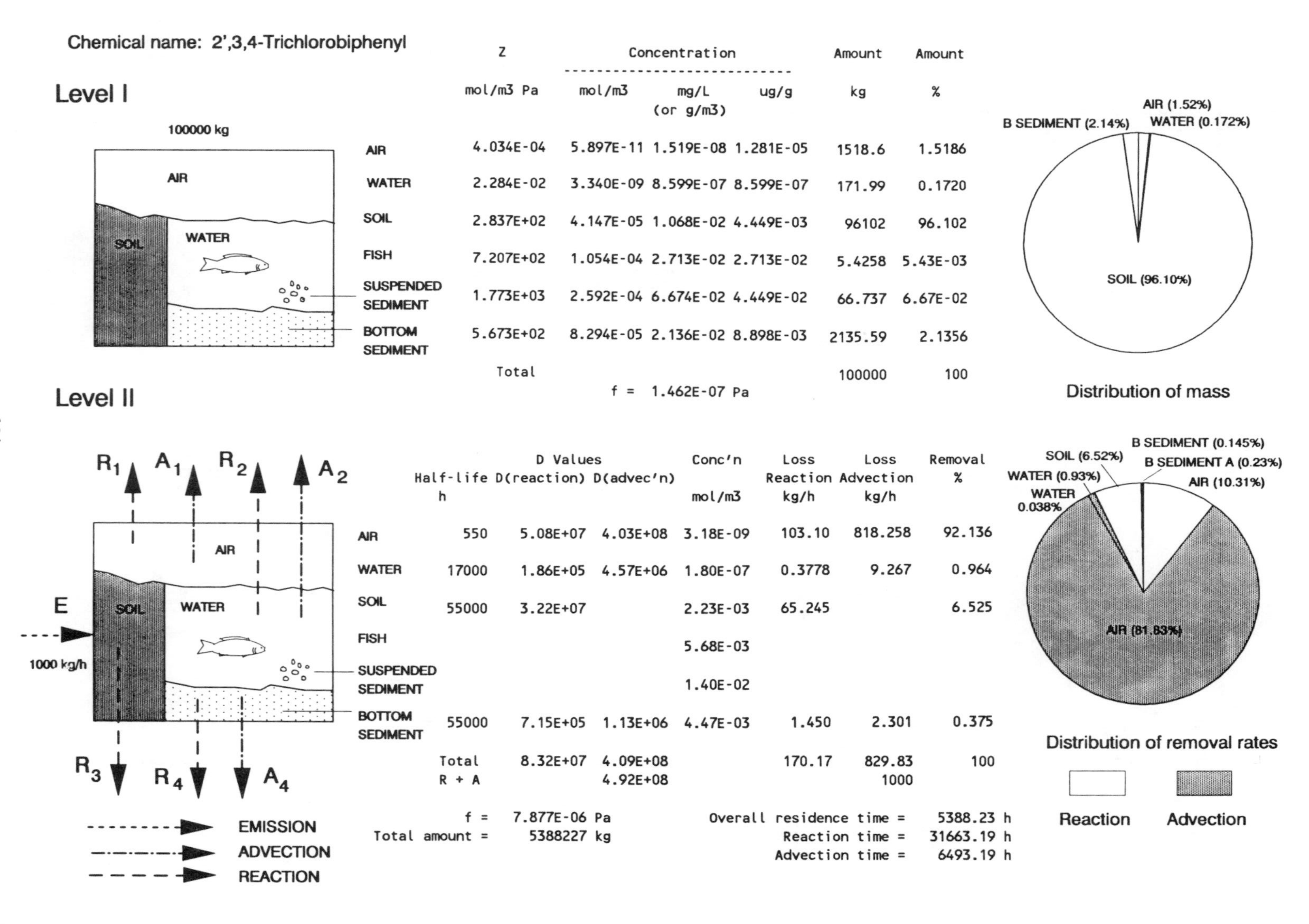

	Z	Concentration			Amount	Amount
	mol/m3 Pa	mol/m3	mg/L (or g/m3)	ug/g	kg	%
AIR	4.034E-04	5.897E-11	1.519E-08	1.281E-05	1518.6	1.5186
WATER	2.284E-02	3.340E-09	8.599E-07	8.599E-07	171.99	0.1720
SOIL	2.837E+02	4.147E-05	1.068E-02	4.449E-03	96102	96.102
FISH	7.207E+02	1.054E-04	2.713E-02	2.713E-02	5.4258	5.43E-03
SUSPENDED SEDIMENT	1.773E+03	2.592E-04	6.674E-02	4.449E-02	66.737	6.67E-02
BOTTOM SEDIMENT	5.673E+02	8.294E-05	2.136E-02	8.898E-03	2135.59	2.1356
	Total				100000	100

f = 1.462E-07 Pa

Level II

	Half-life	D Values		Conc'n	Loss	Loss	Removal
	h	D(reaction)	D(advec'n)	mol/m3	Reaction kg/h	Advection kg/h	%
AIR	550	5.08E+07	4.03E+08	3.18E-09	103.10	818.258	92.136
WATER	17000	1.86E+05	4.57E+06	1.80E-07	0.3778	9.267	0.964
SOIL	55000	3.22E+07		2.23E-03	65.245		6.525
FISH				5.68E-03			
SUSPENDED SEDIMENT				1.40E-02			
BOTTOM SEDIMENT	55000	7.15E+05	1.13E+06	4.47E-03	1.450	2.301	0.375
	Total	8.32E+07	4.09E+08		170.17	829.83	100
	R + A		4.92E+08			1000	

f = 7.877E-06 Pa
Total amount = 5388227 kg

Overall residence time = 5388.23 h
Reaction time = 31663.19 h
Advection time = 6493.19 h

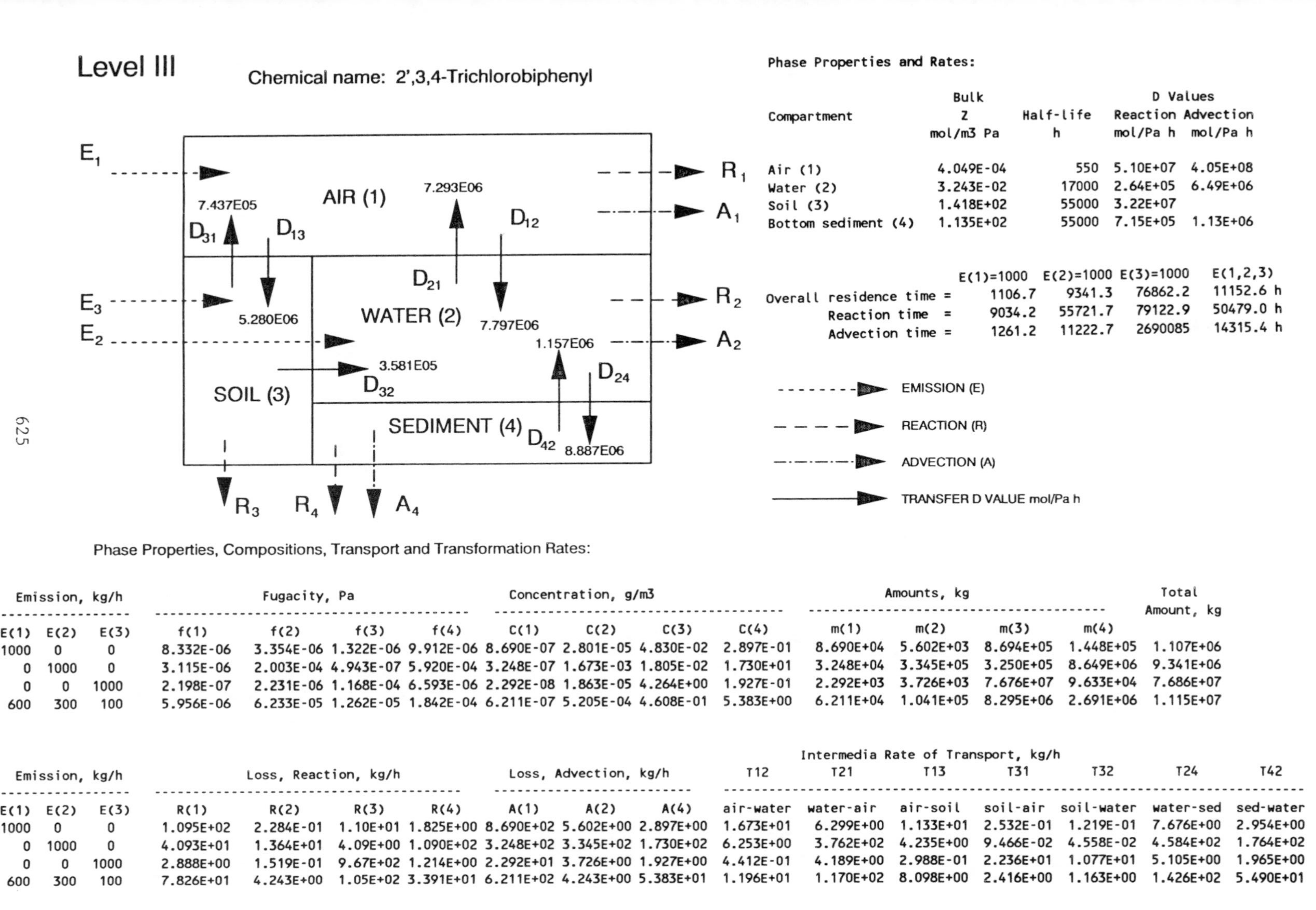

Phase Properties and Rates:

Compartment	Bulk Z mol/m3 Pa	Half-life h	D Values Reaction mol/Pa h	D Values Advection mol/Pa h
Air (1)	4.049E-04	550	5.10E+07	4.05E+08
Water (2)	3.243E-02	17000	2.64E+05	6.49E+06
Soil (3)	1.418E+02	55000	3.22E+07	
Bottom sediment (4)	1.135E+02	55000	7.15E+05	1.13E+06

	E(1)=1000	E(2)=1000	E(3)=1000	E(1,2,3)
Overall residence time =	1106.7	9341.3	76862.2	11152.6 h
Reaction time =	9034.2	55721.7	79122.9	50479.0 h
Advection time =	1261.2	11222.7	2690085	14315.4 h

Phase Properties, Compositions, Transport and Transformation Rates:

Emission, kg/h			Fugacity, Pa				Concentration, g/m3				Amounts, kg				Total Amount, kg
E(1)	E(2)	E(3)	f(1)	f(2)	f(3)	f(4)	C(1)	C(2)	C(3)	C(4)	m(1)	m(2)	m(3)	m(4)	
1000	0	0	8.332E-06	3.354E-06	1.322E-06	9.912E-06	8.690E-07	2.801E-05	4.830E-02	2.897E-01	8.690E+04	5.602E+03	8.694E+05	1.448E+05	1.107E+06
0	1000	0	3.115E-06	2.003E-04	4.943E-07	5.920E-04	3.248E-07	1.673E-03	1.805E-02	1.730E+01	3.248E+04	3.345E+05	3.250E+05	8.649E+06	9.341E+06
0	0	1000	2.198E-07	2.231E-06	1.168E-04	6.593E-06	2.292E-08	1.863E-05	4.264E+00	1.927E-01	2.292E+03	3.726E+03	7.676E+07	9.633E+04	7.686E+07
600	300	100	5.956E-06	6.233E-05	1.262E-05	1.842E-04	6.211E-07	5.205E-04	4.608E-01	5.383E+00	6.211E+04	1.041E+05	8.295E+06	2.691E+06	1.115E+07

Emission, kg/h			Loss, Reaction, kg/h				Loss, Advection, kg/h			Intermedia Rate of Transport, kg/h						
										T12	T21	T13	T31	T32	T24	T42
E(1)	E(2)	E(3)	R(1)	R(2)	R(3)	R(4)	A(1)	A(2)	A(4)	air-water	water-air	air-soil	soil-air	soil-water	water-sed	sed-water
1000	0	0	1.095E+02	2.284E-01	1.10E+01	1.825E+00	8.690E+02	5.602E+00	2.897E+00	1.673E+01	6.299E+00	1.133E+01	2.532E-01	1.219E-01	7.676E+00	2.954E+00
0	1000	0	4.093E+01	1.364E+01	4.09E+00	1.090E+02	3.248E+02	3.345E+02	1.730E+02	6.253E+00	3.762E+02	4.235E+00	9.466E-02	4.558E-02	4.584E+02	1.764E+02
0	0	1000	2.888E+00	1.519E-01	9.67E+02	1.214E+00	2.292E+01	3.726E+00	1.927E+00	4.412E-01	4.189E+00	2.988E-01	2.236E+01	1.077E+01	5.105E+00	1.965E+00
600	300	100	7.826E+01	4.243E+00	1.05E+02	3.391E+01	6.211E+02	4.243E+00	5.383E+01	1.196E+01	1.170E+02	8.098E+00	2.416E+00	1.163E+00	1.426E+02	5.490E+01

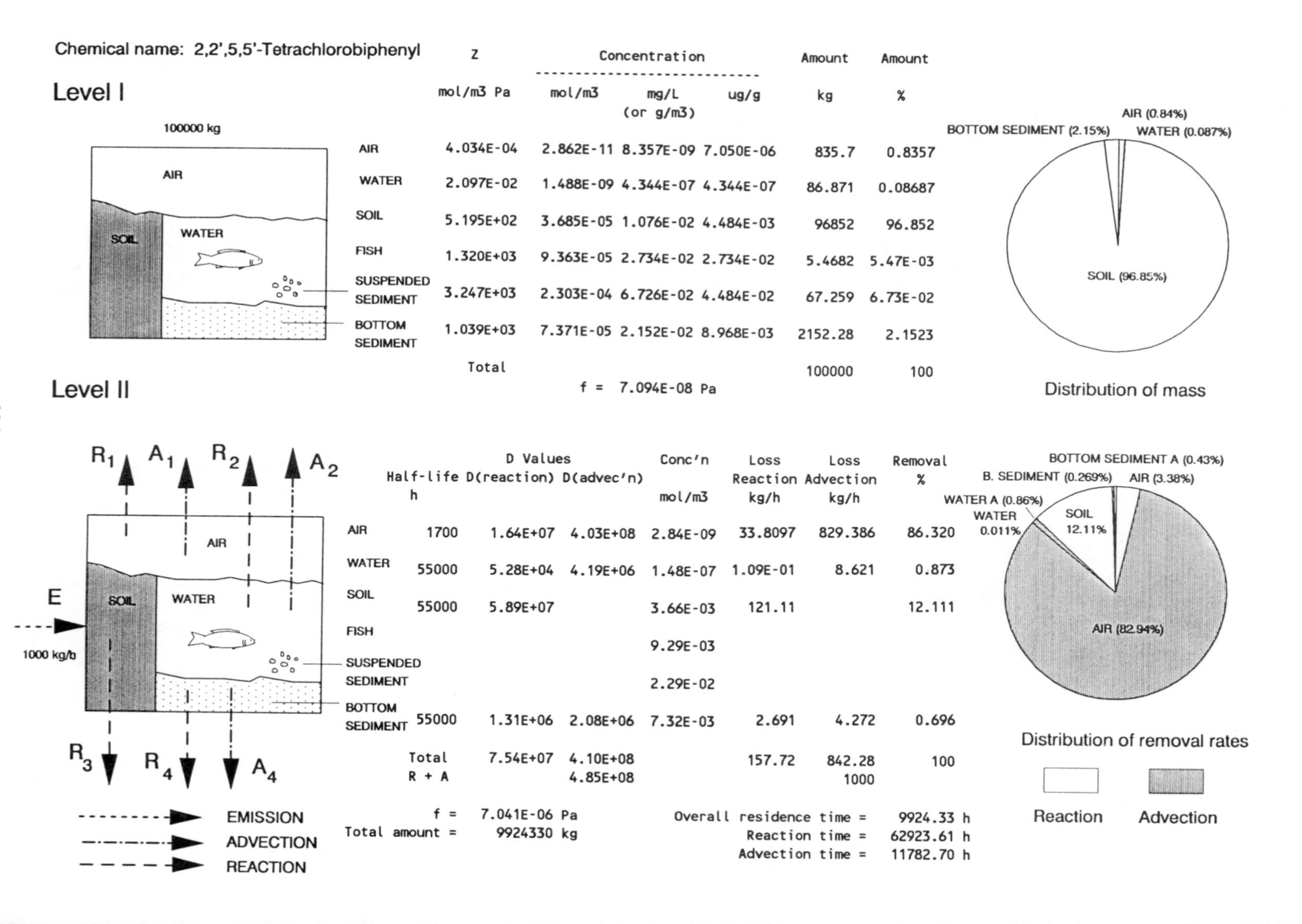

Chemical name: 2,2',5,5'-Tetrachlorobiphenyl

Level I

	Z	Concentration			Amount	Amount
	mol/m3 Pa	mol/m3	mg/L (or g/m3)	ug/g	kg	%
AIR	4.034E-04	2.862E-11	8.357E-09	7.050E-06	835.7	0.8357
WATER	2.097E-02	1.488E-09	4.344E-07	4.344E-07	86.871	0.08687
SOIL	5.195E+02	3.685E-05	1.076E-02	4.484E-03	96852	96.852
FISH	1.320E+03	9.363E-05	2.734E-02	2.734E-02	5.4682	5.47E-03
SUSPENDED SEDIMENT	3.247E+03	2.303E-04	6.726E-02	4.484E-02	67.259	6.73E-02
BOTTOM SEDIMENT	1.039E+03	7.371E-05	2.152E-02	8.968E-03	2152.28	2.1523
	Total				100000	100

f = 7.094E-08 Pa

Level II

	Half-life h	D Values D(reaction)	D(advec'n)	Conc'n mol/m3	Loss Reaction kg/h	Loss Advection kg/h	Removal %
AIR	1700	1.64E+07	4.03E+08	2.84E-09	33.8097	829.386	86.320
WATER	55000	5.28E+04	4.19E+06	1.48E-07	1.09E-01	8.621	0.873
SOIL	55000	5.89E+07		3.66E-03	121.11		12.111
FISH				9.29E-03			
SUSPENDED SEDIMENT				2.29E-02			
BOTTOM SEDIMENT	55000	1.31E+06	2.08E+06	7.32E-03	2.691	4.272	0.696
	Total	7.54E+07	4.10E+08		157.72	842.28	100
	R + A		4.85E+08			1000	

f = 7.041E-06 Pa
Total amount = 9924330 kg

Overall residence time = 9924.33 h
Reaction time = 62923.61 h
Advection time = 11782.70 h

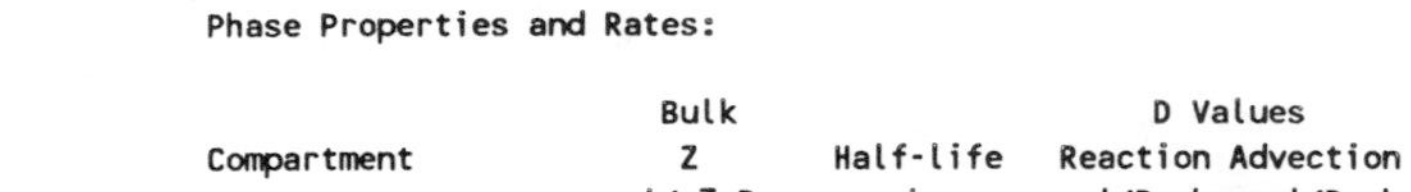

Level III

Chemical name: 2,2'5,5'-Tetrachlorobiphenyl

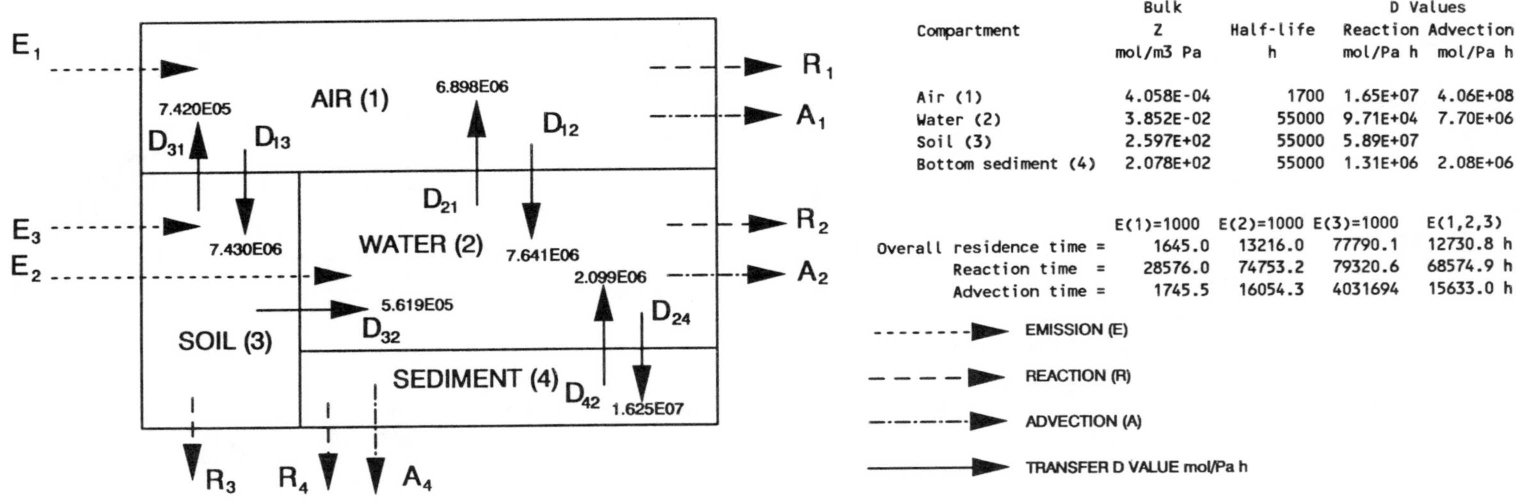

Phase Properties and Rates:

Compartment	Bulk Z mol/m3 Pa	Half-life h	D Values Reaction mol/Pa h	D Values Advection mol/Pa h
Air (1)	4.058E-04	1700	1.65E+07	4.06E+08
Water (2)	3.852E-02	55000	9.71E+04	7.70E+06
Soil (3)	2.597E+02	55000	5.89E+07	
Bottom sediment (4)	2.078E+02	55000	1.31E+06	2.08E+06

	E(1)=1000	E(2)=1000	E(3)=1000	E(1,2,3)
Overall residence time =	1645.0	13216.0	77790.1	12730.8 h
Reaction time =	28576.0	74753.2	79320.6	68574.9 h
Advection time =	1745.5	16054.3	4031694	15633.0 h

EMISSION (E)
REACTION (R)
ADVECTION (A)
TRANSFER D VALUE mol/Pa h

Phase Properties, Compositions, Transport and Transformation Rates:

Emission, kg/h			Fugacity, Pa				Concentration, g/m3				Amounts, kg				Total Amount, kg
E(1)	E(2)	E(3)	f(1)	f(2)	f(3)	f(4)	C(1)	C(2)	C(3)	C(4)	m(1)	m(2)	m(3)	m(4)	
1000	0	0	7.869E-06	2.453E-06	9.710E-07	7.268E-06	9.325E-07	2.759E-05	7.365E-02	4.410E-01	9.325E+04	5.518E+03	1.326E+06	2.205E+05	1.645E+06
0	1000	0	2.195E-06	1.391E-04	2.708E-07	4.122E-04	2.601E-07	1.565E-03	2.054E-02	2.501E+01	2.601E+04	3.130E+05	3.697E+05	1.251E+07	1.322E+07
0	0	1000	1.174E-07	1.329E-06	5.689E-05	3.936E-06	1.392E-08	1.494E-05	4.315E+00	2.389E-01	1.392E+03	2.989E+03	7.767E+07	1.194E+05	7.779E+07
600	300	100	5.392E-06	4.335E-05	6.353E-06	1.284E-04	6.389E-07	4.876E-04	4.818E-01	7.793E+00	6.389E+04	9.751E+04	8.673E+06	3.896E+06	1.273E+07

Emission, kg/h			Loss, Reaction, kg/h				Loss, Advection, kg/h			Intermedia Rate of Transport, kg/h T12	T21	T13	T31	T32	T24	T42
E(1)	E(2)	E(3)	R(1)	R(2)	R(3)	R(4)	A(1)	A(2)	A(4)	air-water	water-air	air-soil	soil-air	soil-water	water-sed	sed-water
1000	0	0	3.801E+01	6.953E-02	1.67E+01	2.778E+00	9.325E+02	5.518E+00	4.410E+00	1.756E+01	4.941E+00	1.707E+01	2.104E-01	1.593E-01	1.164E+01	4.454E+00
0	1000	0	1.060E+01	3.944E+00	4.66E+00	1.576E+02	2.601E+02	3.130E+02	2.501E+02	4.897E+00	2.803E+02	4.762E+00	5.867E-02	4.443E-02	6.604E+02	2.526E+02
0	0	1000	5.673E-01	3.766E-02	9.79E+02	1.505E+00	1.392E+01	2.989E+00	2.389E+00	2.621E-01	2.676E+00	2.548E-01	1.233E+01	9.334E+00	6.306E+00	2.412E+00
600	300	100	2.605E+01	1.229E+00	1.09E+02	4.909E+01	6.389E+02	1.229E+00	7.793E+01	1.203E+01	8.731E+01	1.170E+01	1.376E+00	1.042E+00	2.057E+02	7.871E+01

Chemical name: 2,2',4,5,5'-Pentachlorobiphenyl

Level I

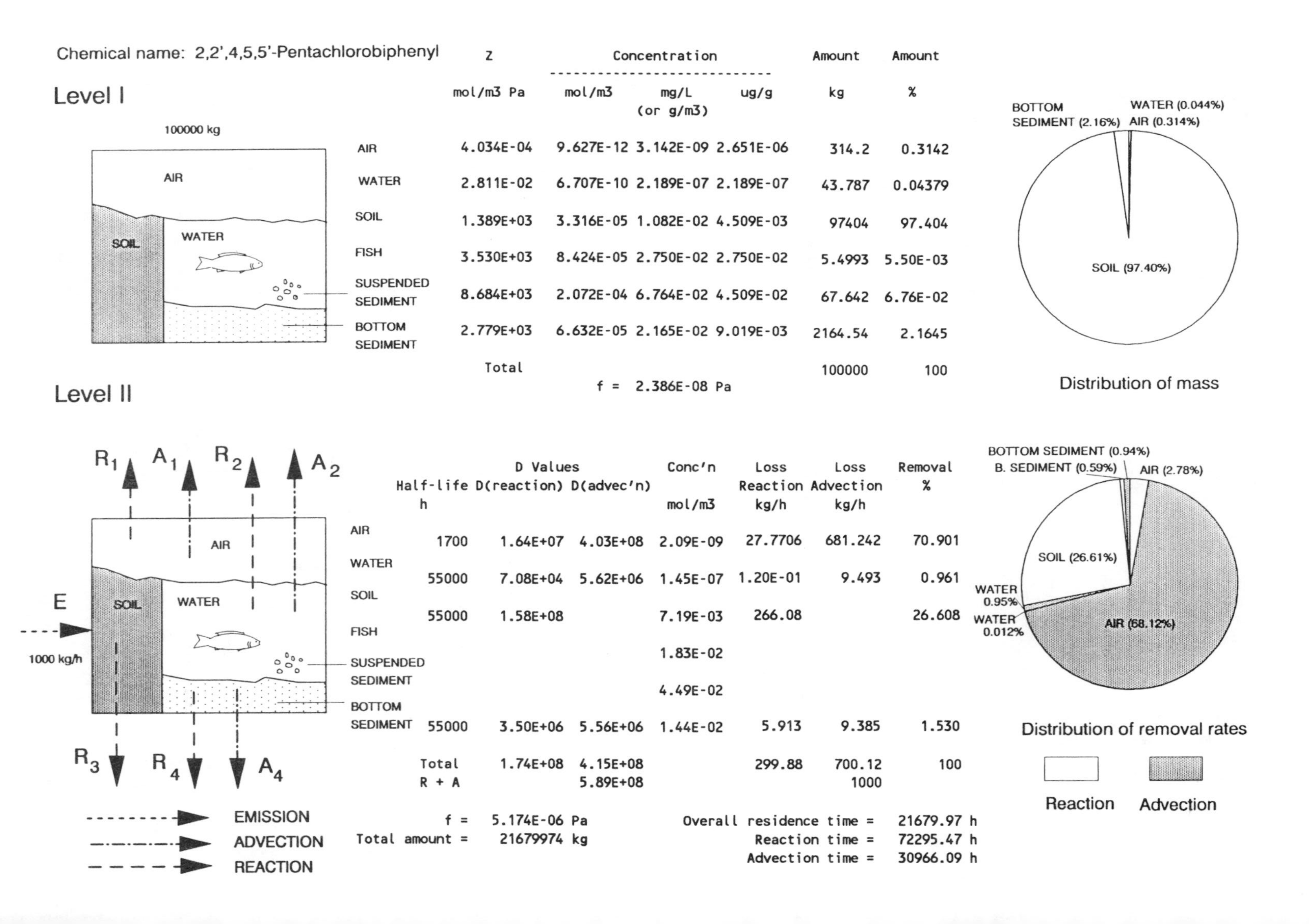

	Z	Concentration			Amount	Amount
	mol/m3 Pa	mol/m3	mg/L (or g/m3)	ug/g	kg	%
AIR	4.034E-04	9.627E-12	3.142E-09	2.651E-06	314.2	0.3142
WATER	2.811E-02	6.707E-10	2.189E-07	2.189E-07	43.787	0.04379
SOIL	1.389E+03	3.316E-05	1.082E-02	4.509E-03	97404	97.404
FISH	3.530E+03	8.424E-05	2.750E-02	2.750E-02	5.4993	5.50E-03
SUSPENDED SEDIMENT	8.684E+03	2.072E-04	6.764E-02	4.509E-02	67.642	6.76E-02
BOTTOM SEDIMENT	2.779E+03	6.632E-05	2.165E-02	9.019E-03	2164.54	2.1645
	Total				100000	100

f = 2.386E-08 Pa

Level II

	Half-life	D Values		Conc'n	Loss	Loss	Removal
	h	D(reaction)	D(advec'n)	mol/m3	Reaction kg/h	Advection kg/h	%
AIR	1700	1.64E+07	4.03E+08	2.09E-09	27.7706	681.242	70.901
WATER	55000	7.08E+04	5.62E+06	1.45E-07	1.20E-01	9.493	0.961
SOIL	55000	1.58E+08		7.19E-03	266.08		26.608
FISH				1.83E-02			
SUSPENDED SEDIMENT				4.49E-02			
BOTTOM SEDIMENT	55000	3.50E+06	5.56E+06	1.44E-02	5.913	9.385	1.530
	Total	1.74E+08	4.15E+08		299.88	700.12	100
	R + A		5.89E+08			1000	

f = 5.174E-06 Pa
Total amount = 21679974 kg

Overall residence time = 21679.97 h
Reaction time = 72295.47 h
Advection time = 30966.09 h

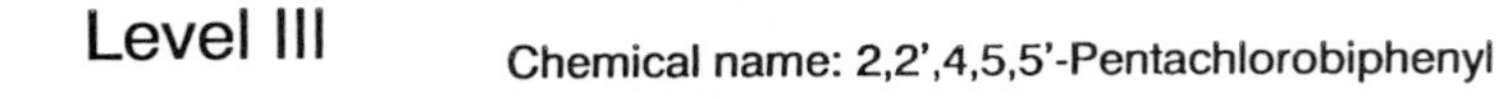

Level III

Chemical name: 2,2',4,5,5'-Pentachlorobiphenyl

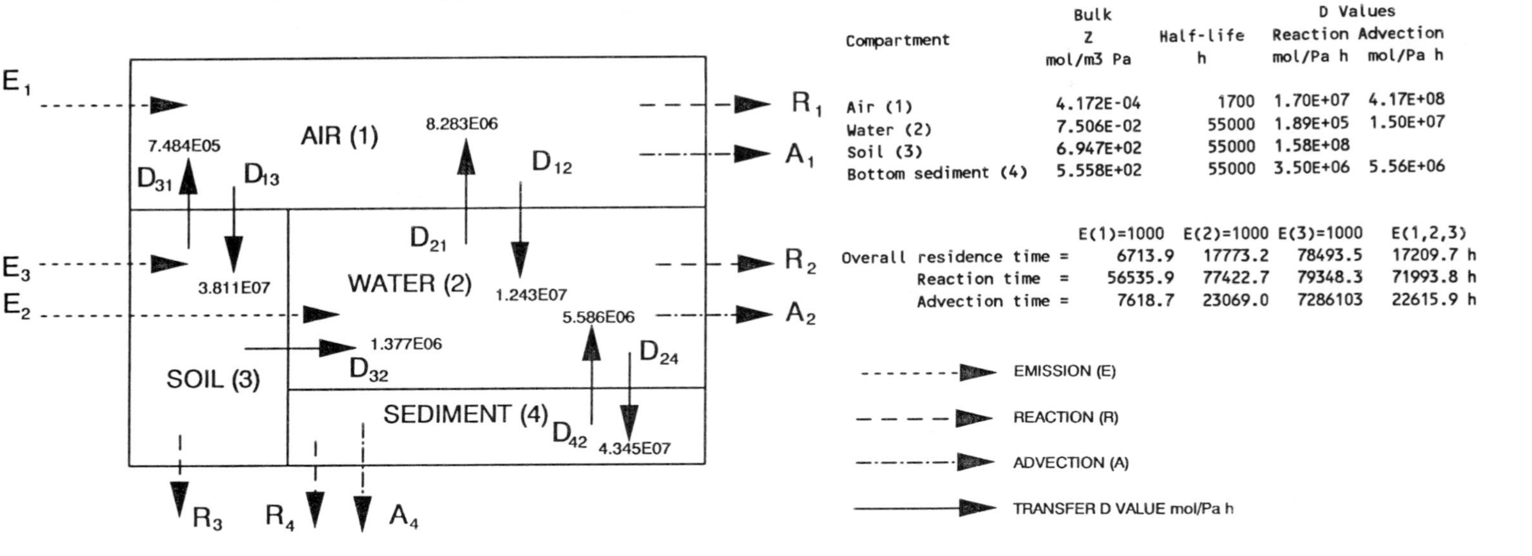

Phase Properties and Rates:

Compartment	Bulk Z mol/m3 Pa	Half-life h	D Values Reaction mol/Pa h	D Values Advection mol/Pa h
Air (1)	4.172E-04	1700	1.70E+07	4.17E+08
Water (2)	7.506E-02	55000	1.89E+05	1.50E+07
Soil (3)	6.947E+02	55000	1.58E+08	
Bottom sediment (4)	5.558E+02	55000	3.50E+06	5.56E+06

	E(1)=1000	E(2)=1000	E(3)=1000	E(1,2,3)
Overall residence time =	6713.9	17773.2	78493.5	17209.7 h
Reaction time =	56535.9	77422.7	79348.3	71993.8 h
Advection time =	7618.7	23069.0	7286103	22615.9 h

EMISSION (E)
REACTION (R)
ADVECTION (A)
TRANSFER D VALUE mol/Pa h

Phase Properties, Compositions, Transport and Transformation Rates:

Emission, kg/h			Fugacity, Pa				Concentration, g/m3				Amounts, kg				Total
E(1)	E(2)	E(3)	f(1)	f(2)	f(3)	f(4)	C(1)	C(2)	C(3)	C(4)	m(1)	m(2)	m(3)	m(4)	Amount, kg
1000	0	0	6.351E-06	1.609E-06	1.516E-06	4.775E-06	8.647E-07	3.943E-05	3.437E-01	8.662E-01	8.647E+04	7.886E+03	6.186E+06	4.331E+05	6.714E+06
0	1000	0	1.044E-06	6.110E-05	2.493E-07	1.813E-04	1.422E-07	1.497E-03	5.653E-02	3.288E+01	1.422E+04	2.994E+05	1.017E+06	1.644E+07	1.777E+07
0	0	1000	3.877E-08	5.344E-07	1.919E-05	1.585E-06	5.278E-09	1.309E-05	4.353E+00	2.876E-01	5.278E+02	2.618E+03	7.835E+07	1.438E+05	7.849E+07
600	300	100	4.128E-06	1.935E-05	2.904E-06	5.740E-05	5.620E-07	4.740E-04	6.584E-01	1.041E+01	5.620E+04	9.481E+04	1.185E+07	5.207E+06	1.721E+07

Emission, kg/h			Loss, Reaction, kg/h				Loss, Advection, kg/h			Intermedia Rate of Transport, kg/h T12	T21	T13	T31	T32	T24	T42
E(1)	E(2)	E(3)	R(1)	R(2)	R(3)	R(4)	A(1)	A(2)	A(4)	air-water	water-air	air-soil	soil-air	soil-water	water-sed	sed-water
1000	0	0	3.525E+01	9.936E-02	7.79E+01	5.457E+00	8.647E+02	7.886E+00	8.662E+00	2.577E+01	4.351E+00	7.900E+01	3.702E-01	6.812E-01	2.282E+01	8.705E+00
0	1000	0	5.797E+00	3.772E+00	1.28E+01	2.072E+02	1.422E+02	2.994E+02	3.288E+02	4.239E+00	1.652E+02	1.299E+01	6.089E-02	1.120E-01	8.665E+02	3.305E+02
0	0	1000	2.152E-01	3.299E-02	9.87E+02	1.812E+00	5.278E+00	2.618E+00	2.876E+00	1.573E-01	1.445E+00	4.823E-01	4.688E+00	8.627E+00	7.579E+00	2.891E+00
600	300	100	2.291E+01	1.195E+00	1.49E+02	6.561E+01	5.620E+02	1.195E+00	1.041E+02	1.675E+01	5.231E+01	5.135E+01	7.092E-01	1.305E+00	2.744E+02	1.047E+02

Chemical name: 2,2',3,3',4,4'-Hexachlorobiphenyl

Level I

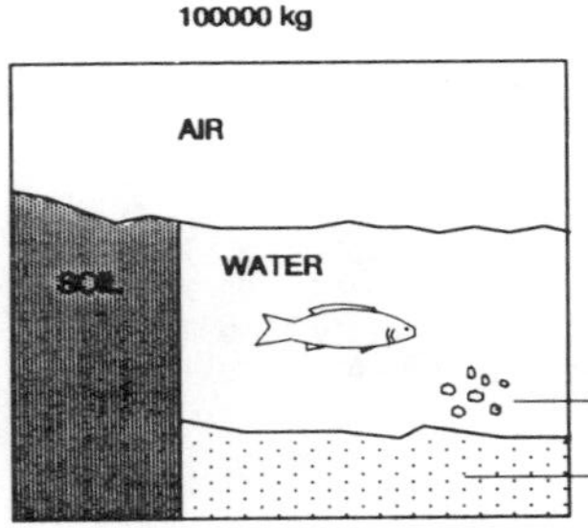

100000 kg

	Z	Concentration			Amount	Amount
	mol/m3 Pa	mol/m3	mg/L (or g/m3)	ug/g	kg	%
AIR	4.034E-04	7.345E-13	2.651E-10	2.236E-07	26.5	0.0265
WATER	8.397E-02	1.529E-10	5.517E-08	5.517E-08	11.034	0.01103
SOIL	1.652E+04	3.008E-05	1.086E-02	4.524E-03	97718	97.718
FISH	4.198E+04	7.643E-05	2.759E-02	2.759E-02	5.5170	5.52E-03
SUSPENDED SEDIMENT	1.033E+05	1.880E-04	6.786E-02	4.524E-02	67.859	6.79E-02
BOTTOM SEDIMENT	3.305E+04	6.017E-05	2.172E-02	9.048E-03	2171.50	2.1715
	Total				100000	100

f = 1.821E-09 Pa

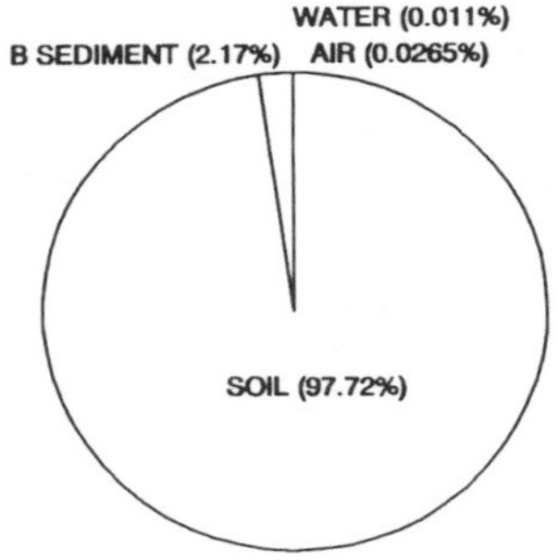

Distribution of mass

Level II

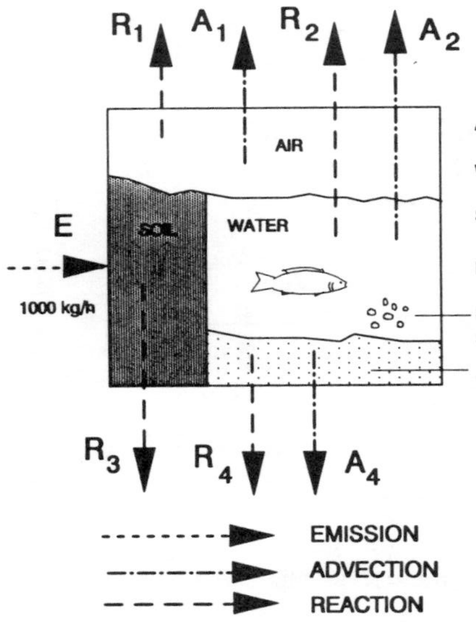

	Half-life h	D Values D(reaction)	D(advec'n)	Conc'n mol/m3	Loss Reaction kg/h	Loss Advection kg/h	Removal %
AIR	5500	5.08E+06	4.03E+08	4.64E-10	2.1117	167.595	16.971
WATER	55000	2.12E+05	1.68E+07	9.67E-08	8.79E-02	6.976	0.706
SOIL	55000	1.87E+09		1.90E-02	778.47		77.847
FISH				4.83E-02			
SUSPENDED SEDIMENT				1.19E-01			
BOTTOM SEDIMENT	55000	4.16E+07	6.61E+07	3.80E-02	17.299	27.459	4.476
Total		1.88E+09	4.86E+08		797.97	202.03	100
R + A			2.37E+09			1000	

f = 1.151E-06 Pa
Total amount = 63226482 kg

Overall residence time = 63226.48 h
Reaction time = 79234.19 h
Advection time = 312955.35 h

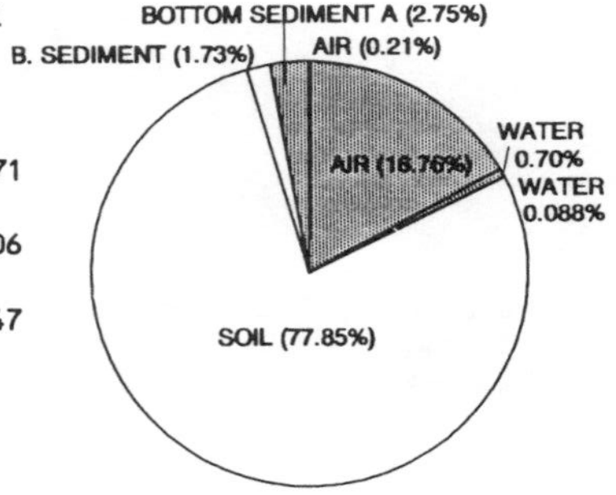

Distribution of removal rates

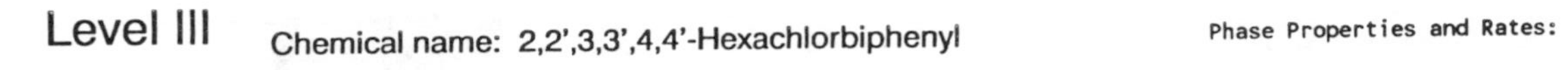

Level III

Chemical name: 2,2',3,3',4,4'-Hexachlorbiphenyl

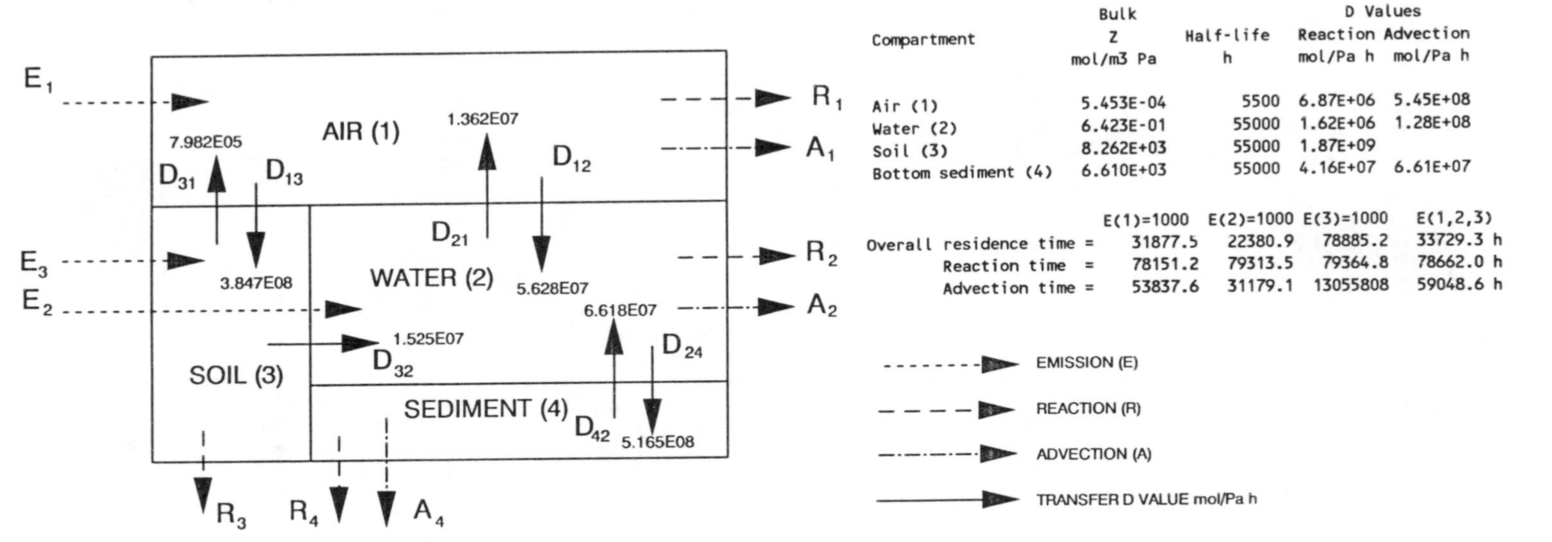

Phase Properties and Rates:

Compartment	Bulk Z mol/m3 Pa	Half-life h	D Values Reaction mol/Pa h	D Values Advection mol/Pa h
Air (1)	5.453E-04	5500	6.87E+06	5.45E+08
Water (2)	6.423E-01	55000	1.62E+06	1.28E+08
Soil (3)	8.262E+03	55000	1.87E+09	
Bottom sediment (4)	6.610E+03	55000	4.16E+07	6.61E+07

	E(1)=1000	E(2)=1000	E(3)=1000	E(1,2,3)
Overall residence time =	31877.5	22380.9	78885.2	33729.3 h
Reaction time =	78151.2	79313.5	79364.8	78662.0 h
Advection time =	53837.6	31179.1	13055808	59048.6 h

Phase Properties, compositions, Transport and Transformation Rates:

Emission, kg/h			Fugacity, Pa				Concentration, g/m3				Amounts, kg				Total Amount, kg
E(1)	E(2)	E(3)	f(1)	f(2)	f(3)	f(4)	C(1)	C(2)	C(3)	C(4)	m(1)	m(2)	m(3)	m(4)	
1000	0	0	2.795E-06	3.580E-07	5.690E-07	1.063E-06	5.501E-07	8.299E-05	1.697E+00	2.536E+00	5.501E+04	1.660E+04	3.054E+07	1.268E+06	3.188E+07
0	1000	0	8.215E-08	5.987E-06	1.672E-08	1.778E-05	1.617E-08	1.388E-03	4.985E-02	4.241E+01	1.617E+03	2.776E+05	8.974E+05	2.120E+07	2.238E+07
0	0	1000	1.843E-09	4.846E-08	1.467E-06	1.439E-07	3.628E-10	1.123E-05	4.373E+00	3.433E-01	3.628E+01	2.247E+03	7.871E+07	1.716E+05	7.889E+07
600	300	100	1.702E-06	2.016E-06	4.930E-07	5.986E-06	3.350E-07	4.673E-04	1.470E+00	1.428E+01	3.350E+04	9.345E+04	2.646E+07	7.139E+06	3.373E+07

Emission, kg/h			Loss, Reaction, kg/h				Loss, Advection, kg/h			Intermedia Rate of Transport, kg/h T12	T21	T13	T31	T32	T24	T42
E(1)	E(2)	E(3)	R(1)	R(2)	R(3)	R(4)	A(1)	A(2)	A(4)	air-water	water-air	air-soil	soil-air	soil-water	water-sed	sed-water
1000	0	0	6.932E+00	2.091E-01	3.85E+02	1.598E+01	5.501E+02	1.660E+01	2.536E+01	5.678E+01	1.760E+00	3.881E+02	1.639E-01	3.131E+00	6.673E+01	2.539E+01
0	1000	0	2.037E-01	3.497E+00	1.13E+01	2.672E+02	1.617E+01	2.776E+02	4.241E+02	1.668E+00	2.944E+01	1.140E+01	4.816E-03	9.202E-02	1.116E+03	4.246E+02
0	0	1000	4.571E-03	2.831E-02	9.92E+02	2.163E+00	3.628E-01	2.247E+00	3.433E+00	3.744E-02	2.383E-01	2.559E-01	4.225E-01	8.071E+00	9.032E+00	3.437E+00
600	300	100	4.221E+00	1.178E+00	3.33E+02	8.996E+01	3.350E+02	1.178E+00	1.428E+02	3.457E+01	9.911E+00	2.363E+02	1.420E-01	2.714E+00	3.757E+02	1.430E+02

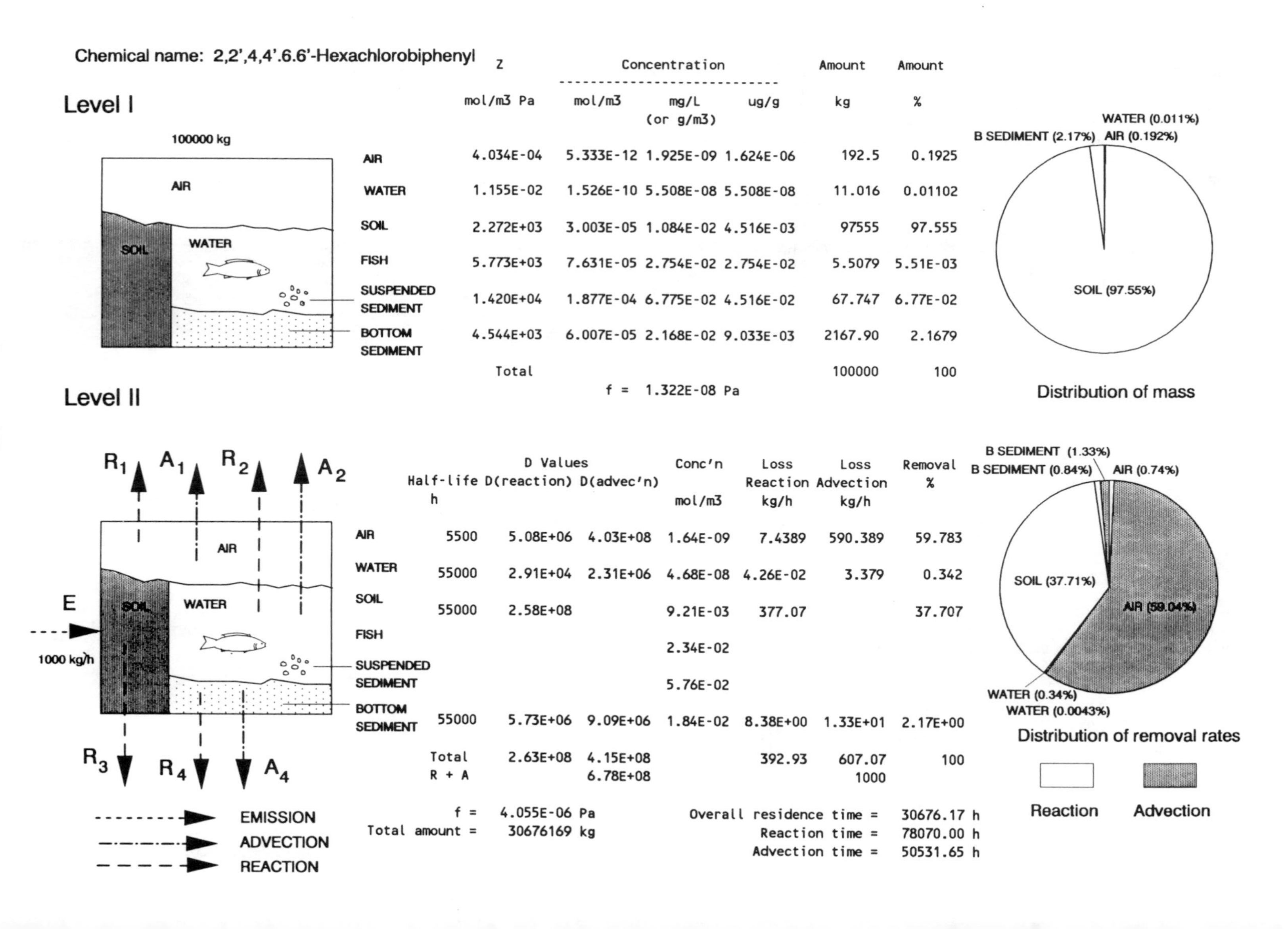

Chemical name: 2,2',4,4'.6.6'-Hexachlorobiphenyl

Level I

	Z	Concentration			Amount	Amount
	mol/m3 Pa	mol/m3	mg/L (or g/m3)	ug/g	kg	%
AIR	4.034E-04	5.333E-12	1.925E-09	1.624E-06	192.5	0.1925
WATER	1.155E-02	1.526E-10	5.508E-08	5.508E-08	11.016	0.01102
SOIL	2.272E+03	3.003E-05	1.084E-02	4.516E-03	97555	97.555
FISH	5.773E+03	7.631E-05	2.754E-02	2.754E-02	5.5079	5.51E-03
SUSPENDED SEDIMENT	1.420E+04	1.877E-04	6.775E-02	4.516E-02	67.747	6.77E-02
BOTTOM SEDIMENT	4.544E+03	6.007E-05	2.168E-02	9.033E-03	2167.90	2.1679
	Total				100000	100

f = 1.322E-08 Pa

Distribution of mass

Level II

	Half-life h	D Values D(reaction)	D(advec'n)	Conc'n mol/m3	Loss Reaction kg/h	Loss Advection kg/h	Removal %
AIR	5500	5.08E+06	4.03E+08	1.64E-09	7.4389	590.389	59.783
WATER	55000	2.91E+04	2.31E+06	4.68E-08	4.26E-02	3.379	0.342
SOIL	55000	2.58E+08		9.21E-03	377.07		37.707
FISH				2.34E-02			
SUSPENDED SEDIMENT				5.76E-02			
BOTTOM SEDIMENT	55000	5.73E+06	9.09E+06	1.84E-02	8.38E+00	1.33E+01	2.17E+00
	Total R + A	2.63E+08	4.15E+08 6.78E+08		392.93	607.07 1000	100

f = 4.055E-06 Pa
Total amount = 30676169 kg

Overall residence time = 30676.17 h
Reaction time = 78070.00 h
Advection time = 50531.65 h

Distribution of removal rates

Level III

Chemical name: 2,2',4,4',6,6'-Hexachlorobiphenyl

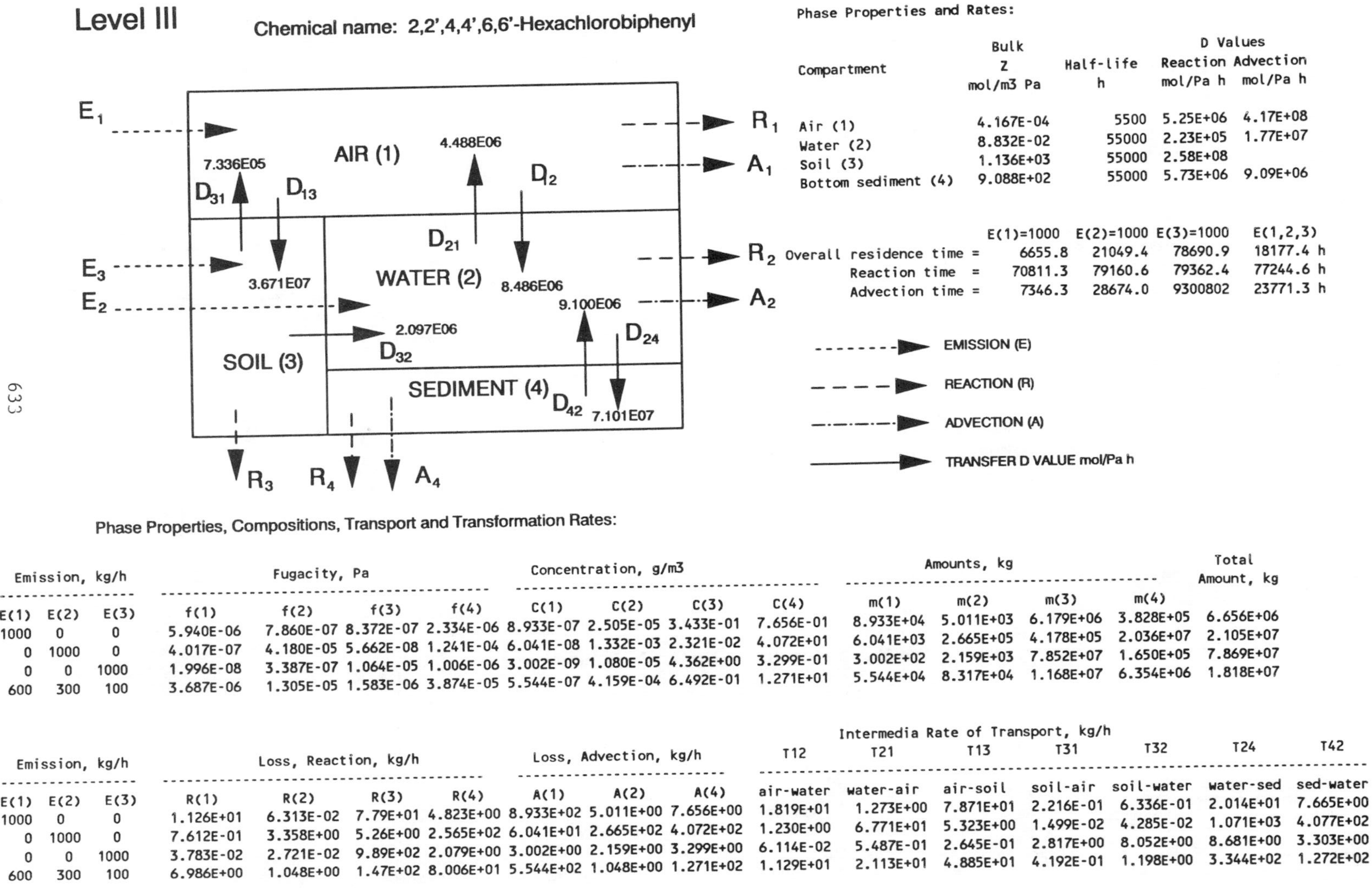

Phase Properties and Rates:

Compartment	Bulk Z mol/m3 Pa	Half-life h	D Values Reaction mol/Pa h	D Values Advection mol/Pa h
Air (1)	4.167E-04	5500	5.25E+06	4.17E+08
Water (2)	8.832E-02	55000	2.23E+05	1.77E+07
Soil (3)	1.136E+03	55000	2.58E+08	
Bottom sediment (4)	9.088E+02	55000	5.73E+06	9.09E+06

	E(1)=1000	E(2)=1000	E(3)=1000	E(1,2,3)
Overall residence time =	6655.8	21049.4	78690.9	18177.4 h
Reaction time =	70811.3	79160.6	79362.4	77244.6 h
Advection time =	7346.3	28674.0	9300802	23771.3 h

Phase Properties, Compositions, Transport and Transformation Rates:

Emission, kg/h			Fugacity, Pa				Concentration, g/m3				Amounts, kg				Total
E(1)	E(2)	E(3)	f(1)	f(2)	f(3)	f(4)	C(1)	C(2)	C(3)	C(4)	m(1)	m(2)	m(3)	m(4)	Amount, kg
1000	0	0	5.940E-06	7.860E-07	8.372E-07	2.334E-06	8.933E-07	2.505E-05	3.433E-01	7.656E-01	8.933E+04	5.011E+03	6.179E+06	3.828E+05	6.656E+06
0	1000	0	4.017E-07	4.180E-05	5.662E-08	1.241E-04	6.041E-08	1.332E-03	2.321E-02	4.072E+01	6.041E+03	2.665E+05	4.178E+05	2.036E+07	2.105E+07
0	0	1000	1.996E-08	3.387E-07	1.064E-05	1.006E-06	3.002E-09	1.080E-05	4.362E+00	3.299E-01	3.002E+02	2.159E+03	7.852E+07	1.650E+05	7.869E+07
600	300	100	3.687E-06	1.305E-05	1.583E-06	3.874E-05	5.544E-07	4.159E-04	6.492E-01	1.271E+01	5.544E+04	8.317E+04	1.168E+07	6.354E+06	1.818E+07

Emission, kg/h			Loss, Reaction, kg/h				Loss, Advection, kg/h			Intermedia Rate of Transport, kg/h T12	T21	T13	T31	T32	T24	T42
E(1)	E(2)	E(3)	R(1)	R(2)	R(3)	R(4)	A(1)	A(2)	A(4)	air-water	water-air	air-soil	soil-air	soil-water	water-sed	sed-water
1000	0	0	1.126E+01	6.313E-02	7.79E+01	4.823E+00	8.933E+02	5.011E+00	7.656E+00	1.819E+01	1.273E+00	7.871E+01	2.216E-01	6.336E-01	2.014E+01	7.665E+00
0	1000	0	7.612E-01	3.358E+00	5.26E+00	2.565E+02	6.041E+01	2.665E+02	4.072E+02	1.230E+00	6.771E+01	5.323E+00	1.499E-02	4.285E-02	1.071E+03	4.077E+02
0	0	1000	3.783E-02	2.721E-02	9.89E+02	2.079E+00	3.002E+00	2.159E+00	3.299E+00	6.114E-02	5.487E-01	2.645E-01	2.817E+00	8.052E+00	8.681E+00	3.303E+00
600	300	100	6.986E+00	1.048E+00	1.47E+02	8.006E+01	5.544E+02	1.048E+00	1.271E+02	1.129E+01	2.113E+01	4.885E+01	4.192E-01	1.198E+00	3.344E+02	1.272E+02

Chemical name: 2,2',3,3',4,4',6-Heptachlorobiphenyl

Level I

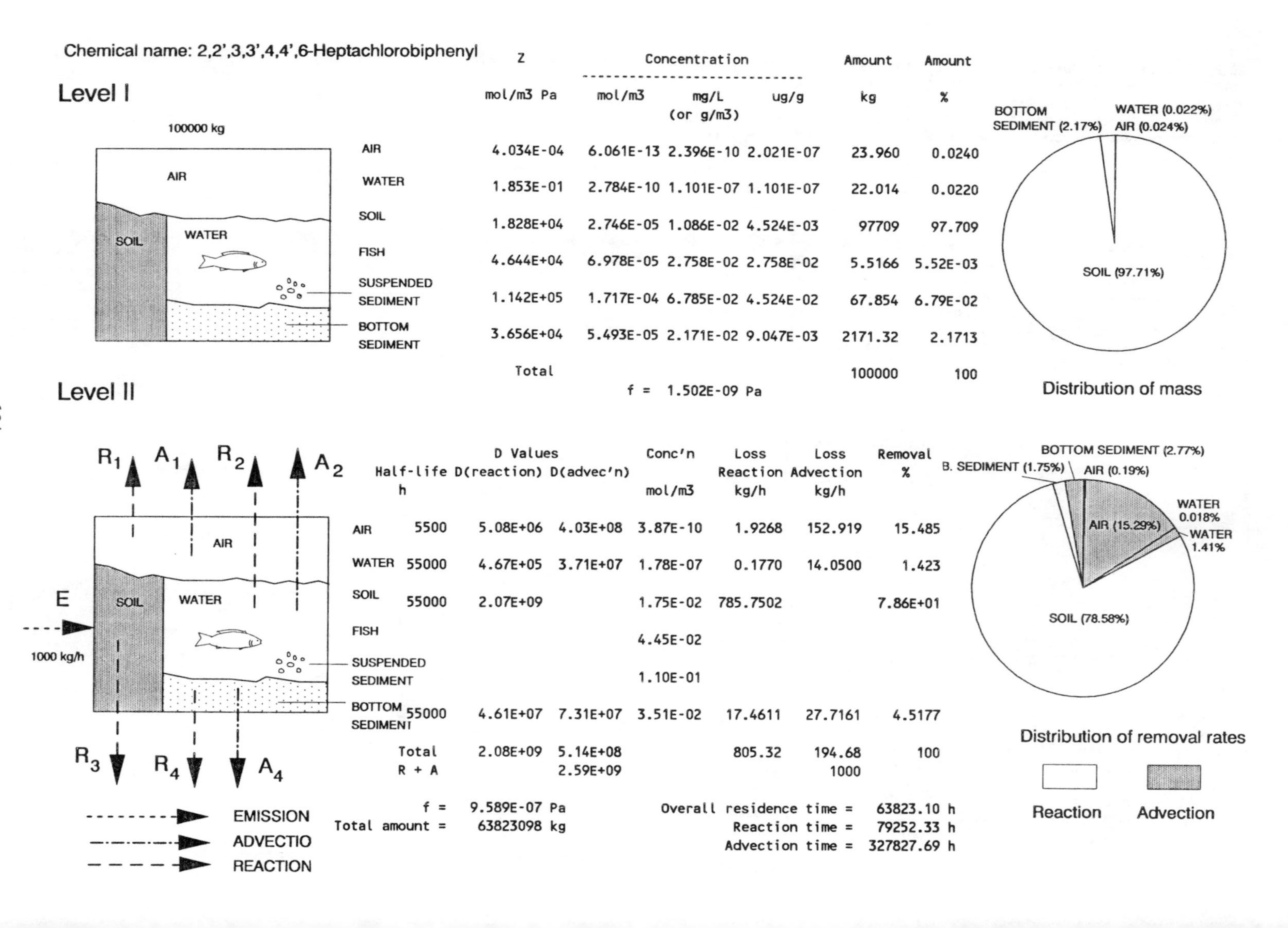

	Z	Concentration			Amount	Amount
	mol/m3 Pa	mol/m3	mg/L (or g/m3)	ug/g	kg	%
AIR	4.034E-04	6.061E-13	2.396E-10	2.021E-07	23.960	0.0240
WATER	1.853E-01	2.784E-10	1.101E-07	1.101E-07	22.014	0.0220
SOIL	1.828E+04	2.746E-05	1.086E-02	4.524E-03	97709	97.709
FISH	4.644E+04	6.978E-05	2.758E-02	2.758E-02	5.5166	5.52E-03
SUSPENDED SEDIMENT	1.142E+05	1.717E-04	6.785E-02	4.524E-02	67.854	6.79E-02
BOTTOM SEDIMENT	3.656E+04	5.493E-05	2.171E-02	9.047E-03	2171.32	2.1713
	Total				100000	100

f = 1.502E-09 Pa

Level II

	Half-life h	D Values D(reaction)	D(advec'n)	Conc'n mol/m3	Loss Reaction kg/h	Loss Advection kg/h	Removal %
AIR	5500	5.08E+06	4.03E+08	3.87E-10	1.9268	152.919	15.485
WATER	55000	4.67E+05	3.71E+07	1.78E-07	0.1770	14.0500	1.423
SOIL	55000	2.07E+09		1.75E-02	785.7502		7.86E+01
FISH				4.45E-02			
SUSPENDED SEDIMENT				1.10E-01			
BOTTOM SEDIMENT	55000	4.61E+07	7.31E+07	3.51E-02	17.4611	27.7161	4.5177
Total		2.08E+09	5.14E+08		805.32	194.68	100
R + A			2.59E+09			1000	

f = 9.589E-07 Pa
Total amount = 63823098 kg

Overall residence time = 63823.10 h
Reaction time = 79252.33 h
Advection time = 327827.69 h

Level III

Chemical name: 2,2'3,3',4,4',6-Heptachlorobiphenyl

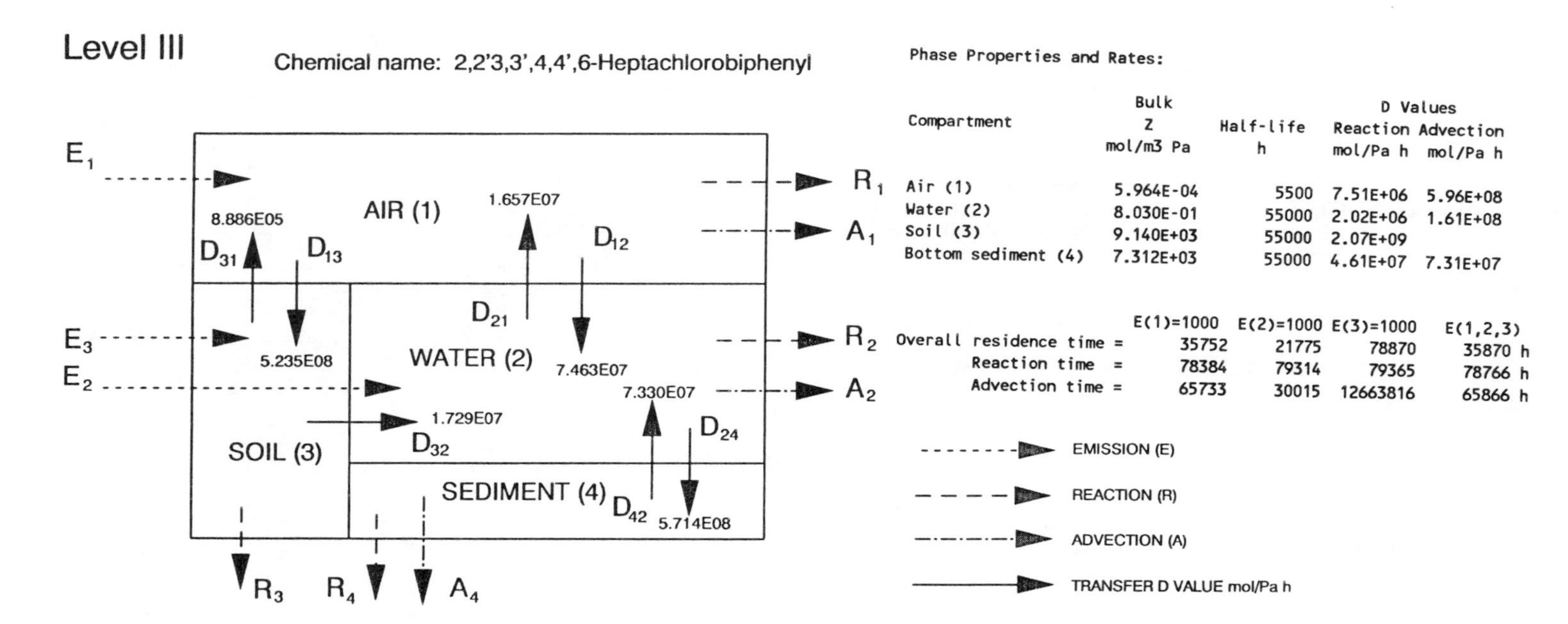

Phase Properties and Rates:

Compartment	Bulk Z mol/m3 Pa	Half-life h	D Values Reaction mol/Pa h	D Values Advection mol/Pa h
Air (1)	5.964E-04	5500	7.51E+06	5.96E+08
Water (2)	8.030E-01	55000	2.02E+06	1.61E+08
Soil (3)	9.140E+03	55000	2.07E+09	
Bottom sediment (4)	7.312E+03	55000	4.61E+07	7.31E+07

	E(1)=1000	E(2)=1000	E(3)=1000	E(1,2,3)
Overall residence time =	35752	21775	78870	35870 h
Reaction time =	78384	79314	79365	78766 h
Advection time =	65733	30015	12663816	65866 h

Phase Properties, Compositions, Transport and Transformation Rates:

Emission, kg/h			Fugacity, Pa				Concentration, g/m3				Amounts, kg				Total Amount, kg
E(1)	E(2)	E(3)	f(1)	f(2)	f(3)	f(4)	C(1)	C(2)	C(3)	C(4)	m(1)	m(2)	m(3)	m(4)	
1000	0	0	2.109E-06	3.125E-07	5.281E-07	9.276E-07	4.972E-07	9.919E-05	1.908E+00	2.681E+00	49724	19838	34341532	1340597	35751691
0	1000	0	6.555E-08	4.756E-06	1.641E-08	1.412E-05	1.545E-08	1.510E-03	5.929E-02	4.081E+01	1545.4	301932	1067295	20404018	21774791
0	0	1000	1.438E-09	3.945E-08	1.210E-06	1.171E-07	3.390E-10	1.252E-05	4.372E+00	3.385E-01	33.90	2504.3	78698714	169235	78870487
600	300	100	1.285E-06	1.618E-06	4.428E-07	4.804E-06	3.030E-07	5.137E-04	1.600E+00	1.388E+01	30301	102732	28794978	6942468	35870479

Emission, kg/h			Loss, Reaction, kg/h				Loss, Advection, kg/h			Intermedia Rate of Transport, kg/h T12	T21	T13	T31	T32	T24	T42
E(1)	E(2)	E(3)	R(1)	R(2)	R(3)	R(4)	A(1)	A(2)	A(4)	air-water	water-air	air-soil	soil-air	soil-water	water-sed	sed-water
1000	0	0	6.265E+00	2.500E-01	4.33E+02	1.689E+01	4.972E+02	1.984E+01	2.681E+01	6.223E+01	2.046E+00	4.365E+02	1.855E-01	3.608E+00	7.058E+01	2.688E+01
0	1000	0	1.947E-01	3.804E+00	1.34E+01	2.571E+02	1.545E+01	3.019E+02	4.081E+02	1.934E+00	3.114E+01	1.357E+01	5.765E-03	1.121E-01	1.074E+03	4.091E+02
0	0	1000	4.272E-03	3.155E-02	9.92E+02	2.132E+00	3.390E-01	2.504E+00	3.385E+00	4.243E-02	2.583E-01	2.976E-01	4.251E-01	8.269E+00	8.910E+00	3.393E+00
600	300	100	3.818E+00	1.294E+00	3.63E+02	8.748E+01	3.030E+02	1.294E+00	1.388E+02	3.792E+01	1.060E+01	2.660E+02	1.555E-01	3.025E+00	3.655E+02	1.392E+02

Chemical name: 2,2',3,3',5,5',6,6'-Octachlorobiphenyl

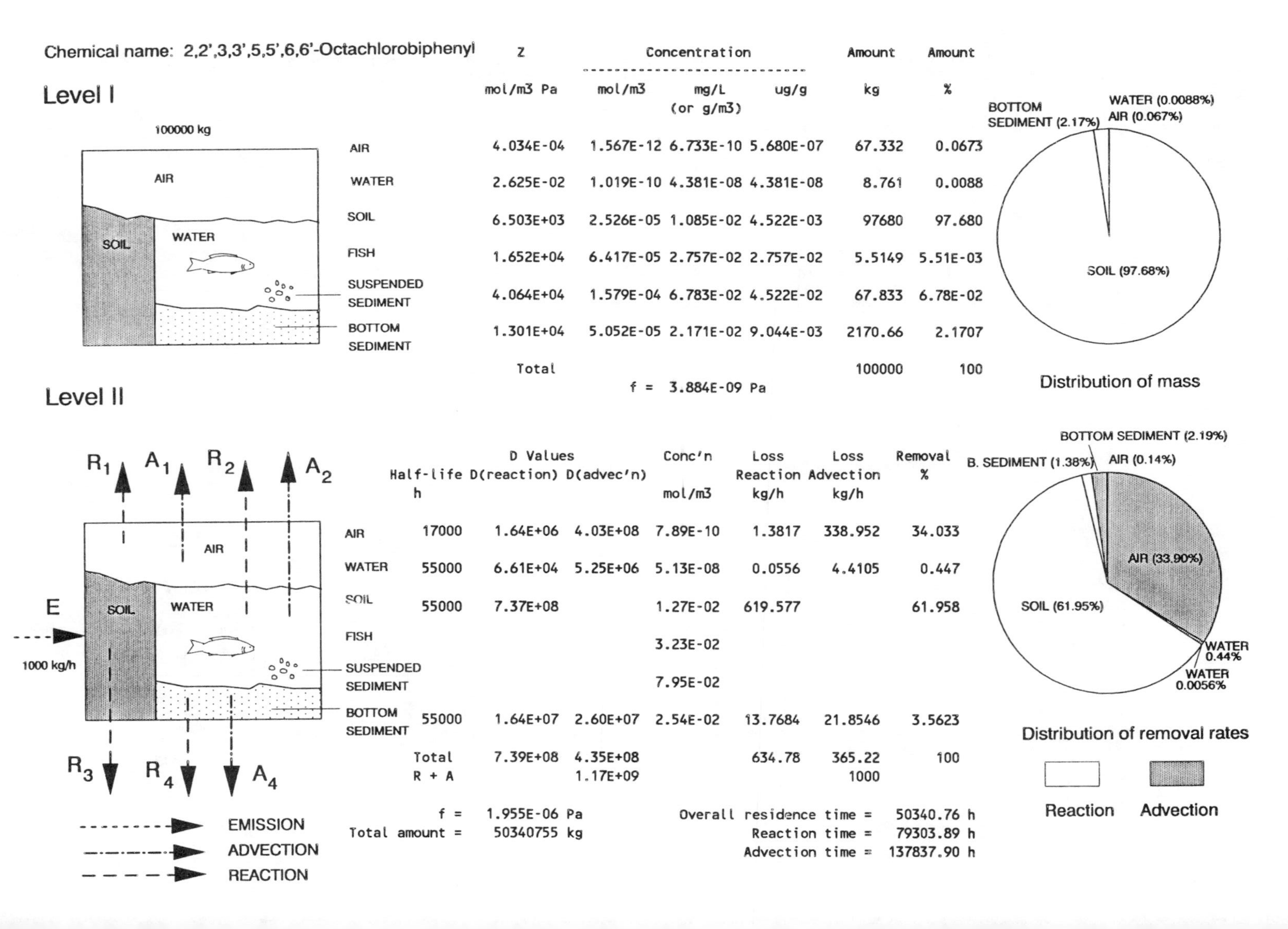

Level I

	Z	Concentration			Amount	Amount
	mol/m3 Pa	mol/m3	mg/L (or g/m3)	ug/g	kg	%
AIR	4.034E-04	1.567E-12	6.733E-10	5.680E-07	67.332	0.0673
WATER	2.625E-02	1.019E-10	4.381E-08	4.381E-08	8.761	0.0088
SOIL	6.503E+03	2.526E-05	1.085E-02	4.522E-03	97680	97.680
FISH	1.652E+04	6.417E-05	2.757E-02	2.757E-02	5.5149	5.51E-03
SUSPENDED SEDIMENT	4.064E+04	1.579E-04	6.783E-02	4.522E-02	67.833	6.78E-02
BOTTOM SEDIMENT	1.301E+04	5.052E-05	2.171E-02	9.044E-03	2170.66	2.1707
	Total				100000	100

f = 3.884E-09 Pa

Level II

	Half-life h	D Values D(reaction)	D(advec'n)	Conc'n mol/m3	Loss Reaction kg/h	Loss Advection kg/h	Removal %
AIR	17000	1.64E+06	4.03E+08	7.89E-10	1.3817	338.952	34.033
WATER	55000	6.61E+04	5.25E+06	5.13E-08	0.0556	4.4105	0.447
SOIL	55000	7.37E+08		1.27E-02	619.577		61.958
FISH				3.23E-02			
SUSPENDED SEDIMENT				7.95E-02			
BOTTOM SEDIMENT	55000	1.64E+07	2.60E+07	2.54E-02	13.7684	21.8546	3.5623
	Total	7.39E+08	4.35E+08		634.78	365.22	100
	R + A		1.17E+09			1000	

f = 1.955E-06 Pa
Total amount = 50340755 kg

Overall residence time = 50340.76 h
Reaction time = 79303.89 h
Advection time = 137837.90 h

Level III

Chemical name: 2,2',3,3',5,5',6,6'-Octachlorobiphenyl

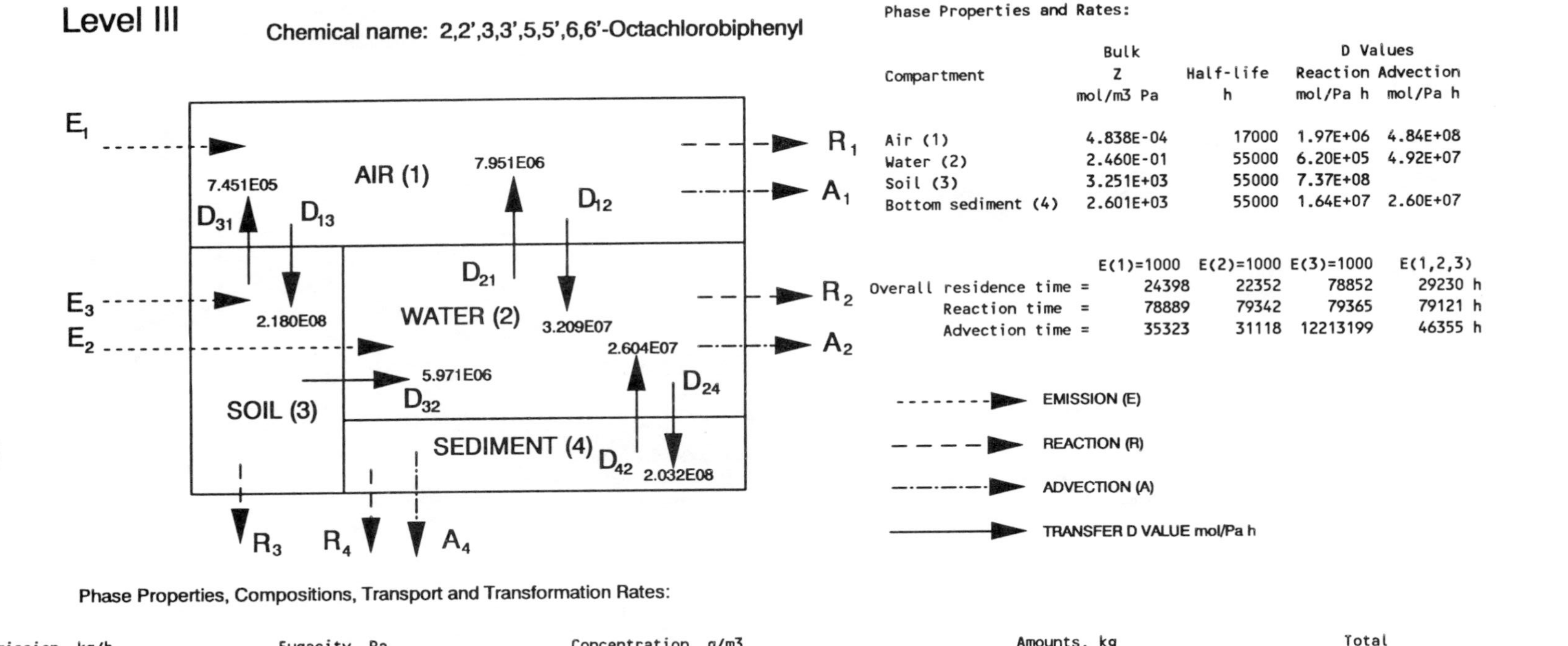

Phase Properties and Rates:

Compartment	Bulk Z mol/m3 Pa	Half-life h	D Values Reaction mol/Pa h	D Values Advection mol/Pa h
Air (1)	4.838E-04	17000	1.97E+06	4.84E+08
Water (2)	2.460E-01	55000	6.20E+05	4.92E+07
Soil (3)	3.251E+03	55000	7.37E+08	
Bottom sediment (4)	2.601E+03	55000	1.64E+07	2.60E+07

	E(1)=1000	E(2)=1000	E(3)=1000	E(1,2,3)
Overall residence time =	24398	22352	78852	29230 h
Reaction time =	78889	79342	79365	79121 h
Advection time =	35323	31118	12213199	46355 h

Phase Properties, Compositions, Transport and Transformation Rates:

Emission, kg/h			Fugacity, Pa				Concentration, g/m3				Amounts, kg				Total Amount, kg
E(1)	E(2)	E(3)	f(1)	f(2)	f(3)	f(4)	C(1)	C(2)	C(3)	C(4)	m(1)	m(2)	m(3)	m(4)	
1000	0	0	3.170E-06	5.839E-07	9.285E-07	1.734E-06	6.590E-07	6.172E-05	1.297E+00	1.938E+00	6.590E+04	1.234E+04	2.335E+07	9.691E+05	2.440E+07
0	1000	0	1.372E-07	1.270E-05	4.019E-08	3.770E-05	2.852E-08	1.342E-03	5.615E-02	4.214E+01	2.852E+03	2.684E+05	1.011E+06	2.107E+07	2.235E+07
0	0	1000	4.282E-09	1.024E-07	3.129E-06	3.042E-07	8.901E-10	1.083E-05	4.371E+00	3.401E-01	8.901E+01	2.166E+03	7.868E+07	1.700E+05	7.885E+07
600	300	100	1.944E-06	4.169E-06	8.820E-07	1.238E-05	4.040E-07	4.407E-04	1.232E+00	1.384E+01	4.040E+04	8.813E+04	2.218E+07	6.919E+06	2.923E+07

Emission, kg/h			Loss, Reaction, kg/h				Loss, Advection, kg/h			Intermedia Rate of Transport, kg/h T12	T21	T13	T31	T32	T24	T42
E(1)	E(2)	E(3)	R(1)	R(2)	R(3)	R(4)	A(1)	A(2)	A(4)	air-water	water-air	air-soil	soil-air	soil-water	water-sed	sed-water
1000	0	0	2.686E+00	1.555E-01	2.94E+02	1.221E+01	6.590E+02	1.234E+01	1.938E+01	4.371E+01	1.995E+00	2.969E+02	2.979E-01	2.382E+00	5.100E+01	1.940E+01
0	1000	0	1.163E-01	3.381E+00	1.27E+01	2.655E+02	2.852E+01	2.684E+02	4.214E+02	1.892E+00	4.337E+01	1.285E+01	1.290E-02	1.031E-01	1.109E+03	4.218E+02
0	0	1000	3.629E-03	2.729E-02	9.91E+02	2.142E+00	8.901E-01	2.166E+00	3.401E+00	5.904E-02	3.500E-01	4.010E-01	1.004E+00	8.027E+00	8.947E+00	3.404E+00
600	300	100	1.647E+00	1.110E+00	2.79E+02	8.718E+01	4.040E+02	1.110E+00	1.384E+02	2.680E+01	1.424E+01	1.820E+02	2.830E-01	2.263E+00	3.641E+02	1.385E+02

Chemical name: 2,2',3,3',4,4',5,5',6-Nonachlorobiphenyl

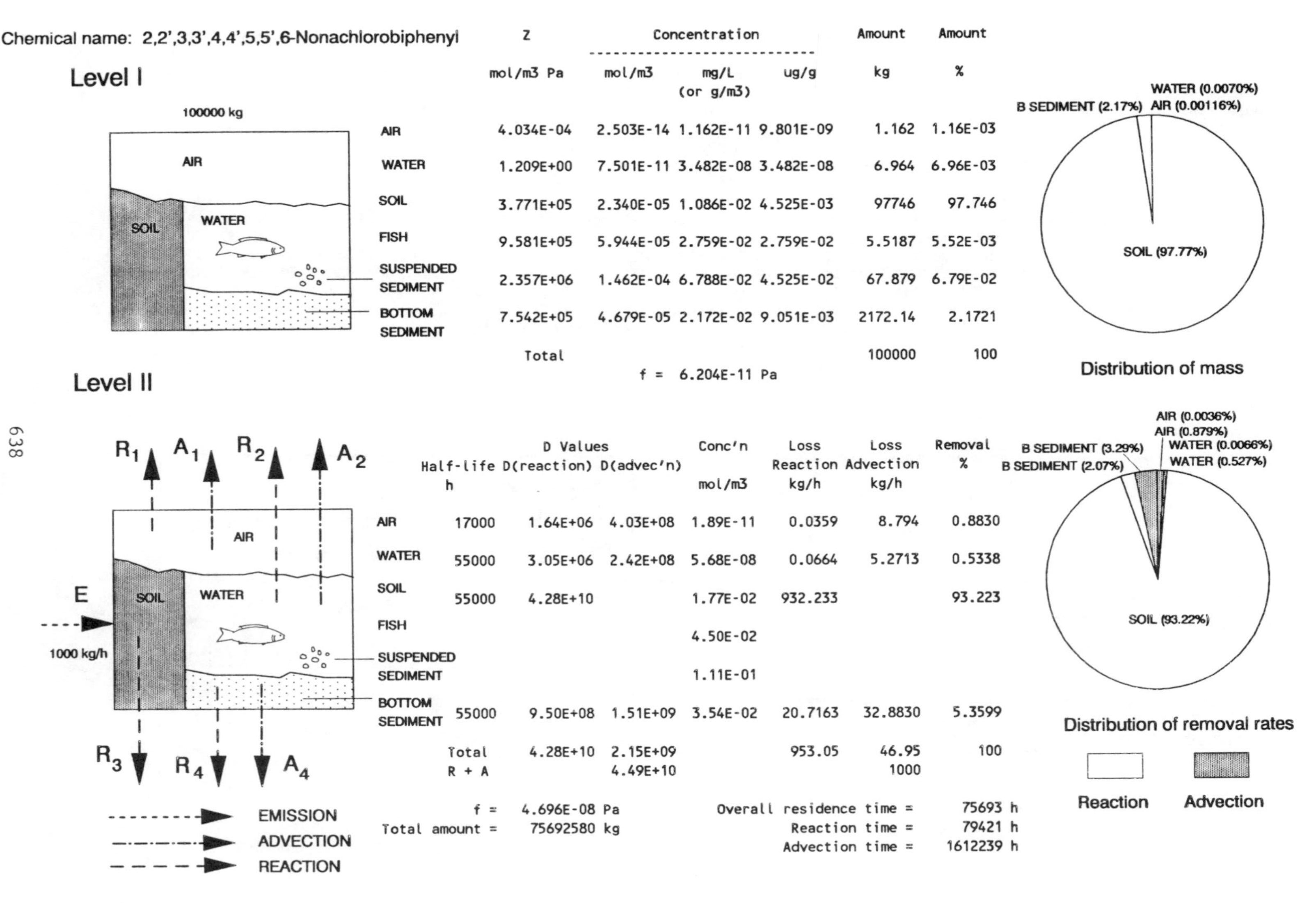

Level I

	Z	Concentration			Amount	Amount
	mol/m3 Pa	mol/m3	mg/L (or g/m3)	ug/g	kg	%
AIR	4.034E-04	2.503E-14	1.162E-11	9.801E-09	1.162	1.16E-03
WATER	1.209E+00	7.501E-11	3.482E-08	3.482E-08	6.964	6.96E-03
SOIL	3.771E+05	2.340E-05	1.086E-02	4.525E-03	97746	97.746
FISH	9.581E+05	5.944E-05	2.759E-02	2.759E-02	5.5187	5.52E-03
SUSPENDED SEDIMENT	2.357E+06	1.462E-04	6.788E-02	4.525E-02	67.879	6.79E-02
BOTTOM SEDIMENT	7.542E+05	4.679E-05	2.172E-02	9.051E-03	2172.14	2.1721
Total					100000	100

f = 6.204E-11 Pa

Level II

	Half-life h	D Values D(reaction)	D(advec'n)	Conc'n mol/m3	Loss Reaction kg/h	Loss Advection kg/h	Removal %
AIR	17000	1.64E+06	4.03E+08	1.89E-11	0.0359	8.794	0.8830
WATER	55000	3.05E+06	2.42E+08	5.68E-08	0.0664	5.2713	0.5338
SOIL	55000	4.28E+10		1.77E-02	932.233		93.223
FISH				4.50E-02			
SUSPENDED SEDIMENT				1.11E-01			
BOTTOM SEDIMENT	55000	9.50E+08	1.51E+09	3.54E-02	20.7163	32.8830	5.3599
Total		4.28E+10	2.15E+09		953.05	46.95	100
R + A			4.49E+10			1000	

f = 4.696E-08 Pa
Total amount = 75692580 kg

Overall residence time = 75693 h
Reaction time = 79421 h
Advection time = 1612239 h

Level III

Chemical name: 2,2',3,3',4,4',5,5',6-Nonachlorobiphenyl

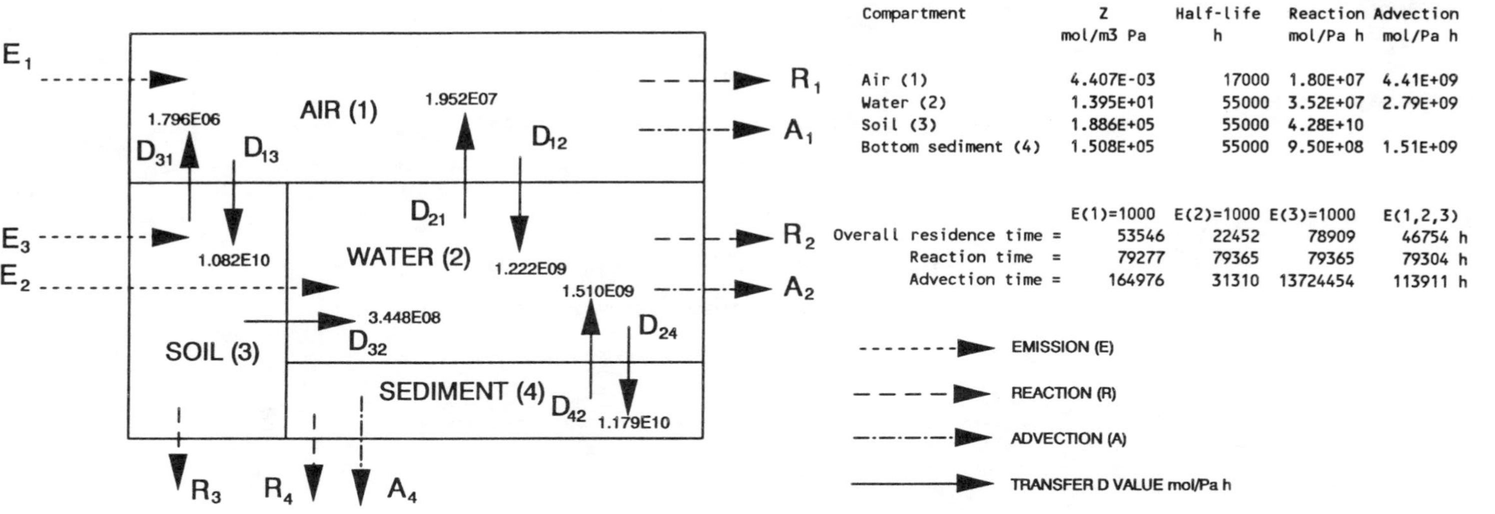

Phase Properties and Rates:

Compartment	Bulk Z mol/m3 Pa	Half-life h	D Values Reaction mol/Pa h	D Values Advection mol/Pa h
Air (1)	4.407E-03	17000	1.80E+07	4.41E+09
Water (2)	1.395E+01	55000	3.52E+07	2.79E+09
Soil (3)	1.886E+05	55000	4.28E+10	
Bottom sediment (4)	1.508E+05	55000	9.50E+08	1.51E+09

	E(1)=1000	E(2)=1000	E(3)=1000	E(1,2,3)	
Overall residence time =	53546	22452	78909	46754	h
Reaction time =	79277	79365	79365	79304	h
Advection time =	164976	31310	13724454	113911	h

EMISSION (E)

REACTION (R)

ADVECTION (A)

TRANSFER D VALUE mol/Pa h

Phase Properties, Compositions, Transport and Transformation Rates:

Emission, kg/h			Fugacity, Pa				Concentration, g/m3				Amounts, kg				Total Amount, kg
E(1)	E(2)	E(3)	f(1)	f(2)	f(3)	f(4)	C(1)	C(2)	C(3)	C(4)	m(1)	m(2)	m(3)	m(4)	
1000	0	0	1.308E-07	1.687E-08	3.284E-08	5.010E-08	2.676E-07	1.092E-04	2.875E+00	3.508E+00	2.676E+04	2.185E+04	5.174E+07	1.754E+06	5.355E+07
0	1000	0	2.516E-10	2.123E-07	6.318E-11	6.306E-07	5.148E-10	1.375E-03	5.530E-03	4.416E+01	5.148E+01	2.750E+05	9.954E+04	2.208E+07	2.245E+07
0	0	1000	7.464E-12	1.699E-09	4.997E-08	5.046E-09	1.527E-11	1.100E-05	4.374E+00	3.533E-01	1.527E+00	2.201E+03	7.873E+07	1.767E+05	7.891E+07
600	300	100	7.857E-08	7.399E-08	2.472E-08	2.197E-07	1.607E-07	4.792E-04	2.164E+00	1.539E+01	1.607E+04	9.584E+04	3.895E+07	7.693E+06	4.675E+07

Emission, kg/h			Loss, Reaction, kg/h				Loss, Advection, kg/h			Intermedia Rate of Transport, kg/h T12	T21	T13	T31	T32	T24	T42
E(1)	E(2)	E(3)	R(1)	R(2)	R(3)	R(4)	A(1)	A(2)	A(4)	air-water	water-air	air-soil	soil-air	soil-water	water-sed	sed-water
1000	0	0	1.091E+00	2.753E-01	6.52E+02	2.210E+01	2.676E+02	2.185E+01	3.508E+01	7.420E+01	1.529E-01	6.573E+02	2.739E-02	5.257E+00	9.229E+01	3.511E+01
0	1000	0	2.099E-03	3.465E+00	1.25E+00	2.782E+02	5.148E-01	2.750E+02	4.416E+02	1.427E-01	1.924E+00	1.264E+00	5.268E-05	1.011E-02	1.162E+03	4.419E+02
0	0	1000	6.225E-05	2.773E-02	9.92E+02	2.226E+00	1.527E-02	2.201E+00	3.533E+00	4.234E-03	1.540E-02	3.750E-02	4.167E-02	7.999E+00	9.296E+00	3.536E+00
600	300	100	6.552E-01	1.208E+00	4.91E+02	9.694E+01	1.607E+02	1.208E+00	1.539E+02	4.456E+01	6.704E-01	3.947E+02	2.061E-02	3.957E+00	4.048E+02	1.540E+02

Chemical name: Decachlorobiphenyl

Level I

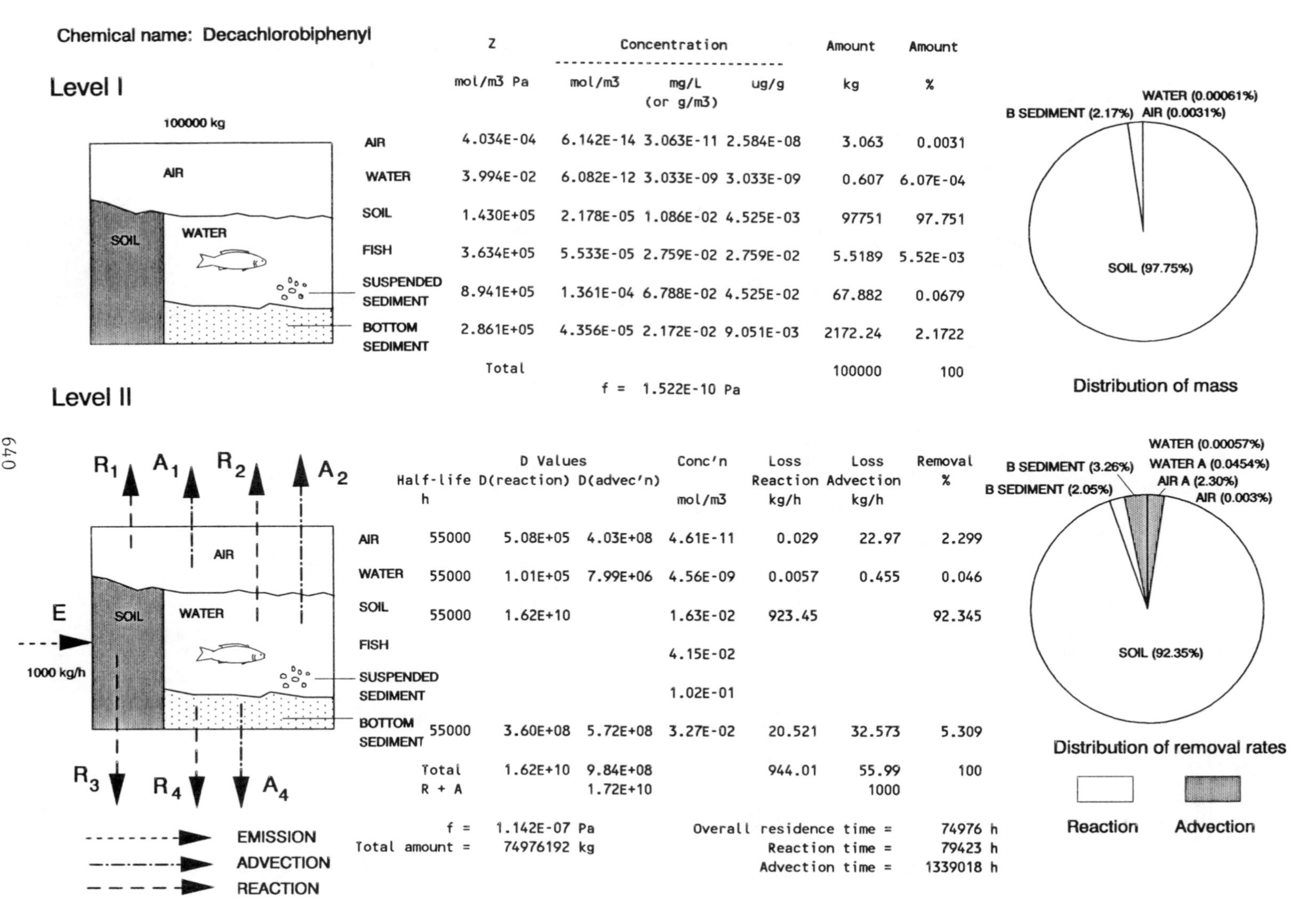

	Z	Concentration			Amount	Amount
	mol/m3 Pa	mol/m3	mg/L (or g/m3)	ug/g	kg	%
AIR	4.034E-04	6.142E-14	3.063E-11	2.584E-08	3.063	0.0031
WATER	3.994E-02	6.082E-12	3.033E-09	3.033E-09	0.607	6.07E-04
SOIL	1.430E+05	2.178E-05	1.086E-02	4.525E-03	97751	97.751
FISH	3.634E+05	5.533E-05	2.759E-02	2.759E-02	5.5189	5.52E-03
SUSPENDED SEDIMENT	8.941E+05	1.361E-04	6.788E-02	4.525E-02	67.882	0.0679
BOTTOM SEDIMENT	2.861E+05	4.356E-05	2.172E-02	9.051E-03	2172.24	2.1722
	Total				100000	100

f = 1.522E-10 Pa

Distribution of mass

Level II

	Half-life h	D Values D(reaction)	D(advec'n)	Conc'n mol/m3	Loss Reaction kg/h	Loss Advection kg/h	Removal %
AIR	55000	5.08E+05	4.03E+08	4.61E-11	0.029	22.97	2.299
WATER	55000	1.01E+05	7.99E+06	4.56E-09	0.0057	0.455	0.046
SOIL	55000	1.62E+10		1.63E-02	923.45		92.345
FISH				4.15E-02			
SUSPENDED SEDIMENT				1.02E-01			
BOTTOM SEDIMENT	55000	3.60E+08	5.72E+08	3.27E-02	20.521	32.573	5.309
Total		1.62E+10	9.84E+08		944.01	55.99	100
R + A			1.72E+10			1000	

f = 1.142E-07 Pa
Total amount = 74976192 kg

Overall residence time = 74976 h
Reaction time = 79423 h
Advection time = 1339018 h

Distribution of removal rates

Level III   Chemical name: Decachlorobiphenyl

E1, E2, E3
AIR (1)
WATER (2)
SOIL (3)
SEDIMENT (4)
D31 7.589E05
D13 4.340E09
D21 1.004E07
D12 4.921E08
D32 1.289E08
D42 5.772E08
D24 4.470E09
R1, A1, R2, A2, R3, R4, A4

EMISSION (E)
REACTION (R)
ADVECTION (A)
TRANSFER D VALUE mol/Pa h

Phase Properties and Rates:

Compartment	Bulk Z mol/m3 Pa	Half-life h	D Values Reaction mol/Pa h	D Values Advection mol/Pa h
Air (1)	2.010E-03	55000	2.53E+06	2.01E+09
Water (2)	4.874E+00	55000	1.23E+07	9.75E+08
Soil (3)	7.152E+04	55000	1.62E+10	
Bottom sediment (4)	5.722E+04	55000	3.60E+08	5.72E+08

	E(1)=1000	E(2)=1000	E(3)=1000	E(1,2,3)
Overall residence time =	51716	22953	78919	45808 h
Reaction time =	79365	79365	79365	79365 h
Advection time =	148452	32293	14041530	108338 h

Phase Properties, Compositions, Transport and Transformation Rates:

Emission, kg/h			Fugacity, Pa				Concentration, g/m3				Amounts, kg				Total Amount, kg
E(1)	E(2)	E(3)	f(1)	f(2)	f(3)	f(4)	C(1)	C(2)	C(3)	C(4)	m(1)	m(2)	m(3)	m(4)	
1000	0	0	2.930E-07	4.094E-08	7.777E-08	1.216E-07	2.938E-07	9.950E-05	2.774E+00	3.470E+00	2.938E+04	1.990E+04	4.993E+07	1.735E+06	5.172E+07
0	1000	0	7.805E-10	5.323E-07	2.072E-10	1.581E-06	7.825E-10	1.294E-03	7.389E-03	4.512E+01	7.825E+01	2.588E+05	1.330E+05	2.256E+07	2.295E+07
0	0	1000	1.975E-11	4.199E-09	1.226E-07	1.247E-08	1.980E-11	1.021E-05	4.374E+00	3.559E-01	1.980E+00	2.041E+03	7.874E+07	1.780E+05	7.892E+07
600	300	100	1.761E-07	1.847E-07	5.899E-08	5.486E-07	1.765E-07	4.489E-04	2.104E+00	1.565E+01	1.765E+04	8.978E+04	3.787E+07	7.827E+06	4.581E+07

Emission, kg/h			Loss, Reaction, kg/h				Loss, Advection, kg/h			Intermedia Rate of Transport, kg/h T12	T21	T13	T31	T32	T24	T42
E(1)	E(2)	E(3)	R(1)	R(2)	R(3)	R(4)	A(1)	A(2)	A(4)	air-water	water-air	air-soil	soil-air	soil-water	water-sed	sed-water
1000	0	0	3.702E-01	2.507E-01	6.29E+02	2.186E+01	2.938E+02	1.990E+01	3.470E+01	7.192E+01	2.049E-01	6.342E+02	2.943E-02	5.000E+00	9.127E+01	3.470E+01
0	1000	0	9.860E-04	3.261E+00	1.68E+00	2.843E+02	7.825E-01	2.588E+02	4.512E+02	1.916E-01	2.664E+00	1.689E+00	7.840E-05	1.332E-02	1.187E+03	4.513E+02
0	0	1000	2.495E-05	2.572E-02	9.92E+02	2.242E+00	1.980E-02	2.041E+00	3.559E+00	4.848E-03	2.102E-02	4.275E-02	4.642E-02	7.885E+00	9.361E+00	3.560E+00
600	300	100	2.224E-01	1.131E+00	4.77E+02	9.862E+01	1.765E+02	1.131E+00	1.565E+02	4.321E+01	9.243E-01	3.810E+02	2.233E-02	3.793E+00	4.117E+02	1.566E+02

4.4 COMMENTARY ON THE PHYSICAL-CHEMICAL PROPERTIES AND ENVIRONMENTAL FATE

Summary Tables

As discussed in Chapter 1, there have been reports of calculated values of aqueous solubility, vapor pressure (Burkhard 1984, Burkhard et al. 1985), octanol-water partition coefficient (Hawker and Connell 1988), melting point and aqueous solubility (Abramowitz and Yalkowsky 1990) for all 209 PCB congeners based on the concept of Quantitative-Structure-Activity Relationships (QSARs). When these calculated quantities are plotted against a molecular descriptor, a linear or near-linear relationship is expected. Therefore, the values selected for the summary tables and included in the QSPR plots are largely based on experimental rather than calculated values.

QSPR Plots

The QSPR plots of the PCBs given in Figures 4.1 to 4.6 show consistently near-linear relationships. The various commonly used molecular descriptors are shown in Figure 4.1.

An increase of 20 cm^3/mol (approximately one chlorine atom) causes (approximately):

(i) a decrease in log solubility by 0.52 units (a factor of 3.3);

(ii) a decrease in log vapor pressure by 0.6 units (a factor of 4.0);

(iii) an increase in log K_{OW} by 0.44 units (a factor of 2.7);

(iv) a decrease in log Henry's law constant by 0.08 units (a factor of 1.2). The data are, however, quite scattered.

The slope of the log K_{OW} versus log C_L line in Figure 4.6 is thus approximately 0.44/0.52 or 0.84, however, depending on the range selected lower values are possible.

It must be noted that the solubilities and vapor pressures of most of these PCB congeners are those of the subcooled liquids, not the actual solid phase values.

No attempt is made here to discriminate between difference in properties of isomers. As more data become available it will be desirable to use molecular descriptors such as total surface area which can explain and predict the effect of chlorine placement on these properties within isomer groups.

Evaluative Calculation

Level I, II and III calculations are shown for biphenyl and selected isomers ranging from a monochlorobiphenyl to decachlorobiphenyl.

The Level I calculations show that increased chlorination results in a tendency to partition from air to soil and sediment with only minor amounts in water. Even the relatively involatile trichlorobiphenyls show significant partitioning into air.

The Level II calculations show that atmospheric removal processes are significant, but as partitioning into air becomes negligible, reaction in soil becomes dominant as the major removal process. These results must be treated with caution because the half-lives of 55000 h (6 years) in soil of the more chlorinated PCBs are speculative. Clearly PCBs are highly persistent with residence times ranging up to nearly 10 years.

The Level III calculations also show long persistences and a tendency to accumulate to high concentrations.

If discharged to air, the dominant removal mechanism is usually advection from air, but there is appreciable deposition of the more chlorinated homologs to soil and water. For example, if 1000 kg of hexa-PCB is discharged to air, 550 kg advects from air, 7 kg reacts in air and 388 kg deposits on soil and 57 kg on water. This is consistent with the observed tendency for the atmosphere to control PCB concentrations in remote lakes and soils. The residence time of the more chlorinated PCBs is several years. It is striking that the water, soil and sediment fugacities are similar to those in the air, i.e., these media tend to become equilibrated with the air because of slow reaction rates.

If discharged to water there is appreciable evaporation and sedimentation, indeed for the higher PCBs the sediment becomes a major sink with appreciable rates of loss by reaction (even with the long half-life) and by burial (sediment advection).

Very high sediment concentrations are encountered for the more chlorinated PCBs. Residence times tend to be about 20000 h (2 years) when chlorine number exceeds three with advection from water and sediment, and sediment reaction dominating. Very high concentrations of the higher homologs are encountered in sediments, and thus in fish.

If discharged to soil most PCB remains and reacts in the soil but with some evaporation. The overall residence time increases with chlorine number to about 80000 h or 9 years as dictated by the assumed 6 year half life in soil. Evaporation is fairly slow because of the high sorption tendency.

The calculated behavior of PCBs is consistent with their observed behavior. They tend to accumulate and persist in soils and sediments. They are subject to appreciable atmospheric deposition, to sedimentation, bioaccumulation and to long range transport. Their physical-chemical and reactivity properties combine to render them among the environmental contaminants of greatest concern.

4.5 REFERENCES

Abramowitz, R., Yalkowsky, S.H. (1990) Estimation of aqueous solubility and melting point of PCB congeners. *Chemosphere* 21(10-11), 1221-1229.

Aihara A. (1959) Estimation of the energy of hydrogen bonds formed in crystals. I. Sublimation pressures of some organic molecular crystals and the additivity of lattice energy. *Bull. Chem. Soc. Japan* 32, 1242-1248.

Akiyoshi, M., Deguchi, T., Sanemasa, I. (1987) The vapor saturation method for preparing aqueous solutions of solid aromatic hydrocarbons. *Bull. Chem. Soc. Jpn.* 60, 3935-3939

Andrews, L.J., Keefer, R.M. (1949) Cation complexed of compounds containing carbon-carbon double bonds. IV. The argentation of aromatic hydrocarbons. *J. Am. Chem. Soc.* 71, 3644-3647.

Arbuckle, W.B. (1983) Estimating acitivity coefficients for use in calculating environmental parameters. *Environ. Sci. Technol.* 17, 537-542.

Arbuckle, W.B. (1986) Using UNIFAC to calculate aqueous solubilities. *Environ. Sci. Technol.* 20, 1060-1064.

Armstrong, R.W., Sloan, R.J. (1985) PCB patterns in Hudson River fish. I. Residue/fresh water species. *Proc. Hudson River Environ. Soc.* Norrie Point, New York.

Atkinson, R. (1987) Estimation of OH radicals reaction rate constants and atmospheric lifetimes for polychlorobiphenyls, dibenzo-p-dioxins and dibenzofurans. *Environ. Sci. Technol.* 21, 305-307.

Atkinson, R., Aschmann, S.M. (1985) Rate constants for the gas-phase reaction of hydroxyl radicals with biphenyl and the monochlorobiphenyls at 295°K. *Environ. Sci. Technol.* 19, 462-464.

Atlas, E., Foster, R., Giam, C.S. (1982) Air-sea exchange of high molecular weight organic pollutants: laboratory studies. *Environ. Sci. Technol.* 16, 283-286.

Atlas, E., Giam, C.S. (1986) Sea-air exchange of high-molecular weight synthetic organic compounds. In: *The Role of Air-Sea Exchange in Geochemical Cycling*. pp.295-329, P.Buat-Ménard Ed., D. Reidel Publishing Company.

Augood, D.R., Hey D.H., Williams, G.H. (1953) Homolytic aromatic substitution. Part III. Ratio of isomerides formed in phenylation of chlorobenzene. Competitive experiments on the phenylation of p-dichlorobenzene and 1,3,5-trichlorobenzene. Partial rate factors for chlorobenzene. *J. Chem. Soc.(London),* pp.44-50.

Bahnick, D.A., Doucette, W.J. (1988) Use of molecular connectivity indices to estimate soil sorption coefficients for organic chemicals. *Chemosphere* 17, 1703-1715.

Bailey, R.E., Gonsior S.J., Rhinehart, W.L. (1983) Biodegradation of the monochlorobiphenyls and biphenyl in river water. *Environ. Sci. Technol.* 17(10), 617-621.

Baker, J.E., Capel, P.D., Eisenreich, S.J. (1986) Influence of colloids on sediment and water partition coefficients of polychlorobiphenyl congeners in natural water. *Environ. Sci. Technol.* 20, 1136-1143.

Ballschmiter, K., M. (1980) Analysis of polychlorinated biphenyls (PCB) by glass capillary gas chromatography. Composition of technical Aroclor and Clopen-PCB mixtures. *Zell, Fres. Z. Anal. Chem.* 302, 20-31.

Ballschmiter, K., Wittlinger, R. (1991) Interhemisphere exchange of hexachlorocyclohexanes, hexachlorobenzenes, polychlorobiphenyls, and 1,1,1-trichloro-2,2-bis(p-chlorophenyl)ethane in the lower troposphere. *Environ. Sci. Technol.* 25(6), 1103-1111.

Banerjee, S. (1985) Calculation of water solubility of organic compounds with UNIFAC-derived parameters. *Environ. Sci. Technol.* 19, 369-370.

Banerjee, S., Baughman, G.L. (1991) Bioconcentration factors and lipid solubility. *Environ. Sci. Technol.* 25, 536-539.

Banerjee, S., Howard, P.H. (1988) Improved estimation of solubility and partitioning through correction of UNIFAC-derived activity coefficients. *Environ. Sci. Technol.* 22, 839-841.

Banerjee, S. Howard, P.H., Lande, S.S. (1990) General structure vapor pressure relationships for organics. *Chemosphere* 21(10-11), 1173-1180.

Banerjee, S., Yalkowsky, S.H., Valvani, S.C. (1980) Water solubility and octanol/water partition coefficients of organics. Limitations of the solubility-partition coefficient correlation. *Environ. Sci. Technol.* 14, 1227-1229.

Baxter, R.A., Gilbert, P.E., Lidgett, R.A., Mainprize, J.H., Kooden, H.A. (1975) The degradation of polychlorinated biphenyls by microorganisms. *Sci. Total Environ.* 4, 53.

Baughman, G.L., Paris, D.F. (1981) Microbial bioconcentration of organic pollutants from aquatic systems - a critical review. *CRC Critical Reviews in Microbiology,* pp. 205-228.

Beaven, G.H., Hassan de la Mare, P.B.D.M., Johnson, E.A., Klassen, N.V. (1961) The kinetics and mechanisms of aromatic halogen substitution. Part X. Products in chlorination of biphenyl in acetic acid. *J. Chem. Soc.* 2749.

Bellavita, V. (1935) Biphenyl series. VI. The halogenation of biphenyline (2,4'-diaminobiphenyl). *Gazz. Chim. Ital.* 65, 632-646.

Bidleman, T.F. (1984) Estimation of vapor pressures for nonpolar organic compounds by capillary gas chromatography. *Anal. Chem.* 56, 2490-2496.

Bidleman T.F., Christinsen E.J. (1979) Atmospheric removal process for high molecular weight organochlorines. *J. Geophysical Res.* 84(c12), 7857-7862.

Billington, J.W. (1982) *The solubility of polynuclear aromatic hydrocarbons and polychlorinated biphenyls in aqueous system.* M.A.Sc. Thesis, University of Toronto, Toronto, Ontario.

Billington, J.W., Huang, G.L., Szeto, F., Shiu, W.Y., Mackay, D. (1988) Preparations of aqueous solutions of sparingly soluble organic substances: I. Single component systems. *Environ. Toxicol. & Chem.* 7, 117-124.

Binns, F., Suschitzky, H. (1971) Polyhalogenoaromatic compounds. Part XX. Some reactions of decachlorobiphenyl. *J. Chem. Soc.* (C) 1913.

Bohon, R.L., Claussen, W.F. (1951) The solubility of aromatic hydrocarbons in water. *J. Am. Chem. Soc.* 72, 1571-1578.

Bopp, R.F. (1983) Revised parameters for modeling the transport of PCB components across the air-water interface. *J. Geophys. Res.* 88, 2521-2529.

Boublik, T., Fried, V., Hala, E. (1973) *The Vapour Pressure of Pure Substances.* Elsevier, Amsterdam.

Boublik, T., Fried, V., Hala, E. (1984) *The Vapour Pressures of Pure Substances.* (second revised edition), Elsevier, Amsterdam.

Boublik T. et al. (1984) *Dynamics, Exposure, Hazard Assessment Toxic Chemicals.* p.379-392.

Bradley, R.S., Cleasby, T.G. (1953) The vapor pressure and lattice energy of some aromatic ring compounds. *J. Chem. Soc.* 1953, 1690-1692.

Branson, D.R. (1977) A new capacity fluid: a case study in product stewardship. In: *Aquatic Toxicology and Hazard Evaluation.* ASTM, STP 634, Am. Soc. for Testing Materials, Philadelphia, pp.44-61.

Branson, D.R., Blau, G.E., Alexander, H.C., Neely, W.B. (1975) Bioconcentration of 2,2',4,4'-tetrachlorobiphenyl in rainbow trout as measured by accelerated test. *Trans. Am. Fish Soc.* 104, 785-792.

Briggs, G.G. (1981) Theoretical and experimental relationships between soil adsorption, octanol-water partition coefficients, water solubilities, bioconcentration factors, and the Parachor. *J. Agric. Food Chem.* 29, 1050-1059.

Bright, N.F.H. (1951) The vapour pressure of diphenyl, dibenzyl, and diphenylmethane. *J. Chem. Soc.* 624-625.

Brinkman, U.A.T., De Kock, A. (1980) Production, properties and usage. In: *Halogenated Biphenyls, Terphenyls, Naphthalenes, Dibenzodioxins and Other Products.* pp.1-36, R.D. Kimbrough, Ed., Elsevier/North Holland Biomedical Press, Amsterdam.

Brooke, D.N., Dobbs, A.J., Williams, N. (1986) Octanol:water partition coefficients (P): Measurement, estimation, and interpretation, particularly for chemicals with $P > 10^5$. *Ecotoxicol. Environ. Safety* 11, 251-260.

Brodsky, J., Ballschmiter, K. (1988) Reversed phase liquid chromatography of PCBs as a basis for the calculation of water solubility and log K_{OW} for polychlorobiphenyls. *Fresenius Z. Anal. Chem.* 331, 295-301.

Bruggeman, W.A., Martron, L.B.J.M., Kooiman, D., Hutzinger, O. (1981) Accumulation and elimination kinetics of di-, tri-, and tetrachlorobiphenyls by goldfish after dietary and aqueous exposure. *Chemosphere* 10(8), 811-832.

Bruggeman, W.A., Van Der Steen, J., Hutzinger, O. (1982) Reversed-phase thin-layer chromatography of polynuclear aromatic hydrocarbons and chlorinated biphenyls. Relationship with hydrophobicity as measured by aqueous solubility and octanol-water partition coefficient. *J. Chromatogr.* 238, 335-346.

Bruggeman, W.A., Opperhuizen, A. Wizbeuga, A., Hutzinger, O. (1984) Bioaccumulation of super-lipophilic chemicals in fish. *Toxicol. Environ. Chem.* 7, 173-189.

Brunner, S., Hornung, E., Santl, H., Wolff, E., Pringer, O.G., Altschuh, J., Bruggemann, R. (1990) Henry's law constants for polychlorinated biphenyls: Experimental determination and structure-property relationships. *Environ. Sci. Technol.* 24, 1751-1754.

Burkhard, L.P. (1984) *Physical-chemical properties of the polychlorinated biphenyls: measurement, estimation, and application to environmental systems.* Ph.D. Thesis, University of Wisconsin-Madison.

Burkhard, L.P., Kuehl, D.W., Veith, G.D. (1985c) Evaluation of reversed phase LC/MS for estimation of n-octanol/water partition coefficients of organic chemicals. *Chemosphere* 14, 1551-1560.

Burkhard, L.P., Kuehl, D.W. (1986) n-Octanol/water partition coefficients by reversed phase liquid chromatography/mass spectrometry for eight tetrachlorinated planar molecules. *Chemosphere* 15, 163-167.

Burkhard, L.P., Armstrong, D.E., Andren, A.W. (1984) Vapor pressures for biphenyl, 4-chlorobiphenyl, 2,2',3,3',5,5',6,6'-octachlorobiphenyl, and decachlorobiphenyl. *J. Chem. Eng. Data* 29, 248-250.

Burkhard, L.P., Andren, A.W., Armstrong, D.E. (1985a) Estimation of vapor pressures for polychlorinated biphenyls: A comparison of eleven predictive methods. *Environ. Sci. Technol.* 19, 500-507.

Burkhard, L.P., Armstrong, D.E., Andren, A.W. (1985b) Henry's law constants for the polychlorinated biphenyls. *Environ. Sci. Technol.* 19, 590-596.

Bysshe, S.E. (1982) Bioconcentration factor in aquatic organisms. In: *Handbook of Chemical Property Estimation Methods.* Lyman, W.J., Reehl, W.F., Rosenblatt, D.H. Editors, Chapter 5, Ann Arbor Sci., Ann Arbor, Michigan.

Callahan, M.A., Slimak, M.W., Gabel, N.W., May, I.P., Fowler, C.F., Freed, J.R., Jennings, P., Durfee, R.L., Whitmore, F.C., Maestri, B., Mabey, W.R., Holt, B.R., Gould, C. (1979) *Water-Related Environmental Fate of 129 Priority Pollutants.* Vol. I, EPA Report No. 440/4-79-029ab. Versar, Inc., Springfield, Virginia.

Capel, P.D., Eisenreich, S.J. (1990) Relationship between chlorinated hydrocarbons and organic carbon in sediment and porewater. *J. Great Lakes Res.* 16(2), 245-257.

Capel, P.D., Leuenberger, C., Giger, W. (1991) Hydrophobic organic chemicals in urban fog. *Atmos. Environ.* 25A(7), 1355-1346.

Carlberg, G.E., Martinsen, K., Kringstad, A., Gjessing, E., Grande, M., Käleqvist, T., Skåre, J.U. (1986) Influence of aqueous humus on the bioavailability of chlorinated micropollutants in Atlantic salmon. *Arch. Environ. Contam. Toxicol.* 15, 543-548.

Chin, Y.-P., Weber, Jr., W.J. (1989) Estimating the effects of dispersed organic polymers on the sorption of contaminants by natural solids. 1. A predictive thermodynamic humic substance-organic solute interaction model. *Environ. Sci. Technol.* 23, 978-984.

Chin, Y.-P., Weber, Jr., W.J., Eadie, B.J. (1990) Estimating the effects of dispersed organic polymers on the sorption of contaminants by natural solids. 2. Sorption in the presence of humic and orther natural maccromolecules. *Environ. Sci. Technol.* 24, 837-842.

Chiou, C.T. (1985) Partition coefficients of organic compounds in lipid-water systems and correlations with fish bioconcentration factors. *Environ. Sci. Technol.* 19, 57-62.

Chiou, C.T., Block, J.B. (1986) Parameters affecting the partition coefficient of organic compounds in solvent-water and lipid-water systems. In: *Partition Coefficient, Determination and Estimation.* W.J. Dunn III, J.H. Block, R.S. Pearlman, Eds., pp.36-60. Pergamon Press, New York.

Chiou, C.T., Freed, V.H., Schmedding, D.W., Kohnert, R.L. (1977) Partition coefficient and bioaccumulation of selected organic chemicals. *Environ. Sci. Technol.* 11, 475-478.

Chiou, C.T., Peters, L.J., Freed, V.H. (1979) A physical concept of soil-water equilibria for nonionic organic compounds. *Science* 206, 831-832.

Chiou C.T., Schmedding, D.W., Manes, M. (1982) Partitioning of organic compounds in octanol-water system. *Environ. Sci. Technol.* 16, 4-10.

Chiou, C.T., Porter, P.E., Schmedding, D.W. (1983) Partition equilibria of nonionic organic compounds between soil organic matter and water. *Environ. Sci. Technol.* 17, 227-231.

Chiou, C.T., Malcolm, R.L., Brinton, T.I., Kile, D.E. (1986) Water solubility enhancement of some organic pollutants and pesticides by dissolved humic and fulvic acids. *Environ. Sci. Technol.* 20, 502-508.

Chiou, C.T., Dile, D.E., Brinton, T.I., Malcolm, R.L., Leenheer, J.A., MacCarthy, P. (1987) A comparison of water solubility enhancements of organic solutes by aquatic humic materials and commercial humic acids. *Environ. Sci. Technol.* 21, 1231-1234.

Chiou, C.T., Kile, D.E., Rutherford, D.W. (1991) The natural oil in commercial linear alkylbenzenesulfonate and its effect on organic solute solubility in water. *Environ. Sci. Technol.* 25(4), 660-665.

Chou, S.F.J., Griffin, R.A. (1987) Solubility and soil mobility of polychlorinated biphenyls. In: *PCBs and the Environment.* Waid, Ed., Chapter 5, pp.101-120. CRC Press, Inc., Boca Raton, Florida.

Ciam, C.S., Murray, H.E., Ray, L.E., Kira, S. (1980) Bioaccumulation of hexachlorobenzene in killifish. *Bull. Environ. Contam. Toxic.* 25, 891-897.

Clark, K.E., Gobas, F.A.P.C., Mackay, D. (1990) Model of organic chemical uptake and clearance by fish from food and water. *Environ. Sci. Technol.* 24(8), 1203-1213.

Coates, J.T. (1984) *Sorption equilibria and kinetics for selected polychlorinated biphenyls on river sediments.* Ph.D. Thesis, Clemson University.

Coates, J.T., Elzerman, A.W. (1986) Desorption kinetics for selected PCB congeners from river sediments. In: *Transport and Transformation of Organic Contaminants.* D.L. Macalady, Ed., *J. Contam. Hydrology* 1, 191-210.

Connell, D.W., Hawker, D.W. (1986) Bioconcentration of lipophilic compounds by some aquatic organisms. *Ecotoxicol. Environ. Safety* 11, 184-197.

Connell, D.W., Hawker, D.W. (1988) Use of polynomial expressions to describe the bioconcentration of hydrophobic chemicals by fish. *Ecotoxicol. Envrion. Safety* 16, 242-257.

Coover, M.P., Sims, R.C.C. (1987) The effects of temperature on polyaromatic hydrocarbon persistence in an un-acclimated agricultural soil. *Haz. Waste Haz. Mat.* 4, 69-82.

De Bruijn, J., Busser, F., Seinen, W., Hermens, J. (1989) Determination of octanol/water partition coefficients for hydrophobic organic chemicals with the "slowing-stirring" method. *Environ. Toxicol. Chem.* 8, 499-512.

De Bruijn, J., Hermens, J. (1990) Relationships between octanol/water partition coefficients and total molecular surface area and total molecular volume of hydrophobic organic chemicals. *Quant. Struct.-Act. Relat.* 9, 11-21.

De Kock, A.C., Lord, D.A. (1987) A simple procedure for determining octanol-water partition coefficients using reversed phase high performance liquid chromatography (RPHPLC). *Chemosphere* 16(1), 133-142.

De Kock, A.C., Lord, D.A. (1988) Kinetics of the uptake and elimination of polychlorinated biphenyls by an estuarine fish species (rhabdosargus holubi) after aqueous exposure. *Chemosphere* 17(12), 2381-2390.

Dean, J.D., Ed. (1979) *Lange's Handbook of Chemistry.* 12th ed. McGraw-Hill, Inc., New York.

Dean, J.D., Ed. (1985) *Lange's Handbook of Chemistry.* 13th ed., McGraw-Hill, Inc., New York.

DeFoe, D.L., Veith, G.D., Carlson, R.W. (1978) Effects of Aroclor 1248 and 1260 on fathead minnows (*pimephales promelas*). *J. Fish Res. Board Can.* 35, 997.

Dexter, R.N., Pavlou, S.P. (1978) Mass solubility and aqueous activity coefficients of stable organic chemicals in the marine environment: polychlorinated biphenyls. *Mar. Chem.* 6, 41-53.

Dickerman, S.C., Weiss, K. (1957) Arylation of aromatic compounds by the Meerwein reaction. Evidence for aryl radicals from orientation studies. *J. Org. Chem.* 22, 1070.

Dickhut, R.M. (1985) *Investigation of the effects of temperature and organic co-solvents on the solubilties of selested polychlorinted biphenyls.* M.S. Thesis, University of Wisconsin, Madison.

Dickhut, R.M., Andren, A.W., Armstrong, D.E. (1986) Aqueous solubilities of six polychlorinated biphenyl congeners at four temperatures. *Environ. Sci. Technol.* 20, 807-810.

Dilling, W.L., Miracle, G.E., Boggs, G.U. (1983) Organic photochemistry. XVIII. Tropospheric phototransformation rates of 2-, 3-, and 4-chlorobiphenyl. *Preprints, Div. of Environ. Chem. ACS Natl Meeting,* Washington, D.C., pp 343-346.

Di Toro, D.M., Jeris, J.S., Ciarcia D. (1985) Diffusion & partitioning of hexachlorobiphenyl in sediments. *Environ. Sci. Technol.* 19, 1169-1176.

Dobbs, A.J., Cull, M.R. (1982) Volatilization of chemicals-relative loss rates and the estimation of vapor pressures. *Environ. Pollut.* (series B) 3, 289-298.

Doskey, P.V., Andren, A.W. (1981) Modelling the flux of atmospheric polychlorinated biphenyls across the air/water interface. *Environ. Sci. Technol.* 15, 705.

Doucette, W.J. (1985) *Measurement and estimation of octanol/water partition coefficients and aqueous solubility for halogenated aromatic hydrocarbons.* Ph.D. Thesis, University of Wisconsin, Madison, Wisconsin.

Doucette, W.J., Andren, A.W. (1987) Correlation of octanol/water partition coefficients and total molecular surface area for highly hydrophobic aromatic compounds. *Environ. Sci. Technol.* 21, 821-824.

Doucette, W.J., Andren, A.W. (1988) Estimation of octanol/water partition coefficients: Evaluation of six methods for highly hydrophobic aromatic hydrocarbons. *Chemosphere* 17, 345-359.

Doucette, W., Andren, A.W. (1988) Aqueous solubility of biphenyl, furan, and dioxin congeners. *Chemosphere* 17, 243-252.

Dow Chemical Co. (1982) *Private Communication.*

Dulin, D., Drossman, H., Mill, T. (1986) Products and quantum yields for photolysis of chloroaromatics in water. *Environ. Sci. Technol.* 20, 72-77.

Dunnivant, F.M., Coates, J.T., Elzerman, A.W. (1988) Experimentally determined Henry's law constants for 17 polychlorobiphenyl congeners. *Environ. Sci. Technol.* 22, 448-453.

Dunnivant, F.M., Elzerman, A.W. (1988) Aqueous solubility and Henry's law constant data for PCB congeners for evaluation of quantitative structure-property relationships (QSARs). *Chemosphere* 17, 525-541.

Eadie, B.J., Morehead, N.R., Landrum, P.F. (1990) Three phase partitioning of hydrophobic organic compounds in Great Lakes waters. *Chemosphere* 20, 161-178.

Eadsforth, C.V., Moser, P.(1983) Assessment of reversed phase chromatographic methods for determining partition coefficients. *Chemosphere* 12, 1459-1475.

Eganhouse, R.P., Calder, J.A. (1976) The solubility of medium molecular weight aromatic hydrocarbons and the effects of hydrocarbon co-solutes and salinity. *Geochim. Cosmochim. Acta* 40, 555-561.

Eisenreich, S.J. (1987) The chemical liminology of nonpolar organic contaminants: polychlorinated biphenyl in Lake Superior. In: *Sources and Fates of Aquatic Pollutants.* Hites, R.A., Eisenreich, S.J., Eds. pp. 393-469. *Advances in Chemistry Series* 216. Am. Chem. Soc., Washington D.C.

Eisenreich, S.J., Looney, B.B., Hollod, G.J. (1983) PCBs in the Lake Superior Atmosphere 1978-1980. In: *Physical Behavior of PCBs in the Great Lakes.* D. Mackay, S Paterson, S.J. Eisenreich and M.S. Simmons, Eds., p.115-125, Ann Arbor Sci. Publ., Ann Arbor, Michigan.

Eisenreich, E.J., Looney, B.B., Thornton, J.D. (1981) Airborne organic contaminants in the Great Lakes ecosystem. *Environ. Sci. Technol.* 15, 30-38.

Eitzer, B.D., Hite, R.A. (1988) Vapor pressures of chlorinated dioxins and dibenzofurans. *Environ. Sci. Technol.* 22, 1362-1364.

Erickson, M.D. (1986) *Analytical Chemistry of PCB's.* Ann Arbor Sci. Book, Butterworth Publishers, Stoneham, MA.

Ernst, W. (1977) Determination of the bioconcentration potential of marine organisms-a steady-state approach. *Chemosphere* 6, 731-740.

Evans, H.E. (1988) The binding of three PCB congeners to dissolved organic carbon in fresh waters. *Chemosphere* 17(12), 2325-2338.

Evans, M.S., Landrum, P.F. (1989) Toxicokinetics of DDE, Benzo(a)pyrene, 2,4,5,2',4',5'-Hexachlorobiphenyl in *Pontoporela hoyi* and *Mysis relicta*. *J. Great Lakes Res.* 15(4), 589-600.

Fendinger, N.J., Glotfelty, D.E. (1990) Henry's law constants for selected pesticides, PAHs, and PCBs. *Environ. Toxicol. Chem.* 9, 731-735.

Fichter, F. Adler, M. (1927) Electrochemical oxidation of nuclear chlorinated hydrocarbons. *Helv. Chim. Acta.* 9, 287.

Foreman, W.T., Bidleman, T.F. (1985) Vapor pressure estimates of individual polychlorinated biphenyls and commercial fluids using gas chromatographic retention data. *J. Chromatogr.* 330, 203-216.

Formica, S.J., Baron, J.A., Thibodeaux, L.J., Valsaraj, K.T. (1988) PCB transport into lake sediments. Conceptual model and laboratory simulation. *Environ. Sci. Technol.* 22, 1435-1440.

Freed, V.H., Chiou, C.T., Haque, R. (1977) Chemodynamics: transport and behaviour of chemicals in the environment-a problem in environmental health. *Environ. Health Perspec.* 20, 55-70.

Freitag, D., Lay, J.P. Korte, F. (1984) Environmental hazard profile-test results as related to structures and translation into the environment. In: *QSAR in Environmental Toxicology,* Kaiser, K.L.E., Ed., D. Reidel Publ. Co., Dordrecht, Netherlands.

Freitag, D., Ballhorn, L., Geyer, H., Korte, F. (1985) Environmental hazard profile of organic chemicals. An experimental method for the assessment of the behaviour of chemicals in the ecosphere by simple laboratory tests with C-14 labelled chemicals. *Chemosphere* 14, 1589-1616.

Furukawa, K., Matsumara, F. (1976) Microbial metabolism of polychlorinated biphenyls: Studies of th relative degradability of polychlorinated biphenyls by Alkaligenes Sp. *J. Agri. Food Chem.* 24, 251.

Furukawa, K., Tonomura, K., Kamibayashi, A. (1978) Effects of chlorine substitution on the biodegradability of polychlorinated biphenyls. *Appl. Environ. Microbiol.* 35, 223-227.

Garst, J.E. (1984) Accurate, wide-range, automated, high-performance liquid chromatographic method for the estimation of octanol/water partition coefficients. II: Equilibrium in partition coefficient measurements, additivity of substituent constants, and correlation of biological data. *J. Pharm. Sci.* 73, 1623-1629.

Garst, J.E., Wilson, W.C. (1984) Accurate, wide-range, automated, high-performance chromatographic method for the estimation of octanol/water partition coefficients. I: Effect of chromatographic conditions and procedure variables on accuracy and reproducibility of the method. *J. Pharm. Sci.* 73, 1616-1623.

Garten, Jr., C.T., Trabalka, J.R. (1983) Evaluation of models for predicting terrestrial food chain behavior of xenobiotics. *Environ. Sci. Technol.* 17, 590-595.

General Aniline and Film Corp. (1942) Compounds of the biaryl series. by W. Zerweek & K. Schütz. *Chem. Abstr.* 36, 5658., U.S. Patent 2,280,504, 1939.

Gerstl, Z., Mingelgrin, U. (1984) Sorption of organic substances by soils and sediments. *J. Environ. Sci. Health* B19(3), 297-312.

Getzen, F.W., Ward, T.M. (1971) Influence of water structure on aqueous solubility. *Ind. Eng. Chem. Prod. Res. Develop.* 10, 122-132.

Geyer, H., Politzki, G.R., Freitag, D. (1984) Prediction of ecotoxicological behaviour of chemicals: relationship between n-octanol/water partition coefficient and bioaccumulation of organic chemicals by Alga Chlorella. *Chemosphere* 13, 269-184.

Geyer, H. J., Scheunert, I., Korte, F. (1987) Correlation between the bioconcentration potential of organic environmental chemicals in humans and their n-octanol/water partition coefficients. *Chemosphere* 16(1), 239-252.

Geyer, H., Kraus, A.G., Klein, W., Richter, E., Korte, F. (1980) Relationship between water solubility and bioaccumulation potential of organic chemicals in rats. *Chemosphere* 9, 277-291.

Giam, C.S., Atlas, E., Chan, H.S., Neff, G.S. (1980) Phthalate esters, PCB and DDT residues in the gulf of Mexico atmosphere. *Atmos. Environ.* 14, 65-69.

Gibson, D.T. (1976) Microbial degradation of carcinogenic hydrocarbons and related compounds. In: *Proceedings of Symposium on Sources, Effects and Sinks of Hydrocarbons in the Aquatic Environment,* pp.224-238, American Institute of Biological Sciences, Washington, D.C., August 1976.

Glew, D.N., Moelwyn-Hughes, E.A. (1987) Dynamics of hydrophobic organic chemicals bioconcentration in fish. *Environ. Toxicol. Chem.* 6, 495-504.

Gobas, F.A.P.C., Bedard, D.C., Ciborowski, J.J.H. (1989) Bioaccumulation of chlorinated hydrocarbons by the mayfly (*hexagenia limbata*) in Lake St. Clair. *J. Great Lakes Res.* 15(4), 581-588.

Gobas, F.A.P.C., Mackay, D. (1989) *Biosorption, Bioaccumulation and Food Chain Transfer of Organic Chemicals.* Report prepared for Ontario Ministry of Environment.

Gobas, F.A.P.C., Clark, K., Shiu, W.Y., Mackay, D. (1989) Bioconcentration of polybromainated benzenes and biphenyls and related superhydrophobic chemicals in fish: role of bioavailability and elimination into the feces. *Environ. Toxicol. Chem.* 8, 231-245.

Gobas, F.A.P.C., Mackay, D. (1987) Dynamics of hydrophobic chemicals bioconcentration in fish. *Environ. Toxicol. Chem.* 6, 495-504.

Gobas, F.A.P.C., Shiu, W.Y., Mackay, D. (1987) Factors determining partitioning of hydrophobic organic chemicals in aquatic organisms. In: *QSAR in Environmental Toxicology* - II. Kaiser, K.L.E., Ed., D. Reidel Publ. Co., Dordrecht, Holland.

Goerke, H., Ernst, W. (1977) Fate of ^{14}C-labelled di-, tri-, and pentachlorobiphenyl in the marine annelid nereis virens. I.Accumulation and elimination after oral administration. *Chemosphere* 9, 551.

Gomberg, M., Bachman, W.E. (1924) Synthesis of biaryl compounds by means of the diazo reaction. *J. Am. Chem. Soc.* 46, 2339-2343.

Gooch, J.A., Hamdy, M.K. (1982) Depuration and biological half-life of ^{14}C-PCB in aquatic organisms. *Bull. Environ. Contam. Toxicol.* 28, 305.

Goto, M. et al. (1978) Accumulation of polychlorinated biphenyls and polybrominated biphenyls in fish: Limitation of "correlation between partition coefficients and accumulation factors". *Chemosphere* 7, 731.

Griffin, R.A., Clark, R., Lee, M.C., Chian, E.S.K. (1978) Disposal and removal of polychlorinated biphenyls in soil. In: *Land Disposal of Hazardous Waste.* David Schultz Ed., EPA-600/9-78-016, pp. 169-181. U.S. Environmental Protection Agency, Cincinnati, Ohio.

Griffin, R.A., Chian E.S.K. (1980) *Attenuation of water soluble polychlorinated biphenyl by earth materials.* Final Report, EPA 600/2-80-027, *PB* 80-219652, 104pp. Environmental Protection Agency, Washinton DC.

Guiney, P.D., Peterson, R.E., Melancon, M.J.Jr., Lech, J.J. (1977) The distribution and elimination of 2,5,2',5'-^{14}C tetrachlorobiphenyl in rainbow trout (salmo gairdneri). *Toxicol. Appl. Pharmacol.* 39, 329.

Guiney, P.D., Lech, J.J., Peterson, R.E. (1980) Distribution and elimination of a polychlorinated biphenyl during early life stages of rainbow trout (salmo gairdneri). *Toxicol. Appl. Pharm.* 53, 521-529.

Hafkenscheid, T.L., Tomlinson, E. (1981) Estimation of aqueous solubilities of organic nonelectrolytes using liquid chromatographic retention data. *J. Chromatog.* 218, 409-425.

Hafkenscheid, T.L., Tomlinson, E. (1983) Correlation between alkane/water and octanol/water distribution coefficients and isocratic reversed-phase liquid chromatographic capacity factor of acids, bases and neutrals. *Int. J. Pharm.* 16, 225-240.

Haky, J.E., Young, A.M. (1984) Evaluation of a simple HPLC correlation method for the estimation of the octanol-water partition coefficients of organic compounds. *J. Liq. Chromatogr.* 7, 675-689.

Halfon, E., Reggiani, M.G. (1986) On ranking chemicals for environmental hazard. *Environ. Sci. Technol.* 20, 1173-1179.

Hall, D.M., Minhaj, F. (1957) Relation between configuration and conjugation in biphenyl derivatives. IX. Some tetrachloro-2,2'-bridged compounds. *J. Chem. Soc.* 4585.

Halter, M.T., Johnson, H.E. (1977) In: *Aquatic Toxicology and Hazard Evaluation.* F.L. Mayer, J.L. Hamelink Eds., Am. Soc. Testing Materials, ASTM STP 634, pp. 178-195., Philadelphia.

Hammers, W.E., Meurs, G.J., De Ligny, C.L. (1982) Correlations betweeen liquid chromatographic capacity ratio data on Lichrosorb RP-18 and partition coefficients in the octanol-water system. *J. Chromatogr.* 247, 1-13.

Hansch, C., Nabomoto, K., Gorin, M., Denisevich, P., Garrett, E.R., Herman-Acbah, S.M., Won, C.H. (1973) Structure-activity relationship of chloramphenicols. *J. M. Chem.* 16, 917-922.

Hansch, C., Leo, A. (1985) *Medchem Project* Issue No. 26.Pomona College, Claremont, Calofornia.

Hansch, C., Quinlan, J.E., Lawrence, G.L (1968) The linear free-energy relationship between partition coefficients and the aqueous solubility of organic liquids. *J. Org. Chem.* 33, 347-350.

Hansch, C., Leo, A.J. (1979) *Substituent Constants for Correlation Analysis in Chemistry and Biology.* John Wiley & Sons, New York.

Hansen, D.J., Parrish, P.R., Lowe, J.I., Wilson, A.J., Wilson, P.D. (1971) Chronic toxicity, uptake and retention of Aroclor 1254 in two estuarine fishes. *Bull. Environ. Contam. Toxicol.* 6, 113.

Hansen (1976), L.G., Wiekhorst, W.B., Simon, J. (1976) Effects of dietary aroclor 1242 on channel catfish (ictalurus punctatus) and the selective accumulation of PCB congeners. *J. Fish Res. Board Can.* 33, 1343-1352.

Haque, R., Falco, J., Cohen, S., Riordan, C. (1980) Role of transport and fate studies in the exposure, assessment and screening of toxic chemicals. In: *Dynamics, Exposure, and Hazard Assessment of Toxic Chemicals.* R. Haque Ed., Ann Arbor Science Publishers, Ann Arbor, Michigan.

Haque, R., Schmedding, D. (1975) A method of measuring the water solubility of hydrophobic chemicals: Solubility of five polychlorinated biphenyls. *Bull.Environ. Contam.* Toxicol. 14, 13-18.

Haque, R., Schmedding, D. (1976) Studies on the adsorption of selected polychlorinated biphenyl isomers on several surfaces. *J. Environ. Sci. Health* B11(2), 129-137.

Hashimoto, Y., Tokura, K., Ozaki, K., Strachan, W.M.J. (1982) A comparison of water solubility by the flask and micro-column methods. *Chemosphere* 11(10), 991-1001.

Hassett, J.P., Milicic, E. (1985) Determination of equilibrium and rate constants of a polychlorinated congener by dissolved humic substances. *Environ. Sci. Technol.* 19, 638-643.

Hawker, D.W. (1989a) Vapor pressures and Henry's law constants of polychlorinated biphenyls. *Environ. Sci. Technol.* 23, 1250-1253.

Hawker, D.W. (1989b) The relationship between octan-1-ol/water partition coefficient and aqueous solubility in terms of solvatochromic parameters. *Chemosphere* 19, 1585-1593.

Hawker, D.W. (1990) Description of fish bioconcentration factors in terms of solvatochromic parameters. *Chemosphere* 20, 467-477.

Hawker, D.W., Connell, D.W. (1985) Relationships between partition coefficient uptake rate constant, clearance rate constant and time to equilibration for bioaccumulation. *Chemosphere* 14, 1205-1219.

Hawker, D.W., Connell, D.W. (1986) Bioconcentration of lipophilic compounds by some aquatic organisms. *Ecotox. Environ. Safety* 11, 184-197.

Hawker D.W., Connell, D.W. (1988a) Octanol-water partition coefficients of polychlorinated biphenyl congeners. *Environ. Sci. Technol.* 22, 382-387.

Hawker, D.W., Connell, D.W. (1988b) Influence of partition coefficient of lipophilic compounds on bioconcentration kinetics with fish. *Water Res.* 22, 701-707.

Hawker, D.W., Connell, D.W. (1989) A simple water/octanol partition system for bioconcentraiton investigations. *Environ. Sci. Technol.* 23, 961-965.

Hetling, L., Horn, E., Toftlemire, J. (1978) *Summary of Hudson river PCB study results.* New York State Department of Environmental Conservation, Tech. Report No. 51, Albany, New York.

Hinckley, D.A., Bidleman, T.F., Foreman, W.T. (1990) Determination of vapor pressures for nonpolar and semipolar organic compounds from gas chromatographic retention data. *J.Chem. Eng. Data* 35, 232-237.

Hinkel, L.E., Hay, D.H. (1928) Conversion of hydroaromatic into aromatic compounds. III. 3,5-dichloro-1-phenyl-$\Delta^{2,4}$-cyclohexadiene and its behavior with chlorine. *J. Chem. Soc.* 2786.

Hodson, J., Williams, N.A. (1988) The estimation of the adsorption coefficient (K_{oc}) for soils by high performance liquid chromatography. *Chemosphere* 17, 67-77.

Hollifield H.C. (1979) Rapid nephelometric estimate of water solubility of highly insoluble organic chemicals of environmental interests. *Bull. Environ. Contam. Toxicol.* 23, 579-586.

Hoover, T.B. (1971) Water solubility of PCB isomers. *PCB Newsletter* No. 3.

Horvath, A.L. (1982) *Halogenated Hydrocarbons, Solubility-Miscibility with Water.* Marcel Dekker, Inc., New York, N.Y.

Huang, G.L. (1983) The aqueous solubility and partitioning of hydrophobic organic solutes. *M.A.Sc. Thesis, University of Toronto.*

Hutzinger, O., Safe, S., Zitko, V. (1971) *Polychlorobiphenyls. Synthesis of some individual chlorobiphenyls. Bull. Environ. Contam. Toxicol.* 6, 209-219.

Hutzinger, O., Safe, S., Zitko, V. (1974) *The Chemistry of PCBs.* CRC Press, Inc., Cleveland, Ohio.

Isnard, P., Lambert, S. (1988) Estimating bioconcentration factors from octanol-water partition coefficient and aqueous solubility. *Chemosphere* 17, 21-34.

Isnard, P., Lambert, S. (1989) Aqueous solubility/n-octanol water partition coefficient correlations. *Chemosphere* 18, 1837-1853.

IUPAC Solubility Data Series (1985) Vol. 20: *Halogenated Benzenes, Toluenes and Phenols with Water.* Horvath, A.L., Getzen, F.W. Eds., Pergamon Press, Oxford, England.

Jaber, H.M., Smith, J.H., Cwirla, A.N. (1982) Evaluation of gas saturation methods to measure vapor pressure. *(EPA Contract* No. 68-01-5117) SRI International, Menlo Park, CA.

Johnstone, G.J., Ecobichon, D.J., Hutzinger, O. (1974) Effects of pure polychlorinated biphenyl compounds on hepatic function in the rat. *Toxicol. Appl. Pharmacol.* 28, 66-81.

Jota, M.A.T., Hassett, J.P. (1991) Effects of environmental variables on binding of a PCB congener by dissolved humic substances. *Environ. Toxicol. & Chem.* 10, 483-491.

Kaiser, K.L.E. (1983) A non-linear function for the approximation of octanol/water partition coefficients of aromatic compounds with multiple chlorine substitution. *Chemosphere* 12(9/10), 1159-1167.

Kamlet, M.J., Doherty, R.M., Abraham, M.H., Carr, P.W., Doherty, R.F., Taft, R.W. (1987) Linear solvation energy relationships. 42. Important differences between aqueous solubility relationships for aliphatic and aromatic solutes. *J. Phys. Chem.* 91, 1996.

Kamlet, M.J., Doherty, R.M., Carr, P.W., Mackay, D., Abraham, M.H., Taft, R.W. (1988) Linear solvation energy relationship. 44. Parameter estimation rules that allow accurate prediction of octanol/water partition coefficients and other solubility and toxicity properties of polychlorinated biphenyls and polycyclic aromatic hydrocarbons. *Environ. Sci. Technol.* 22, 503-509.

Kapoor, I.P., Metcalf, R.L., Hirwe, A.S., Coats, J.R., Khaisa, M.S. (1973) Structure activity correlations of biodegradability of DDT analogs. *J. Agric. Food Chem.* 21(2), 310-315.

Karickhoff, S.W. (1981) Semiempirical estimation of sorption of hydrophobic pollutants on natural sediments and soils. *Chemosphere* 10, 833-846.

Karickhoff, S.W., Brown, D.S., Scott, T.A. (1979) Sorption of hydrophobic pollutants on natural water sediments. *Water Res.* 13, 241-248.

Kenaga, E.E. (1980) Predicted bioconcentration factors and soil sorption coefficients of pesticides and other chemicals. *Ecotoxicol. Environ. Safety* 4, 26-38.

Kenaga, E.E., Goring, C.A.I. (1980) Relationship between water solubility, soil sorption, octanol-water partitioning and bioconcentration of chemicals in biota. In: *Aquatic Toxicology.* Eaton, J.G., Parrish, P.R., Hendricks, A.C. Eds., Am. Soc. for Testing and Materials, STP 707, pp. 78-115.

Kilzer, L., Scheunert, I., Geyer, H., Klein, W., Korte, F. (1979) Laboratory screening of the volatilization rates of organic chemicals from water & soil. *Chemosphere* 8, 751-761.

Kishi, H., Kogure, N., Hashimoto, Y. (1990) Contribution of soil constituents in adsorption coefficient of aromatic compounds, halogenated alicyclic and aromatic compounds to soil. *Chemosphere* 21(7), 867-876.

Koch, R. (1983) Molecular connectivity index for assessing ecotoxicological behaviour of organic chemicals. *Toxicol. Environ. Chem.* 6, 87-96.

Könemann, H. (1981) Quantitative structure-activity relationships in fish toxicity studies. Part 1: Relationship for 50 industrial pollutants. *Toxicology* 19, 209-221.

Könemann, W.H. (1979) *Quantitative Structure Activity Relationship for Kinetics and Toxicity of Aquatic Pollutants and Their Mixtures in Fish.* Ph.D. Thesis, University Utrecht, Netherlands.

Kong, H-L, Sayler, G.S. (1983) Degradation and total mineralization of monohalogenated biphenyls in natural sediment and mixed bacterial culture. *Appl. Environ. Microbiol.* 46(3), 666-672.

Korte, F., Freitag, D., Geyer, H., Klein, W., Kraus, A.G., Lahaniatis, E. (1978) Ecotoxicologic profile analysis-a concept for establishing ecotoxicologic priority lists for chemicals. *Chemosphere* No. 1, 79-102.

Lande, S.S., Banerjee, S. (1981) Predicting aqueous solubility of organic nonelectrolytes from molar volume. *Chemosphere* 10, 751-759.

Landrum P.F., Nihart, S.R., Eadle, B.J., Gardner, W.S. (1984) Reversed-phase separation method to determining pollutant binding to Aldrich humic acid and dissolved organic carbon of natural waters. *Environ. Sci. Technol.* 18, 187-192.

Landrum, P.F., Poore, R. (1988) Toxicokinetics of selected xenobiotics in hexagenia limbata. *J. Great Lakes Res.* 14(4), 427-437.

Lara, A., Ernst, W. (1989) Interaction between polychlorinated biphenyls and marine humic substances, determination of association coefficients. *Chemosphere* 19, 1655-1664.

Lawrence J., Tosine, H.M. (1976) Adsorption of polychlorinated biphenyls from aqueous solutions and sewage. *Environ. Sci. Technol.* 10, 381-383.

Leahy, D.E. (1986) Intrinsic molecular volume as a measure of the cavity term in linear solvation energy relationships: octanol-water partition coefficients and aqueous solubilities. *J. Pharm. Sci.* 75, 629-636.

Lee, M.C., Chian S.K., Griffin, R.A. (1979) Solubility of polychlorinated biphenyls and capacitor fluid in water. *Water Res.* 13, 1249-1258.

Lee, R.F., Ryan, C. (1976) Biodegradation of petroleum hydrocarbons by marine microbes. In: *Proc. Int'l Biodegradation Symp.* 3rd. 1975, pp.119-125.

Leo, A. (1985) *Medchem. Project.* Issue No.26, Pomona College, Claremont, CA.

Leo, A., Hansch, C., Elkins, D. (1971) Partition coefficients and their uses. *Chem. Rev.* 71, 525-616.

Lindenberg, M.A.B. (1956) Physicochimie des solutions-sur une relation simple entre le volume moliéculaire et la solubilité dans l'eau des hydrocarbures et dérivés halogénés. *Séance Du* 17, 2057-2060.

LOGP and Related Computerized Data Base, *Pomona College Med-Chem. Project,* Pomona College, Claremont, CA.. Technical Data Base (TDS) Inc.

Lyman, W.J., Reehl, W.F., Rosenblatt, D.H., Eds. (1982) *Handbook on Chemical Property Estimation Methods.* Environmental Behavior of Organic Compounds. McGraw-Hill, New York.

Lyman, W.J. (1982) Adsorption coefficient for soils and sediments. In: *Handbook of Chemical Property Estimation Methods.* Lyman, W.J., Reehl, W.F., Rosenblatt, D.H. Editors, Chapter 4, Ann Arbor Sci., Ann Arbor, Michigan.

Lynch, T.R., Johnson, H.E., Adams, W.J. (1982) The fate of atrazine and a hexachlorobiphenyl isomer in naturally-derived model stream ecosystem. *Environ. Toxicol. Chem. 1, 179-192.*

Mabey, W., Smith, J.H., Podoll, R.T., Johnson, H.L., Mill, T., Chou, T.W., Gate, J., Waight-Partridge, I., Jaber, H., Vandenberg, D. (1982) *Aquatic Fate Process for Organic Priority Pollutants.* EPA Report, No. 440/4-81-14.

Mabey, W., Mill, T. (1978) Critical review of hydrolysis of organic compounds in water under environmental conditions. *J. Phys. Chem. Ref Data* 7, 383-415.

Mackay, D. (1982) Correlation of bioconcentration factors. *Environ. Sci. Technol.* 16, 274-278.

Mackay, D. (1986) *Personal Communication.*

Mackay, D. (1989) Modeling the long term behavior of an organic contaminant in a large lake: Application to PCBs in Lake Ontario. J. Great Lakes Res. 15(2), 283-297.

Mackay, D., Bobra, A.M., Chan, D.W., Shiu, W.Y. (1982) Vapor pressure correlation for low-volatility environmental chemicals. *Environ. Sci. Technol.* 16, 645-649.

Mackay, D., Bobra, A.M., Shiu, W.Y., Yalkowsky, S.H. (1980) Relationships between aqueous solubility and octanol-water partition coefficient. *Chemosphere* 9, 701-711.

Mackay, D., Hughes, A.I. (1984) Three-parameter equation describing the uptake of organic compounds by fish. *Environ. Sci. Technol.* 18, 439-444.

Mackay, D., Leinonen, P.J. (1975) Rate of evaporation of low-solubility contaminants from water to atmosphere. *Environ. Sci. Technol.* 7, 1178-1180.

Mackay, D., Mascarenhas, R., Shiu, W.Y., S.C. Valvani, S.C., Yalkowsky, S.H. (1980) Aqueous solubility of polychlorinated biphenyls. *Chemosphere* 9, 257-264.

Mackay, D., Paterson, S. (1991) Evaluating the multimedia fate of organic chemicals: a level III fugacity model. *Environ. Sci. Technol.* 25(3), 427-436.

Mackay, D., Paterson, S., Schroeder, W.H. (1986) Model describing the rates of transfer processes of organic chemicals between atmosphere and water. *Environ. Sci. Technol.* 20, 810-816.

Mackay, D., Shiu, W.Y. (1977) Aqueous solubility of polynuclear aromatic hydrocarbons. *J. Chem. Eng. Data* 22, 399-402.

Mackay, D., Shiu, W.Y. (1981) A critical review of Henry's law constants for chemicals of environmental interest. *J. Phys. Chem. Ref. Data* 10, 1175-1199.

Mackay, D., Shiu, W.Y, Billington, J.W., Haung, G.L. (1983) Physical chemical properties of polychlorinated biphenyls. In: *Physical Behavior of PCBs in the Great Lakes.* Mackay, D., Paterson, S., Eisenreich, S.J., Simmons, M.S. Eds., pp. 59-69, Ann Arbor Sci. Publ., Ann Arbor, Michigan.

Mackay, D., Shiu, W.Y., Bobra, A., Billington, J., Chau, E., Yeun, A., Ng, C.,Szeto, F. (1980) *Volatilization of organic pollutants from water.* EPA 600/3-82-019.

Mackay, D., Shiu, W.Y., Sutherland, R.P. (1979) Determination of air-water Henry's law constants for hydrophobic pollutants. *Environ. Sci. Technol.* 13, 333-337.

Mackay, D., Wolkoff, A.W. (1973) Rate of evaporation of low-solubility contaminants from water bodies to atmosphere. *Environ. Sci. Technol.* 7, 611-614.

Mailhot, H. (1987) Prediction of algae bioaccumulation and uptake rate of nine organic compounds by ten physicochemical properties. *Environ. Sci. Technol.* 21, 1009-1013.

Mascarelli, L. Gatti, D. (1933) *Biphenyls and its derivatives; assymetric 2,2'-disubstituted derivatives of biphenyl. Gazz. Chim. Ital.* 63, 654.

Mayer, F.L., Mehrle, P.M., Sanders, H.O. (1977) Residue dynamics and biological effects of polychlorinated biphenyls in aquatic organisms. *Arch. Environ. Contam. Toxicol.* 5, 501.

McCall, P.J., Laskowski, D.A., Swann, R.L., Dishburger, H.J. (1983) Estimation of environmental partitioning of organic chemicals in model ecosystems. *Residue Reviews* 85, 231-243.

McBee, E.T., Divelet, W.R., Burch, J.E. (1955) The Diels-Alder reaction with 5,5-dimethoxy-1,2,3,4-tetrachlorocyclopentadiene. *J. Am. Chem. Soc.* 77, 385-387.

McDuffie, D. (1981) Estimation of octanol/water partition coefficients for organic pollutants using reversed-phase HPLC. *Chemosphere* 10, 73-83.

McKillop, A., Elsom, L.F., Taylor, E.C. (1968) Thallium in organic synthesis. III. Coupling of aryl and alkyl Grignard reagents. *J. Am. Chem. Soc.* 90, 2423-2424.

McKim, J., Schnieder, P., Veith, G. (1985) Absorption dynamics of organic chemical transport across trout gills as related to octanol-water partition coefficient. *Toxicol. Appl. Pharmacol.* 77, 1-10.

Metcalf, R.L., Sanborn, J.R., Lu, P-Y, Nye, D. (1975) Laboratory model ecosystem studies of the degradation and fate of radiolabelled tri-, tetra-, and pentachloro-biphenyls compared with DDE. *Arch. Environ. Contam. Toxicol.* 3, 151-165.

Metcalfe, D.E., Zukova, G., Mackay, D., Paterson S. (1988) Polychlorinated biphenyls (PCBs), physical and chemical property data. In: *Hazards, Decontamination and Replacement of PCB. A Comprehensive Guide.* Crine, J.P. Ed., pp. 3-33, Plenum Press, New York, N.Y.

Miller, M.M., Ghodbane, S., Wasik, S.P., Tewari, Y.B., Martire, D.E. (1984) Aqueous solubilities, octanol/water partition coefficients and entropies of melting of chlorinated benzenes and biphenyls. *J. Chem. Eng. Data* 29, 184-190.

Miller, M.M., Wasik, S.P., Huang, G.L., Shiu, W.Y., Mackay, D. (1985) Relationships between octanol-water partition coefficient and aqueous solubility. *Environ. Sci. Technol.* 19, 522-529.

Mills, W.B., Dean, J.D., Porcella, D.B., Gherini, S.A., Hudson, R.J.M., Frick, W.E., Rupp, G.L., Bowie, G.L. (1982). *Water Quality Assessment: A Screening Procedure for Toxic and Conventional Pollutants.* Part 1, EPA-600/6-82-004a.

Monsanto Co. (1972) Presentation to the interdepartmental task force on PCB, May 15, 1972, Washington, DC.

Muir, D.C.G., Marshall, W.K., Webster, G.R.B. (1985) Bioconcentration of PCDDs by fish: effects of molecular structure and water chemistry. *Chemosphere* 14(6/7), 829-833.

Mullin, M.D., Pochini,C.M., McCrindle, S., Romkes, M., Safe, S., Safe, L.M. (1984) High resolution PCB analysis: Synthesis and chromatographic properties of all 209 PCB congeners. *Environ. Sci. Technol.* 18, 468.

Mullin M.D., Filins, J.C. (1981) Analysis of polychlorinated biphenyls by glass capillary and packed-column chromatography. In: *Advances in the Identification and Analysis of Organic Pollutants in Water.* Vol. 1, L.H. Keith, Ed. chapter 11, p.187-196. Ann Arbor Publ.,Inc., Ann Arbor, Michigan.

Murphy, T.J. (1984) Atmospheric inputs of chlorinated hydrocarbons to the Great Lakes. In: *Toxic Contaminants in the Great Lakes.* J.D. Nriagu and M.S. Simmons, Eds., p.53-79, John Wiley & Sons, Inc., N.Y.

Murphy, T.J., Pokojowczyk, J.C., Mullin, M.D. (1983) Vapor exchange of PCBs with Lake Michigan: The Atmosphere as a Sink for PCBs. In: *Physical Behavior of PCBs in the Great Lakes.* D. Mackay, S. Paterson, S.J. Eisenreich and M.S. Simmons, Eds., pp.49-58. Ann Arbor Sci. Publ., Ann Arbor, Michigan.

Murphy, T.J., Rezszutko, C.P. (1977) Precipitation inputs of PCBs to Lake Michigan. *J. Great Lake Res.* 3, 305.

Murphy, T.J., Mullin, M.D., Meyer, J.A. (1987) Equilibration of polychlorinated biphenyls and toxaphene with air and water. *Environ. Sci. Technol.* 21(2), 155-162.

NAS (1979) *National Academy of Sciences, Polychorinated Biphenyls*: A Report prepared by the Committee on the Assessment of Polychlorinated-biphenyls in the Environment of Environmental Studies Board Commission on Natural Resources of National Research Council, Washington, D.C.

Neely, W.B. (1976) Predicting the flux of organics across the air/water interface. In: *National Conference on Control of Hazardous Material Spills*, New Orleans.

Neely, W.B. (1979) Estimating rate constants for the uptake and clearance by fish. *Environ. Sci. Technol.* 13, 1506-1510.

Neely, W.B. (1980) A method for selecting the most appropriate environmental experiments on a new chemical. In: *Dynamics, Exposure and Hazard Assessment of Toxic Chemicals.* R. Haque, Ed., Ann Arbor Science Publishers, Ann Arbor, Michigan.

Neely, W.B. (1981) Complex problems-simple solutions. *Chemtech.* 11, 249.

Neely, W.B. (1982) Organizing data for environmental studies. *Environ. Toxicol. & Chem.* 1, 259-266.

Neely, W.B. (1983) Reactivity and Environmental Persistence of PCB Isomers. In: *Physical Behavior of PCBs in the Great Lakes.* D. Mackay, S. Paterson, S.J. Eisenreich and M.S. Simmons, Eds., p.71-88, Ann Arbor Sci. Publ., Ann Arbor, Michigan.

Neely, W.B., Branson, D.R. Blau, G.E. (1974) Partition coefficient to measure bioconcentration potential of organic chemicals in fish. *Environ. Sci. Technol.* 8, 1113-1115.

Nelson, N., Hammond, P.B., Nisbet, I.C.T., Sarofim, A.F., Drury, W.H. (1972) Polychlorinated biphenyls: Environmental impact. *Environ. Res.* 5, 249-362.

Niimi, A.J., Oliver, B. (1983) Biological half-lives of polychlorinated biphenyl (PCB) congeners in whole fish and muscle of rainbow trout (salmo gairdneri). *Can. J. Fish Aquatic Sci.* 40, 1388.

Nirmalakhandan, N.N., Speece, R.E. (1988a) Prediction of aqueous solubility of organic chemicals based on molecular structure. *Environ. Sci. Technol.* 22, 328-338.

Nirmalakhandan, N.N., Speece, R.E. (1988b) QSAR model for predicting Henry's law constant. *Environ. Sci. Technol.* 22, 1349-1357.

Nirmalakhanden, N.N., Speece, R.E. (1989) Prediction of aqueous solubility of organic chemicals based on molecular structure. 2. Application to PNAs, PCBs, PCDDs, etc. *Environ. Sci. Technol.* 23(6), 708-713.

Nisbet I.C.T., Sarofim, A.F. (1972) Rates and routes of transport of PCBs in the environment. *Environ. Health Perspectives* 1, 21-38.

OECD (1981) *OECD Guidelines for Testing of Chemicals.* Organization for Economic Co-Operation and Development. OECD, Paris.

Oliver, B.G. (1984) The relationship between bioconcentration factor in rainbow trout and physical-chemical properties for some halogenated compounds. In: *QSAR in Environmental Toxicology.* Kaiser, K.L.E. (Ed.), pp. 300-317. D. Reidel Publishing Company.

Oliver, B.G. (1985) Desorption of chlorinated hydrocarbons from spiked and anthropogenically contaminated sediments. *Chemosphere* 14, 1087-1106.

Oliver, B.G. (1987a) Partitioning relationships for chlorinated organics between water and particulates in the St. Clair, Detroit and Niagara Rivers. In: *QSAR in Environmental Toxicology - II.* Kaiser, K.L.E. ED., pp.251-260. D. Reidel Publ. Co., Dordrecht, Holland.

Oliver, B.G. (1987b) Fate of some chlorobenzenes from Niagara River in Lake Ontario. In: *Sources and Fates of Aquatic Pollutants.* Hite, R.A., Eisenreich, S.J. Eds. pp.471-489. Advances in Chemistry Series 216, Am. Chem. Soc., Washington D.C.

Oliver, B.G. (1987c) Bio-uptake of chlorinated hydrocarbons from laboratory-spiked and field sediments by oligochaete worms. *Environ. Sci. Technol.* 21, 785-790.

Oliver, B.G., Nimii, A.J. (1983) Bioconcentration of chlorobenzenes from water by rainbow trout: correlations with partition coefficients and environmental residues. *Environ. Sci. Technol.* 17, 287-291.

Oliver, B.G., Charlton, M.N. (1984) Chlorinated organic contaminants on settling particulates in the Niagara River vicinity of Lake Ontario. *Environ. Sci. Technol.* 18, 903-908.

Oliver, B.G., Niimi, A.J. (1984) Rainbow trout bioconcentration of some halogenated aromatics from water at environmental concentrations. *Environ. Toxicol. & Chem.* 3, 271-277.

Oliver, B.G., Niimi, A.J. (1985) Bioconcentration factors of some halogenated organics for rainbow trout: limitations in their use for prediction of environmental residues. *Environ. Sci. Technol.* 19, 842-849.

Oloffs, P.C., Abright, L.J., Szeto, S.Y. (1972) Fate and behaviour of five chlorinated hydrocarbons in three natural waters. *Can. J. Microbiol.* 18, 1393.

Opperhuizen, A. (1986) Bioconcentration of hydrophobic chemicals in fish. In: *Aquatic Toxicology and Environmental Fate.* Nineth Volume. *ASTM STP* 921. Poston, T.M., Purdy, Eds., pp. 304-315, Am. Soc. for Testing and Materials. Philadelphia.

Opperhuizen, A., Van Develde, E.W., Gobas, F.A.P.C., Liem, D.A.K., Van der Steen, J.M., Hutzinger, O. (1985) Relationship between bioconcentration in fish and steric factors of hydrophobic chemicals. *Chemosphere* 14, 1871-1896.

Opperhuizen A., Gobas, F.A.P.C., Van der Steen, J.M.D., Hutzinger, O. (1988) Aqueous solubility of polychlorinated biphenyls related to molecular structure. *Environ. Sci. Technol.* 22, 638-646.

Pal, D., Weber, J.B., Overcash, M.R. (1980) Fate of polychlorinated biphenyls (PCBs) in soil plant systems. *Pesticide Reviews* 74, 45-98.

Paris, D.F., Steen, W.C., Baughman, G.E. (1978) Role of physico-chemical properties of Aroclors 1016 and 1242 in determining their fate and transport in aquatic environments. *Chemosphere* 7, 319-325.

Patil, G.S. (1991) Correlation of aqueous solubility and octanol-water partition coefficient based on molecular structure. *Chemosphere* 22(8), 723-738.

Pavlou, S.P. (1987) The use of the equilibrium partitioning approach in determining safe levels of contaminants in marine sediments. In: *Fate and Effects of Sediment-Bound Chemicals in Aquatic systems.* Dickson, K.L., Maki, A.W., Brungs, W.A., Eds., pp 395-412, Pergamon Press, New York.

Pavlou, S.P., Weston, D.P. (1983, 1984) *Initial evaluation of alternatives for development of sediment related criteria for toxic contaminants in marine waters (Puget Sound) phase I & II.* EPA Contract No. 68-01-6388.

Paya-Perez, B., Riaz, M., Larsen, B.R. (1991) Soil sorption of 20 PCB congeners and six chlorobenzenes. *Ecotoxicol. Environ. Safety* 21, 1-17.

Pearlman, R.S., Yalkowsky, S.H., Banerjee, S. (1984) Water solubilities of polynuclear aromatic and heteroaromatic compounds. *J. Phys. Chem. Ref. Data* 13, 555-562.

Prausnitz, J.M. (1969). *Molecular Thermodynamic of Fluid-Phase Equilibria.* Prentic Hall, Englewood Cliffs, N.J.

Prausnitz, J.M., Lichtenthaler, R.N., de Azevedo, E.G. (1986) *Molecular Thermodynamics of Fluid Plase Equilibria,* 2nd ed., Prentice Hall, Englewood Cliffs, N.J.

Radchenco, L.G., Kitiagorodskii, A.I. (1974) Vapor pressure and heat of sublimation of naphthalene, biphenyl, octafluoronaphthalene, decafluorobiphenyl, acenaphthene and α-nitronaphthalene. *Zhur. Fiz. Khim.* 48, 2702-2704.

Radding, S.B., Liu, D.H., Johnson, H.L., Mill, T. (1977) *Review of the Environmental Fate of Selected Chemicals*. U.S. Environmental Protection Agency Report No. EPA-560/5-77-003.

Rapaport, R.A., Eisenreich, S.J. (1984) Chromatographic determination of octanol-water partition coefficients (K_{OW}'s) for 58 polychlorinated biphenyl congeners. *Environ. Sci. Technol.* 18, 163-170.

Reichardt, P.B., Chadwick, B.L., Cole, M.A., Robertson, B.R., Button, D.K. (1981) Kinetic study of the biodegradation of biphenyl and its monochlorinated analogues by a mixed marine microbial community. *Environ. Sci. Technol.* 15(1), 75-79.

Reischl, A., Reissinger, M., Thoma, H., Hutzinger, O. (1989) Uptake and accumulation of PCDD/F in terrestial plants: basic considerations. *Chemosphere* 19(1-6), 467-474.

Rekker, R.F. (1977) *The Hydrophobic Fragmental Constants. Its Derivation and Application, a Means of Characterizing Membrane Systems*. Elsevier Sci. Publ. Co., Oxford, England.

Renberg, L., Sundstrom, G., Sundh-Nygard, K. (1980) Partition coefficients of organic chemicals derived from reversed phase thin layer chromatography, evaluation of methods and application on phosphate esters, polychlorinated paraffins and some PCB-substitutes. *Chemosphere* 9, 683-691.

Richardson, W.L., Smith, V.E., Wethington, R. (1983) Dynamic mass balance of PCB and suspended solids in Saginaw Bay- a case study. In: *Physical Behavior of PCBs in the Great Lakes*. D. Mackay, S. Patterson, S.J. Eisenreich, Eds., Chapter 18, pp.329-366. Ann Arbor Science Publishers, Inc., Ann Arbor, Michigan.

Rippen, G., Frank, R. (1986) In: *Hexachlorobenzene: Proceedings of an International Symposium*. Morris, C.R., Cabral, J.R.P. (Eds.), pp. 45-52. IARC, Lyon, France.

Rogers, K.S., Cammarata, A. (1969) Superdelocalizability and charge density. A correlation with partition coefficients. *J. Med. Chem.* 12, 692.

Ryan, J.A., Bell, R.M., Davidson, J.M., O'Connor, G.A. (1988) Plant uptake of non-ionic organic chemicals from soils. *Chemosphere 17*, 2299-2323.

Sabljic, A. (1984) Predictions of the nature and strength of soil sorption of organic pollutants by molecular topology. *J. Agric. Food Chem.* 32, 243-246.

Sabljic, A. (1987a) On the prediction of soil sorption coefficients of organic pollutants from molecular structure: application of molecular topology model. *Environ. Sci. Technol.* 21, 358-366.

Sabljic, A. (1987b) Nonempirical modeling of environmental distribution and toxicity of major organic pollutants. In: *QSAR in Environmental Toxicology - II*. Kaiser, K.L.E., Ed., pp 309-322, D. Reidel Publ. Co., Dordrecht, Netherlands.

Sabljic, A., Gusten, H. (1989) Predicting Henry's law constants for polychlorinated biphenyls. *Chemosphere* 19, 1503-1511.

Sabljic A., Lara, R., Ernst, W. (1989) Modelling association of highly chlorinated biphenyls with marine humic substances. *Chemosphere* 19(10-11), 1665-1676.

Saeki, S., Tsutsui, A., Oguri, K., Yoshimura, H., Hamana, M. (1971) Isolation and structure elucidation of the amine component of KC-400 (chlorobiphenyls). *Fukuoka Igaku Zhasshi 62, 20. Chem. Abstr.* 74, 146294.

Safe, S. (1980) Metabolism, Uptake, Storage and Bioaccumulation. In: *Halogenated Biphenyls, Terphenyls, Naphthalenes, Dibenzo-dioxins and Related Products*, R.D. Kimbrough, Ed., Chapter 4. Elsevier/North- Holland Biomedical Press, Amsterdam.

Safe, S., Hutzinger, O. (1972) The mass spectra of polychlorinated biphenyls. *J. Chem. Soc. Perkin Trans.* I, 686-691.

Sahyun, M.R. (1966) Binding of aromatic compounds to bovine serum albumin. *Nature* (London) 209, 613-614.

Sanborn, J.R., Childers, W.F., Metcalf, R.L. (1975) Uptake of three polychlorinated biphenyls, DDT and DDE by green sunfish, *Lepomis cyanellus Raf. Bull. Environ. Contam. Toxicol.* 13, 209.

Sangster, J. (1989) Octanol-water partition coefficients of simple organic compounds. *J. Phys. Chem. Ref. Data* 18, 1111-1230.

Sarna, L.P., Hodge, P.E., Webster, G.R.B. (1984) Octanol-water partition coefficients of chlorinated dioxins and dibenzofurans by reversed-phase HPLC using several C_{18} columns. *Chemosphere* 13, 975-983.

Sawhney, B.L. (1987) Chemistry and properties of PCBs in relation to environmental effects. In: *PCBs and the Environment.* Waid, Ed., Chapter 2, pp.48-64. CRC Press Inc., Boca Raton, Florida.

Schwarzenbach, R.P., Westall, J. (1981) Transport of nonpolar compounds from surface water to groundwater. Laboratory sorption studies. *Environ. Sci. Technol.* 11, 1360-1367.

Seki & Suzuki (1953) Physical chemical studies on molecular compounds. III. Vapor pressures of diphenyl, 4,4'-dinitrophenyl and molecular compound between them. *Bull. Chem. Soc. Jpn.* 26, 209-213.

Seip, H.M., Alstad, J., Carlberg, G.E., Martinsen, K., Skaane, P. (1986) Measurement of solubility of organic compounds in soils. *Sci. Total Environ.* 50, 87-101.

Sengupta, S.K. (1966) Studies on fluorine compounds: Part I.-Halofluorenes. *Ind. J. Chem.* 4, 235-239.

Shiu, W.Y., Ma, K.C., Mackay, D., Seiber, J.N., Wauchope, R.D. (1990) Solubilities of Pesticide Chemicals in Water. Part I. Environmental Physical Chemistry, Part II. Data Compilation. *Review Environ. Contam. Toxicol.* 116, 1-187.

Shiu, W.Y., Gobas, F.A.P.C., Mackay, D. (1987) Physical-chemical properties of three congeneric series of chlorinated aromatic hydrocarbons. In: *QSAR in Environmental Toxicology - II*, K.L.E. Kaiser, Ed., pp. 347-362. D. Reidel Publishing Company.

Shiu, W.Y., Mackay, D. (1986) A critical review of aqueous solubilities, vapor pressures, Henry's law constants and octanol-water partition coefficients of the polychlorinated biphenyls. *J. Phys. Chem. Data* 15, 911-929.

Skea, J.C., Simonin, H.A., Dean, H.J., Colquhoun, J.R., Spagnoli, J.J., Veith, G.D. (1979) Bioaccumulation of Aroclor 1016 in Hudson River fish. *Bull. Environ. Contam. Toxicol.* 22, 332.

Sklarew, D.S., Girvin, D.C. (1987) Attenuation of polychlorinated biphenyls in soils. *Rev. Environ. Contam. & Toxicol.* 98, 1-41.

Slinn, W.G.N., Hasse, L., Hicks, B.B., Hogan, A.W., Lal, D., Liss, P.S., Munnich, K.O., Sehmel, G.A., Vittori, O. (1978) Some aspects of the transfer of atmospheric trace constituents past the air-sea interface. *Atmos. Environ.* 12, 2055-2087.

Smith, J.H., Mabey, W.R., Bahonos, N., Holt, B.R., Lee, S.S., Chou, T.W., Venberger, D.C., Mill, T. (1978) *Environmental pathways of selected chemicals in fresh water systems: Part II. Laboratory Studies.* Interagency Energy-Environment Research Program Report. EPA-600/7-78-074. Environmental Research Laboratory Office of Research and Development. U.S. Environment Protection Agency, Athens, Georgia 30605, p. 304.

Smith, N.K., Gorin, G., Good, W.D., McCullough, J.P. (1964) The heats of combustion, sublimation and formation of four dichlorobiphenyls. *J. Phys. Chem.* 68, 940.

Södergren, A. (1987) Solvent filled dialysis membranes simulate uptake of pollutants by aquatic organisms. *Environ. Sci. Technol.* 21, 855-859.

Stalling, D.L., Mayer, F.L. (1972) Toxicities of PCBs to fish and environmental residues. *Environ. Health Prospect* 1, 159-164.

Stolzenburg, T.R., Andren, A.W. (1983) Determination of the aqueous solubility of 4-chlorobiphenyl. *Anal. Chim. Acta*, 151, 271-274.

Stull, D.R. (1947) Vapor pressure of pure substances. Organic compounds. *Ind. Eng. Chem.* 39, 517-540.

Sugiura, K., Ito, Y., Matsumoto, N., Mihara, Y., Murata, K., Tsukakoshi, Y., Goto, M. (1978) Accumulation of polychlorinated biphenyls and polybrominated biphenyls in fish: limitation of correlation between partition coefficients and accumulation factors. *Chemosphere* 7, 731-736.

Sugiura, K., Washino, T., Hattori, M., Sato, E., Goto, M. (1979) Accumulation of organochlorines in fishes-difference of accumulation factors by fishes. *Chemosphere* No.6, 359-364.

Swackhamer, D.L., Armstrong, D.E. (1987) Distribution and characterization of PCBs in Lake Michigan water. *J. Great Lakes Res.* 13(1), 24-36.

Swann, R.L., Laskowski, D.A., McCall, P.J., Vander Kuy, K., Dishburger, H.J. (1983) A rapid method for the estimation of the environmental parameters octanol/water partition coefficient, soil sorption constant, water to air ratio, and water solubility. *Residue Rev.* 85, 17-28.

Tabak, H.H., Quave, S.A., Mashni, C.I., Barth, E.F. (1981) Biodegradability studies with organic priority pollutant compounds.. *J. Water Pollut. Control. Fed.* 53, 1503-1518.

Tadokoro, H., Tomita, Y. (1987) The relationship between bioaccumulation and lipid content of fish. In: *QSAR in Environmental Toxicology - II.* Kaiser, K.L.E., Ed., pp. 363-373, D. Reidel Publ. Co., Dorhrect, Holland.

Tas, A.C., DeVos, R.H. (1971) Characterization of four major compounds in a technical polychlorinated biphenyl mixture. *Environ. Sci. Technol.* 5, 1217.

The Merck Index. *An Encyclopedia of Chemicals, Drugs and Biologicals.* (1989). Budavari, S., Ed., Merck & Co., Inc., Rahway, N.J., 11th edition.

The Merck Index. *An Encyclopedia of Chemicals, Drugs and Biologicals* (1983). Windholz, M., Ed., Merck and Co., Inc., Rahway, N.J., U.S.A., 10th edition.

Thibodeaux, L.J. (1979) *Chemodynamics*. John Wiley & Sons, New York.

Thomann, R.V. (1989) Bioaccumulation of organic chemical distribution in aquatic food chains. *Environ. Sci. Technol.* 23(6), 699-707.

Tomlinson, E., Hafkenscheid, T.L. (1986) Aqueous solution and partition coefficient estimation from HPLC data. pp.101-141. In: *Partition Coefficient Determination and Estimation*. W.J. Dunn III, J.H. Block, R.S. Pearlman Eds., Pergamon Press.

Travis, C.C., Arms, A.D. (1988) Bioconcentration of organics in beef, milk, and vegetation. *Environ. Sci. Technol.* 22, 271-174.

Tsonopoulos, C., Prausnitz, J.M. (1971) Activity coefficients of aromatic solutes in dilute aqueous solutions. *Ind. Eng. Chem. Fundam.* 10, 593-600.

Tucker, E.S., Litschgi, W.J., Mees, W.M. (1975) Migration of polychlorinated biphenyls in soil induced by percolating water. *Bull. Environ. Contam. Toxicol.* 13, 86.

Tulp, M.T.M., Hutzinger, O. (1978) Some thoughts on aqueous solubilities and partition coefficients of PCB, and the mathematical correlation between bioaccumulation and physico-chemical properties. *Chemosphere* 7, 849-860.

Ullmann, F. (1904) *Ann. Chem.* 332, 38.

Vaishnav, D.D., Babeu, L. (1987) Comparison of occurrence and rates of chemical biodegradation in natural waters. *Bull. Environ. Contam. Toxicol.* 39, 237-244.

Valvani S.C., Yalkowsky, S.H. (1980) Solubility and partitioning in drug design. In: *Physical Chemical Properties of Drug. Med. Res. Ser.* Vol. 10. Yalkowsky, S.H., Sinkinla, A.A., Valvani, S.C. Eds., pp. 201-229. Marcel Dekker Inc., New York, N.Y.

Van der Linden, A.C. (1978) Degradation of oil in the marine environment. *Dev. Biodegradation Hydrocarbons* 1, 165-200.

Van Roosmalen, F.L.W. (1934) Chloro and bromo derivatives of biphenyl. *Rec. Trav. Chim.* 53, 359.

Veith, G.D., Kuehl, D.W., Puglish, F.A., Glass, G.E., Eaton, J.G. (1977) Residues of PCB's and DDT in the western Lake Superior ecosystem. *Arch. Environ. Contam. Toxicol.* 5, 487-499.

Veith, G.D., DeFoe, D.L. Bergstedt, B.V. (1979b) Measuring and estimating the bioconcentration factor of chemicals in fish. *J. Fish Res. Board Can.* 26, 1040-1048.

Veith, G.D., Austin, N.M., Morris, R.T. (1979a) A rapid method for estimating log P for organic chemicals. *Water Res.* 13, 43-47.

Veith, G.D., Kosian, P. (1983) Estimating bioconcentration potential from octanol/water partition coefficients. In: *Physical Behaviour of PCBs in the Great Lakes*. D. Mackay, S. Patterson, S.J. Eisenreich, M.S. Simmons Eds., Ann Arbor Science Publishers, Ann Arbor, Michigan.

Verlag Chemie (1983) *Deutsche Forschungsgemeinschaft Hexachlocyclohexanals Schadstoff in Lebensmitten.* p.13, Verlag Chemie, Weinheim, Germany.

Versar Inc. (1979) *Water related environmental fate of 129 priority pollutants. A literature search. V. Polycyclic aromatic hydrocarbons, PCBs and related compounds.* pp. 36-1 to 36-12. Office Water and Waste Management, U.S. Environmental Protection Agency, Washington DC.

Verschueren,K. (1977) *Handbook of Environmental Data on Organic Chemicals.* Van Nostrands Reinhold, New York.

Verschueren, K. (1983) *Handbook of Environmental Data on Organic Chemicals*, 2nd ed.. Van Nostrand Reinhold Co., New York.

Vesala, A. (1974) Thermodynamics of transfer of nonelectrolytes from light to heavy water. I. Linear free energy correlation of free energy of transfer with solubility and heat of melting of nonelectrolytes. *Acta Chem. Scand.* 28A(8), 839-845.

Voice, T.C., Rice, C.P., Weber Jr., W.J. (1983) Effects of solid concentration on the sorptive partitioning of hydrophobic pollutants in organic systems. *Environ. Sci. Technol.* 17(9), 513-518.

Voice, T.C., Weber Jr., W.J. (1983) Sorption of hydrophobic compounds by sediments, soils and suspended solids-I. *Water Res.* 17(10), 1433-1441.

Voice, T.C., Weber Jr., W.J. (1985) Sorbent concentration effects in liquid/solid partitioning. *Environ. Sci. Technol.* 19(9), 789-796.

Vreeland V. (1974) Uptake of chlorinated biphenyls by oysters. *Environ. Pollution* 6, 135-140.

Waid, J.S., Ed. (1986) *PCBs in the Environment.* CRC Press, Inc., Boca Raton, Florida.

Wallnöfer, P.R., M. Koniger, M., Hutzinger, O. (1973) The solubilities of twenty-one chlorobiphenyls in water. *Analab Res. Notes* 13(3), 14-19.

Warner, H. et al. (1980) Determination of Henry's law constants of selected priority pollutants. *MERL*, Cincinnati.

Watts, C.D., Moore, K. (1987). Fate and transport of organic compounds in river. In: Organic Micropollutants in the Aquatic Environment. Angelletti, G., Bjorseth, A., Eds., *Proceedings of the 5th European Symposium,* Rome, Italy, pp.154-169, Kluwer Academic Publ.

Wauchope, R.D., Getzen, F.W. (1972) Temperature dependence of solubilities in water and heats of fusion of solid aromatic hydorcarbons. *J. Chem. Eng. Data* 17, 38-41.

Weast, R.C., Ed. (1972-73) *Handbook of Chemistry and Physics*, 53th ed. CRC Press, Cleveland.

Weast, R. (1976-77) *Handbook of Chemistry and Physics.* 57th ed., CRC Press, Boca Raton, Florida.

Weast, R.C., Ed. (1983-84). *Handbook of Chemistry and Physics,* 64th ed., CRC Press, Boca Raton, Florida.

Webb, R.G. (1970) *PCB Newsletter* NO.1.

Webb, R.G., McCall, A.C. (1972) Identities of polychlorinated biphenyl isomers in Aroclors. *J. Assoc. Offic. Anal. Chem.* 55, 746.

Weber Jr., W.J., Voice, T.C., Pirbazari, M., Hunt, G.E., Ulanoff, D.M. (1983) Sorption of hydrophobic compounds by sediments, soils and suspended solids-II. *Water Res.* 17(10), 1443-1452.

Weil, L., Dure, G., Quentin, K.L. (1974) Solubility in water of insecticide chlorinated hydrocarbons and polychlorinated biphenyls in view of water pollution. *Z. Wasser Abwasser Forsch.* 7, 169-175.

Weingarten, H. (1961) Chlorination of biphenyl. *J. Org. Chem.* 26, 4347-4350.

Weingarten, H. (1962) Aluminum chloride induced isomerization of chlorinated biphenyls. *J. Org. Chem.* 27, 2024-2026.

Wiese, C.S., Griffin, D.A. (1978) The solubility of Aroclor 1254 in seawater. *Bull. Environ. Contamin. Toxicol.* 19, 403-411.

Westcott, J.W., Bidleman, T.F. (1981) Determination of polychlorinated biphenyl vapor pressures by capillary gas chromatography. *J. Chromatogr.* 210, 331-336.

Westcott, J.W., Simon, C.G., Bidleman, T.F. (1981) Determination of polychlorinated biphenyl vapor pressures by a semimicro gas saturation method. *Environ. Sci. Technol.* 15, 1375-1378.

Whitehouse,B.G., Cooke, R.C. (1982) Estimating the aqueous solubility of aromatic hydrocarbons by high performance liquid chromatography. *Chemosphere* 11, 689-699.

Wildish, D.J., Metcalfe, C.D., Akagi, H.M., McLeese, D.W. (1980) Flux of Aroclor 1254 between estuarine sediments and water. *Bull. Environ. Contam. Toxicol.* 24, 20-26.

Wingender, R.J., Williams, R.M. (1984) Evidence for the long-distance atmospheric transport of polychlorinated terphenyl. *Environ. Sci. Technol.* 18, 625-628.

Wittlinger, R., Ballschmiter, K. (1990) Studies of the global baseline pollution XIII C_6 -C_{14} organohalogens (α and g-HCH, HCB, PCB, 4,4'-DDT, 4,4'-DDE, cis- and trans-chlordane, trans-nonachlor, anisols) in lower troposphere of southern Indian ocean. *Fresenius J. Anal. Chem.* 336, 193-200.

Wolfe, D.A. (1987) Interactions of spilled oil with suspended materials and sediments in aquatic systems. In: *Fate and Effects of Sediment-Bound Chemicals in Aquatic Systems.* Dickson, K.L., Maki, A.W., Brungs, W.A., Eds., pp. 299-316. Pergamon Press, New York.

Wong, P.T.S., Kaiser, K.L.E. (1975) Bacterial degradation of polychlorinated biphenyls. II Rate studies. *Bull. Contam. Toxicol.* 3, 249.

Woodburn, K.B. (1982) *M.S. Thesis, University of Wisconsin*, Madison, Wisconsin.

Woodburn, K.B., W.J. Doucette, W.J., Andren, A.W. (1984) Generator column determination of octanol/water partition coefficients for selected polychlorinated biphenyl congeners. *Environ. Sci. Technol.* 18, 457-459.

Yalkowsky, S.H. (1979) Estimation of entropies of fusion of organic compounds. *Ind. Eng. Chem. Fundam.* 18, 108-111.

Yalkowsky, S.H., Valvani, S.C. (1979) Solubilities and partitioning. 2. Relationships between aqueous solubilities, partition coefficients, and molecular surface areas of rigid aromatic hydrocarbons. *J. Chem. Eng. Data* 24, 127-129.

Yalkowsky, S.H., Valvani, S.C. (1980) Solubility and partitioning. I. Solubility of nonelectrolytes in water. *J. Pharm. Sci.* 69, 912-922.

Yalkowsky, S.H., Valvani, S.C., Mackay, D. (1983) Estimation of the aqueous solubility of some aromatic compounds. *Residue Rev.* 85, 43-55.

Yoshida, K., Shigeoka, T., Yamauchi, F. (1983) Relationship between molar refraction and n-octanol/water partition coefficient. *Ecotox. Environ. Safety* 7, 558-565.

Zaroogian, G.E., Heltshe, J.F., Johnson, M. (1985) Estimation of bioconcentration in marine species using structure-activity models. *Envriron. Toxicol. Chem.* 4, 3-12.

Zimmerli, B., Marek, B. (1974) Moldelversuche Zur kontamination von lebensmitten mit pestiziden via gasphase. *Mitt. Gebiete Lebensm. Hyg.* 65, 55-64.

Zitko, V. (1970) Polychlorinated biphenyls (PCB) solubilized in water by monoionic surfactants for study of toxicity to aquatic animals. *Bull. Contam. Toxicol.* 5(3), 279-285.

Zitko, V. (1971) Polychlorinated biphenyls and organochlorine pesticides in some fresh water and marine fishes. *Bull. Contam. Toxicol.* 6(5), 464-470.

List of Symbols and Abbreviations:

A_i	area of phase i, m^2
BCF	bioconcentration factor
C	molar concentration, mol/m^3
C^S	saturated aqueous concentration, mol/m^3
C_L	subcooled liquid concentration, mol/m^3
C_S	solid molar concentration, mol/m^3
C_A	concentration in air phase, mol/m^3
C_W	concentration in water phase, mol/m^3
^{14}C	radioactive labelled carbon-14 compound
D	D values, mol/Pa·h
D_A	D values for advection, mol/Pa·h
D_{Ai}	D values for advective loss in phase i, mol/Pa·h
D_R	D value for reaction, mol/Pa·h
D_{Ri}	D value for reaction loss in phase i, mol/Pa·h
D_{ij}	intermedia D values, mol/Pa·h
D_{VW}	intermedia D value for air-water diffusion (absorption), mol/Pa·h
D_{RW}	intermedia D value for air-water dissolution, mol/Pa·h
D_{QW}	D value for total particle transport (dry and wet), mol/Pa·h
D_{RS}	D value for rain dissolution (air-soil), mol/Pa·h
D_{QS}	D value for wet and dry deposition (air-soil), mol/Pa·h
D_{VS}	D value for total soil-air transport, mol/Pa·h
D_S	D value for air-soil boundary layer diffusion, mol/Pa·h
D_{SW}	D value for water transport in soil, mol/Pa·h
D_{SA}	D value for air transport in soil, mol/Pa·h
D_{Ti}	total transport D value in bulk phase i, mol/Pa·h
DOC	dissolved organic carbon
E	emission rate, mol/h or kg/h
EPICS	Equilibrium Partitioning In Closed System

F	Fugacity ratio
f	fugacity, Pa
f_i	fugacity in pure phase i, Pa
f-const.	fragmental constants
fluo.	fluorescence method
G	advective inflow, mol/h
G_B	advective inflow to bottom sediment mol/h
ΔG_v	Gibbs's free energy of vaporization kJ/mol or kcal/mol
GC	gas chromatography
GC/FID	GC analysis with flame ionization detector
GC/ECD	GC analysis with electron capture detector
GC-RT	GC retention time
gen. col.	generator-column
H, HLC	Henry's law constant, Pa m^3/mol
ΔH_{fus}	enthalpy of fusion, kcal/mol
ΔH_v	enthalpy of vaporization, kJ/mol or kcal/mol
HPLC	high pressure liquid chromatography
HPLC/UV	HPLC with UV detector
HPLC-k'	HPLC capacity factor
HPLC-RI	HPLC retention index
HPLC-RT	HPLC retention time
HPLC-RV	HPLC retention volume
IP	ionization potential
J	intermediate quantities for fugacity calculation
k	first order rate constant, h_{-1} (hour^{-1})
k_i	first order rate constant in phase i, h^{-1}
k_A	air-water mass transfer coefficient, air-side, m/h
k_W	air-water mass transfer coefficient, water-side, m/h
K_{AW}	dimensionless air/water partition coefficient

K_B	bioconcentration factor
K_h	association coefficient
K_{OC}	organic-carbon sorption partition coefficient
K_{OM}	organic-matter sorption partition coefficient
K_{OW}	octanol-water partition coefficient
K_p	sorption coefficient
k_1	uptake/accumulation rate constant, d^{-1} (day^{-1})
k_2	elimination/clearance/depuration rate constant, d^{-1}
k_b	biodegradation rate constant, d^{-1}
k_h	hydrolysis rate constant, d^{-1}
k_p	photolysis rate constant, d^{-1}
L	lipid content of fish
LSC	Liquid Scintillation Counting
m_i	amount of chemical in phase i, mol or kg
M	total amount of chemical, mol or kg
MO	molecular orbital calculation
MR	molar refraction
MW	molecular weight, g/mol
n_C	number of carbon atoms
n_{Cl}	number of chlorine atoms
P	vapor pressure, Pa (Pascal)
P_L	liquid or subcooled liquid vapor pressure, Pa
P_S	solid vapor pressure, Pa
Q	scavenge ratio
QSPR	Quantitative Structure-Property Relationship
QSAR	Quantitative Structural-Activity Relationship
S	water solubility, mg/L or mg/m^3
ΔS_{fus}	entropy of fusion, J/mol • K or cal/mol • K (e.u.)
t	residence time, h

t_o	overall residence time, h
t_A	advection persistence time, h
t_B	sediment burial residence time, h
t_R	reaction persistence time, h
$t_{1/2}$	half-life, h
T_{ij}	intermedia transport rate, mol/h or kg/h
T	system temperature, K
T_M	melting point, K
TLV	thin-layer chromatography
TMV	total molecular volume per molecule, $Å^3$ (Angstrom^3)
TSA	total surface area per molecule, $Å^2$
U_1	air side, air-water MTC (same as k_A), m/h
U_2	water side, air-water MTC (same as k_W), m/h
U_3	rain rate (same as U_R), m/h
U_4	aerosol deposition rate, m/h
U_5	soil-air phase diffusion MTC, m/h
U_6	soil-water phase diffusion MTC, m/h
U_7	soil-air boundary layer MTC, m/h
U_8	sediment-water MTC, m/h
U_9	sediment deposition rate, m/h
U^{10}	sediment resuspension rate, m/h
U_{11}	soil-water run-off rate, m/h
U_{12}	soil-solids run-off rate, m/h
U_R	rain rate, m/h
U_Q	dry deposition velocity, m/h
U_B	sediment burial rate, m/h
UV	UV spectrometry
UNIFAC	UNIQUAC Functional Group Activity Coefficients
V_i	volume of pure phase i, m^3

V_S	volume of bottom sediment, m^3
V_{Bi}	volume of bulk phase i, m^3
V_I	intrinsic molar volume, cm^3/mol
V_M	molar volume, cm^3/mol
v_i	volume fraction of phase i
v_Q	volume fraction of aerosol
W	molecular mass, g/mol
Z_i	fugacity capacity of phase i, mol/m^3 Pa
Z_{Bi}	fugacity capacity of bulk phase i, mol/m^3 Pa

Greek characters:

π-const.	substituent constants
γ	solute activity coefficient
γ_o	solute activity coefficient in octanol phase
γ_W	solute activity coefficient in water phase
ρ_i	density of pure phase i, kg/m^3
ρ_{Bi}	density of bulk phase i, kg/m^3
χ	molecular connectivity indices
ϕ_{OC}	organic carbon fraction
ϕ_i	organic carbon fraction in phase i

APPENDICES:

A1 BASIC COMPUTER PROGRAM FOR FUGACITY CALCULATIONS

```
10 REM Fugacity Level I,II and III program, 6 compartments,(LEWIS)
20 REM Select condensed print
30 WIDTH "lpt1:",250
40 LPRINT CHR$(15)
50 DIM N$(9),V(9),Z(9),C(9),F(9),M(9),P(9),CG(9),CU(9),DEN(9),ORG(9),VZ(9),
DR(9),DA(9),CB(9),PA(9),PR(9),RK(9),GA(9),VB(9),DENB(9)
60 DIM NR(9),NA(9),I(9),GD(9,9),D(9,9),N(9,9),GRA(9),TD(9,9),HL(9),U(20),
Y(20),U$(20)
70 REM N$  = six phases : air, water, soil, sediment, susp sedt and fish
80 REM V   = volume of the six phases  (m3)
85 REM VB  = bulk volumes in Level III (m3)
90 REM DEN = density of the six phases (kg/m3)
95 REM DENB = bulk density in Level III
100 REM HT  = depth of air, water, soil and sediment (m)
110 REM AR  = area of air, water, soil and sediment  (m2)
120 REM ORG = the fraction of organic carbon in sediment and susp sedt
130 REM Z   = Z values for each phase (mol/m3.Pa)
140 REM VZ  = VZ values for each phase (mol/Pa)
150 REM F   = fugacity for each phase (Pa)
155 REM TD  = transport half times (h)
160 REM C   = concentration of chemical in each phase (mol/m3)
170 REM CG  = concentration of chemical in each phase (g/m3)
180 REM CU  = concentration of chemical in each phase (ug/g)
190 REM M   = the total amount of chemical in each phase (mol)
200 REM MK  = the total mass of chemical in each phase (kg)
210 REM DR  = reaction D values
214 REM DA  = advection D values
216 REM U   = transport velocities
220 N$(1)="Air      "
230 N$(2)="Water    "
240 N$(3)="Soil     "
250 N$(4)="Sediment "
260 N$(5)="Susp sedt"
270 N$(6)="Fish     "
275 N$(7)="Aerosols"
280 ART=100000!*1000000!:FAR(2)=.1:FAR(3)=1-FAR(2)'totalarea and fractions
290 AR(2)=FAR(2)*ART:AR(3)=FAR(3)*ART:AR(1)=ART:AR(4)=AR(2)'areas m2
300 HT(1)=1000:HT(2)=20:HT(3)=.1:HT(4)=.01
310 V(1)=AR(1)*HT(1):V(2)=AR(2)*HT(2):V(3)=AR(3)*HT(3):V(4)=AR(4)*HT(4)
320 V(5)=.000005*V(2):V(6)=.000001*V(2)
330 REM input properties
```

```
340 PRINT "Select chemical ,user-spec =1, benzene= 2, HCB= 3, 123TCB= 4, Type 5 to
exit program"
350 INPUT QC
360 ON QC GOTO 370,510,520,530,4360
370 INPUT "Name of chemical ",CHEM$
380 INPUT "Temperature eg 25 deg C";TC
390 INPUT "Melting point temperature or data temperature if chemical is liquid eg 80 deg
C";TM
400 INPUT "Molecular mass eg 200 g/mol";WM
410 INPUT "Vapor pressure eg 2 Pa ";P
420 INPUT "Water solubility eg 50 g/m3 ";SG
430 INPUT "Log octanol water coefficient eg 4.0 ";LKOW
440 PRINT "Input overall reaction rate half-lives eg 100 h "
450 PRINT "For zero reaction enter a fictitiously long half life eg 1E11 h"
460 INPUT "Half life in air         ";HL(1)
470 INPUT "Half life in water       ";HL(2)
480 INPUT "Half life in soil        ";HL(3)
490 INPUT "Half life in sediment    ";HL(4)
500 GOTO 590
510 CHEM$="Benzene":TC=25!:WM=78.11:P=12700:SG=1780:LKOW=2.13:TM=5.53:
HL(1)=17:HL(2)=170:HL(3)=550:HL(4)=1700:GOTO590
520 CHEM$="Hexachlorobenzene(HCB)":TC=25!:WM=284.8:P=.0023:SG=.005:
LKOW=5.5:TM=230!:HL(1)=17000:HL(2)=55000!:HL(3)=55000!:HL(4)=55000!:GOTO
590
530 CHEM$="1,2,3-Trichlorobenzene(123TCB)":TC=25!:WM=181.45:P=28:SG=21:
LKOW=4.1:TM=53:HL(1)=550:HL(2)=1700!:HL(3)=5500!:HL(4)=17000!:GOTO590
590 MTK=100000!
600 MT=MTK*1000/WM
610 REM    Input for Fugacity Level II program
620 EK=1000
630 E=EK*1000/WM
640 PRINT "Input emission rates of chemical for Level III calculation kg/h"
650 INPUT "Emission into air       ";IK(1)
660 INPUT "Emission into water     ";IK(2)
670 INPUT "Emission into soil      ";IK(3)
680 GRA(1)=100'advection residence times (h)
690 GRA(2)=1000
700 GRA(4)=50000!
710 S=SG/WM 'solubility mol/m3
720 H=P/S    'Henry's law constant Pa.m3/mol
730 KOW=10^LKOW  'Octanol-water partition coefficient
740 KOC=.41*KOW  'Organic carbon-water partition coefficient
750 KFW=.05*KOW  'Fish-water bioconcentration factor
```

```
760 TK=TC+273.15 'Temperature K
770 RG=8.314     'Gas constant
780 IF TM>TC GOTO 790 ELSE GOTO 810
790 FR=EXP(6.79*(1-(TM+273.15)/TK))
800 GOTO 820
810 FR=1
820 PL=P/FR'subcooled liquid vapor pressure
830 ORG(3)=.02:ORG(4)=.04:ORG(5)=.2'Organic carbon contents g/g
840 DEN(1)=.029*101325!/RG/TK:DEN(2)=1000:DEN(3)=2400'Densities kg/m3
850 DEN(4)=2400:DEN(5)=1500:DEN(6)=1000:DEN(7)=2000'Densities kg/m3
860 REM calculate Z values
870 Z(1)=1/RG/TK
880 Z(2)=1/H
890 Z(3)=Z(2)*DEN(3)*ORG(3)*KOC/1000
900 Z(4)=Z(2)*DEN(4)*ORG(4)*KOC/1000
910 Z(5)=Z(2)*DEN(5)*ORG(5)*KOC/1000
920 Z(6)=Z(2)*DEN(6)*KFW/1000
930 K71=6000000!/PL
940 Z(7)=Z(1)*K71
950 K12=Z(1)/Z(2) 'Partition coefficients
960 K32=Z(3)/Z(2)
970 K42=Z(4)/Z(2)
980 K52=Z(5)/Z(2)
990 K62=Z(6)/Z(2)
1000 REM calculate Level I distribution
1010 VZT=0
1020 FOR N= 1 TO 6
1030 VZ(N)=V(N)*Z(N)
1040 VZT=VZT+VZ(N)
1050 NEXT N
1060 F1=MT/VZT 'fugacity
1070 FOR N=1 TO 6
1080 F(N)=F1
1090 C(N)=F(N)*Z(N) 'concentration mol/m3
1100 M(N)=C(N)*V(N) 'amount mol
1110 MK(N)=M(N)*WM/1000
1120 P(N)=100*M(N)/MT 'percentages
1130 CG(N)=C(N)*WM  'concentration g/m3
1140 CU(N)=CG(N)*1000/DEN(N)'concentration ug/g
1150 NEXT N
1160 REM print out results
1170 LPRINT " PROGRAM 'LEWIS':SIX COMPARTMENT FUGACITY LEVEL I
CALCULATION "
```

```
1180 LPRINT " "
1190 LPRINT "Properties of "CHEM$
1200 LPRINT " "
1210 LPRINT "Temperature deg C                ";TC
1220 LPRINT "Molecular mass g/mol             ";WM
1230 LPRINT "Melting point deg C              ";TM
1240 LPRINT "Fugacity ratio                   ";FR
1250 LPRINT "Vapor pressure Pa                ";P
1260 LPRINT "Sub-cooled liquid vapor press Pa ";PL
1270 LPRINT "Solubility g/m3                  ";SG
1280 LPRINT "Solubility mol/m3                ";S
1290 LPRINT "Henry's law constant Pa.m3/mol   ";H
1300 LPRINT "Log octanol-water p-coefficient  ";LKOW
1310 LPRINT "Octanol-water partn-coefficient  ";KOW
1320 LPRINT "Organic C-water ptn-coefficient  ";KOC
1330 LPRINT "Fish-water partition coefficient ";KFW
1340 LPRINT "Air-water partition coefficient  ";K12
1350 LPRINT "Soil-water partition coefficient ";K32
1360 LPRINT "Sedt-water partition coefficient ";K42
1370 LPRINT "Susp sedt-water partn coeffnt    ";K52
1380 LPRINT "Aerosol-air partition coeff      ";K71
1390 LPRINT "Aerosol Z value                  ";Z(7)
1400 LPRINT "Aerosol density kg/m3            ";DEN(7)
1410 LPRINT " "
1420 LPRINT "Amount of chemical moles         ";MT
1430 LPRINT "Amount of chemical kilograms     ";MTK
1440 LPRINT "Fugacity Pa                      ";F1
1450 LPRINT "Total of VZ products             ";VZT
1460 LPRINT " "
1470 LPRINT "Phase properties and compositions"
1480 LPRINT " "
1490 LPRINT "Phase        "TAB(15) N$(1) TAB(30) N$(2) TAB(45) N$(3) TAB(60) N$(4)
TAB(75) N$(5) TAB(90) N$(6)
1500 LPRINT "Volume m3    "TAB(15) V(1) TAB(30) V(2) TAB(45) V(3) TAB(60) V(4)
TAB(75) V(5) TAB(90) V(6)
1510 LPRINT "Density kg/m3"TAB(15) DEN(1) TAB(30) DEN(2) TAB(45) DEN(3)
TAB(60) DEN(4) TAB(75) DEN(5) TAB(90) DEN(6)
1520 LPRINT "Depth m      "TAB(15) HT(1) TAB(30) HT(2) TAB(45) HT(3) TAB(60) HT(4)
1530 LPRINT "Area  m2     "TAB(15) AR(1) TAB(30) AR(2) TAB(45) AR(3) TAB(60)
AR(4)
1540 LPRINT "Frn org carb " TAB(45) ORG(3) TAB(60) ORG(4) TAB(75) ORG(5)
1550 LPRINT "Z mol/m3.Pa "TAB(15) Z(1) TAB(30) Z(2) TAB(45) Z(3) TAB(60) Z(4)
TAB(75) Z(5) TAB(90) Z(6)
```

```
1560 LPRINT "VZ mol/Pa   "TAB(15) VZ(1) TAB(30) VZ(2) TAB(45) VZ(3) TAB(60) VZ(4
) TAB(75) VZ(5) TAB(90) VZ(6)
1570 LPRINT "Fugacity Pa "TAB(15) F(1) TAB(30) F(2) TAB(45) F(3) TAB(60) F(4)
TAB(75) F(5) TAB(90) F(6)
1580 LPRINT "Conc mol/m3 "TAB(15) C(1) TAB(30) C(2) TAB(45) C(3) TAB(60) C(4)
TAB(75) C(5) TAB(90) C(6)
1590 LPRINT "Conc g/m3   "TAB(15) CG(1) TAB(30) CG(2) TAB(45) CG(3) TAB(60)
CG(4 ) TAB(75) CG(5) TAB(90) CG(6)
1600 LPRINT "Conc ug/g   "TAB(15) CU(1) TAB(30) CU(2) TAB(45) CU(3) TAB(60)
CU(4) TAB(75) CU(5) TAB(90) CU(6)
1610 LPRINT "Amount mol  "TAB(15) M(1) TAB(30) M(2) TAB(45) M(3) TAB(60) M(4)
TAB(75) M(5) TAB(90) M(6)
1620 LPRINT "Amount kg   "TAB(15) MK(1) TAB(30) MK(2) TAB(45) MK(3) TAB(60)
MK(4) TAB(75) MK(5) TAB(90) MK(6)
1630 LPRINT "Amount %    "TAB(15) P(1) TAB(30) P(2) TAB(45) P(3) TAB(60) P(4)
TAB(75) P(5) TAB(90) P(6)
1640 LPRINT CHR$(12)
1650 REM Fugacity Level II program, 6 compartments
1660 REM calculate total flows, m3/h
1670 GA(1)=V(1)/GRA(1)
1680 GA(2)=V(2)/GRA(2)
1690 GA(4)=V(4)/GRA(4)
1700 REM calculate D values
1710 NRT=0:NRTK=0:NAT=0:NATK=0:MT=0:DT=0:DTA=0:DTR=0 'set totals to
zero
1720 FOR N= 1 TO 4
1730 RK(N)=.693/HL(N) 'rate constants from half lives
1740 DR(N)=V(N)*Z(N)*RK(N):DA(N)=GA(N)*Z(N) 'reaction and advection D values
1750 DTR=DTR+DR(N):DTA=DTA+DA(N) 'total D values
1760 NEXT N
1770 DT=DTR+DTA 'total D value
1780 F2=E/DT 'fugacity
1790 FOR N=1 TO 6
1800 F(N)=F2
1810 C(N)=F(N)*Z(N) 'concentration mol/m3
1820 M(N)=C(N)*V(N) 'amount mol
1830 MK(N)=M(N)*WM/1000
1840 MT=MT+M(N) 'total amount
1850 CG(N)=C(N)*WM 'concentration g/m3
1860 CU(N)=CG(N)*1000/DEN(N) 'concentration ug/g
1870 NR(N)=V(N)*C(N)*RK(N):NRK(N)=NR(N)*WM/1000 'reaction rates mol/h and kg/h
1880 NA(N)=GA(N)*C(N):NAK(N)=NA(N)*WM/1000 'advection rates mol/h and kg/h
1890 NRT=NRT+NR(N):NAT=NAT+NA(N) 'total rates mol/h
```

```
1900 NRTK=NRTK+NRK(N):NATK=NATK+NAK(N) 'total rates kg/h
1910 NEXT N
1920 NT=NRT+NAT:NTK=NRTK+NATK
1930 MTK=MT*WM/1000
1940 FOR N=1 TO 6
1950 P(N)=100*M(N)/MT 'percentages of amount
1960 PR(N)=100*NR(N)/NT 'percentages of reaction rate
1970 PA(N)=100*NA(N)/NT 'percentages of advection rate
1980 NEXT N
1990 IF NRT=0 THEN TR=0 ELSE TR=MT/NRT
2000 IF NAT=0 THEN TA=0 ELSE TA=MT/NAT
2010 TOV=MT/NT 'overall residence time h
2020 REM print out results
2030 LPRINT
2040 LPRINT "SIX COMPARTMENT FUGACITY LEVEL II CALCULATION  ";CHEM$
2050 LPRINT " "
2060 LPRINT "Emission rate of chemical mol/h  ";E
2070 LPRINT "Emission rate of chemical  kg/h  ";EK
2080 LPRINT "Fugacity Pa                      ";F2
2090 LPRINT "Total amount of chemical mol     ";MT
2100 LPRINT "Total amount of chemical kg      ";MTK
2110 LPRINT " "
2120 LPRINT "Phase properties,compositions and rates"
2130 LPRINT " "
2140 LPRINT "Phase      "TAB(15) N$(1) TAB(30) N$(2) TAB(45) N$(3) TAB(60) N$(4)
TAB(75) N$(5) TAB(90) N$(6)
2150 LPRINT "Adv.flow m3/h"TAB(15) GA(1) TAB(30) GA(2) TAB(45) GA(3) TAB(60)
GA(4)
2160 LPRINT "Adv.restime h"TAB(15) GRA(1) TAB(30) GRA(2) TAB(45) GRA(3) TAB(60)
GRA(4)
2170 LPRINT "Rct halflife h"TAB(15) HL(1) TAB(30) HL(2) TAB(45) HL(3) TAB(60)
HL(4)
2180 LPRINT "Rct rate c.h-1"TAB(15) RK(1) TAB(30) RK(2) TAB(45) RK(3) TAB(60)
RK(4)
2190 LPRINT "Fugacity Pa "TAB(15) F(1) TAB(30) F(2) TAB(45) F(3) TAB(60) F(4)
TAB(75) F(5) TAB(90) F(6)
2200 LPRINT "Conc mol/m3 "TAB(15) C(1) TAB(30) C(2) TAB(45) C(3) TAB(60) C(4)
TAB(75) C(5) TAB(90) C(6)
2210 LPRINT "Conc g/m3   "TAB(15) CG(1) TAB(30) CG(2) TAB(45) CG(3) TAB(60)
CG(4 ) TAB(75) CG(5) TAB(90) CG(6)
2220 LPRINT "Conc ug/g   "TAB(15) CU(1) TAB(30) CU(2) TAB(45) CU(3) TAB(60)
CU(4) TAB(75) CU(5) TAB(90) CU(6)
```

```
2230 LPRINT "Amount mol  "TAB(15) M(1) TAB(30) M(2) TAB(45) M(3) TAB(60) M(4)
TAB(75) M(5) TAB(90) M(6)
2240 LPRINT "Amount kg   "TAB(15) MK(1) TAB(30) MK(2) TAB(45) MK(3) TAB(60)
MK(4) TAB(75) MK(5) TAB(90) MK(6)
2250 LPRINT "Amount %    "TAB(15) P(1) TAB(30) P(2) TAB(45) P(3) TAB(60) P(4)
TAB(75) P(5) TAB(90) P(6)
2260 LPRINT "D rct mol/Pa.h"TAB(15) DR(1) TAB(30) DR(2) TAB(45) DR(3) TAB(60)
DR(4)
2270 LPRINT "D adv mol/Pa.h"TAB(15) DA(1) TAB(30) DA(2) TAB(45) DA(3) TAB(60)
DA(4)
2280 LPRINT "Rct rate mol/h"TAB(15) NR(1) TAB(30) NR(2) TAB(45) NR(3) TAB(60)
NR(4)
2290 LPRINT "Adv rate mol/h"TAB(15) NA(1) TAB(30) NA(2) TAB(45) NA(3) TAB(60)
NA(4)
2300 LPRINT "Rct rate kg/h "TAB(15) NRK(1) TAB(30) NRK(2) TAB(45) NRK(3) TAB(60)
NRK(4)
2310 LPRINT "Adv rate kg/h "TAB(15) NAK(1) TAB(30) NAK(2) TAB(45) NAK(3)
TAB(60) NAK(4)
2320 LPRINT "Reaction %    "TAB(15) PR(1) TAB(30) PR(2) TAB(45) PR(3) TAB(60)
PR(4)
2330 LPRINT "Advection %   "TAB(15) PA(1) TAB(30) PA(2) TAB(45) PA(3) TAB(60)
PA(4)
2340 LPRINT " "
2350 LPRINT "Total advection D value     ";DTA
2360 LPRINT "Total reaction D value      ";DTR
2370 LPRINT "Total D value               ";DT
2380 LPRINT " "
2390 LPRINT "Output by reaction    mol/h ";NRT
2400 LPRINT "Output by advection   mol/h ";NAT
2410 LPRINT "Total output by reaction and advection mol/h ";NT
2420 LPRINT" "
2430 LPRINT "Output by reaction    kg/h  ";NRTK
2440 LPRINT "Output by advection   kg/h  ";NATK
2450 LPRINT "Total output by reaction and advection kg/h  ";NTK
2460 LPRINT" "
2470 LPRINT "Overall residence time   h  ";TOV
2480 LPRINT "Reaction residence time  h  ";TR
2490 LPRINT "Advection residence time h  ";TA
2500 LPRINT CHR$(12)
2510 LPRINT
2520 REM Fugacity Level III Program
2530 REM Set bulk phase volumes, densities and Z values
2540 VB(1)=V(1):VB(2)=V(2):VB(3)=1.8E+10:VB(4)=5E+08
```

```
2550 VA(1)=1:VQ(1)=2E-11 'volume fractions
2560 VW(2)=1:VP(2)=.000005:VF(2)=.000001
2570 VA(3)=.2:VW(3)=.3:VE(3)=.5
2580 VW(4)=.8:VS(4)=.2
2590 DENB(1)=VA(1)*DEN(1)+VQ(1)*DEN(7)
2600 DENB(2)=VW(2)*DEN(2)+VP(2)*DEN(5)+VF(2)*DEN(6)
2610 DENB(3)=VW(3)*DEN(2)+VA(3)*DEN(1)+VE(3)*DEN(3)
2620 DENB(4)=VW(4)*DEN(2)+VS(4)*DEN(4)
2630 ZB(1)=VA(1)*Z(1)+VQ(1)*Z(7)
2640 ZB(2)=VW(2)*Z(2)+VP(2)*Z(5)+VF(2)*Z(6)
2650 ZB(3)=VA(3)*Z(1)+VW(3)*Z(2)+VE(3)*Z(3)
2660 ZB(4)=VW(4)*Z(2)+VS(4)*Z(4)
2670 KB12=ZB(1)/ZB(2)
2680 KB32=ZB(3)/ZB(2)
2690 KB42=ZB(4)/ZB(2)
2700 REM Parameters
2710 U(1)=5         :U$(1)="air side air-water MTC              "
2720 U(2)=.05       :U$(2)="water side air-water MTC            "
2730 U(3)=.0001     :U$(3)="rain rate                           "
2740 U(4)=6E-10     :U$(4)="aerosol deposition velocity         "
2750 U(5)=.02       :U$(5)="soil air phase diffusion MTC        "
2760 U(6)=.00001    :U$(6)="soil water phase diffusion MTC      "
2770 U(7)=5         :U$(7)="soil air boundary layer MTC         "
2780 U(8)=.0001     :U$(8)="sediment-water diffusion MTC        "
2790 U(9)=.0000005 :U$(9)="sediment deposition velocity        "
2800 U(10)=.0000002:U$(10)="sediment resuspension velocity     "
2810 U(11)=.00005  :U$(11)="soil water runoff rate              "
2820 U(12)=1E-08   :U$(12)="soil solids runoff rate             "
2830 'Calculate D values
2840 DRW=AR(2)*U(3)*Z(2)
2850 DQW=AR(2)*U(4)*Z(7)
2860 DVWA=AR(2)*U(1)*Z(1)
2870 DVWW=AR(2)*U(2)*Z(2)
2880 DVW=1/(1/DVWA+1/DVWW)
2890 D(2,1)=DVW
2900 D(1,2)=DVW+DQW+DRW
2910 DVSB=AR(3)*U(7)*Z(1)
2920 DVSA=AR(3)*U(5)*Z(1)
2930 DVSW=AR(3)*U(6)*Z(2)
2940 DRS=AR(3)*U(3)*Z(2)
2950 DQS=AR(3)*U(4)*Z(7)
2960 DVS=1/(1/DVSB+1/(DVSW+DVSA))
2970 D(3,1)=DVS
```

```
2980 D(1,3)=DVS+DRS+DQS
2990 DSWD=AR(2)*U(8)*Z(2)
3000 DSD=AR(2)*U(9)*Z(5)
3010 DSR=AR(4)*U(10)*Z(4)
3020 D(2,4)=DSWD+DSD
3030 D(4,2)=DSWD+DSR
3040 DSWW=AR(3)*U(11)*Z(2)
3050 DSWS=AR(3)*U(12)*Z(3)
3060 D(3,2)=DSWW+DSWS
3070 D(2,3)=0
3080 REM calculate total chemical inflows
3090 IN=0:INK=0
3100 FOR N=1 TO 4
3110 I(N)=IK(N)*1000/WM
3120 IN=IN+I(N):INK=INK+IK(N)
3130 NEXT N
3140 REM calculate reaction and advection D values for bulk phases
3150 GAB(1)=VB(1)/GRA(1)
3160 GAB(2)=VB(2)/GRA(2)
3170 GAB(4)=VB(4)/GRA(4)
3180 VZBT=0
3190 FOR N= 1 TO 4
3200 RK(N)=.693/HL(N)
3210 DR(N)=VB(N)*ZB(N)*RK(N):DA(N)=GAB(N)*ZB(N)
3220 VZB(N)=VB(N)*ZB(N)
3230 VZBT=VZBT+VZB(N)
3240 NEXT N
3250 FOR N=1 TO 4
3260 FOR NN=1 TO 4
3270 GD(N,NN)=D(N,NN)/ZB(N)
3280 IF GD(N,NN)=0 GOTO 3300 ELSE GOTO 3290
3290 TD(N,NN)=.693*VB(N)/GD(N,NN)
3300 NEXT NN
3310 NEXT N
3320 DT(1)=DR(1)+DA(1)+D(1,2)+D(1,3)
3330 DT(2)=DR(2)+DA(2)+D(2,1)+D(2,3)+D(2,4)
3340 DT(3)=DR(3)+DA(3)+D(3,1)+D(3,2)
3350 DT(4)=DR(4)+DA(4)+D(4,2)
3360 J1=I(1)/DT(1)+I(3)*D(3,1)/DT(3)/DT(1)
3370 J2=D(2,1)/DT(1)
3380 J3=1-D(3,1)*D(1,3)/DT(1)/DT(3)
3390 J4=D(1,2)+D(3,2)*D(1,3)/DT(3)
```

```
3400
F(2)=(I(2)+J1*J4/J3+I(3)*D(3,2)/DT(3)+I(4)*D(4,2)/DT(4))/(DT(2)-J2*J4/J3-D(2,4)*D(4,2)
/DT(4))
3410 F(1)=(J1+F(2)*J2)/J3
3420 F(3)=(I(3)+F(1)*D(1,3))/DT(3)
3430 F(4)= (I(4)+F(2)*D(2,4))/DT(4)
3440 NRT=0:NAT=0:MT=0
3450 FOR N=1 TO 4
3460 C(N)=F(N)*ZB(N)
3470 M(N)=C(N)*VB(N)
3480 MK(N)=M(N)*WM/1000
3490 MT=MT+M(N)
3500 CG(N)=C(N)*WM
3510 CU(N)=CG(N)*1000/DENB(N)
3520 NR(N)=F(N)*DR(N):NRK(N)=NR(N)*WM/1000
3530 NA(N)=F(N)*DA(N):NAK(N)=NA(N)*WM/1000
3540 NRT=NRT+NR(N):NAT=NAT+NA(N)
3550 NEXT N
3560 MTK=MT*WM/1000
3570 NRTK=NRT*WM/1000:NATK=NAT*WM/1000
3580 NT=NRT+NAT:NTK=NT*WM/1000
3590 FOR N=1 TO 4
3600 P(N)=100*M(N)/MT
3610 PR(N)=100*NR(N)/NT
3620 PA(N)=100*NA(N)/NT
3630 NEXT N
3640 N(1,2)=D(1,2)*F(1):NK(1,2)=N(1,2)*WM/1000
3650 N(1,3)=D(1,3)*F(1):NK(1,3)=N(1,3)*WM/1000
3660 N(2,1)=D(2,1)*F(2):NK(2,1)=N(2,1)*WM/1000
3670 N(2,4)=D(2,4)*F(2):NK(2,4)=N(2,4)*WM/1000
3680 N(3,1)=D(3,1)*F(3):NK(3,1)=N(3,1)*WM/1000
3690 N(3,2)=D(3,2)*F(3):NK(3,2)=N(3,2)*WM/1000
3700 N(4,2)=D(4,2)*F(4):NK(4,2)=N(4,2)*WM/1000
3710 TR=MT/(NRT+.0000001)
3720 TA=MT/(NAT+.0000001)
3730 TOV=MT/NT
3740 TOVD=TOV/24
3750 REM print out results
3760 LPRINT  " FOUR COMPARTMENT FUGACITY LEVEL III
CALCULATION",CHEM$
3770 LPRINT
3780 LPRINT "Bulk phase properties,compositions and rates"
3790 LPRINT " "
```

```
3800 LPRINT "Phase        "TAB(15) N$(1) TAB(30) N$(2) TAB(45) N$(3) TAB(60) N$(4)
TAB(75) "Total"
3810 LPRINT "Bulk vol m3 "TAB(15) VB(1) TAB(30) VB(2) TAB(45) VB(3) TAB(60)
VB(4)
3820 LPRINT "Density kg/m3"TAB(15) DENB(1) TAB(30) DENB(2) TAB(45) DENB(3)
TAB(60) DENB(4)
3830 LPRINT "Bulk Z value"TAB(15) ZB(1) TAB(30) ZB(2) TAB(45) ZB(3) TAB(60) ZB(4
)
3840 LPRINT "Bulk VZ      "TAB(15) VZB(1) TAB(30) VZB(2) TAB(45) VZB(3) TAB(60)
VZB(4) TAB(75) VZBT
3850 LPRINT "Emission mol/h"TAB(15) I(1) TAB(30) I(2) TAB(45) I(3) TAB(60) I(4)
TAB(75) IN
3860 LPRINT "Emission kg/h "TAB(15) IK(1) TAB(30) IK(2) TAB(45) IK(3) TAB(60) IK(4)
TAB(75) INK
3870 LPRINT "Fugacity Pa "TAB(15) F(1) TAB(30) F(2) TAB(45) F(3) TAB(60) F(4)
3880 LPRINT "Conc mol/m3 "TAB(15) C(1) TAB(30) C(2) TAB(45) C(3) TAB(60) C(4)
3890 LPRINT "Conc g/m3    "TAB(15) CG(1) TAB(30) CG(2) TAB(45) CG(3) TAB(60)
CG(4 )
3900 LPRINT "Conc ug/g    "TAB(15) CU(1) TAB(30) CU(2) TAB(45) CU(3) TAB(60)
CU(4)
3910 LPRINT "Amount mol   "TAB(15) M(1) TAB(30) M(2) TAB(45) M(3) TAB(60) M(4)
TAB(75) MT
3920 LPRINT "Amount kg    "TAB(15) MK(1) TAB(30) MK(2) TAB(45) MK(3) TAB(60)
MK(4) TAB(75) MTK
3930 LPRINT "Amount %     "TAB(15) P(1) TAB(30) P(2) TAB(45) P(3) TAB(60) P(4)
3940 LPRINT "Adv.flow m3/h"TAB(15) GAB(1) TAB(30) GAB(2) TAB(45) GAB(3)
TAB(60) GAB(4)
3950 LPRINT "D rct mol/Pa.h"TAB(15) DR(1) TAB(30) DR(2) TAB(45) DR(3) TAB(60)
DR(4 )
3960 LPRINT "D adv mol/Pa.h"TAB(15) DA(1) TAB(30) DA(2) TAB(45) DA(3) TAB(60)
DA(4 )
3970 LPRINT "Rct rate mol/h"TAB(15) NR(1) TAB(30) NR(2) TAB(45) NR(3) TAB(60)
NR(4 ) TAB(75) NRT
3980 LPRINT "Rct rate kg/h "TAB(15) NRK(1) TAB(30) NRK(2) TAB(45) NRK(3) TAB(60)
NRK(4 ) TAB(75) NRTK
3990 LPRINT "Adv rate mol/h"TAB(15) NA(1) TAB(30) NA(2) TAB(45) NA(3) TAB(60)
NA(4 ) TAB(75) NAT
4000 LPRINT "Adv rate kg/h "TAB(15) NAK(1) TAB(30) NAK(2) TAB(45) NAK(3)
TAB(60) NAK(4 ) TAB(75) NATK
4010 LPRINT "Reaction %     "TAB(15) PR(1) TAB(30) PR(2) TAB(45) PR(3) TAB(60) PR(4
)
4020 LPRINT "Advection %   "TAB(15) PA(1) TAB(30) PA(2) TAB(45) PA(3) TAB(60)
PA(4 )
```

```
4030 LPRINT " "
4040 LPRINT "Overall residence time    h   ";TOV
4050 LPRINT "Reaction residence time  h   ";TR;
4060 LPRINT "    Advection residence time h   ";TA
4070 LPRINT
4080 LPRINT "Intermedia Data.   Half times   Equiv flows   D values    Rates  of transport "
4090 LPRINT "                              h            m3/h          mol/Pa.h    mol/h        kg/h"
4100 LPRINT "Air to water      ";:LPRINT USING " ##.####^^^^ ";TD(1,2);GD(1,2);D(1,2)
;N(1,2);NK(1,2)
4110 LPRINT "Air to soil       ";:LPRINT USING " ##.####^^^^ ";TD(1,3);GD(1,3);D(1,3)
;N(1,3);NK(1,3)
4120 LPRINT "Water to air      ";:LPRINT USING " ##.####^^^^ ";TD(2,1);GD(2,1);D(2,1)
;N(2,1);NK(2,1)
4130 LPRINT "Water to sediment";:LPRINT USING " ##.####^^^^
";TD(2,4);GD(2,4);D(2,4) ;N(2,4);NK(2,4)
4140 LPRINT "Soil to air       ";:LPRINT USING " ##.####^^^^ ";TD(3,1);GD(3,1),D(3,1)
,N(3,1);NK(3,1)
4150 LPRINT "Soil to water     ";:LPRINT USING " ##.####^^^^ ";TD(3,2);GD(3,2);D(3,2)
;N(3,2);NK(3,2)
4160 LPRINT "Sediment to water";:LPRINT USING " ##.####^^^^
";TD(4,2);GD(4,2),D(4,2) ,N(4,2);NK(4,2)
4170 LPRINT "   Transport velocity parameters              m/h              m/year   "
4180 FOR I=1 TO 12
4190 UY(I)=U(I)*8760
4200 LPRINT TAB(5) I TAB(10) U$(I) TAB(45) U(I) TAB(60) UY(I)
4210 NEXT I
4220 LPRINT "Individual process D values "
4230 LPRINT "Air-water diffusion (air-side)     ";DVWA TAB(50);
4240 LPRINT "Air-water diffusion (water-side)  ";DVWW
4250 LPRINT "Air-water diffusion (overall)     ";DVW
4260 LPRINT "Rain dissolution to water          ";DRW  TAB(50);
4270 LPRINT "Aerosol deposition to water        ";DQW
4280 LPRINT "Rain dissolution to soil           ";DRS  TAB(50);
4290 LPRINT "Aerosol deposition to soil         ";DQS
4300 LPRINT "Soil-air diffusion (air-phase)     ";DVSA TAB(50);
4310 LPRINT "Soil-air diffusion (water-phase)  ";DVSW
4320 LPRINT "Soil-air diffusion (bndry layer)  ";DVSB TAB(50);
4330 LPRINT "Soil-air diffusion (overall)      ";DVS
4340 LPRINT "Water-sediment diffusion           ";DSWD
4350 LPRINT "Water-sediment deposition          ";DSD TAB(50);
4360 LPRINT "Sediment-water resuspension        ";DSR
4370 LPRINT "Soil-water runoff (water)          ";DSWW TAB(50);
4380 LPRINT "Soil-water runoff (solids)         ";DSWS
```

PROGRAM 'LEWIS':SIX COMPARTMENT FUGACITY LEVEL I CALCULATION

Properties of Benzene

Temperature deg C	25
Molecular mass g/mol	78.11
Melting point deg C	5.53
Fugacity ratio	1
Vapor pressure Pa	12700
Sub-cooled liquid vapor press Pa	12700
Solubility g/m3	1780
Solubility mol/m3	22.78838
Henry's law constant Pa.m3/mol	557.3017
Log octanol-water p-coefficient	2.13
Octanol-water partn-coefficient	134.8964
Organic C-water ptn-coefficient	55.30751
Fish-water partition coefficient	6.744818
Air-water partition coefficient	.2248255
Soil-water partition coefficient	2.65476
Sedt-water partition coefficient	5.309521
Susp sedt-water partn coeffnt	16.59225
Aerosol-air partition coeff	472.441
Aerosol Z value	.1905911
Aerosol density kg/m3	2000

Amount of chemical moles	1280246
Amount of chemical kilograms	100000
Fugacity Pa	3.14213E-05
Total of VZ products	4.074452E+10

Phase properties and compositions

Phase	Air	Water	Soil	Sediment	Susp sedt	Fish
Volume m3	1E+14	2E+11	8.999999E+09	1E+08	999999.9	200000
Density kg/m3	1.185413	1000	2400	2400	1500	1000
Depth m	1000	20	.1	.01		
Area m2	1E+11	1E+10	9E+10	1E+10		
Frn org carb			.02	.04	.2	
Z mol/m3.Pa	4.034179E-04	1.79436E-03	4.763596E-03	9.527192E-03	2.977247E-02	1.210263E-02
VZ mol/Pa	4.034179E+10	3.58872E+08	4.287237E+07	952719.2	29772.47	2420.526
Fugacity Pa	3.14213E-05	3.14213E-05	3.14213E-05	3.14213E-05	3.14213E-05	3.14213E-05
Conc mol/m3	1.267591E-08	5.638113E-08	1.496784E-07	2.993568E-07	9.354898E-07	3.802804E-07
Conc g/m3	9.901156E-07	4.40393E-06	1.169138E-05	2.338276E-05	7.307111E-05	2.970371E-05
Conc ug/g	8.352493E-04	4.40393E-06	4.871408E-06	9.742816E-06	4.871407E-05	2.970371E-05
Amount mol	1267591	11276.23	1347.105	29.93568	.9354897	7.605609E-02
Amount kg	99011.56	880.786	105.2224	2.338276	7.307111E-02	5.940741E-03
Amount %	99.01157	.880786	.1052224	2.338276E-03	7.307111E-05	5.940741E-06

SIX COMPARTMENT FUGACITY LEVEL II CALCULATION Benzene

Emission rate of chemical mol/h 12802.46
Emission rate of chemical kg/h 1000
Fugacity Pa 6.245664E-06
Total amount of chemical mol 254476.6
Total amount of chemical kg 19877.16

Phase properties,compositions and rates

Phase	Air	Water	Soil	Sediment	Susp sedt	Fish
Adv.flow m3/h	1E+12	2E+08	0	2000		
Adv.restime h	100	1000	0	50000		
Rct halflife h	17	170	550	1700		
Rct rate c.h-1	4.076471E-02	4.076471E-03	.00126	4.076471E-04		
Fugacity Pa	6.245664E-06	6.245664E-06	6.245664E-06	6.245664E-06	6.245664E-06	6.245664E-06
Conc mol/m3	2.519613E-09	1.120697E-08	2.975182E-08	5.950364E-08	1.859489E-07	7.558897E-08
Conc g/m3	1.968069E-07	8.753765E-07	2.323915E-06	4.64783E-06	1.452447E-05	5.904255E-06
Conc ug/g	1.660239E-04	8.753765E-07	9.682979E-07	1.936596E-06	9.682976E-06	5.904255E-06
Amount mol	251961.3	2241.394	267.7664	5.950364	.1859488	1.511779E-02
Amount kg	19680.69	175.0753	20.91523	.464783	1.452446E-02	1.180851E-03
Amount %	99.01158	.8807861	.1052224	2.338276E-03	7.307111E-05	5.940741E-06
D rct mol/Pa.h	1.644521E+09	1462931	54019.18	388.3732		
D adv mol/Pa.h	4.034179E+08	358872	0	19.05438		
Rct rate mol/h	10271.13	9.136978	.3373856	2.425649E-03		
Adv rate mol/h	2519.613	2.241394	0	1.190073E-04		
Rct rate kg/h	802.2778	.7136893	2.635319E-02	1.894674E-04		
Adv rate kg/h	196.8069	.1750753	0	9.295659E-06		
Reaction %	80.22778	7.136893E-02	2.63532E-03	1.894674E-05		
Advection %	19.6807	1.750753E-02	0	9.295659E-07		

Total advection D value 4.037768E+08
Total reaction D value 1.646039E+09
Total D value 2.049815E+09

Output by reaction mol/h 10280.6
Output by advection mol/h 2521.854
Total output by reaction and advection mol/h 12802.46

Output by reaction kg/h 803.018
Output by advection kg/h 196.982
Total output by reaction and advection kg/h 1000

Overall residence time h 19.87717
Reaction residence time h 24.75308
Advection residence time h 100.9085

FOUR COMPARTMENT FUGACITY LEVEL III CALCULATION   Benzene

Bulk phase properties,compositions and rates

Phase	Air	Water	Soil	Sediment	Total
Bulk vol m3	1E+14	2E+11	1.8E+10	5E+08	
Density kg/m3	1.185413	1000.009	1500.237	1280	
Bulk Z value	4.034179E-04	1.794521E-03	3.00079E-03	3.340927E-03	
Bulk VZ	4.034179E+10	3.589042E+08	5.401422E+07	1670463	4.075638E+10
Emission mol/h	12802.46	0	0	0	12802.46
Emission kg/h	1000	0	0	0	1000
Fugacity Pa	6.249393E-06	2.0235E-06	5.78145E-06	1.555557E-06	
Conc mol/m3	2.521117E-09	3.631213E-09	1.734892E-08	5.197001E-09	
Conc g/m3	1.969245E-07	2.836341E-07	1.355124E-06	4.059378E-07	
Conc ug/g	1.661231E-04	2.836317E-07	9.032731E-07	3.171389E-07	
Amount mol	252111.7	726.2427	312.2805	2.598501	253152.8
Amount kg	19692.45	56.72682	24.39223	.2029689	19773.77
Amount %	99.58874	.2868792	.1233565	1.026455E-03	
Adv.flow m3/h	1E+12	2E+08	0	10000	
D rct mol/Pa.h	1.644521E+09	1463063	68057.93	680.9596	
D adv mol/Pa.h	4.034179E+08	358904.3	0	33.40927	
Rct rate mol/h	10277.26	2.960507	.3934735	1.059271E-03	10280.62
Rct rate kg/h	802.7568	.2312452	3.073421E-02	8.273968E-05	803.019
Adv rate mol/h	2521.117	.7262427	0	5.197001E-05	2521.843
Adv rate kg/h	196.9244	5.672682E-02	0	4.059378E-06	196.9812
Reaction %	80.27568	2.312452E-02	3.073421E-03	8.273966E-06	
Advection %	19.69244	5.672681E-03	0	4.059377E-07	

Overall residence time h 19.77377
Reaction residence time h 24.62429   Advection residence time h 100.3841

Intermedia Data.	Half times h	Equiv flows m3/h	D values mol/Pa.h	Rates of transport mol/h	kg/h
Air to water	3.2479E+04	2.1337E+09	8.6077E+05	5.3793E+00	4.2018E-01
Air to soil	3.7728E+04	1.8369E+09	7.4102E+05	4.6309E+00	3.6172E-01
Water to air	2.8956E+02	4.7866E+08	8.5897E+05	1.7381E+00	1.3577E-01
Water to sediment	1.2799E+05	1.0829E+06	1.9432E+03	3.9321E-03	3.0714E-04
Soil to air	5.1640E+01	2.4156E+08	7.2486E+05	4.1907E+00	3.2734E-01
Soil to water	4.6333E+03	2.6923E+06	8.0789E+03	4.6708E-02	3.6483E-03
Sediment to water	6.3837E+02	5.4279E+05	1.8134E+03	2.8209E-03	2.2034E-04

	Transport velocity parameters	m/h	m/year
1	air side air-water MTC	5	43800
2	water side air-water MTC	.05	438
3	rain rate	.0001	.876
4	aerosol deposition velocity	6E-10	5.256E-06
5	soil air phase diffusion MTC	.02	175.2
6	soil water phase diffusion MTC	.00001	.0876
7	soil air boundary layer MTC	5	43800
8	sediment-water diffusion MTC	.0001	.876
9	sediment deposition velocity	.0000005	.00438
10	sediment resuspension velocity	.0000002	.001752
11	soil water runoff rate	.00005	.438
12	soil solids runoff rate	1E-08	.0000876

Individual process D values

Air-water diffusion (air-side)	2.017089E+07	Air-water diffusion (water-side)	897180.1
Air-water diffusion (overall)	858973.8		
Rain dissolution to water	1794.36	Aerosol deposition to water	1.143547
Rain dissolution to soil	16149.24	Aerosol deposition to soil	10.29192
Soil-air diffusion (air-phase)	726152.2	Soil-air diffusion (water-phase)	1614.924
Soil-air diffusion (bndry layer)	1.815381E+08	Soil-air diffusion (overall)	724861.2
Water-sediment diffusion	1794.36		
Water-sediment deposition	148.8624	Sediment-water resuspension	19.05438
Soil-water runoff (water)	8074.621	Soil-water runoff (solids)	4.287237

Fugacity calculations: Benzene

```
LEVEL I, II and III                                                                                  * input
*  Amount of chemicals, moles                                  1280245.8 moles     100000 kg      Adv. flow, G(air)       1.000E+12 mol/hr
*  Emission rate of chemicals, E =                             12802.458 mol/h       1000 kg/h    Adv. flow G(water)=     5.00E+08 mol/hr
   Gas constant, Pa m3/mol K, R=                                   8.314                          Adv. flow, G(sed.)        10000 mol/hr
   System temperature, K, T= (t + 273.15)                         298.15                          Half-lives, hours
*  Molecular weight, g/mol                             MW =        78.11  * input data          *  t(air)                    17 h
*  Melting point, t C                                M.P., C        5.53  * input data          *  t(water)                 170 h
#  If MP > 25 C enter Tm (mp+273.15) or else system temp.,K       298.15  * input               *  t(soil)                  550 h
   Fugacity ratio = exp(6.79(1-Tm/T)) for solid comp'ds           1.0000                        *  t(sediment)             1700 h
*  Solublitiy, g/m3 or mg/L                            S =          1780  * input data             t(s. sediment)     1.000E+11 h
   molar solubility, mol/m3,  c=S/MW                            2.28E+01                           t(biota)           1.000E+11 h
*  Vapor pressure, Pa                                  P =      1.27E+04  * input data          Emission, kg/h :
   Vap. pressure, subcooled liquid, Pa                          1.27E+04                                    air       water      soil
   Henry's law constant, Pa m3/mol, H=p/c                      557.30168                        E            600        300       100
*  Octanol/water partition coefficient,              log Kow        2.13  * input data          E(A)        1000          0         0
   Kow =                                                             135                        E(B)           0       1000         0
   Partition coefficient, organic C, Koc = 0.41*Kow*y          55.307478                        E(C)           0          0      1000
    for soil (mole fraction organic C), y(3) =                      0.02
    suspended sediment, y(5) =                                       0.2
    bottom sediment, y(6) =                                         0.04
   Kp(3) =0.41*Kow*y(3)                                                1
   Kp(4) =0.41*Kow*y(4)                                               11
   Kp(5) = 0.41*Kow*y(5)                                               2
   Bioconcentraion factor, K(6) = 0.050*Kow                            7
   Air/water partition coeff., Z(air)/Z(water)                 0.2248254
   Soil/water partition coeff., Z(soil)/Z(water)               2.6547589
   Sediment/water partition coeff., Z(sediment)/Z(water)       5.3095179
   Sus. sediment/water partition coeff., Z(ss)/Z(water)        16.592243
   Aerosol/water partition coeff., Z(aerosol)/Z(air)           472.44094
   Densities, g/cm3 or kg/L
    air, d(1) = (0.029*101325/RT)                              0.0011854
    water, d(2)                                                        1
    soil, d(3) =                                                     2.4
    bottom sediment, d(4) =                                          2.4
    suspended sediment, d(5) =                                       1.5
    biota, d(6) =                                                      1
    aerosol, d(7)                                                      2
  Fugacity capacities, Z:
    Z(1) or Z(air) = 1/RT                                      4.034E-04
    Z(2) or Z(water) = 1/H = c/p                               1.794E-03
    Z(3) or Z(soil) = Kp(s)*Z(water)*d(s)                      4.764E-03
    Z(4) or Z(bottom sediment) = Kp(bs)*Z(water)*d(bs)         9.527E-03
    Z(5) or Z(suspended sediment) = Kp(ss)*Z(water)*d(ss)      2.977E-02
    Z(6) or Z(biota) = K *Z(water)*d(B)                        1.210E-02
    Z(7) or Z(aerosol) = 6*E6/p(L)RT                           1.906E-01
  Fugacity (Level I), f = total no. of moles/sum(ViZi)         3.142E-05
  Fugacity (Level II), f = emission/sum(D values)              6.246E-06
```

Tansport parameters:

	m/h k's	m/yr k/8760
air-water MTC, air side, U(1)	5	43800
air-water MTC, water side, U(2)	0.05	438
rain rate, U(3) = 0.85/8760	0.0001	0.876
aerosol deposition velocity, U(4)	6.00E-10	0.0000052
soil air phase diffusion MTC, U(5)	0.02	175.2
soil water phase diffusion MTC, U(6)	1.00E-05	0.0876
soil air boundary layer MTC, U(7)	5	43800
sediment-water MTC, U(8)	0.0001	0.876
sediment deposition rate, U(9)	5.00E-07	0.00438
sediment resuspended rate, U(10)	2.00E-07	0.001752
soil water runoff rate, U(11)	5.00E-05	0.438
soil solids runoff rate, U(12)	1.00E-08	0.0000876
sediment burial rate, U(13)	0	0
diffusion to stratosphere, U(14)	0	0

Define unit world:

Compartment	Volume, Vi, m3	Depth, h, m	Area, A, m2	Density d, kg/m3	Fugacity cap., Zi mol/Pa m3	VZ	*input Advective flow, G mol/h	Residence time, V/G t(R),h	*input Reaction half-life t(1/2),h	*input Rate const. 0.693/t k, 1/hr
Air (1)	1.00E+14	1000	1.000E+11	1.1854132	4.034E-04	4.03E+10	1.000E+12	100	17	0.04076470
Water (2)	2.00E+11	20	1.000E+10	1000	1.794E-03	3.59E+08	2.000E+08	1000	170	0.00407647
Soil (3)	9.00E+09	0.1	9.000E+10	2400	4.764E-03	4.29E+07	0		550	0.00126
bottom sediment (4)	1.00E+08	0.01	1.000E+10	2400	9.527E-03	9.53E+05	2000	50000	1700	0.00040764
Susp. sediment (5)	1.00E+06			1500	2.977E-02	2.98E+04			1.0000E+11	6.930E-12
Biota (fish) (6)	2.00E+05			1000	1.210E-02	2.42E+03			1.0000E+11	6.930E-12
Aerosol (7)	2000			2000	1.906E-01	4.07E+10				

Level I calculation:

Compartment	Volume, Vi, m3	Fugacity capacity Zi	VZ	Conc'n, c C = f*Z mol/m3	Amount m= CiVi mol	Amount w=m*MW/1E3 kg	Amount %	Conc'n,S mg/L (or g/m3)	Conc'n (S/d)*1000 ug/g
Air (1)	1.00E+14	4.034E-04	4.034E+10	1.268E-08	1.27E+06	9.901E+04	9.90E+01	9.90E-07	8.35E-04
Water (2)	2.00E+11	1.794E-03	3.589E+08	5.638E-08	1.13E+04	8.808E+02	8.81E-01	4.40E-06	4.40E-06
Soil (3)	9.00E+09	4.764E-03	4.287E+07	1.497E-07	1.35E+03	1.052E+02	1.05E-01	1.17E-05	4.87E-06
Bottom sediment (4)	1.00E+08	9.527E-03	9.527E+05	2.994E-07	2.99E+01	2.338E+00	2.34E-03	2.34E-05	9.74E-06
Sus. sediment (5)	1.00E+06	2.977E-02	2.977E+04	9.355E-07	9.35E-01	7.307E-02	7.31E-05	7.31E-05	4.87E-05
Biota (fish) (6)	2.00E+05	1.210E-02	2.421E+03	3.803E-07	7.61E-02	5.941E-03	5.94E-06	2.97E-05	2.97E-05
			4.074E+10		1280245.8	100000	100		

Level II phase properties and rates:

Compartment	Rate const. k, 1/hr	D(reaction) VZk mol/Pa h	D(advec'n) GZ mol/Pa h	conc'n C =f*Z mol/m3	Amount m= CiVi mol	Amount m*MW/1000 kg	conc'n mg/L (or g/m3)	Conc'n ug/g (S/d)*1000	Loss Reaction mol/h VCk	Loss Advection mol/h GC	% Loss reaction	% Loss advection	Removal %
Air (1)	0.040764705	1.645E+09	4.034E+08	2.520E-09	2.52E+05	1.97E+04	1.968E-07	1.66E-04	1.027E+04	2.520E+03	8.02E+01	19.68	99.91
Water (2)	0.004076470	1.463E+06	3.589E+05	1.121E-08	2.24E+03	1.75E+02	8.754E-07	8.75E-07	9.137E+00	2.241E+00	7.14E-02	0.018	0.09
Soil (3)	0.00126	5.402E+04		2.975E-08	2.68E+02	2.09E+01	2.324E-06	9.68E-07	3.374E-01		2.64E-03		0.00
Bottom sediment (4)	0.000407647	3.884E+02	1.905E+01	5.950E-08	5.95E+00	4.65E-01	4.648E-06	1.94E-06	2.426E-03	1.190E-04	1.89E-05	9.30E-07	1.99E-05
Sus. sediment (5)	6.9300E-12	2.063E-07		1.859E-07	1.86E-01	1.45E-02	1.452E-05	9.68E-06	1.289E-12		1.01E-14		1.01E-14
Biota (fish) (6)	6.9300E-12	1.677E-08		7.559E-08	1.51E-02	1.18E-03	5.904E-06	5.90E-06	1.048E-13		8.18E-16		8.18E-16
	Total	1.646E+09	4.038E+08		254476.58	19877.165			10280.60	2521.85	80.30	19.70	100
	R + A		2.050E+09							12802.4580			

Total amount of chemicals,	254476.58	moles	19877	kg
Total reaction D value	1.65E+09			
Total advection D value	403776792			
Total D value	2.05E+09			
Fugacity, E/sum D values	6.246E-06			
Output by reaction, mol/h	10280.603		803.01796	kg/h
Output by advection, mol/h	2521.8542		13.945853	kg/h
Total output, mol/h	12802.458		3817.0528	kg/h
Overall resistence time, h	19.877165			
Reaction resistence time, h	24.753077			
Advection resistence time, h	100.90852			

Compartment	Subcomp't	volume fraction fvi	Fugacity capacity Zi	Bulk vol. VB, m3	Bulk ZB(i) sum(viZi)	partition coeff. (Zi/Zw)	Bulk VZ=VB*Zi	Bulk den. sum(vidi) kg/m3	Bulk Adv. flow GAB=VB/Gi
Air (1)	Air	1	0.0004034	1.00E+14	0.0004034	2.25E-01	4.034E+10	1.18541328	1.00E+12
	Aerosol	2.0000E-11	0.19						
Water (2)	Water	1	0.0017943	2.00E+11	0.0017945	2.6547589	3.589E+08	1000.0085	2.00E+08
	Particulate	0.000005	0.0297724						
	Biota(fish)	0.000001	0.0121026						
Soil (3)	Air	0.2	0.0004034	1.80E+10	3.00E-03	5.3095179	5.401E+07	1500.23708	
	Water	0.3	0.0017943						
	Solids	0.5	0.0047635						
Bottom sediment (4)	Water	0.8	0.0017943	5.00E+08	0.0033409	6.7448144	1.670E+06	1280	10000
	Solids	0.2	0.0095271						
					0.0085396		4.076E+10		

Level III Intermedia Data:

	D values Dij mol/Pa h	Eq. flows Dij/Z(i) GDij, m3/h	half-life .693Vi/G t(1/2), h	Rate of transport Dij*f(i) N, mol/h	N*MW/1000 Nk, kg/h Tij
Air to water (D12)	8.6077E+05	2.134E+09	3.248E+04	4.233E+00	3.31E-01
Air to soil (D13)	7.4102E+05	1.837E+09	3.773E+04	3.644E+00	2.85E-01
Water to air (D21)	8.5897E+05	4.787E+08	2.896E+02	1.236E+03	9.65E+01
Water to sed. (D24)	1.9432E+03	1.083E+06	1.280E+05	2.796E+00	2.18E-01
Soil to air (D31)	7.2486E+05	2.416E+08	2.582E+01	1.162E+03	9.08E+01
Soil to water (D32)	8.0789E+03	2.692E+06	2.317E+03	1.295E+01	1.01E+00
Sed. to water (D42)	1.8134E+03	5.428E+05	1.277E+02	2.006E+00	1.57E-01
Water to soil (D23)	0				
Sed. burial DL(4)	0.0000E+00				
Stratosphere DL(1)	0.0000E+00				

	D values	Eq. flows	half-life	E(A)		E(B)		E(C)	
	Dij mol/Pa h	Dij/Z(i) GDij, m3/h	.693Vi/G t(1/2), h	Rate of transport Dij*f(i) N, mol/h	N*MW/1000 Nk, kg/h Tij	Rate of transport Dij*f(i) N, mol/h	N*MW/1000 Nk, kg/h Tij	Rate of transport Dij*f(i) N, mol/h	N*MW/1000 Nk, kg/h Tij
Air to water (D12)	8.6077E+05	2.134E+09	3.248E+04	5.379E+00	4.20E-01	1.723E+00	1.35E-01	4.885E+00	3.816E-01
Air to soil (D13)	7.4102E+05	1.837E+09	3.773E+04	4.631E+00	3.62E-01	1.483E+00	1.16E-01	4.206E+00	3.285E-01
Water to air (D21)	8.5897E+05	4.787E+08	2.896E+02	1.738E+00	1.36E-01	4.102E+03	3.20E+02	4.293E+01	3.353E+00
Water to sed. (D24)	1.9432E+03	1.083E+06	1.280E+05	3.932E-03	3.07E-04	9.279E+00	7.25E-01	9.712E-02	7.586E-03
Soil to air (D31)	7.2486E+05	2.416E+08	2.582E+01	4.191E+00	3.27E-01	1.342E+00	1.05E-01	1.159E+04	9.052E+02
Soil to water (D32)	8.0789E+03	2.692E+06	2.32E+03	4.671E-02	3.65E-03	1.496E-02	1.17E-03	1.292E+02	1.009E+01
Sed. to water (D42)	1.8134E+03	5.428E+05	1.277E+02	2.821E-03	2.20E-04	6.657E+00	5.20E-01	6.967E-02	5.442E-03
Water to soil (D23)	0								
Sed. burial DL(4)	0.0000E+00								
Stratosphere DL(1)	0.0000E+00								

Phase properties and rates:

Compartment	Rate const. k, 1/hr	D Values Reaction DRi VB*ZB*k mol/Pa h	Advection DAi GAB*ZB mol/Pa h	DTs	Js,(E)	Js, E(A)	Js, E(B)	Js, E(C)
Air (1)	0.040764705	1.645E+09	4.034E+08	2.050E+09	4.313E-06	6.247E-06	0.000E+00	5.653E-06
Water (2)	0.004076470	1.463E+06	3.589E+05	2.683E+06	4.191E-04	4.191E-04	4.191E-04	4.191E-04
Soil (3)	0.00126	6.806E+04	0	8.010E+05	9.997E-01	9.997E-01	9.997E-01	9.997E-01
Bottom sediment (4)	0.000407647	6.810E+02	3.341E+01	2.528E+03	8.682E+05	8.682E+05	8.682E+05	8.682E+05
Total	Total	1.646E+09	4.038E+08					
	R + A		2.050E+09					

For E(1,2,3):

Compartment	fugacity f's Pa	Concentration C mol/m3	S mg/L	ug/g	amount m, mol C*VBi	amount %	Loss Reaction mol/h	Loss Advection mol/h	Loss rate mol/h	Reaction %	Advection %	% Loss	Removal %
Air (1)	4.918E-06	1.984E-09	1.550E-07	1.307E-04	1.98E+05	24.70073	8.09E+03	1983.9130	0.00E+00	63.1704	15.4963	0.00E+00	78.67
Water (2)	1.439E-03	2.582E-06	2.017E-04	2.017E-04	5.16E+05	64.28991	2.10E+03	516.3636	0	16.4417	4.0333	0	20.48
Soil (3)	1.603E-03	4.810E-06	3.757E-04	2.504E-04	8.66E+04	10.77932	1.09E+02	0.0000	0	0.8521	0.0000	0	0.85
Bottom sediment (4)	1.106E-03	3.695E-06	2.886E-04	2.255E-04	1.85E+03	0.23003	7.53E-01	0.0370	0.00E+00	0.0059	0.0003	0.00E+00	6.17E-03
Total					8.03E+05	100	1.03E+04	2500.3135	0.0000				100
								1.28E+04					

For E(A), i.e., E(1) = 1000 kg/L conditions:

Compartment	fugacity f, Pa	Concentration mol/m3	mg/L	ug/g	amount m, mol	amount %	Loss Reaction mol/h	Loss Advection mol/h	Loss rate mol/h	Reaction %	Advection %	% Loss	Removal %
Air (1)	6.249E-06	2.521E-09	1.969E-07	1.661E-04	2.52E+05	99.58874	1.03E+04	2.52E+03	0.00E+00	8.03E+01	19.6924	0.00E+00	1.00E+02
Water (2)	2.023E-06	3.631E-09	2.836E-07	2.836E-07	7.26E+02	0.28688	2.96E+00	7.26E-01	0	2.31E-02	0.0057	0	2.88E-02
Soil (3)	5.781E-06	1.735E-08	1.355E-06	9.033E-07	3.12E+02	0.12336	3.93E-01	0.00E+00	0	3.07E-03	0.0000	0	3.07E-03
Bottom sediment (4)	1.556E-06	5.197E-09	4.059E-07	3.171E-07	2.60E+00	1.026E-03	1.06E-03	5.20E-05	0.00E+00	8.27E-06	4.06E-07	0.00E+00	8.68E-06
Total					2.53E+05		1.03E+04	2521.8434	0.0000				100
								12802					

For E(B), i.e., E(2) = 1000 kg/L conditions:

Compartment	fugacity f, Pa	Concentration mol/m3	mg/L	ug/g	amount m, mol	amount %	Reaction mol/h	Advection mol/h	Loss rate mol/h	Reaction %	Advection %	% Loss	Removal %
Air (1)	2.002E-06	8.076E-10	6.308E-08	5.321E-05	80760	4.48475	3.29E+03	8.08E+02	0.00E+00	2.57E+01	6.31E+00	0.00E+00	3.20E+01
Water (2)	4.775E-03	8.569E-06	6.693E-04	6.693E-04	1.71E+06	95.16918	6.99E+03	1.71E+03	0	5.46E+01	1.34E+01	0	6.80E+01
Soil (3)	1.852E-06	5.557E-09	4.341E-07	2.893E-07	100	0.00556	1.26E-01	0.00E+00	0	9.85E-04	0.00E+00	0	9.85E-04
Bottom sediment (4)	3.671E-03	1.226E-05	9.579E-04	7.484E-04	6132	0.34052	2.50E+00	1.23E-01	0.00E+00	0.0195	9.58E-04	0.00E+00	2.05E-02
Total					1800772.1		10280.96	2521.5043	0.0000				100
								12802					

For E(C) i.e., E(3) = 1000 kg/L conditions:

Compartment	fugacity f, Pa	Concentration mol/m3	mg/L	ug/g	amount m, mol	amount %	Reaction mol/h	Advection mol/h	Loss rate mol/h	Reaction %	Advection %	% Loss	Removal %
Air (1)	5.676E-06	2.290E-09	1.788E-07	1.509E-04	2.29E+05	20.61676	9.33E+03	2.29E+03	0.00E+00	7.29E+01	1.79E+01	0.00E+00	9.08E+01
Water (2)	4.998E-05	8.969E-08	7.006E-06	7.005E-06	1.79E+04	1.61518	7.31E+01	1.79E+01	0	5.71E-01	1.40E-01	0	7.11E-01
Soil (3)	1.599E-02	4.798E-05	3.75E-03	2.498E-03	8.64E+05	77.76228	1.09E+03	0.00E+00	0	8.50E+00	0.00E+00	0	8.50E+00
Bottom sediment (4)	3.842E-05	1.284E-07	1.003E-05	7.833E-06	6.42E+01	0.00578	2.62E-02	1.28E-03	0.00E+00	2.04E-04	1.00E-05	0.00E+00	2.14E-04
Total					1110563.6		10494.86	2307.5609	0.0000				100
								12802					

Level III input, output summary:	E(1,2,3)	E(A)	E(B)	E(C)
Total emission rate mol/h	12802.458	12802.458	12802.458	12802.458
Total VZ products	4.076E+10	4.076E+10	4.076E+10	4.076E+10
Total amount of chemicals	803179.80	253152.83	1800772.1	1110563.6
Total advection D value, mol/Pa h	1.65E+09	1.65E+09	1.65E+09	1.65E+09
Total reaction D value, mol/Pa h	403776843	403776843	403776843	403776843
Total D value, mol/Pa h	2.05E+09	2.05E+09	2.05E+09	2.05E+09
Output by reaction, mol/h	10302.144	10280.614	10280.963	10494.860
Output by advection, mol/h	2500.3135	2521.8433	2521.5042	2307.5609
Output by losses	0	0	0	0
Overall residence time, h	62.736374	19.773767	140.65820	86.746373
Reaction residence time, h	77.962389	24.624289	175.15596	105.81976
Advection residence time, h	321.23163	100.38404	714.16582	481.27164

Output for Fugacity Level I and II diagram

Benzene

Level I calculation:

Compartment	Z	Concentration			Amount	Amount
	mol/m3 Pa	mol/m3	mg/L (or g/m3)	ug/g	kg	%
Air	4.034E-04	1.268E-08	9.901E-07	8.352E-04	99012	99.012
Water	1.794E-03	5.638E-08	4.404E-06	4.404E-06	880.79	0.881
Soil	4.764E-03	1.497E-07	1.169E-05	4.871E-06	105	0.105
Biota (fish)	1.210E-02	3.803E-07	2.970E-05	2.970E-05	0.0059	5.94E-06
Suspended sediment	2.977E-02	9.355E-07	7.307E-05	4.871E-05	0.073	7.31E-05
Bottom sediment	9.527E-03	2.994E-07	2.338E-05	9.743E-06	2.34	0.0023
	Total				100000	100

f = 3.142E-05 Pa

Level II Calculation:

Compartment	Half-life h	D Values D(reaction)	D(advec'n)	Conc'n mol/m3	Loss Reaction kg/h	Loss Advection kg/h	Removal %
Air	17	1.64E+09	4.03E+08	2.52E-09	802.28	196.81	99.908
Water	170	1.46E+06	3.59E+05	1.12E-08	0.714	0.175	0.089
Soil	550	5.40E+04		2.98E-08	0.026		2.64E-03
Biota (fish)				7.56E-08			
Suspended sediment				1.86E-07			
Bottom sediment	1700	3.88E+02	1.91E+01	5.95E-08	1.89E-04	9.30E-06	1.99E-05
	Total	1.65E+09	4.04E+08		803.02	196.98	100
	R + A		2.05E+09			1000	

f = 6.246E-06 Pa

Total amount = 19877 kg

Overall residence time = 19.88 h

Reaction time = 24.75 h

Advection time = 100.91 h

Output for Fugacity Level III diagram

Level III Calculation: Benzene

Phase Properties and Rates:

Compartment	Bulk Z mol/m3 Pa	Half-life h	D Values Reaction mol/Pa h	D Values Advection mol/Pa h
Air (1)	4.034E-04	17	1.64E+09	4.03E+08
Water (2)	1.795E-03	170	1.46E+06	3.59E+05
Soil (3)	3.001E-03	550	6.81E+04	
Bottom sediment (4)	3.341E-03	1700	6.81E+02	3.34E+01

	E(1)=1000	E(2)=1000	E(3)=1000	E(1,2,3)	
Overall residence time =	19.77	140.66	86.75	62.74	h
Reaction time =	24.62	175.16	105.82	77.96	h
Advection time =	100.38	714.17	481.27	321.23	h

Phase Properties, Compositions, Transport and Transformation Rates:

Emission, kg/h			Fugacity, Pa				Concentration, g/m3				Amounts, kg				Total amount, kg
E(1)	E(2)	E(3)	f(1)	f(2)	f(3)	f(4)	C(1)	C(2)	C(3)	C(4)	m(1)	m(2)	m(3)	m(4)	
1000	0	0	6.249E-06	2.023E-06	5.781E-06	1.556E-06	1.969E-07	2.836E-07	1.355E-06	4.059E-07	19692	56.73	24.39	0.203	1.977E+04
0	1000	0	2.002E-06	4.775E-03	1.852E-06	3.671E-03	6.308E-08	6.693E-04	4.341E-07	9.579E-04	6308	133863	7.81	478.96	1.407E+05
0	0	1000	5.676E-06	4.998E-05	1.599E-02	3.842E-05	1.788E-07	7.006E-06	3.748E-03	1.003E-05	17884	1401	67456	5.013	8.675E+04
600	300	100	4.918E-06	1.439E-03	1.603E-03	1.106E-03	1.550E-07	2.017E-04	3.757E-04	2.886E-04	15496	40333	6763	144.31	6.274E+04

Emission, kg/h			Loss, Reaction, kg/h				Loss, Advection, kg/h			Intermedia Rate of Transport, kg/h T12	T21	T13	T31	T32	T24	T42
E(1)	E(2)	E(3)	R(1)	R(2)	R(3)	R(4)	A(1)	A(2)	A(4)	air-water	water-air	air-soil	soil-air	soil-water	water-sed	sed-water
1000	0	0	8.028E+02	2.312E-01	3.07E-02	8.274E-05	1.969E+02	5.673E-02	4.059E-06	4.202E-01	1.358E-01	3.617E-01	3.273E-01	3.648E-03	3.071E-04	2.203E-04
0	1000	0	2.572E+02	5.457E+02	9.85E-03	1.952E-01	6.308E+01	1.339E+02	9.579E-03	1.346E-01	3.204E+02	1.159E-01	1.049E-01	1.169E-03	7.248E-01	5.200E-01
0	0	1000	7.290E+02	5.712E+00	8.50E+01	2.044E-03	1.788E+02	1.401E+00	1.003E-04	3.816E-01	3.353E+00	3.285E-01	9.052E+02	1.009E+01	7.586E-03	5.442E-03
600	300	100	6.317E+02	1.644E+02	8.52E+00	5.883E-02	1.550E+02	1.644E+02	2.886E-03	3.306E-01	9.653E+01	2.846E-01	9.075E+01	1.011E+00	2.184E-01	1.567E-01